Volkswagen Cabriolet, Scirocco

Service Manual
1985, 1986, 1987, 1988, 1989, 1990, 1991, 1992, 1993

Including Scirocco 16V

BentleyPublishers®
.com

Selected Books and Repair Information From Bentley Publishers

Motorsports

Alex Zanardi: My Sweetest Victory
Alex Zanardi and Gianluca Gasparini
ISBN 978-0-8376-1249-2

The Unfair Advantage
Mark Donohue and Paul van Valkenburgh
ISBN 978-0-8376-0069-7

Equations of Motion - Adventure, Risk and Innovation *William F. Milliken*
ISBN 978-0-8376-1570-7

Engineering

Hack Mechanic Guide to European Automotive Electrical Systems
Rob Siegel ISBN 978-0-8376-1751-0

Bosch Automotive Handbook *Robert Bosch GmbH* ISBN 978-0-8376-1732-9

Bosch Fuel Injection and Engine Management *Charles O. Probst, SAE*
ISBN 978-0-8376-0300-1

Maximum Boost: Designing, Testing, and Installing Turbocharger Systems
Corky Bell ISBN 978-0-8376-0160-1

Supercharged! Design, Testing and Installation of Supercharger Systems
Corky Bell ISBN 978-0-8376-0168-7

Scientific Design of Exhaust and Intake Systems *Phillip H. Smith & John C. Morrison* ISBN 978-0-8376-0309-4

Physics for Gearheads
Randy Beikmann ISBN 978-0-8376-1615-5

Audi

Audi A4 Service Manual: 2002-2008, 1.8L Turbo, 2.0L Turbo, 3.0L, 3.2L
Bentley Publishers
ISBN 978-0-8376-1574-5

Audi A4 Service Manual: 1996-2001, 1.8L Turbo, 2.8L *Bentley Publishers*
ISBN 978-0-8376-1675-9

Audi TT Service Manual: 2000-2006, 1.8L turbo, 3.2 L *Bentley Publishers*
ISBN 978-0-8376-1625-4

Audi A6 (C5 platform) Service Manual: 1998-2004 *Bentley Publishers*
ISBN 978-0-8376-1670-4

BMW

Memoirs of a Hack Mechanic
Rob Siegel ISBN 978-0-8376-1720-6

BMW 3 Series (F30, F31, F34) Service Manual: 2012-2015 *Bentley Publishers*
ISBN 978-0-8376-1752-7

BMW X3 (E83) Service Manual: 2004-2010 *Bentley Publishers*
ISBN 978-0-8376-1731-2

BMW X5 (53) Service Manual: 2000-2006 *Bentley Publishers*
ISBN 978-0-8376-1643-8

BMW 3 Series (E90, E91, E92, E93) Service Manual: 2006-2011
Bentley Publishers ISBN 978-0-8376-1723-7

BMW 3 Series (E46) Service Manual: 1999-2005 *Bentley Publishers*
ISBN 978-0-8376-1657-5

BMW 3 Series (E36) Service Manual: 1992-1998 *Bentley Publishers*
ISBN 978-0-8376-1709-1

BMW Z3 (E36/7) Service Manual: 1996-2002 *Bentley Publishers*
ISBN 978-0-8376-1617-9

BMW 5 Series (E60, E61) Service Manual: 2004-2010 *Bentley Publishers*
ISBN 978-0-8376-1689-6

BMW 5 Series (E39) Service Manual: 1997-2003 *Bentley Publishers*
ISBN 978-0-8376-1672-8

Corvette

Corvette: America's Star-Spangled Sports Car *Karl Ludvigsen*
ISBN 978-0-8376-1659-9

Corvette by the Numbers: The Essential Corvette Parts Reference 1955-1982 *Alan Covin*
ISBN 978-0-8376-0228-2

Zora Arkus-Duntov: The Legend Behind Corvette *Jerry Burton*
ISBN 978-08376-0858-7

Porsche

Porsche 911 (996) Service Manual: 1999-2005 *Bentley Publishers*
ISBN 978-0-8376-1710-7

Porsche 911 (993) Service Manual: 1995-1998 *Bentley Publishers*
ISBN 978-0-8376-1719-0

Porsche Boxster Service Manual: 1997-2004 *Bentley Publishers*
ISBN 978-0-8376-1645-2

Porsche 911 Carrera Service Manual: 1984-1989 *Bentley Publishers*
ISBN 978-0-8376-1696-4

Porsche 911 SC Service Manual: 1978-1983 *Bentley Publishers*
ISBN 978-0-8376-1705-3

Porsche: Excellence Was Expected
Karl Ludvigsen
ISBN 978-0-8376-0235-6

Porsche — Origin of the Species
Karl Ludvigsen
ISBN 978-0-8376-1331-4

Volkswagen

Volkswagen Rabbit, GTI Service Manual: 2006-2009 *Bentley Publishers*
ISBN 978-0-8376-1664-3

Volkswagen Jetta Service Manual: 2005-2010 *Bentley Publishers*
ISBN 978-0-8376-1616-2

Volkswagen Jetta, Golf, GTI Service Manual: 1999-2005 *Bentley Publishers*
ISBN 978-0-8376-1678-0

Volkswagen Jetta, Golf, GTI: 1993-1999, Cabrio: 1995-2002 Service Manual *Bentley Publishers*
ISBN 978-0-8376-1660-5

Volkswagen GTI, Golf, Jetta Service Manual: 1985-1992 *Bentley Publishers*
ISBN 978-0-8376-1637-7

Volkswagen Corrado Repair Manual: 1990-1994 *Bentley Publishers*
ISBN 978-0-8376-1699-5

Volkswagen Passat, Passat Wagon Service Manual: 1998-2005
Bentley Publishers
ISBN 978-0-8376-1669-8

MINI Repair Manuals

MINI Cooper Service Manual: 2007-2013 *Bentley Publishers*
ISBN 978-0-8376-1730-5

MINI Cooper Service Manual: 2002-2006 *Bentley Publishers*
ISBN 978-0-8376-1639-1

Mercedes-Benz

Mercedes-Benz C-Class (W202) Service Manual 1994-2000
Bentley Publishers
ISBN 978-0-8376-1692-6

Mercedes Benz E-Class (W124) Owner's Bible: 1986-1995
Bentley Publishers
ISBN 978-0-8376-0230-1

Mercedes-Benz Technical Companion
Staff of The Star and members of Mercedes-Benz Club of America
ISBN 978-0-8376-1033-7

Bentley Publishers®
Automotive Reference

Bentley Publishers has published service manuals and automobile books since 1950. For more information, please contact Bentley Publishers at 1734 Massachusetts Avenue, Cambridge, MA 02138 USA, or visit our web site at
BentleyPublishers.com

Volkswagen Cabriolet, Scirocco

Service Manual
1985, 1986, 1987, 1988, 1989, 1990, 1991, 1992, 1993

Including Scirocco 16V

BentleyPublishers.com

BENTLEY PUBLISHERS® | Automotive Reference™

Bentley Publishers, a division of Robert Bentley, Inc.
1734 Massachusetts Avenue
Cambridge, MA 02138 USA
800-423-4595 / 617-547-4170

Information that makes the difference®

BentleyPublishers®.com

Technical contact information

We welcome your feedback. Please submit corrections and additions to our Volkswagen technical discussion forum at:

`http://www.BentleyPublishers.com`

Updates and corrections

We will evaluate submissions and post appropriate editorial changes online as updates or tech discussion. Appropriate updates and corrections will be added to the book in future printings. Check for updates and corrections for this book before beginning work on your vehicle. See the following web address for additional information:

`http://www.BentleyPublishers.com/updates/`

WARNING—Important Safety Notice

Do not use this manual for repairs unless you are familiar with automotive repair procedures and safe workshop practices. This manual illustrates workshop procedures for some maintenance and service work. It is not a substitute for full and up-to-date information from the vehicle manufacturer or for proper training as an automotive technician. Note that it is not possible for us to anticipate all of the ways or conditions under which vehicles may be serviced or to provide cautions as to all of the possible hazards that may result.

We have endeavored to ensure the accuracy of the information in this manual. Please note, however, that considering the vast quantity and the complexity of the service information involved, we cannot warrant the accuracy or completeness of the information contained in this manual.

FOR THESE REASONS, NEITHER THE PUBLISHER NOR THE AUTHOR MAKES ANY WARRANTIES, EXPRESS OR IMPLIED, THAT THE INFORMATION IN THIS MANUAL IS FREE OF ERRORS OR OMISSIONS, AND WE EXPRESSLY DISCLAIM THE IMPLIED WARRANTIES OF MERCHANTABILITY AND OF FITNESS FOR A PARTICULAR PURPOSE, EVEN IF THE PUBLISHER OR AUTHOR HAVE BEEN ADVISED OF A PARTICULAR PURPOSE, AND EVEN IF A PARTICULAR PURPOSE IS INDICATED IN THE MANUAL. THE PUBLISHER AND AUTHOR ALSO DISCLAIM ALL LIABILITY FOR DIRECT, INDIRECT, INCIDENTAL OR CONSEQUENTIAL DAMAGES THAT RESULT FROM ANY USE OF THE EXAMPLES, INSTRUCTIONS OR OTHER INFORMATION IN THIS MANAUL. IN NO EVENT SHALL OUR LIABILITY WHETHER IN TORT, CONTRACT OR OTHERWISE EXCEED THE COST OF THIS MANUAL.

Before attempting any work on your Volkswagen, read the **WARNINGS** and **CAUTIONS** on page ix and any **WARNING** or **CAUTION** that accompanies a procedure in the manual. Review all **WARNINGS** and **CAUTIONS** each time you prepare to work on your Volkswagen.

Your common sense and good judgment are crucial to safe and successful service work. Read procedures through before starting them. Think about whether the condition of your car, your level of mechanical skill, or your level of reading comprehension might result in or contribute in some way to an occurrence which might cause you injury, damage your car, or result in an unsafe repair. If you have doubts for these or other reasons about your ability to perform safe repair work on your car, have the work done at an authorized Volkswagen dealer or other qualified shop.

Part numbers listed in this manual are for identification purposes only, not for ordering. Always check with your authorized Volkswagen dealer to verify part numbers and availability before beginning service work that may require new parts.

Special tools and safety equipment are required to perform certain service operations are identified in the manual and are recommended for use. Use of improper tools may be detrimental to the car's safe operation as well as the safety of the person servicing the car.

The vehicle manufacturer will continue to issue service information updates and parts retrofits after the editorial closing of this manual. Some of these updates and retrofits may apply to procedures and specifications in this manual. We regret that we cannot supply updates to purchasers of this manual.

This manual is prepared, published and distributed by Bentley Publishers, 1734 Massachusetts Avenue, Cambridge, Massachusetts 02138 USA. All information contained in this manual is based on information available to the publisher from Volkswagen at the time of editorial closing. Volkswagen has not reviewed and does not vouch for the accuracy or completeness of the technical specifications and work procedures described and given in this manual.

ISBN 978-0-8376-1636-0 **Job code: VS93-15** editorial closing 08 / 1992.

Library of Congress Catalog Card No. for the paperback version: 84-70138

VWoA Inc. Part No. LPV 800 113

VWC Inc. Part No. MAN 101 171 F DSL

The paper used in this publication is acid free and meets the requirements of the National Standard for Information Sciences-Permanence of Paper for Printed Library Materials. ∞

Manufactured in the United States of America.

202009R005

Contents:

Foreword

Service to Volkswagen owners is of top priority to the Volkswagen organization and has always included the continuing development and introduction of new and expanded services. This manual has been prepared with the Volkswagen owner in mind. The aim throughout has been simplicity, clarity and completeness, with practical explanations, step-by-step procedures and accurate specifications.

This Volkswagen Cabriolet, Scirocco service manual covers the model years 1985, 1986, 1987, 1988, 1989, 1990, 1991, 1992 and 1993, and includes Wolfsburg editions and Scirocco 16V. This manual was created specifically to cover Cabriolet and Scirocco models built for sale in the United States and Canada. Information in this manual applies to all Cabriolet and Scirocco models unless particular model names, engine codes or other characteristics are used to denote differences.

For the Volkswagen owner with basic mechanical skills and for independent auto service professionals, this manual includes the same specifications and procedures that were available in an authorized Volkswagen dealer Service department as this manual went to press. The Volkswagen owner who has no intention of working on his or her car will find that owning and referring to this manual will make it possible to be better informed and to more knowledgeably discuss repairs with a professional automotive service technician.

This manual has been prepared from the information that the Volkswagen organization provides to its own factory-trained technicians and has also been developed with the do-it-yourself Volkswagen owner in mind. The aim throughout has been clarity and completeness, with step-by-step procedures and accurate specifications.

The Volkswagen owner intending to do maintenance and repair should have screwdrivers and a set of metric wrenches, sockets and hex wrenches, since these basic hand tools are needed for most of the work described in this manual. Most procedures will also require a torque wrench to ensure that fasteners are tightened properly and in accordance with specifications. In some cases, the text refers to special tools that are recommended or required to accomplish adjustments or repairs. These tools are identified by their Volkswagen special tool number and illustrated. A thorough review of each procedure is recommended to determine in advance the need for particular tools and for essential replacement parts such as gaskets.

Some of the information in this manual applies only to cars of a particular model year or range of years. For example, "1990 m.y." refers to the 1990 model year. The model year does not necessarily match the calender year in which the car was manufactured or sold. To be sure of the model year of a particular car, check the vehicle identification number (VIN) on the car. Technical changes made in production within a model year are identified in this manual by listing the VIN for the first car produced with this change.

The VIN is a unique sequence of 17 characters assigned by the manufacturer to identify each car. WVWCA015ZHK005670 is an example of a Volkswagen VIN. Each of the 17 letters and numbers indicates certain facts about the car and its manufacture. Information in this manual may refer to VINs to identify information that applies only to specific cars, usually using only the last eleven characters: 15ZHK005670 or 15_H_005670 in the example above.

Your Volkswagen's VIN can be found on a plate mounted on the top of the instrument panel, on the driver's side where the number can be seen through the windshield, and on an identification label in the luggage compartment. The 10th character refers to model year: "F" for 1985 models, "G" for 1986, "H" for 1987 and so on. The letters "I" and "O" are not used, so the sequence skips to "J" for 1988 and skips again to "P" for 1993. This manual covers Volkswagen Cabriolet and Scirocco from model years 1985, 1986, 1987, 1988, 1989, 1990, 1991, 1992 and 1993, with these VINs:

1985:	______30_F_ 000 001	to	______30_F_ 999 999
1986:	______30_G_000 001	to	______30_G_999 999
1987:	______30_H_000 001	to	______30_H_999 999
1988:	______30_J_ 000 001	to	______30_J_ 999 999
1989:	______30_K_000 001	to	______30_K_999 999
1990:	______30_L_ 000 001	to	______30_L_ 999 999
1991:	______30_M_000 001	to	______30_M_999 999
1992:	______30_N_000 001	to	______30_N_999 999
1993:	______30_P_000 001	to	______30_P_999 999

We have endeavored to ensure the accuracy of the information in this manual. When the vast array of data presented in the manual is taken into account, however, no claim to infallibility can be made. We therefore cannot be responsible for the result of any errors that may have crept into the text. The Publisher encourages comments from the readers of this manual with regard to errors, and also suggestions for improvement in the presentation of the technical material. These communications have been and will be carefully considered in the preparation of this and other manuals. Please write to Robert Bentley, Inc., Cambridge, Massachusetts 02138.

Volkswagen offers extensive warranties, especially on components of the fuel delivery and emission control systems. Before deciding to repair a car that may be covered wholly or in part by any warranties issued by Volkswagen United States, Inc., consult your authorized Volkswagen dealer. You may find that the dealer can make the repair either without charge or at minimal cost. Regardless of its age, or whether it is covered by warranty, your Volkswagen is an easy car to service and an easy car to get serviced. If at any time a repair is needed that you feel is too difficult to do yourself, a trained Volkswagen technician is ready to do the job for you.

Robert Bentley

Please read these warnings and cautions before proceeding with maintenance and repair work.

WARNING—

● Some repairs may be beyond your capability. If you lack the skills, tools and equipment, or a suitable workplace for any procedure described in this manual, we suggest you leave such repairs to an authorized Volkswagen dealer service department, or other qualified shop.

● Volkswagen is constantly improving its cars. Sometimes these changes, both in parts and specifications, are made applicable to earlier models. Therefore, before starting any major jobs or repairs to components on which passenger safety may depend, consult your authorized Volkswagen dealer about Technical Bulletins that may have been issued since the editorial closing of this manual.

● Do not re-use any fasteners that are worn or deformed in normal use. Some fasteners are designed to be used only once, and are unreliable and may fail if used a second time. This includes, but is not limited to, nuts, bolts, washers, circlips and cotter pins. Always follow the recommendations in this manual to replace these fasteners with new parts where indicated, and any other time it is deemed necessary by inspection.

● Never work under a lifted car unless it is solidly supported on stands designed for the purpose. Do not support a car on cinder blocks, hollow tiles or other props that may crumble under continuous load. Never work under a car that is supported solely by a jack. Never work under the car while the engine is running.

● If you are going to work under a car on the ground, make sure that the ground is level. Block the wheels to keep the car from rolling. Disconnect the battery negative (–) terminal (Ground strap) to prevent others from starting the car while you are under it.

● Never run the engine unless the work area is well ventilated. Carbon monoxide kills.

● Finger rings, bracelets and other jewelry should be removed so that they cannot cause electrical shorts, get caught in running machinery, or be crushed by heavy parts.

● Tie long hair behind your head. Do not wear a necktie, a scarf, loose clothing, or a necklace when you work near machine tools or running engines. If your hair, clothing, or jewelry were to get caught in the machinery, severe injury could result.

● Do not attempt to work on your car if you do not feel well. You increase the danger of injury to yourself and others if you are tired, upset or have taken medication or any other substance that may keep you from being fully alert.

● Illuminate your work area adequately but safely. Use a portable safety light for working inside or under the car. Make sure the bulb is enclosed by a wire cage. The hot filament of an accidentally broken bulb can ignite spilled fuel or oil.

● Catch draining fuel, oil, or brake fluid in suitable containers. Do not use food or beverage containers that might mislead someone into drinking from them. Store flammable fluids away from fire hazards. Wipe up spills at once, but do not store the oily rags, which can ignite and burn spontaneously.

● Always observe good workshop practices. Wear goggles when you operate machine tools or work with battery acid. Gloves or other protective clothing should be worn whenever the job requires working with harmful substances.

● Friction materials such as brake or clutch discs may contain asbestos fibers. Do not create dust by grinding, sanding, or by cleaning with compressed air. Avoid breathing asbestos fibers and asbestos dust. Breathing asbestos can cause serious diseases such as asbestosis or cancer, and may result in death.

● Disconnect the battery negative (–) terminal (Ground strap) whenever you work on the fuel system or the electrical system. Do not smoke or work near heaters or other fire hazards. Keep an approved fire extinguisher handy.

● Batteries give off explosive hydrogen gas during charging. Keep sparks, lighted matches and open flame away from the top of the battery. If hydrogen gas escaping from the cap vents is ignited, it will ignite gas trapped in the cells and cause the battery to explode.

● Connect and disconnect battery cables, jumper cables or a battery charger only with the ignition switched off, to prevent sparks. Do not disconnect the battery while the engine is running.

● Do not quick-charge the battery (for boost starting) for longer than one minute. Wait at least one minute before boosting the battery a second time.

● Do not allow battery charging voltage to exceed 16.5volts. If the battery begins producing gas or boiling violently, reduce the charging rate. Boosting a sulfated battery at a high charging rate can cause an explosion.

● The air-conditioning system is filled with chemical refrigerant, which is hazardous. The A/C system should be serviced only by trained technicians using approved refrigerant recovery/recycling equipment, trained in related safety precautions, and familiar with regulations governing the discharging and disposal of automotive chemical refrigerants.

● Do not expose any part of the A/C system to high temperatures such as open flame. Excessive heat will increase system pressure and may cause the system to burst.

● Some aerosol tire inflators are highly flammable. Be extremely cautious when repairing a tire that may have been inflated using an aerosol tire inflator. Keep sparks, open flame or other sources of ignition away from the tire repair area. Inflate and deflate the tire at least four times before breaking the bead from the rim. Completely remove the tire from the rim before attempting any repair.

● Some cars covered by this manual are equipped with a supplemental restraint system (SRS), that automatically deploys an airbag in the event of a frontal impact. The airbag is inflated by an explosive device. Handled improperly or without adequate safeguards, it can be accidently activated and cause serious injury.

● To prevent personal injury or airbag system failure, **only trained Volkswagen dealer service technicians** should test, disassemble or service the airbag system.

● Disconnect the power supply before working on the airbag system, or when doing repairs that require removing airbag system components. Disconnect the battery negative (–) terminal and cover the battery. Also disconnect the red single-pin harness connector (marked "airbag") behind the instrument panel.

continued on next page

Please read these warnings and cautions before proceeding with maintenance and repair work.

WARNING (continued) —

- The airbag unit should not be removed, nor should any other work be performed on the airbag system, until at least 20 minutes after the battery is disconnected.

- On airbag-equipped cars, do not use a computer memory-saver tool to maintain battery power to an anti-theft radio. It will keep the airbag system charged, which may cause accidental activation of the airbag.

- On airbag-equipped cars, never apply stickers or any other type of covering on the steering wheel. Do not let chemical cleaners, oil or grease come into contact with the vinyl covering of the airbag unit.

- When driving or riding in an airbag-equipped vehicle never hold test equipment in your hands or lap while the vehicle is in motion. Objects between you and the airbag can increase the risk of injury in an accident.

- Never open or otherwise attempt to repair airbag system parts. Always use new parts. Never leave airbag parts or the partially disassembled airbag system unattended.

- Never use a test light to conduct electrical tests on the airbag system. The system must only be tested by trained Volkswagen Service technicians using the Volkswagen VAG 1551 Scan Tool (ST) or an approved equivalent. The airbag unit must never be electrically tested while it is not installed in the car.

- Do not expose the airbag unit to temperature above 194°F (90°C), even for brief periods. Keep clear of heat sources such as hot plates, soldering irons, heat lamps and welding equipment.

- Greases, lubricants and other automotive chemicals contain toxic substances, many of which are absorbed directly through the skin. Read manufacturer's instructions and warnings carefully. Use hand and eye protection. Avoid direct skin contact.

CAUTION—

- If you lack the skills, tools and equipment, or a suitable workshop for any procedure described in this manual, we suggest you leave such repairs to an authorized Volkswagen dealer or other qualified shop.

- Volkswagen offers extensive warranties, especially on components of fuel delivery and emission control systems. Therefore, before deciding to repair a Volkswagen that may still be covered wholly or in part by any warranties issued by Volkswagen United States, Inc., consult your authorized Volkswagen dealer. You may find that he can make the repair for free, or at minimal cost.

- Volkswagen part numbers listed in this manual are for identification purposes only, not for ordering. Always check with your authorized Volkswagen dealer to verify part numbers and availability before beginning service work that may require new parts.

- Before starting a job, make certain that you have all the necessary tools and parts on hand. Read all the instructions thoroughly, do not attempt shortcuts. Use tools appropriate to the work and use only replacement parts meeting Volkswagen specifications. Makeshift tools, parts and procedures will not make good repairs.

- Use pneumatic and electric tools only to loosen threaded parts and fasteners. Never use these tools to tighten fasteners, especially on light alloy parts. Always use a torque wrench to tighten fasteners to the tightening torque specification listed.

- Be mindful of the environment and ecology. Before you drain the crankcase, find out the proper way to dispose of the oil. Do not pour oil onto the ground, down a drain, or into a stream, pond or lake. Consult local ordinances that govern the disposal of wastes.

- On cars equipped with anti-theft radios, make sure you know the correct radio activation code before disconnecting the battery or removing the radio. If the wrong code is entered into the radio when power is restored, that radio may lock up and be rendered inoperable, even if the correct code is then entered.

- Connect and disconnect a battery charger only with the battery charger switched off.

- Do not quick-charge the battery (for boost starting) for longer than one minute. Wait at least one minute before boosting the battery a second time.

- Sealed or "maintenance free" batteries should be slow-charged only, at an amperage rate that is approximately 10% of the battery's ampere-hour (Ah) rating.

- Do not allow battery charging voltage to exceed 16.5 volts. If the battery begins producing gas or boiling violently, reduce the charging rate. Boosting a sulfated battery at a high charging rate can cause an explosion.

Section 1

FUNDAMENTALS

Contents

Fundamentals

Introduction

Although Volkswagens are sophisticated and complex machines, nearly all of their basic maintenance and most repairs can be accomplished by any interested owner with basic mechanical skills and the right information. While some of the repairs covered in this manual are complicated and require special knowledge and equipment, most of the care that is required in the lifetime of the average Volkswagen is well within the capabilities of the do-it-yourselfer.

This section of the manual is dedicated to helping the beginner get started smartly and safely with Volkswagen maintenance and repair. Of interest to all owners, the section begins with a **General Description** of the car, broken down into its individual systems so that it can be more easily understood, and a discussion on **How To Use This Manual**—a simple directory of the kind of information you can expect to find and where to find it.

The section continues with extensive information on **Safety** and **General Advice For The Beginner**, including a number of tips on mechanic's skills and workshop techniques which can help the beginner do a faster, complete, and more thorough job. Also for the beginner, this section includes a discussion of **Tools** with information on the tools needed to do 90% of the jobs in this manual, and advice on how to buy them wisely and use them effectively.

Finally, and once again of interest to any owner, this section ends with a quick reference guide to emergencies—what to do when the car won't start or when a warning light comes on, including basic troubleshooting and information on how to gauge the seriousness of the problem.

1. GENERAL DESCRIPTION

Volkswagen's Scirocco and Cabriolet models are sophisticated examples of the state of today's automotive engineering art, blending advanced design and manufacturing to provide an outstanding combination of performance, economy, roadholding, and reliability at very reasonable cost. In spite of this sophistication, much of the necessary maintenance and repair can be accomplished by the average owner using this manual. While the complexity of the car may seem to make this a difficult challenge for the novice mechanic, it can be simplified and more easily understood by viewing the car as a well-designed assembly of simpler systems, each performing its own independent functions.

1.1 Body

The body is the basic building block. The two models covered in this manual feature unitized body construction, meaning that they do not have a separate frame. A complex body shell, shown in Fig. 1-1, is the main structural platform to which all the other systems are attached.

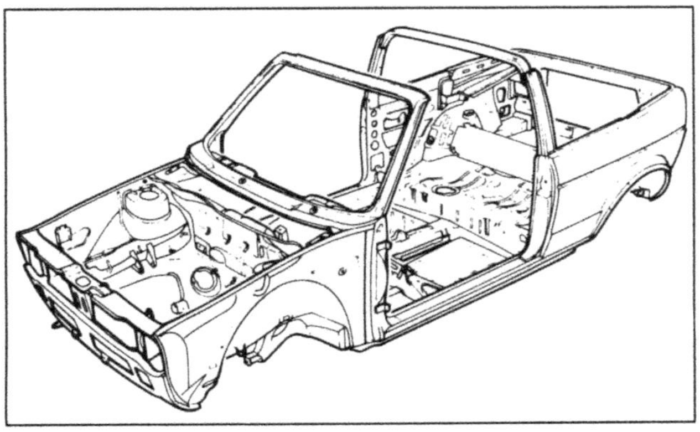

Fig. 1-1. Modern Volkswagen unitized body.

For comfort, convenience, and safety, the body itself has several additional subsystems and substructures. The doors, the instrument panel, the seats, and other interior trim pieces are all added to the basic body shell. Other parts of the body shell function as mounting points for the other major and minor subsystems, such as the engine, drivetrain, suspension, and steering. For more information, see **BODY AND INTERIOR**.

1.2 Engine

The engine is the system which produces the power to move the car. It does so by burning a precise mixture of fuel and air, converting the fuel's stored energy into mechanical work, and delivering that mechanical work in a useful form.

All of the Volkswagen engines covered in this manual are of reciprocating-piston design, and operate on the four-stroke cycle. The combustion of the air/fuel mixture creates tremendous pressure in a closed space above a piston, forcing the piston downward in its cylinder, and translating the energy of combustion into a mechanical force. The crankshaft converts each piston's motion into rotating motion, in much the same way that the up-and-down motion of a person's legs rotates the pedals of a bicycle. The power, transmitted in this rotary form, can then be used to move the car. The four-stroke cycle, the heart of how and why this all happens, is illustrated in Fig. 1-2.

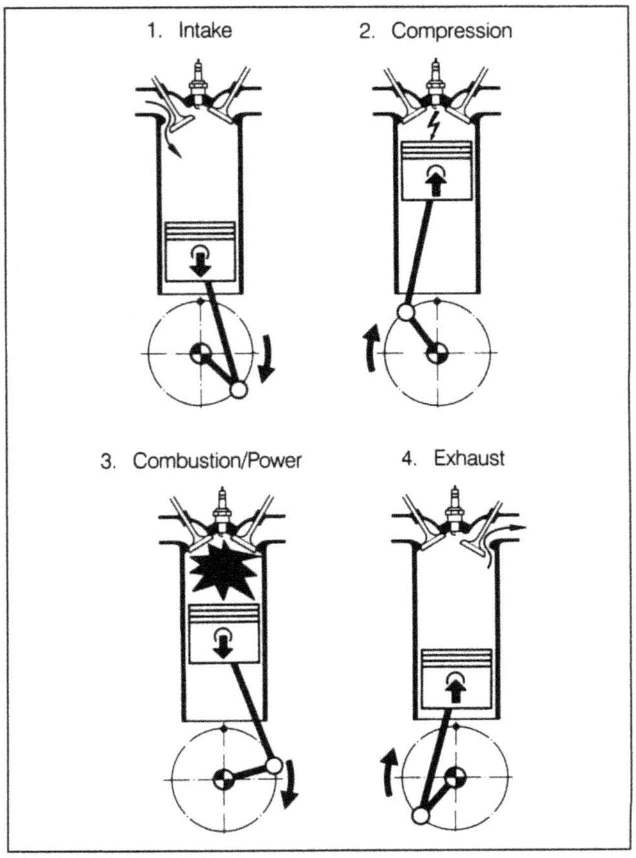

Fig. 1-2. The four-stroke cycle.

Intake Stroke. The piston, traveling downward, creates low pressure inside the cylinder. With the intake valve open, this low pressure causes the fresh air/fuel mixture to rush in. When the piston is near the bottom of its travel, the intake valve closes, leaving the cylinder sealed, full of air and vaporized fuel.

Compression Stroke. As the piston begins its upward travel in the sealed cylinder, the air/fuel mixture is compressed to a small percentage of its original volume, creating a very flammable mixture in a very small space, referred to as the combustion chamber. Just before the piston reaches the top of its travel, the air/fuel mixture is ignited with a precisely timed spark and begins to burn very rapidly.

Combustion or **Power Stroke**. As the confined air/fuel mixture burns, temperature and pressure rise very rapidly, forcing the piston downward, turning combustion energy into work. Generally, the faster an engine runs and the more often this combustion cycle happens, the more power is produced.

Exhaust Stroke. At the end of the power stroke, the piston is near the bottom of its travel and the cylinder is filled with the waste products of combustion which have little energy value and must be expelled to make way for a fresh load of air/fuel mixture. The exhaust valve opens and the piston, now traveling back upward, pushes the burned gasses out into the exhaust system. Near the top of the piston's travel, the exhaust valve closes, the intake valve opens, and the process begins anew with another intake stroke.

Engine Systems

The engine, which seems so complex, is a collection of simpler systems whose sole purpose is to efficiently perform the engine's power-making and power-transmitting functions. Virtually all automobile engines are multi-cylinder designs, in which a number of individual pistons and cylinders are joined together in a common housing or cylinder block, transmitting their power to a common crankshaft.

A camshaft, driven by the crankshaft, precisely times and controls the opening and closing of the intake and exhaust valves. Since each valve must cycle open and closed once for every two turns of the crankshaft, camshafts always turn at one-half crankshaft speed. The rest of the engine assembly is made up of systems which supply the essential fuel, air, and ignition, and provide for continuous operation of the pistons, crankshaft, and valves. A cutaway view of the engine is shown in Fig. 1-3. For more information, see **ENGINE**.

Ignition System

In the case of gasoline-fueled, spark-ignition engines, the ignition system creates the high-voltage spark necessary to ignite the combustible mixture in the cylinders. An ignition distributor, synchronized to the rotating parts of the engine, delivers the spark to the right cylinder at precisely the right time. Since each cylinder has to have a spark once for every two revolutions of the crankshaft, the distributor always turns at one-half crankshaft speed. The basic system is shown schematically in Fig. 1-4. For more information, see **IGNITION**.

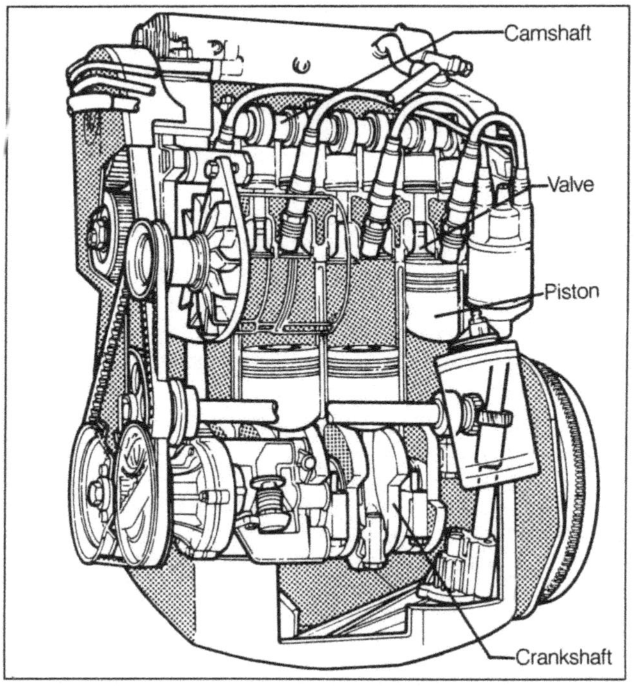

Fig. 1-3. Cutaway view of 4-cylinder overhead-cam engine.

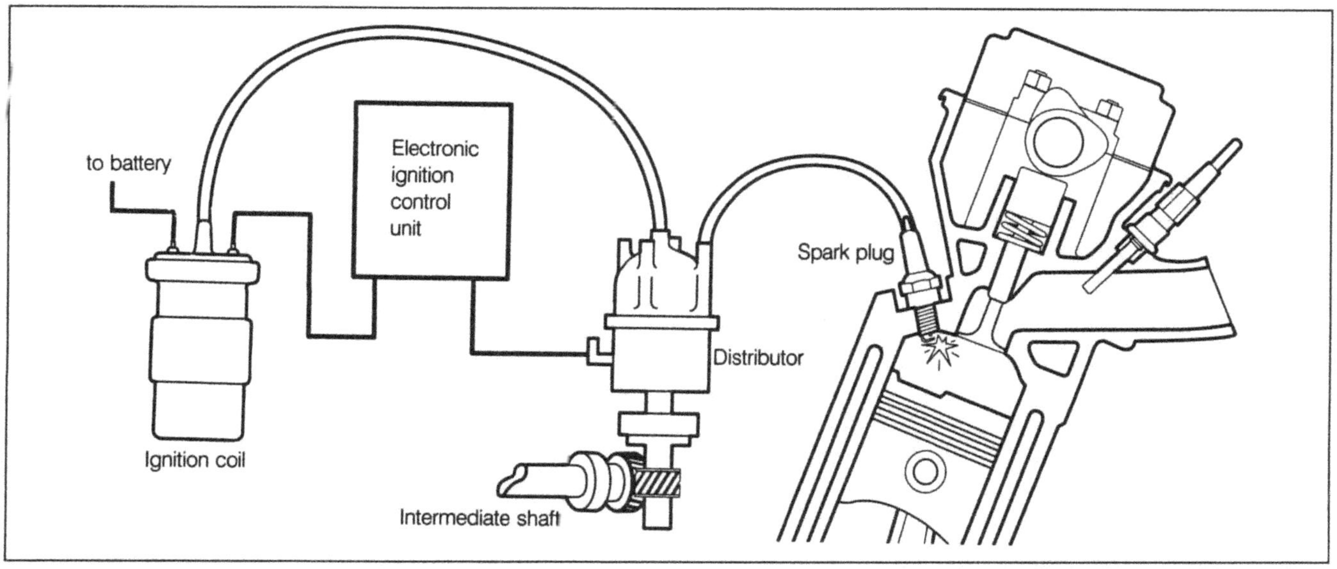

Fig. 1-4. Schematic representation of basic spark-ignition system. Ignition coil boosts voltage, electronic control unit and distributor control timing, and distributor delivers spark to spark plugs.

Fuel System

To run smoothly and produce power most efficiently, the engine requires the proper mixture of air and fuel—the proper air/fuel ratio. Depending on conditions, the optimum ratio for gasoline-fueled engines is about 15:1, fifteen parts of air for every one part of fuel. The throttle controls the amount of air entering the engine. The fuel system's job is to deliver and disperse fuel in the proper ratio to the incoming air.

Traditionally, the job of fuel delivery has been handled by a carburetor, a device carefully calibrated to dispense and atomize fuel in proportion to the amount of air passing through it. To meet the increasing demand for performance with economy and reduced exhaust emissions, many modern engines, including the Volkswagen engines covered in this manual, use a more sophisticated fuel injection system. It measures the incoming air more precisely, and in turn meters fuel more precisely for better control of the air/fuel ratio. This precise control means greater efficiency over a wider variety of operating conditions. In spite of all this sophistication, the fundamental task of the fuel injection system is still simply to control the ratio of fuel and air entering the engine's combustion chambers.

The fuel system of the automobile also includes a means of fuel storage, the fuel tank, and a network of pump and lines to transfer the fuel from the tank to the injection system.

For more information on the fuel system, see **FUEL SYSTEM**.

Cooling System

Even the most advanced engines lose some of their combustion energy as heat, which must be dissipated to prevent damage to the engine parts. Some heat is carried away in the exhaust, but much of it is absorbed by the valves, the pistons, and the rest of the combustion chamber. Most modern automobile engines are liquid-cooled, using a network of passages around the cylinders and combustion chambers, filled with circulating water-based coolant, to carry away heat. Circulation is provided by an engine-driven coolant pump, often called the water pump. The heat which the coolant absorbs from the hot engine is eventually dissipated to the surrounding atmosphere by the radiator at the front of the car. A smaller radiator-like heater core, located near the interior of the car, radiates heat to warm the passenger compartment. A basic cooling system layout, similar to that used on cars covered by this manual, is shown in Fig. 1-5.

Since some heat is necessary for the engine to run most efficiently, a thermostat in the cooling system restricts the flow of coolant through the radiator until the engine has reached normal operating temperature. For more information on the entire engine cooling system, see **COOLING SYSTEM**.

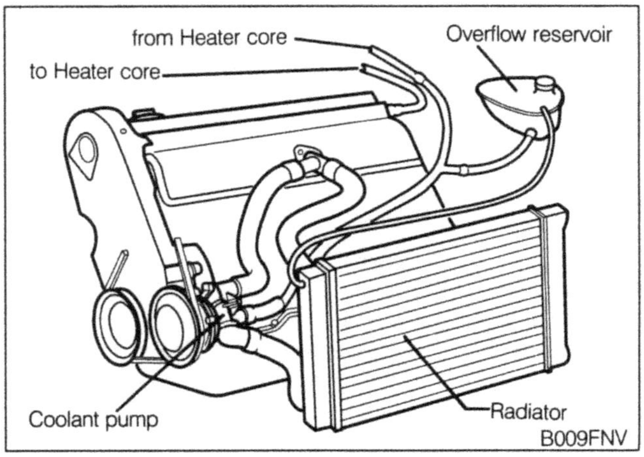

Fig. 1-5. Cooling system. Heat from engine is absorbed by coolant which is then circulated by coolant pump and cooled by radiator.

Lubrication System

The crankshaft and camshaft rotate at speeds up to several thousand revolutions per minute (rpm). Valves and pistons accelerate at tremendous rates, abruptly changing direction between velocities of hundreds of feet per second. In order to endure these harsh conditions, the parts are manufactured to exact dimensions, assembled with precision clearances, and lubricated by a pressurized oiling system. An engine-driven oil pump supplies oil under pressure to the engine where it is routed through a network of small passages which deliver it to each critical bearing surface.

A filter to clean the oil and a system to warn the driver of low oil pressure are part of the oil system. A secondary function of the oil in the lubrication system is to help carry away excess heat, and some lubrication systems contain an oil cooler for reducing oil temperature before it is recirculated back to the engine. More information on the lubrication system is found in **ENGINE**. For information on oil and oil filter replacement intervals, see **LUBRICATION AND MAINTENANCE**.

Exhaust System

The exhaust system serves several functions, but the primary one is to carry spent gasses from the engine and to route them away from the passenger compartment. Modern automobiles' exhaust systems include muffling devices to reduce noise, chemically reactive components to reduce harmful emissions, and sensors which can evaluate the exhaust gasses and provide information on engine efficiency to the fuel injection system. Fig. 1-6 shows a typical Volkswagen exhaust system. For more information, see **EXHAUST SYSTEM AND EMISSION CONTROLS**.

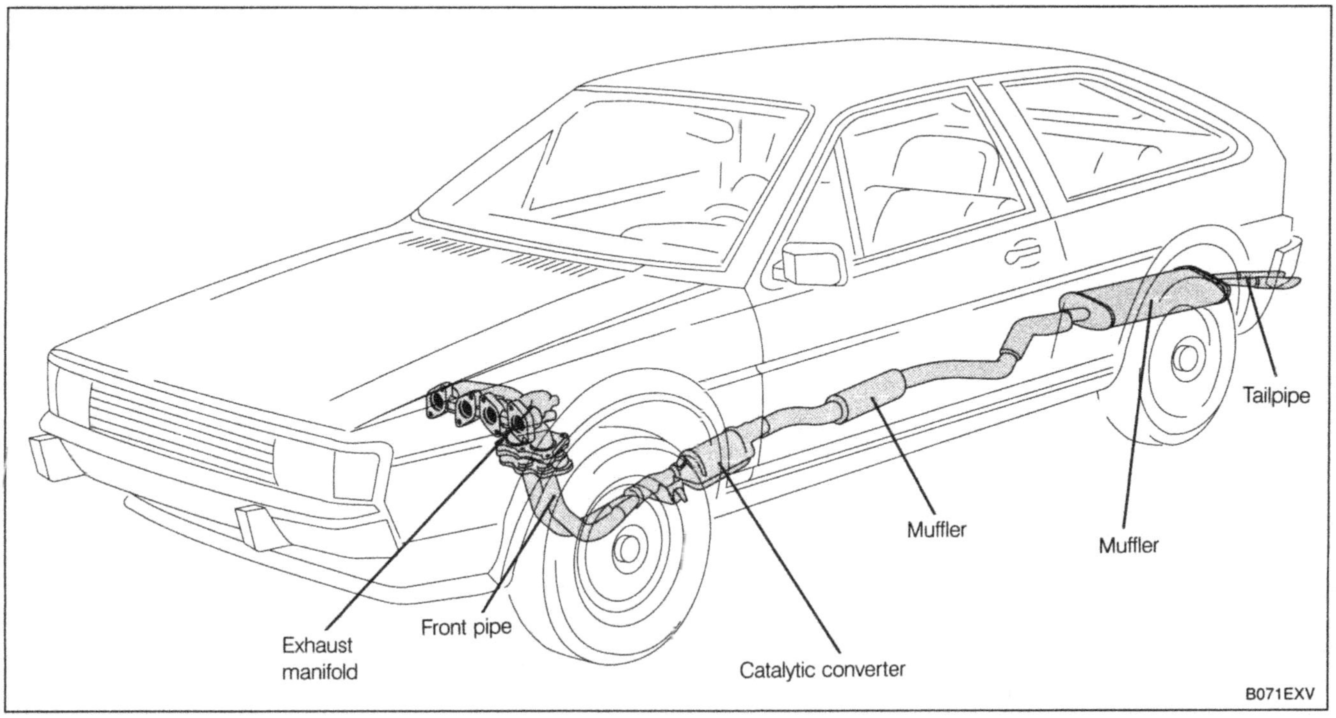

Fig. 1-6. Typical exhaust system. (Scirocco 16V shown; Cabriolet and Scirocco systems are similar).

1.3 Drivetrain

The drivetrain is a series of mechanisms which takes the power developed by the engine and delivers it to the wheels in order to actually move the car. It consists of the clutch or torque converter, the transmission, the final drive, the drive axles, and the wheels and tires. The Volkswagens covered by this manual, like most front-wheel-drive cars, combine the transmission and the final drive in one unit called a transaxle. A manual transaxle is shown in Fig. 1-7.

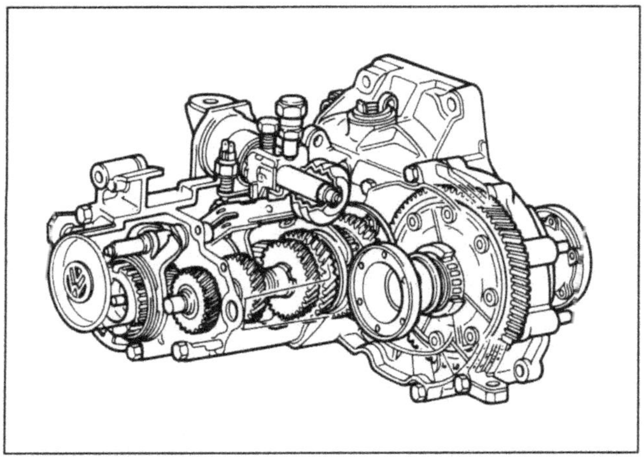

Fig. 1-7. Cutaway view of manual transaxle. Transmission and final drive are combined in one major assembly.

Clutch or Torque Converter

The clutch provides a way to connect or disconnect the engine and the rest of the drivetrain. A friction disc, attached to the input shaft of the transmission, gets squeezed between the heavily spring-loaded clutch and the flywheel, both positively attached to the engine. Normally, the friction between the clutch, friction disc, and flywheel, aided by the heavy spring, makes the transmission input shaft turn at the same speed as the engine. For stopping, starting, and shifting gears, depressing the clutch pedal works against the clutch spring, reducing the friction bond and disconnecting the engine from the drivetrain. In normal use the clutch and flywheel wear, much like brakes, and need periodic maintenance or replacement. More information on the clutch is found in **MANUAL TRANSMISSION AND CLUTCH**.

Cars with automatic transmission achieve the same result with a sophisticated fluid clutch called a torque converter. With one part attached to the engine and another to the transmission, power is transmitted between the two by the action of a viscous fluid. The design of the torque converter is such that it allows the engine to turn at idle speeds without transmitting much driving force to the transmission. A slight forward tug may be noticed when the car is in a forward gear, but at idle speed it is barely enough to drive the car. Above idle speeds, the torque converter becomes increasingly resistant to slip, and transmits power through the fluid coupling to the transmission, delivering power to the wheels. Fig. 1-8 is a cutaway view of the automatic transaxle showing the torque converter. For more information, see **AUTOMATIC TRANSMISSION**.

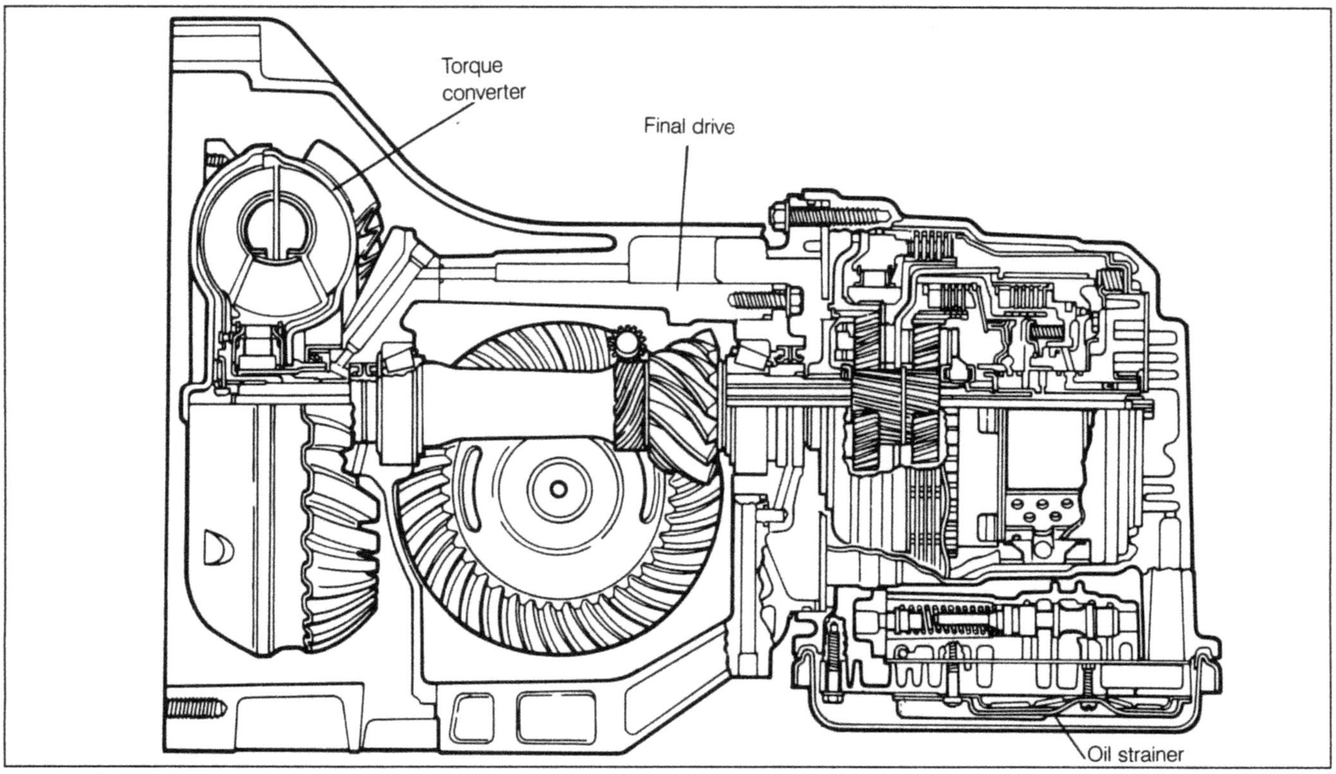

Fig. 1-8. Cutaway view of automatic transaxle showing torque converter, automatic transmission, and final drive.

Transmission

Although the engine develops a substantial amount of power, it does so best at relatively high revolutions per minute (rpm). To handle all driving conditions, it is necessary to use gearing to change the ratio of engine rpm to vehicle speed. A manual transmission is simply a mechanism which contains several sets of gears in a compact assembly, each suited to a particular range of driving speeds, and a mechanism to allow the driver to change from one gear ratio to the next on the move as vehicle speed requirements change. For information on manual transmission maintenance and adjustments, see **MANUAL TRANSMISSION AND CLUTCH**.

An automatic transmission has the same purpose, but uses hydraulic fluid under pressure in a complex network of passages, valves, and hydraulically-operated control mechanisms to engage and disengage constantly meshed planetary gear sets. Hydraulic controls select the appropriate gear based on vehicle speed, engine load, throttle position, and gear shift position. For information on automatic transmission maintenance and adjustments, see **AUTOMATIC TRANSMISSION**.

Final Drive

The final drive is another gearset which delivers transmission output to the drive axles, and includes the differential. When a car turns, the wheels on the outside of the turn have to turn slightly faster than those on the inside, since they have to travel a larger arc in an equal amount of time. The drivetrain must be able to transmit power to the wheels and still allow for these variations in wheel speed when cornering. The device which allows this, called the differential, uses a system of gears to allow a speed differential between wheels on opposite sides of the car.

Drive Axles

The final step in the transfer of power from the engine to the wheels is the drive axles, which provide a connection between the differential and the wheel hubs. Information on the drive axles is found in **SUSPENSION AND STEERING**.

1.4 Suspension and Steering

The suspension and steering systems are what allow the wheels to move independently of the body structure for improved ride, stability, and directional control. In addition, these systems limit and control that movement.

The suspension system is the combination of springs, shock absorbers, and other stabilizing devices which separate the wheels from the body structure and support its weight. The springs cushion the effects of bumps. For added control, the suspension system also includes dampers, or shock absorbers, to resist excessive movement of the springs. Volkswagen's strut-type suspension, like that of many modern cars, combines the spring and shock absorber into a single unit, performing the same jobs in less space with fewer individual components. Stabilizer bars aid stability by causing the forces acting on the suspension to be more uniformly distributed.

The remainder of the suspension system are the parts that link it all together, designed with bushings, bearings, and joints which purposefully allow or restrict movement.

The steering system is an assembly of gearbox mechanism and linkage which translates the rotating motion of the steering wheel into the side-to-side motion of the front wheels. Cars covered by this manual use a rack-and-pinion type of steering mechanism. A pinion gear, connected to the steering column, moves a toothed rack from side to side. The rack can be thought of as a long, straight gear. The entire assembly is also referred to as the steering rack. Power-assisted steering uses hydraulic fluid under pressure to do some of the work normally done by the driver turning the steering wheel.

Fig. 1-9 and Fig. 1-10 show the typical compact system of steering and front and rear suspension, and their proximity to other systems in the car. For more information, see **SUSPENSION AND STEERING**.

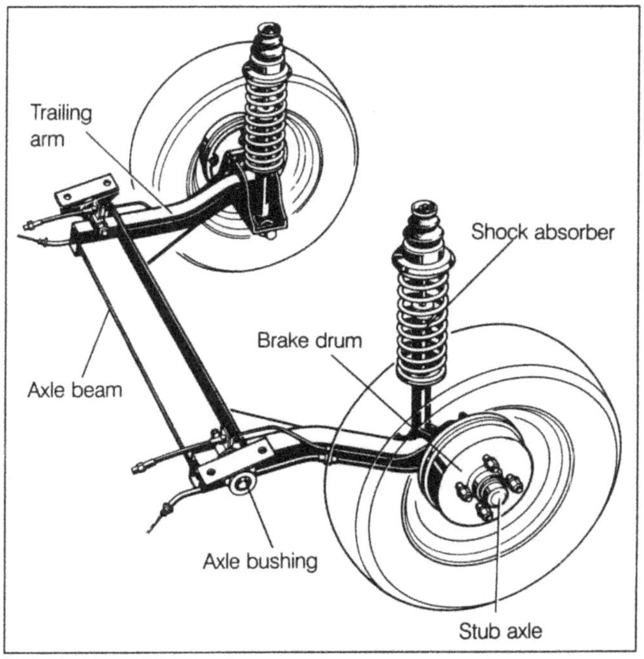

Fig. 1-9. Rear suspension of a front-wheel-drive car showing integral axle and trailing arms.

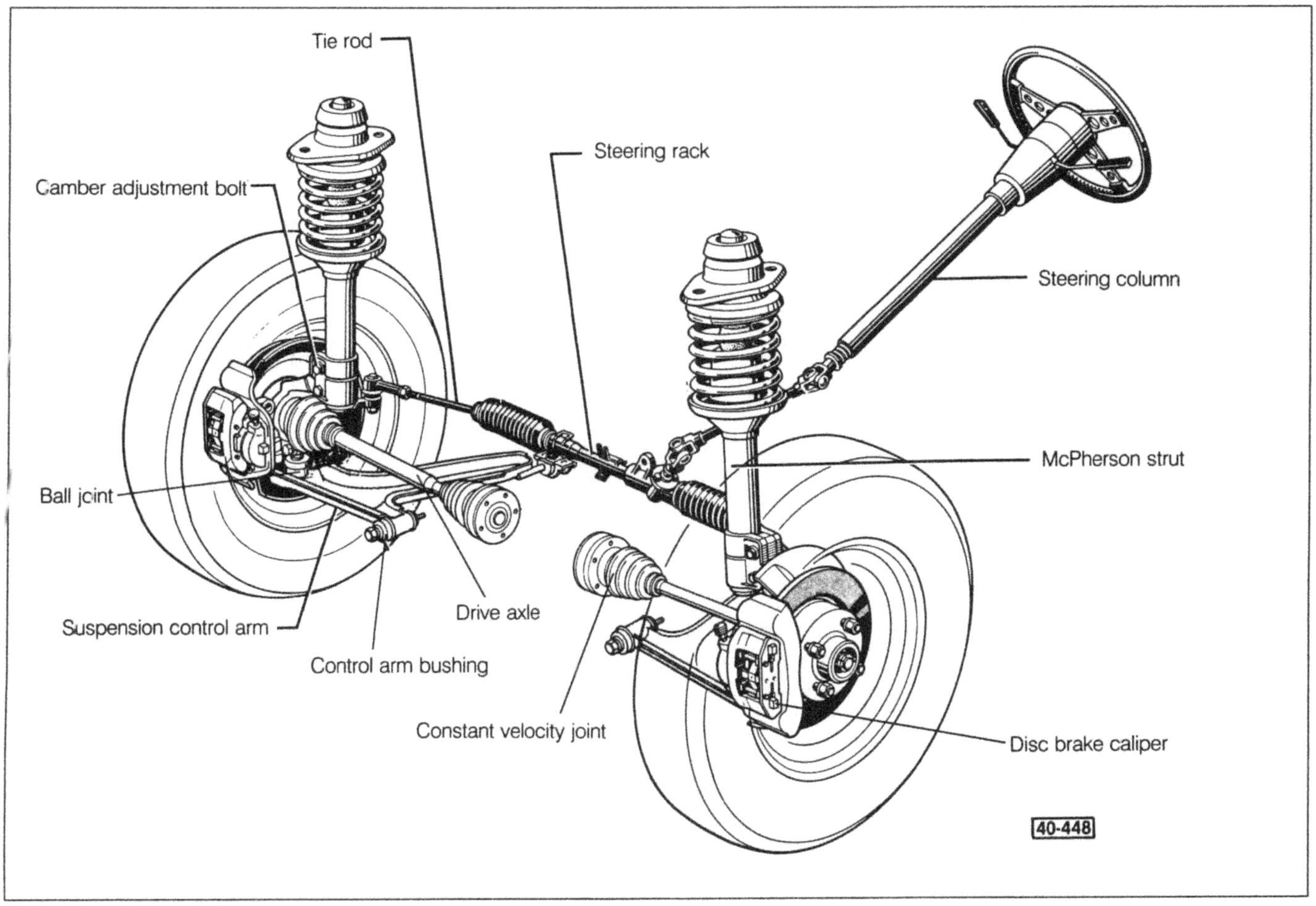

Fig. 1-10. Front suspension and steering of a front-wheel-drive car, showing drive axles and brakes.

1.5 Brakes

The system for slowing and stopping the car is, not surprisingly, completely independent of the systems which make it go. Although the brakes are located at the wheels where they can act most directly, and they mount to parts of the suspension system, the brakes and the suspension systems are separate systems.

The brakes act to slow or stop the car by causing friction. Since cars are relatively heavy, the amount of friction required to stop it safely and effectively is quite high, and requires considerable force. Virtually all modern cars use a system of hydraulics to multiply the force applied to the brake pedal and to distribute it uniformly to the four wheels.

The brake pedal is connected by a mechanical linkage to the first major hydraulic component, the master cylinder, usually mounted on the firewall at the back of the engine compartment. A piston in the master cylinder creates hydraulic pressure in the brake lines going to the wheels. Because the brakes are located at the wheels and move relative to the body, the final length of brake line at each wheel is flexible, so that it follows the movement. At each wheel, the hydraulic pressure acts on the brake mechanism to cause friction and slow the wheel. The hydraulic components are sized in such a way that the driver's force applied to the brake pedal is highly multiplied by the time it acts on the wheels.

Two types of brake mechanisms are in common use in modern cars. Both create braking friction by forcing a stationary friction material against a larger, rotating member attached to the wheel. A drum brake forces semi-circular shoes, lined with friction material, against the inside of a round brake drum. A disc brake squeezes pads lined with friction material against both sides of a flat, round brake disc. Both types are shown below in Fig. 1-11 and Fig. 1-12. A disc brake is generally capable of generating higher braking forces for a given size, is self-adjusting and self-cleaning, and dissipates heat more easily. Heat is a major enemy of brake efficiency, reducing the friction material's ability to grip. Under extreme conditions, excess heat from repeated heavy braking can cause the brake fluid to boil, resulting in severely diminished braking capability, called brake fade. Disc brakes, because of their greater ability to dissipate heat, are more resistant to brake fade.

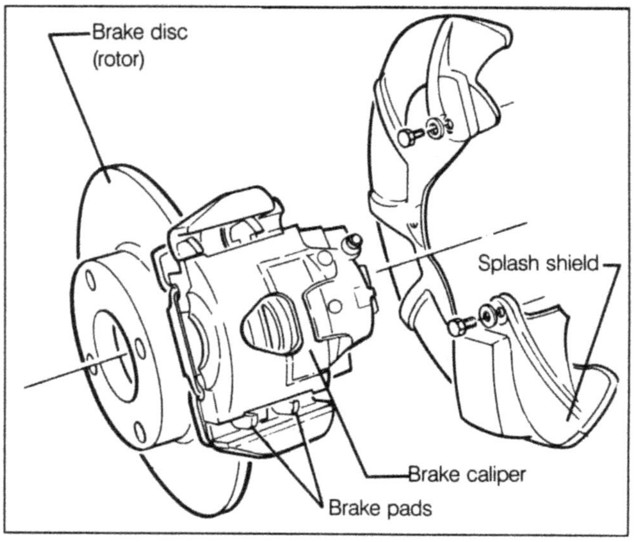

Fig. 1-11. Disc brake assembly showing disc, caliper, and splash shield. Caliper holds pads with friction material.

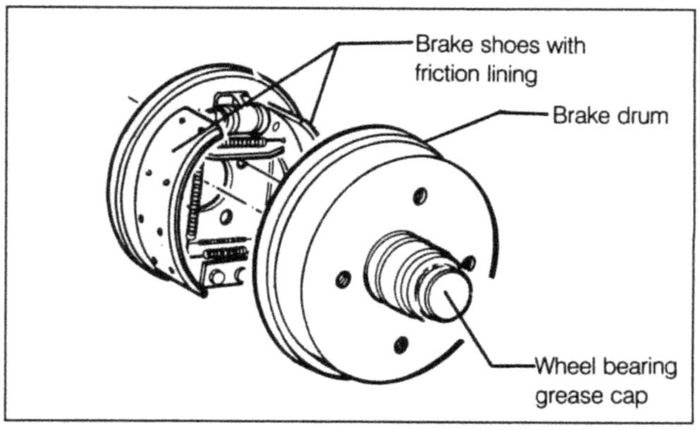

Fig. 1-12. Drum brake assembly with brake drum removed to show brake shoes. Friction material on brake shoes contacts inner surface of brake drum.

1.6 Electrical System

Many components, including accessory items, are powered electrically. The power comes from the car's electrical system. The electrical system uses a battery to store energy, an engine-driven alternator to generate electricity and recharge the battery, and various wiring harnesses and other circuits to distribute electric power to the rest of the car. The electrical system is represented in Fig. 1-13.

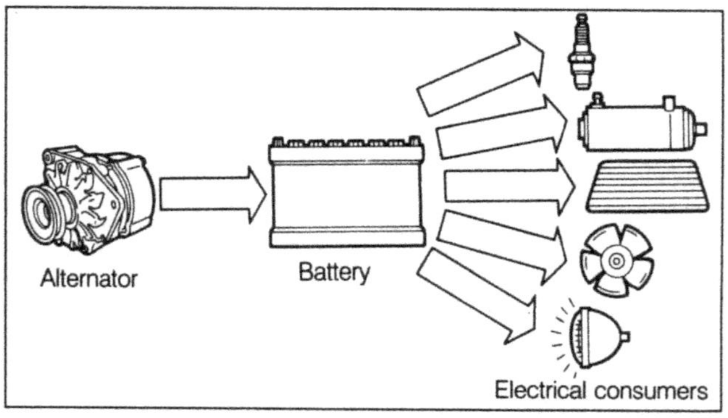

Fig. 1-13. In electrical system, alternator generates electricity to recharge battery and power other electrical consumers.

Battery

Almost every electrical component in the car operates from 12 volt direct current (DC). The battery converts electrical energy into chemical energy for storage, and converts its stored chemical energy back into electrical energy on demand.

Alternator

Left alone to meet all the electrical demands of an automobile, the battery would soon be completely discharged, so the electrical system includes an alternator. Once the running engine is turning the alternator, it takes over supplying electrical energy to the various electrical components. When the power generating capability of the alternator exceeds demand, as it usually does at driving speeds, the extra energy recharges the battery to its full potential.

Wiring Harness and Circuits

The flow of electricity depends upon a closed-loop path—a complete circuit. Electrical current flows through wires to the consumer, a light bulb for example, and back to the battery in a complete circuit. The electrical route back to the source, which completes the circuit, is called a path to ground. Every consumer of electrical power in the car must have a source of power and a path to ground in order to operate. In this manual, the electrical circuits are represented by current flow diagrams which illustrate electrical current flow from voltage source to ground for each circuit.

Commonly, the electrically conductive metal structure of the automobile is used as a ground path. The negative (–) terminal of the battery connects to the car body, and all of the electrical consumers in the car make a ground connection to the car body, thus eliminating the need for many feet of additional wire. Electrical components near the engine are often grounded directly to the engine, which is then grounded to the body. Some components are grounded through their housings which are bolted to a ground. Electrically, the effect is the same.

2. How To Use This Manual

The manual is divided into 13 sections: **FUNDAMENTALS**, **LUBRICATION AND MAINTENANCE**, and 11 repair sections, each covering a particular system or portion of the car. Thumb-tabs on the page margins help locate each section. A page listing section titles and showing their thumb-tab locations is near the front of the manual. An index is located at the back of the manual.

Each section has a Table of Contents listing the major subject headings within the section, and the pages on which they begin. Page numbers always refer to pages within the same section. References to numbered headings always refer to headings in the same section. Reference to a procedure in another section is by section title only.

2.1 Fundamentals

This first section is **FUNDAMENTALS.** It contains basic information on equipment and safety which is important to any do-it-yourselfer, regardless of experience, as well as information on getting started and helpful suggestions for the novice. Anyone can use this manual. This section helps show how.

2.2 Lubrication and Maintenance

LUBRICATION AND MAINTENANCE is the section dedicated to taking care of the car and preventing future problems. Volkswagen specifies certain periodic maintenance to prevent trouble and keep the car at its best. This section details those maintenance tasks, shows how they are done, and tells what is needed to do them.

2.3 Repair Sections

The 11 repair sections give more involved and more detailed information about system function, troubleshooting, and repair. For clarity and ease of use, each repair section begins with **1. General Description**, **2. Maintenance**, and **3. Troubleshooting.**

General Description

The General Description is an overview of the system's technical features. It describes the general layout and function of the system, discusses unique aspects of different versions, and gives information on identifying each version and the repair information which applies to it.

Maintenance

Maintenance is a brief checklist of all the routine maintenance tasks which pertain to the system being discussed. Each listed maintenance item also includes a reference to the part of the manual which covers the work procedure.

Troubleshooting

A systematic approach to problem solving, based on observing trouble symptoms and isolating their causes, is called troubleshooting. The Troubleshooting heading of each repair section begins with a discussion of the system's basic operating principles. Following that general discussion is a more specific list of symptoms—particular problems which affect the car—and their probable causes. Suggested corrective actions include references to the numbered heading or section where the repair information can be found.

2.4 Index

A comprehensive index is found at the back of the manual. Each index entry is followed by a page reference giving the section and the section page number. For example, **3-16** refers to section three, **ENGINE**, and page 16 of that section.

2.5 Notes, Cautions, and Warnings

Throughout this manual are many passages with the headings **NOTE**, **CAUTION**, or **WARNING**. These very important headings have different meanings.

> **NOTE —**
> A note contains helpful information about the job, tips which will help in doing a better job and completing it more easily.

> ***CAUTION —***
> *A caution calls attention to important precautions to be observed during the repair work which will prevent accidently damaging the car.*

> ***WARNING —***
> *A warning is the most serious of the three. It warns of unsafe practices which will very possibly cause injury, either by direct threat to the person(s) doing the work or by increased risk of accident or mechanical failure while driving.*

Please read every **NOTE, CAUTION**, and **WARNING** at the front of the manual and as they appear in repair procedures. They are very important. Read them before you begin any maintenance or repair job. Some **CAUTION**s and **WARNING**s are repeated wherever they apply. Read them all and do not skip any. These messages are important, even to the owner who never intends to work on the car.

3. GETTING STARTED

Most of the necessary maintenance and minor repair that a Volkswagen will need through the years can be accomplished with ordinary tools, even by owners with little or no experience in car repair. Below is some important information on how to work safely, a discussion of what tools will be needed and how to use them, and a series of mechanic's tips on methods and workmanship.

3.1 Safety

Although an automobile presents many hazards, common sense and good equipment can ensure safety. Accidents happen because of carelessness. Pay attention, and stick to these few important safety rules.

> ***WARNING —***
> - *Never run the engine in the work area unless it is well-ventilated. The exhaust should be ducted to the outside. Carbon Monoxide (CO) in the exhaust kills.*

> ***WARNING —***
> - *Remove all neckties, scarfs, loose clothing, or jewelry when working near running engines or power tools. Tuck in shirts. Tie long hair and secure it under a cap. Severe injury can result from these things being caught in rotating parts.*
> - *Remove rings, watches, and bracelets. Aside from the dangers of moving parts, metallic jewelry conducts electricity and may cause shorts, sparks, burns, or damage to the electrical system when accidently contacting the battery or other electrical terminals.*
> - *Disconnect the battery negative (–) terminal whenever working on the fuel system or anything that is electrically powered. Accidental electrical contact may damage the electrical system or cause fire.*
> - *Never work under a lifted car unless it is solidly supported on stands designed for the purpose. Do not support a car on cinder blocks, hollow tiles, or other props that may crumble under continuous load. Never work under a car that is supported solely by a jack.*
> - *If you are going to work under a car on the ground, make sure that the ground is level. Block the wheels to keep the car from rolling. Never work under the car while the engine is running. Disconnect the battery ground strap to prevent others from starting the car while you are under it.*
> - *The fuel system is designed to retain pressure even when the ignition is off. When working with the fuel system, loosen the fuel lines very slowly to allow the residual pressure to dissipate gradually. Avoid spraying fuel.*
> - *Fuel is highly flammable. When working around fuel, do not smoke or work near heaters or other fire hazards. Keep an approved fire extinguisher handy.*
> - *Illuminate the work area adequately and safely. Use a portable safety light for working inside or under the car. A fluorescent type is best because it gives off less heat. If using a light with a normal incandescent bulb, use rough service bulbs to avoid breakage. The hot filament of an accidently broken bulb can ignite spilled fuel or oil.*
> - *Keep sparks, lighted matches, and open flame away from the top of the battery. Hydrogen gas emitted by the battery is highly flammable. Any nearby source of ignition may cause the battery to explode.*
> - *Never lay tools or parts in the engine compartment or on top of the battery. They may fall into confined spaces and be difficult to retrieve, become caught in belts or other rotating parts when the engine is started, or cause electrical shorts and damage to the electrical system.*
> - *Do not attempt to work on your car if you do not feel well. You increase the danger of injury to yourself and others if you are tired, upset or have taken medicine or any other substance that may impair you from being fully alert.*

Lifting The Car

For those repairs that require raising the car, the proper jacking points should be used to raise the car safely and avoid damage. To use the jack supplied with the car by Volkswagen for changing wheels, there are four jack points—two on each side of the car—marked by dimples in the lower panel sheet metal just behind the front wheel or just in front of the rear wheel. See Fig. 3-1.

Fig. 3-1. Jacking point for use with Volkswagen-supplied jack. Dimple in lower panel marks exact point just behind front wheel or just ahead of rear wheel.

To lift the car with a floor jack or hydraulic lift, use the lifting points shown in Fig. 3-2 and Fig. 3-3. The lifting points are located on the underbody between the wheel wells and the dimples in the lower panel sheet metal, just behind the front wheel or just in front of the rear wheel.

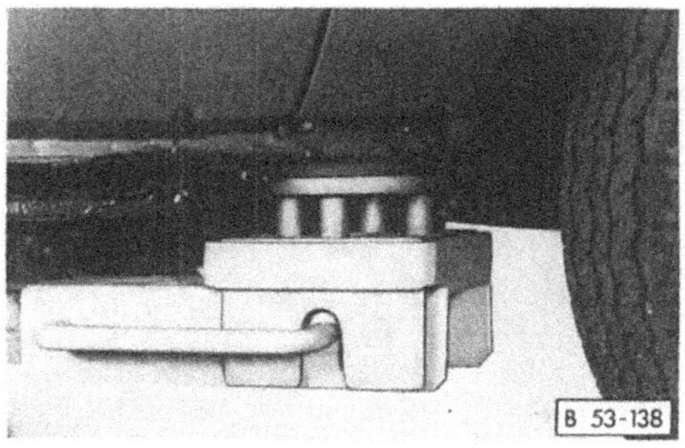

Fig. 3-2. Front lifting point for use with floor jack or hydraulic lift. Lift on underbody under front door hinge pillar. (Scirocco shown; Cabriolet is similar).

Fig. 3-3. Rear lifting point for use with floor jack or hydraulic lift. Lift on point in front of rear wheel well. (Scirocco shown; Cabriolet is similar).

To raise the car safely:

1. Park the car on a flat, level surface.
2. Place the jack in position. Make sure the jack is resting on flat, solid ground. Use a board or other support to provide a firm surface for the jack, if necessary.
3. Raise the car slowly.

> ***WARNING —***
> *Watch the jack closely. Make sure that it stays stable and does not shift or tip. As the car is raised, the car, or the jack if it is on wheels, will want to roll slightly.*

4. Once the car is raised, block the wheel that is opposite and farthest from the jack to prevent the car from unexpectedly rolling.

> ***WARNING —***
> *Do not rely on the transmission or the emergency brake to keep the car from rolling. While they will help, they are not a substitute for blocking the opposite wheel.*

> ***WARNING —***
> *Never work under car that is supported only by a jack. Use jack stands which are properly designed to support the car. See* **4. Tools**.

To work safely under a car:

1. Disconnect the battery negative (-) terminal so that no one else can start the car. Let others know what you will be doing.
2. Use at least two jack stands to support the car. A jack is a temporary lifting device and should not be used alone to support the car while you are under it. Use positively locking jack stands which are designed for the purpose of supporting a car. For more information on jack stands, see **4. Tools**.

> ***WARNING —***
> *Do not use wood, concrete blocks, or bricks to support a car. Wood may split. Blocks or bricks, while strong, are not designed for that kind of load, and may break or collapse.*

3. Place jack stands on a firm, solid surface, just like the jack. If necessary, use a flat board or similar solid object to provide a firm footing.

4. After placing the jack stands, lower the car slowly until its weight is fully supported by the jack stands. Watch to make sure that the jack stands do not tip or lean as the car settles on them, and that they are placed solidly and will not move.

5. Observe all jacking precautions again when raising the car to remove the jack stands.

3.2 General Advice For The Beginner

The tips in the paragraphs which follow are general advice to help any do-it-yourself Volkswagen owner perform repairs and maintenance tasks more easily and more professionally.

Planning Ahead

Most of the repairs and maintenance tasks described in this manual can be successfully completed by anyone with basic tools and abilities. Some cannot. To prevent getting in too deep, know what the whole job requires before starting. Read the procedure thoroughly, from beginning to end, in order to know just what to expect and what parts will have to be replaced.

Cleanliness

Keeping things organized, neat, and clean is essential to doing a good job, and a more satisfying way to work. When working under the hood, use fender covers to protect the finish from scratches and other damage. Make sure the car is relatively clean so that dirt under the cover does not scratch.

Avoid getting tools or clothing near the battery. Battery electrolyte is a corrosive acid. Be careful with brake fluid. If it comes in contact with paint, it will cause permanent damage. Finally, keep rubber parts such as hoses and belts free from oil or gasoline, as they will cause the material to soften and fail prematurely.

Tightening Bolts

When tightening the bolts or nuts which attach a component, it is always good practice to tighten the bolts gradually and evenly to avoid overstressing any one portion of the component. For components sealed with gaskets, this method helps to ensure that the gasket will seal properly and completely. With several fasteners, tighten them in a sequence alternating between opposite sides of the component. Fig. 3-4 shows such a sequence for tightening six bolts attaching an imaginary component.

For some repairs, a specific tightening sequence is necessary, or a particular order of assembly is required. Such special conditions are noted in the text, and the necessary sequence is described or illustrated.

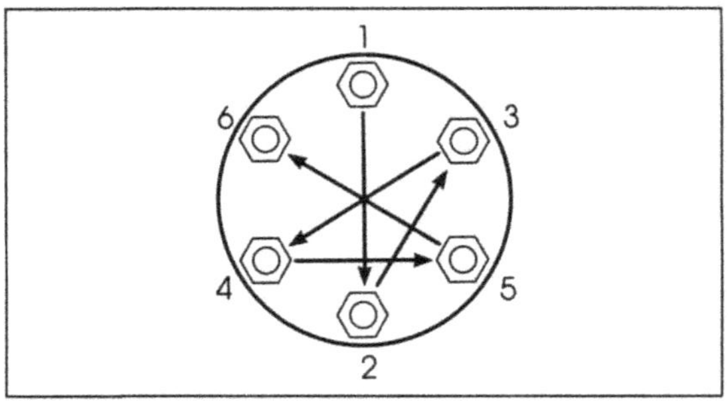

Fig. 3-4. Sequence for tightening alternating opposite fasteners attaching component. Repeat sequence until all are evenly tightened to desired specification.

Bolt Torque

Tightening bolts and nuts to a specified torque value, as measured with a torque wrench, is a way to ensure that they are tight enough without being in danger of going too far and breaking the fastener or stripping the threads.

Vibration of assembled parts causes stress. A fastener which is simply holding things together will be subjected to stress alternating in opposite directions which will eventually cause it to become loose. To counter this loosening, fasteners are tightened more, and actually stretched, in order to pre-stress them. When tightened this way, they are always stressed in one direction, in spite of vibration, and will not work loose.

The proper torque for a fastener is related to the amount of stretch necessary to prevent the fastener from working loose in normal use. To do the best work, use a torque wrench and follow Volkswagen's torque specifications whenever possible. See **4. Tools** for more information on torque wrenches.

Gaskets

The smoothest metal mating surfaces still have imperfections which can allow leakage. As an aid to preventing leakage at critical joints, gaskets of soft, form-fitting material are used to fill in the imperfections.

To be most effective at their task, gaskets are designed to "crush", to become thinner as they are pressed together between mating parts. Once a gasket has been used and crushed, it is no longer capable of making as good a seal as when new, and is much more likely to leak. For this reason, gaskets should not be reused. Always plan to use new gaskets for any reassembly.

This same logic applies to any part used for sealing, including rubber O-rings and copper sealing washers.

Seals

In places where a shaft must pass through a housing, flexible lip seals are used to keep the lubricating oil or grease from leaking out past the rotating shaft. With a little consideration for the seals and the job they have to do, future leaks can be avoided.

Seals are designed to be installed in the housing only once, and should never be reused once they have been removed. As long as they are not removed from the housing and not leaking, they need not be replaced. However, there is no easier time to replace seals than when the car is already apart for some other repair.

When doing repairs that require removing a seal, be very careful not to scratch or otherwise damage the metal sealing surfaces. Even minor damage can cause seal damage and leakage.

The key to seal installation is to get the seal in straight and without damage. Use an object that is the same diameter as the seal to gently and evenly drive it into place. A socket of the right size works well. Coat the entire seal with a little grease or oil to help it go in more easily. Seals are directional. Make sure that it is being installed with the lip facing the correct way. Normally the lip faces the inside. Note the installation direction of the old seal before removing it.

Wire Repairs

Repairs to a wiring harness to reconnect broken wires or correct shorts to ground deserve special care to make the repair permanent.

The wire ends must be clean. If frayed or otherwise damaged, cut off the end. If it is necessary to maintain proper length, splice in a new piece of wire of the same size and make two connections.

Use connectors which are designed for the purpose. Crimped-on or soldered-on connectors are best. Crimp connectors and special crimping pliers can be found at most electronics supply or hardware stores. If soldering, use needlenose pliers to hold the wire near the solder joint and create a "heat dam", keeping the solder from "wicking" up the wire. Use a solder specifically for electronics, without acid flux which will promote corrosion. Twisting wires together is a temporary repair at best, since corrosion will eventually spoil the connection.

Insulate the finished connection. Electronics stores can supply heat-shrinkable insulating tubing which can be placed onto the wire before connecting, slid over the finished joint, and shrunk to a tight fit with a heat gun or hair dryer. The next best alternative is electrical tape. Make sure the wire is clean and free of solder flux or other contamination. Wrap the joint tightly and completely to seal out moisture.

Cleaning

Any repair job is more enjoyable and less likely to be troublesome if the parts are clean. For cleaning old parts, there are any number of solvents and parts cleaners available commercially.

For cleaning parts prior to assembly, commercially available aerosol cans of carburetor cleaner or brake cleaner are handy to use, and the cleaner will evaporate completely, leaving no residue.

WARNING —
Virtually all solvents used for cleaning parts are highly flammable, especially in aerosol form. Use with extreme care. Do not smoke. Do not use these products near any source of sparks or flame.

Let any solvent or cleaning product dry completely. Low-pressure, dry compressed air is helpful if available. Also, use only lint-free rags for cleaning and drying.

WARNING —
When drying roller or ball bearings with compressed air, do not allow them to spin. Unlubricated, they may fail and come apart, causing injury.

Electrical Testing

A great many electrical problems can be understood and solved with only a little fundamental knowledge of how electrical circuits function.

Electric current only flows in a complete circuit. To operate, every electrical device in the car requires a complete circuit including a voltage source and a path to ground. The positive (+) side of the battery is the original voltage source, and ground is any return path to the negative (−) side of the battery, whether through the wiring harness or the car body. Except for portions of the charging system, all electrical current in the car is direct current (DC) and flows from positive (+) to negative (−). Switches are used to turn components on or off by completing or interrupting the circuit. A switch is "open" when the circuit is interrupted, and "closed" when the circuit is completed. Fig. 3-5 shows a complete circuit schematically.

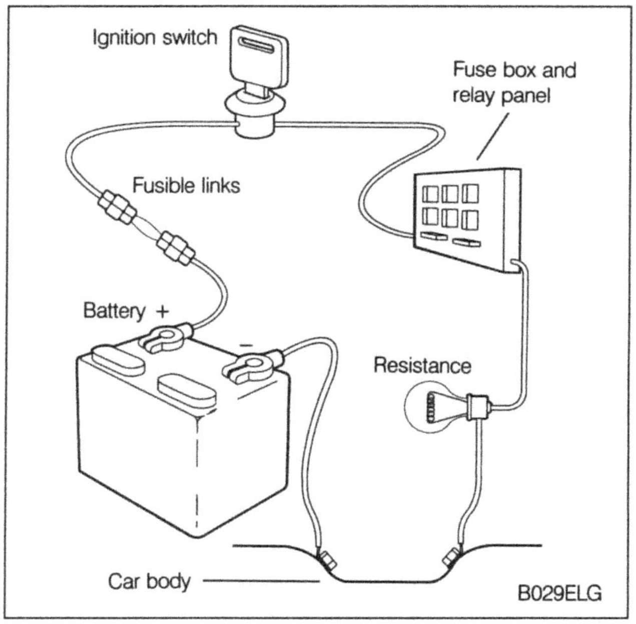

Fig. 3-5. Schematic representation of simple circuit for light bulb.

The first step in tracing an electrical problem is to check with a test light, a voltmeter, or a multimeter (DC volts scale) to see that voltage is reaching the component. If so, then the circuit is sound as far as that point. If voltage is not reaching the component, that indicates that the circuit is open (interrupted) somewhere between the battery positive (+) terminal and the component. Look for a blown fuse, an open switch, a broken wire, or a failed component earlier in the same circuit. Isolate the location of the problem by doing more voltage measurements at different points in the circuit. Voltage found at any point in the circuit means that the circuit is good at least up to that point. Look for problems after the point in the circuit where voltage is found.

To test for voltage:

1. Start with an electrical connection closest to the component in question. If necessary, remove the terminal cover or disconnect the harness connector.

2. Connect the clip lead of a test light or the black (−) probe of a voltmeter or multimeter to ground (any clean, unpainted metal part of the engine or car).

3. Touch the probe of the test light or the red (+) probe of the meter to the terminal being tested. A meter reading or the test light lighting indicates that voltage is present in the circuit up to that point.

4. Double check a no-voltage indication by wiggling the probes and connections to make sure adequate contact is being made.

Once it is confirmed that voltage is reaching the component, check the remainder of the circuit by testing for continuity to ground. If the component is grounded through its physical mounting, make sure that the contact area is clean, dry, and free of corrosion.

To check component ground:

1. If the component is grounded through the wiring harness (usually a brown or brown striped wire), disconnect the harness connector or the ground wire. Connect the clip lead of a test light or the black (−) probe of a meter to the removed ground terminal.

2. If the component is grounded by its mounting to the car, connect one end of a test light or the black (−) probe of a meter to the clean metal surface of the component.

3. Briefly touch the remaining probe to a known source of battery voltage. If the previous test for voltage reaching the component was successful, that terminal is a suitable connection.

4. A meter reading or the test light lighting indicates current flow in a complete circuit and, therefore, a good connection to ground.

A continuity test, performed with an ohmmeter, is a universal test of any wire, connection, or component which will tell if current can flow through it. It can be used to check wires for breaks, to find out whether switches are open or closed, to find poor connections, and many other things. Continuity is a measure of resistance. A complete circuit has continuity—resistance is nearly zero. An open circuit due to a broken wire or an open switch has infinite resistance—there is no continuity.

To test the continuity of any conductor between any two points, connect one ohmmeter probe to each test point. If the path between points is uninterrupted, the ohmmeter should read nearly zero. If it reads significantly above zero, there may be a component in the circuit which is supposed to have some resistance. If not, there may be a poor connection or a damaged wire somewhere between the test points.

Making An LED Test Light

Many of the electrical tests in this manual require the use of a special LED test light, since the use of a more conventional test light with incandescent bulb can damage sensitive electronic circuits in the ignition, fuel injection, and emission control systems. One suitable test light is Volkswagen special tool no. US 1115 (order no. TU1 115 000 28 ZEL).

A lower-cost alternative is to make an LED test light, using parts available from an electronics supply outlet. Assemble the components as shown in Fig. 3-6. Use a needlenose plier to hold the parts and to act as a heat dam while soldering, as described above in **Wire Repairs**. Insulate all connections with heat-shrinkable tubing or electrical tape.

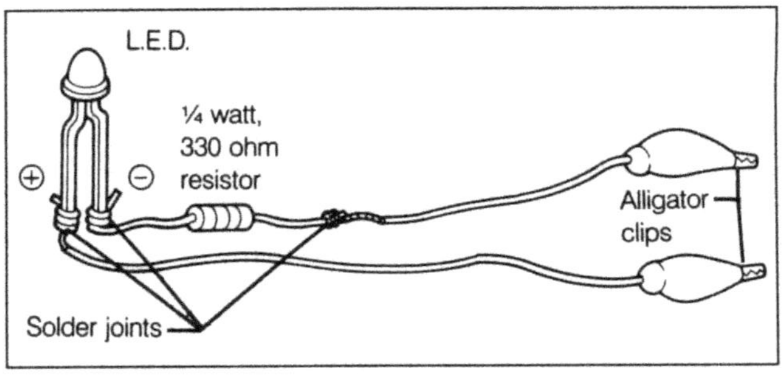

Fig. 3-6. Do-it-yourself LED test light for safe testing of ignition, fuel injection, and emission control circuits.

Parts:

1. (1) LED

2. (1) ¼ watt, 330 ohm resistor

3. wire and two alligator clips. (Purchase a jumper wire with an alligator clip on each end, and cut it in half)

4. solder and soldering iron

5. heat-shrinkable tubing or electrical tape

Disconnecting Wiring Harness Connectors

Volkswagen harness connectors used throughout the car are designed to positively lock into place to prevent coming loose. One type, shown in Fig. 3-7, is equipped with an easy disconnect feature. To disconnect this type of connector, press on the wire clip to release the lock and carefully pull the connector loose.

CAUTION —
Always pull only on the connector body to disconnect it. Never pull on the wires themselves.

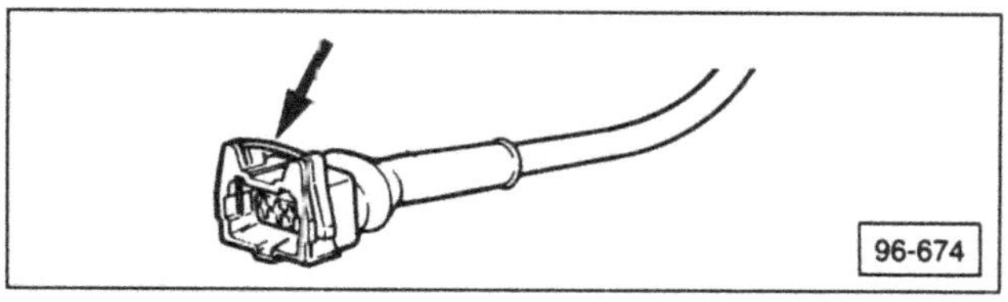

Fig. 3-7. Harness connector with quick disconnect feature. Push on wire lock (arrow) and gently pull on connector to release.

4. BUYING PARTS

Many of the maintenance and repair tasks in this manual call for the installation of new parts, or the use of new gaskets and other materials when reinstalling parts. Most often, the parts which will be needed can be on hand before beginning the job. Read the introductory text and the complete procedure to determine which parts will be needed.

NOTE —
For some bigger jobs, partial disassembly and inspection is required in order to determine a complete parts list. Read the procedure carefully and, if necessary, make other arrangements to get the necessary parts while your car is disassembled.

Genuine Volkswagen Parts

Without any doubt, the best parts for maintenance and repair of Volkswagens are genuine Volkswagen parts, from an authorized Volkswagen dealer.

Genuine Volkswagen replacement parts from an authorized Volkswagen dealer are designed and manufactured to the same high standards as the original parts. They will be the correct material, manufactured to the same specifications, and guaranteed to fit and work as intended by the engineers who designed the car. Volkswagen is constantly updating and improving their cars, often making improvements during a given model year. So, when Volkswagen recommends a newer, improved part as a replacement, your authorized dealer's parts department will know about it and provide it.

The Volkswagen parts organization is best equipped to deal with any Volkswagen parts need. Most parts, if not immediately available at an authorized dealer, are available through Volkswagen's highly computerized parts distribution system in less than 48 hours.

Most reputable independent repair shops also make a point of using genuine Volkswagen parts. They know the value of doing the job right with the right parts. So, even if someone else is buying parts for your Volkswagen, make sure they are genuine Volkswagen parts from an authorized Volkswagen dealer.

Volkswagen Remanufactured Parts

If major repairs become necessary, one attractive alternative to expensive new parts is to install genuine Volkswagen remanufactured parts. They are remanufactured—not just rebuilt—to the same exact standards as new parts and usually covered by a warranty, but are available at lower cost than new through authorized Volkswagen dealers. In most cases, the old assembly, called a "core", can be turned in for partial credit against the cost of the remanufactured unit, sometimes even if it is severely damaged.

Volkswagen remanufactured parts which are available from an authorized Volkswagen dealer include:

- Alternators
- Brake Calipers
- Coolant Pumps
- Crankshafts
- Cylinder Heads
- Distributors
- Drive Axles
- Engines
- Flywheels
- Fuel Distributors
- Fuel Injectors
- Power Steering Pumps
- Starters
- Steering Gears
- Torque Converters
- Transaxles
- Windshield Wiper Motors

Non-returnable Parts

Some parts cannot be returned for credit, even if it is the wrong part for the car or it was not required. The best example is electrical parts, which are almost universally considered non-returnable because they are so easily damaged internally.

Buy electrical parts carefully, and be as sure as possible that a replacement is needed, especially for expensive parts such as control units. It may be wise to let an authorized dealer confirm your diagnosis before replacing a part which cannot be returned.

What Information You Need To Know

Model Year. This is not necessarily the same as date of manufacture or date of sale. A 1986 model could have been manufactured in late 1985, or perhaps sold as late as early 1987. It would still be a 1986 model. If in doubt, consult the vehicle identification number (VIN), explained below.

Model. Scirocco and Cabriolet are model names. In some cases, buying trim parts for example, it may be helpful to know more details, such as Wolfsburg Edition, etc.

Vehicle Identification Number (VIN). This is a combination of 17 letters and numbers which identify the particular vehicle. The VIN appears on the state vehicle registration document, and on the car itself. One location is shown in Fig. 4-1.

Fig. 4-1. Location (arrow) of vehicle identification number (VIN) on Volkswagen Scirocco and Cabriolet. Number, on driver's side of instrument panel, is most easily viewed from outside, through windshield.

Copy down the VIN and keep it handy when you buy parts. Because of mid-year production changes, the correct replacement parts are often identified by the car's VIN.

Each of the 17 characters indicates facts about the car's manufacture. The 10th character is a letter that indicates model year, which is not necessarily the calender year the car was sold. The letter "F" signifies a 1985 model. "G" is for 1986, "H" for 1987, and so on. The letters "I" and "O" are not used, so the sequence skips to "J" for 1988 and to "P" for 1993.

Engine Code. The Scirocco and Cabriolet engines covered by this manual are all similar, based on one basic design, but some parts of the engines and engine management systems are different. Each engine type is identified by a two-character code, part of the engine number stamped on the cylinder block, as shown in Fig. 4-2. For a complete explanation of the engine codes and how to find them, see the section titled **ENGINE.**

Transmission Number. Although most internal repairs to the transaxle are beyond the scope of this manual, the transmission number with its identifying code may be important when buying clutch parts, seals, gaskets, and other transmission-related parts for repairs which are covered. The location of the transmission number is shown in Fig. 4-3. The number consists of two or three transmission code letters followed by a number signifying date of manufacture. For example, "AGB18117" identifies an AGB transaxle built on the 18th of November, 1987. More information on transmission codes and their meanings can be found in **MANUAL TRANSMISSION AND CLUTCH** and **AUTOMATIC TRANSMISSION**.

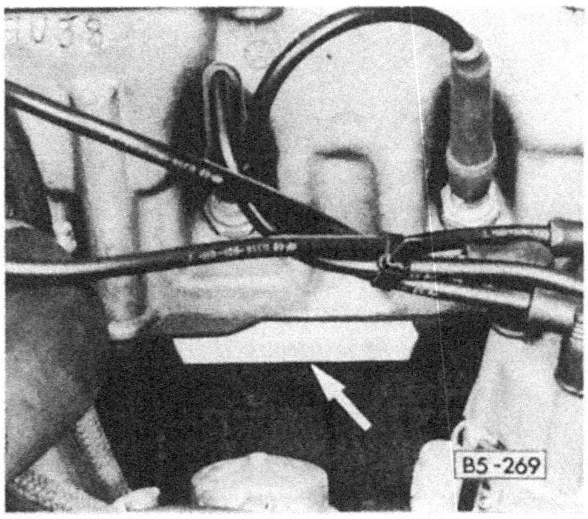

Fig. 4-2. Location (arrow) of stamped engine number on engine just below cylinder head. First two characters are engine identifying code.

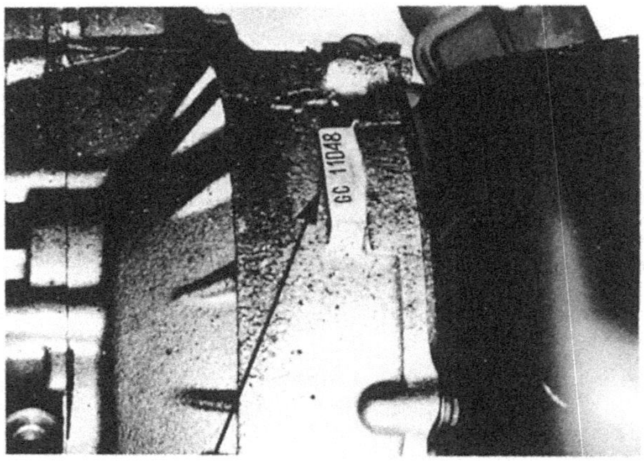

Fig. 4-3. Location (arrow) of transmission number. First characters are transmission code letters.

Volkswagen dealers are uniquely qualified to provide the best possible service for Volkswagen cars. Their ongoing relationship with the large Volkswagen service organization means that they are constantly receiving new tools and equipment, together with the latest and most accurate repair information.

The Volkswagen dealer's service technicians are highly trained and very capable. In addition to periodic Volkswagen Service Training courses which keep them up to date on all the latest technical features and service methods, Volkswagen dealers recognize and actively promote professional technician certification by the National Institute for Automotive Service Excellence (ASE).

Unlike independent repair shops, authorized Volkswagen dealers are intensely committed to supporting the Volkswagen product. They share every owner's interest in Volkswagen value, performance, and reliability.

5. TOOLS

Most maintenance can be accomplished with a small selection of the right tools. Tools range in quality from inexpensive junk, which may break at first use, to very expensive and well-made tools which, to the professional, are worth every bit of their high cost. The best tools for most do-it-yourself Volkswagen owners lie somewhere in between.

Cheap tools are no bargain. They often do not hold up to even casual use, and they present a greater risk of personal injury. If they fit poorly, they can actually damage the fasteners they are intended to remove, making it that much harder to use a good tool the next time around.

Many reputable tool manufacturers offer good quality, moderately priced tools with a lifetime guarantee. A broken tool can be exchanged for a new one, for the life of the tool. These are your best buy. They cost a little more, but they are good quality tools which will do what is expected of them. Sears' Craftsman™ line is one such source of good quality, reasonably priced, and guaranteed tools.

Basic Tool Requirements

The basic hand tools described below can be used to accomplish most of the simple maintenance and repair tasks.

Screwdrivers. Two types, the common flat-blade type and the Phillips type, will handle 99% of all screws used on Volkswagens. Two or three different sizes of each type will be best, since a screwdriver of the wrong size will damage the screw head. Screwdrivers are for screws. Do not use them for anything else, such as prying or chiseling. A complete set of screwdrivers can often be purchased for about the same money as the four or six individual ones that are really necessary. See Fig. 5-1.

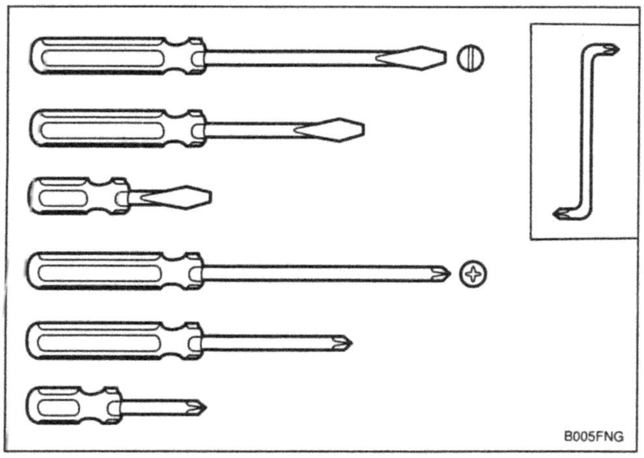

Fig. 5-1. Common flat-blade (top) and Phillips (bottom) screwdrivers. Offset screwdriver (inset) is used for screws with limited access.

For the more complete tool box, include "stubby" screwdrivers or offset screwdrivers for use in tight spots where a normal length screwdriver will not easily fit.

Wrenches. Wrenches come in different styles for different uses. Fig. 5-2 shows several. The basic open-end wrench is most widely useful, but grips on only two sides. It can spread apart and slip off more easily. The box wrench has better grip, on all six sides of a nut or bolt, and is much less prone to slip. A 12-point box end can loosen a nut or bolt with less movement, while a 6-point box end provides greater grip for tight fasteners. For hex fasteners on fluid lines, like brake lines and fuel lines, a flare-nut wrench offers better grip and less chance of fastener damage, but will still fit over the line.

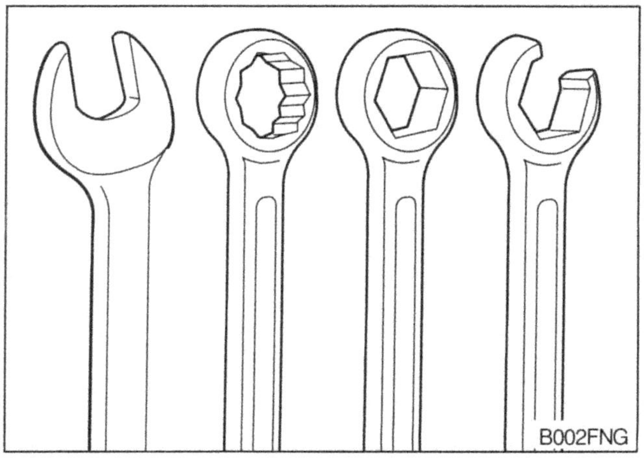

Fig. 5-2. Types of wrench heads. From left, open-end, 12-point box, 6-point box, flare nut.

Most universal is the combination wrench, shown in Fig. 5-3, with one open-end and one 12-point box end. For Volkswagens, 10mm and 13mm wrenches are the most common sizes needed. A 19mm wrench is needed to loosen and tighten the engine oil drain plug. A complete set should also include 6mm, 7mm, 8mm, 9mm, 11mm, 12mm, 14mm, 15mm, and 17mm.

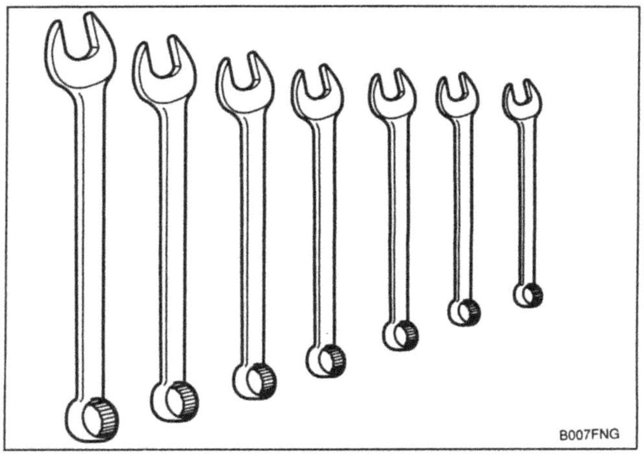

Fig. 5-3. Combination wrenches with one open-end and one 12-point box end. 12-point box end is more universal than 6-point.

Sockets. Sockets perform the same job as box wrenches, but offer greater flexibility. Normally used with a ratchet handle for speed and convenience, they can be combined with extensions to reach fasteners more easily. Standard sockets come in 6-point and 12-point styles. For use with a ratchet, the universal nature of the 12-point is unnecessary, and the 6-point type offers better grip on tight nuts and bolts. Sockets are also characterized by their drive size, the size of the connection to a handle or extension. To start, choose 6-point sockets with a ⅜ in. square drive, two or three extensions of different lengths, and a ⅜ in. drive ratchet handle. As with wrenches, 6mm to 15mm, 17mm, and 19mm are the most needed sizes. See Fig. 5-4.

For the more complete tool box, add deep sockets and a greater variety of handles and extensions. A universal joint extension can allow access from an angle where a straight extension will not quite fit.

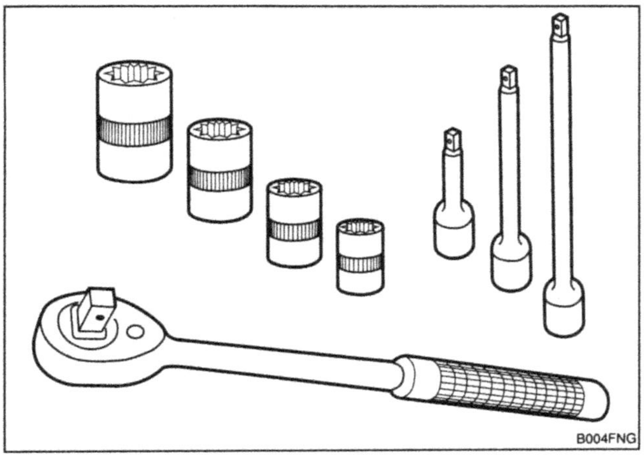

Fig. 5-4. Sockets, handles, and extensions.

Spark Plug Socket. A special socket for spark plugs is the correct size, is deep enough to accommodate a spark plug's length, and contains a rubber insert to both protect the spark plug from damage and grip it for easy removal. This gripping feature is especially necessary on 16-valve engines. A spark plug socket is shown in Fig. 5-5.

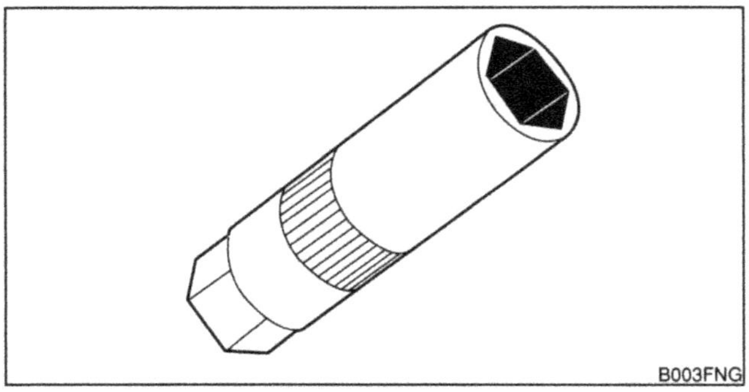

Fig. 5-5. Spark plug socket.

The spark plugs used in all Volkswagen engines except 16-valve require a 13/16 in. socket. The spark plugs used in 16-valve engines require a ⅝ in. socket. Get one with a drive size to match the socket handle and extensions.

Pliers. A few of the many types of pliers are shown in Fig. 5-6. Most all are used for holding irregular objects, bending, or crimping. Some have special applications. A needlenose plier is used for gripping small and poorly accessible objects, and is useful for wiring and other electrical work. A locking plier, such as the famous Vise-Grip™ is useful because it can grip tightly all by itself. Snap-ring pliers with special tipped jaws are used to remove and install snap-rings or circlips. A channel lock or water pump plier has adjustable jaws which can be quickly changed to match the size of the object being held to give greater leverage. There are many different types and sizes of pliers. Start with a small selection of different types of medium size.

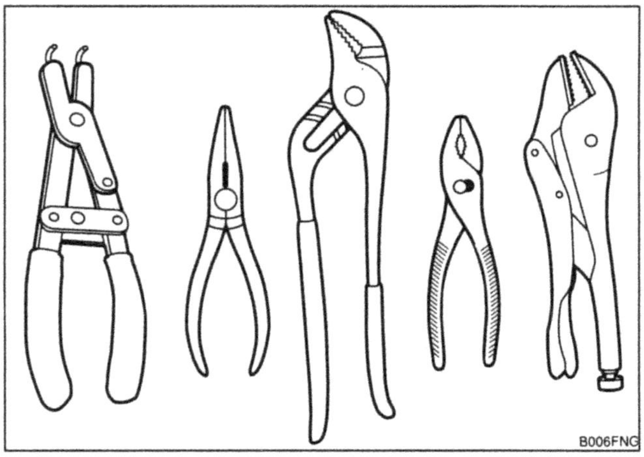

Fig. 5-6. Pliers. From left, snap-ring, needlenose, channel lock, common, locking.

Adjustable Wrench. An adjustable wrench, shown in Fig. 5-7, can be a useful addition to a small tool kit. It can substitute in a pinch if, for example, two wrenches of the same size are needed to remove a nut and bolt. Use extra care with adjustable wrenches, as they especially tend to loosen, slip, and damage fasteners if not used carefully. Compared to a wrench of the correct size, an adjustable wrench is always second best. Choose one of average size range, about 6 to 8 inches in length.

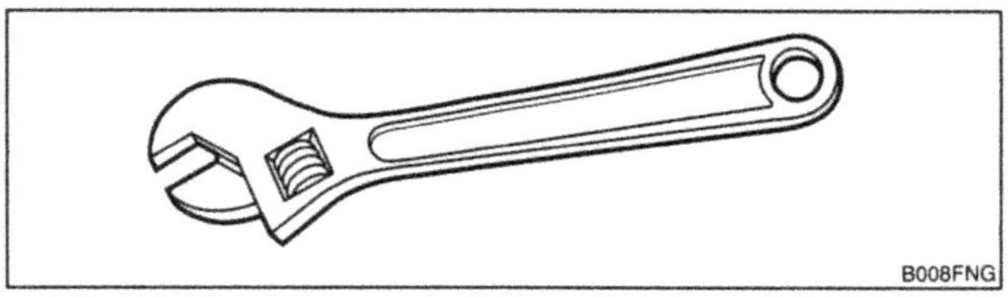

Fig. 5-7. Adjustable wrench is useful if used with care. Adjust for best possible fit. Use only when correct size wrench is not available.

Jack Stands

Strong jack stands are extremely important for any work that is done under the car. A jack should never, ever be used alone to support the car while working underneath.

Use only jack stands which are designed for the purpose. Blocks of wood, concrete, bricks, etc. are not safe or suitable substitutes.

Jack stands are available in several styles. A typical jack stand is shown in Fig. 5-8. The best ones are made of heavy material for strength, have a wide base for stability, and are equipped to positively lock in their raised positions. They also cost more. Think about their purpose. We recommend the best ones available.

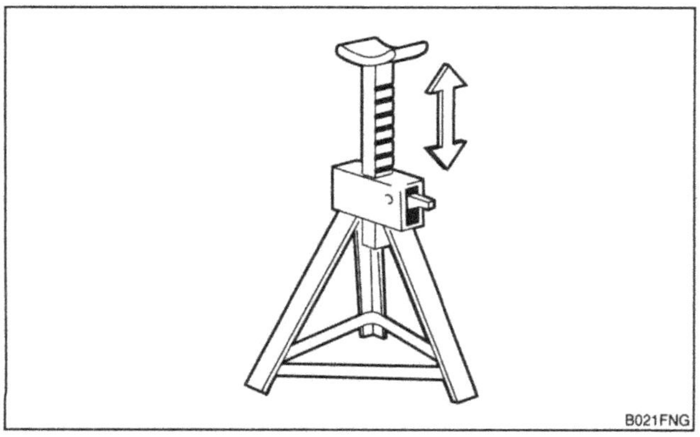

Fig. 5-8. Jack stand for safely supporting car to work underneath. Get the best available.

Oil Change Equipment

Changing oil requires a box wrench or socket to loosen and tighten the drain plug (19mm or ¾ in.), a drain pan (at least 5 qt.), and an oil filter wrench. These items are shown in Fig. 5-9. A wide, low drain pan will fit more easily under the car. Use a funnel to pour the new oil into the engine.

An oil filter wrench is used to remove the oil filter. Volkswagen oil filters are of smaller diameter than many others. Be sure to get a filter wrench which will grip the smaller Volkswagen filter tightly.

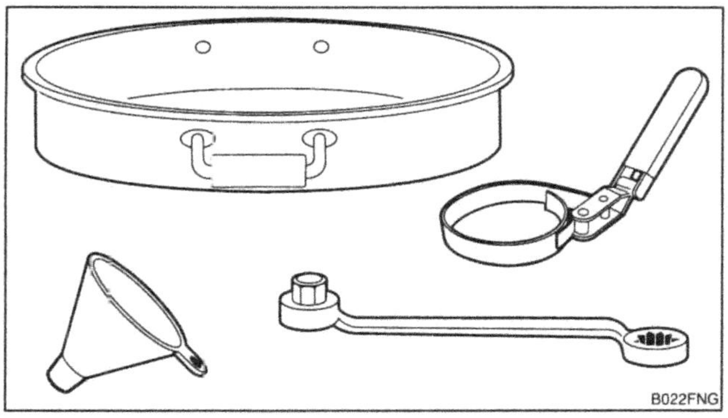

Fig. 5-9. Oil change equipment includes drain plug wrench (19mm or ¾ in.), 5 qt. drain pan, oil filter wrench, and funnel.

Torque Wrench

A torque wrench is used to precisely tighten threaded fasteners to a predetermined value. Nearly all of the repair procedures in this manual include Volkswagen's recommended torque values, in Newton-meters (Nm) and foot-pounds (ft. lb.), to be used during reassembly.

Several types of torque wrenches are widely available. They all do the same job, but offer different convenience features at different prices. Some typical torque wrenches are shown in Fig. 5-10. The most convenient ones have a built-in ratchet, and can be preset to indicate when a specific torque value has been reached. Follow the wrench manufacturer's directions for use to achieve the greatest accuracy.

A torque wrench with a range up to about 250 Nm (185 ft. lb.) has adequate capacity for all of the repairs covered in this manual. For recommended torque values of 10 Nm or below, the English system equivalent is given in inch-pounds (in. lb.). These small values may be most easily reached using a torque wrench calibrated in inch-pounds. To convert foot-pounds to inch-pounds, multiply by 12. To convert inch-pounds to foot-pounds, divide by 12.

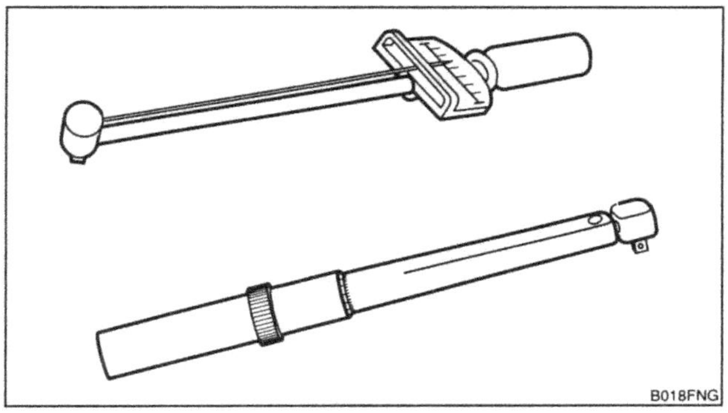

Fig. 5-10. Torque wrenches. Inexpensive beam-type (top) is adequate but must be read visually. Ratchet-type (bottom) can be preset to indicate when torque value has been reached.

Timing Light

A timing light connects to the battery and to the #1 spark plug wire to check ignition timing. A bright strobe light flashes in sequence with the firing of the #1 spark plug. One example is shown in Fig. 5-11. Guard against less expensive models with lights which are too dim to be seen in daylight.

An attractive feature of the more expensive models is an inductive pickup which just clamps on over the spark plug wire to get its signal, rather than needing a more positive connection.

An adjustable timing light is a more sophisticated instrument used to actually measure the timing and the advance and retard characteristics in engine crankshaft degrees. This type is only necessary for detailed analysis of the ignition system's function.

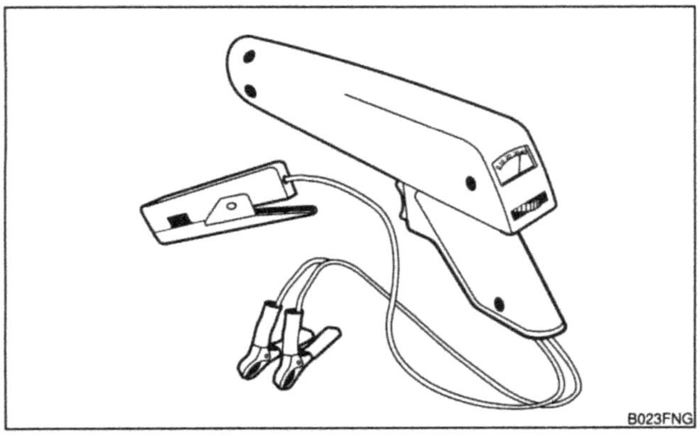

Fig. 5-11. Timing light and electrical connections, including inductive pickup. Dial shown is found on sophisticated adjustable timing lights.

Tachometer

An external tachometer is used to precisely measure engine speed (rpm) for various tests and adjustments. Most tachometers are powered by connection to the battery, and measure engine rpm through a connection to terminal 1 of the ignition coil. To locate terminal 1, see **IGNITION**.

Duty Cycle Meter or Dwell Meter

The duty cycle meter is used on today's Volkswagens to measure the functions of certain components of the fuel injection system and emission controls. A dwell meter, the same type used on older cars to measure ignition breaker point dwell, can also be used for this purpose. Some tachometers can also be used as a dwell meter.

Duty cycle is measured on a 0 to 100 % scale. Using a dwell meter (0° to 90° scale), multiply duty cycle by ⁴⁄₅ to get the dwell equivalent, and multiply dwell readings by ⁵⁄₄ to convert to duty cycle percentage.

Feeler Gauges

Feeler gauges are thin metal strips of precise thickness, used to measure small clearances. They are normally available as a set, covering a range of sizes. For Volkswagens, metric feeler gauges (in millimeters) are the best choice. Fig. 5-12 shows a set of feeler gauges.

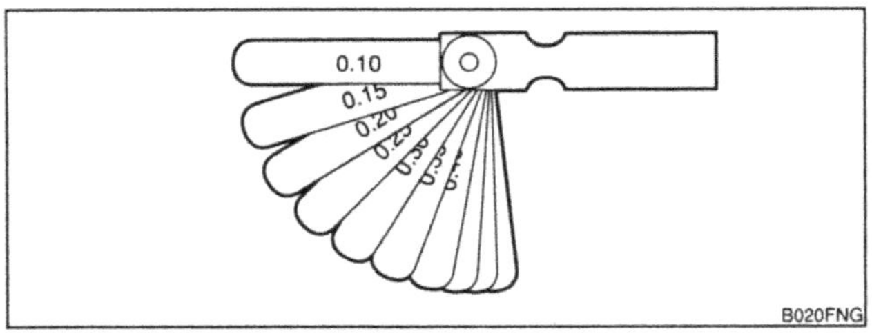

Fig. 5-12. Feeler gauge set, used for precise measurement of clearances between parts.

Micrometers

Precision measurements of internal engine parts and other critical dimensions are made with micrometers, some of which can accurately measure to within thousandths of a millimeter. These are expensive instruments, and are only recommended for those who plan to be repeatedly involved in engine overhauls or other similar work requiring detailed measurement. If such measurements are necessary on a one-time basis, a qualified machine shop can be called upon to make these measurements, particularly if they are also going to be doing the necessary machine work.

Test Light

A test light, shown in Fig. 5-13, is a simple tool used to check electrical circuits for voltage or continuity to ground when actual voltage values are unimportant. A bulb in the handle will light whenever current is flowing through the circuit. The use of a test light is described in **3.2 General Advice For The Beginner**.

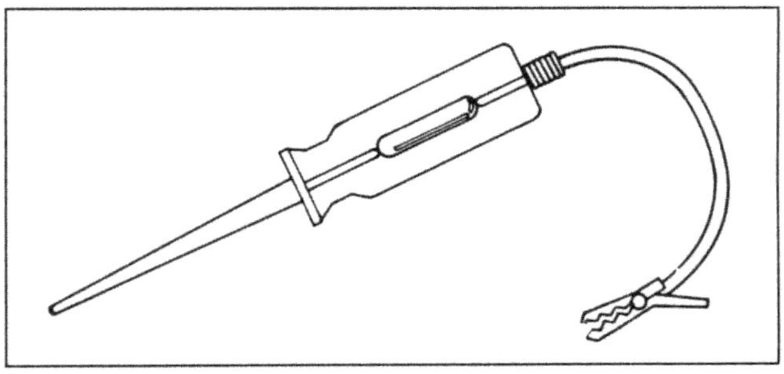

Fig. 5-13. Test light with alligator clip test lead.

CAUTION —
*The ignition, fuel injection, and emission control electronics may be damaged by a test light with a normal incandescent bulb. For these applications, use a high input impedance tester such as Volkswagen special tool no. US 1115 (order no. TU1 115 000 28 ZEL), or build one as described in **3.2 General Advice For The Beginner** under **Making An LED Test Light**.*

Volt-Ohm Meter or Multimeter

Many of the electrical tests in this manual call for the measurement of resistance (ohms) or voltage values. For tests of ignition, fuel injection, and emission control systems, the multimeter should be a digital, high input impedance type. One good example of this type of meter is Volkswagen special tool no. US 1119 (order no. TU1 119 000 00 KTM), shown in Fig. 5-14.

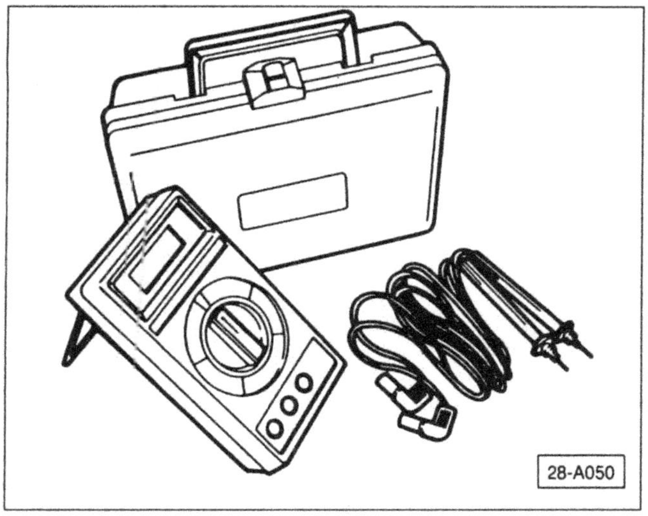

Fig. 5-14. Multimeter with test probes and case. Volkswagen special tool no. US 1119 (order no. TU1 119 000 00 KTM).

Jumper Wires

Some of the electrical tests in this manual require the use of extra jumper wires to bypass a component or a portion of the wiring harness. For most basic electrical tests, jumper wires with alligator clips at each end (made or purchased) are sufficient. For tests involving harness connectors, hookup of jumper wires may damage the connector and cause inferior connections later on. To avoid this damage, jumper connections to harness connectors should be made using a small, flat-blade terminal which will mate properly with the connector. Use Volkswagen part no. N 17 457 2. See Fig. 5-15.

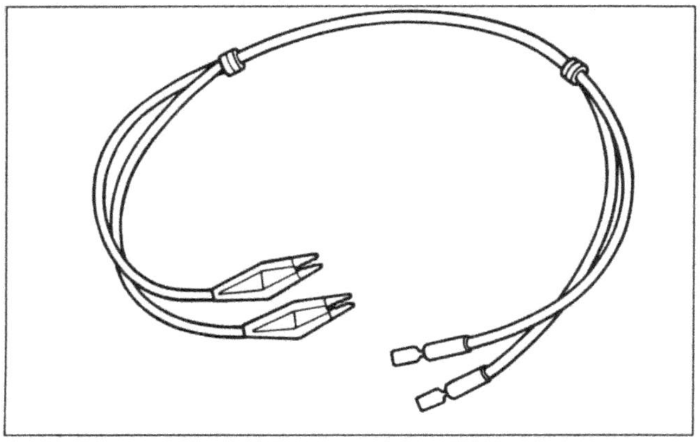

Fig. 5-15. Jumper wires with alligator clips. Flat connectors shown (Volkswagen part no. N 17 457 2) are for electrical testing at harness connectors.

Volkswagen Special Tools

Many parts of this manual refer to the use of Volkswagen special tools. These are tools used by Volkswagen Service technicians in an authorized Volkswagen dealership.

The reference to special tools does not automatically mean that the job is too complicated or out of reach of the novice. Many are inexpensive and are recommended because they are the best way or the only way to do a job properly.

There are some jobs for which expensive special tools are essential, but these may be too costly to purchase for a one-time do-it-yourself repair. These difficult repairs are included in this manual for the benefit of those with the necessary experience and access to tools. The do-it-yourselfer, with full knowledge of the need for special tools, can plan accordingly.

Volkswagen special tools can usually be purchased by special-order through your authorized Volkswagen dealer Parts department. To do so, refer to the tool order number that is listed in the accompanying text. Do not try to order tools from the dealer using the shorter, factory-assigned tool number.

Many Volkswagen special tools are also available directly from their manufacturers using the Volkswagen special tool number. Check with:

Assenmacher Specialty Tools
6440 Odell Place
Boulder, CO 80301
telephone: (303) 530-2424
toll free (except CO): (800) 525-2943
fax: (303) 530-4720

Baum Tools Unlimited
876 Tarawitt Drive
P.O. Box 87
Longboat Key, FL 34228
toll free: (800) 848-6657

Schley Products
5350 E. Hunter Avenue
Anaheim Hills, CA
telephone: (714) 693-7666
fax: (714) 693-8558

Zelenda Machine and Tools Corp.
66-02 Austin Street
Forest Hills, NY 11374
telephone: (718) 896-2288

6. TROUBLESHOOTING FUNDAMENTALS

Troubleshooting is a systematic approach to identifying and solving a problem. The exact approach will depend on the individual circumstances, but it usually relies on carefully observing what is happening, and under what conditions.

The basic rule of troubleshooting is to never overlook the obvious. Always start with the basics and work toward the more complex. Lots of time and money can be wasted on exotic testing only to find, eventually, that the problem is a loose wire or an empty fuel tank.

6.1 Starting

There are three main requirements for starting the engine:

1. The starting system (battery and starter) must provide adequate engine cranking speed.

2. The ignition system must function properly.

3. The fuel system must correctly deliver fuel.

Observing the symptoms of a starting problem will give clues to its cause. Too low cranking speed indicates problems with the electrical system—probably the battery or starter. Further troubleshooting should focus on the electrical system—tests of the ignition or fuel systems at this point would be meaningless. See **ELECTRICAL SYSTEM**.

An engine that cranks normally is confirmation that the battery and starter are fine, so starting problems suggest an ignition or fuel system problem. If there is no sign whatsoever of starting, make sure there is adequate fuel in the tank. Check for loose wires around the coil and distributor. Check to see that the distributor cap and spark plug wires are dry.

Unless this basic inspection turns up a cause, the cause could be in either the ignition system or the fuel system. The ignition system is the more likely culprit and is also easier to evaluate. The condition and function of the ignition system should always be confirmed before suspecting the fuel system. See **IGNITION**. If troubleshooting of the ignition system indicates that all is well, proceed to **FUEL SYSTEM**.

6.2 Running

Problems with the way the engine runs may be caused by either ignition system or fuel system faults. The fuel injection system is far more likely to be influenced by temperature. Symptoms which are present only when the engine is cold, or only when it is warm, tend to suggest fuel system problems. The ignition system should be investigated and confirmed to be in good condition before beginning any work on the fuel system.

For an engine which has recorded a high number of miles, the general mechanical condition of the engine may also be a factor. Particularly in cases where running problems have developed slowly over time, troubleshooting should include evaluation of the engine's mechanical condition with a compression test. See **ENGINE**.

When attempting to evaluate noise or vibration problems which exist when the engine is running, try to eliminate other possible causes. With a manual transaxle, if the symptoms change depending on whether or not the clutch is engaged, the problem may be in the clutch disc, the clutch release mechanism, or the transaxle. See **MANUAL TRANSMISSION AND CLUTCH**. With an automatic transaxle, check for the symptoms in different shift positions. If the symptoms differ, the problem may be in the torque converter or the transaxle. See **AUTOMATIC TRANSMISSION**.

6.3 Driving

To track down noise and vibration problems which occur while driving, first try to learn more about the symptom. Does it occur all the time, or only at certain speeds? If the symptom is speed-dependent, the first step is to find out whether it is related to engine speed or vehicle speed. Compare driving in different gears at the same approximate engine speed. A symptom which persists at a certain engine speed or speed range regardless of gear selection suggests an engine or exhaust system problem. See **ENGINE** or **EXHAUST SYSTEM AND EMISSION CONTROLS**.

For more analysis, try driving the car at the speed which creates the symptom, and then briefly shifting into neutral and coasting. If the symptom continues unchanged, then it is the car's speed which is a factor, and not engine speed. Symptoms which are related to vehicle speed suggest problems with running gear. See **SUSPENSION AND STEERING** for more detailed troubleshooting information.

7. EMERGENCIES

7.1 Changing a Tire

If the tire goes flat while driving, pull well off the road. Changing a tire on a busy street or highway can be dangerous. If necessary, drive a short distance on the flat tire to get to a safe place. It is much better to ruin a tire or rim than to risk being hit.

Stop the car on as flat a surface as possible, turn on the emergency flashers, and set out flares or emergency markers well behind the car. Passengers should get out of the car and stand well away from the road. Take the jack, tools, and spare wheel from the trunk. If you have wheel chocks, chock the wheel diagonally opposite to the one being changed. If not, use whatever is available, including large rocks or pieces of wood.

Remove the hubcap from the wheel using either a puller clip and wheel wrench supplied with the tool kit or, on 16V models, a special wrench as shown in Fig. 7-1, which should be in the glove compartment. On cars without separate hubcaps, use a screwdriver to pry off the black plastic protective caps which cover the wheel bolts. If using the puller clip, insert the clip into the holes at the hubcap rim. Put the wheel wrench on the wheel rim and pull lightly on the other end.

Loosen the wheel bolts about one turn, but do not remove them. Place the jack under the lifting point nearest the wheel being changed (lifting points are described in **3.1 Safety**). Use a board to provide a firm footing for the jack if the ground is soft. Raise the jack until it is just touching the notch at the lifting point, and adjust the jack so that its base is slightly under the car. Raise the car only far enough so that the wheel is off the ground, and then remove the wheel bolts and the wheel.

Fig. 7-1. Special wrench being used to remove hubcap. Insert wrench into locking bolt and remove bolt counterclockwise.

Install the spare wheel and the wheel bolts; tighten them by hand, then lower the car. With all wheels on the ground, fully tighten the bolts in a cross-wise pattern. If a torque wrench is available, torque the wheel bolts to 110 Nm (80 ft. lb.), and on 16V models torque the hubcap bolt to 6 Nm (4.5 ft. lb.).

Check the inflation pressure of the spare tire. The compact spare should be inflated to 60 psi. Inflation pressures for other tires are given in **LUBRICATION AND MAINTENANCE**.

> ***WARNING —***
> *The compact spare supplied with the car is for temporary use only. Do not drive at speeds in excess of 50 mph (80 km/h), and avoid rapid acceleration, hard cornering, and hard braking. Do not use two or more compact spares at one time. Failure to heed these warnings may result in tire failure and loss of car control.*

7.2 Car Will Not Start

If the engine turns over slowly or not at all, especially on cold mornings, it may be because the battery is not sufficiently charged. Jump-starting the battery from another car may help. Jump-starting is described below in **6.3 Jump-Starting**.

If the starter seems to be operating but the engine does not turn over (indicated by a high-pitched whine or grinding when the ignition key is turned to START), then there is a problem with the starter; jump-starting will not help.

> ***CAUTION —***
> *Volkswagen does not recommend "bump starting" the engine by pushing the car or driving it down a hill. Damage to the catalytic converter and other parts of the car may result.*

If the engine is turning over at normal speed, it indicates that the battery and starter are fine. Check to make sure that there is fuel in the tank. Don't rely on the fuel gauge, it may be faulty. Instead, remove the gas filler cap and rock the car. If there is gas in the tank, you should hear a sloshing sound from the filler neck. If so, turn the ignition on and listen for the sound of the fuel pump. It should run for a few seconds, and then go off. If not, then fuel is not reaching the engine.

The engine may have difficulty starting because it has too much fuel, because the fuel system is vapor-locked on a hot day, or because the ignition system is wet on a very damp day. There will be a strong smell of gas if the engine is flooded. Repeatedly cranking the engine with the gas pedal floored may damage the catalytic converter. Instead, wait for a few minutes, and then try starting the engine again. If you suspect vapor lock, raise the hood, let the engine cool, and then try to start the engine. On damp days, check the distributor cap and spark plug wires for condensation. If they are wet, remove and replace the wires one at a time and dry them off with a clean dry cloth, then remove the distributor cap and wipe it dry inside and out.

7.3 Jump-Starting

Cars with partially discharged or completely dead batteries can be jump-started using the good battery from another car. When jump-starting the engine, always heed the following warnings.

> ***WARNING —***
> *• Battery acid (electrolyte) can cause severe burns, and will damage the car and clothing. If electrolyte is spilled, wash the surface with large quantities of water. If it gets into eyes, flush them with water for several minutes and call a doctor.*
>
> *• Batteries produce explosive gasses. Keep sparks and flames away. Do not smoke near batteries.*
>
> *• Do not jump-start the engine if you suspect that the battery is frozen; trapped gas may explode. Allow the battery to thaw first.*

To jump-start the engine, place the cars close together, but do not allow the cars to touch. With the engine of the car with the good battery off, remove the battery cover from the dead battery and connect the jumper cables as shown in Fig. 7-2. First connect the end of one cable to the positive post of the good battery, and the other end of the same cable to the positive post of the dead battery. The positive post is the fatter of the two posts and is usually marked with a plus (+) sign. Connect one end of the other cable to the negative (−) post of the good battery, and connect the other end of the same cable to the engine block of the car with the dead battery. Make the connection as far away from the battery as possible, as there may be sparks.

Have a helper start the car with the good battery and rev the engine slightly, then start the car with the dead battery. Leave the cars running and disconnect the cables in the reverse order in which they were installed. The car with the dead battery should be run for at least a half an hour to recharge the battery.

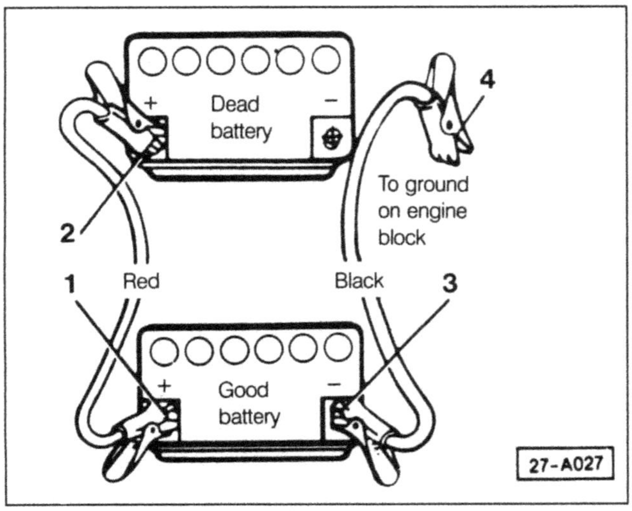

Fig. 7-2. Battery jumper cables connections. Numbers indicate correct sequence for cable attachment.

7.4 Coolant Temperature Warning Light - Overheating

If the coolant warning light on the instrument panel flashes, the engine may be overheated. Even if the temperature gauge is not indicating a high temperature, find a safe place to stop, stop the engine, and inspect the cooling system. Open the hood to help cool the engine. Continuing to drive an overheated car can cause expensive damage.

> ***WARNING —***
> *Do not remove the coolant reservoir or radiator caps with the engine hot. Undoing either could spray hot coolant, and cause burns, or damage the engine.*

> ***WARNING —***
> *Do not touch the radiator fan blades. They will rotate spontaneously when the thermoswitch turns the fan on, even with the ignition off, until the engine has cooled sufficiently. This may take ten minutes or more.*

> ***CAUTION —***
> *Do not add cold water or coolant to a hot engine. Severe engine damage could result from the sudden temperature change.*

The radiator fan should be running. If not, then this may be the cause of overheating, especially if the warning light came on while the car was in slow traffic. Let the engine cool, and check the coolant level as described in **LUBRICATION AND MAINTENANCE**. The car can still be driven, but it should only be as far as is necessary until repairs are made. Watch the temperature gauge closely, and stop to cool the engine if the warning light comes on again.

If the fan is running and everything seems normal in the engine compartment, then the overheating may be caused by the driving conditions, such as operating the air conditioner in slow traffic, or by a low coolant level or damaged V-belt. Turn off the air conditioner. Let the engine cool and check the coolant and V-belt as described in **LUBRICATION AND MAINTENANCE**.

If no coolant is visible in the coolant reservoir, then remove the radiator cap to check the level and add coolant. The car can be driven, but have the cooling system thoroughly checked as soon as possible. If replacement coolant is not available, then plain water can be used, but the coolant should eventually be drained and refilled with coolant of the proper anti-freeze/water ratio.

If steam comes from the engine compartment then there most likely is a burst coolant hose or a large leak in the cooling system. To find the hole, look for signs of coolant leakage on hoses, at hose connections, or on the radiator. Let the engine cool thoroughly, then add coolant or water to fill the system and start the engine. If a great deal of water or coolant flows out of the hole, then the car should not be driven until repairs are made. If there is a slight seepage, then it may be possible to drive a short distance, adding coolant as needed.

If none of the above conditions apply, and the coolant gauge does not indicate a hot engine, then the warning light system may be faulty. The car can be driven, but closely watch the temperature gauge.

7.5 Oil Pressure Warning Light

If the oil pressure warning light comes on while driving the car, stop the engine immediately to prevent severe engine damage. Check the oil level as described in **LUBRICATION AND MAINTENANCE**. If the level is low, add oil to the correct level and start the engine. If the light still flashes, do not operate the car at all. Have it towed to the nearest authorized Volkswagen dealer or qualified repair shop.

7.6 Dim Lights

Headlights that are dim or gradually getting dimmer generally indicate a problem with the battery or charging system. The red alternator warning light may come on as the lights are dimming. In either case, the engine and accessories are running off of the battery alone, and will soon discharge it altogether.

If possible, do not turn the engine off, unless you have the capability to jump start it. There may not be enough power in the starting system to restart the engine. Instead, turn off as many electrical consumers as possible. This will reduce the current drain and will allow the car to be driven further before you lose all battery power.

With the engine and ignition off, check to see if the battery cables are firmly attached, or if there are any loose wires leading to the battery or to the alternator. If any of the wires look

7.7 Towing

Some restrictions apply to towing the cars covered by this manual. Scirocco models and 1985-1987 Cabriolet models can be towed either flat, on all four wheels, or by a tow truck using sling-type or wheel lift towing equipment. 1988 and later Cabriolet models must be towed with wheel lift towing equipment to avoid damage to bumpers and body panels.

> ***WARNING —***
> - *Never work under a vehicle that is lifted by towing equipment.*
> - *Never allow passengers to ride in a vehicle that is being towed.*

> ***WARNING —***
> - *When towing with the rear wheels lifted, always secure front wheels with a steering wheel clamping device that is designed for towing service.*
> - *Never use the ignition/steering lock to secure front wheels.*

> ***CAUTION —***
> - *Tow by lifting the front (drive) wheels whenever possible.*
> - *Automatic transmission cars must never be towed with the front (drive) wheels on the ground.*

If flat-towing the car, use the towing eyes at the front of the car under the bumper as shown in Fig. 6-3. Set the transmission in neutral.

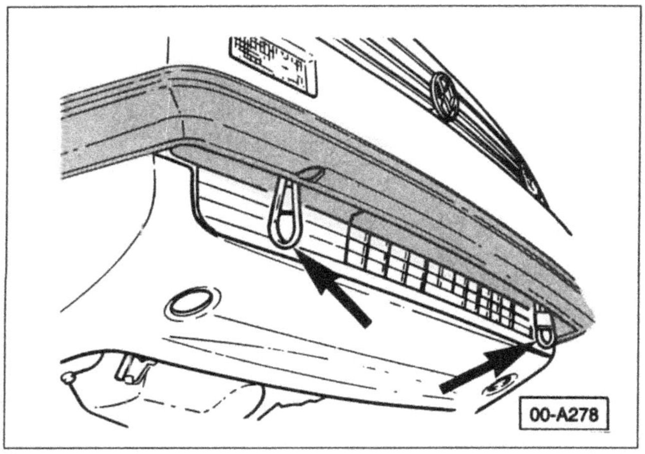

Fig. 7-3. Front towing eyes (arrows) used when flat-towing cars. Model shown is not covered by this manual. Models covered by this manual are similar.

If using sling-type towing equipment, attach grab hooks to the towing eyes under the bumper, position a four-foot-long piece of 4 x 4 against the eyes and under the bumper, and raise the car four inches off the ground as shown in Fig. 7-4. Attach safety chains at the front to the lower control arms, and at the rear to the rear axle. Set the transmission in neutral.

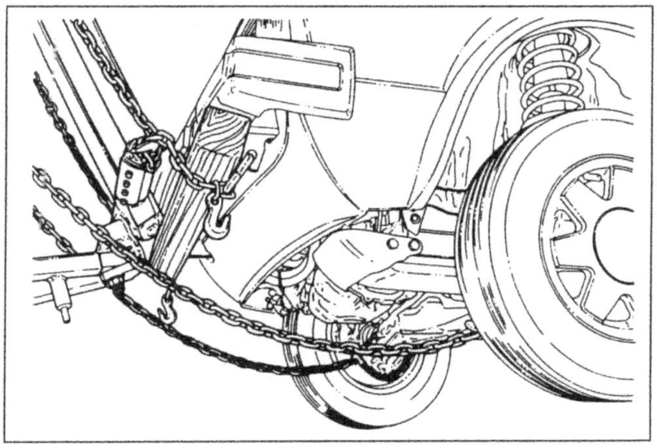

Fig. 7-4. Correct installation of sling-type towing equipment at front of car. Rear hook-up is similar.

> ***CAUTION —***
> *1988 and later Cabriolet models cannot be towed using conventional sling-type towing equipment or dollies. Bumper and body panel damage will result.*

If using wheel lift towing equipment, attach wheel lift equipment and safety straps to wheels, and attach safety chains to lower control arms. See Fig. 7-5. Maintain a towing clearance of 6–12 inches between tires and ground.

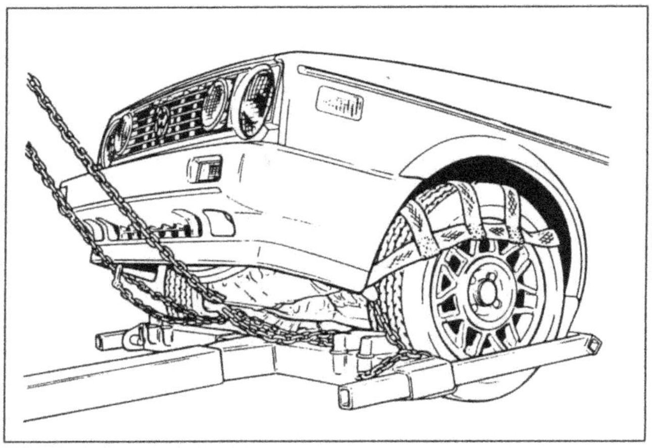

Fig. 7-5. Correct installation of wheel lift towing equipment at front of car. Rear hook-up is similar.

Cars with manual transmission should not be towed faster than 50 mph or further than a distance of 50 miles (80 km/h, 80 km) if the front wheels are on the ground. When towing a car with automatic transmission flat (four wheels on the ground) do not exceed a speed of 30 mph or a distance of 30 miles (48 km/hr, 48 km). Damage to the transmission could result. Do not tow a car with automatic transmission backwards with the front wheels on the ground. If exceeding the limits in either case, use wheel dollies, or remove the drive axles and support the front wheel bearings as described in **SUSPENSION AND STEERING.**

7.8 Spare Parts Kit

Carrying a basic set of spare parts with you in your car can prevent a minor breakdown from turning into a major annoyance. Many of the following items won't allow you to do major repair work on the car, but they will help in the event of the failure of something which can disable the car or compromise its safety.

Spare Parts Kit - Basic Contents:

1. V-belt(s) for the alternator and/or coolant pump (on some models there is one belt, on others two separate belts)
2. one or two quarts of engine oil
3. a gallon container of engine coolant (pre-mixed anti-freeze and water)
4. spare fuel pump relay
5. a new, unopened bottle of brake fluid
6. 10 amp, 15 amp, and 20 amp fuses
7. upper and lower radiator hoses

Spare Parts Kit - Additional Contents:

1. replacement headlight (sealed beam or bulb)
2. brake light, turn signal light, and tail light bulbs
3. other relays such as headlight, turn signal, or load reduction
4. wiper blades
5. distributor cap and rotor

Section 2

LUBRICATION AND MAINTENANCE

Contents

TABLES

Lubrication and Maintenance

Introduction

The service life of your car depends on the kind of maintenance it receives. The procedures described in this section of the manual include all periodic checks and maintenance steps necessary for the long and reliable operation of your car, as well as instructions for basic car care.

The Owner's Manual, the Maintenance Record, and the Warranty Booklet originally supplied with the car contain the maintenance schedules that apply to your Volkswagen. Following these schedules will ensure safe and dependable operation. In addition, many of the maintenance items are necessary to maintain warranty protection. Volkswagen is constantly updating their recommended maintenance procedures and requirements. The information contained here is as accurate as possible at the time of publication. If there is any doubt about what procedures apply to a specific model or model year, or what intervals should be followed, always take the Owner's Manual, the Maintenance Record, and the Warranty Booklet as your guide, and remember that an authorized Volkswagen dealer always has the latest maintenance information.

Some maintenance procedures, such as oil change service, require no special tools and can be carried out by almost all Volkswagen owners, regardless of mechanical experience. Certain other diagnostic and maintenance tasks require special tools and equipment. Cylinder compression tests, idle speed and idle mixture checks, wheel alignment, and ignition timing are some examples. If you lack the skills, the tools, or a suitable workplace for performing any of the maintenance described in this section, we suggest you leave this work to an authorized Volkswagen dealer or other qualified shop. We especially urge you to consult an authorized Volkswagen dealer before beginning any repairs on a car that is still subject to Volkswagen's extensive warranty coverage.

Although the Volkswagens covered by this manual generally require less maintenance than those of five or ten years ago, all of the maintenance work described here is important and should be performed promptly and correctly. Your Volkswagen should not be thought of as a maintenance-free machine. Correct care will protect your investment and help you to get many years of reliability and enjoyable driving from your Volkswagen.

1. GENERAL DESCRIPTION

Lubrication and maintenance is the performance of those routine procedures which are necessary to keep a car operating at its peak and to maintain the requirements for full warranty coverage.

This section of the manual contains information about all of the routine maintenance that is specified for the Volkswagens covered by this manual.

1.1 Scheduled Maintenance

The scheduled maintenance tables list all of the routine maintenance required for a particular model or model year, and the mileage and time intervals at which particular maintenance tasks should be performed.

1.2 Fluid and Lubricant Specifications

The fluids and lubricants used in Volkswagens have been carefully chosen for their ability to perform and adequately protect. To maintain these high standards of performance, and to ensure that Volkswagen's warranty coverage remains in effect, use only the fluids and lubricants which meet the standards set forth by Volkswagen and listed under this heading.

1.3 Engine Oil Change

Regular changing of the engine lubricating oil and the engine oil filter is perhaps the single most important kind of maintenance that a car can receive. It is also simple and easy.

1.4 Tune-up

Much of what has traditionally been considered part of a tune-up has been rendered obsolete by sophisticated Volkswagen technology. Therefore, tune-ups have become a less frequent and simpler part of Volkswagen maintenance.

The heading **5. Tune-up**, covers those few parts of the traditional tune-up which are still included by Volkswagen as scheduled maintenance.

1.5 Engine Compartment Maintenance

Many of the most important routine maintenance tasks are performed under the hood within easy reach. They are grouped together so that complete maintenance can be planned and carried out most efficiently.

1.6 Under-car Maintenance

Thorough maintenance requires periodic inspection and servicing of parts which are only accessible by raising the car. Since a suitable level workspace and equipment to raise the car are required, this maintenance requires a little more planning. As a convenient alternative, you may wish to leave these items to an authorized Volkswagen dealer or other qualified and suitably equipped repair shop.

1.7 Body and Interior Maintenance

Periodic service and inspection of certain safety-related body and interior equipment is specified by Volkswagen and covered under this heading.

1.8 Cleaning and Preserving

Keeping a car clean not only improves its appearance. It can also reduce harmful effects of dirt and other contaminants and actually preserve the car's appearance. Information on recommended cleaning materials and methods is covered under this heading.

2. SCHEDULED MAINTENANCE

The maintenance schedules below list all of the routine maintenance tasks specified by Volkswagen, as well as the mileage and time intervals at which they should be performed. Maintenance should be performed according to the specified time **or** mileage interval, whichever comes first. In any instance where the publisher has recommended maintenance or a maintenance interval which differs from that specified by Volkswagen, such recommendations are more conservative, and still meet Volkswagen's maintenance requirements

Except where noted, the maintenance items listed apply to all models covered by this manual. The boldface numbers which appear after each listing are the headings in this section where the maintenance procedure is discussed in more detail.

CAUTION —
Aside from keeping Volkswagens in the best possible condition, the scheduled maintenance plays a role in maintaining full coverage under Volkswagen's extensive warranties. If in doubt about the terms and conditions of your car's warranty, an authorized Volkswagen dealer can explain them.

NOTE —
The maintenance schedules published by Volkswagen list what Volkswagen considers to be the minimum maintenance requirements, based on typical expected use at reasonable speeds and in moderate climates. If the car is subjected to frequent short trips or extended use under conditions such as temperature extremes or excessive dust, more frequent maintenance intervals are necessary. See an authorized Volkswagen dealer for advice.

Table a. Maintenance Schedules

First 1000 miles (1500 km) (1985, 1986, and 1987 models, except 16-valve)
Brakes: inspect hoses, check for damage and leaks. Check brake pad and brake shoe wear. **7.2** Clutch: check freeplay and adjust as necessary (except models with self-adjusting clutch cable). **6.6** Cooling system: check coolant level and add coolant as necessary. **6.3** Engine: inspect for leaks around seals and gaskets. Exhaust system: check for damage and exhaust leaks. **7.3** Oil level: Check and add oil if necessary. **4.1** Valve clearance: Check and adjust if necessary, replace valve cover gaskets. **5.7**
First 7,500 miles (12,000 km) or 6 months:
Oil Change Engine oil: drain and replace. **4.1** Engine oil filter: replace. **4.1** **Engine Compartment** Clutch: check freeplay and adjust as necessary (except models with self-adjusting clutch cable). **6.6** **Body & Chassis** Automatic shiftlock: check operation including neutral safety switch (1990 and later models with automatic transmission). See **AUTOMATIC TRANSMISSION**
Every 7,500 miles (12,000 km) or 6 months thereafter:
Oil Change Engine oil: drain and replace. **4.1** Engine oil filter: replace*. **4.1** **Engine Compartment** Clutch: check freeplay and adjust as necessary (except models with self-adjusting clutch cable). **6.6** **Body & Chassis** Automatic shiftlock: check operation including neutral safety switch (1990 and later models with automatic transmission). See **AUTOMATIC TRANSMISSION**

*** Publisher's recommendation—meets or exceeds Volkswagen requirements**

continued on next page

Table a. Maintenance Schedules (continued)

Every 15,000 miles (24,000 km) or 12 months:
All items listed under the previous mileage category, plus: **Oil Change** Engine oil filter: replace. **4.1** **Tune-up** Air cleaner filter element: clean. **5.1** Fuel system: remove mini fuel filter (if applicable) and replace with standard banjo fitting (first 15,000 mile service). **5.4** Valve clearance: Check and adjust if necessary, replace cylinder head cover gaskets. **5.7** **Engine Compartment** Brakes: check brake fluid level and add if necessary. **7.2** Cooling system: check coolant level and add coolant as necessary. **6.3** On-board diagnosis system: check fault memory, purge if necessary (1990 and later models). See **FUEL SYSTEM** Power steering: check fluid level and add as necessary. **6.4** **Body and Chassis** Automatic transmission: check automatic transmission fluid (ATF) level and add as necessary. **7.6** Automatic transaxle final drive: check for leaks, check oil level if leaks are present. **7.6** Brakes: inspect hoses, check for damage and leaks. Check brake pad and brake shoe wear. **7.2** Door check rods: lubricate (1985 models). **8.2** Drive axle shafts: check CV joint protective boots for damage. **7.4** Exhaust system: check for damage and exhaust leaks. **7.3** Front suspension: inspect ball joint dust seals, drive axles and constant velocity (CV) joint boots. **7.4, 7.7** Headlights: check aiming and adjust as necessary. See **BODY AND INTERIOR** Manual transmission: check for leaks, check oil level if leaks are present. **7.5** Sliding sunroof: clean guide rails, lubricate with silicone spray. See **BODY AND INTERIOR** Steering: inspect steering rack, tie rod ends and tie rods. **7.7** Wheels and tires: inspect. **7.1** **During Road Test** Check: braking, steering, heating, ventilation, air conditioning, power accessories, electrical system and automatic transmission kickdown **After Road Test:** Engine: check for leaks, especially around seals and gaskets Idle speed: check and adjust if necessary (except California models). **5.5** Exhaust system: check for damage and exhaust leaks. **7.3**
Every 30,000 miles (48,000 km) or 2 years
All items listed under the previous mileage category, plus: **Tune-up** Air cleaner filter element: replace. **5.1** Engine: check compression. See **ENGINE** Fuel filter: replace*. **5.4** Ignition system: inspect distributor cap, rotor and spark plug wires, and replace as necessary*. **5.3** Oxygen sensor: replace sensor and reset OXS indicator light every 30,000 miles or 48,000 km regardless of time interval (1985 through 1987 models, except 16-valve). **6.5** Spark plugs: replace. **5.2**

*** Publisher's recommendation—meets or exceeds Volkswagen requirements**

continued on next page

Table a. Maintenance Schedules (continued)

Every 30,000 miles (48,000 km) or 2 years (continued)
Engine Compartment
Cooling system: drain and flush system, replace coolant with new anti-freeze and water mixture. See **COOLING SYSTEM**
V-belt(s): check tension and condition, adjust or replace as necessary. **6.2**
Body and Chassis
Automatic transmission: drain and replace automatic transmission fluid (ATF), replace filter (recommended for transmission subjected to severe service). **7.6**
Brakes: replace brake fluid (every 2 years, regardless of mileage). **7.2**
Front suspension: check dust seals on ball joints, steering tie rod ends, and tie rods for wear and damage. **7.7**
Every 60,000 miles (96,000 km) or 4 years
All items listed under the previous mileage category, plus:
Engine Compartment
Cooling system: replace hoses (every four years regardless of mileage)*. **6.3**
Engine: replace camshaft timing belt*. See **ENGINE**
Oxygen sensor: replace every 60,000 miles or 96,000 km regardless of time interval (1987 16-valve, and 1988 and later models) **6.5**
V-belt(s): replace (every four years regardless of mileage)*. **6.2**

*** Publisher's recommendation—meets or exceeds Volkswagen requirements**

3. FLUID AND LUBRICANT SPECIFICATIONS

The fluids and lubricants listed in the tables which follow are those specified by Volkswagen for use in the cars covered by this manual.

> ***CAUTION —***
> *The use of fluids which do not meet Volkswagen's specifications may impair performance and reliability, and may void warranty coverage.*

Table b and **Table c** give engine oil viscosity (SAE grade) vs. operating temperature range for the different Volkswagen engine types covered in this manual. **Table d** lists fluid and lubricant specifications.

> ***CAUTION —***
> *Avoid high-speed long distance driving when using SAE 5W-20 or SAE 10W oil, especially if the outside temperature rises above the indicated limits. If maximum loads on the engine or continuous speeds above 60 mph (100 km/h) are expected, use an oil with the next higher viscosity rating.*

Table b. Oil Viscosity Requirement vs. Temperature
(All engines except 16-valve)

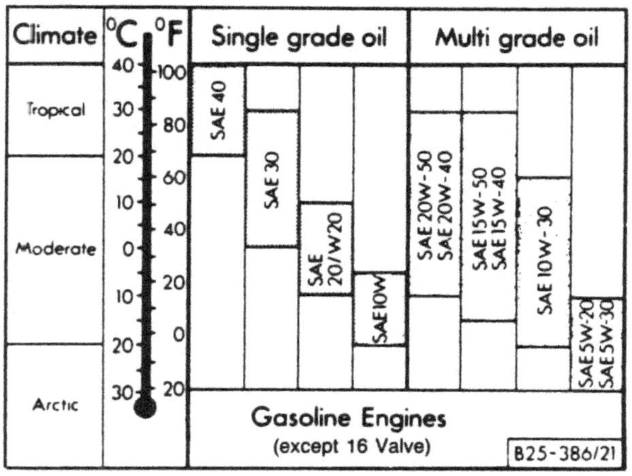

Table c. Oil Viscosity Requirement vs. Temperature

(16-valve engines)

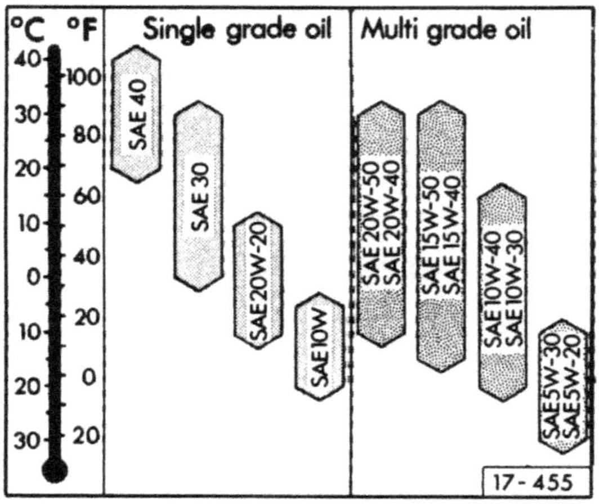

Table d. Fluids and Lubricants

Fluid	Specification	Approximate capacity
Engine oil	API Service SF	4.0 L (4.2 US qt.) with filter change 3.5 L (3.7 US qt.) without filter change
Manual transmission and final drive	SAE 80W hypoid oil API service GL-4, Mil-L-2105	1.9 L (2.1 US qt.)
Final drive (with automatic transmission)	SAE 90W hypoid oil, API service GL-5, Mil-L-2105B	0.75 L (0.8 US qt.)
Automatic transmission fluid (ATF)	Dexron® or Dexron II®	refill: 3.0 L (3.2 US qt.) dry fill: 6.1L (6.4 US qt.)
Power steering fluid	Dexron® or Dexron II® ATF or equivalent	0.77 L (0.81 US qt.)
Brake fluid	DOT 4, SAE J 1703, FMVSS 116	—
Engine coolant	Mixture of 50% phosphate-free ethylene glycol anti-freeze ("G11" Volkswagen Part No. ZVW 237 104) with 50% distilled water	4.6 L (4.9 US qt.)

Engine Oil

Oil is essential to the engine. It provides a lubricating film between all moving parts, and also has a cooling function. Maintaining an adequate supply of clean oil is one of the best ways of making an engine last. Some engine oil is consumed during normal operation, making it necessary to regularly check and "top up" the oil supply. Since oil also becomes contaminated and breaks down over time, regular oil changes are necessary.

Engine oil requirements are defined by the oil's American Petroleum Institute (API) service rating, and by its Society of Automotive Engineers (SAE) viscosity rating. This information can be found on the oil can or bottle, often on a standard label as shown in Fig. 3-1.

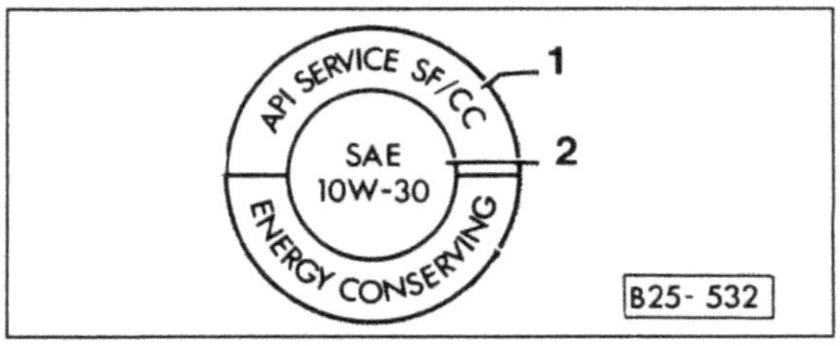

Fig. 3-1. Engine oil label showing API service rating **1** and SAE viscosity number **2**.

The API service rating designates the type of use for which the oil is suited. It is based on the additives that are used to resist oil break-down and carbon formation, inhibit corrosion, resist foaming, neutralize acids, and help remove deposits and keep contaminants suspended in the oil. For gasoline engines, the service ratings range from SA to SF, with SF rated for the most demanding applications.

The SAE viscosity rating indicates resistance to flow. An oil designated SAE 40 has a higher viscosity, greater resistance to flow, than an oil designated SAE 30. While higher viscosity oils will generally offer greater engine protection, they may be too thick and resistant to flow and may inhibit starting during cold weather.

The correct engine oil viscosity depends on the operating temperature range. See the viscosity vs. temperature tables elsewhere in this section. Select a viscosity rating for the lowest anticipated temperature at which the engine must start.

Multi-viscosity oils have additives that make them suitable for use over a wider range of temperatures. For example, an oil rated SAE 10W-30 offers the flow characteristics of SAE 10 at low temperatures, but the protection capability of SAE 30 at engine operating temperature. The "W" in the SAE rating indicates that the oil is suitable for winter use.

Oils of different viscosity ratings can be mixed, but mixing oils of different API service ratings or brands is not recommended.

Gear Oil - Manual Transmission and Final Drive

Gear oil requirements are also defined by API service rating and SAE viscosity rating, as described in **Engine Oil** above. Manual transmissions and final drives share a common oil supply. Automatic transmissions and final drives are separate. With an automatic transmission, only the final drive uses gear oil.

Brake Fluid

Brake fluid deserves special consideration because it absorbs moisture easily, and moisture in the fluid affects brake performance and reliability. When replacing or adding brake fluid, use only new fluid from previously unopened containers. Do not use brake fluid that has been bled from the system, even if it is brand new. Volkswagen does not recommend the use of DOT 5 synthetic brake fluids.

Engine Coolant (Anti-freeze)

Volkswagen recommends coolant that is a mixture of water and phosphate-free anti-freeze containing ethylene glycol. Anti-freeze raises the boiling point and lowers the freezing point of the coolant. It also contains additives which help prevent cooling system corrosion.

Gasoline Additive

Because many gasolines do not contain the necessary additives to help prevent carbon deposits in the fuel injection system, Volkswagen recommends the periodic use of Autobahn Gasoline Additive, Volkswagen part no. ZVW 239 003, which is available from any authorized Volkswagen dealer.

CAUTION —
Follow all label directions. Do not use more than once every 2500 mi. (4000 km).

Greases

Two different types of grease are used for lubrication of drive train and brake components. Multipurpose grease (lithium grease) has a wider temperature tolerance range than ordinary grease and should be used for most general lubrication purposes, including roller bearings.

Molybdenum grease is lithium grease with a friction-reducing molybdenum disulfide additive. This special grease, Volkswagen part no. G 000 602, is recommended for certain applications including lubrication of the drive axle constant velocity (CV) joints.

4. ENGINE OIL CHANGE

The engine oil level is checked with a dipstick, located in the engine block behind the alternator. To check the level, pull out the dipstick, wipe it clean, reinsert it all the way, and withdraw it again. The oil level is correct if it is between the two marks near the end of the stick. The location of the dipstick and the level marks are shown in Fig. 4-1. The upper (max) mark indicates full, the highest acceptable oil level. The lower mark (min) indicates the minimum acceptable level. On 1985 through 1988 models, the dipstick is a round rod with 1.0 L (1 US qt.) difference between marks. On 1989 and later models, the dipstick is a flat rod with 0.75 L (0.79 US qt.) difference between marks.

Always check the oil with the car on a level surface, after the engine has been stopped for at least a few minutes. For the most accurate check, especially on 16-valve engines, wait a few hours or overnight.

Add oil through the filler cap on the top of the cylinder head cover shown in Fig. 4-2. Add only the amount needed to bring the oil level to the "max" mark on the dipstick, using an oil of the correct viscosity and grade as described above in **Engine Oil**. Too much oil is just as harmful as too little.

The amount of oil which needs to be added between oil changes varies from one engine to another. Generally, a new engine or an engine operated routinely at high speeds will consume more oil. It is helpful to become familiar with the rate at which a particular engine requires oil. A sudden increase may be an early warning of engine mechanical problems.

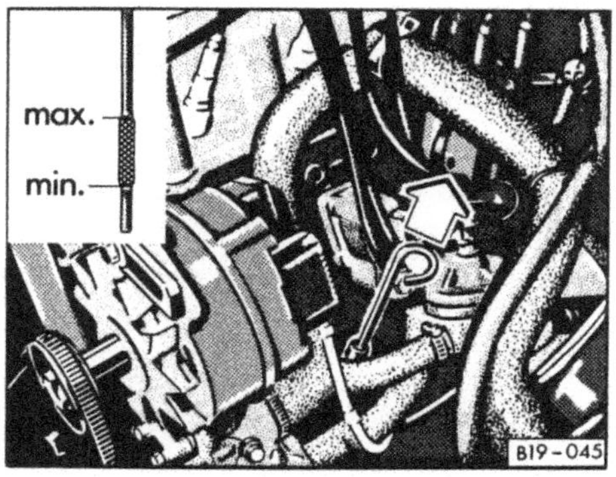

Fig. 4-1. Engine oil dipstick location in engine compartment. Withdraw in direction of arrow. Inset shows maximum (**max**) and minimum (**min**) marks on dipstick used in determining correct oil level.

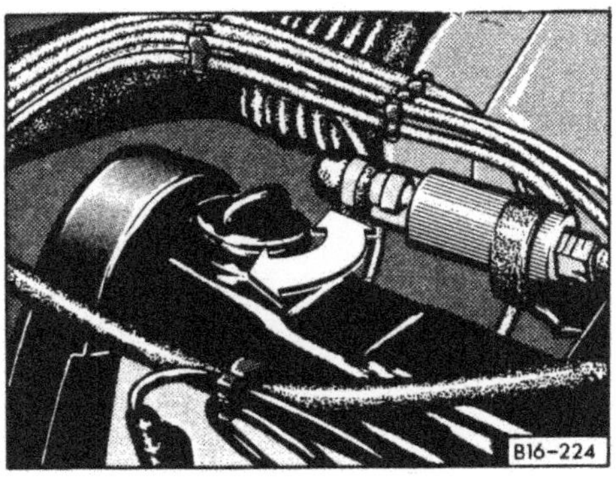

Fig. 4-2. Location of oil filler cap. Remove and install by turning (arrows). 16-valve engine (not shown) is similar.

4.1 Changing Engine Oil and Filter

The 7,500-mile or 6-month oil change interval specified in the maintenance schedule should be the basic guide to scheduling oil changes. Do not rely on the color of the oil on the dipstick to indicate when a change is needed. Because of the detergent additives in the oil, fresh oil will look dark after only a few hundred miles.

If the car is used primarily for short trips, or routinely operated in dusty conditions or slow-moving traffic, the oil should be changed more frequently. In general, changing the oil at more frequent intervals (two or three times as often) will help better protect the engine, and promote longer engine life.

Although it is an acceptable practice to change the oil filter at every other oil change, reusing the old filter leaves about ½ quart of old oil in the lubrication system. The publisher recommends also changing the oil filter with every oil change, no matter what the interval.

A complete oil change requires approximately 5 US qt. of new oil (see **3. Fluid and Lubricant Specifications**), a new oil filter, and a new drain plug sealing washer. Part numbers and oil capacities are given in **Table e**. The equipment needed, a drain plug socket or box wrench (19 mm), a drain pan of at least 5 US qt. (4.7 l) capacity, and an oil filter wrench, is described in **FUNDAMENTALS**.

Table e. Engine Oil Change Specifications

Engine oil capacity (approximate) (with filter change) (without filter change)	 4.0 liters (4.2 US qt.) 3.5 liters (3.7 US qt.)
Oil filter	Volkswagen part no. 056 115 561 G
Oil pan drain plug sealing washer	Volkswagen part no. N 013 849 2

To change oil and filter:

1. Run the car for a few minutes to slightly warm the engine and the oil, then shut the engine off.

2. With the car on level ground, place a drain pan under the oil drain plug shown in Fig. 4-3.

   NOTE —
   If a shallow enough drain pan is used, the car will not need to be raised off the ground.

3. Using a 19 mm socket or box wrench, loosen the drain plug. By hand, remove the plug and let the oil drain into the pan.

Fig. 4-3. Engine oil drain plug (arrow) in oil pan underneath engine (shown from rear).

CAUTION —
Pull the loose plug away from the hole quickly to avoid being burned by the hot oil. It will run out quickly when the plug is removed.

4. When the oil flow has diminished to an occasional drip, reinstall the drain plug with a new metal sealing washer and torque the plug to 30 Nm (22 ft. lb.).

5. Move the drain pan slightly forward so that it is directly under the oil filter, shown in Fig. 4-4. Using an oil filter wrench, loosen the filter by turning it counterclockwise (as viewed from below), then remove it by hand.

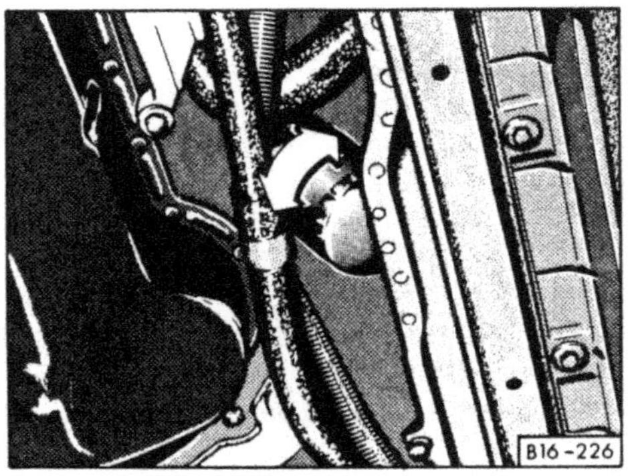

Fig. 4-4. Engine oil filter located in front of engine, behind radiator. (Shown viewed from below).

6. After the oil stops dripping, wipe clean the oil filter gasket surface on the filter mounting flange. Lubricate the rubber gasket of the new oil filter with a light coating of clean engine oil.

7. Install the filter by hand until the gasket contacts the mounting flange, then turn the filter another ¾ turn to tighten it.

CAUTION —

Overtightening the oil filter will make the next change much more difficult, and may deform the gasket, causing leaks.

8. Refill the crankcase with oil. Approximate oil capacity is given in **Table e.** above. Use the dipstick to check for the correct oil level. Oil specifications are found in **3. Fluid and Lubricant Specifications**.

9. Start the engine and allow it to run for three to five minutes to circulate the new oil, then check for leaks at the drain plug and around the oil filter. Stop the engine and recheck the oil level.

NOTE —

Dispose of the used oil properly. Use tight-sealing containers and mark them clearly. Check with a local service station or oil retailer about proper disposal.

5. TUNE-UP

A tune-up is regular maintenance of the ignition and fuel systems to compensate for normal wear. Modern Volkswagen electronic ignition and fuel injection systems have eliminated much work involved in a tune-up. For the Volkswagens covered by this manual, only limited tune-up maintenance is necessary to maintain peak performance and economy.

5.1 Air Filter

The specified maintenance intervals for the air filter are based on "normal" use. If the car is operated primarily in dusty conditions, the air filter should be serviced more frequently. A dirty air filter element starves the engine for air, reducing power output and increasing fuel consumption. Fig. 5-1 shows the air filter housing for the engines covered by this manual.

Fig. 5-1. Air filter housing, showing air filter being removed (black arrow) after mixture control unit is lifted up (white arrow). The replacement filter element for all engines except 16-valve is Volkswagen part no. 049 133 843. The element for 16-valve engines is Volkswagen part no. 027 133 843.

The upper and lower parts of the air filter are fastened together with wire spring-clips around the outside edge. To replace the air filter element, unfasten the clips and lift the upper part of the air filter housing. Notice the filter's installed position, then remove the old filter element and wipe the inside of the air cleaner housing using a lint-free cloth. The filter element is installed with the gasket side facing up. Make sure that the two halves of the air filter housing are mated correctly, and refasten the spring-clips.

On California models, clean the element at the specified interval by tapping it lightly, dirty-side down, on a hard surface to remove the dirt. Do not blow it clean with compressed air, as this can cause small tears which will lead to engine contamination. Do not clean it in gasoline or cleaning solvents, and do not coat it with oil.

CAUTION —

It is especially important when reinstalling a used air filter element that it be in the same position as when it was removed. Reversed installation will allow accumulated dirt to be drawn into the engine.

5.2 Spark Plugs

High temperature and high voltage sparks eventually wear out the spark plug electrodes, and the spark plugs must be replaced. To replace a spark plug, gently remove the spark plug wire by pulling on the protective boot, and blow or brush away any dirt from around the base of the plug to prevent it from entering the engine when the plug is removed.

CAUTION —

Label the spark plug wires so that each one can be reinstalled on the correct spark plug.

Use a 13/16 in. spark plug socket to remove spark plugs on all except 16-valve engines. On 16-valve engines, use a 5/8 in. spark plug wrench with a rubber insert that will hold the plugs for installation and removal in the deep spark plug recesses.

The correct spark plugs for the different engines covered by this manual are listed in **Table f.** Use a spark plug gap gauge to check the gap. If necessary, bend the outer electrode slightly to adjust the gap to meet the specification. Do not bend or file the center electrode.

NOTE —

The electrode gap on the triple-electrode spark plugs used on 16-valve engines cannot be adjusted.

Lightly lubricate the new spark plug threads with a little oil or grease, and thread the plugs in the engine by hand, to prevent cross-threading. Torque the spark plugs to no more than 20 Nm (15 ft. lb.).

Table f. Spark Plug Specifications

Engine code*	Spark plug			Electrode gap
	Bosch	**Beru**	**Champion**	
JH	WR 7 DS W7 DTC	RS 35	N8GY	0.7±0.1 mm (.028±.004 in.)
PL	F6 DTC	–	–	0.8±0.1 mm (.032±.004 in.)
2H	WR 7 DS	–	–	0.7 to 0.9 mm (.028 to .035 in.)
*See **ENGINE** for more information on engine codes				

For California models, it is recommended for emission control that the spark plugs be removed and cleaned every 15,000 miles (24,000 km). Checking the plugs at this interval on all models can be informative, as the condition of the spark plug is a good indicator of combustion quality which can help diagnose engine faults. Fig. 5-2 shows some examples of spark plug condition and what they mean.

Normal

Normal spark plug has gray or light tan color which indicates proper combustion.

Oil-fouled

Oil-fouled spark plug has wet, oily black deposits caused by excess engine oil getting into combustion chamber, probably due to worn piston rings or valve guide oil seals.

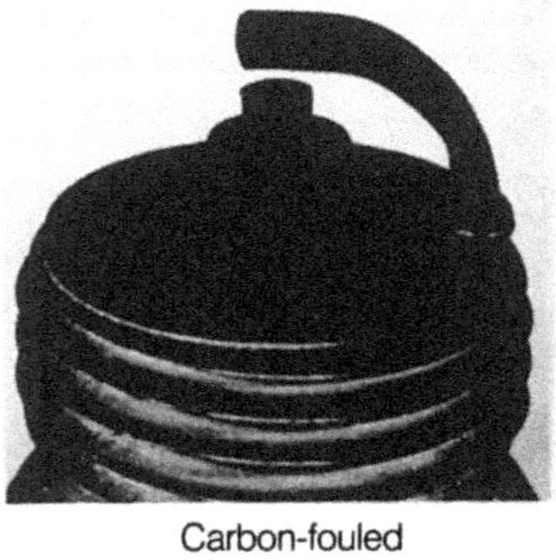

Carbon-fouled

Carbon-fouled spark plug has dry, sooty black deposits caused by too much fuel which may indicate fuel injection or ignition problems.

Worn out

Worn out spark plug may have correct gray or light tan color, but shows physical deterioration (enlarged gap, eroded electrodes).

Fig. 5-2. Spark plug appearance which may indicate engine condition. For more information on interpreting spark plug condition, see **ENGINE**. Photos courtesy of Champion Spark Plug Co.

NOTE —

Any of the abnormal spark plug conditions described above could also result from spark plugs of the wrong specification being installed. Check replacement plugs carefully and follow the spark plug manufacturer's recommendations.

Clean lightly-fouled spark plugs with a light wire brush, and remove all debris from around the electrode. Do not chip the ceramic insulator. Badly fouled spark plugs should be replaced, and the cause of the fouling should be investigated and corrected.

5.3 Distributor Cap, Rotor, and Spark Plug Wires

The distributor cap, the rotor, and the spark plug wires deliver high-voltage spark to the spark plugs. They are subject to insulation breakdown, corrosion fouling, and electrode wear and damage. Inspection every 30,000 miles (48,000 km) or 2 years, and replacement as necessary, is recommended to ensure maximum ignition system efficiency. Guidelines for visual inspection and testing, and instructions for replacement are found in **IGNITION**.

5.4 Fuel Filters

Main Fuel Filter

Because of the varying quality of available gasoline, the fuel filter may become clogged enough during normal operation to restrict fuel flow. Periodic replacement of the fuel filter will help to prevent any such problems, and ensure continued good performance. The main fuel filter is located in the engine compartment next to the air flow sensor, as shown in Fig. 5-3.

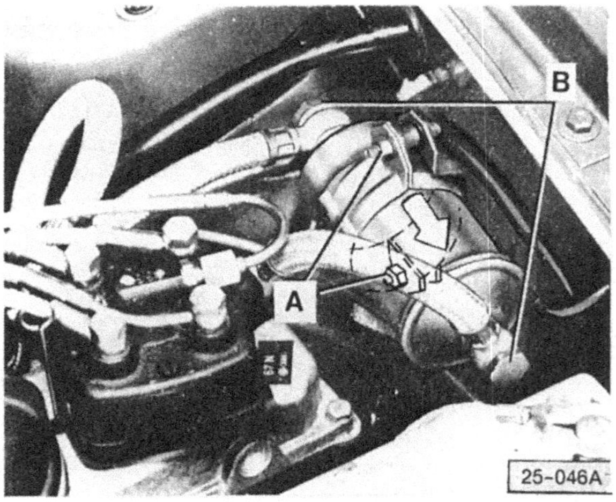

Fig. 5-3. Main fuel filter. Direction of flow (arrow) is indicated by arrow on filter housing. Mounting nuts are at **A**, fuel line union bolts at **B**. Mounting bracket on some models may vary from one shown.

When replacing the fuel filter, disconnect the battery negative (–) terminal, and clamp the filter inlet and outlet hoses to minimize fuel spillage. Torque the inlet bolt to 25 Nm (18 ft. lb.) and the outlet bolt to 20 Nm (15 ft. lb.).

WARNING —
Fuel will be expelled when the filter is removed. Do not smoke or work near heaters or other fire hazards. Keep a fire extinguisher handy.

CAUTION —
Clean thoroughly around the filter connections before removing them, and make sure that no dirt gets into the fuel lines.

Mini Fuel Filter

Many Volkswagens equipped with CIS or CIS-E fuel injection have a mini fuel filter installed in the system to collect any contaminants which may have been trapped in the system during manufacturing. The filter, shown in Fig. 5-4, is a small mesh screen built into the fuel inlet union bolt on the fuel distributor. The bolt with the mini filter is identified by a disc shape on its head. Removal of the mini filter is a one-time procedure. The old bolt/filter is replaced with a new union bolt (Volkswagen part no. N 021 071 2) and two new sealing washers (Volkswagen part no. N 013 812 2). Torque the new union bolt to 20 Nm (15 ft. lb.).

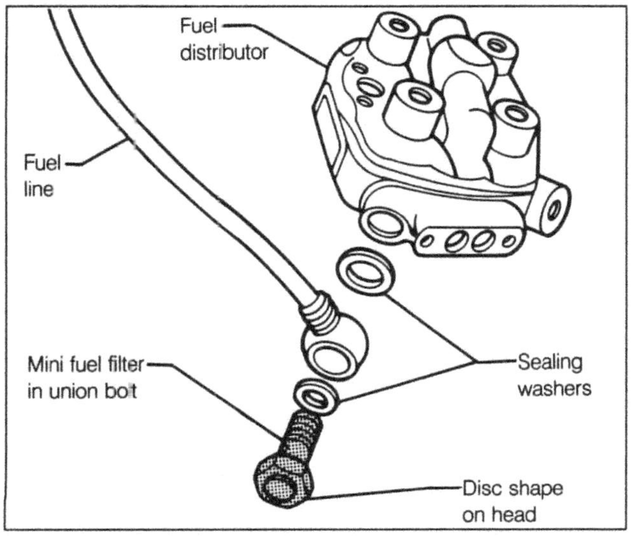

Fig. 5-4. Fuel line union bolt with mini fuel filter. (CIS-E fuel injection shown, CIS fuel injection is similar).

WARNING —
Fuel will be expelled when the filter is removed. Do not smoke or work near heaters or other fire hazards. Keep a fire extinguisher handy.

CAUTION —
Thoroughly clean the area around the union bolt before removing it, and make sure that no dirt enters the fuel lines.

5.5 Idle Speed

Engine idle speed can change due to a number of factors, including normal wear of engine and fuel injection components. In all of the cars covered by this manual, engine idle speed is controlled electronically, through the fuel injection system. Usually, a check of idle speed simply confirms that the electronic idle speed stabilizer system is functioning correctly. See **FUEL SYSTEM** for detailed information on checking engine idle speed.

5.6 Compression Test

A check of engine cylinder compression is recommended for California models. However, a compression test is a good indicator of overall valve and piston ring condition on any engine. In some cases it can pinpoint specific problems. See **ENGINE** for detailed information on engine compression testing.

5.7 Valve Adjustment
(1985 through 1987 except 16-valve)

All 1985 through 1987 models except those with 16-valve engines have mechanical cam followers which require periodic valve clearance adjustment. All other models covered in this manual have hydraulic cam followers which do not require adjustment.

For more information on identifying engines which require valve adjustment, and for the complete valve adjustment procedure, see **ENGINE**.

6. ENGINE COMPARTMENT MAINTENANCE

6.1 Battery

Simple maintenance of the battery and its terminal connections will ensure maximum starting performance, especially in winter when colder temperatures reduce battery power. For a more detailed discussion of the battery and charging system, see **ELECTRICAL SYSTEM**.

Checking and Cleaning Battery

Terminal clamps should be tight. The terminals, the terminal clamps, and the battery case should be clean and free of the white deposits which indicate corrosion and acid salts. Even a thin layer of dust containing conductive acid salts can cause the battery to discharge.

To remove corrosion, disconnect the battery terminals, negative (−) terminal first. Clean the terminal posts and the terminal clamps with a wire brush. Corrosion can be washed away with a baking soda and water solution which will neutralize the acid. Avoid getting the solution into the battery cells through vent holes, as it will also neutralize the acid inside the battery. Reconnect the terminal clamps, positive (+) terminal first, then lightly coat the outsides of the terminals and clamps with petroleum jelly, grease, or a commercial battery terminal corrosion inhibitor.

WARNING —

Battery acid is extremely corrosive. Take care to keep it from contacting eyes, skin, or clothing. Wear eye protection.

CAUTION —

Disconnecting the battery cables with the engine running, or reconnecting the cables to the incorrect posts will damage the electrical system.

Battery electrolyte should be maintained at the correct level just above the battery plates and their separators. On conventional and low-maintenance batteries, the level can be checked using the indicator marks on the side of the battery, or through a filler hole, as shown in Fig. 6-1. The correct level is approximately 5 mm (¼ in.) above the battery plates. If the electrolyte level is low, replenish it by adding distilled water only.

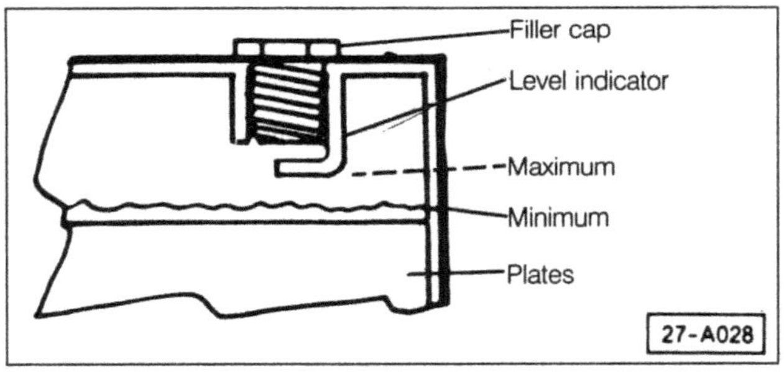

Fig. 6-1. Proper level for battery electrolyte.

On the low-maintenance batteries which are factory-installed in the cars covered by this manual, the filler caps are removed using the tip of a flat-bladed screwdriver, as shown in Fig. 6-2. Maintenance-free batteries, because of the sealed case, cannot have their electrolyte levels adjusted. For additional information on batteries, see **ELECTRICAL SYSTEM**.

Replacing Battery

Batteries are rated by ampere hours (Ah), the number of hours they can sustain a specific current drain before complete discharge, or by cold cranking amps (CCA), the number of amps they produce to crank the engine in cold weather conditions. They may be rated according to European (DIN) standards, by Society of Automotive Engineers (SAE) standards, or both. **Table g** lists approximate rating equivalents. Replacement batteries should always be rated equal to or higher than the original battery.

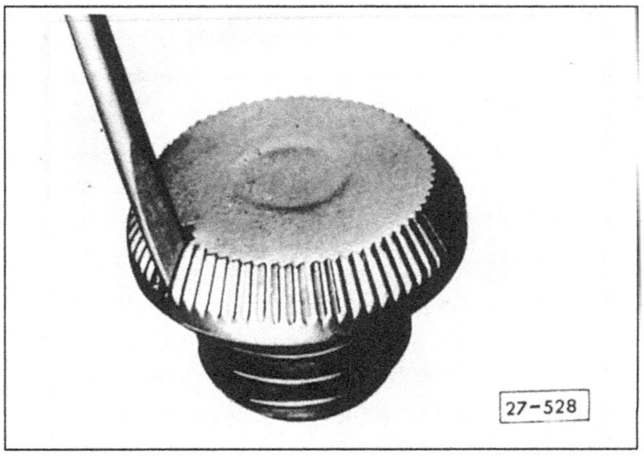

Fig. 6-2. Screwdriver being used to open filler cap on low-maintenance batttery. Turn upper part of filler cap counterclockwise to its stop. Install cap by hand.

Table g. DIN and SAE Battery Rating Equivalents

Cold cranking amps DIN	SAE	Ampere hours
220A	365A	45 Ah
265A	440A	54 Ah
300A	500A	63 Ah
320A	535A	50 Ah
380A	635A	63 Ah
395A	660A	88 Ah

The battery is held in place by a single hold-down bolt and plate. When replacing the battery, always disconnect the negative (−) terminal first, and reconnect it last. Clean away any corrosion in or around the battery tray using a baking soda and water solution to neutralize the acid. Make sure that the battery hold-down is secure in order to prevent vibrations and road shock from damaging the battery's internal plates.

Charging Battery

A discharged battery is not necessarily faulty. It may be restored by recharging, using a battery charger. There are some limitations on the rate at which low-maintenance and maintenance-free batteries may be charged. Frozen batteries should be recharged only after they have thawed. For more information on battery charging, as well as important cautions and warnings, see **ELECTRICAL SYSTEM**.

6.2 V-Belts

V-belts and pulleys transfer power from the engine crankshaft to various accessories. Cars covered by this manual have at least one V-belt, and may have as many as four, depending on the number of accessories.

Inspecting and Adjusting

Proper V-belt maintenance, by correcting loose or overtight belts, can extend the life of the belt and the component it drives. Inspect belts with the engine off, and twist the belt to see its sidewalls and bottom. Belt structural damage, glazed or shiny sidewalls caused by a loose belt, or separation caused by oil contamination are all reasons to replace a belt. Some of these faults are illustrated in Fig. 6-3.

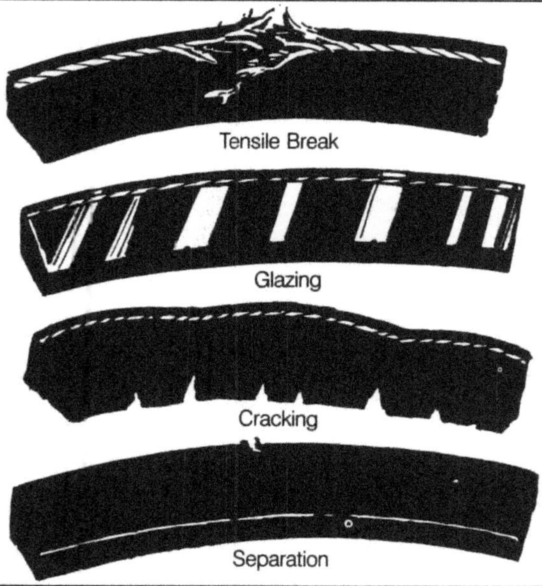

Fig. 6-3. Examples of belt failure.

V-belt squealing is normally caused by incorrect belt tension (too loose) or by dirt between the belt and pulley. Extremely loud squealing may only be corrected by replacing the belt. Belt dressings should not be used to correct the problem. Many dressings contain oil-based compounds which can soften the rubber and reduce belt life.

Check V-belt tension by checking the amount of deflection when the belt is pressed midway between two pulleys. Specifications are given in **Table h**. Alternator V-belt adjustment is described in **ELECTRICAL SYSTEM**.

Table h. V-Belt Deflection Specifications

Belt	Approximate deflection
Air conditioning compressor	5 to 10 mm ($^3/_{16}$ to $^3/_8$ in.)
Alternator	2 mm ($^1/_8$ in.) (checking and adjusting, see **ELECTRICAL SYSTEM**)
Power steering	10 mm ($^3/_8$ in.)

The power steering pump and air conditioning compressor V-belts are each adjusted using a tensioning bolt that is part of the components' respective mounting brackets. To adjust the tension, loosen the mountings and turn the tensioning bolt, clockwise to tighten the belt or counterclockwise to loosen it.

Replacing V-Belts

To reduce the chance of V-belt failure while driving, replacement of the belts every four years is recommended. Loosen the mounting bolts and adjust until the belt tension is very loose, then remove the belt by slipping it over the pulleys. In some cases it may be necessary to remove one V-belt to replace another. Cross section and length determine belt size. Use the old belt for comparison, or make sure that the new belt fits into the pulley groove as shown in Fig. 6-4.

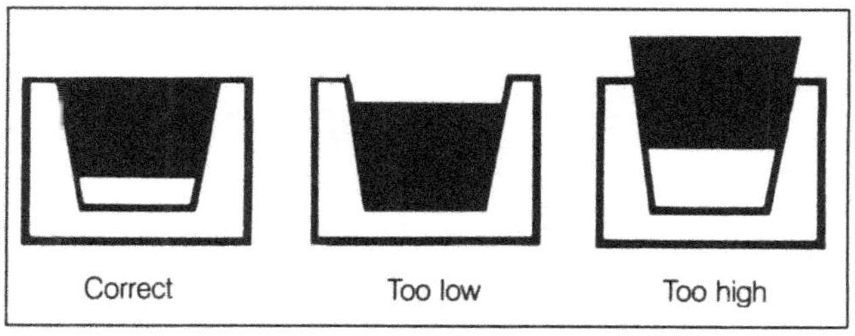

Fig. 6-4. Cross-section of correct V-belt position in pulley.

With the belt off, remove dirt and grease from the belt pulleys using a suitable solvent. Inspect the pulleys for wear or damage which may cause premature failure of the new belt. This is also a good opportunity to inspect the belt-driven accessory, checking for bearing wear and excess play, for example. When installing the new belt, gently pry it over the pulleys. Too much force may damage the belt. Tension the belt(s), run the engine for about 15 minutes, then recheck the tension.

6.3 Cooling System

Cooling system maintenance includes maintaining the coolant level and inspecting the hoses. Because the coolant's anti-corrosion and anti-freeze additives gradually lose their effectiveness, replacement of the coolant every 2 years is recommended. As a preventive measure, replacement of the cooling system hoses every 4 years is recommended.

CAUTION —

Use only phosphate-free anti-freeze when filling the cooling system. Use of anti-freeze containing phosphates is considered to be harmful to the cooling system and may void warranty coverage.

Checking Coolant Level

Because the expansion tank is translucent, the coolant level can be checked visually without opening the system. With the engine cold, the coolant level should be between the **MIN** and **MAX** marks on the expansion tank, as shown in Fig. 6-5. If it is necessary to add coolant, remove the cap and add it to the expansion tank only.

Fig. 6-5. Correct coolant level indicated by **MAX** and **MIN** marks on expansion tank.

Inspecting Hoses

Connections should be tight and dry. Coolant seepage indicates either that the hose clamp is loose, that the hose is damaged, or that the connection is dirty or corroded. Hoses should be firm and springy. Replace any hose that is cracked, that has become soft and limp, or has been contaminated by oil. See Fig. 6-6.

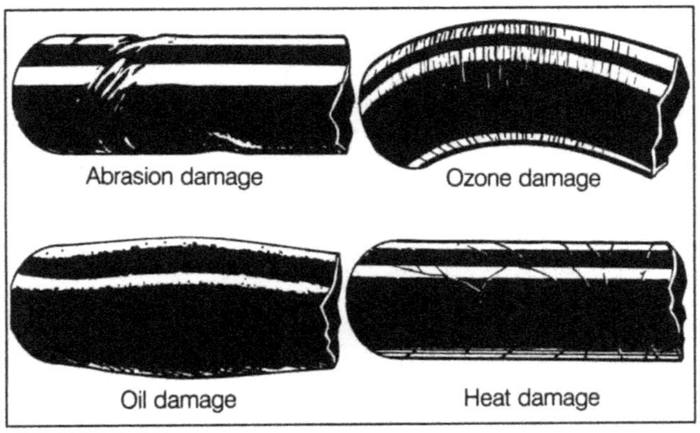

Fig. 6-6. Examples of damage to coolant hoses. Any of conditions shown is cause for replacement.

6.4 Power Steering

Check the power steering fluid level in the fluid reservoir, just to the right of the radiator as shown in Fig. 6-7. Park the car on level ground and start the engine. The fluid level is correct if it is between the **MAX** and **MIN** marks on the fluid reservoir. If the level is below the **MIN** mark, add fluid to the reservoir to bring the level up. Hand-tighten the reservoir cap.

NOTE —
The power steering fluid does not routinely need to be changed.

Fig. 6-7. Power steering fluid reservoir showing **MIN** and **MAX** marks which indicate correct fluid level.

6.5 Oxygen Sensor

The oxygen sensor monitors engine combustion efficiency by measuring the oxygen content of the exhaust gasses. That information in turn helps to control the fuel injection system. Any problems with the oxygen sensor will directly affect how well the engine runs and how well emissions are controlled.

Replacement of the oxygen sensor at the specified interval ensures that the engine and emission control system will continue to perform correctly with no interruption. Failure to replace the oxygen sensor when scheduled may void the emission control warranty.

On JH and 2H engines, the sensor is mounted in the exhaust manifold and is accessible from inside the engine compartment. On PL (16-valve) engines, the sensor is mounted in the exhaust system at the inlet to the catalytic converter. See **EXHAUST SYSTEM AND EMISSION CONTROLS** for more information on the oxygen sensor system.

Replacing Oxygen Sensor

The oxygen sensor connects to the wiring harness with an in-line, detachable connector. The sensor itself is threaded into place. When installing, apply a light coat of anti-seize compound to the threads, but be careful not to get any into the slits of the probe. Torque the sensor to 50 Nm (37 ft. lb.) and reconnect the wiring.

Resetting OXS Warning Light Mileage Counter

The mileage counter is located in the engine compartment, in-line with the speedometer cable, behind the coolant overflow reservoir. To reset the counter, first remove the reservoir by pulling it straight up from its mounting bracket, and then push in the counter reset button as shown in Fig. 6-8. It is not necessary to disconnect any of the coolant hoses to reach the counter.

NOTE —
Beginning with the 1988 models, the mileage counter and warning light have been eliminated. For these models, the oxygen sensor has a 60,000 mile (96,000 km) replacement interval.

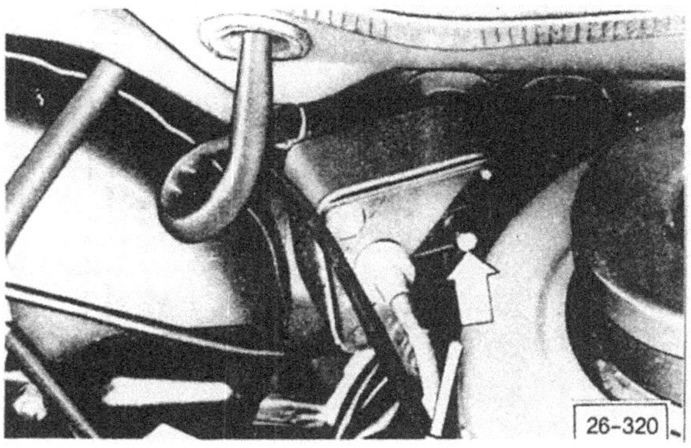

Fig. 6-8. Reset button (arrow) on mileage counter for OXS warning light. (Car shown is slightly different from those covered by this manual).

6.6 Clutch Adjustment

Those cars which are not equipped with a self-adjusting clutch cable require periodic adjustment to compensate for the wear of the clutch friction material. See **MANUAL TRANSMISSION AND CLUTCH** for information on identifying the two cable types and for the clutch cable adjustment procedure.

7. UNDER-CAR MAINTENANCE

7.1 Tires and Wheels

For stability and car control, the wheels and tires must be of the correct size and in good condition. Tires must be inflated to the correct recommended air pressures, and the wheels must be in proper alignment. For maximum safety and best all-around handling, always install replacement radial tires that are of the same specifications. When possible, all four tires should be replaced at once, or at least in pairs on the front or rear. New tires do not provide maximum traction, and should be broken in gently for the first 100 miles (160 kilometers) or so. Volkswagen warns against mixing tires of different design on the same car, i.e., bias ply with radial ply.

Tire Inflation Pressure

Correct tire inflation pressures are important to handling and stability, fuel economy, and tire wear. Tire pressures change with temperature. Pressures should be checked often during seasonal temperature changes. The correct inflation pressures can be found on a sticker in the glove box or on the rear edge of the right door. Notice that the pressures should be higher when the car is more heavily loaded.

WARNING —
Do not inflate any tire to a higher pressure than the tire's maximum inflation pressure listed on the sidewall.

All inflation pressures given are for cold inflation—that is, when the car has not been driven for at least three hours, or for more than one mile after sitting for at least three hours. Inflate radial snow tires (M + S) to 3 psi above normal pressure, and inflate all-season radials to normal pressures.

Tire Rotation

Volkswagen recommends tire rotation every 7,500 miles (12,000 km) to promote even wear and maximum tire life. Though it is a minor added expense, it is prudent to also have the tires balanced at the same time, to correct any imbalance due to wear. Rotate the tires as shown in Fig. 7-1.

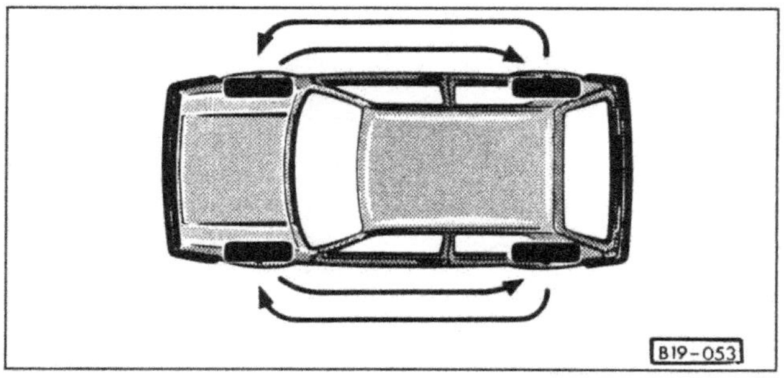

Fig. 7-1. Recommended tire rotation pattern, showing tires rotated from front to rear and rear to front on same side of car.

Wheel Alignment

Volkswagen makes no specific recommendation of how often the car should have a wheel alignment. It is a good idea to have the wheels aligned whenever they are rotated, or at least once a year. See **SUSPENSION AND STEERING** for a more detailed discussion of alignment requirements and specifications.

7.2 Brakes

Routine maintenance of the brake system includes maintaining an adequate level of brake fluid in the reservoir, checking brake pads and brake shoes for wear, and inspecting the system for fluid leaks or other damage.

WARNING —
Friction materials such as brake linings may contain asbestos fibers. Do not create dust by grinding, sanding, or cleaning the pads with compressed air. Avoid breathing asbestos fibers and asbestos dust, as it may result in serious diseases such as asbestosis and cancer, or in death.

WARNING —
Brake fluid is poisonous. Wear gloves when working with brake fluid to prevent contamination of cuts.

CAUTION —
Immediately wipe up any brake fluid that spills on painted surfaces. Brake fluid is very damaging to paint.

Checking Brake Fluid Level

The level of the brake fluid will drop slightly as the brakes wear. The level is checked at the brake fluid reservoir, located to the left of the coolant reservoir as shown in Fig. 7-2. Use only new brake fluid from previously unopened containers. See **3. Fluid and Lubricant Specifications** for brake fluid specifications.

Fig. 7-2. Level indicators on brake fluid reservoir. Correct level is between **MIN** and **MAX** marks.

Inspecting Brake Hoses and Lines

Gently bend the hoses to check for cracks, and make sure that they are routed correctly to avoid chafing or kinking. Inspect the unions and the brake calipers for signs of fluid leaks. Inspect the lines for corrosion, dents, or other damage. Replace faulty hoses or lines as described in **BRAKES**.

WARNING —
Incorrect installation or overtightening hoses, lines, and unions may cause chafing or leakage, leading to partial or complete brake system failure.

Checking Disc Brake Pad Wear

In most cases, brake pads can be inspected by looking through a hole in the wheel as shown in Fig. 7-3 to inspect the outer pad, and using a small mirror to inspect the inner pad. For a more thorough inspection, or to view the pads when the wheels have no suitable holes, raise the car and remove the wheel. Minimum brake pad thickness specifications are given in **Table i**.

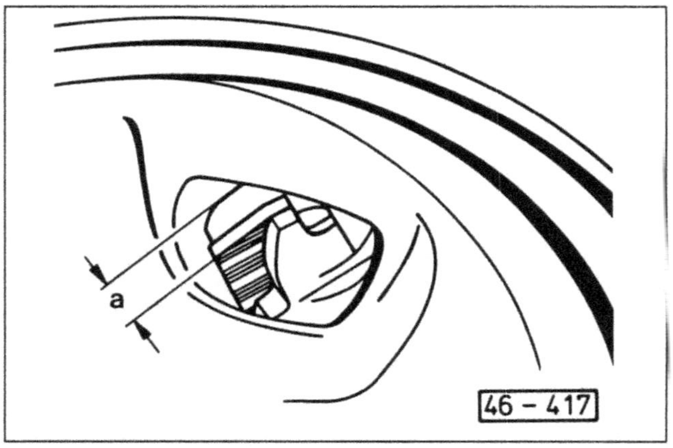

Fig. 7-3. Outer disc brake pad wear being checked through opening in wheel.

Table i. Brake Pad and Lining Minimum Thickness (Including brake pad backing plate or brake shoe)

Front disc brake pads	7 mm (.276 in.)
Rear disc brake pads	7 mm (.276 in.)
Rear Drum brake linings	2.5 mm (.098 in.)

Checking Rear Drum Brake Lining Wear

Drum brake lining wear can be checked from behind the brake backing plate after removing a rubber plug, as shown in Fig 7-4. Minimum thickness specifications are given in **Table i** above.

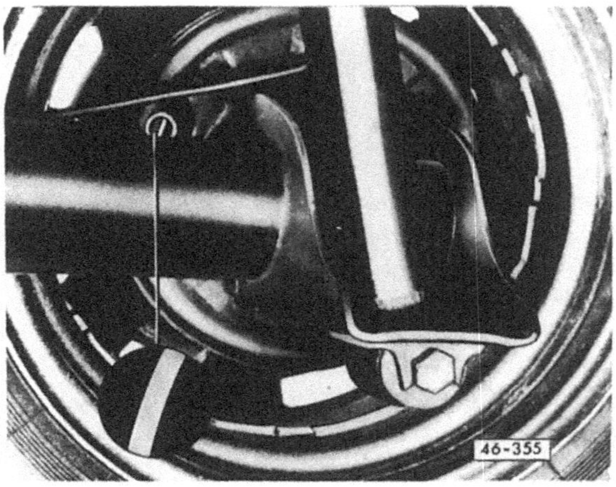

Fig. 7-4. Rear brake lining thickness being checked through hole in backing plate after removing rubber plug.

Replacing Brake Fluid

Volkswagen recommends replacement of the brake fluid every 2 years, to protect against corrosion and the effects of moisture absorbed into the brake fluid. The procedure is described in detail in **BRAKES**. If you lack suitable equipment to perform this job properly, service by an authorized Volkswagen dealer or other qualified repair shop is recommended.

7.3 Exhaust System

Exhaust system life varies widely according to driving habits and environmental conditions. If short-distance driving predominates, the moisture and condensation in the system will not fully dry out, leading to corrosion damage and more frequent replacement of systems or system components. Scheduled maintenance of the exhaust system is limited to inspection. Check for restrictions due to dents or kinks. Check for weakness or perforation due to rust. Check to see that all the hangers are in place and properly supporting the system and that the system does not strike the body. Alignment of the system and the location of the hangers are described in **EXHAUST SYSTEM AND EMISSION CONTROLS**.

7.4 Constant Velocity Joints

The constant velocity (CV) joints' protective boots must be closely inspected for cracks and any other damage which will allow contaminants to get into the joint. If the rubber boots fail, the water and dirt which enter the joint will quickly damage it. Replacement of the CV joint boots and inspection of the CV joints are described in **SUSPENSION AND STEERING**.

7.5 Manual Transaxle Service

The areas where transmission oil leaks are most likely to occur are around the drive axle mounting flanges on the final drive, and at the bottom of the bellhousing, between the transmission and the engine. For more information on identifying oil leaks and their causes, see **MANUAL TRANSMISSION**.

Checking and Filling Transaxle Lubricant

Because the manual transaxle oil is not contaminated by combustion by-products, the lubricant does not need to be *routinely* changed. There is no dipstick to check the oil level, and there should be no need to check the level or fill the transaxle unless leaks are detected. The oil filler and checking hole is shown in Fig. 7-5.

If leaks are present, check and fill the transaxle with the car on level ground. Use a 17 mm hex wrench to remove the plug. Add gear lubricant until it is level with the edge of the hole. When the level is correct, reinstall and torque the oil filler plug to 25 Nm (18 ft. lb.).

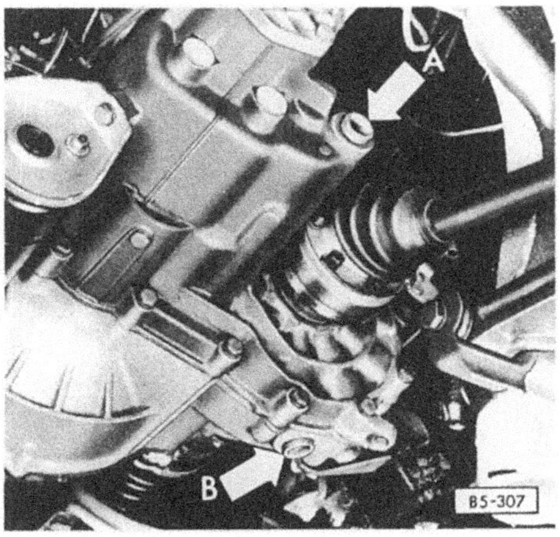

Fig. 7-5. Oil filler plug (**A**) and oil drain plug (**B**) on manual transmissions. Use 17 mm hex wrench to remove plugs.

7.6 Automatic Transaxle Service

Smooth and efficient operation of the automatic transmission relies on the automatic transmission fluid (ATF). Many automatic transmission problems can be traced to an incorrect fluid level, a clogged ATF strainer, or contaminated fluid. With regular preventative maintenance, expensive and unnecessary automatic transmission repair may be avoided.

Before checking the ATF level, inspect for leaks. ATF leaks are most likely to occur around the ATF pan gasket and at the bottom of the bellhousing, where the transaxle joins the engine. Final drive lubricant leaks will most likely occur around the drive flanges. All leaks should be corrected. If necessary, replace a leaky ATF pan gasket as described below. For more information, see **AUTOMATIC TRANSMISSION**.

CAUTION —
Extreme cleanliness is important when working on the automatic transmission. Use lint-free rags to check the level, and use a clean funnel when adding fluid.

Checking and Filling ATF

The location of the dipstick for checking the ATF is shown in Fig. 7-6. Two types of dipsticks are used on cars covered by this manual, as shown in Fig. 7-7. The level should be checked with the car on a level surface, the ATF warm, the engine idling, the parking brake firmly set, and the transmission selector lever in park. The ATF level is correct if it is between the **Min** and **Max** marks on the dipstick. The ATF is sufficiently warm if the car is driven approximately 6 mi. (10 km) prior to checking the level.

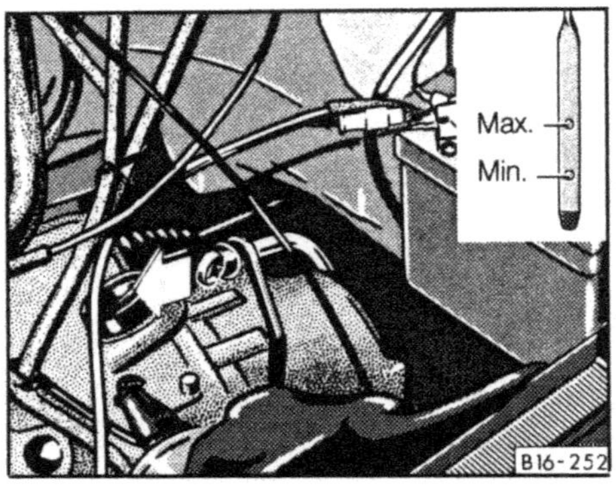

Fig. 7-6. ATF dipstick removed in direction of arrow. Inset shows marks on ATF dipstick. Level should be between **Min** and **Max** marks.

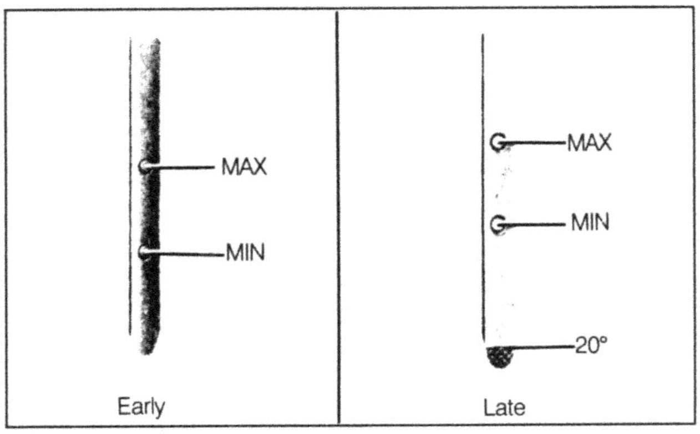

Fig. 7-7. Two versions of automatic transmission dipsticks used on earlier and later transmissions. Difference between marks is 0.33 L (.35 US qt.) on early version and 0.23 L (.24 US qt.) on later version.

If the level is too low, use a clean funnel to add ATF as specified in **3. Fluid and Lubricant Specifications** until the level falls between the two marks on the dipstick.

NOTE —
If the ATF level is incorrect, and a visual inspection does not indicate any external leaks, then an internal oil seal between the transmission and final drive may be faulty. Correcting the ATF level will not eliminate the problem. See **AUTOMATIC TRANSMISSION** for more information on faulty oil seals.

Check the condition of the ATF by rubbing some between fingers and sniffing it. The ATF should not be foamy or gritty, or have a burnt odor. Contaminated ATF should be drained and replaced to prevent further damage, but doing so will not repair any internal transmission damage that has already occurred.

NOTE —
Because Dexron II® ATF is a red/brown color that discolors to black/brown during normal use, ATF color may not be a good indicator of its condition.

Draining and Replacing ATF, and Cleaning ATF Strainer

With the car raised and supported securely on jack stands, place a drain pan of at least 6 US qt. capacity under the transmission and remove the two rear pan bolts. See Fig. 7-8. Loosen the two front pan bolts, then pull the ATF pan down at the rear and allow as much fluid as possible to drain. Then remove the front bolts and remove the pan and gasket. Pour out the fluid remaining in the pan.

CAUTION —
Towing the car or running the engine with no fluid in the transmission will ruin the transmission bearings.

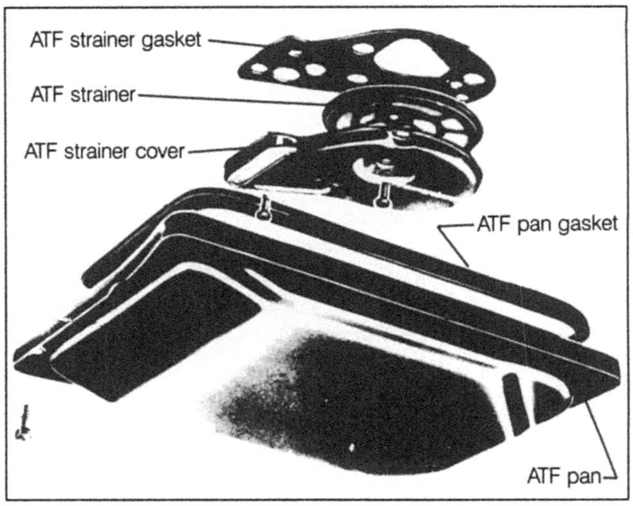

Fig. 7-8. Exploded view of ATF pan and strainer components. Always use new gasket without sealer.

Clean the pan and dry it completely. Remove the two screws retaining the strainer and remove the cover and the strainer. The strainer can be cleaned if it is dirty or clogged. Replace a damaged strainer or one that cannot be thoroughly cleaned. Remount the strainer using a new gasket. Torque the screws to 4 Nm (35 in. lb.). Using a new gasket without any sealer, install the ATF pan. Working diagonally, tighten the pan bolts to 20 Nm (15 ft. lb.). Refill the transmission with fluid according to the type and amount specified in **3. Fluid and Lubricant Specifications**. Then check the fluid level as described above.

NOTE —
On 1985 and 1986 models, transmission slippage during cornering or at stops may only be caused by a low ATF level. If, however, the level is correct and the problem still occurs, there may be another cause. A redesigned strainer cover (Volkswagen part no. 010 325 429A) and gasket (Volkswagen part no. 010 325 443) are available to correct the problem.

Checking and Filling Final Drive Lubricant

Because the oil is not contaminated by combustion by-products, the final drive lubricant does not need to be routinely changed. Checking the lubricant level should only be necessary if hypoid oil leaks are discovered.

Check the lubricant level with the car on a level surface. Remove the oil filler plug, shown in Fig. 7-9, using a 17mm hex wrench. The level is correct when the fluid just reaches the edge of the filler hole. Torque the oil filler plug to 25 Nm (18 ft. lb.).

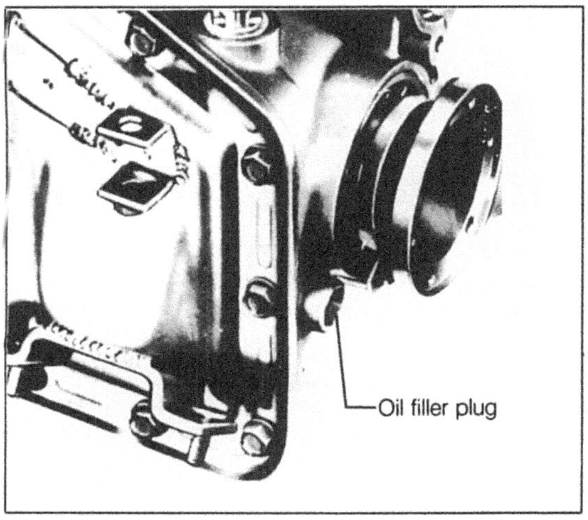

Fig. 7-9. Oil filler plug location on automatic transaxle final drive, behind right-side drive flange. Remove with 17 mm hex wrench.

Hypoid oil leaks from the final drive housing cover may be corrected by replacing the final drive housing cover gasket with a new metal gasket, Volkswagen part no. 010 409 597A. Apply sealing compound (Volkswagen D3 or equivalent) to the cover mounting bolts. Torque to 75 Nm (55 ft. lb.). Refill the final drive gear oil with the type and amount specified in **3. Fluid and Lubricant Specifications**. Then recheck the fluid level.

Checking Kickdown Operation

Kickdown operation is a feature designed into the transmission which gives maximum acceleration when the accelerator pedal is depressed to the full-throttle position. Depending on road speed and engine speed, kickdown either delays the upshift or downshifts into the next lower gear. The kickdown specifications vary with the exact type of transmission installed. See **AUTOMATIC TRANSMISSION** for information on how to identify a particular transmission.

Table j lists kickdown shift points. Shifts should be smooth and positive. Incorrect shift points simply mean that the accelerator cable is misadjusted. Slow shifting may mean that the bands or clutches are slipping. See **AUTOMATIC TRANSMISSION** for more troubleshooting information.

Table j. Kickdown Shift Points

Shift points in mph (km/h)	
Shift	Transmission code TN, TNA
1–2	40–43 (65–69)
2–3	69–71 (112–114)
3–2	66–67 (106–108)
2–1	31–33 (50–53)

7.7 Front Suspension and Steering

Inspection of the front suspension and steering includes a check of all moving parts for wear and excessive play, as well as checks of the condition of rubber seals and boots which are designed to keep the joints free of dirt, water, and other wearing contaminants. Complete front suspension inspection and troubleshooting information can be found in **SUSPENSION AND STEERING**.

8. BODY AND INTERIOR MAINTENANCE

8.1 Windshield Wiper Blades

Common problems with the windshield wipers include streaking or sheeting, water drops after wiping, and blade chatter. Streaking, illustrated in Fig. 8-1, is usually caused when wiper blades are coated with road film or car wash wax. The blades can be cleaned with alcohol or an alcohol-based cleaning solution, using a lint-free cloth or hard-bristle brush. If cleaning the blades does not cure the problem then they should be replaced.

Fig. 8-1. Streaking caused by dirty, worn, or damaged wiper blades.

Drops which remain behind after wiping, as shown in Fig. 8-2, are caused by oil, road film, or diesel exhaust coating the windshield. Again, use an alcohol or ammonia solution, or a non-abrasive cleanser to clean the windshield.

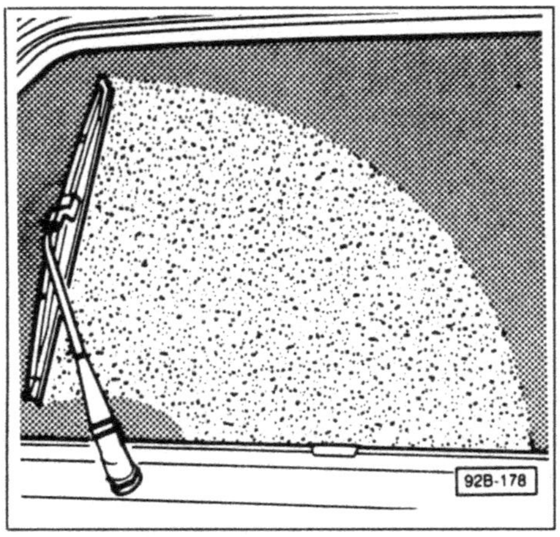

Fig. 8-2. Water beading after wiping caused by oil and dirt contaminating windshield.

Wiper blade chatter, illustrated in Fig. 8-3, may be caused by dirty or worn blades, by a dirty windshield, or by bent or twisted wiper arms. Clean the blades and windshield as described above. Adjust the wiper arm so that there is even pressure along the blade, and so that the blade is perpendicular to the windshield at rest. If the problem persists, the blades are excessively aged or worn and should be replaced.

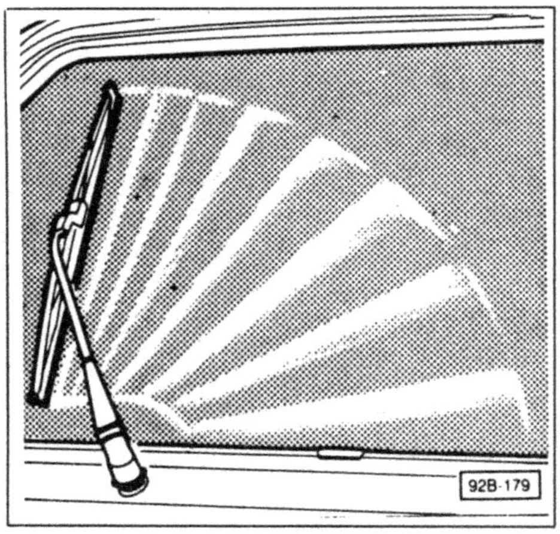

Fig. 8-3. Blade chatter caused by dirty or worn blade, or by twisted or bent wiper arm.

8.2 Body Lubrication

The door locks and lock cylinders can be lubricated with an oil that contains graphite. The body and door hinges, the hood latch, and the door check rods should be lubricated with SAE 30 or SAE 40 engine oil. Lubricate the seat runners with multipurpose grease. Do not apply any oil to rubber parts. Lubricate the sunroof guide rails with silicone spray. If door weatherstripping is sticking, lubricate it with silicone spray or talcum powder.

8.3 Seat Belts

Dirt and other abrasive particles will damage seat belt webbing. If it is necessary to clean seat belts, use a mild soap solution. Bleach and other strong cleaning agents may weaken the belt webbing.

The condition of the belt webbing and the function of the retractor mechanisms should be inspected. See **BODY AND INTERIOR** for seat belt inspection information.

9. CLEANING AND PRESERVING

9.1 Care of Exterior Finish

The longer dirt is left on the paint, the greater the risk of damaging the glossy finish, either by scratching or by the chemical effect dirt particles may have on the painted surface.

Washing

Washing the car in direct sunlight is not recommended. Beads of water not only leave spots when dried rapidly by the sun, but also act as tiny magnifying glasses that can burn spots into the finish. Use plenty of water, a mild soap, and a soft sponge.

Begin by spraying water over the dry finish to remove all loose dirt. Wash with lukewarm soapy water. Special car wash soap can be used, but a gentle soap such as Ivory® can do the job just as well at less expense. The idea is to get rid of the dirt without removing the wax underneath. Rinse using plenty of clear water under as little pressure as possible. Wipe the body dry with a soft terry-cloth towel or chamois to prevent water-spotting.

Waxing

For a long-lasting, protective, and glossy finish, apply a hard wax after the car has been washed and dried. Waxing is not needed after every washing. You can tell when waxing is required by looking at the finish when it is wet. If the water coats the paint in smooth sheets instead of forming beads that roll off, the old wax is gone and a new coat of wax is needed.

Polishing

Use paint polish only if the finish assumes a dull look after long service. Polish can be used to remove tar spots and tarnish, but afterwards a coat of wax should be applied to protect the clean finish. Do not use abrasive polish or cleaners on aluminum trim or accessories.

Washing Chassis

Periodic washing of the underside of the car, especially in winter, will help prevent accumulation of road salt and rust. The best time to wash the underside is just after the car has been driven in wet conditions. Spray the chassis with a powerful jet of water to remove dirt and de-icing salt that may have accumulated there. Commercial or self-service car washes may not be best for this, as they may recycle the salt-contaminated water.

Special Cleaning

Tar spots can be removed with tar remover. Never use gasoline, kerosene, nail polish remover, or other unsuitable solvents. Insect spots also respond to tar remover. A bit of baking soda dissolved in the wash water will facilitate their removal. This method can also be used to remove spotting from tree sap.

9.2 Care of Interior

Clean the carpet with a vacuum cleaner or whisk broom. Dirt spots can usually be removed with lukewarm soapy water. Use spot remover for grease and oil spots. Do not pour the liquid directly on the carpet, but dampen a clean cloth and rub carefully, starting at the edge of the spot and working inward. Do not use gasoline, naptha, or other flammable substances to clean the carpeting.

Leatherette Upholstery and Trim

Use a dry foam cleaner. Grease or paint spots can be removed by wiping with a cloth soaked with this cleaner. Use the same cleaner, applied with a soft cloth or brush, on the headliner and side trim panels. For cloth-covered seat areas, use the techniques described above for cleaning the carpeting.

ENGINE

Contents

TABLES

Engine

Introduction

The Cabriolet and Scirocco engines are all four-cylinder, front-mounted and water-cooled. The cast iron cylinder block supports the crankshaft in five main bearings. The overhead camshaft running in the cylinder head is driven from the crankshaft by a reinforced toothed belt. Displacement is 1781 cc (109 cu. in.), commonly referred to as 1.8 liters. Horsepower is 94 (SAE net). The Scirocco 16V engine is also 1.8 liters, but its cross-flow cylinder head design and dual overhead camshafts with four valves per cylinder (two intake and two exhaust) raises rated horsepower to 123 (SAE net).

The engine is transverse-mounted and inclined toward the rear of the car—a design that permits a lower hood line and improves weight distribution. The engine and transaxle are supported as a single unit by a combination of bonded rubber mounts that isolate the drivetrain and help reduce the transmission of noise and vibration to the rest of the car.

The engine's transverse placement takes up less overall space than other designs and makes possible a roomier passenger compartment. It also concentrates drivetrain weight over the front driving wheels where it helps improve traction under slippery conditions. Despite a slight forward weight bias when the car is empty, total vehicle weight is distributed about equally on the front and rear wheels when the car is loaded. Near-equal weight distribution increases stability and assures precise handling.

The information in this section of the manual is intended both as a guide to Cabriolet and Scirocco owners and as a reference for professional mechanics. Some of the procedures and repairs described require special equipment and some prior knowledge of general engine repair and assembly practices. If you lack the skills, tools, or a suitable workplace for servicing or repairing the engine, we suggest you leave these repairs to an authorized Volkswagen dealer or other qualified automotive repair shop. We especially urge you to consult your authorized Volkswagen dealer before beginning any repairs on a car that may be covered wholly or in part by any of the extensive warranties issued by Volkswagen United States, Inc.

1. GENERAL DESCRIPTION

The in-line four-cylinder overhead cam engine is water-cooled and transmits power through a piston-driven crankshaft. The cylinder block is made of cast iron with integral cylinders completely exposed on all sides to the coolant that circulates through the water jacket. A separate cast aluminum alloy cylinder head contains the belt-driven camshaft and the entire valve train. A cutaway view of the engine appears in Fig. 1-1.

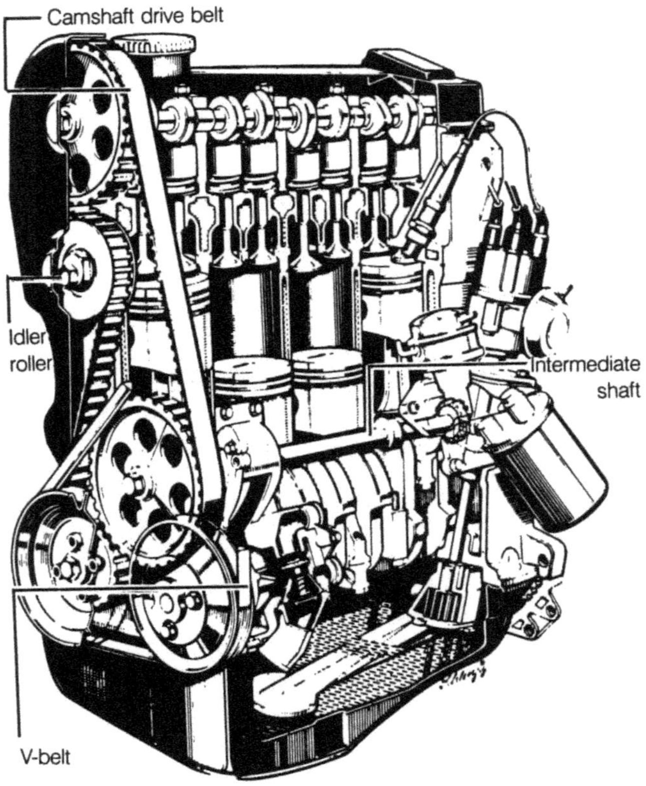

Fig. 1-1. Cutaway view of engine.

1.1 Engine Components

Crankshaft and Bearings

The fully counterweighted crankshaft rotates in five replaceable split-shell main bearings. A 6-piece center main bearing (upper shell, lower shell, and four thrust bearings) controls crankshaft end thrust. Flexible lip seals, pressed into light alloy seal carriers, are used at both ends of the crankshaft to prevent oil leakage.

Connecting Rods and Pistons

The connecting rods are steel forgings. Replaceable split-shell bearings are used at the crankshaft end and lead-bronze coated steel bushings at the piston pin end. The pistons are of the three-ring type with two upper compression rings and a lower one-piece oil scraper ring. Full-floating piston pins are retained at each end by circlips.

Cylinder Head

The cylinder head is an aluminum alloy casting. The overhead camshaft bearing surfaces and the bores for the cam followers (valve lifters) are machined directly into the cylinder head casting. The valve guides are press-fit and replaceable.

Valve Train

The single overhead camshaft is driven from the crankshaft by a toothed, steel-reinforced belt. The cam lobes and bucket-style cam followers directly actuate the intake and the exhaust valves. Valve clearance is adjusted by fitting the cam followers with shims of different sizes. Starting with the 1988 model year, 8-valve engines have hydraulic cam followers. These use engine oil pressure to maintain correct valve clearances. Intake and exhaust valves are equipped with dual valve springs and both upper and lower spring seats.

Intermediate Shaft

The intermediate shaft, turning in the cylinder block, is located above and parallel to the crankshaft. It is driven at half crankshaft speed by the camshaft drive belt, and in turn drives the oil pump and the ignition distributor shaft.

Lubrication System

A gear-type oil pump driven by the intermediate shaft draws oil through a strainer in the bottom of the oil pan and forces it through a spin-on replaceable filter and into the engine's oil passages. A pressure relief valve limits the pressure in the system, and a filter bypass valve assures lubrication even if the filter is plugged. An oil cooler is attached to the filter housing. Engine coolant circulates through the oil cooler to help moderate oil temperature. In addition to a warning system for low oil pressure, a more sensitive dynamic oil pressure warning system operates at elevated engine rpm.

1.2 16-valve Engine

The 16-valve engine is a highly developed version of the basic 1.8 liter engine, sharing most of its major design features. There are several differences, in the cylinder head as well as other related systems, which distinguish the 16-valve version.

Cylinder Block

The cylinder block used for the 16-valve engine has several modifications to help make it more suitable to the higher power output, primarily in the lubrication system. The passageways connecting the crankcase and the upper cylinder head are larger, to allow better ventilation and oil return, and an additional oil passage in the block supplies oil to a series of spray nozzles, one below each cylinder. At oil pressure above 3.5 bar (50 psi), the nozzles open and small streams of oil are sprayed at the underside of each piston to provide additional cooling. The cylinder block casting is also modified to accept a larger crankcase breather.

Pistons

The 16-valve engine, because of its different cylinder head design and valve arrangement, uses a different piston than the other 1.8 liter engines to achieve the necessary combination of compression ratio and valve clearance.

Cylinder Head

The cylinder head is, of course, the 16-valve engine's major departure from Volkswagen's other 1.8 liter engines. Its cross-flow design features four valves per cylinder, versus the usual two, operated by two overhead camshafts. The result is more efficient air flow into and out of the combustion chambers, and an engine with greater efficiency and power over a wide rpm range.

The cylinder head is an aluminum alloy casting with sintered steel valve seats. The press-fit valve guides are replaceable, while the bearing surfaces for the dual camshafts and the bores for the cam followers are machined directly into the cylinder head casting. Fig. 1-2 shows the 16-valve cylinder head and its four-valve combustion chambers.

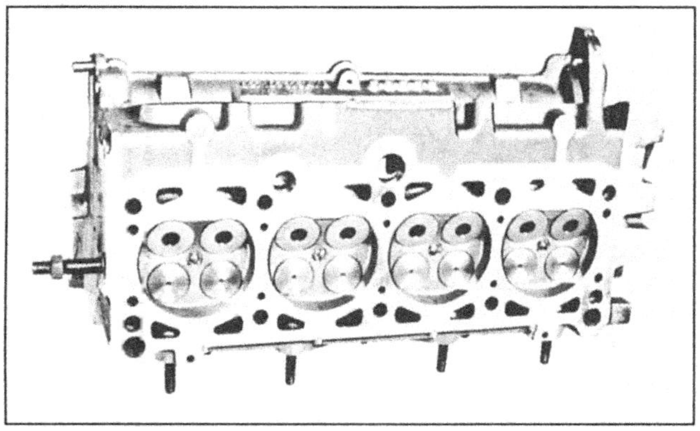

Fig. 1-2. 16-valve cylinder head showing orientation of intake valves (top) and exhaust valves (bottom).

Valve Train

The camshafts are driven by the crankshaft via a single toothed, steel-reinforced belt. The belt drives a single sprocket attached to the exhaust camshaft. The exhaust camshaft in turn drives the intake camshaft through a roller chain and sprockets at the other end of the cylinder head. The drive belt and sprockets, although similar to those of the other 1.8 liter engines, are modified to handle the extra loads of the 16-valve valve train. Fig. 1-3 shows the orientation of the camshafts and drive mechanisms in the assembled cylinder head. The cam lobes operate the intake and exhaust valves directly through hydraulic cam followers which use engine oil, supplied under pressure by the lubrication system, to maintain correct valve clearances. A spray-jet assembly, similar to that used to cool the pistons, is used in the cylinder head to prevent oil from draining out of the supply gallery for the cam followers. Sodium-filled exhaust valves aid heat transfer and reduce valve seat temperature. Dual valve springs and both upper and lower spring seats are used on the intake and exhaust valves.

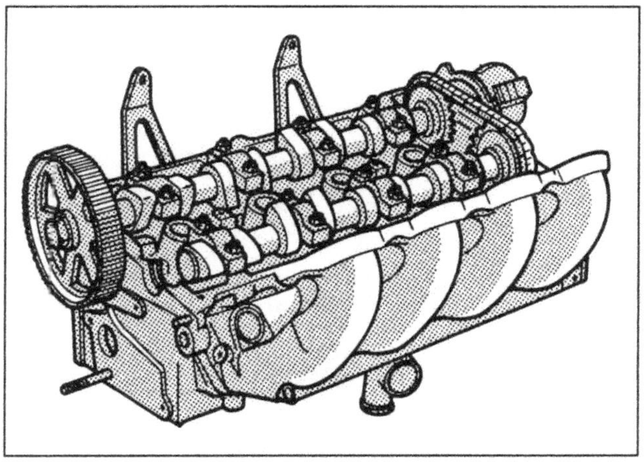

Fig. 1-3. 16-valve cylinder head and valve train.

Intermediate Shaft

The ignition distributor on the 16-valve engine is driven directly by the exhaust camshaft. The intermediate shaft, which drives only the oil pump, is driven at 60% of crankshaft speed by the camshaft drive belt.

1.3 Engine Identification Codes and Specifications

Different versions of the basic four-cylinder engine are used in the cars covered by this manual. Volkswagen uses a unique engine code to identify each version and to distinguish between their different parts and specifications. The engine code is the single most important piece of information for you to know about your engine. Throughout this manual, where different procedures or specifications may apply, the engine codes are used to identify the correct information for each engine.

Finding Engine Codes

The engine code can be viewed in the engine compartment. The code, JH for example, appears at the beginning of the engine number. The number is stamped on a flat area of the top of the engine block, just below the lower edge of the cylinder head, between cylinders No. 3 and No. 4. The engine number stamp is shown in Fig. 1-4. **Table a** lists Cabriolet and Scirocco engine codes, application information, and major engine specifications.

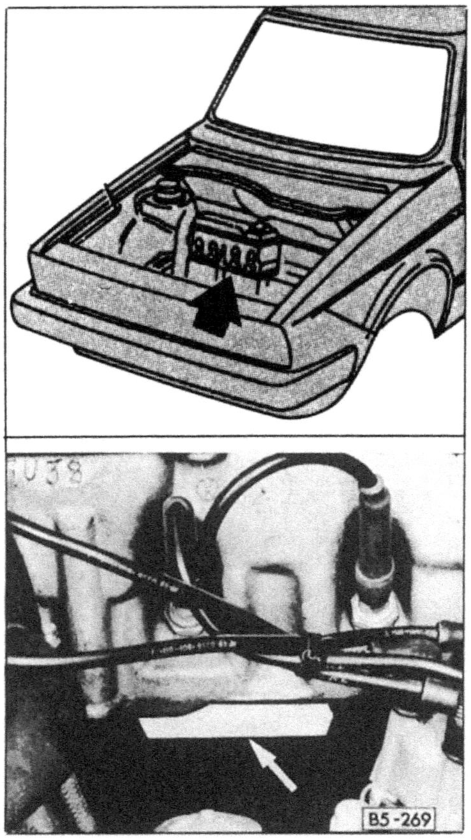

Fig. 1-4. Engine code and production number stamped on cylinder block between cylinders three and four.

Table a. Engine Codes and Specifications

Engine code	**JH**	**2H**	**PL**
Number of cylinders	4		
Bore – mm (in.)	81.0 (3.19)		
Stroke – mm (in.)	86.4 (3.40)		
Displacement – cc (cu. in.)	1781 (109)		
Compression ratio	8.5:1	10:1	10:1
Horsepower – SAE net @ rpm	94 @ 5500	94 @ 5400	123 @ 5800
Torque – lbs-ft @ rpm	102 @ 3000	100 @ 3000	120 @ 4250
Fuel injection / Engine management	CIS	Digifant	CIS-E
Application notes	–	from 1990	Scirocco 16V

2. MAINTENANCE

Volkswagen specifies the maintenance steps below, to be carried out at specific time and mileage intervals. A number in bold type indicates that the procedure is in this section, under that numbered heading. See **LUBRICATION AND MAINTENANCE** for other listed items, and for the prescribed intervals.

1. Checking engine oil level
2. Changing engine oil and oil filter
3. Replacing spark plugs
4. Checking engine compression pressure. **3.2**
5. Adjusting valve clearance (with mechanical cam followers, 1985-1987 8-valve engine only). **4.4**
6. Replacing cylinder head cover gasket (only with valve adjustment listed above). **4.1**

3. TROUBLESHOOTING

The principles of internal combustion can be very complex, but the basic engine functions are pretty simple. The potential for problems is limited.

Wear is the main cause of any problem in an engine that sees normal use and receives proper maintenance. Wear affects the ability of the pistons, piston rings and valves to seal the combustion chambers—to create and maintain compression. Wear increases clearances between moving parts. Finally, wear affects the components that keep the crankshaft and pistons, valve train and ignition properly timed.

This troubleshooting section deals with the basic engine functions. To investigate an engine oil pressure warning, or to troubleshoot the lubrication system, see **7. Lubrication System** in this section. To troubleshoot other, more general problems with engine cooling or the way the engine starts or runs, see **COOLING SYSTEM, FUEL SYSTEM** and **IGNITION**.

3.1 Basic Troubleshooting Principles

To isolate and identify engine problems, always begin with careful observation of the symptoms. Some important things to look for are:

How quickly did the problem develop? Problems that occur suddenly are probably caused by a failure that can be corrected by some sort of repair. A problem that has developed gradually, especially after years and many tens of thousands of miles, is more likely brought on by wear. It may mean that more comprehensive overhaul work is required.

Does the symptom change with engine speed? If so, the problem is most likely an engine problem. Noises which repeatedly occur only in a certain rpm range suggest a vibration problem. Noises or other symptoms that vary with the speed of the car are more likely caused by problems somewhere else in the drivetrain or running gear.

Is the symptom load dependent? Forces at work inside a running engine vary as the demand for power varies. Symptoms which are more severe during hard acceleration indicate certain kinds of problems. Symptoms which are more apparent at no load or high vacuum (example: coasting at high rpm) point to other problems. Note that higher engine loads also affect the fuel and ignition systems, which may be responsible for high-load performance problems.

Is the symptom temperature dependent? Does it only occur when the engine is cold? Does it change as the engine warms up? How? Metal parts expand and contract with changes in temperature. Clearances change. Oil viscosity and cooling system pressure change. In troubleshooting symptoms which change as the engine gets warm, look for an engine characteristic that changes with temperature.

The paragraphs which follow describe particular types of problems and some of the basic troubleshooting principles associated with them. **Table b** lists symptoms of problems commonly associated with engine trouble, their probable causes, and suggested corrective actions. The numbers in bold type in the corrective action column refer to numbered headings in this section where the suggested repairs are described.

Noise

In order to run reliably and smoothly under harsh conditions, the internal engine parts are manufactured to precise dimensions, assembled with precision clearances between moving parts, and lubricated by a pressurized oiling system.

Most unidentified engine noises result from clearances which have become too large due to worn or failed parts, lack of adequate lubrication, or both. The importance of lubrication cannot be over-emphasized. For best results, troubleshooting engine noises should only be done when the oil and filter have been recently changed and the oil level is correct.

High-pitched metallic tapping noises are caused by relatively small, lightweight parts and are most likely an indication of excessive clearances in the valve train. Valve train noise accompanied by burning oil (blue-gray smoke in the exhaust), particularly at startup or when decelerating from high rpm, is an indication of worn valve guides which can only be remedied by overhaul or replacement of the cylinder head. In a high-mileage engine, a light metallic rattle or chatter under acceleration, accompanied by increased oil consumption and smoking, may indicate severely worn or broken piston rings. Since this diagnosis means overhaul or replacement of the engine, the problem should be further investigated with a compression or cylinder leakage test. See **3.2 Diagnostic Testing**.

Deep, metallic knocking sounds are caused by excessive clearances between heavier components. Closer analysis of the noise will often help identify the problem. Piston slap, caused by excessive piston skirt to cylinder wall clearance, is worse when the engine is cold and may be accompanied by increased oil consumption and reduced compression due to accelerated piston ring wear. A double knock, most pronounced at idle or low load, is due to excessive clearance at the piston pin and upper connecting rod bushing.

Crankshaft bearing problems produce a deep, hollow knock that is worst when the engine is warm. A noise that is very pronounced under load, perhaps louder during the transition from acceleration to coasting, is most likely caused by a damaged connecting rod bearing. Crankshaft main bearings produce a lower, dull knock. An intermittent knock, indicating excessive crankshaft end play, may be most apparent when depressing or releasing the clutch. These problems seldom occur as isolated failures. They are almost always an indication of the overall engine condition which can only be properly corrected by complete engine overhaul or replacement.

Rumbling or groaning from the engine compartment may not indicate engine problems at all, but rather a worn bearing or bushing in an engine-driven accessory. They include the coolant pump, alternator, and may include a power steering pump and air conditioning compressor. The air conditioning compressor is equipped with an electrically-switched clutch-type pulley, so a bad compressor will only be noisy when the air conditioning is on. To check other accessories, run the engine briefly with the drive belt disconnected and see if the noise has stopped. Once the drive belt is removed, turning the pulley and shaft by hand may also reveal a bad bearing or bushing. A properly functioning accessory should turn smoothly.

Fluid Leaks

Fluid leaking from and around the engine is most likely either oil, coolant, or brake fluid. Look for wet spots on the engine to help pinpoint the source. It may be helpful to start by cleaning the suspected area.

The most likely sources of engine oil leaks are the oil filter gasket, the crankcase oil seals, the cylinder head cover gaskets, or the oil pan gaskets. See **6. Cylinder Block and Pistons** for more information on the gaskets and seals.

The power steering system is another possible source of oil leaks near the engine. For repairs to the power steering system, see **SUSPENSION AND STEERING**.

Coolant is a mixture of water and anti-freeze, yellow-green in color or perhaps brown if the cooling system is corroded. A pressure test of the cooling system is the best way to discover and pinpoint leaks. See **COOLING SYSTEM**.

Brake fluid is clear, perhaps slightly purple, and a little slippery. Look for wet spots around the master cylinder or brake lines. Especially check the flexible hoses near the wheels. See **BRAKES** for repair information.

Smoking

Smoke which is visible under the hood will be either blue-gray smoke from burning oil, or white steam from the cooling system. Both symptoms indicate a leak. See **Fluid Leaks** above.

Smoke in the exhaust indicates something getting into the combustion chamber and being burned which does not belong there. The color of the smoke identifies the contaminant.

Blue-gray smoke is from oil. Oil smoke, probably accompanied by increased oil consumption and oil residue on spark plugs, indicates that engine oil is getting past piston rings, valve guides, the cylinder head gasket, or some combination of the three. Use a compression test for diagnosis. See **3.2 Diagnostic Testing**. Compression pressures in an older engine which are even, but below specifications, point to piston ring and cylinder wall wear and the need for engine overhaul or replacement. If the smoking is most obvious under high engine vacuum, such as while coasting at high rpm, and compression pressures are within specifications, leaking valve guide oil seals or valve guides are a probable cause. See **4. Cylinder Head** for repair information.

Oil smoke or steam appearing suddenly in the exhaust, along with low compression pressure in one cylinder or two adjoining cylinders, is very probably due to a failed cylinder head gasket. Look also for coolant loss, oil in the radiator, or water in the oil (which turns the oil an opaque, creamy brown). See **4.8 Removing and Installing Cylinder Head** for repair procedures.

Black smoke is caused by the engine getting too much fuel. See **FUEL SYSTEM** for more troubleshooting information.

Excessive Oil Consumption

Some oil consumption is normal and indicates healthy flow and distribution of the vital lubricant in the engine. This is why oil level must be checked, and occasionally corrected, between oil changes. Aside from leaks, increased oil consumption will usually be accompanied by some smoking, however slight, and the causes of excessive oil consumption are the same as those for oil smoke in the exhaust. As with smoking symptoms, gradual increases are caused by worn piston rings and/or valve guides. Sudden high oil consumption suggests broken rings or a failed cylinder head gasket. See **Smoking** above for more troubleshooting information.

Poor Fuel Consumption and Low Power

Poor fuel consumption and low power can, of course, suggest problems with the fuel or ignition systems, particularly on a low-mileage engine. On an engine with high mileage, suffering the effects of wear, low compression may be the cause.

Normal wear of the valves, piston rings, and cylinder walls decreases their ability to seal. The intake and compression of the air/fuel mixture becomes less efficient, and the engine has to work harder, using more fuel, to produce the same amount of power. Engine condition can be evaluated with a compression test. See **3.2 Diagnostic Testing**.

Engine Not Running

An engine problem which affects timing may prevent the engine from starting or running. The camshaft drive belt and sprockets are responsible for timing the actions of the valves and the ignition system relative to the pistons and crankshaft. A worn belt and sprockets may jump teeth, throwing off all the engine's timing functions, and still appear to be perfectly normal. To check camshaft and ignition timing, see **4.2 Camshaft Drive Belt** and **IGNITION**. Other troubleshooting information for an engine which fails to start can be found in **ELECTRICAL SYSTEM**, **FUEL SYSTEM**, and **IGNITION**.

Table b. Engine Troubleshooting

Symptom	Probable cause	Corrective action
1. Pinging or rattling noise under load, uphill or accelerating, especially from low speeds. Indicates detonation or pre-ignition	**a.** Ignition timing too far advanced **b.** Fuel does not meet manufacturer's octane requirements **c.** Overheating **d.** Spark plugs damaged or wrong heat range **e.** Air/fuel mixture too lean	**a.** Correct ignition timing. See **IGNITION** **b.** Switch to higher octane fuel. See **FUEL SYSTEM** for fuel octane requirements **c.** See **COOLING SYSTEM** **d.** Replace spark plugs. See **IGNITION** **e.** See **FUEL SYSTEM**
2. Screeching or squealing noise under load. Goes away when coasting. Indicates slipping V-belt	**a.** Loose, worn, or damaged V-belt(s) **b.** Excessive belt loads due to failed engine-driven component	**a.** Inspect, replace, or tighten belt(s). See **LUBRICATION AND MAINTENANCE** **b.** Locate and replace failed component. **3.2**

continued on next page

Table b. Engine Troubleshooting (continued)

Symptom	Probable cause	Corrective action
3. Growling or rumbling, varies with engine rpm. Bad bearing or bushing in an engine-driven accessory	**a.** Coolant pump **b.** Alternator **c.** Power steering pump **d.** Camshaft drive belt tensioner bearing **e.** Air conditioning compressor	**a.** See **COOLING SYSTEM** **b.** See **ELECTRICAL SYSTEM** **c.** See **SUSPENSION AND STEERING** **d.** Replace belt tensioner. **4.2** **e.** Replace compressor
4. Light metallic tapping noise, varies directly with engine speed. Oil pressure warning light **not** illuminated	**a.** Low oil pressure and defective warning light circuit **b.** Valve lash out of adjustment (1985–1987, engine code JH) **c.** Defective cam follower(s)	**a.** Check oil pressure and/or warning system. **7.** **b.** Adjust valve lash. **4.4** **c.** Check cam followers and replace as required. **4.6**
5. Light metallic knock, varies directly with engine speed. Oil pressure warning light blinking or fully illuminated (may be most noticeable after hard stops or during hard cornering) indicates lack of sufficient oil supply	**a.** Low oil level **b.** Restricted (dirty) oil filter **c.** Insufficient oil pressure	**a.** Check and correct oil level. See **LUBRICATION AND MAINTENANCE** **b.** Change engine oil and filter. See **LUBRICATION AND MAINTENANCE** **c.** Check oil pressure. **7.**
6. Blue-gray exhaust smoke, oily spark plugs. Indicates oil burning in combustion chamber	**a.** Leaking valve stem oil seals **b.** Worn valve guides **c.** Worn or broken pistons or piston rings	**a.** Replace valve stem oil seals. **4.7** **b.** Overhaul or replace cylinder head. **4.** **c.** Overhaul or replace engine
7. Blue-gray smoke and/or white steam in exhaust	**a.** Failed cylinder head gasket (probably accompanied by low compression readings). See **3.2** **b.** Warped or cracked cylinder head (probably accompanied by low compression readings). See **3.2** **c.** Cracked cylinder block	**a.** Replace cylinder head gasket. **4.8** **b.** Resurface or replace cylinder head gasket. **4.8** **c.** Replace engine or short block. **5.**
8. Black exhaust smoke	**a.** Rich air/fuel mixture	**a.** See **FUEL SYSTEM**
9. Engine runs badly, pops and backfires	**a.** Spark plug wires installed incorrectly **b.** Incorrect valve timing	**a.** Install wires correctly. See **IGNITION** **b.** Check camshaft drive belt and camshaft timing. **4.2**
10. Engine will not start or run. Starter operates, engine turns over at normal speed	**a.** Failed ignition system **b.** Broken camshaft drive belt **c.** Incorrect camshaft timing due to jumped belt or incorrect assembly	**a.** See **IGNITION** **b.** Check cam sprocket rotation as engine turns over. Install camshaft drive belt as necessary. **4.2** **c.** Check camshaft timing. Replace belt and sprockets as necessary. Adjust belt tension. **4.2**

3.2 Diagnostic Testing

The tests that follow can be used to help isolate engine problems, to better understand a problem before starting expensive repairs, or just to periodically check engine condition.

Compression Test

A test of compression pressures in the individual cylinders will tell a lot about the condition of the engine without the need for taking for taking it apart. The test is relatively simple and requires a compression tester, a spark plug wrench, a screwdriver, and a jumper wire to disable the ignition system.

The battery and starter must be capable of turning the engine at normal cranking speed to achieve meaningful results. The area around the spark plugs should be clean, to avoid getting debris inside the engine when they are removed. Because engine temperature may affect compression, the most accurate results are obtained when the engine is at normal operating temperature.

To test compression:

1. With the ignition off, disconnect the coil wire from the center of the distributor cap (terminal 4) and connect it to ground on the engine block using a jumper wire. This is to disable the secondary circuit of the ignition system.
2. Disconnect one end of the duct between the air flow sensor and the throttle body to disable the fuel injection system.
3. Remove the spark plug wires from the spark plugs. Use care to pull on both the wire and the boot at the same time to avoid damage to the connectors. Label the wires so that they can be reattached to the correct spark plugs.
4. Remove the spark plugs and set them aside, in order corresponding to the cylinders from which they were removed.
5. Thread the compression tester into the first cylinder's spark plug hole, just tight enough to seal around the spark plug hole. Use care not to damage the seal on the gauge line.
6. With the transmission in neutral and the throttle held wide open, crank the engine with the starter. The gauge reading should increase with each engine revolution. Continue cranking until the gauge reading stops increasing (about 4 to 5 revolutions). Record the highest value indicated by the gauge.
7. Release the pressure, either with the gauge valve or by slowly removing the gauge, allowing the pressure to bleed off while threading it out of the spark plug hole.
8. Repeat the test for each of the other cylinders. Record the data and compare with **table c**.

Table c. Compression Pressure Specifications

	Engine code	
Compression pressure – psi (bar)	**JH**	**2H, PL**
new engines	131–174 (9.0–12.0)	145–189 (10.0–13.0)
wear limit	109 (7.5)	109 (7.5)
maximum difference between cylinders	44 (3.0)	44 (3.0)

9. Reinstall the spark plugs and the spark plug wires in their original locations. Reinstall the air duct. Reconnect the coil wire.

Low compression is evidence of poorly sealed combustion chambers. The characteristics of the test results help isolate the cause or causes. Generally, compression pressures which are relatively even but below acceptable specifications indicate worn piston rings and/or cylinder walls. Low but erratic values tend to indicate valve leakage. Dramatic differences, such as acceptable values in some cylinders and very low values in one or two cylinders are the sign of a localized failure, probably of a head gasket. There are two more tests which can further isolate the problem.

Wet Compression Test

To analyze poor compression and further identify the source of the leakage, repeat the compression test, this time with about a tablespoon of oil squirted into each cylinder. The oil will temporarily help seal between the piston rings and the cylinder wall, practically eliminating leakage past the rings for a short time. If this test yields higher compression readings than the "dry" compression test, the difference can be attributed to leakage between the piston rings and cylinder walls, due either to wear or to broken piston rings. Little or no change in compression readings indicates other leakage, probably from the valves or a failed cylinder head gasket.

Leak-down Test

The most conclusive diagnosis of low compression symptoms requires a leak-down test. Using a special tester and a supply of compressed air, each cylinder is pressurized. The rate at which the air leaks out of the cylinder, as well as the sound of the air escaping, can more accurately pinpoint the magnitude and source of the leakage. Any engine compression diagnosis which will require major disassembly should first be confirmed by the more accurate leak-down test. Because the test requires special equipment and experience, it may be desirable to have it performed by a Volkswagen dealer or other qualified repair shop.

4. CYLINDER HEAD

The cylinder head can be removed from the engine for repairs without first removing the engine from the car, and many cylinder head repairs can be accomplished without removing the cylinder head from the engine. The cylinder head cover gasket, the camshaft(s), the camshaft drive belt, the camshaft oil seal, the valve guide oil seals, and the cam followers are all accessible with the cylinder head installed.

Reconditioning the cylinder head is not overly complicated, but requires time and an extensive tool selection. If good machine shop services are not available in your area, or time is a factor, installation of a remanufactured cylinder head is an alternative. Remanufactured cylinder heads are available from an authorized Volkswagen dealer.

4.1 Cylinder Head Cover and Gasket

Gaskets that have been compressed will never seal as well if used a second time. They should be replaced any time the cover is removed. Any sign of oil leakage around or below the cover is also cause to replace the gaskets. The 8-valve engines use a two-piece gasket set. Notice that the half-round plug at the back of the head is also an important seal. The two-piece gasket set for 16-valve engines has built-in plugs.

To remove and install cylinder head cover and gaskets (8-valve engines):

1. Remove the eight 10 mm nuts and the reinforcing strips.
2. Remove the socket-head flanged nut from the camshaft drive belt upper cover.
3. Lift off the cylinder head cover and its gasket. Raise the camshaft drive belt upper cover slightly for clearance. If the gasket is stuck to the cylinder head, use a gasket removing tool to separate the gasket from the head. Remove the sealing plug from the rear of the cylinder head and remove the semi-circular gasket piece from inside the front of the cover.

> ***CAUTION —***
> *Remove all of the old gasket. Work carefully if the gasket is stuck. Damage to either surface will cause leaks. Only use a gasket remover tool designed for the purpose.*

4. Installation is the reverse of removal. Make sure the tabs on the semi-circular piece correctly mate the slots on the gasket to ensure a good seal. Torque the eight nuts and the socket head flanged nut to 10 Nm (87 in. lb.).

To remove and install cylinder head cover and gasket (16-valve engines):

1. Remove the spark plug wire connectors. Remove the intake air boot from the throttle housing and remove the small rubber hose between the intake air boot and the idle air valve.
2. Disconnect the electrical connector for the charcoal canister control valve, and the large vacuum hose from the back of the upper intake manifold.
3. Remove the support bolt from the back of the upper intake manifold, then remove the five nuts and washers that join the upper intake manifold to the lower intake manifold.
4. Rotate the upper manifold up and away from the engine. Remove and label electrical connections or vacuum hoses as required. Then secure the manifold out of the way.
5. Remove the two inner bolts and the six outer bolts from the cover, and lift the cover from the cylinder head. If necessary, use a soft hammer to loosen the cover. Remove the center gasket from the cylinder head or the cover.

> ***CAUTION —***
> *Remove all of the old gasket. Work carefully if the gasket is stuck. Damage to either surface will cause leaks. Only use a gasket remover tool designed for the purpose.*

6. Installation is the reverse of removal. Torque the eight cover bolts to 10 Nm (87 in. lb.). Torque the five manifold nuts and the manifold support bolt to 20 Nm (15 ft. lb.). Torque the support bolt last.

4.2 Camshaft Drive Belt

The camshaft drive belt and its related parts are shown in Fig. 4-1. The belt is installed for the life of the engine. However, to guard against unexpected failure the publisher recommends periodic inspection of the belt, as well as the additional precaution of belt replacement at 60,000 to 75,000 miles (96,000 to 120,000 km) or every 4 to 5 years.

Removal of the belt is the best way to perform a thorough inspection. For PL (16-valve) engines, a special wrench is required to adjust the camshaft drive belt tensioner. On all engines, the drive belt can be replaced without removing the drive belt sprockets.

Whenever the drive belt is installed, the camshaft and intermediate shaft timing must be adjusted.

To remove and inspect camshaft drive belt:

1. Remove the nut and the socket-head shouldered nut securing the upper camshaft drive belt cover, and remove the cover.
2. Loosen the coolant pump pulley bolts, and remove the V-belts as described in **LUBRICATION AND MAINTENANCE**. Remove the coolant pump pulley.
3. Using a socket wrench on the crankshaft vibration damper bolt, rotate the engine by hand to set the No. 1 piston at Top Dead Center (TDC), and remove the vibration damper from the crankshaft.

> ***CAUTION —***
> *Rotating the crankshaft or the camshaft with the drive belt removed may cause interference which can damage pistons and valves.*

4. Remove the nuts and bolt that hold the lower camshaft drive belt cover to the coolant pump and to the front of the engine, and remove the lower cover.

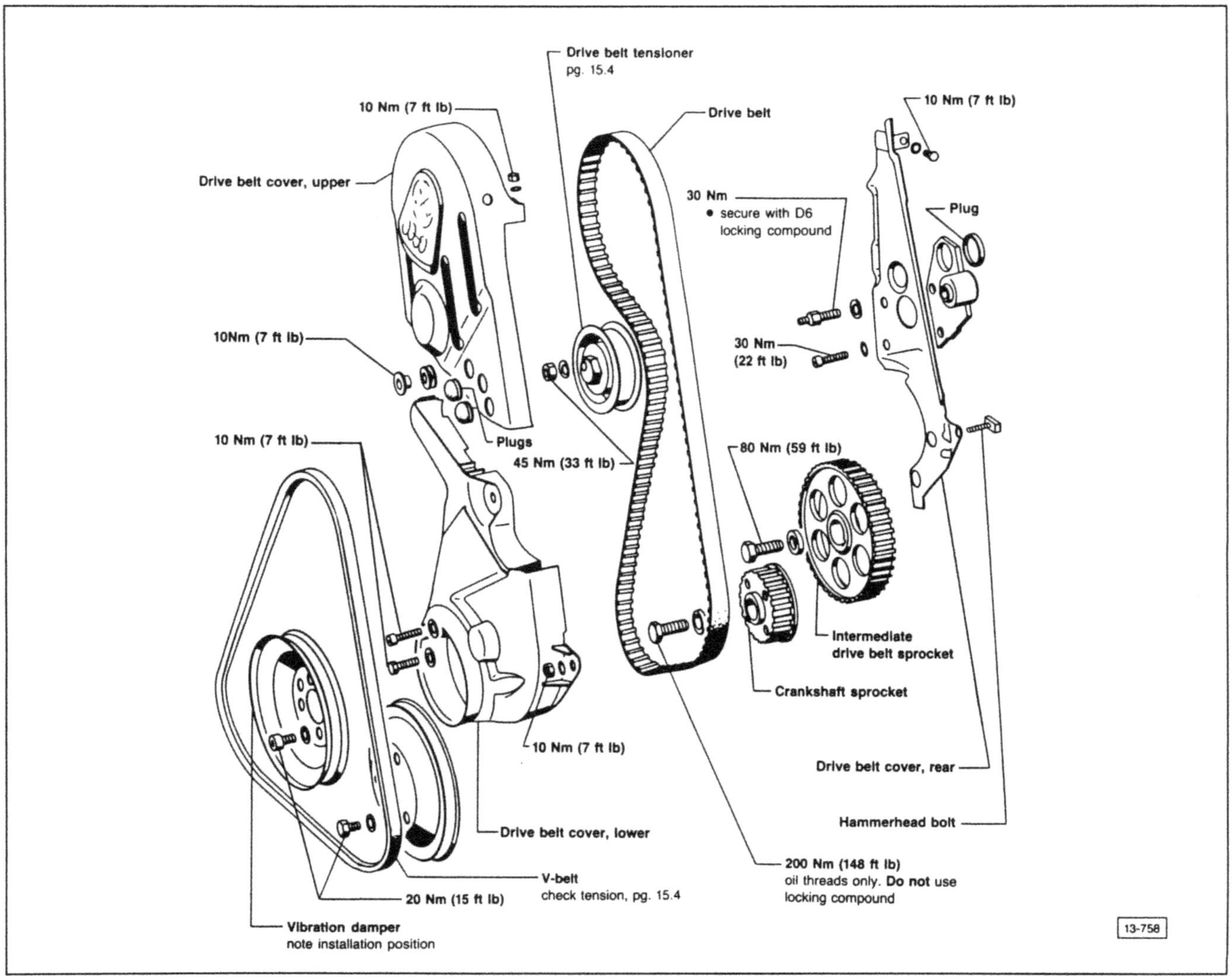

Fig. 4-1. Camshaft drive belt and cover.

5. Loosen the camshaft drive belt tensioner locknut to relieve belt tension. Then remove the drive belt by working it off the sprockets.

6. Inspect the tensioner bearing and the belt. Spin the bearing to check that it runs smoothly. Inspect the drive belt for visible damage such as exposed threads and missing teeth. Damaged components should be replaced.

> **NOTE —**
> Failure of the sealed tensioner bearing is a possibility. On a high-mileage engine, consider replacing the tensioner when replacing the belt.

To install and adjust camshaft drive belt:

1. Use a socket on the center bolt to rotate the camshaft sprocket by hand until the timing marks are aligned. The camshaft timing marks for PL (16-valve) engines are shown in Fig. 4-2. The camshaft timing marks for the 8-valve engines are shown in Fig. 4-3.

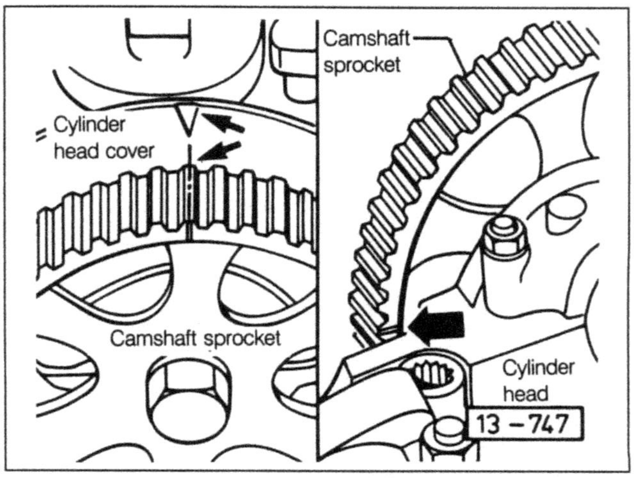

Fig. 4-2. Camshaft timing marks on PL (16-valve) engine with cylinder head cover on (left) and off (right).

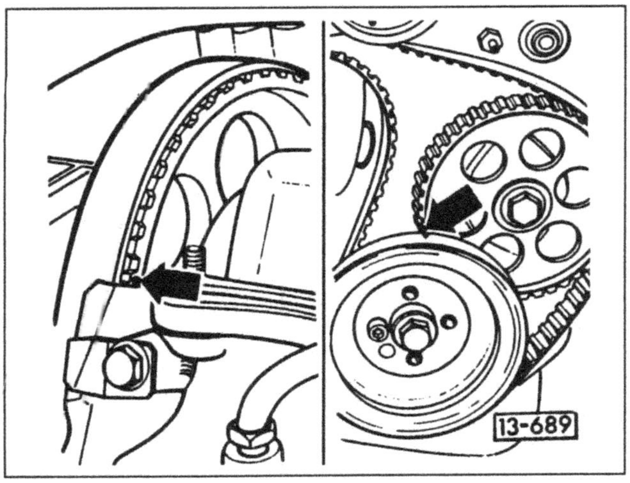

Fig. 4-3. Drive belt timing marks (arrows) for camshaft and intermediate shaft on 8-valve engines.

2. Loosely install the camshaft drive belt over the crankshaft and intermediate shaft sprockets. Then temporarily install the crankshaft vibration damper onto the crankshaft.
3. Rotate the crankshaft and the intermediate shaft by hand until the timing marks line up. The crankshaft and the intermediate shaft timing marks for 8-valve engines are shown above in Fig. 4-3. The crankshaft timing marks for PL (16-valve) engines are shown in Fig. 4-4.

NOTE —
On 8-valve engines only, the intermediate shaft timing is included because it drives the ignition distributor and determines basic ignition timing. It is not necessary to time the intermediate shaft on 16-valve engines.

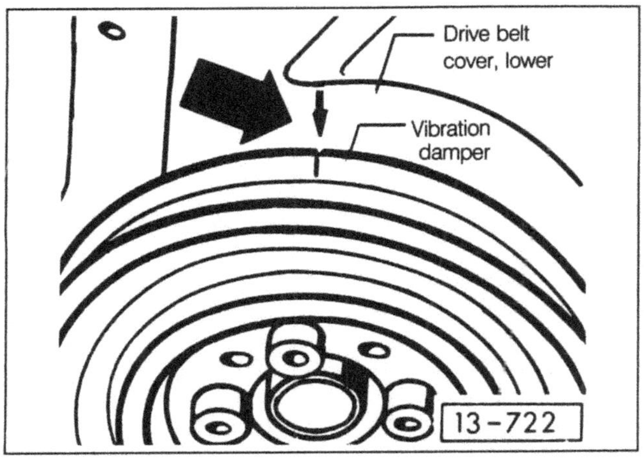

Fig. 4-4. Crankshaft timing marks on vibration damper and lower drive belt cover on PL (16-valve) engine.

4. Install the camshaft drive belt. Start on the crankshaft and intermediate shaft sprockets. Install the belt so that it is as tight as possible between the crankshaft and the intermediate shaft sprockets, and between the intermediate shaft and the camshaft sprockets.
5. Adjust belt tension by turning the tensioner clockwise. The direction is important to avoid contact with the belt cover. Lock the tensioner in position with the locknut. Check the tension by twisting the belt, as shown in Fig. 4-5. On 8-valve engines, you should be able to twist the belt no more than 90°. On PL (16-valve) engines, you should be able to twist the belt no more than 45°.

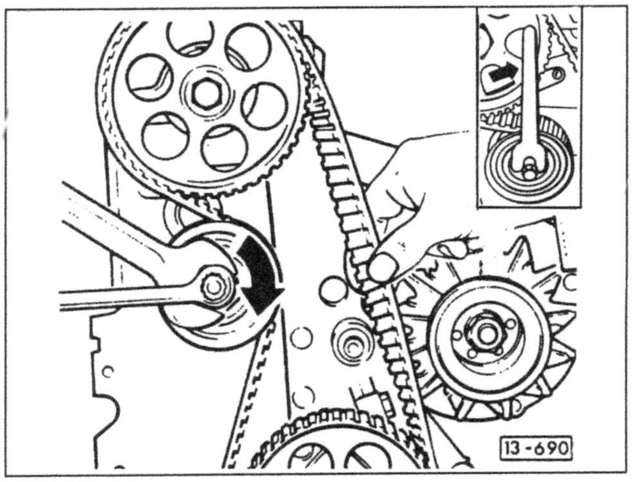

Fig. 4-5. Camshaft drive belt tension being adjusted. Check tension halfway between the camshaft and the intermediate shaft sprockets. For later style tensioner use spanner (inset) to engage holes in tensioner.

NOTE —
Only the later-style tensioner is available as a replacement part.

6. Torque the tensioner locknut to 45 Nm (33 ft. lb.). Turn the crankshaft two full revolutions and recheck the tension and the timing marks.

NOTE —
Slight movement of the sprockets and marks is to be expected as belt tension is adjusted. Small movement is OK. Keep in mind that the smallest possible increment of adjustment is one whole tooth of the belt or sprocket.

7. The remaining installation is the reverse of removal. Remove the vibration damper to install the lower drive belt cover. On PL (16-valve) engines, torque the upper belt cover bolt to 6 Nm (53 in. lb.). Torque 8-valve belt cover nuts and bolts to 10 Nm (87 in. lb.). Torque the coolant pump pulley bolts and the crankshaft vibration damper bolts to 20 Nm (15 ft. lb.). Install the V-belts as described in **LUBRICATION AND MAINTENANCE**.

4.3 Camshaft Oil Seal

To save disassembly, the camshaft oil seal can be removed using a seal extractor such as Volkswagen special tool no. 2085, shown in Fig. 4-6. The procedure that follows involves more disassembly and does not require special tools.

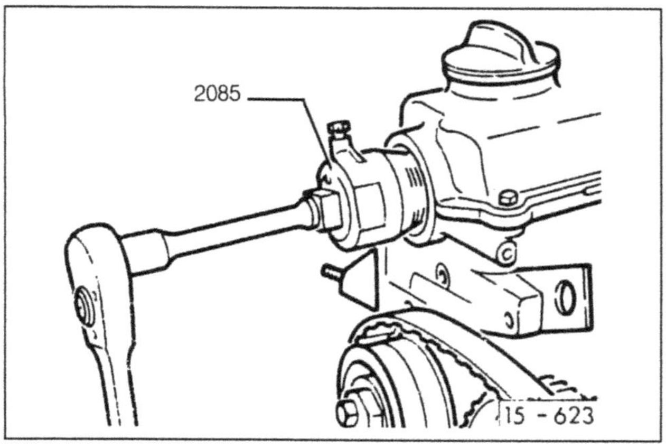

Fig. 4-6. Oil seal extractor being used to remove camshaft oil seal.

To replace camshaft oil seal:

1. Remove the camshaft drive belt as described in **4.2 Camshaft Drive Belt**.
2. Remove the cylinder head cover as described in **4.1 Cylinder Head Cover and Gasket**.
3. Remove the camshaft sprocket. Loosen the center sprocket bolt $^1/_2$ turn and, using a soft-faced hammer, tap the sprocket loose. On 8-valve engines, take care also to remove the woodruff key from the shaft.
4. On PL (16-valve) engines only, remove the ignition distributor from the cylinder head, as described in **IGNITION**.
5. Loosen the camshaft bearing caps in exact order as described in **4.5 Removing and Installing Camshaft**, so that the camshaft is loose and free to move slightly.

CAUTION —

Failure to loosen and tighten the bearing caps in the order prescribed may damage the bearing caps or warp the camshaft.

6. Lift slightly on the sprocket end of the camshaft and remove the oil seal. A small screwdriver may help to pry it loose. On PL (16-valve) engines, use care not to disturb the camshaft drive chain at the other end of the cylinder head.
7. Tighten the camshaft bearing caps in exact order as described in **4.5 Removing and Installing Camshaft**.
8. Install the cylinder head cover, with a new gasket, as described in **4.1 Cylinder Head Cover and Gasket**. On PL (16-valve) engines, first check to see that the markings on the camshaft chain sprockets are correctly aligned. See **4.5 Removing and Installing Camshaft**.
9. Lightly oil the seal lip and camshaft's sealing surface. Fit the seal into position, and carefully drive it into place until it is flush with the front of the cylinder head. Use care not to distort the seal as it is installed. For best results, use a seal driver such as Volkswagen special tool no. 10-203, shown in Fig. 4-7.

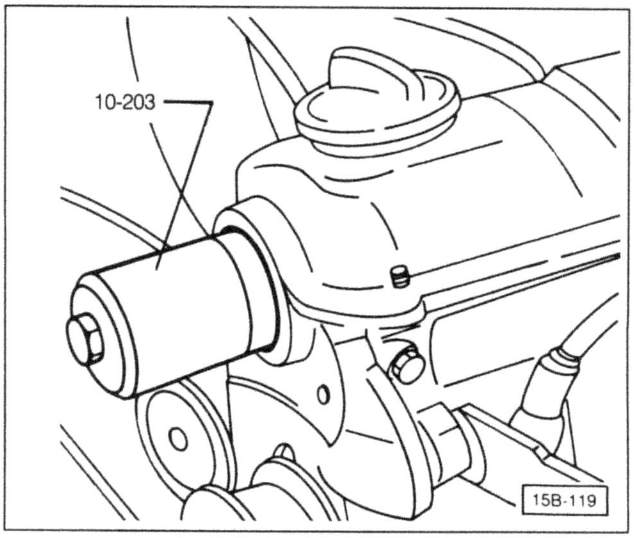

Fig. 4-7. Camshaft oil seal being installed.

10. Reinstall the camshaft drive belt sprocket. On 8-valve engines, the woodruff key should be a tight fit. If not, replace it. Torque the sprocket bolt to 80 Nm (59 ft. lb.). On PL (16-valve) engines, the curved side of the integral key faces the engine. Torque the sprocket bolt to 65 Nm (48 ft. lb.) on PL (16-valve) engines.
11. Reinstall the camshaft drive belt as described in **4.2 Camshaft Drive Belt**.

4.4 Valve Adjustment

(8-valve engine code JH, 1985-1987 only)

1985, 1986 and 1987 8-valve JH engines have mechanical cam followers (valve lifters) that require periodic valve lash or valve clearance adjustment. The 16-valve engines and the later 8-valve engines do not. For more information on those engines see **4.6 Hydraulic Cam Followers**.

The **LUBRICATION AND MAINTENANCE** section specifies particular valve adjustment intervals. Valve clearance should also be rechecked 1000 miles (1500 km) after any repairs to the cylinder head.

Valve clearance is determined by the thickness of an adjusting disc on the top of each cam follower. Clearance is adjusted by replacing the disc with one of a different thickness. Fig. 4-8 shows a typical cam lobe, cam follower and adjusting disc.

The discs are available in a range of thickness from 3.30 to 4.25 mm (0.130 to 0.167 in.) in increments of 0.05 mm (about 0.002 in.). **Table d** lists thicknesses and part numbers of the available adjusting discs. The thickness of each disc is etched on its underside. The procedure which follows requires two Volkswagen special tools. These are no. 2078, used to compress the valve spring, and no. 10-208, used to replace the adjusting disc.

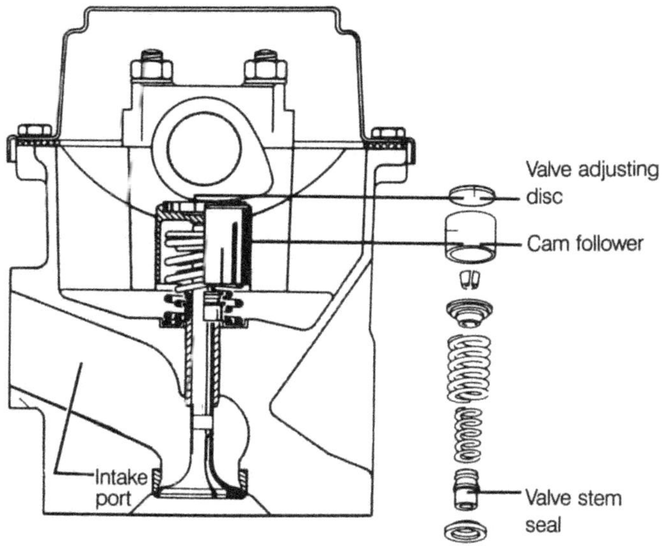

Fig. 4-8. Cross-section of cylinder head showing cam lobe and cam follower. Adjusting disc fits in recess in follower.

Table d. Valve Adjusting Discs

Thickness mm (in.)	Part No.
3.30 (.1299)	.056 109 561
3.35 (.1319)	.056 109 562
3.40 (.1339)	.056 109 563
3.45 (.1358)	.056 109 564
3.50 (.1378)	.056 109 565
3.55 (.1398)	.056 109 566*
3.60 (.1417)	.056 109 567*
3.65 (.1437)	.056 109 568*
3.70 (.1457)	.056 109 569*
3.75 (.1476)	.056 109 570*
3.80 (.1496)	.056 109 571*
3.85 (.1516)	.056 109 572
3.90 (.1535)	.056 109 573
3.95 (.1555)	.056 109 574
4.00 (.1575)	.056 109 575
4.05 (.1594)	.056 109 576
4.10 (.1614)	.056 109 577
4.15 (.1634)	.056 109 578
4.20 (.1654)	.056 109 579
4.25 (.1673)	.056 109 580

*** most commonly used adjusting discs**

Valve clearance should be measured with the engine warm. Coolant temperature should be at least 35°C (95°F) or the engine should be at least warm to the touch. The clearance changes with engine temperature, so more precise adjustment is possible when the engine is warm. Because it may be necessary to adjust the valves when the engine is cold, such as after cylinder head repairs, specifications for both hot and cold engines are given.

To adjust valve clearance:

1. Remove the cylinder head cover, as described in **4.1 Cylinder Head Cover and Gasket**.

2. Using a socket wrench on the crankshaft vibration damper bolt, hand-turn the crankshaft clockwise until both the No. 1 cylinder's cam lobes are pointing approximately upward (valves fully closed).

NOTE —

No. 1 cylinder is the one closest to the camshaft drive belt.

3. Measure the valve clearances. Insert feeler gauges between the cam lobe and the adjusting disc on the top of the cam follower, as shown in Fig. 4-9. Compare the measured clearance values with the specifications given in **Table e**.

Table e. Valve Clearance Specifications (1985–1987, engine code JH)

engine warm	
intake	0.20–0.30 mm (.008–.012 in.)
exhaust	0.40–0.50 mm (.016–.020 in.)
engine cold	
intake	0.15–0.25 mm (.006–.010 in.)
exhaust	0.35–0.45 mm (.014–.018 in.)

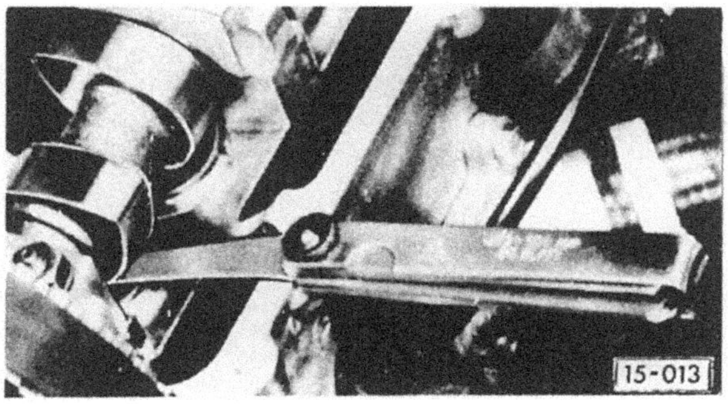

Fig. 4-9. Valve clearance being measured. Both valves are closed. Both cam lobes are pointing up, away from cam followers. Feeler gauge is inserted between cam lobe and adjusting disc on top of cam follower.

4. If valve clearance is not within the specifications, replace the adjusting disc. Depress the cam follower and lift out the disc as shown in Fig. 4-10.

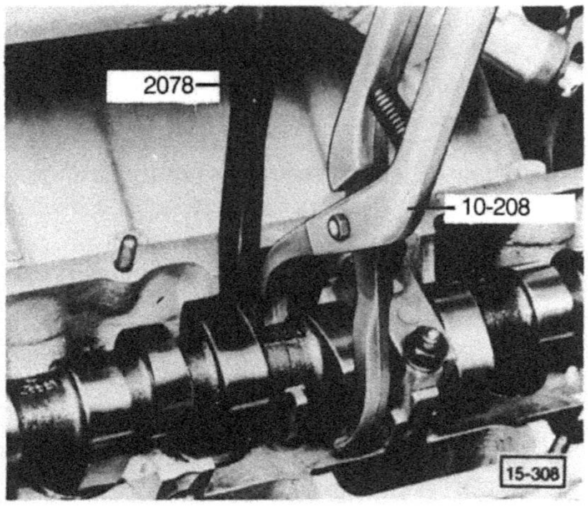

Fig. 4-10. Adjusting disc being removed from cylinder head. Volkswagen tool no. 2078 is used to depress cam followers. Tool no. 10-208 is used to remove adjusting disc.

NOTE —
To calculate the correct adjusting disc thickness, first determine the thickness of the old disc. Read the thickness in millimeters that is etched on its underside, or measure it with a micrometer. The change in thickness required is the same as the difference between the measured clearance and the specification given in **table e** above. A thicker disc will reduce valve clearance. A thinner disc will increase valve clearance.

5. Install the correct adjusting disc. Lightly oil the contact surfaces. Depress the cam follower and install the disc the same way it was removed, as shown in Fig. 4-10, above. The etched disc thickness numbers should face down.

6. Using a socket wrench, rotate the crankshaft clockwise by hand for two complete turns, to ensure that the adjusting disc is properly seated, and recheck the clearance.

7. Rotate the crankshaft by hand clockwise 180° (1/2 turn) and repeat the procedure on the valves for the No. 3 cylinder. Turn another 180° to adjust the No. 4 cylinder, and another 180° to adjust the No. 2 cylinder.

8. Install the cylinder head cover with new gaskets, as described in **4.1 Cylinder Head Cover and Gasket**.

4.5 Removing and Installing Camshaft

The camshaft can be removed without removing the cylinder head. Removing the camshaft allows access to the cam followers, the valve springs, and the valve stem oil seals. When removed, the camshaft should be checked for wear and other visible damage as described in **4.9 Disassembly, Assembly, and Reconditioning**.

To remove and install camshaft (8-valve engines):

1. Turn the engine over by hand until the No. 1 cylinder is at Top Dead Center (TDC), and remove the camshaft drive belt from the camshaft sprocket as described in **4.2 Camshaft Drive Belt**.

2. Remove the cylinder head cover and gasket as described in **4.1 Cylinder Head Cover and Gasket**.

3. Loosen the nuts on each of the bearing caps, a little at a time so that valve spring tension is relieved evenly. Then, remove the bearing caps.

4. To install, place the bearing caps in position as shown in Fig. 4-11. Install the washers and nuts, gradually tightening all the nuts until the camshaft is drawn down fully and evenly into the bearing saddles. Torque all the nuts to 20 Nm (15 ft. lb.).

CAUTION —
To avoid uneven and accelerated wear, the bearing caps must be reinstalled in their exact original positions. Although each bearing cap is numbered, the numbers are not always marked in the same place on the cap.

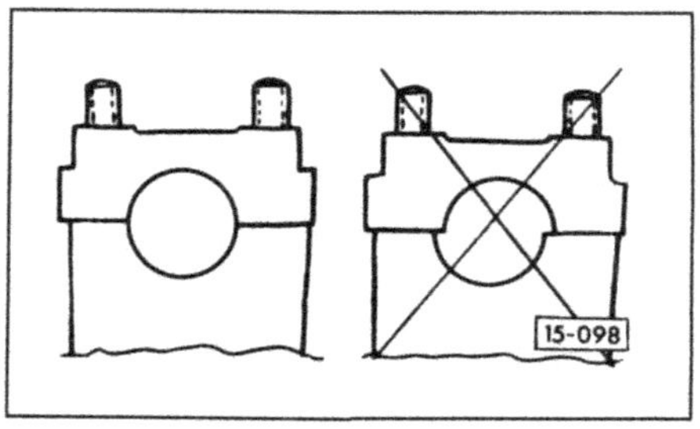

Fig. 4-11. Cross-section of camshaft bearing, showing cap installed correctly (left) and incorrectly (right).

5. Reinstall the camshaft drive belt as described in **4.2 Camshaft Drive Belt**. Reinstall the cylinder head cover with a new gasket, as described in **4.1 Cylinder Head Cover and Gasket**.

To remove and install camshafts (16-valve engines):

1. Turn the engine over by hand until the No. 1 cylinder is at Top Dead Center (TDC), and remove the camshaft drive belt from the camshaft sprocket as described in **4.2 Camshaft Drive Belt**.

2. Remove the cylinder head cover and gasket as described in **4.1 Cylinder Head Cover and Gasket**.

3. Remove the ignition distributor from the rear of the cylinder head as described in **IGNITION**.

4. Remove bearing caps 1, 3, 5 and 7, as illustrated in Fig. 4-12. Also remove the three end caps at this time.

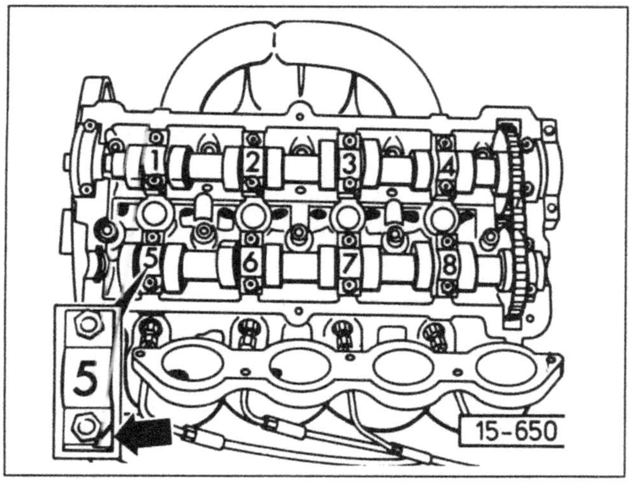

Fig. 4-12. Identification of 16-valve engine camshaft bearing caps. Notice also three camshaft bearing end caps (not numbered). Inset shows proper installation (beveled corners face intake side).

5. Gradually and evenly loosen each of the eight nuts from bearing caps 2, 4, 6 and 8, a little at a time so that valve spring tension and cam chain tension are relieved evenly.

> ***CAUTION —***
> *Removal of camshaft bearing caps by any other method may damage the camshafts, the camshaft chain, or the bearing caps.*

6. Installation is the reverse of removal. Install the camshafts and the drive chain so that the markings on the chain sprockets face each other and align with the top of the cylinder head, as shown in Fig. 4-13.

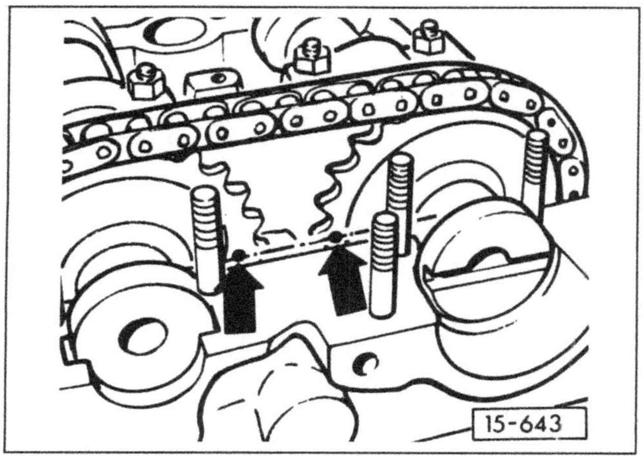

Fig. 4-13. Camshaft drive chain sprockets correctly aligned in 16-valve cylinder head.

7. Torque the bearing cap nuts to 15 Nm (11 ft. lb.). Reinstall the cylinder head cover with a new gasket, as described in **4.1 Cylinder Head Cover and Gasket**.

8. Install the camshaft drive belt and adjust the camshaft timing as described in **4.2 Camshaft Drive Belt**.

9. Install the ignition distributor and adjust ignition timing as described in **IGNITION**.

4.6 Hydraulic Cam Followers

Hydraulic cam followers (valve lifters) maintain proper valve clearance and do not require adjustment. They are installed on 8-valve engines starting with 1988 models, and on all PL (16-valve) engines. The cylinder head cover label shown in Fig. 4-14 indicates an engine equipped with hydraulic cam followers.

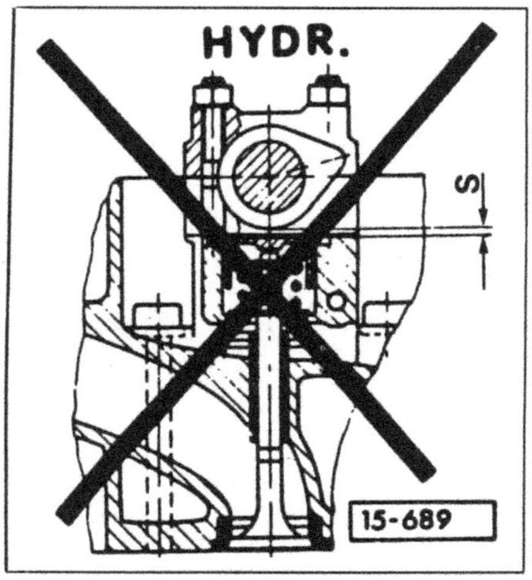

Fig. 4-14. Label on cylinder head cover identifies engines with hydraulic cam followers that maintain proper valve clearance and do not require adjustment.

Hydraulic cam followers are pumped up by engine oil pressure. They expand as necessary to close the gap between the valve stem and the camshaft lobe. Oil pressure inside keeps the follower at the proper height. Excess oil pressure is allowed to bleed off through a small orifice.

Some valve noise at start-up is normal, especially when the engine has not run for some time and the hydraulic cam followers have bled down from lack of oil pressure. Allow a minute or so with the engine running for the lubrication system to properly pump up the cam followers. Before checking noisy cam followers, make sure that the engine oil level is adequate.

> ***CAUTION —***
> *After installing new cam followers, the engine should not be started for at least 30 minutes. New cam followers are at full height and must be allowed to bleed down to their proper height. Failure to do this may cause valve or piston damage.*

Checking Hydraulic Cam Followers

Cam followers should only be checked when the engine is fully warm. Run the engine, preferably drive the car, for 20 to 30 minutes. Shut the engine off and proceed immediately while the engine is still warm.

To check the hydraulic cam followers, remove the cylinder head cover as described in **4.1 Cylinder Head Cover and Gasket**. Turn the engine by hand until both the camshaft lobes of one cylinder are pointing approximately up. Using a non-metal object such as wood or plastic, lightly apply pressure to the top of the cam follower, as shown in Fig. 4-15.

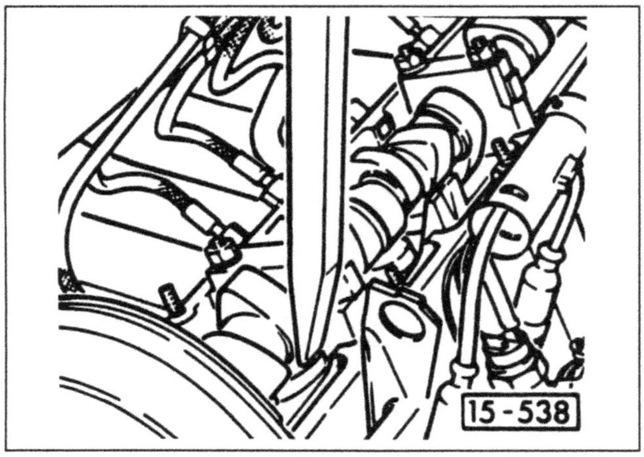

Fig. 4-15. Hydraulic cam follower being checked with a non-metal object.

If the follower can be pushed down more than 0.1 mm (.004 in.) with hand pressure, it is faulty and should be replaced. Repeat the test for the other cylinders. Replace a faulty cam follower by removing the camshaft, as described in **4.5 Removing and Installing Camshaft**, and pulling the follower from the cylinder head. Hydraulic cam followers are non-adjustable and non-repairable, and are replaced only as complete assemblies.

NOTE —
Place removed hydraulic cam followers on a clean surface with the camshaft contact surface facing down, to prevent bleed down, and cover them.

4.7 Valve Stem Oil Seals

The sign of faulty valve stem oil seals are excessive oil consumption and blue-gray oil smoke from the exhaust after starting and during sudden deceleration. For more information on excessive oil consumption and smoking, see **3. Troubleshooting**.

Replacement of the valve stem oil seals requires that the camshaft(s), the cam followers, and the valve springs be removed. Valve springs can be removed while the cylinder head is installed with the use of compressed air to hold the valves closed, or with the cylinder head removed. Either method requires the use of a valve spring compressor. An additional special tool is highly recommended for installation of the new seals. If the special tools are not available, an alternative is to remove and partially dismantle the cylinder head, following the procedures in this section, and have the valve spring and valve stem oil seal work performed by an authorized Volkswagen dealer or other qualified repair shop.

To replace valve stem oil seals:

1. If working with the cylinder head installed, remove the cylinder head cover, as described in **4.1 Cylinder Head Cover and Gasket**. Otherwise, remove the cylinder head as described below in **4.8 Removing and Installing Cylinder Head**.

2. Remove the camshaft, as described in **4.5 Removing and Installing Camshaft**, and remove the cam followers. Keep all parts so that they can be reinstalled in their original locations.

3. Remove the valve springs as shown in Fig. 4-16. The valve spring compressors shown are Volkswagen special tools, designed specifically for use on Volkswagen engines.

NOTE —
To support the valves with compressed air and remove the valve springs with the cylinder head installed, install an air hose adapter in the spark plug hole. Position the piston for the cylinder being worked on at top dead center (TDC). Apply a constant air pressure of at least 6.0 bar (87 psi).

4. Remove the valve stem oil seals. Volkswagen recommends the use of a slide hammer, special tool no. 3047A, to remove the seals without damage to the valve stems.

5. Install each new valve stem oil seal by temporarily fitting the protective plastic cap over the valve stem end, and then hand-pressing the seal onto the valve stem as shown in Fig. 4-17. Remove the protective cap for use on the next valve.

NOTE —
The protective cap protects the seal from being damaged by the keeper grooves in the valve stem. In the absence of a protective cap, temporarily wrap the valve stem with plastic tape.

NOTE —
The use of an inexpensive hand-held tool will guard against damaging the new seals on installation. Such a tool is Volkswagen special tool no. US5042 (order no. TU5 042 000 15 ZEL).

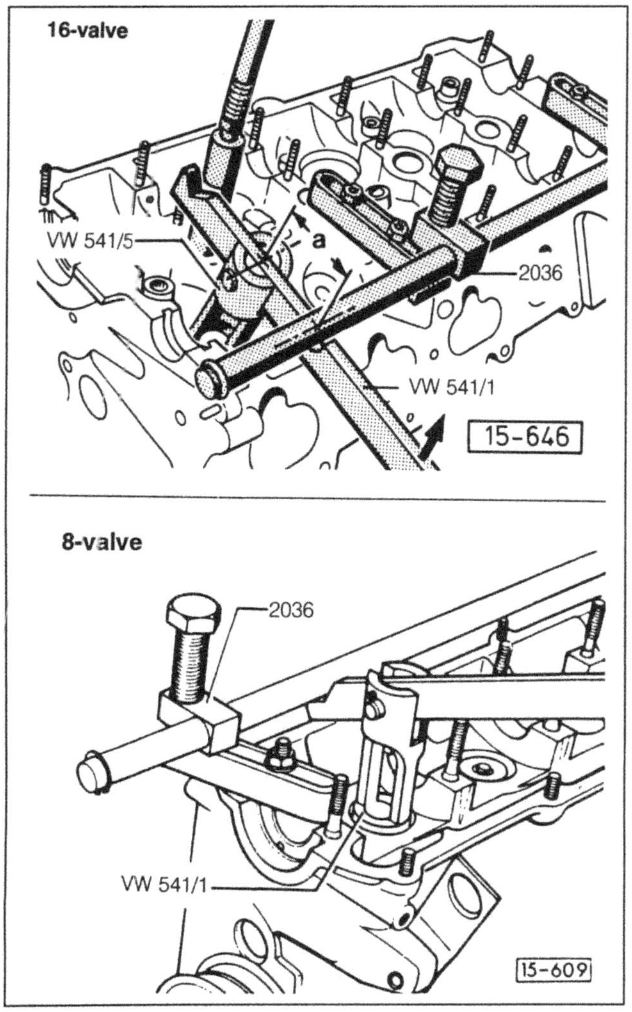

Fig. 4-16. Valve spring compressor being used to remove valve springs from 16-valve engine (top view) and 8-valve engine (bottom view). Numbers identify Volkswagen special tools. For 16-valve engines, notch the handle of special tool VW 541/1 as shown to keep it from slipping. Dimension **a** to notch is 50 mm ($^{31}/_{32}$ in.).

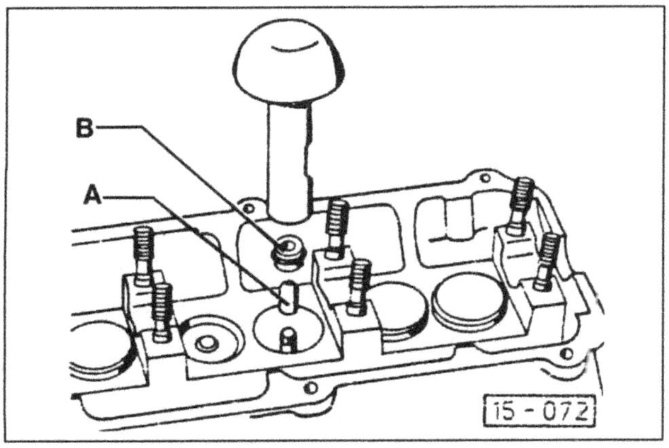

Fig. 4-17. Protective cap **A**, new valve stem oil seal **B**, and installing tool.

6. Reinstall the valve springs. The closely spaced coils of the outer valve springs face down, toward the lower spring seats.

7. Repeat the procedure for each pair of valves. Assembly is the reverse of disassembly. See **4.5 Removing and Installing Camshaft, 4.8 Removing and Installing Cylinder Head**, and **4.1 Cylinder Head Cover and Gasket**. Torque spark plugs to 20 Nm (15 ft. lb.).

4.8 Removing and Installing Cylinder Head

The cylinder head can be removed without removing the engine from the vehicle, and without removing the camshaft(s).

If a failed head gasket or warped head is suspected, a compression test, as described in **3.2 Diagnostic Testing**, may aid diagnosis and should be performed before the cylinder head is removed. A failed head gasket may be caused by a warped cylinder head. When replacing a failed head gasket, always check the cylinder head for straightness. Specifications for maximum permissible warpage can be found in **4.9 Disassembly, Assembly, and Reconditioning**.

To remove cylinder head:

1. Disconnect the battery negative (–) terminal from the battery. Disconnect the ground strap from the cylinder head or intake manifold.

2. Disconnect the intake air duct from the throttle body on the intake manifold.

3. Drain the coolant as described in **COOLING SYSTEM**.

4. Remove the camshaft drive belt as described in **4.2 Camshaft Drive Belt**.

5. Pull out the fuel injectors and remove the cold start valve from the intake manifold, leaving the fuel lines attached. Cover the injectors and the cold start valve to keep them clean, and plug the openings. Disconnect the throttle cable and throttle switches from the throttle body. Disconnect and label the hoses and wires connected to the throttle body and the intake manifold. For more information on fuel injection components, see **FUEL SYSTEM**.

6. Disconnect the exhaust pipe from the exhaust manifold, as described in **EXHAUST SYSTEM AND EMISSION CONTROLS**.

7. Loosen the alternator mounting bolts and remove the upper alternator mounting bracket.

8. Disconnect the coolant hoses from the outlet on the front of the cylinder head and from the heater outlet on the left (driver's side) end of the cylinder head. Disconnect the wires from the temperature and oil pressure sending units, and label them.

9. Disconnect the spark plug connectors from the spark plugs. On PL (16-valve) engines, also disconnect the high voltage lead from the coil, and then remove the distributor as described in **IGNITION**.

10. Remove the cylinder head cover, and, on PL (16-valve engines), the upper intake manifold, as described in **4.1 Cylinder Head Cover and Gasket**.

11. Gradually and evenly loosen the cylinder head bolts, beginning with the outer bolts and working toward the center, and remove the cylinder head. If the head is stuck, use a soft-faced mallet, or pry gently with a wooden stick to loosen it.

To install cylinder head:

1. Clean the cylinder head and the gasket surface of the cylinder block. Clean the threads of the head bolts and bolt holes with thread chasers, and remove all foreign matter from the holes. Avoid letting debris into the cylinders or oil passages in the cylinder block.

> ***CAUTION —***
> *Do not use a metal scraper or wire brush to clean the aluminum cylinder head. These may damage the cylinder head. Instead, use a solvent to soften carbon deposits and old sealing materials. If necessary, use a wooden or plastic scraper.*

2. Check the gasket surfaces of the cylinder head and the cylinder block for warpage. See **4.9 Disassembly, Assembly, and Reconditioning**.

3. Place a new cylinder head gasket on the surface of the cylinder block. The word "OBEN", found imprinted on the gasket near the part number, should face up.

> ***CAUTION —***
> *Cylinder head gaskets will give a reliable seal only once. Always use a new gasket that has not previously been compressed by tightening the cylinder head bolts.*

4. Place the cylinder head in position on the cylinder block. Loosely install all the head bolts, with the special wa ers, then thread them in until they are finger-tight.

> **NOTE —**
> To be sure of proper alignment of the installed cylinder head, place two 200 mm long, 9 mm diameter (8 in. x 3/8 in.) wooden dowels into two of the outermost head bolt holes to hold the gasket and serve as guides as the cylinder head is installed. Thread in several bolts, then remove the dowels and install the remaining bolts.

5. Torque the bolts to 40 Nm (30 ft. lb.), in the sequence shown in Fig. 4-18. Repeat the sequence, torquing the bolts to 60 Nm (43 ft. lb.). Repeat the sequence a third time, tightening each bolt an additional 1/2 turn. Alternatively, this last step may be done as two 1/4 turns.

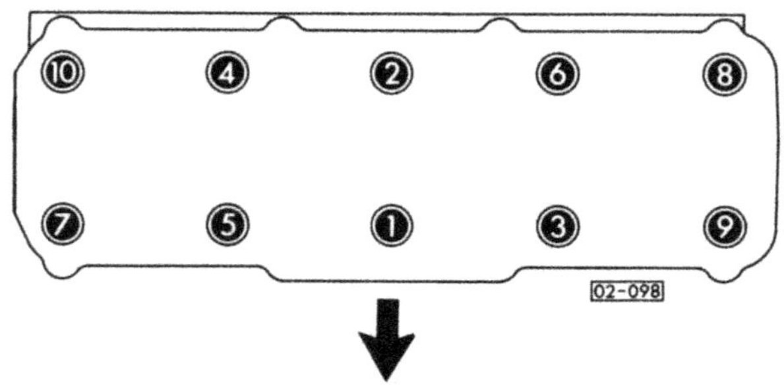

Fig. 4-18. Cylinder head bolt tightening sequence. 8-valve engine shown. 16-valve sequence is the same.

6. Installation of the remaining parts is the reverse of removal. See **4.2 Camshaft Drive Belt**, **4.1 Cylinder Head Cover and Gasket**, **FUEL SYSTEM**, and **EXHAUST SYSTEM AND EMISSION CONTROLS**. For installation of the ignition distributor on PL (16-valve) engines, see **IGNITION**.

7. Refill the cooling system as described in **COOLING SYSTEM**.

4.9 Disassembly, Assembly, and Reconditioning

Disassembly, assembly, and reconditioning procedures for the cylinder heads used on the 1.8 liter Volkswagen engines covered in this manual are similar to those for most other modern 4-cylinder, water-cooled engines. For anyone with the proper tools and equipment and basic experience in cylinder head reconditioning, this section provides the specifications and special reconditioning information necessary to repair these Volkswagen engines.

Those who are without the necessary tools, or unfamiliar with reconditioning procedures, will still find this information to be important to the qualified machinist who does the work. If machine shop services are not readily available, one alternative is to install a Volkswagen remanufactured cylinder head, available from an authorized Volkswagen dealer parts department.

> ***CAUTION —***
> *Volkswagen advises that engine coolant must not be reused when replacing the cylinder head, the cylinder head gasket or the engine.*

Cylinder Head Assembly
(8-valve engines)

Fig. 4-19 shows the 8-valve cylinder head and valve train detail. Later engines with hydraulic cam followers are similar.

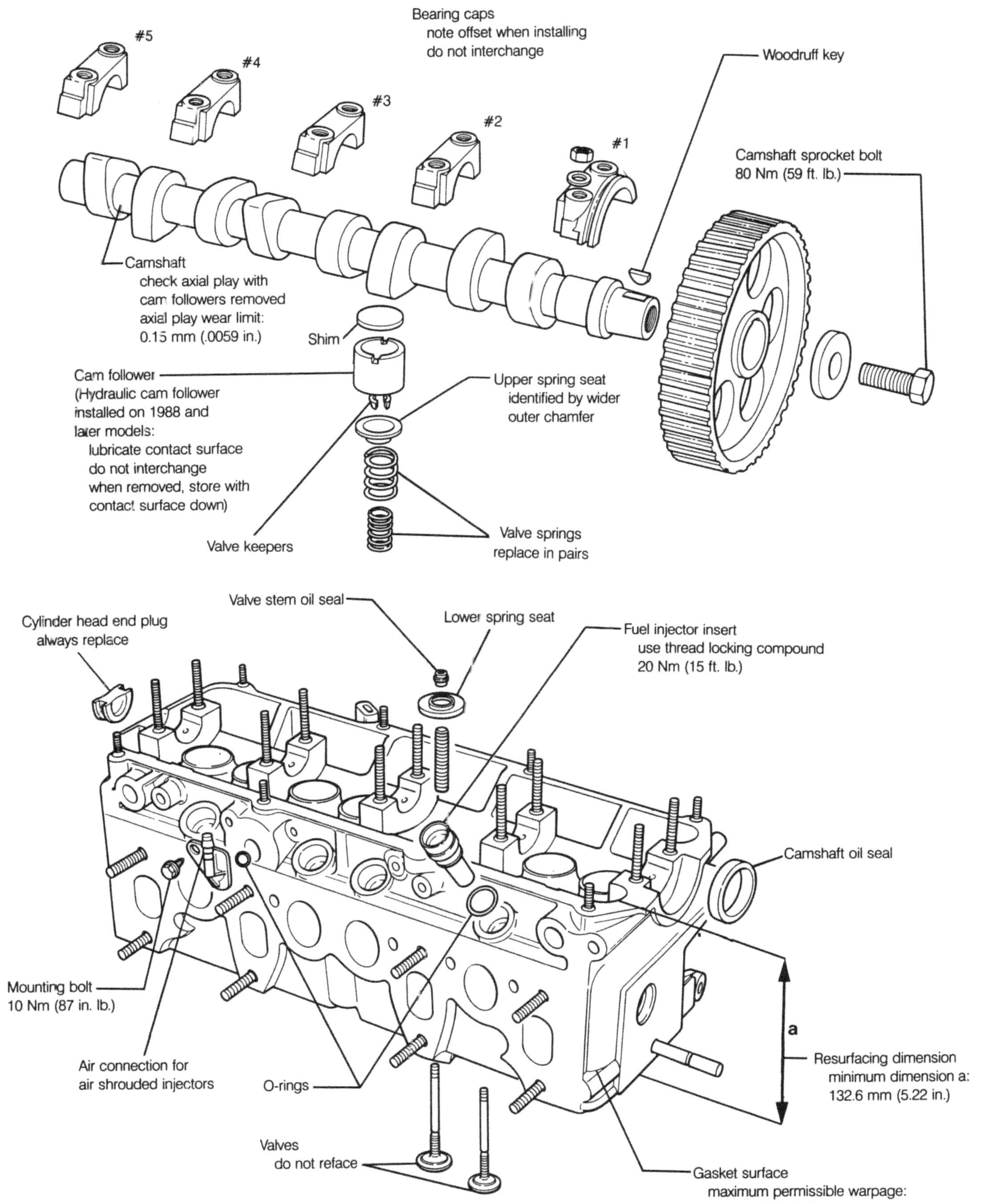

Fig. 4-19. Exploded view of 8-valve cylinder head. Look for small, fine cracks between valve seats and spark plug threads. Heads with cracks can be reused provided cracks are not more than 0.5 mm (0.02 in.) wide, and do not extend into more than the first few spark plug threads.

Cylinder Head Assembly
(16-valve engine, code PL)

Fig. 4-20 shows the cylinder head and valve train assembly used on PL (16-valve) engines.

Camshaft and Cam Followers

To measure camshaft axial play, relieve the tension on the cam lobes by first removing the cam followers. Store hydraulic cam followers with the camshaft contact surface face down. Lubricate the cam follower's contact surfaces before installing.

Do not interchange camshaft bearing caps or cam followers. Note the bearing cap offset when installing. On PL (16-valve) engines, the beveled corners on the bearing caps face the intake side of the cylinder head. On PL (16-valve) engines, align the camshaft chain sprockets as shown in Fig. 4-21.

CAUTION —
After installing new hydraulic cam followers, the engine should not be started for at least 30 minutes. New cam followers are at full height and must be allowed to bleed down to their proper height. Failure to do this may cause valve or piston damage.

Fig. 4-21. Camshaft chain sprocket timing marks on PL (16-valve) engine. Marks should align with each other and cylinder head top surface.

Valves and Valve Springs

All engine valves should be hand-lapped only. Valve specifications are listed in **table f**.

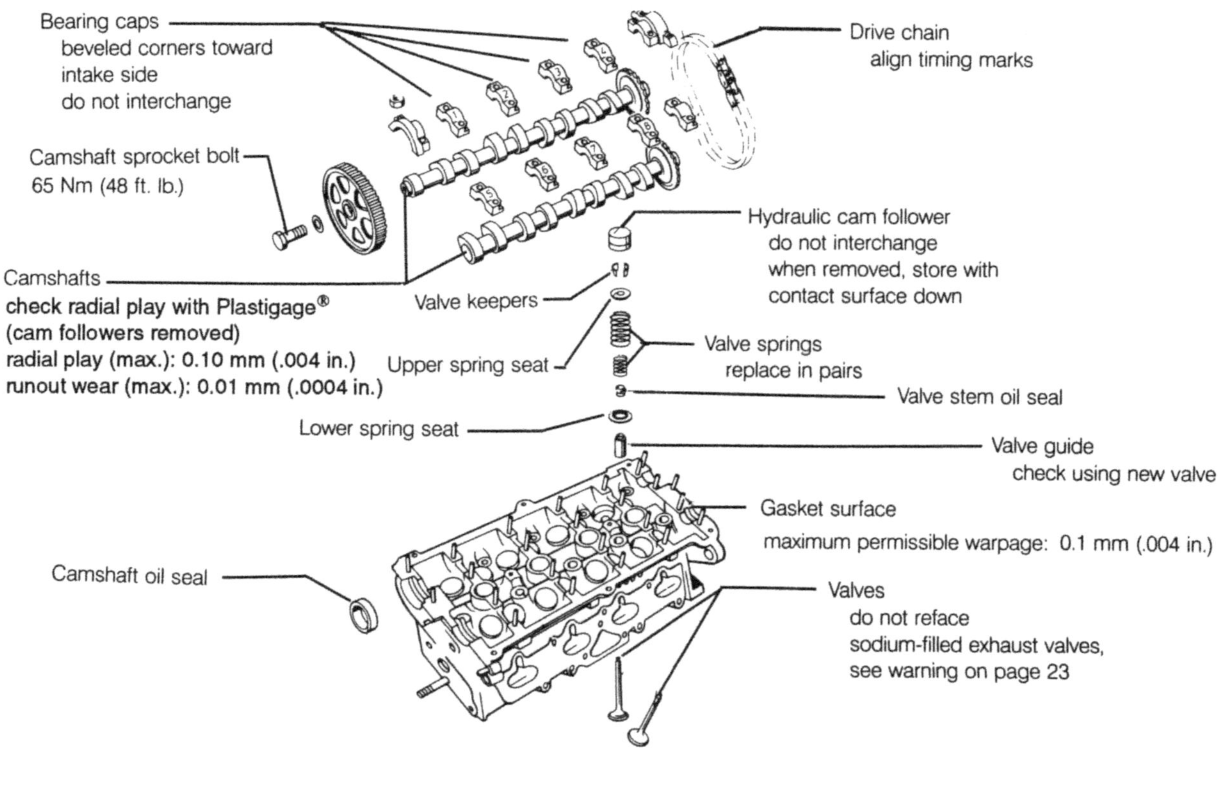

Fig. 4-20. Exploded view of 16-valve cylinder head. Cylinder heads with small, fine cracks between valve seats and plug threads are usable provided the cracks are not more than 0.5 mm (.02 in.) wide and do not extend into more than the first few spark plug threads.

CAUTION —
16-valve cylinder heads must not be resurfaced.

Table f. Valve Specifications

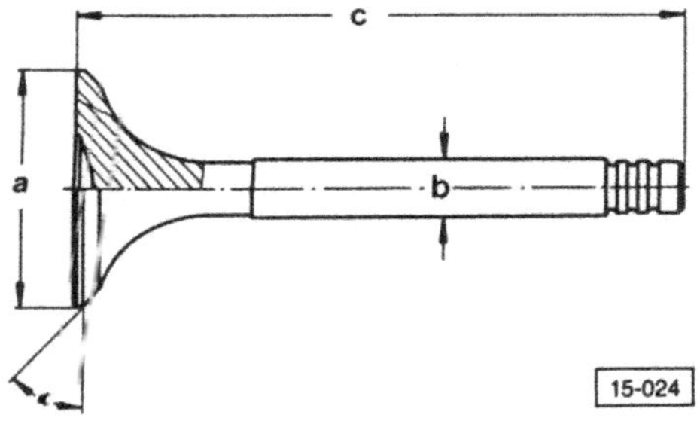

Engine code	JH (1985-1987)	JH (from 1988)	2H (from 1990)	PL (16-valve)
Cam followers	mechanical	hydraulic		
Head diameter (a)				
intake	40.0 mm (1.58 in.)			32.0 mm (1.34 in.)
exhaust	33.0 mm (1.30 in.)			28.0 mm (1.10 in.)
Stem diameter (b)				
intake	7.97 mm (0.314 in.)			6.97 mm (0.274 in.)
exhaust	7.95 mm (0.313 in.)			6.94 mm (0.273 in.)
Length (c)				
intake	98.7 mm (3.89 in.)	91.0 mm (3.58 in.)		95.5 mm (3.76 in.)
exhaust	98.5 mm (3.88 in.)	90.8 mm (3.56 in.)		98.2 mm (3.87 in.)
Valve face angle (d)				
intake	45°			
exhaust	45°			

Exhaust valves used in PL (16-valve) engines are sodium-filled. Disposal of used valves is dangerous and requires special care to avoid personal injury.

WARNING —

Dispose of 16-valve engine sodium-filled exhaust valves properly. By hand, cut off the valve stem near the head of each valve. Use only a hack saw. Do not use a power saw. Do not let water contact the valve while cutting. Throw the valve parts (no more than 10 valves at a time) into a bucket of water and stand clear. Discard the valves when the reaction has ceased.

Valve Guides

Special tools and a press are required to replace valve guides. Check valve guide wear with a new valve, as shown in Fig. 4-22. Inspect the valve seats to ensure that the cylinder head can be reconditioned before installing new valve guides.

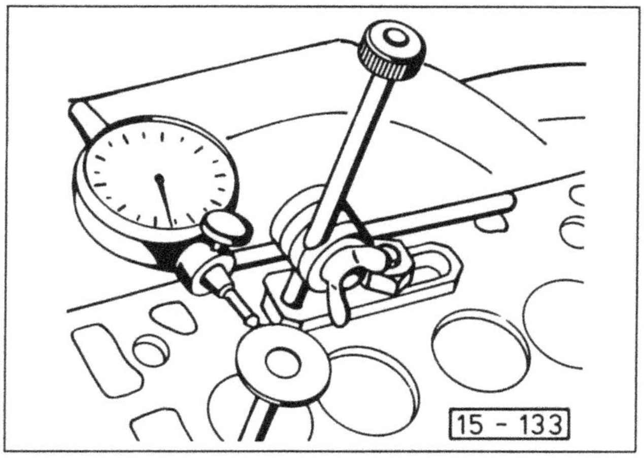

Fig. 4-22. Valve guide wear being checked with a new valve. Insert valve until stem end is flush with end of guide. Maximum play for all intake valves is 1.0 mm (.039 in.). Maximum play for exhaust valves is 1.3 mm (.051 in.).

Valve guides are pressed out from the combustion chamber side of the cylinder head. Lubricate new valve guides with oil and press them in from the camshaft side. For 8-valve engines, remove and install the valve guides using a drift, Volkswagen special tool 10-206 (order no. T10 206 000 15 ZEL).

On PL (16-valve) engines, remove and install guides with Volkswagen special tool no. 3121 (order no. T03 121 000 15 ZEL), and ream guides with Volkswagen special tool no. 3120 (order no. T03 120 000 15 ZEL). To remove and install intake valve guides, support the cylinder head at an angle as shown in Fig. 4-23, using Volkswagen special tool no. 3123 (order no. T03 123 000 15 ZEL).

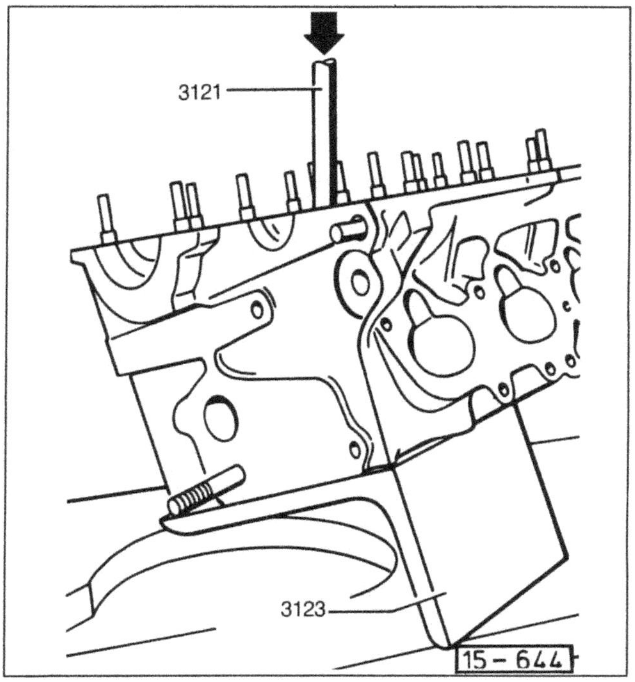

Fig. 4-23. Angle plate (Volkswagen special tool no. 3123) used to support 16-valve cylinder head when removing and installing intake valve guides.

Valve Seats

When resurfacing valve seats on engines with hydraulic cam followers, there is a limit to the amount of material which can be removed. If too much material is removed, the final assembly will leave too little space for the hydraulic cam follower to function correctly. The maximum refacing dimension, the maximum amount of material that can be removed from the valve seat, is calculated from the measurement shown in Fig. 4-24.

Measure dimension **a**, and subtract the minimum dimension, as given in **Table g**. The difference is the maximum amount of material that can be removed from the valve seat.

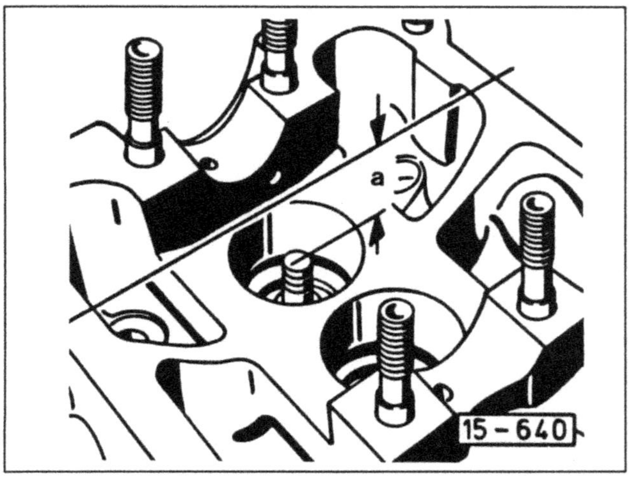

Fig. 4-24. Top view of cylinder head. Dimension **a**, distance between top of valve stem and upper surface of cylinder head, is used to calculate maximum refacing dimension.

Table g. Minimum Dimensions for Calculating Valve Seat Refacing Dimensions

Engine Code	Intake	Exhaust
JH , 2H (8-valve)	33.8 mm (1.331 in.)	34.8 mm (1.370 in.)
PL (16-valve)	34.4 mm (1.354 in.)	34.7 mm (1.366 in.)

Valve seat dimensions are listed in the tables that follow. **Table h** lists dimensions for 8-valve engines. **Table i** lists dimensions for 16-valve engines.

Table h. Valve Seat Dimensions (8-valve engines)

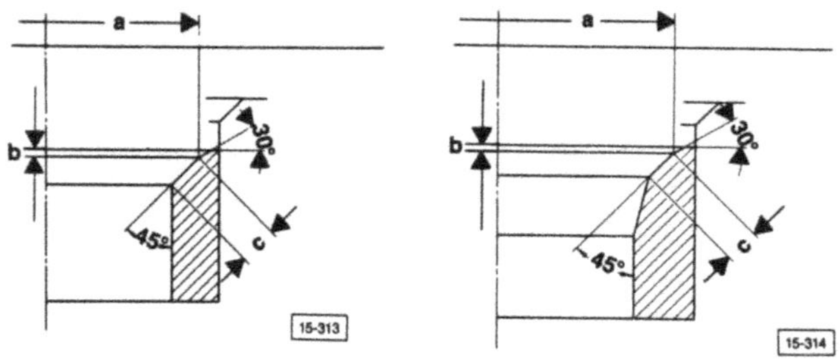

Engine Code	JH (all years)
diameter **a**	
intake	39.2 mm (1.543 in.)
exhaust	32.4 mm (1.276 in.)
Maximum permissible refacing depth **b**	calculate See Fig. 4-24, **Table g.**
Seat width **c**	
intake	approx. 2.0 mm (.080 in.)
exhaust	approx. 2.4 mm (.094 in.)
Valve seat angle	45°
Upper correction angle	30°

Table i. Valve Seat Dimensions (16-valve engines)

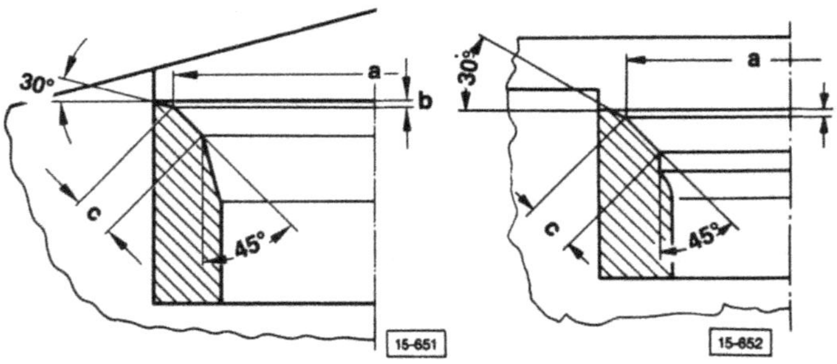

Engine Code	PL
diameter **a**	
intake	31.2 mm (1.228 in.)
exhaust	27.6 mm (1.087 in.)
Maximum permissible refacing depth **b**	calculate See Fig. 4-24, **Table g**
Seat width **c**	
intake	1.5 to 1.8 mm (.06 to .07 in.)
exhaust	approx. 1.8 mm (.07 in.)
Valve seat angle	45°
Upper correction angle	30°
Lower correction angle	75°

5. REMOVING AND INSTALLING ENGINE AND TRANSAXLE

The engine and transaxle are removed as a unit, and separated from each other for additional work once they are removed. The assembly is lifted out from the top, but it is not necessary to remove the hood, unless it will interfere with the type of lifting equipment being used.

Removing

On air-conditioned cars, the air conditioning compressor and condenser are removed and set aside without disconnecting the refrigerant lines. On cars with power steering, remove the power steering pump and fluid reservoir. Set them aside without disconnecting the power steering fluid lines.

To remove:

1. Disconnect the negative (–) battery terminal, and then the positive (+) terminal. Remove the battery.
2. Loosen the fuel filler cap in order to relieve fuel tank pressure.
3. Remove the rubber duct that connects the throttle valve assembly to the fuel injection system's mixture control unit.
4. Drain the cooling system as described in **COOLING SYSTEM**, and remove the coolant hoses connected to the engine.

   CAUTION —
   Volkswagen advises that engine coolant must not be reused when replacing the cylinder head, the cylinder head gasket or the engine.

5. Disconnect the electrical connectors for the radiator cooling fan and the radiator thermoswitch, then remove the radiator and fan assembly as described in **COOLING SYSTEM**.
6. Remove the wiring from the alternator, and then remove the alternator. See **ELECTRICAL SYSTEM**.
7. At the front of the engine, disconnect wiring from the oil pressure switch, the coolant thermoswitch and, on engines with CIS fuel injection, the control pressure regulator. Remove the intake air preheat pipe.
8. On cars with air conditioning, remove the compressor drive belt. Without disconnecting any refrigerant lines, remove the air conditioning compressor from the engine. Set it aside on the fender, using care to avoid kinking the refrigerant lines. Tie the compressor down to hold it.
9. At the left side of the engine, disconnect the coolant temperature sensor wires, the heater hoses, and the starter wiring harness. Disconnect the vacuum hoses, the center coil wire, and the Hall sender connection on the distributor.
10. Detach the fuel lines from the control pressure regulator and the cold start valve. Disconnect the wiring from the cold start valve, the auxiliary air regulator, and the full-load enrichment switch. Disconnect the vacuum hoses for the brake vacuum servo and the vacuum amplifier. Remove the PCV hose.

    WARNING —
    Fuel will be expelled. Do not smoke or work near heaters or other fire hazards. Have a fire extinguisher handy.

11. On PL (16-valve) engines, remove the upper intake manifold, as described in **4.1 Cylinder Head Cover and Gasket**, and disconnect the oxygen sensor connector behind the cylinder head. Remove all engine ground connections.
12. Remove the four fuel injectors.
13. Remove the speedometer drive cable from the transaxle housing.
14. On models with manual transmission, detach the clutch cable and shift linkage as described in **MANUAL TRANSMISSION AND CLUTCH**. Remove the accelerator cable from the engine as described in **FUEL SYSTEM**.
15. On models with automatic transmissions, place the selector lever in the "Park" position. Remove the accelerator cable and the selector lever cable as shown in **AUTOMATIC TRANSMISSION**.
16. Disconnect all remaining electrical connections to the engine and transaxle, including the wires for the starter solenoid and the back-up light switch. Remove the horn.
17. Detach the exhaust front pipe from the exhaust manifold. See **EXHAUST AND EMISSION CONTROLS** if necessary.
18. Working under the car, remove the drive axles as described in **SUSPENSION AND STEERING**, and then reinstall the ball joints into the wheel bearing housings so that the car can be lowered onto its wheels.
19. On cars with manual transmission, remove the starter as described in **ELECTRICAL SYSTEM**.
20. Remove the engine/transaxle front mounting.
21. Prepare to lift the engine by connecting the Volkswagen engine sling or its equivalent to the lifting points shown in Fig. 5-1 (for 8-valve engines) or Fig. 5-2 (for 16-valve engines).
22. Raise the hoist slightly, so that the weight of the engine and transaxle assembly is supported by it.
23. Remove the rear transaxle mount as shown in Fig. 5-3.

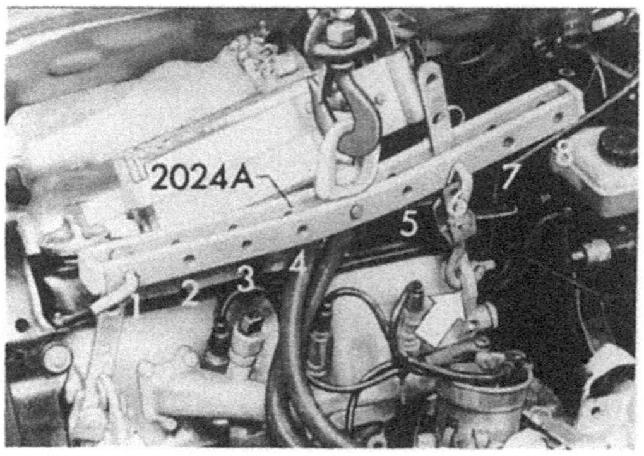

Fig. 5-1. Engine sling (Volkswagen special tool no. 2024A) correctly attached for lifting 8-valve engines with proper weight distribution. Note specific connecting points. Always secure connecting pins and hooks with cotter pins as shown.

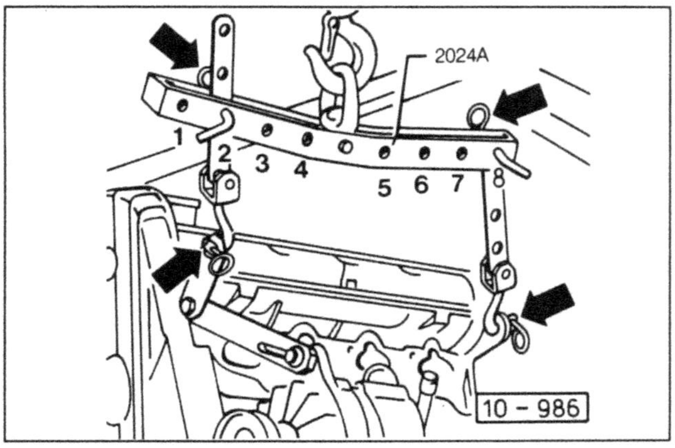

Fig. 5-2. Engine sling (Volkswagen special tool no. 2024A) correctly attached for lifting 16-valve engines with proper weight distribution. Note specific connecting points. Always secure connecting pins and hooks with cotter pins as shown.

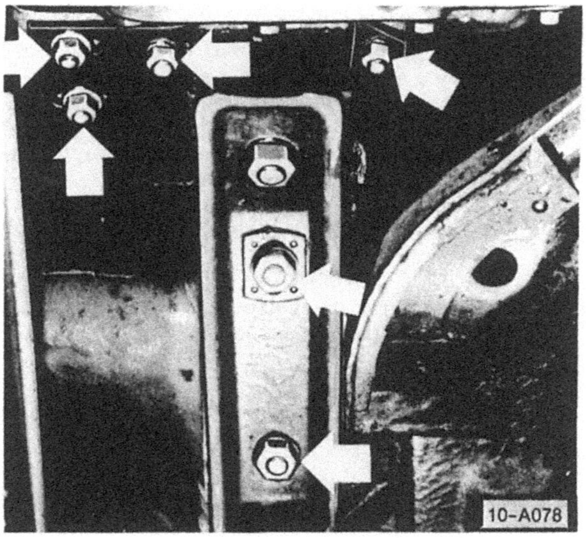

Fig. 5-3. Nuts (arrows) to be removed to remove rear transaxle mount.

24. Remove the bolt that holds the bonded rubber part of the right transaxle mounting to the brackets on the body. Similarly remove the bolt for the right mounting.

25. Raise and remove the engine and transaxle assembly from the car. Proceed slowly. Check frequently to make sure no hoses or wires are interfering with engine removal. If the car has been raised on jack stands, check often see that it remains stable and adequately supported.

Separating Engine and Transaxle

Both the engine and the transaxle should be supported. On cars with automatic transmission, remove the three bolts holding the torque converter to the drive plate. These bolts are near the outside diameter of the torque converter, and are accessible from the bottom as shown in Fig. 5-4.

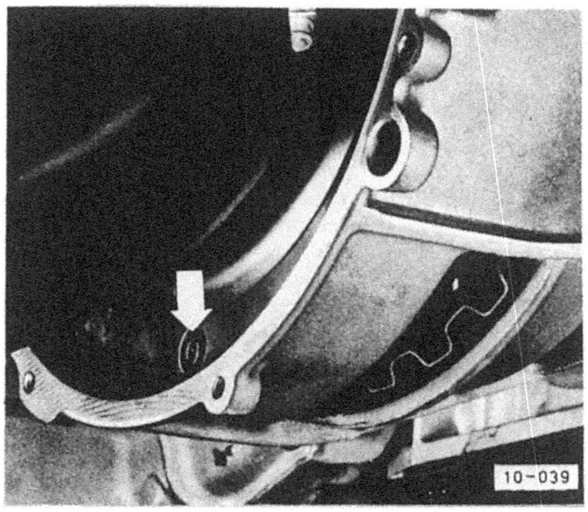

Fig. 5-4. One of three bolts (arrow) that hold torque converter to drive plate. Use crankshaft pulley bolt to turn engine for access to bolts.

On all models, remove the cover plate shown in Fig. 5-5, and remove the bolts which hold the transaxle to the engine. Separate the engine and transaxle using care not to place any strain on the transmission mainshaft.

CAUTION —
On models with automatic transmission, be sure that the drive plate separates cleanly from the torque converter without pulling the torque converter off its support. As soon as the engine and transaxle are separated, install a bar across the open bellhousing to keep the torque converter from falling.

CAUTION —
On models with manual transmission, at no time should the weight of either the engine or the transaxle be supported by the transmission mainshaft. If it is, the clutch, the clutch pushrod, or the mainshaft may be damaged.

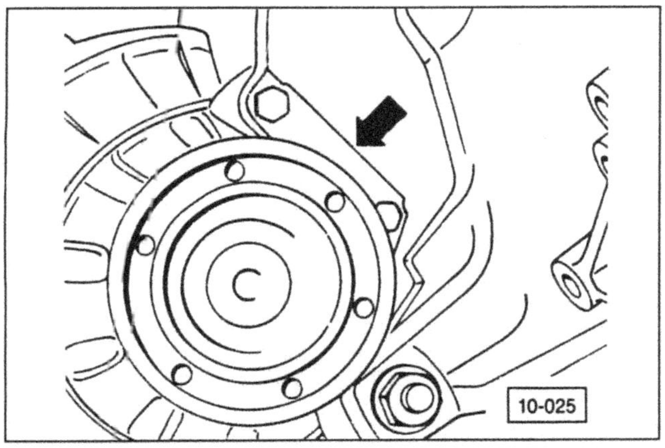

Fig. 5-5. Cover plate (arrow) to be removed when separating engine and transaxle.

Installing

Engine and transaxle installation is the reverse of removal. On models with automatic transmission, make certain during reassembly that the torque converter has not slipped from its support. See Fig. 5-6.

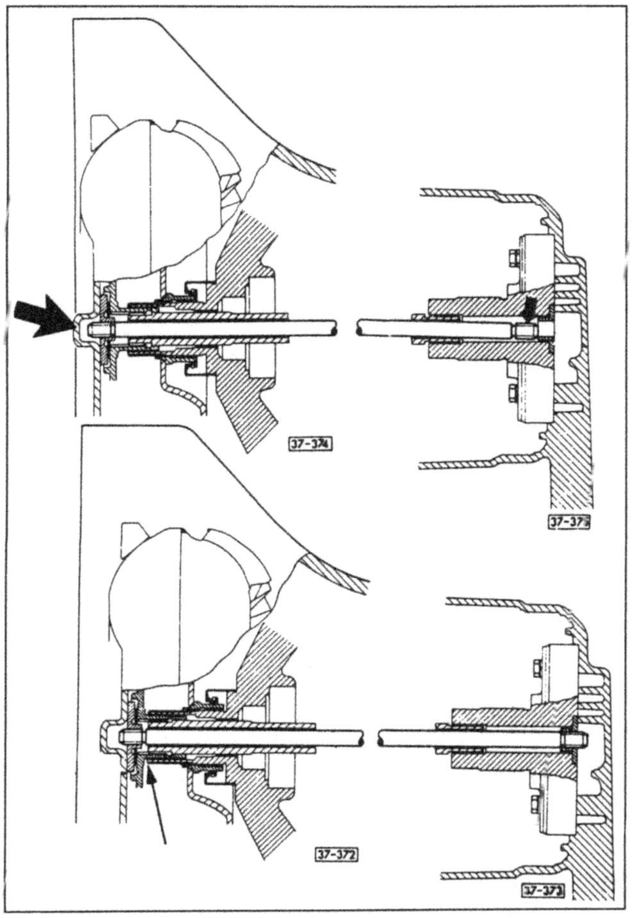

Fig. 5-6. Torque converter positions. Top: converter slipped off support; pilot projects at left arrow, pump driveshaft may be disengaged, as at right arrow. Bottom: converter correctly positioned on support (arrow).

The engine-to-transaxle bolts are two sizes, 10 mm and 12 mm. Torque the larger (12 mm) bolts to 75 Nm (55 ft. lb.) and the smaller (10 mm) bolts to 45 Nm (33 ft. lb.). On models with automatic transmission, torque the bolts holding the engine drive plate to the torque converter to 30 Nm (22 ft. lb.). Torque the starter mounting bolts to 45 Nm (33 ft. lb.), and the cover plate mounting bolts to 15 Nm (11 ft. lb.).

With the engine and transaxle in position, install the left transaxle mount first. Roughly align the engine/transaxle assembly, then install the remaining mounts. Do not fully torque the mounting bolts. Torque the horn mounting bolts to 18 Nm (13 ft. lb.).

On cars with air conditioning, adjust the tension of the compressor drive belt as described in **LUBRICATION AND MAINTENANCE**.

Following installation of the remaining engine and transaxle parts and fluids, the assembly should be aligned in its mounts as described in **Aligning Engine and Transaxle Assembly**.

Aligning Engine and Transaxle Assembly

It is normally necessary to align the engine and transaxle mounts only after the assembly has been installed, or after the mounts have been loosened as part of another repair. However, a misaligned engine/transaxle can be one of the causes of excessive engine vibration, indicated by buzzing noises in the passenger compartment.

To align:

1. Loosen the mounting bolts and nuts as indicated in Fig. 5-7.

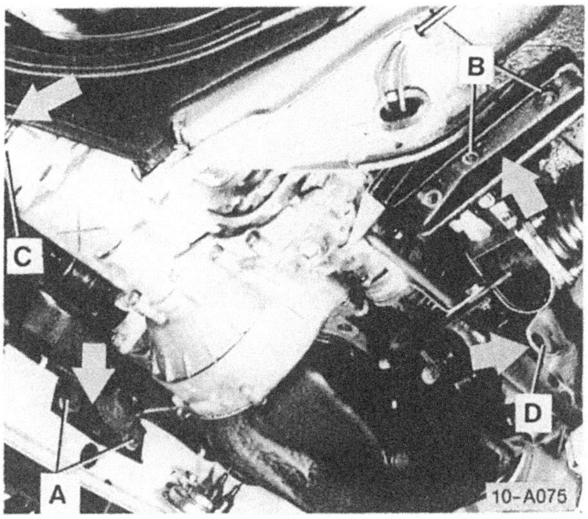

Fig. 5-7. Bolts and nuts to be loosened when aligning engine and transaxle. Loosen front mount bolts at **A**, rear mount nuts at **B**, left mount center bolt at **C**, and right mount center bolt at **D**.

2. With the car sitting on the ground, move the engine and transaxle assembly until the rear transaxle mount is straight as shown in Fig. 5-8, and the left and right engine and transaxle mounts are centered as shown in Fig. 5-9. Then torque the rear mount nuts and the left and right center bolts to 35 Nm (26 ft. lb.). Recheck the mounts to make sure that there is no twisting or strain.

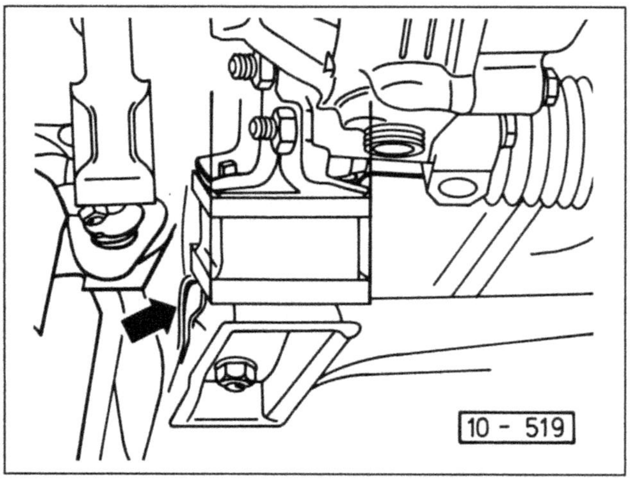

Fig. 5-8. Rear transaxle mount correctly aligned (lines).

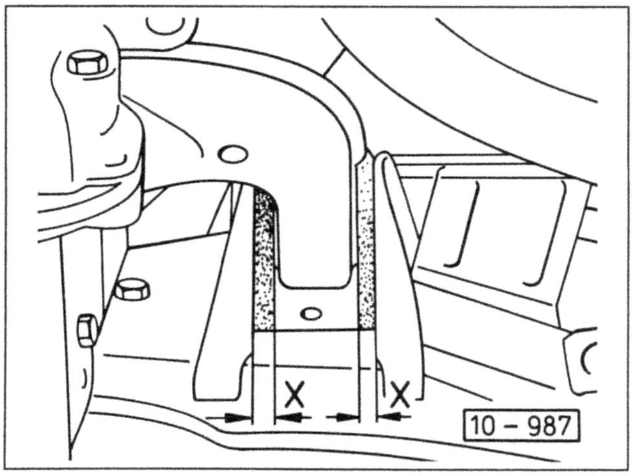

Fig. 5-9. Correct alignment (lines) of left and right engine and transaxle mounts.

3. Move the front mount on the body until the rubber core is centered in the mount as shown in Fig. 5-10, then torque the nuts. For 8-valve engines, torque the mount nuts to 52 Nm (38 ft. lb.). For 16-valve engines, torque the nuts to 40 Nm (30 ft. lb.).

> **NOTE —**
> After the alignment, check gearshift function and exhaust system alignment. If necessary, adjust the shift mechanism as described in either **MANUAL TRANSMISSION AND CLUTCH** or **AUTOMATIC TRANSMISSION**, and align the exhaust system as described in **EXHAUST AND EMISSION CONTROLS**.

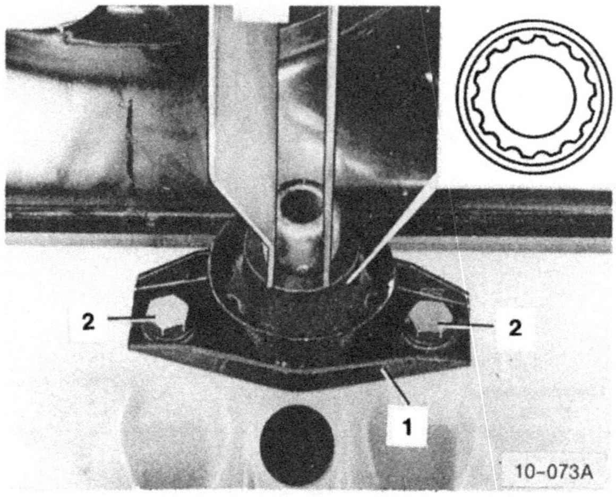

Fig. 5-10. Front engine mount being aligned. Move housing (**1**) on bolts (**2**) until rubber core is centered (inset).

6. CYLINDER BLOCK AND PISTONS

The cylinder block and pistons are normally affected only by wear due to high mileage. Low compression and noise caused by worn clearances between parts are signs of wear. For more information, see **3. Troubleshooting**.

> ***CAUTION —***
> *Replacement blocks may not include the oil pump shaft/distributor drive bushing. Assembly without the bushing will result in little or no oil pressure.*

6.1 Cylinder Block Oil Seals

The front crankshaft and intermediate shaft oil seals can be replaced with the engine installed. Replacing of the rear crankshaft oil seal requires separating the engine and transaxle.

Replacing Front Crankshaft Oil Seal

Remove the drive belt as described in **4.2 Camshaft Drive Belt**. Then remove the belt sprocket from the crankshaft. To remove the seals, Volkswagen special tools shown in Fig. 6-1 are recommended. Alternatively, remove the seal carrier from the cylinder block and the oil pan. Always use care to prevent marring the sealing surfaces. Be careful during reassembly to seat the seal properly. Use a new gasket.

Lubricate the new seal with clean engine oil. Press it all the way into the carrier, being careful not to distort the seal. The sealing lip faces inward. For best results, use a seal driver such as the one shown in Fig. 6-2. Lubricate the threads of the crankshaft sprocket bolt with engine oil.

Tightening torques:
- seal carrier (small M6 bolt) 10 Nm (87 in. lb.)
- seal carrier (all other bolts) 20 Nm (15 ft. lb.)

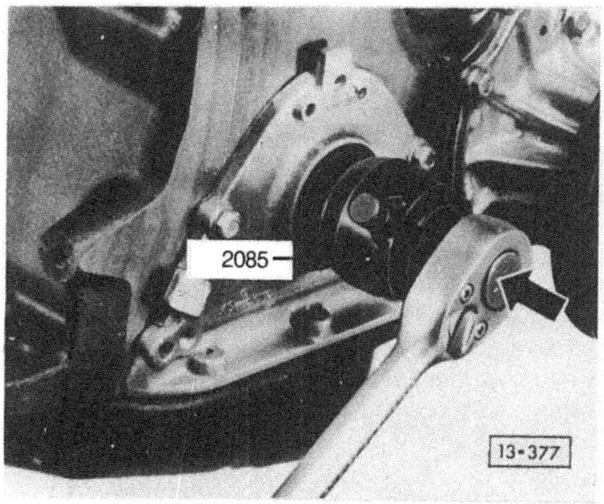

Fig. 6-1. Crankshaft front oil seal being removed using Volkswagen special tool no. 2085. Rotate tool clockwise to remove seal.

Tightening torques:

- hex bolt (with washer) 150 Nm (111 ft. lb.)
- 12-point bolt (without washer) 90 Nm (66 ft. lb.) plus $^1/_2$ turn (180°)

NOTE —
Only 12-point bolts are supplied as replacement parts. Install without washer. Always replace.

Fig. 6-2. Front crankshaft oil seal being installed, using Volkswagen special tool no. 3083.

Replacing Intermediate Shaft Oil Seal

Remove the camshaft drive belt, as described in **4.2 Camshaft Drive Belt**. Loosen the intermediate shaft sprocket bolt and, using a soft faced hammer, tap the sprocket loose. Remove the bolt and the sprocket. Remove the two bolts holding the oil seal carrier to the cylinder block and pull out the seal carrier. Pry or hand-press the seal out of the carrier.

Reinstall the seal carrier with a new O-ring. Torque the bolts to 25 Nm (18 ft. lb.). To install the seal, lightly oil the seal and shaft's sealing surface. Fit the seal into position, and carefully drive it into place. Use care not to distort the seal as it is installed. For best results, use a seal driver such as Volkswagen special tool no. 10-203.

6.2 Disassembly, Assembly, and Reconditioning

Disassembly, assembly, and reconditioning procedures for the 1.8 liter Volkswagen engines covered in this manual are similar to those for most other modern 4-cylinder, water-cooled engines. For anyone with the proper tools and equipment and basic experience in engine reconditioning, this section provides the specifications and special reconditioning information necessary to repair these Volkswagen engines.

Those who are without the necessary tools, or unfamiliar with reconditioning procedures, will still find this information to be important to the qualified mechanic who does the work. If machine shop services are not readily available, one alternative is to install a Volkswagen remanufactured engine or short block, available from an authorized Volkswagen dealer parts department.

To minimize wear during engine start-up, clean engine oil should be used to lubricate all friction surfaces during engine assembly.

Pistons and Connecting Rods

Pistons, piston pins, piston rings, connecting rods, and bearings, if they are to be reused, should never be interchanged. Mark the cylinder numbers on connecting rods and connecting rod caps before removal. Mark the piston tops as shown in Fig. 6-3. (avoid scratching the piston). The components of one piston and connecting rod assembly are shown in Fig. 6-4.

Fig. 6-3. Cylinder numbers and arrows marked on piston crowns. Used pistons should always be installed in their original orientation.

The piston pin should fit such that, with the piston heated to approximately 60°C (140°F) in an oil bath, a light push will move the pin. Replace the piston and the pin if the fit is excessively loose. Connecting rods should always be replaced in complete sets.

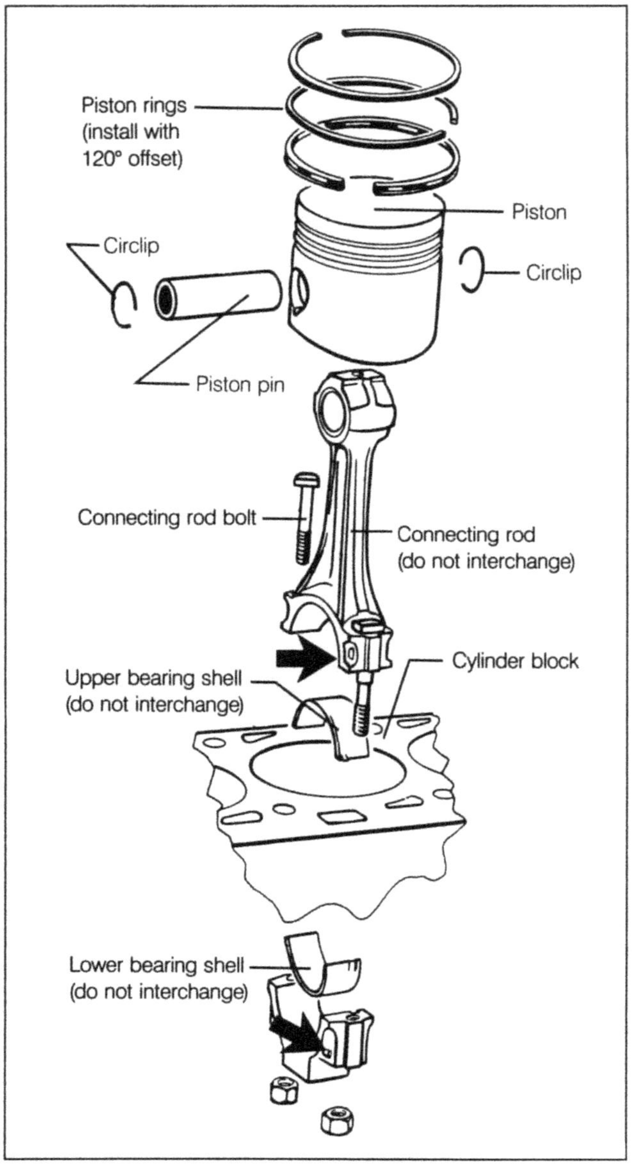

Fig. 6-4. Components of piston and connecting rod assembly. Connecting rod markings (arrows) face camshaft drive belt.

Measure the cylinder bores at approximately the top, the middle, and the bottom of piston travel, and at right angles (90°), as shown in Fig. 6-5. The top and bottom measurements should be made approximately 10 mm (3/8 in.) from the ends of the cylinder. Nominal piston and cylinder bore diameter specifications are given in **Table j**.

NOTE —

Nominal piston diameters are marked on the piston crowns.

Maximum out-of-round variation in any one cylinder is 0.04 mm (.0016 in.). Minor irregularities may be corrected by honing. If the diameter of any one cylinder exceeds the nominal dimension by more than 0.08 mm (.003 in.), all four cylinders should be rebored and honed to accept new oversized pistons. Visually inspect the bores for scoring and other imperfections which will also prevent good piston ring to cylinder wall sealing. When fitting pistons, the piston-to-cylinder clearance should be 0.03 mm (.0012 in.). Maximum piston-to-cylinder clearance is 0.07 mm (.0028 in.).

CAUTION —

Mounting the bare cylinder block to an engine stand will distort its shape and cause inaccurate cylinder bore measurements.

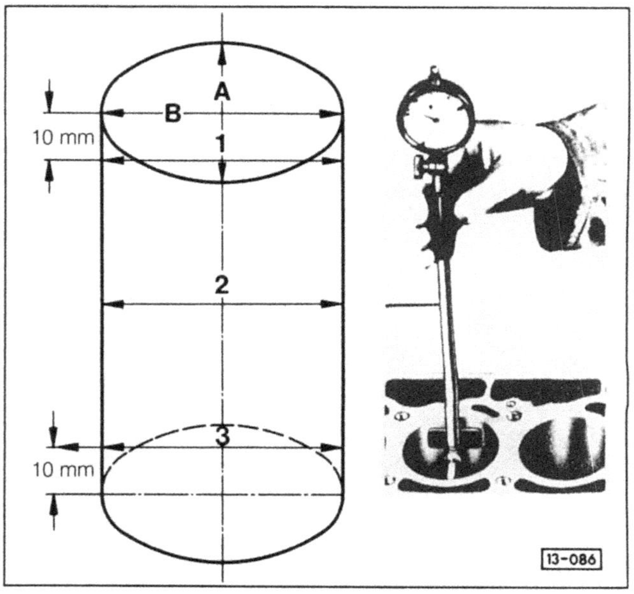

Fig. 6-5. Cylinder bore measuring points. Top **1** and bottom **3** measurements should be made at least 10 mm (3/8 in.) from ends of cylinder, first in direction **A** and then in direction **B**.

Table j. Cylinder and Piston Diameter Specifications

Engine Code	Repair Stage	Piston Diameter	Cylinder Bore
JH, 2H	Basic Dimension (standard)	80.98 mm (3.1882 in)	81.01 mm (3.1894 in.)
	1st oversize	81.23 mm (3.1980 in.)	81.26 mm (3.1992 in.)
	2nd oversize	81.48 mm (3.2079 in.)	81.51 mm (3.2091 in.)
PL	Basic Dimension	80.98 mm (3.1882 in.)	81.01 mm (3.1894 in.)

NOTE —
At the time of printing, only standard size pistons are available for PL (16-valve) engines. Check with an authorized Volkswagen dealer parts department for the latest information on parts availability.

Measure connecting rod bearing clearances using Plastigage®. Inspect the connecting rods for any bending, distortion, or other visible damage. Connecting rod specifications are listed in **Table k**.

Table k. Connecting Rod Specifications

Radial Clearance Wear Limit	0.12 mm (.0047 in.)
Side Clearance Wear Limit	0.37 mm (.0145 in.)
Checking Torque	30 Nm (22 ft. lb.)
Assembly Torque	30 Nm (22 ft. lb.) + ¼ turn (90°)

NOTE —
When checking radial clearance, lubricate the contact surface of the nut before tightening.

If connecting rod radial clearance is excessive, the crankshaft connecting rod journals should be checked as described in **Crankshaft and Intermediate Shaft**. If the crankshaft journal diameters are within specifications, recheck radial clearance using new bearing shells.

When installing each connecting rod to the crankshaft, the two bearing shells' retaining tabs should be on the same side of the journal, and the marks on the connecting rods, shown earlier in Fig. 6-4, should face the drivebelt side of the engine.

Piston Rings

Piston ring end gaps are checked with the piston rings inserted approximately 15mm (5/8 in.) from the top of the cylinder. The piston ring gap specifications are listed in **Table l**.

Table l. Piston Ring End Gap

	Gap	Wear Limit
Compression Rings	0.30 to 0.45 mm (.012 to .018 in.)	1.0 mm (.039)
Oil Scraper Rings (bottom ring)	0.25 to 0.45 mm (.010 to .018 in.)	1.0 mm (.039)

Piston ring side clearance is checked using feeler gauges. Measure each ring in its original groove. Piston ring side clearance specifications are listed in **Table m**.

Table m. Piston Ring Side Clearance

Ring	Clearance (new parts)	Wear Limit
all	0.02–0.05 mm (.0008–.0019 in.)	0.15 mm (.0059 in.)

Install the piston rings with the gaps offset from each other by 120°, as illustrated earlier in Fig. 6-4.

Crankshaft and Intermediate Shaft

Fig. 6-6 shows the assembly of the crankshaft and the intermediate shaft in the cylinder block. To remove the crankshaft, both crankshaft oil seal carriers must be removed from the ends of the cylinder block. Crankshaft main bearing caps must not be interchanged. Crankshaft main bearing shells, if they are to be reused, should only be installed in their original positions. On 8-valve engines, the ignition distributor must be removed before removing the intermediate shaft.

NOTE —
On models with automatic transmission, mark the position of the driveplate on the crankshaft before removing the driveplate.

NOTE —
Replacement No. 3 main bearing shells are one-piece with integral thrust washers.

Crankshaft journal specifications are listed in **Table n**. Tightening torques are listed in **Table o**. Crankshaft and intermediate shaft clearance specifications are listed in **Table p**. Crankshaft bearings are available in three undersizes to fit reconditioned crankshafts. If a crankshaft must be replaced, a Volkswagen remanufactured crankshaft is available from an authorized Volkswagen dealer.

Table n. Crankshaft Journal Diameter

	Main Bearing Journal Diameter	Connecting Rod Journal Diameter
Basic Dimension	53.958–53.978 mm (2.1243–2.1251 in.)	47.758–47.778 (1.8802–1.8810 in.)
1st Undersize (0.25 mm)	53.708–53.728 mm (2.1145–2.1153 in.)	47.508–47.528 mm (1.8704–1.8712 in.)
2nd Undersize (0.50 mm)	53.458–53.478 mm (2.1046–2.1054 in.)	47.258–47.278 mm (1.8606–1.8613 in.)
3rd Undersize (0.75 mm)	53.208–53.228 mm (2.0948–2.0956 in.)	47.008–47.028 mm (1.8507–1.8515 in.)

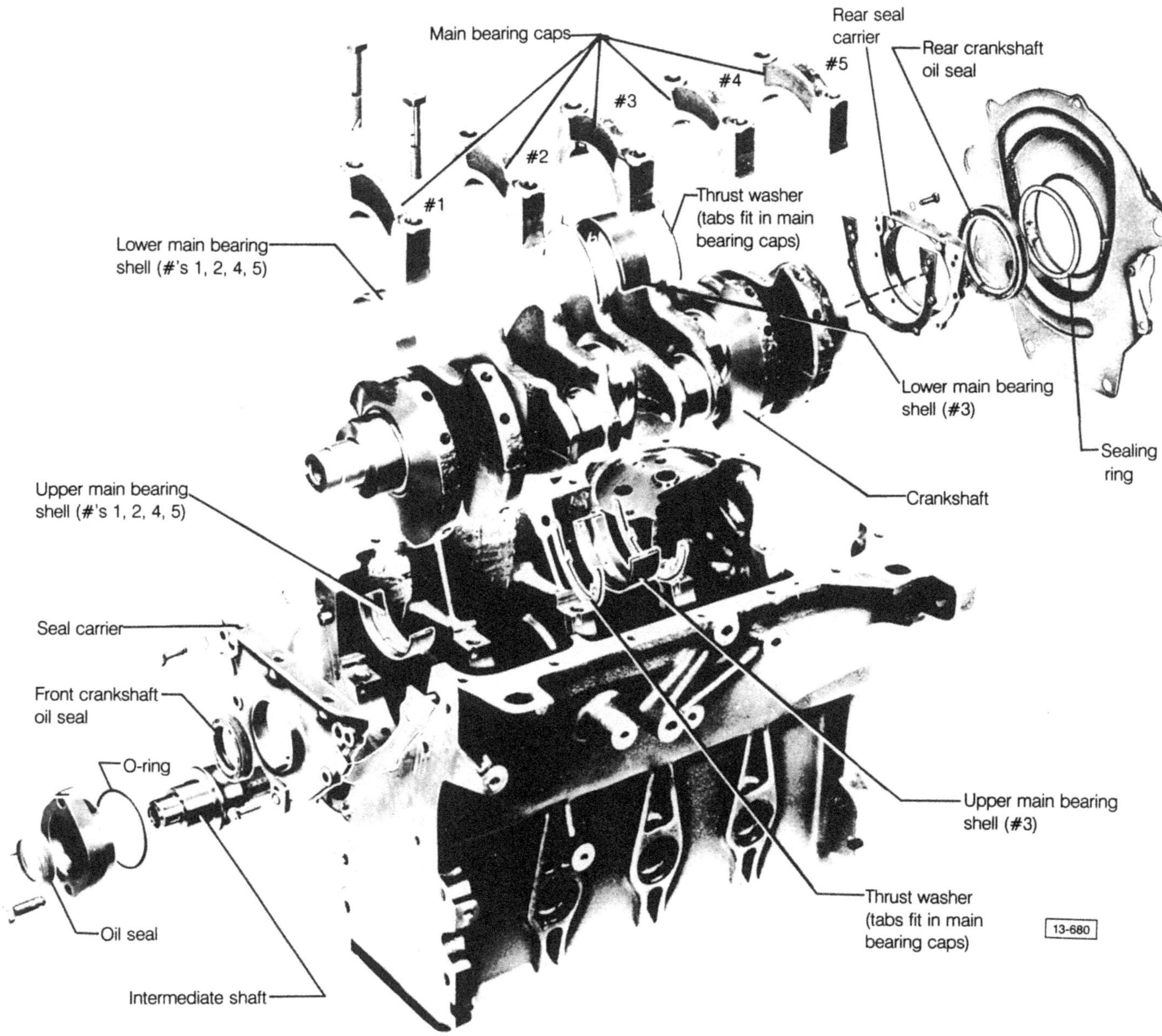

Fig. 6-6. Exploded view of cylinder block, crankshaft, intermediate shaft, and related parts.

Table o. Crankshaft and Cylinder Block Tightening Torques

Crankshaft Oil Seal Carrier Bolts	
M6	10 Nm (87 in. lb.)
M8	20 Nm (15 ft. lb.)
Intermediate Shaft Oil Seal Carrier Bolts	25 Nm (18 ft. lb.)
Main Bearing Cap Nuts	65 Nm (48 ft. lb.)
Oil Pan Bolts	20 Nm (15 ft. lb.)

Table p. Crankshaft and Intermediate Shaft Clearance

	New parts	Wear limit
Crankshaft main bearings		
radial clearance	0.03–0.08 mm (.0012–.0031 in.)	0.17 mm (.0067 in.)
axial (thrust) clearance	0.07–0.17 mm (.0028–.0067 in.)	0.25 mm (.0098 in.)
Connecting rod bearings		
radial clearance	--	0.12 mm (.0047 in.)
axial (side) clearance	--	0.37 mm (.0146 in.)
Intermediate shaft		
axial clearance	--	0.25 mm (.0098 in.)

Flywheel or Drive Plate

Fig. 6-7 and Fig. 6-8 show the tightening sequence and a holding fixture, Volkswagen special tool VW 558. Torque the bolts, with locking compound, to 30 Nm (22 ft. lb.) plus $^1/_4$ turn (90°).

> ***CAUTION —***
> *Always replace flywheel or drive plate mounting bolts with new bolts for final installation.*

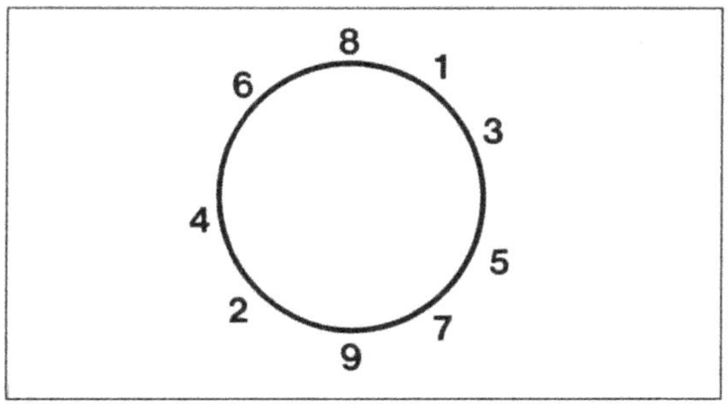

Fig. 6-7. Flywheel or drive plate bolt torque sequence. Loosen and remove in reverse order.

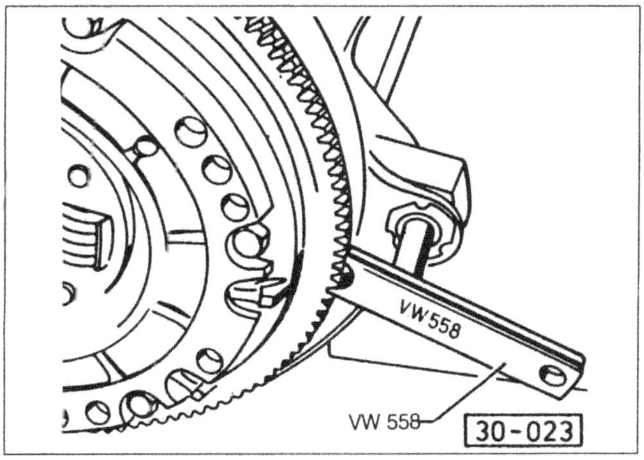

Fig. 6-8. Flywheel secured with holding fixture VW 558.

Replacement flywheels may not have an ignition timing mark. Make a timing mark 12 mm (0.47 in.) arc length to the left of the Top Dead Center (0) mark, as shown in Fig. 6-9.

On automatic transmission cars, mark the drive plate during disassembly and reinstall it with the same orientation. See Fig. 6-10. When replacing the drive plate, first install it with old bolts and a 1 mm shim and check drive plate to cylinder block clearance as shown in Fig. 6-11. Install 4 mm and/or 1 mm shims to achieve a dimension of 31.3 ± 0.7 mm (1.23 ± 0.028 in.). With the correct shims, install the drive plate with new bolts. Use one shim between the drive plate and the bolts. Torque as described above. Torque the torque converter bolts, with D6 locking compound or equivalent, to 35 Nm (26 ft. lb.).

> **NOTE —**
> A modified drive plate, installed in production as of VIN _15MK_005500 (1989), is the only one available as a replacement. It requires a different combination of shims than the earlier drive plate.

Fig. 6-9. Location on replacement flywheel to make new ignition timing mark. Dimension **a** is arc length from Top Dead Center (0) mark.

Fig. 6-10. Drive plate assembly. Use only one shim **1**. Combine shims **2** (1 mm and/or 4 mm) to achieve correct cylinder block clearance dimension.

Fig. 6-11. Drive plate to cylinder block clearance being measured. Dimension **a** should be 31.3 ± 0.7 mm (1.23 ± 0.028 in.). Adjust distance by adding or removing shims behind drive plate.

7. LUBRICATION SYSTEM

The primary function of the lubrication system is to lubricate the internal moving parts of the engine. The circulation of oil also aids engine cooling. Proper lubrication relies on a constant supply of oil, fed to the moving parts under pressure. Pressure is supplied by a gear-type oil pump located inside the oil pan. Engine oil returns to the oil pan where it is stored for pickup by the pump. It is cleaned by circulating through a replaceable filter. Fig. 7-1 is a schematic view of the lubrication system showing the paths of pressurized oil supply to the engine bearings.

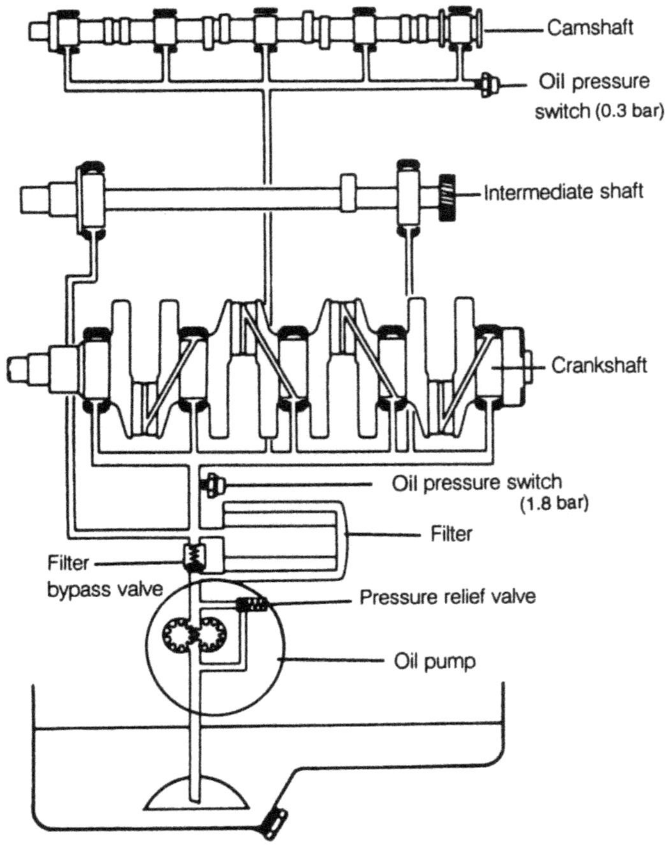

Fig. 7-1. Schematic view of lubrication system. Pressure relief is part of pump. On 16-valve engines, both pressure switches are on oil filter flange.

This section covers inspection, repair and assembly of the lubrication system. Oil and filter change and engine oil specifications are covered in **LUBRICATION AND MAINTENANCE**.

To prevent serious engine damage, a dynamic oil pressure warning system warns the driver of insufficient oil pressure. Other safety features include a filter by-pass, to guard against bursting the filter due to over-pressure, and an oil pump pressure relief valve to prevent excessive system pressure.

On cars equipped with the multi-function indicators, an oil temperature sensor, located on the oil filter flange, monitors the oil temperature. On PL (16-valve) engines the sensor is located on the end of the cylinder head.

Lubrication system problems are caused by an inability to create or sustain adequate oil pressure. In rare cases, indications of trouble may be false alarms caused by faults in the oil pressure warning system.

> ***CAUTION —***
> *Always assume that a warning is valid. Take proper precautions to avoid engine damage until adequate oil supply and proper oil pressure can be verified.*

> **NOTE —**
> To prevent oil being drawn into the air filter through the crankcase breather on 16-valve engines, replace the oil pump splash shield with a baffle, Volkswagen part no. 027 103 623.

Proper lubrication is essential to long engine life. Change the engine oil and oil filter regularly, at least as often as specified by Volkswagen's scheduled maintenance intervals, and periodically check engine oil level between oil changes. See **LUBRICATION AND MAINTENANCE.**

7.1 Dynamic Oil Pressure Warning System

The dynamic oil pressure warning system includes pressure switches, the electronic control unit, the warning indicator and a warning buzzer. A 0.3 bar pressure switch provides warning when the oil pressure falls to near zero at any time. The other (1.8 bar) switch provides earlier warning, at higher rpm, whenever oil pressure falls too low.

The pressure switches are located on the oil filter flange (1.8 bar) and on the end of the cylinder head (.3 bar). On 16-valve engines, both switches are on the oil filter flange. The electronic control unit is behind the instrument cluster.

> ***CAUTION —***
> *If the warning indicator stays on after the engine is started, or flashes on while driving, always assume that the warning is valid. Check oil level and test oil pressure before testing the warning system. See **8. Engine Technical Data**.*

Checking Low Oil Pressure Warning System

With the ignition off, the 0.3 bar switch is closed (completing the circuit to ground). When the ignition is switched on, the indicator will flash. When the engine is started and oil pressure rises, it opens the switch (opening the ground circuit), and the indicator goes out. With too little oil pressure or a stuck switch, the indicator will continue to flash. For switch opening and closing pressures see **Testing Oil Pressure Switches**.

To quick-check the 0.3 bar switch, simulate a closed switch. Remove the blue/black wire and connect it to ground. With the ignition on, the indicator should flash. If not, the problem is in the circuit or the indicator. If the indicator flashes only when tested in this way, replace the switch.

If the indicator flashes with the engine running, and the oil pressure tests OK, remove the blue/black wire from the switch with the engine running. The indicator should go out. If not, the circuit is shorted to ground between the wire and indicator. If the indicator does go out, the switch is stuck closed and should be replaced.

Checking Dynamic Oil Pressure Warning System
(high rpm)

The 1.8 bar pressure switch is normally open (no circuit to ground). When the engine is running above 2000 rpm (ignition distributor input to control unit), oil pressure closes the switch (completing the circuit to ground). If oil pressure is insufficient to close the pressure switch, or the switch is stuck open, the indicator will flash and the buzzer will sound. The pressure switch opens and closes at specific pressures. See **Testing Oil Pressure Switches**.

To quick-check the pressure switch, raise the engine speed above 2000 rpm and disconnect the yellow wire from the switch. The indicator should flash and the buzzer should sound. If the wire is connected to ground, the warning should stop. Disconnect the wire from ground and connect it to the switch. If the indicator does not go out, replace the switch.

Testing Oil Pressure Switches

The oil pressure at which the pressure switches react can be tested by temporarily installing an oil pressure gauge in parallel with the switch with a T-fitting and monitoring switch performance with a multimeter or test light, as shown in Fig. 7-2. See **8. Engine Technical Data**.

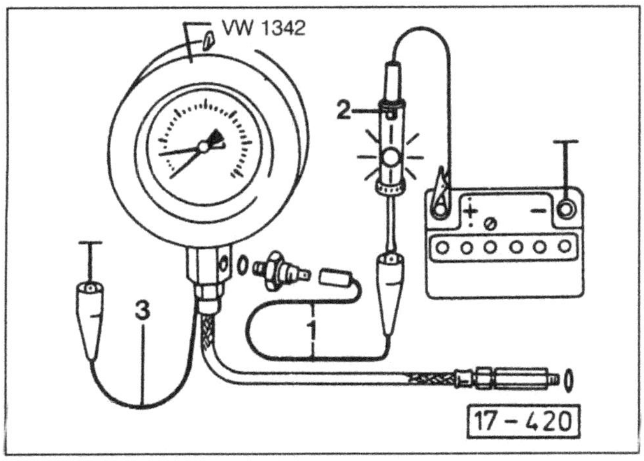

Fig. 7-2. Pressure gauge and test light being used to test oil pressure switches.

7.2 Oil Pump

The oil pump is located inside the engine oil pan, and draws engine oil through a pickup tube from near the bottom of the pan. The pump can be removed for inspection of its internal clearances (a potential source of low oil pressure problems) by removing the oil pan.

There is normally no need to remove and inspect the oil pump unless oil pressure is inadequate. Check the oil pressure by installing a pressure gauge in place of one of the oil pressure switches and run the engine. Engine oil pressure at normal operating temperature should be at least 2.0 bar (29 psi) at 2000 rpm.

To inspect the oil pump, drain the oil and remove the oil pan. Remove the pump and disassemble it. See Fig. 7-3.

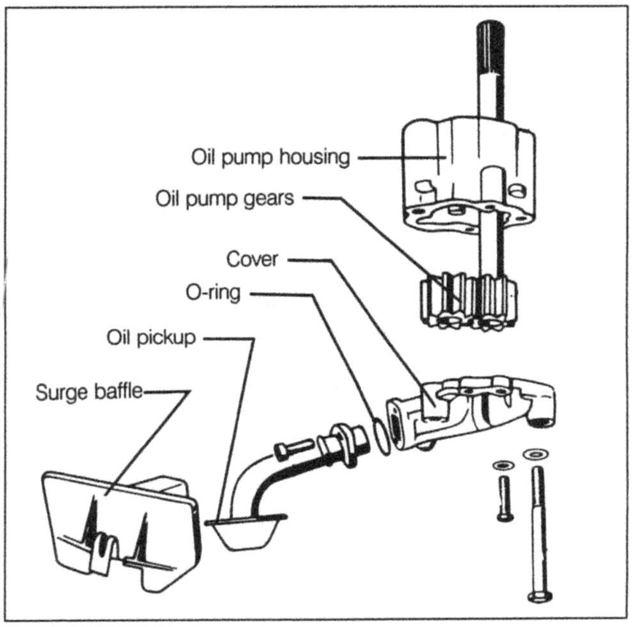

Fig. 7-3. Exploded view of oil pump. Note long mounting bolts and short cover bolts.

With the oil pump removed and disassembled, use feeler gauges to check the clearances. Backlash, shown being measured in Fig. 7-4, should be between 0.05 and 0.20 mm (.002 and .008 in.). If not, replace the gears or the pump. Oil pump gear axial play, shown being measured with a straight-edge in Fig. 7-5, should not exceed 0.15 mm (.006 in.). If it does, replace the pump.

If the oil pickup has been removed from the pump cover, replace the O-ring when installing the pickup. Torque the pickup bolts and cover bolts to 10 Nm (87 in. lb.). Torque the oil pump mounting bolts to 20 Nm (15 ft. lb.).

Replace the oil pan gaskets and install the oil pan. Torque the oil pan bolts to 20 Nm (15 ft. lb.) and the drain plug to 30 Nm (22 ft. lb.). Add engine oil, as described in **LUBRICATION AND MAINTENANCE**.

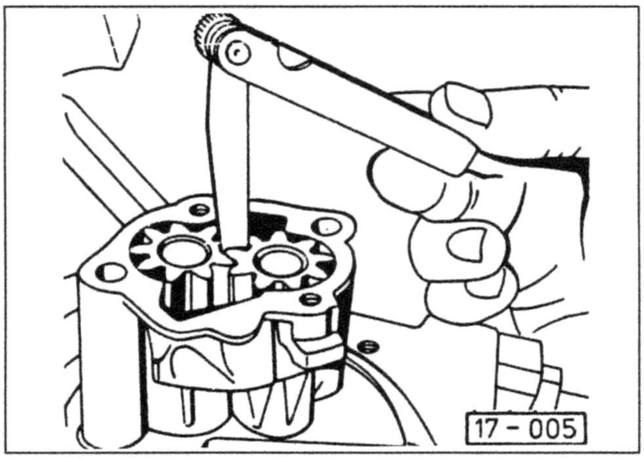

Fig. 7-4. Oil pump gear backlash being measured using feeler gauge.

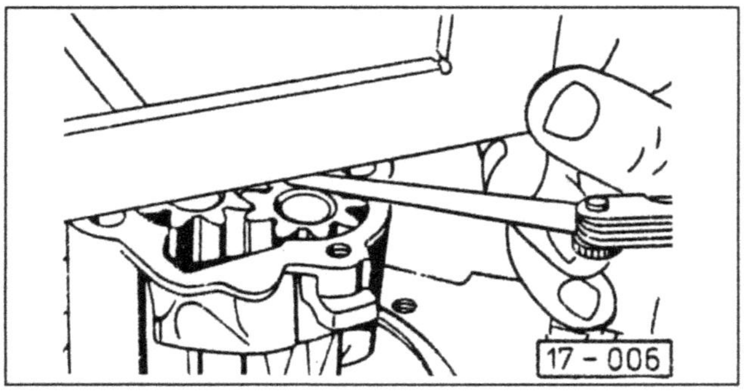

Fig. 7-5. Oil pump gear axial play being measured using straight-edge and feeler gauge.

Oil Spray Nozzles

PL (16-valve) engines are equipped with oil spray nozzles which, above a certain oil pressure threshold, spray oil from the main oil galley against the bottoms of the pistons for added cooling.

Install these nozzles with sealing paste, Volkswagen part no. AMV 188 100 02, and torque the bolts to 10 Nm (87 in. lb.).

NOTE —
Part numbers are subject to change. Always rely on an authorized Volkswagen dealer parts department for the latest information.

7.3 Oil Cooler

The oil cooler is an oil-to-coolant heat exchanger. Engine oil flows through one part of the cooler and gives up heat to the engine coolant flowing through the other part of the cooler. The cooler is a potential source of leakage between the lubrication system and the cooling system, and should be considered whenever such leakage is suspected.

Fig. 7-6 is an exploded view of the oil cooler, the oil filter, and the mounting parts. Remove the oil cooler only if it needs to be cleaned, inspected, or replaced. On installation, use sealing paste, Volkswagen part no. AMV 188 100 02, on the sealing surfaces and check to see that there is adequate space for the coolant hose connections. Torque the oil cooler mounting nut to 25 Nm (18 ft. lb.).

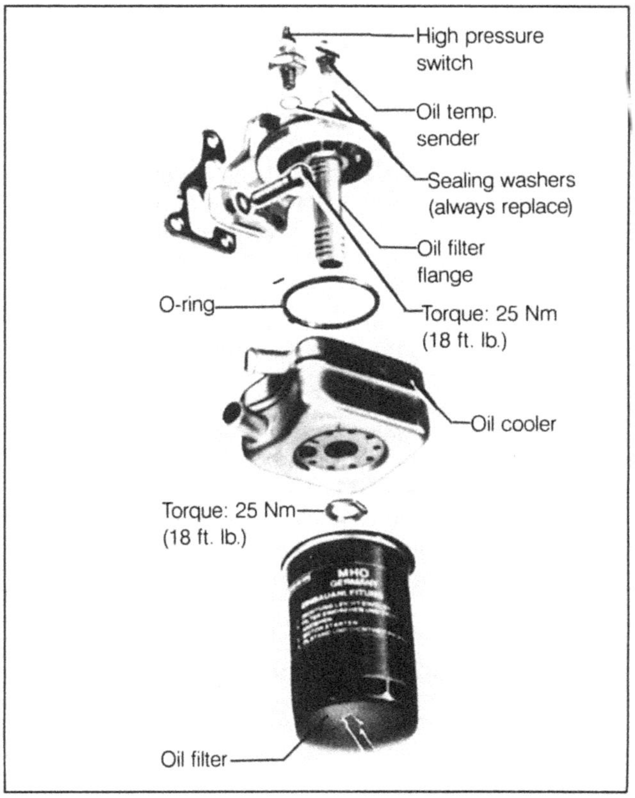

Fig. 7-6. Exploded view of oil cooler and related parts. 8-valve engine components shown. Specifications also apply to similar parts for 16-valve engines.

CAUTION —
If metal shavings are found in the engine oil as a result of engine damage, all engine oil passages must be thoroughly cleaned. The engine oil cooler should be replaced.

8. ENGINE TECHNICAL DATA

I. Tightening Torques

Air connection to cylinder head for 8-valve engines (bolt)	10 Nm (87 in. lb.)
Alternator adjusting bracket to engine (bolt)	20Nm (15 ft. lb.)
Alternator to adjusting bracket of vehicle (bolt)	35 Nm (26 ft. lb.)
Alternator to lower mounting of vehicle (bolt)	35 Nm (26 ft. lb.)
Camshaft bearing caps to cylinder head (nut)	
8-valve engines	20 Nm (15 ft. lb.)
16-valve engines	15 Nm (11 ft. lb.)
Camshaft drive belt sprocket to camshaft or intermediate shaft (bolt)	
8-valve engines	80 Nm (59 ft. lb.)
16-valve engines	65 Nm (48 ft. lb.)
Camshaft drive belt tensioner locknut	45 Nm (33 ft. lb.)
Clutch pressure plate or automatic transmission drive plate to crankshaft	30 Nm (22 ft. lb.) plus 1/4 turn (90°)
Connecting rod cap to connecting rod	30 Nm (22 ft. lb.) plus 1/4 turn (90°)
Cover plate to transaxle	15 Nm (11 ft. lb.)
Crankshaft oil seal carrier to cylinder block	
M 6 bolt	10 Nm (87 in. lb.)
M 8 bolt	20 Nm (15 ft. lb.)
Crankshaft timing belt sprocket to crankshaft	
Hex bolt (with washer)	150 Nm (111 ft. lb.)
12-point bolt (without washer)	90 Nm (66 ft. lb.) plus 1/4 turn (90°)
Cylinder head cover to cylinder head (nut)	10 Nm (87 in. lb.)
Cylinder head to engine block (engine cold)	60 Nm (43 ft. lb.) plus 1/2 turn (180°)
Drive belt rear cover to stud in cylinder head (socket-head bolt)	30 Nm (22 ft.lb.)
Drive belt upper and lower front cover to rear cover (bolt and nut)	
except front bolt on 16-valve engines	10 Nm (87 in. lb.)
front bolt on 16-valve engines	6 Nm (53 in. lb.)
Drive shaft to transaxle drive flange	
8-valve engines	42 Nm (30 ft. lb.)
16-valve engines	45 Nm (33 ft. lb.)
Engine to transaxle bellhousing	
M10	45 Nm (33 ft. lb.)
M12	75 Nm (55 ft. lb.)
Engine front mount to body (bolt)	52 Nm (38 ft. lb.)
Engine right and left mounting to rubber (through bolt)	35 Nm (26 ft. lb.)
Engine/transaxle rear mount to engine (nut)	35 Nm (26 ft. lb.)
Engine/transaxle mount to body	
16-valve engines	40 Nm (30 ft. lb.)
Fuel injector insert to cylinder head (with thread locking compound)	20 Nm (15 ft. lb.)
Flywheel to clutch pressure plate	20 Nm (15 ft. lb.)
Intermediate shaft oil seal carrier to cylinder block	25 Nm (18 ft. lb.)
Manifolds to cylinder head	25 Nm (18 ft. lb.)
Main bearing caps to cylinder block	65 Nm (48 ft. lb.)
Oil cooler to oil filter flange (nut)	25 Nm (18 ft. lb.)
Oil filter mounting flange to engine cylinder block (socket-head bolt)	25 Nm (18 ft. lb.)
Oil pickup to oil pump housing cover (bolt)	10 Nm (87 in. lb.)
Oil pressure switches	25 Nm (18 ft. lb.)

I. Tightening Torques (continued)

Oil pan to cylinder block	20 Nm (15 ft. lb.)
Oil drain plug in oil pan	30 Nm (22 ft. lb.)
Oil pump to cylinder block (M8 bolt)	20 Nm (15 ft. lb.)
Oil pump cover bolt	10 Nm (87 in. lb.)
Power steering pump to mount (bolt)	20 Nm (15 ft. lb.)
Spark plug	20 Nm (15 ft. lb.)
Starter to bellhousing (manual transmission)	46 Nm (34 ft. lb.)
Torque converter to drive plate (with D6 locking compound)	35 Nm (26 ft. lb.)
Upper intake manifold to lower intake manifold (16-valve engines)	20 Nm (15 ft. lb.)
Upper intake manifold to support bracket (16-valve engines)	20 Nm (15 ft. lb.)
V-belt pulleys to coolant pump or crankshaft sprocket	20 Nm (15 ft. lb.)

II. Crankshaft, Intermediate Shaft and Crankshaft Bearing Specifications

Main bearing journal diameter	
basic dimension	53.958–53.978 mm (2.1243–2.1251 in.)
1st undersize (0.25 mm)	53.708–53.728 mm (2.1145–2.1153 in.)
2nd undersize (0.50 mm)	53.458–53.478 mm (2.1046–2.1054 in.)
3rd undersize (0.75 mm)	53.208–53.228 mm (2.0948–2.0956 in.)
Main bearing radial clearance (with Plastigage®)	
new parts	0.03–0.08 mm (.0012–.0031 in.)
wear limit	0.17 mm (.0067 in)
Crankshaft axial play (side clearance)	
new parts	0.07–0.17 mm (.0028–.0067 in.)
wear limit	0.25 mm (.0098 in.)
Crankshaft connecting rod journals diameter	
basic dimension	47.758–47.778 mm (1.8802–1.8810 in.)
1st undersize (0.25 mm)	47.508–47.528 mm (1.8704–1.8712 in.)
2nd undersize (0.50 mm)	47.258–47.278 mm (1.8606–1.8613 in.)
3rd undersize (0.75 mm)	47.008–47.028 mm (1.8507–1.8515 in.)
Connecting rod bearing radial clearance (with Plastigage®)	
wear limit	0.12 mm (.0047 in.)
Connecting rod bearing axial clearance (side clearance)	
wear limit	0.37 mm (.0146 in.)
Intermediate shaft axial clearance (side clearance)	0.25 mm (.0098 in.)

III. Piston, Piston Ring and Cylinder Bore Specifications

Cylinder bore diameter
- 8-valve engines
  - basic dimension .... 81.01 mm (3.189 in.)
  - 1st oversize (0.25 mm) .... 81.26 mm (3.199 in.)
  - 2nd oversize (0.50 mm) .... 81.51 mm (3.209 in.)
  - out-of-round (max.) .... 0.04 mm (.0016 in.)
  - maximum deviation from nominal size .. 0.08 mm (.003 in.)
- 16-valve engines
  - basic dimension .... 81.01 mm (3.189 in.)
  - out-of-round (max.) .... 0.04 mm (.0016 in.)
  - maximum deviation from nominal size .. 0.08 mm (.0031 in.)

Piston diameter
- 8-valve engines
  - basic dimension .... 80.98 mm (3.188 in.)
  - 1st oversize (0.25 mm) .... 81.23 mm (3.198 in.)
  - 2nd oversize (0.50 mm) .... 81.48 mm (3.208 in.)
- 16-valve engines
  - basic dimension .... 80.98 mm (3.188 in.)

Piston-to-cylinder clearance
- with new parts .... 0.03 mm (.0012 in.)
- wear limit .... 0.07 mm (.0028 in.)

Piston rings clearance (ring to ring land) .... 0.02–0.05 mm (.0008–.0019 in.)

Piston ring end gap (measured with ring only in cylinder)
- compression ring .... 0.30–0.45 mm (.012–.018 in.)
- compression ring wear limit .... 1.0 mm (.039 in.)
- oil scraper ring .... 0.25–0.45 mm (.010–.018 in.)
- oil scraper ring wear limit .... 1.00 mm (.039 in.)

IV. Camshaft, Valve and Cylinder Head Specifications

Camshaft
- axial play, maximum .... 0.15 mm (.0059 in.)
- runout, maximum permissible .... 0.01 mm (.0004 in.)
- radial clearance (with Plastigage®)
  - wear limit .... 0.1 mm (.004 in.)

Cylinder head
- warpage, maximum .... 0.1 mm (.004 in.)
- thickness, minimum
  - 8-valve engines .... 132.6 mm (5.220 in.)
  - 16-valve engines (do not resurface) ... 118.1 mm (4.650 in.)

Cylinder block
- deck warpage, maximum .... 0.1 mm (.004 in.)

Valve seat dimensions
- seat angle .... 45°
- upper correction angle .... 30°
- lower correction angle (16-valve engines)
  - intake .... 75°
- seat width, intake
  - 8-valve engines .... 2.00 mm (.079 in.)
  - 16-valve engines .... 1.5-1.8 mm (.059 -.071 in.)
- seat width, exhaust
  - 8-valve engines .... 2.4 mm (.094 in.)
  - 16-valve engines .... approx. 1.8 mm (.071 in.)
- outside seat diameter of 45° face, intake
  - 8-valve engines .... 39.2 mm (1.543 in.)
  - 16-valve engines .... 31.2 mm (1.228 in.)
- outside diameter of 45° face, exhaust
  - 8-valve engines .... 32.4 mm (1.276 in.)
  - 16-valve engines .... 27.6 mm (1.087 in.)

IV. Camshaft, Valve and Cylinder Head Specifications (continued)

Valve guide wear, maximum (measure with new valve)
- intake .... 1.0 mm (.039 in.)
- exhaust .... 1.3 mm (.059 in.)

Intake valve dimensions
- stem diameter
  - 8-valve engines .... 7.97 mm (.314 in.)
  - 16-valve engines .... 6.97 mm (.274 in.)
- length
  - 8-valve engines
    - w/mechanical cam followers ...... 98.70 mm (3.886 in.)
    - w/hydraulic cam followers .... 91.00 mm (3.583 in.)
  - 16-valve engines .... 95.50 mm (3.760 in.)
- head diameter
  - 8-valve engines .... 40.00 mm (1.575 in.)
  - 16-valve enginess .... 32.00 mm (1.259 in.)

Exhaust valve dimensions
- stem diameter
  - 8-valve engines .... 7.95 mm (.313 in.)
  - 16-valve engines .... 6.94 mm (.273 in.)
- length
  - 8-valve engines
    - w/mechanical cam followers ...... 98.50 mm (3.878 in.)
    - w/hydraulic cam followers .... 90.80 mm (3.575 in.)
  - 16-valve engines .... 98.20 mm (3.846 in.)
- head diameter
  - 8-valve engines .... 33.00 mm (1.299 in.)
  - 16-valve engines .... 28.00 mm (1.102 in.)

Valve clearance
- 1985-1987 8-valve (JH) w/adjustable cam followers only
- engine cold
  - intake .... 0.15–0.25 mm (.006–.010 in.)
  - exhaust .... 0.35–0.45 mm (.014–.018 in.)
- engine hot (minimum coolant temperature 95°F (35°C))
  - intake .... 0.20–0.30 mm (.008–.012 in.)
  - exhaust .... 0.40–0.50 mm (.016–.020 in.)

V. Lubrication System Specifications

Oil pump gears
- backlash .... 0.05–0.20 mm (.002–.008 in.)
- axial play, wear limit .... 0.15 mm (.006 in.)

Oil pressure, minimum at 2000 rpm
- (with SAE 10W oil at 140°F (68°C)) .... 2.0 bar (28 psi)

Oil pressure warning system
- low pressure switch
- (nominal 0.3 bar, normally closed)
  - opens .... 0.15–0.45 bar (2.2–6.5 psi)
- dynamic pressure switch
- (nominal 1.8 bar, normally open)
  - closes .... 1.6–2.0 bar (23–29 psi)

Section 4

COOLING SYSTEM

Contents

Cooling System

Introduction

Volkswagen Scirocco and Cabriolet engines are liquid-cooled, and rely on a closed system of circulating coolant to maintain an even engine temperature and help transfer heat away from the engine. To provide adequate cooling system performance over a range of temperature conditions, to supply lubrication to the system's moving parts, and to prevent the buildup of mineral deposits and other contaminants, the coolant recommended for use year-round is a mixture of phosphate-free anti-freeze and clean water. The closed system becomes pressurized as its temperature increases. The pressure in turn raises the boiling point of the coolant and allows engine temperature to exceed the coolant's normal boiling point.

The coolant pump is mechanically driven by the engine, and operates whenever the engine is running. Coolant circulates through the engine to the radiator, the heat exchanger—or heater core—in the passenger compartment heating system, and back to the pump. Coolant is also circulated through a heat exchanger in the engine's lubrication system to help moderate engine oil temperature. Before the engine is up to normal operating temperature, the flow of coolant is controlled by a thermostat. The thermostat forces coolant flow to bypass the radiator, returning directly to the coolant pump, until the engine is warm.

A radiator cooling fan provides auxiliary air flow through the radiator. For compatibility with the transverse engine design, and to minimize power consumption, the fan is electrically operated and thermostatically controlled so that it runs only when the extra air flow is required to maintain proper coolant temperature.

Proper care of the cooling system is easy. Simple preventive maintenance can keep the system operating at its best and help prevent temperature-related problems from shortening engine life. If you lack the tools or a suitable workplace for servicing the cooling system, we suggest you leave this work to an authorized Volkswagen dealer or other qualified shop. We especially urge you to consult your authorized Volkswagen dealer before beginning any repairs on a vehicle still covered by warranty.

1. GENERAL DESCRIPTION

Fig. 1-1 is a schematic view of the cooling system and hose routing. Arrows indicate the direction of coolant flow.

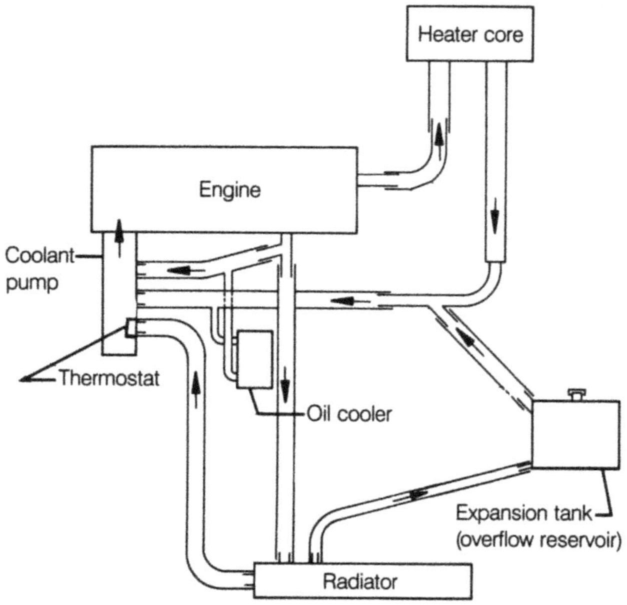

Fig. 1-1. Schematic view of cooling system components and hose routing showing coolant flow (arrows).

Coolant Pump and Thermostat

A centrifugal-type coolant pump and its housing are mounted to the cylinder block. The pump is crankshaft-driven by a V-belt, and circulates coolant through the system whenever the engine is running.

A thermostat is located in the coolant pump housing and controls coolant flow into the pump. When the engine is cold, the thermostat is closed and coolant bypasses the radiator, recirculating from the engine directly back to the pump inlet. When the engine reaches normal operating temperature, the thermostat opens and coolant circulates through the whole system, including the radiator.

Radiator

The radiator is a cross-flow type constructed of an aluminum core and plastic side tanks. A translucent expansion tank, or overflow reservoir, provides for the expansion of the coolant at higher temperatures and easy monitoring of coolant level.

Radiator Cooling Fan

An electric radiator cooling fan, operating independently of the engine, is controlled by a thermoswitch located in the bottom of the radiator. At high coolant temperatures, the switch closes to start the cooling fan. The cooling fan is wired directly to the battery. Any time the coolant temperature is excessive, the fan will start and continue to run until coolant temperature is in the correct range. On models with air conditioning, high refrigerant pressure can also activate the cooling fan.

On some models, a cooling fan after-run system, operated by a control unit on the fuse/relay panel, runs the cooling fan for as long as 15 minutes after the engine is shut off, as a guard against starting problems which may be caused by high engine compartment temperatures.

2. MAINTENANCE

The following routine maintenance steps are covered in detail in **LUBRICATION AND MAINTENANCE**.

1. Checking coolant level
2. Inspecting coolant pump V-belt tension and condition
3. Inspecting coolant hoses
4. Replacing engine coolant

3. TROUBLESHOOTING

This troubleshooting section applies to problems affecting the cooling system, which includes those components which store, pump, and regulate the circulation of engine coolant.

While coolant also circulates through the heater core in the passenger compartment, problems associated with the operation and repair of the heater system are covered in **BODY AND INTERIOR**. Similarly, while this section may help to isolate a problem to a faulty coolant temperature gauge, replacement of the gauge is covered in **ELECTRICAL SYSTEM**.

Overheating problems may also be caused by an engine fault that leaks hot combustion gasses into the cooling system, or by ignition timing that is out of specification. See **ENGINE** or **IGNITION** for additional information on these subjects.

3.1 Basic Troubleshooting Principles

When investigating the cause of overheating or coolant loss, begin with a visual inspection of the system. Check coolant level and for evidence of coolant leaks. Leaks can occur at any place in the cooling system where there is a bolted housing or other connection. An occasional drop of coolant found under the coolant pump is normal, and is not cause for concern.

The system becomes pressurized at normal operating temperatures, and this pressure is important to system function. Leaks may prevent the system from becoming pressurized and allow the coolant to boil at a lower temperature. If visual evidence is inconclusive, a cooling system pressure test will determine whether the system leaks, and may help to indicate the source. See **3.2 Diagnostic Tests**.

If the cooling system is full of coolant and holds pressure, the next most probable cause of overheating is bad coolant circulation caused by a broken V-belt, a failed thermostat, a pinched or restricted hose, or a clogged system. In warm weather, virtually all clogs are caused by neglect of the coolant, or by the addition of substances to the coolant that are not recommended by Volkswagen. In cold weather, a clogged cooling system may also be the result of frozen coolant due to an inadequate amount of anti-freeze.

The engine-driven coolant pump is subject to the same wear as any other rotating engine parts. Complete failure of the pump to circulate coolant is unusual, but excessive wear often results in noise or coolant leaks at the pump shaft.

An otherwise sound cooling system may still have overheating problems, particularly with prolonged idling, if the radiator cooling fan is inoperative. The cooling fan should cycle on and off whenever the coolant temperature is above normal. If the fan does not cycle, test the fan as described in **4.4 Radiator Cooling Fan**.

Table a lists overheating and underheating symptoms, their probable causes, and suggested corrective actions. The bold numbers in the corrective action column refer to headings in this section where the suggested repairs are described.

3.2 Diagnostic Tests

These system tests are used to help isolate and diagnose cooling system problems, and pinpoint their causes.

Pressure Testing Cooling System and Expansion Tank Cap

A pressure test will help find any leaks and show whether the cooling system can maintain pressure. If the system cannot maintain pressure, the boiling point of the coolant is reduced and the engine will overheat more easily. Various kinds of cooling system pressure testers are available. Follow the instructions supplied by the tester's manufacturer. The unique design of the Volkswagen system requires a special adapter. If these tools are not available, a Volkswagen dealer or other qualified repair shop can perform this test inexpensively.

Table a. Cooling System Troubleshooting

Symptom	Probable cause	Corrective action
1. Engine overheats	a. Low coolant level	a. Fill the cooling system. **4.1**. Pressure test for leaks. **3.2**
	b. Burst hose	b. Replace hose. **4.1**
	c. Radiator hose restricted (lower hose may collapse only at highway speeds)	c. Replace hose. **4.1**
	d. V-belt loose or broken	d. Adjust or replace V-belt. See **LUBRICATION AND MAINTENANCE**
	e. Faulty thermostat	e. Remove and test thermostat. Replace if necessary. **4.2, 4.3**
	f. Electric fan not switching on	f. Test thermoswitch and fan. Replace faulty part. **4.4**
	g. Faulty radiator cap	g. Test pressure relief valve in cap. Replace faulty caps. **3.2**
	h. Clogged radiator	h. Replace radiator or have heater core cleaned. **4.5**
	i. Incorrect ignition timing or valve timing	i. Check camshaft drive belt installation. Adjust ignition timing and check spark advance. See **ENGINE**
	j. Coolant pump faulty	j. Test coolant pump. Repair or replace if necessary. **4.3**
2. Temperature gauge reads low, inadequate heater output	a. Faulty thermostat	a. Remove and test thermostat. **4.2**
	b. Electric fan not switching off	b. Replace thermoswitch for fan. **4.4**
3. Temperature gauge reads low, heater output normal	a. Faulty temperature gauge or sending unit	a. Test temperature gauge and sending unit. Replace faulty part. **3.2**
4. Temperature gauge reads normal, inadequate heater output	a. Installed position of heater hoses reversed	a. Install heater hoses. **4., 4.1**
	b. Heater hose restricted	b. Replace hose. **4.1**
	c. Heat exchanger (heater core) clogged	c. Replace heater exchanger or have core cleaned. See **BODY AND INTERIOR**
	d. Heater control out of adjustment	d. Adjust control cables. See **BODY AND INTERIOR**

Pressurize the system to approximately 1.00 bar (14.5 psi). Loss of pressure indicates leaks which should be apparent by the seepage of coolant. If the pressure drops rapidly and there is no sign of coolant leaks, the cylinder head gasket may be faulty. To check for a faulty gasket, see **ENGINE**.

CAUTION —
Do not exceed the specified test pressure. Higher pressure could damage the radiator or other cooling system components.

To test the cap, install it on the pressure tester as shown in Fig. 3-1. Increase pressure to 1.20 to 1.50 bar (18 to 22 psi). The cap's pressure relief valve should open within this range, but not below. Faulty caps should be replaced.

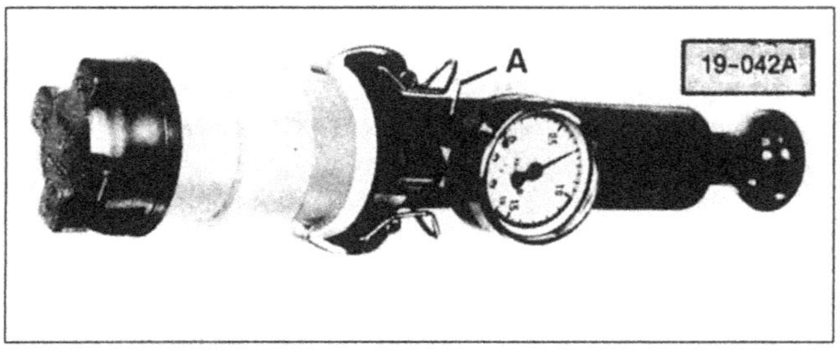

Fig. 3-1. Expansion tank cap installed on pressure tester pump (**A**) using adapter US 4467. Adapter is also used to pressure test the cooling system.

Temperature Gauge and Sending Unit Quick-check

A quick, easy test will determine whether the coolant temperature gauge is correctly representing engine temperature.

If the temperature gauge needle remains at its rest position even though the engine is fully warmed, test the gauge as follows. With the ignition on and the engine not running, disconnect the gauge wire (yellow with red stripe) from the sending unit in the coolant outlet on the cylinder head. On PL (16-valve) engines, the sending unit is on the driver's-side end of the cylinder head. Ground the wire. This simulates a high engine temperature signal to the gauge. If the gauge needle moves upward, the sending unit is faulty and should be replaced. If the needle still does not move, either the wire to the gauge is broken (open circuit) or the gauge is faulty.

If the gauge reads high when the engine is cold and the ignition is switched on, disconnect the temperature sending unit wire from the sending unit. If the needle drops to a lower reading, the sending unit is faulty. If the needle does not drop, the wire or the gauge is shorted to ground. See **ELECTRICAL SYSTEM** for electrical system troubleshooting.

4. COOLING SYSTEM SERVICE

Most repairs to the cooling system are easy and require relatively little time. Always plan to replace gaskets and seals and have them on hand before beginning.

WARNING —
The cooling system at high temperature operates under pressure. If it is necessary to open the hot cooling system, do so very slowly to allow safe release of pressure. Use heavy gloves or other hand protection.

CAUTION —
Avoid adding cold water to the coolant while the engine is hot or overheated. If it is absolutely necessary to add coolant to a hot system, do so only with the engine running and coolant pump turning.

4.1 Coolant and Cooling System Hoses

To guard against cooling system trouble, the coolant level and the hoses should be periodically inspected, as described in **LUBRICATION AND MAINTENANCE**. Hoses deteriorate with time, and periodic inspection will help prevent unexpected failure. When adding or replacing coolant, Volkswagen recommends the use of phosphate-free anti-freeze, formulated to prohibit the formation of harmful, clogging deposits in the cooling system. Always mix anti-freeze with clean water. Distilled water is best for prevention of deposits because of its reduced mineral content.

CAUTION —
- *Use of anti-freeze containing phosphates is considered by Volkswagen to be harmful to the cooling system and may void warranty coverage.*
- *Oil should never be added as a lubricant. Oil in the coolant will encourage the formation of sludge which can clog the system and damage rubber parts.*
- *If using leak sealer as a preventative measure, the system should never contain more than one application. The additives that plug leaks can also plug radiators and heater cores.*
- *Do not reuse coolant when replacing the engine, the cylinder head or cylinder head gasket, the radiator or the heater core. Doing so may damage the cooling system or the engine.*

Draining the coolant is a first step in almost all cooling system repairs. The coolant can be reused provided it is drained into a clean pan. New coolant is recommended every 2 years. Replacing hoses, or draining and filling the coolant, requires only a medium-sized flat bladed screwdriver, a pliers, and a 3-gallon drain pan.

Draining and Filling Coolant

To drain the coolant, disconnect the coolant hoses indicated in Fig. 4-1. To refill the system, first reinstall the lower radiator hoses and their clamps. Then fill the system with the coolant and water mixture until the level reaches the **Max** mark on the coolant reservoir as shown in Fig. 4-2. Mixture proportions are given in **Table b**. Start the engine and let it idle, rechecking the coolant level after it has had a chance to circulate. After the engine has cooled, check the coolant level and add coolant as necessary.

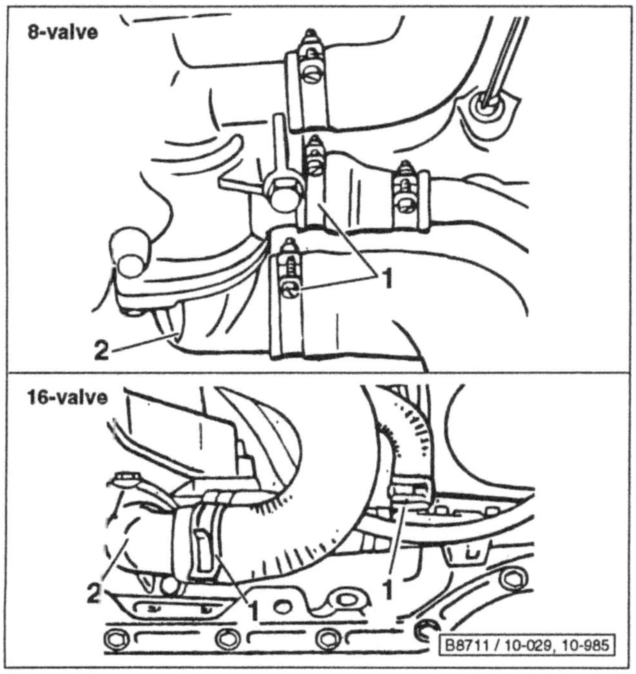

Fig. 4-1. Hoses (**1**) to be removed for draining coolant from coolant pump and thermostat housing (**2**).

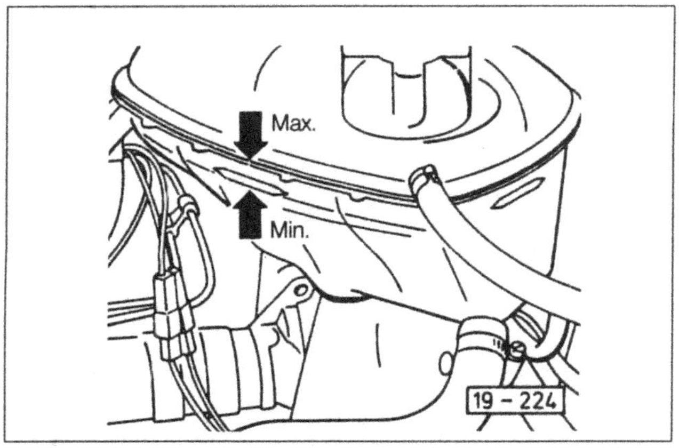

Fig. 4-2. Fill level marks on coolant expansion tank. Level should be between **Max** and **Min** marks (arrows) with engine cold.

Table b. Anti-Freeze-to-Water Proportions

Protection level (outside temperature)	Anti-freeze quarts (liters)	Water quarts (liters)
–13°F (–25°C) –31°F (–35°C)	2.75 (2.6) 3.4 (3.25)	4.1 (3.9) 3.4 (3.25)

To replace a hose:

1. Drain the coolant as described above.
2. Remove the hose. Using a screwdriver, or pliers in the case of spring clamps, loosen each hose clamp and slide the clamps away from the hose ends.

   NOTE —
   If a radiator hose is stuck to the radiator connection by sealer, cut the old hose off the connection, as shown in Fig. 4-3. Prying the hose loose may damage the connection or the radiator.

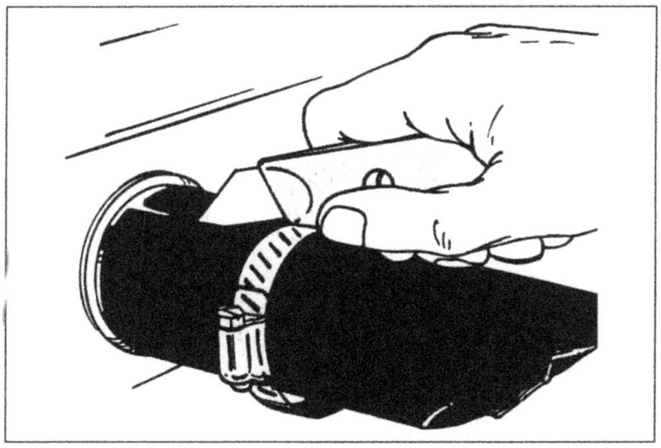

Fig. 4-3. Stuck hose being removed by cutting.

3. Clean the hose connections. Make sure any bits of old hose and sealer are removed. Clean them with a wire brush if necessary.
4. Install the new hose. Lightly coat the connections with water-resistant sealer. Place the loose hose clamps onto the new hose before fitting the hose ends to the connections.

   NOTE —
   Spring clamps should be used only with hoses and connections designed for their use. See Fig. 4-4.

5. Position and tighten the clamps. Place the clamp as near the bead as possible and at least 4 mm (5/32 in.) from the hose end, as shown in Fig. 4-5. Tighten screw-type clamps enough to compress the hose firmly around the connections.

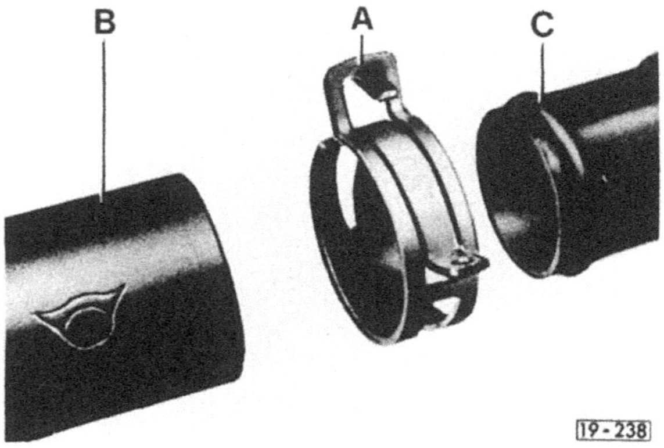

Fig. 4-4. Identification of hose and connection suitable for use with spring clamp (**A**). Hose (**B**) bears spring clamp symbol and is smooth on inside (instead of corrugated). Connection (**C**) has larger bead and is about 5 mm (¼ in.) longer.

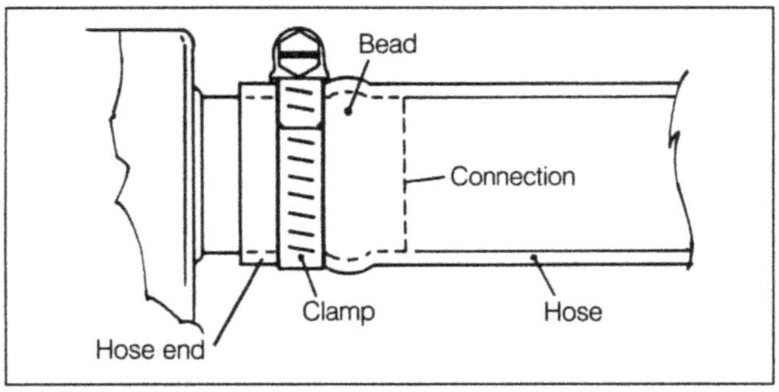

Fig. 4-5. Hose clamp correctly installed on hose end.

CAUTION

Do not overtighten clamps. Tighten just enough to seal. Overtightening may cause hose damage and failure.

6. Refill the radiator as described in **Draining and Filling Coolant**. Run the engine until warm and check for leaks. Check again after the engine has cooled.

4.2 Thermostat

The thermostat controls the coolant temperature by regulating coolant flow to the radiator. A thermostat stuck open will cause the engine to warm up slowly and run below normal temperature at highway speed. A thermostat stuck closed will restrict coolant flow to the radiator and cause overheating.

Replacement is only necessary if the thermostat is faulty. In addition to the tools required for draining the coolant, a wrench or a socket is required to remove the thermostat housing. Use a new O-ring, Volkswagen part no. 059 121 119, when the thermostat housing is reinstalled.

Removing and Installing Thermostat

Drain the coolant as described in **Draining and Filling Coolant**. Remove the two thermostat housing bolts and separate the housing from the coolant pump. See Fig. 4-1, above.

To install the thermostat, position the thermostat in the coolant pump housing with a new O-ring and install the thermostat housing. Torque the bolts to 10 Nm (87 in. lb.). Reinstall any disconnected hoses. Fill the cooling system as described in **Draining and Filling Coolant**. Warm the engine and check for leaks. Check again after the engine has cooled.

Testing Thermostat

Test the thermostat when it is removed from the engine. It is also wise to perform this simple test to a new thermostat before installation.

Measure the thermostat to determine the change in length between cold and hot conditions, as shown in Fig. 4-6. Heat the thermostat in a pan of water as shown in Fig. 4-7 while monitoring temperature with a thermometer.

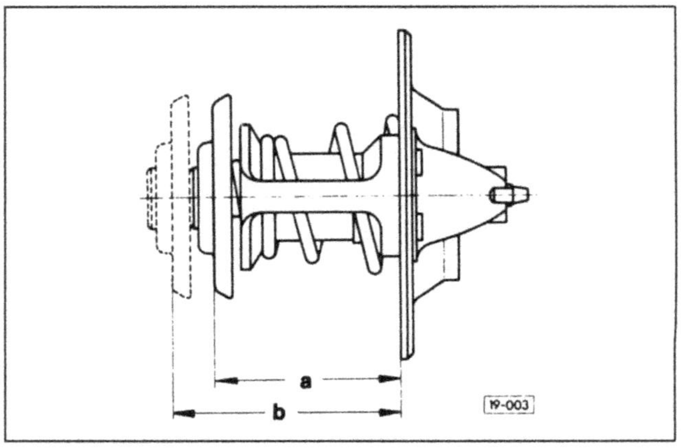

Fig. 4-6. Thermostat length. Dimension **a** is closed (cold) length; dimension **b** is open (hot) length.

Fig. 4-7. Testing thermostat in hot water.

At approximately 189°F (87°C), the length as shown in Fig. 4-6 should be at least 7 mm (9/32 in.) greater than its length when cold, measured earlier. A thermostat which does not open the full amount should not be installed.

4.3 Coolant Pump

Fig. 4-8 is an exploded view of the coolant pump and its related parts. The pump can be removed from the housing and replaced separately, although it is easier to remove the pump and housing from the engine as a unit and separate them later.

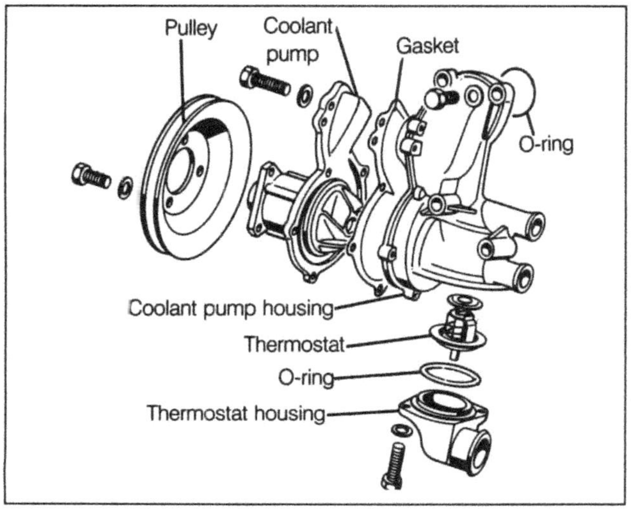

Fig. 4-8. Exploded view of coolant pump and related parts.

Only basic hand tools are required to replace the coolant pump. It is replaced as a unit, as parts for rebuilding are not generally available. A Volkswagen remanufactured pump is available. Always use a new gasket between the pump and housing, and a new O-ring between the housing and engine.

When replacing the coolant pump on cars with power steering, the power steering pump and mounting brackets must be removed first. For more information, see **SUSPENSION AND STEERING**. On cars with air conditioning, replacement of the coolant pump will be considerably easier if the alternator and air conditioning compressor are removed. Remove the alternator as described in **ELECTRICAL SYSTEM**. The air conditioning compressor should be removed from its mountings and set out of the way without loosening or removing the compressor hoses. This procedure is described as part of the engine removal procedure in **ENGINE**.

Inspecting and Replacing Coolant Pump

To inspect the coolant pump, first remove the V-belt. See **LUBRICATION AND MAINTENANCE**. Firmly grasp the coolant pump pulley and check for play in all directions. Rotate the pulley and check that the shaft turns smoothly. Inspect the pump for leaks. An occasional drop of coolant from the pump shaft is acceptable. A larger leak or a shaft that is excessively loose or noisy, indicating a worn bushing, means that the pump is faulty and should be replaced. Also, replace the V-belt if it is worn or damaged.

To remove coolant pump:

1. Drain the cooling system. See **4.1 Coolant and Cooling System Hoses**. Remove the V-belt. See **LUBRICATION AND MAINTENANCE**.
2. Remove the nut and T-bolt that fastens the camshaft drive belt cover to the coolant pump. Disconnect the remaining hoses from the pump housing.
3. Remove pump and housing assembly. Four bolts hold the coolant pump assembly to the engine.
4. Disassemble the pump and housing assembly. Remove the pulley(s) from the coolant pump shaft. Remove the seven bolts and washers and separate the pump from the housing.

Assembly is the reverse of disassembly. Use a new gasket between the coolant pump and housing. Tighten the seven bolts evenly until all are torqued to 10 Nm (87 in. lb.). Install the pulley(s). Torque the bolts to 20 Nm (15 ft. lb.).

To install coolant pump:

1. Clean the surface of the engine block where it will be contacted by the pump housing and the O-ring. Install a new O-ring in the recess that surrounds the water outlet.

   NOTE —
   The O-ring between the coolant pump housing and the engine block does not require sealer.

2. Loosely install the coolant pump on the engine. The two short bolts are used at the top of the pump housing. Torque the four bolts evenly to 20 Nm (15 ft. lb.). Refasten the camshaft drive belt cover to the coolant pump.
3. Install the hoses and refill the cooling system as described in **4.1 Coolant and Cooling System Hoses**. Install and adjust the V-belt as described in **LUBRICATION AND MAINTENANCE**.

4.4 Radiator Cooling Fan

The cooling fan provides additional air flow through the radiator. A faulty cooling fan motor or thermoswitch may be the cause of insufficient air flow and, therefore, overheating. Models covered by this manual have either a single-speed or a two-speed cooling fan.

Some later Scirocco and Cabriolet models have a radiator cooling fan after-run system, designed to run the cooling fan for up to 15 minutes after the engine is shut off, to aid in restarting a hot engine. A thermoswitch mounted on the cylinder head monitors engine compartment temperature.

In either case, the cooling fan can come on at any time, even if the key is out of the ignition. To avoid personal injury, cooling fan and thermoswitch tests should be performed with extreme caution, and preferably with the engine cold. The tests below simulate a warm engine by electrically bypassing the thermoswitch with a jumper wire.

Testing Cooling Fan and Radiator Thermoswitch

Note that the electric cooling fan will operate only when the temperature is high enough to close the radiator thermoswitch or, on cars so equipped, the after-run thermoswitch. If a faulty thermostat is not allowing the coolant to bypass the radiator and warm up to the switching temperature, the thermoswitch will not close and the cooling fan will not run.

To test the cooling fan, disconnect the radiator thermoswitch connector and use a jumper wire (with switch) to make connection between terminals to simulate a closed switch. On three-terminal connectors used with two-speed fans, the fan should run at low speed when the red wire is jumpered to the red/white wire. It should run at high speed when the red wire is jumpered to the red/black wire. Thermoswitch location is shown in Fig. 4-9.

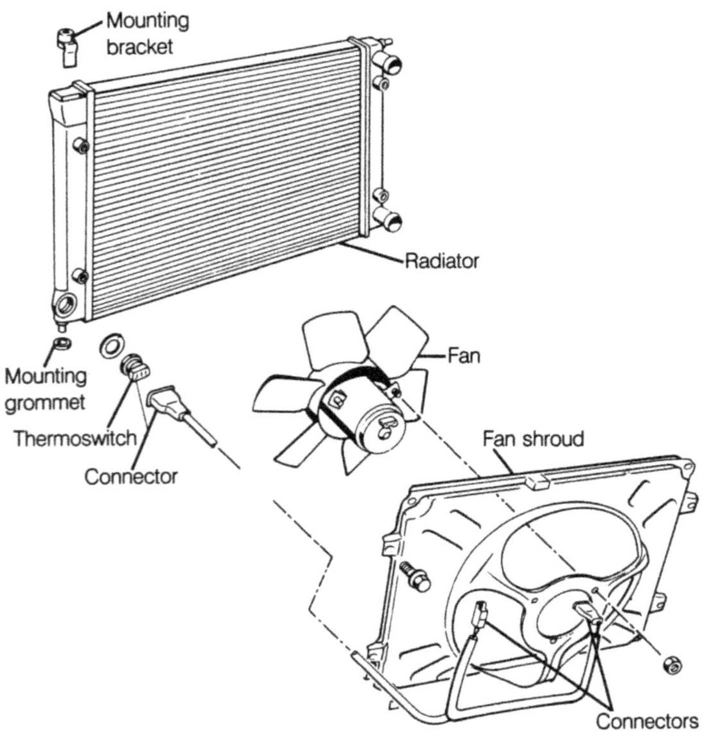

Fig. 4-9. Radiator and cooling fan used on 16-valve models. Other models are similar.

WARNING —
Always keep clear of the fan blades. The cooling fan may run at any time, even with the ignition off.

If the fan does not run, check for voltage at the thermoswitch connector. There should be battery voltage (approx. 12 V) at terminal no. 3 (red wire) whenever the battery is connected. If not, check fuse no.1 in the fuse/relay panel. For more information on the cooling fan circuit, see **ELECTRICAL SYSTEM**.

If the fan runs only when powered directly by the jumpered connector, the radiator thermoswitch should be replaced.

Radiator Cooling Fan After-Run System

A radiator cooling fan after-run system is installed on some later Scirocco and Cabriolet models. A thermoswitch mounted on the cylinder head senses excessive engine compartment temperature. High temperature closes the switch, making a ground connection for the after-run control unit, turning on the cooling fan for up to 15 minutes. The control unit is located on the fuse/relay panel.

For information about the radiator cooling fan after-run system on a specific model, see the current flow diagram for that model in **ELECTRICAL SYSTEM**. 1985 through 1987 models can be retrofitted with the system using a kit which is available from your authorized Volkswagen dealer Parts Department.

Test the system, with the ignition off, by disconnecting the black wire from the thermoswitch and grounding it. The thermoswitch is mounted near the top of the engine, behind the cylinder head cover, between no. 2 and no. 3 cylinders. The fan should run on low speed. If not, check for battery voltage (approx. 12 V) from the control unit (terminal 8/87) to the fan (terminal no. 2, red/white wire). If there is no voltage from the control unit, it is faulty and should be replaced.

If, with the thermoswitch connected normally, the fan runs continuously (more than 15 minutes), disconnect the black wire from the thermoswitch. If the fan stops, the thermoswitch is faulty, permanently shorted to ground, and should be replaced. If the fan continues running with the wire detached, the control unit is faulty and should be replaced.

Removing and Installing Radiator Cooling Fan

To remove the cooling fan without removing the radiator, disconnect the negative (–) battery cable and the fan's electrical connector. See Fig. 4-9 above. On cars with air conditioning, remove the connector leading to the series resistor and the cooling fan relay. See Fig. 4-10. Unbolt the fan shroud from the radiator and remove the shroud and fan together. Installation is the reverse of removal. Torque the mounting bolts to 10 Nm (87 in. lb.).

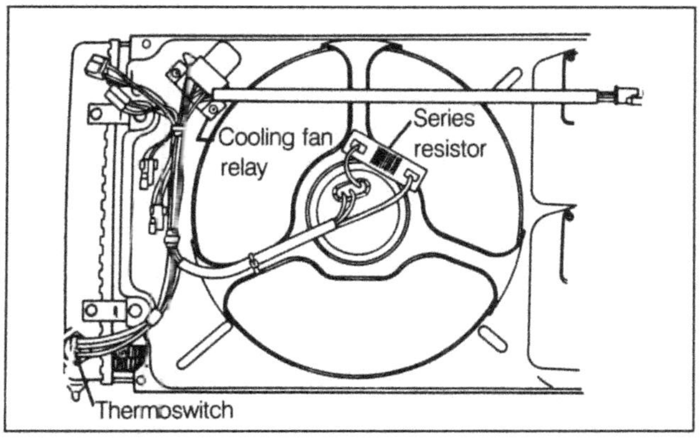

Fig. 4-10. Radiator and related electrical components found on models with air conditioning.

4.5 Radiator

To perform efficiently, the radiator must not be blocked with dirt or debris, and it must be firmly mounted. Excess vibration due to loose, broken, or missing fasteners may damage the radiator.

If the engine overheats and no other cooling system tests indicate trouble, the radiator may have plugged passages restricting coolant flow. Consult a qualified radiator repair shop about repairs.

Removing and Installing Radiator

Drain the cooling system and disconnect the upper and lower radiator hoses, as described in **4.1 Coolant and Cooling System Hoses**. Disconnect the negative battery cable from the battery, and the electrical connectors from the thermoswitch and cooling fan motor. On models with air conditioning, disconnect the connectors to the series resistor and cooling fan relay. See Fig. 4-10 above. Remove the bolts from the upper radiator mounts. Lift the radiator out from the top, complete with the fan and radiator shroud, tilting it toward the engine.

Installation is the reverse of removal. Torque the mounting bolts to 10 Nm (87 in. lb.).

5. COOLING SYSTEM TECHNICAL DATA

I. Cooling System Specifications

Cooling system leakage test maximum test pressure	1.00 bar (14.5 psi)
Expansion tank cap opening pressure	1.20 to 1.50 bar (18 to 22 psi)
Coolant system capacity	6.5 liters (1.7 gal.)
Coolant type	mixture of 50% water and 50% phosphate-free anti-freeze containing ethylene glycol
Thermostat opening temperature	
open	189°F (87°C)
thermostat stroke	7 mm (9/32 in.)
V-belt tension (deflection midway between alternator and crankshaft pulleys)	
without air conditioning	
new belt	2 mm (5/64 in.)
used belt	5 mm (3/16 in.)
with air conditioning	
new belt	10 mm (3/8 in.)
used belt	15 mm (9/16 in.)

II. Tightening Torques

Coolant outlet to cylinder head (bolt)	10 Nm (87 in. lb.)
Coolant pump housing to engine (bolt)	20 Nm (15 ft. lb.)
Coolant pump pulley to coolant pump (bolt)	20 Nm (15 ft. lb.)
Coolant pump to coolant pump housing (bolt)	10 Nm (87 in. lb.)
Coolant temperature sending unit to cylinder head water outlet	10 Nm (87 in. lb.)
Radiator shroud mounting to radiator (bolt)	10 Nm (87 in. lb.)
Thermostat housing to coolant pump housing (bolt)	10 Nm (87 in. lb.)
Upper radiator mounts to body (bolt)	10 Nm (87 in. lb.)

FUEL SYSTEM – GASOLINE

Contents

TABLES

Fuel System—Gasoline

Introduction

The fuel system has several functions. In this manual we refer to fuel supply as those parts of the system that provide storage space for the fuel and deliver it to the engine—the fuel tank, fuel pumps and related components. The remainder of the fuel system is the fuel injection system, or the fuel injection portion of an integrated engine management system. It measures and filters the intake air, mixes fuel and air in precise proportions and delivers the mixture to the cylinders as a combustible vapor.

Unlike a carburetor, which depends on the velocity of the incoming air to vaporize the fuel and draw it into the engine, a fuel injection system injects atomized fuel into the intake air stream under pressure. This method of active fuel metering means that the fuel mixture entering the engine can be controlled more precisely and used more efficiently, yielding improved performance and fuel economy.

Two similar types of gasoline fuel injection systems are used on the 1985-1989 cars covered by this manual. Although at first look these systems appear complicated, they operate on simple principles. This section provides a thorough description of each system, with functional descriptions as well as detailed troubleshooting, testing and repair procedures.

Beginning with 1990 models, the fuel injection is part of a sophisticated engine management system called Digifant. Fuel injection, ignition, idle speed and emission controls are integrated, operated by output signals from a single electronic control unit (ECU). Control of these many functions is more stable and precise because it is based on input from one set of sensors.

Special equipment is necessary for some of the procedures included here, especially for the Digifant engine management system. Please read the information carefully before beginning any test or repair. If you do not have the equipment or the experience necessary to do the job accurately, we suggest leaving those tests or repairs to an authorized Volkswagen dealer or other qualified shop. We especially urge you to consult an authorized Volkswagen dealer before beginning any work on a car which may covered wholly or in part by and of the extensive warranties issued by Volkswagen United States, Inc.

1. GENERAL DESCRIPTION

Following is a brief description of the fuel supply components and the different fuel injection systems fitted to the cars covered by this manual. For help in telling which one is on your car, see **1.4 Applications—Identifying Features**.

Fuel Storage

The fuel tank is located beneath the car, behind the rear seat. The capacity of the fuel tank is 52 liters (13.7 gallons). The fuel tank assembly includes the combined fuel gauge sending unit and transfer pump. Evaporative emission control plumbing and a fuel return line from the fuel reservoir are also connected to the tank.

Fuel Pump and Fuel Circuit

The transfer pump provides a steady supply of fuel to the fuel reservoir for the main fuel pump. The fuel reservoir, the main fuel pump and the main fuel filter are mounted on a common bracket beneath the car in front of the fuel tank.

CIS and CIS-E fuel injection (1987-1989 cars) require relatively high fuel pressure. An accumulator, mounted with the main pump and reservoir, absorbs pressure surges and helps to maintain residual system pressure as an aid to hot restarting.

Air Filter

On all cars covered by this manual, a rectangular-shaped air filter housing on the driver's side of the engine compartment houses a replaceable, dry pleated paper air filter element.

1.1 Continuous Injection System (CIS)

CIS is a fully mechanical fuel injection system. Main components are the air flow sensor, the fuel distributor and the fuel injectors. Fuel metering is determined solely by the mechanical air flow sensor and fuel pressure differentials inside the fuel distributor.

The air flow sensor reacts to changes in intake air volume. It mechanically operates the fuel distributor control plunger, which meters fuel to the injectors. Continuous injection means the injectors deliver fuel constantly when the engine is running.

Another mechanical component, the control pressure regulator, adapts the fuel mixture to different operating conditions such as cold running by altering fuel pressure values within the system. See **5. Continuous Injection System (CIS)**.

1.2 Continuous Injection System-Electronic (CIS-E)

CIS-E is a more advanced version of the CIS system described above. As with CIS, the air flow sensor measures the volume of air entering the engine and the fuel distributor meters fuel to the fuel injectors. Basic fuel metering is determined by fuel pressure differentials inside the fuel distributor.

CIS-E is capable of controlling differential pressures and fuel metering more precisely than CIS. The electronically controlled differential pressure regulator makes small and continuous changes to fuel pressure values within the fuel distributor.

The CIS-E electronic control unit (ECU) monitors engine coolant temperature, engine speed (rpm), throttle position, and oxygen content in the exhaust. It adjusts the differential pressure regulator current and fuel metering to maintain the proper air/fuel mixture. See **6. Continuous Injection System-Electronic (CIS-E)**.

1.3 Digifant Engine Management Systems

The Digifant systems are called engine management systems because fuel injection, ignition, evaporative emission controls and idle speed are controlled by a single integrated system. One array of sensors supplies information for all these engine management functions to the central Digifant electronic control unit (ECU).

Fuel is supplied by electrically operated solenoid valve fuel injectors. The ECU opens and closes the injectors by opening and closing the injectors ground circuits. It controls fuel metering by controlling the length of time that the injectors are open.

1.4 Applications-Identifying Features

This section of the manual includes many specifications and procedures that apply only to certain cars with a particular fuel system. Use the information that follows to determine which fuel system is fitted to a particular car.

1985-1989

For Cabriolet and Scirocco models, the car's fuel injection system depends on the engine type. All 1985-1989 Cabriolets are equipped with the 8-valve engine and CIS fuel injection. Sciroccos with the 8-valve engine are also equipped with CIS fuel injection. The 16-valve engine, fitted to the Scirocco 16V only, is equipped with CIS-E fuel injection.

From 1990

Beginning in 1990, all Cabriolet models are equipped with Volkswagen's Digifant engine management system. Most cars for the United Staes and Canada cars have a version of the system called Digifant II. Cars built for sale in California are a special case.

1990 California cars have a special California-only version of Digifant II. It has some unique features and specifications, but it is still called Digifant II. Beginning with 1991 models, the California cars are equipped with a version called Digifant I.

2. MAINTENANCE

Volkswagen specifies the maintenance steps listed below, to be carried out at particular time or mileage intervals for proper maintenance of the fuel system.

A number in bold type indicates that the procedure or specifications are covered in this **FUEL SYSTEM** section, under that numbered heading. See **LUBRICATION AND MAINTENANCE** for information about the other listed items, and for the prescribed maintenance intervals.

1. Cleaning or replacing the air filter element
2. Inspecting the fuel system for leaks
3. Checking idle speed and the idle mixture adjustment. **5.7** (CIS), **6.6** (CIS-E), **7.4** (Digifant II), **8.4** (Digifant I)
4. Replacing the oxygen sensor. See **EXHAUST SYSTEM AND EMISSION CONTROLS**.

3. TROUBLESHOOTING

This section covers troubleshooting for the CIS and CIS-E fuel injection systems on 1987-1989 cars. Tests of the oxygen sensor are covered in **EXHAUST SYSTEM AND EMISSION CONTROLS**. For information on troubleshooting Digifant cars (from 1990) see **7. Digifant II Fuel Injection**

General performance and driveability problems may also be caused by problems in the ignition system. Make sure that the ignition system has been adequately maintained and that it is delivering a strong and properly timed spark before altering fuel system adjustments to correct a problem.

On an engine that has accumulated high mileage, such problems may be the result of engine wear and low compression. See **ENGINE** for information on engine troubleshooting and compression testing.

3.1 Basic Troubleshooting Principles

The fuel injection system meters fuel in proportion to the amount of air being drawn into the engine, trying always to maintain the ideal air/fuel mixture. Unmeasured air entering the system, from small vacuum leaks for example, does not prompt the system to deliver a corresponding amount of fuel. It will lean the air/fuel mixture and lead to driveability problems.

Inspect the crankcase ventilation and idle speed boost systems. Check for damaged hoses and hose fittings, loose connections, or other air leaks. Leaking gaskets, a loose oil filler cap or a loose dipstick are all potential sources of unmeasured air.

CIS and CIS-E systems supply fuel at relatively high pressures throughout the system. Fuel metering and fuel delivery to the injectors in particular depends on accurate fuel pressure. Check the fuel lines and connections for signs of leaks. Inspect wiring harness connectors and ground connections. Make sure the connecting terminals are clean and dry. Make sure that the connectors are positively attached and well-sealed.

Normal running when the engine is fully warm is the most basic operating condition. Whenever possible, make sure that the car is running well and the system is properly adjusted for normal warm running before troubleshooting cold-start or cold-running (warm-up) problems. Without this starting point as a baseline, other tests and adjustments may just confuse the process and make matters worse.

Certain components alter the basic fuel metering to adapt the system to conditions other than warm running. When troubleshooting specific problems, concentrate on the components that adapt fuel metering for the particular condition. For example, if the engine starts hard when cold, the components that deliver cold-start enrichment are most likely to be at fault. It makes sense to test these functions first.

Please abide by the following cautions and warnings whenever servicing the fuel system.

> ***WARNING —***
> *During many fuel system test procedures, fuel will be discharged. Do not smoke or work near heaters or other fire hazards. Have a fire extinguisher handy.*

> ***CAUTION —***
> - *Before making any electrical tests with the ignition switched on, disconnect the wire from the center tower of the distributor cap and ground it to some clean metal surface to prevent sparks.*
> - *To prevent damage to the ignition system or the electronic fuel system components, connect and disconnect wires and test equipment only with the ignition off.*

3.2 CIS Troubleshooting

This information applies specifically to CIS fuel injection. For troubleshooting CIS-E systems, see **3.3 CIS-E Troubleshooting**.

The basic adjustments of engine idle speed, ignition timing, and idle mixture (% CO) are critical to all other phases of CIS fuel injection performance. Before making any other adjustments to the system, make certain that these parameters are within specifications. See **5.5 Idle Speed**, **5.7 Idle Specifications**, and **IGNITION**.

Table a lists symptoms of CIS fuel system problems, their probable causes, and suggested corrective actions. The boldface numbers in the corrective action column indicate the heading in this section of the manual where the applicable test and repair procedures are found.

Table a. CIS Troubleshooting

Symptom	Probable cause	Corrective action
1. Cold start - Engine starts hard or fails to start when cold	**a.** Cold start valve or thermo-time switch faulty **b.** Fuel pump not running **c.** Air flow sensor plate rest position incorrect **d.** Fuel pressure incorrect	**a.** Test cold start valve and thermo-time switch. **5.2** **b.** Check fuel pump fuse and fuel pump relay. **4.1, 4.3** **c.** Inspect air flow sensor plate rest position and adjust if necessary. **5.1** **d.** Test system pressure and cold control pressure. **5.6**
2. Hot start - Engine starts hard or fails to start when warm	**a.** Hot start pulse relay faulty **b.** Cold start valve leaking or operating continuously **c.** Fuel pressure incorrect **d.** Air flow sensor plate rest position incorrect **e.** Insufficient residual fuel pressure **f.** Fuel leak(s)	**a.** Test hot start pulse relay and replace if necessary. **5.3** **b.** Test cold start valve and thermo-time switch. **5.2** **c.** Test warm control pressure. Replace control pressure regulator if necessary. **5.6** **d.** Inspect air flow sensor plate rest position and adjust if necessary. **5.1** **e.** Test residual fuel pressure. Replace fuel pump check valve or fuel accumulator as necessary. **5.6** **f.** Inspect fuel lines and connections. Correct leaks as required
3. Engine misses and hesitates under load	**a.** Fuel injector(s) clogged **b.** Fuel pressure incorrect **c.** Fuel leak(s)	**a.** Test fuel injectors. Check for clogged injector lines. Replace faulty injectors. **4.7** **b.** Test system pressure and warm control pressure. If necessary, replace control pressure regulator. **5.6** **c.** Inspect fuel lines and connections. Correct leaks as required
4. Engine starts but stalls at idle	**a.** Incorrect fuel pressure **b.** Cold start valve leaking **c.** Auxiliary air regulator faulty **d.** Vacuum (intake air) leak	**a.** Test system pressure and control pressure. **5.6** **b.** Test and, if necessary, replace cold start valve. **5.2** **c.** Test and, if necessary, replace auxiliary air regulator. **5.5** **d.** Inspect intake air components for leaking hoses, hose connections, and cracks or other leaks. Repair as required
5. Engine idles too fast	**a.** Accelerator pedal, cable, or throttle valve binding **b.** Auxiliary air regulator faulty **c.** Air leaking past throttle valve	**a.** Inspect for worn or broken parts, kinked cable, or other damage. Replace faulty parts. **4.6** Also see **BODY AND INTERIOR** **b.** Test and, if necessary, replace auxiliary air regulator. **5.5** **c.** Inspect throttle valve and adjust or replace as required. **5.1**
6. Hesitation on acceleration	**a.** Vacuum (intake air) leak **b.** Fuel injector(s) clogged **c.** Cold start valve leaking **d.** Control plunger in fuel distributor binding or fuel distributor faulty **e.** Air flow sensor plate out of adjustment **f.** Fuel pressure incorrect **g.** Idle mixture (% CO) incorrectly adjusted	**a.** Inspect intake air components for leaking hoses, hose connections, and cracks or other leaks. Repair as required **b.** Test injector spray pattern and quantity. Replace faulty injectors. **4.7** **c.** Test and, if necessary, replace cold start valve. **5.2** **d.** Check air flow sensor plate movement and, if necessary, replace fuel distributor. **5.1** **e.** Inspect air flow sensor plate position and adjust if necessary. **5.1** **f.** Test system pressure and warm control pressure. If necessary, replace control pressure regulator. **5.6** **g.** Check and adjust CO. **5.7**
7. Poor fuel mileage	**a.** Idle speed, ignition timing, and idle mixture (% CO) out of adjustment **b.** Cold start valve leaking **c.** Fuel pressure incorrect	**a.** Check and adjust. **5.7** Also see **IGNITION** **b.** Test and, if necessary, replace cold start valve. **5.2** **c.** Test system pressure and warm control pressure. If necessary, replace control pressure regulator. **5.6**
8. Engine continues to run (diesels) after ignition is turned off	**a.** Incorrect ignition timing or faulty ignition system **b.** Engine overheated	**a.** See **IGNITION** **b.** See **COOLING SYSTEM**

3.3 CIS-E Troubleshooting

This information applies specifically to CIS-E fuel injection fitted to the Scirocco 16V. For troubleshooting CIS systems on all other 1985-1989 cars, see **3.2 CIS Troubleshooting**.

Engine idle speed, ignition timing and idle mixture (differential pressure regulator current) are basic adjustments that affect CIS-E fuel injection performance. Before making any other adjustments to the system to correct running problems, make certain that these basic adjustments are correct. See **6.3 Idle Speed, 6.6 Idle Specifications**, and **IGNITION**.

Table b lists symptoms of CIS-E fuel system problems, their probable causes, and suggested corrective actions. The boldface numbers in the corrective action column indicate the heading in this section of the manual where the applicable test and repair procedures are found.

Table b. CIS-E Troubleshooting

Symptom	Probable cause	Corrective action
1. Cold start - Engine starts hard or fails to start when cold	**a.** Cold start valve or thermo-time switch faulty	**a.** Test cold start valve and thermo-time switch. **6.2**
	b. Fuel pump not running	**b.** Check fuel pump fuse and fuel pump relay. **4.1**, **4.3**
	c. Air flow sensor plate rest position incorrect	**c.** Inspect air flow sensor plate rest position and adjust as necessary. **6.1**
	d. Fuel pressure incorrect	**d.** Check differential pressure regulator, system pressure, and differential pressure. **6.1**, **6.4**, **6.5**
	e. Coolant temperature sensor faulty or wire to sensor broken	**e.** Test coolant temperature sensor and wiring. Repair wiring or replace sensor as required. **6.4**
2. Hot start - Engine starts hard or fails to start when warm	**a.** Cold start valve leaking or operating continuously	**a.** Test cold start valve and thermo-time switch. **6.2**
	b. Fuel pressure incorrect	**b.** Check differential pressure regulator, system pressure, and differential pressure. **6.1**, **6.4**, **6.5**
	c. Insufficient residual fuel pressure	**c.** Test residual fuel pressure. Replace fuel pump check valve or fuel accumulator as necessary **6.5**
	d. Oxygen sensor system faulty	**d.** See **EXHAUST SYSTEM AND EMISSION CONTROLS**
	e. Air flow sensor plate rest position incorrect	**e.** Inspect air flow sensor plate rest position and adjust if necessary. **6.1**
	f. Fuel injector(s) clogged	**f.** Test injector spray patterns and quantity. Replace faulty injectors. **4.7**
	g. Fuel pump delivery inadequate	**g.** Check fuel pump delivery. **4.3**
	h. Fuel leak(s)	**h.** Inspect fuel lines and connections. Correct leaks as required
3. Engine stalls or idles rough (cold or warm)	**a.** Fuel pressure incorrect	**a.** Check differential pressure regulator, system pressure, and differential pressure. **6.1**, **6.4**, **6.5**
	b. Cold start valve faulty	**b.** Test cold start valve. Replace if leaking or otherwise faulty. **6.2**
	c. Fuel injector(s) clogged	**c.** Test injector spray patterns and quantity. Replace faulty injectors. **4.7**
	d. Coolant temperature sensor faulty or wiring to sensor broken	**d.** Test coolant temperature sensor and wiring. Repair wiring or replace sensor as required. **6.4**
	e. Vacuum (intake air) leak	**e.** Inspect intake air components for leaking hoses, hose connections, and cracks or other leaks. Repair as required
	f. Control plunger in fuel distributor binding or fuel distributor faulty	**f.** Check air flow sensor plate movement and, if necessary, replace fuel distributor. **6.2**
	g. Idle air stabilizer valve not operating	**g.** Test idle switch. **6.1**. Test idle air stabilizer valve. **6.3**
	h. Air flow sensor plate rest position incorrect	**h.** Inspect air flow sensor plate rest position and adjust if necessary. **6.1**
4. Engine misses and hesitates under load	**a.** Fuel injector(s) clogged	**a.** Test injector spray patterns and quantity. Replace faulty injectors. **4.7**
	b. Fuel pressure incorrect	**b.** Check differential pressure regulator, system pressure, and differential pressure. **6.1**, **6.4**, **6.5**
	c. Oxygen sensor system faulty	**c.** See **EXHAUST SYSTEM AND EMISSION CONTROLS**
	d. Fuel leak(s)	**d.** Inspect fuel lines and connections. Correct leaks as required
	e. Coolant temperature sensor faulty or wire to sensor broken	**e.** Test coolant temperature sensor and wiring. Repair wiring or replace sensor if faulty. **6.4**

continued on next page

Table b. CIS-E Troubleshooting (continued)

Symptom	Probable cause	Corrective action
5. Engine idles too fast	**a.** Accelerator pedal, cable, or throttle valve binding **b.** Idle air stabilizer valve not operating **c.** Air leaking past throttle valve	**a.** Inspect for worn or broken parts, kinked cable, or other damage. Replace faulty parts. **4.6** Also see **BODY AND INTERIOR** **b.** Test idle switch. **6.1** Test idle air stabilizer valve. **6.3** **c.** Inspect throttle valve and adjust or replace as required. **5.1**
6. Hesitation on acceleration	**a.** Vacuum (intake air) leak **b.** Fuel injector(s) clogged **c.** Cold start valve leaking **d.** Control plunger in fuel distributor binding or fuel distributor faulty **e.** Air flow sensor plate out of adjustment **f.** Fuel pressure incorrect **g.** Air flow sensor plate potentiometer faulty or incorrectly adjusted	**a.** Inspect intake air components for leaking hoses, hose connections, and cracks or other leaks. Repair as required **b.** Test injector spray pattern and quantity. Replace faulty injectors. **4.7** **c.** Test and, if necessary, replace cold start valve. **6.2** **d.** Check air flow sensor plate movement and, if necessary, replace fuel distributor. **6.1** **e.** Inspect air flow sensor plate position and adjust if necessary. **6.1** **f.** Check differential pressure regulator, system pressure, and differential pressure. **6.1**, **6.4**, **6.5** **g.** Test air flow sensor plate potentiometer and adjust if necessary. Replace a faulty potentiometer. **6.4**
7. Low power	**a.** Coolant temperature sensor faulty or wire to sensor broken **b.** Fuel pressure incorrect **c.** Throttle plate not opening fully **d.** Full throttle switch faulty or incorrectly adjusted	**a.** Test coolant temperature sensor and wiring. Repair wiring or replace sensor if faulty. **6.4** **b.** Check differential pressure regulator, system pressure, and differential pressure. **6.1**, **6.4**, **6.5** **c.** Check throttle cable adjustment. **4.6** **d.** Check throttle switch and adjust if necessary. Replace a faulty switch. **6.1**
8. Poor fuel mileage	**a.** Idle speed, ignition timing, and idle mixture out of adjustment **b.** Cold start valve leaking **c.** Fuel pressure incorrect	**a.** Check and adjust. **5.6** Also see **IGNITION** **b.** Test and, if necessary, replace cold start valve. **6.2** **c.** Check differential pressure regulator, system pressure, and differential pressure. **6.1**, **6.4**, **6.5**
9. Engine continues to run (diesels) after ignition is turned off	**a.** Incorrect ignition timing or faulty ignition system **b.** Engine overheated	**a.** See **IGNITION** **b.** See **COOLING SYSTEM**

4. FUEL SUPPLY

Aside from the fuel injection system which controls the amount of fuel metered to the engine, the remainder of the fuel system is dedicated to storing and supplying fuel.

An electric transfer pump, located inside the fuel tank, supplies fuel to the fuel reservoir and the main electric fuel pump. The main pump supplies fuel under pressure to the fuel injection system. If, for any reason, adequate electric power is not available to run these pumps, the engine will not run. The fuel injectors, which deliver fuel from the fuel injection system to the engine, are also covered in this section.

Fig. 4-1 shows the fuel reservoir, the main fuel pump, and the fuel accumulator.

Absolute cleanliness is essential when working with the fuel system. Even a minute particle of dirt can cause trouble if it reaches an injector. Thoroughly clean the unions before disconnecting any of the fuel line connections. Use only clean tools. Keep removed components clean, and seal or cover them with a clean, lint-free cloth, especially if the repair cannot be finished immediately. When replacing parts, install only new, clean components. Do not use compressed air nearby, and do not move the car while the fuel system is open. Always replace seals, O-rings, and clamps.

4.1 Fuses and Relays

Both the transfer pump and the main fuel pump are electric, powered through a fuse and operated by the fuel pump relay. If both pumps are not running, the cause may be a failed fuse or a faulty fuel pump relay. In cold weather, condensation contaminating the fuel may freeze in the pump, causing the circuit to overload and blow the fuse. Any fuel pump troubleshooting should begin with checking the fuel pump fuse and the fuel pump relay. See **Fuel Pump Electrical Tests** later in this section.

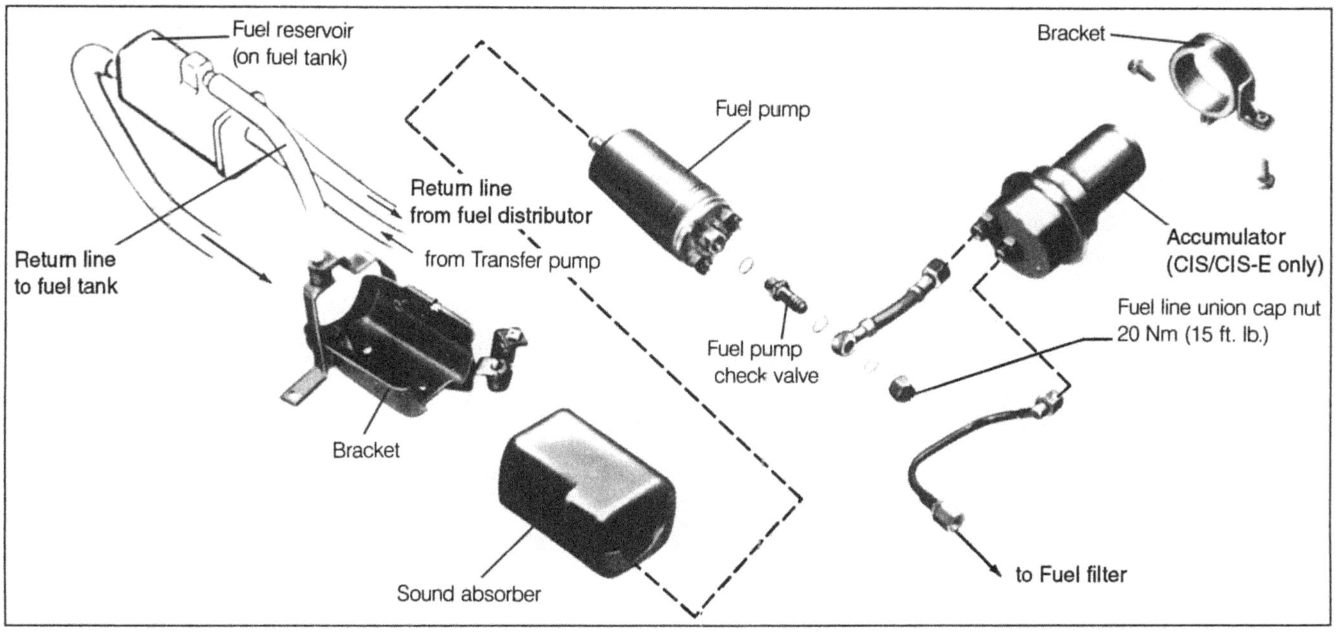

Fig. 4-1. Fuel reservoir, main fuel pump, and accumulator.

The fuel pump relay is activated by the starter circuit. It operates the pumps for starting, but cuts them off after a few seconds if the engine does not start. If the pumps run for only a few seconds after the engine starts, it may be because the relay is not receiving an rpm signal from the ignition system. On all except Scirocco 16V, check the wiring and connections between the ignition coil terminal 1 and the fuel pump relay terminal 5. See the wiring diagrams in **ELECTRICAL SYSTEM.** The Scirocco 16V fuel pump relay gets its ignition signal from the knock sensor control unit. See **IGNITION.**

Operating Fuel Pumps For Testing

The fuel pump relay cuts off the fuel pumps after a few seconds if the engine is not running. To run the pumps longer for fuel system tests, the relay terminals must be bridged to power the pumps directly.

Volkswagen special tool US 4480/3 is a combination jumper wire and switch connected in place of the fuel pump reley. It bridges the relay panel sockets, connecting terminal 30 (battery power) with terminal 87 (switched relay output). See **ELECTRICAL SYSTEM.**

CAUTION —
- *If you construct a jumper to bridge the relay terminals, the wire should be at least 1.5 mm metric size or 16 gauge (awg).*
- *Include an in-line fuse holder with a 20 amp fuse.*
- *Include a toggle switch so that the pumps can be switched on and off without damaging the fuse/relay panel terminals.*

To operate fuel pumps:

1. Disconnect the coil wire from the center of the distributor cap. With a jumper wire, connect it to ground on the engine or body to disable the ignition.
2. With the ignition switched off, remove the fuel pump relay and connect Volkswagen special tool US 4480/3 or equivalent (switched off) in its place to bridge the relay terminals.
3. Switch on the bridging tool to operate the fuel pumps. For more information on fuel pump electrical testing, see **Fuel Pump Electrical Tests** later in this section.

4.2 Transfer Pump

The transfer pump is mounted inside the fuel tank along with the fuel gauge sender. For access to the transfer pump in the top of the fuel tank, remove the rear seat bottom cushion.

WARNING —
The fuel level in the tank must be less than 3/4 full to avoid spilling fuel when the transfer pump and fuel gauge sending unit are removed.

NOTE —
Some 1985 through 1987 Sciroccos and Cabriolets have been recalled for transfer pump replacement. On affected cars, the transfer pump may seize in hot weather, causing the engine to stall. To prevent this, a modified fuel pump and filter are installed. If in doubt about whether a particular car is affected or has been repaired, check with an authorized Volkswagen dealer.

To test transfer pump:

1. Remove the rear seat bottom as described in **BODY AND INTERIOR**. Remove the transfer pump and fuel sender gauge access cover by removing the three mounting screws.

2. With the ignition off, disconnect the coil wire from the center tower of the distributor cap and connect it to ground on the engine block using a jumper wire.

3. With the transmission in neutral, and the parking brake set, operate the starter for 3 to 4 seconds. While the starter is running, and for a few seconds afterward, the running transfer pump should be audible.

4. If the pump is not operating, bypass the fuel pump relay as described in **4.1 Fuses and Relays**.

5. If the transfer pump still does not run, check for voltage at the transfer pump harness connector, as shown in Fig. 4-2. If voltage is reaching the connector, the transfer pump is probably faulty and should be replaced.

> **NOTE —**
> On the harness connector, one wire is for input voltage, one is to ground, and one is for fuel gauge sending unit output. See the current flow diagrams in **ELECTRICAL SYSTEM** for specific terminal and wire identification.

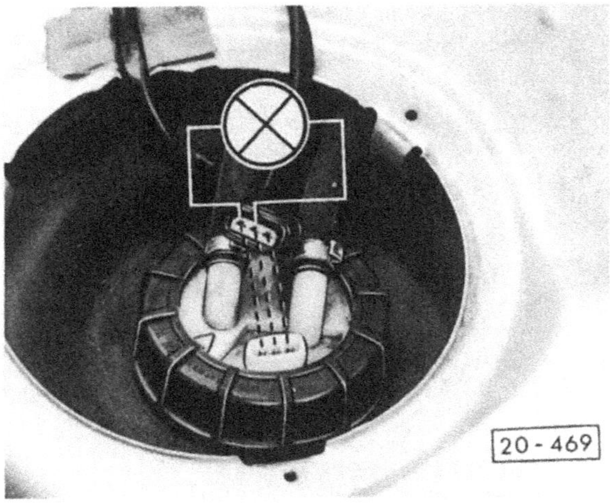

Fig. 4-2. Voltage supply to transfer pump being checked at harness connector using test light (shown schematically). Fuel gauge sender shown is not for model covered by this manual. Scirocco and Cabriolet sender is similar.

6. If the transfer pump runs, check its fuel delivery rate. Remove the fuel filler cap. Remove and plug the black transfer pump output hose.

7. Temporarily connect a hose leading to a measuring container, as shown in Fig. 4-3.

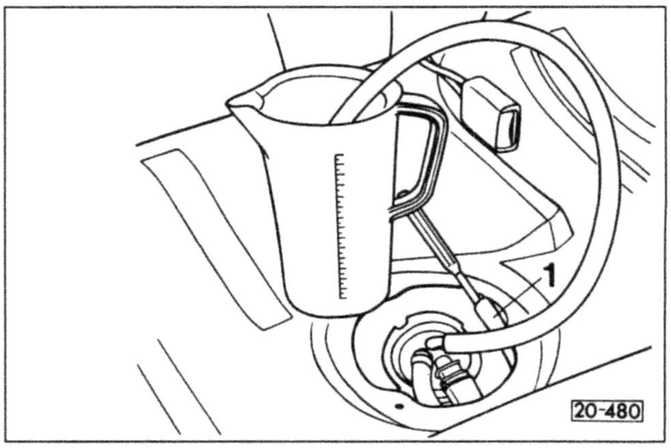

Fig. 4-3. Transfer pump fuel delivery rate being measured. Disconnect and plug output hose (**1**) and temporarily connect test hose.

8. With the fuse/relay panel jumper described in **4.1 Fuses and Relays**, run the pump for exactly 10 seconds. The pump should deliver at least 300 cc (10 oz.) of fuel.

9. If too little fuel is delivered in the preceding step, remove the transfer pump as described below and inspect the pickup strainer/filter for blockage. If when retested the fuel delivery is still too low, the transfer pump is faulty and should be replaced. Reconnect the coil wire when finished.

To remove and install transfer pump:

1. Disconnect the negative (−) battery terminal.

2. Remove the rear seat bottom as described in **BODY AND INTERIOR**. Remove the access cover and the harness connector as described above.

3. Remove the fuel hoses and discard the hose clamps. Label the hoses so they can be reconnected correctly.

> ***WARNING —***
> *Fuel will be exposed. Do not smoke or work near heaters or other fire hazards. Keep a fire extinguisher handy.*

4. Loosen and remove the large lock ring on the transfer pump and fuel gauge sending unit assembly by turning it counterclockwise. See Fig. 4-4. Remove the assembly from the fuel tank.

5. Inspect and, if necessary, clean the strainer/filter. Inspect the sealing gasket on the top of the fuel tank. Replace a damaged gasket. Lubricate the gasket with graphite powder before installing it.

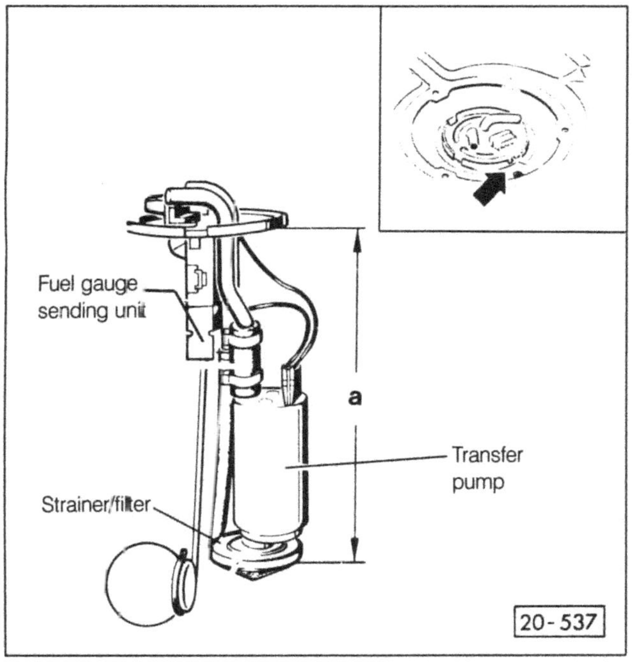

Fig. 4-4. Transfer pump and fuel gauge sending unit assembly shown removed from fuel tank. Inset shows alignment of matching marks (arrow). When installing new pump, dimension **a** should be 175 ± 1 mm (6.89 ± .039 in.).

6. Install the transfer pump assembly, with its gasket in position on the tank, with the matching marks aligned as shown in Fig. 4-4 above.

7. Install the lock ring and tighten clockwise until snug. Using new hose clamps, connect the fuel hoses. Reconnect the harness connector. Install the access cover and its mounting screws. Install the rear seat bottom and reconnect the negative (−) battery terminal.

4.3 Fuel pump

The electric main fuel pump is mounted beneath the car in front of the fuel tank. The pump runs whenever the ignition is switched on, but it stops automatically after a few seconds if the engine stops running or fails to start. Listening under the rear of the car, you should be able to hear the pump. If it does not run, see **4.1 Fuses and Relays.** Before testing the main fuel pump delivery rate, make sure that the transfer pump is supplying adequate fuel. See **4.2 Transfer Pump**.

A fuel pump check valve helps maintain residual pressure in the system after the engine stops, to help prevent vapor lock and hard starting. The higher pressure CIS and CIS-E systems include a fuel pressure accumulator, shown in cross-section in Fig. 4-5 and earlier in Fig. 4-1. It helps maintain residual pressure and damps the pressure pulsations from the pump.

This section covers testing and replacing the fuel pump and replacing the CIS/CIS-E fuel pump check valve. For information on testing residual pressure, see the headings covering pressure tests under **5. Continuous Injection System (CIS)**, **6. Continuous Injection System-Electronic (CIS-E)** or **7. Digifant II Fuel Injection.**

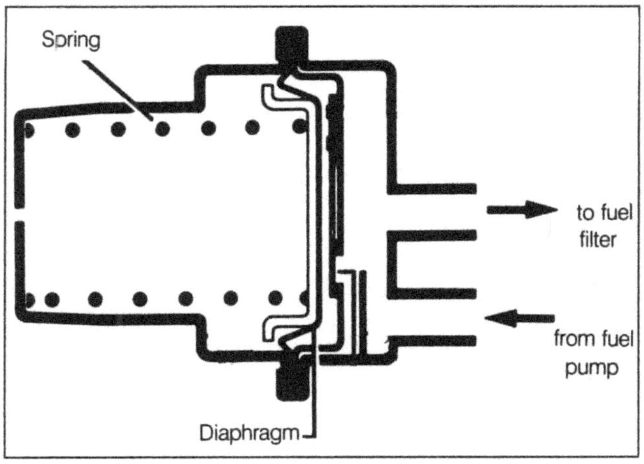

Fig. 4-5. Schematic cross-section of fuel accumulator to damp pressure pulses and maintain residual pressure in CIS and CIS-E fuel systems.

To test fuel pump delivery rate (CIS and CIS-E):

1. Disconnect the fuel return line as shown in Fig. 4-6, and route the end of the fuel return line to a measurement container of at least 1000 ml (1 qt.).

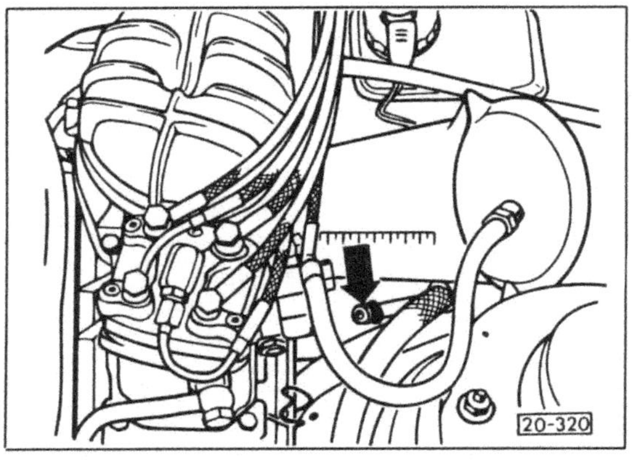

Fig. 4-6. Container being used to measure fuel delivery rate at return line. Return line connection (arrow) is located near left (driver side) front suspension strut tower.

WARNING —

Fuel will be expelled under pressure as the line is disconnected. Do not disconnect wires that could cause sparks. Do not smoke or work near heaters or other fire hazards. Keep a fire extinguisher handy.

2. Run the fuel pump, as described in **4.1 Fuses and Relays**, for exactly 30 seconds.

3. Compare the amount of fuel delivered with the specifications indicated by the chart in Fig. 4-7.

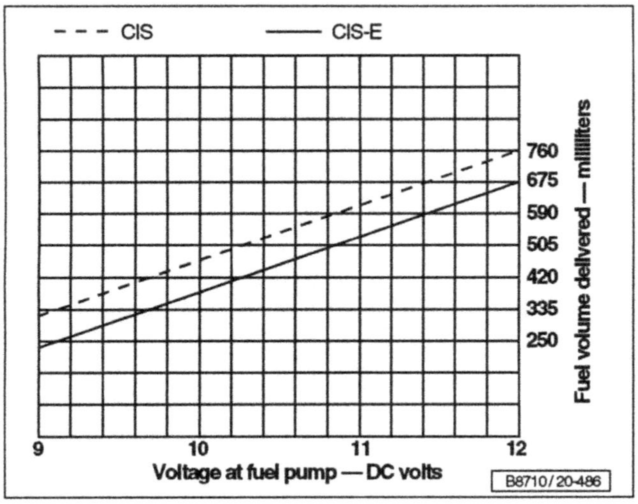

Fig. 4-7. Minimum fuel pump delivery specifications for CIS and CIS-E. Graph shows supply voltage measured at fuel pump (approx. two volts less than battery voltage) vs. volume of fuel delivered in 30 seconds.

If fuel delivered is less than the amount specified, check for leaks, blocked or kinked lines, a clogged filter or fuel reservoir, or a blocked transfer pump strainer/filter. If no such faults are found, the fuel pump is probably faulty and should be replaced.

To test fuel pump delivery rate (Digifant):

1. Clean the fuel line connections at the fuel pressure regulator in the engine compartment. See **7. Digifant II Fuel Injection**.

2. Disconnect the fuel return line running from the fuel pressure regulator back to the fuel reservoir. Prepare to collect fuel from the open line in a suitable container.

> ***WARNING —***
> *Fuel will be expelled as the line is disconnected. Do not disconnect wires that could cause sparks. Do not smoke or work near heaters or other fire hazards. Keep a fire extinguisher handy.*

3. Connect a jumper to bypass the fuel pump relay, as described in **4.1 Fuses and Relays**. Run the pump for 30 seconds and measure how much fuel is collected.

Main fuel pump specification (Digifant):

- delivery rate (minimum) 300 ml (10 fl. oz.) in 30 seconds

If fuel delivered is less than the amount specified, check for leaks, blocked or kinked lines, a clogged filter or fuel reservoir, or a blocked transfer pump strainer/filter. If no such faults are found, the fuel pump is probably faulty and should be replaced.

Fuel Pump Electrical Tests

Use these tests when the fuel pumps do not run at all. First check the fuel pump fuse #5 and, if necessary replace it. Next, disable the ignition and bridge the fuel pump relay terminals to operate the pumps as described in **4.1 Fuses and Relays**.

If the transfer pump and the main pump still do not run, check the fuse/relay panel connections and wiring common to both pumps, and check the fuel pump ground connections. See the wiring diagrams in **ELECTRICAL SYSTEM**. Repair the fuse/relay panel or the connections as necessary.

If only one pump does not run—either the transfer pump or the main fuel pump—check at the pump connections to see whether battery voltage (about 12 volts DC) is reaching the pump. Repair the wiring harness and fuel pump connections as necessary. If voltage is reaching the pump and it does not run, the pump is faulty and should be replaced.

If the pumps run only with the fuel pump relay bypassed, then either the relay or a circuit that operates the relay is faulty. Test at the relay panel as described in the procedure that follows. Fig. 4-8 shows the location of the fuel pump relay on the fuse/relay panel.

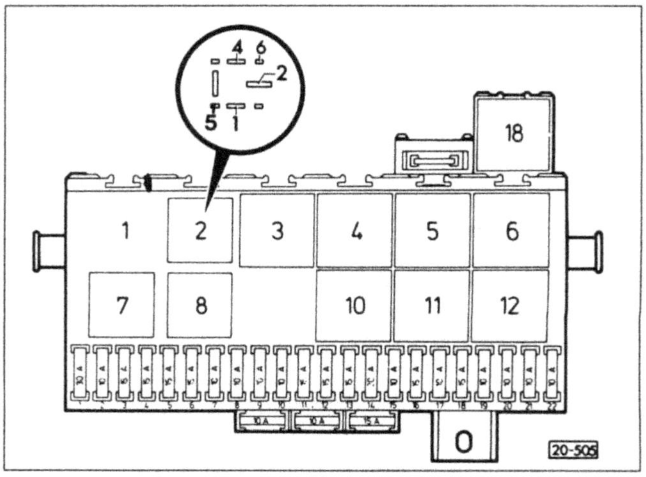

Fig. 4-8. Location of fuel pump relay on typical fuse/relay panel. Inset identifies relay terminals for testing.

To test fuel pump relay circuit:

1. Remove the fuel pump relay. See Fig. 4-8.
2. Switch the ignition on.
3. Using an LED test light or a voltmeter, check for battery voltage (about 12 volts DC) reaching the relay terminals on the fuse/relay panel.
   - terminal **2** (30 circuit) to ground 12 volts
   - terminal **4** (15/86 circuits) to ground 12 volts
   - if not, check and repair fuse/relay panel wiring. See **ELECTRICAL SYSTEM**

4. On all cars except Scirocco 16V, continue checking for battery voltage (about 12 volts DC) to test the relay ground terminal connections.

   - terminal **2** to terminal **1** (31 circuit) 12 volts
   - terminal **4** to terminal **1** 12 volts
   - if not, check fuse/relay panel wiring to ground terminal 1 and check fuse/relay panel ground connections. See **ELECTRICAL SYSTEM**

If these tests reveal that the fuel pump relay is getting power (battery voltage) and has a sound ground connection, but the fuel pumps do not run, then the fuel pump relay is faulty and should be replaced.

Replacing Fuel Pump

The main fuel pump assembly is shown in exploded view earlier in this section in Fig. 4-1. Remove and install it as described below. Always use new sealing washers on the fuel line connections.

> **NOTE —**
> For CIS and CIS-E systems, the fuel pump check valve is available as a separate part.

To replace:

1. Disconnect the negative (–) battery cable from the battery.
2. Working beneath the car, disconnect the wiring harness connections from the fuel pump. The pump and connections are shown in Fig. 4-9. Label the wires to ensure correct installation. Thoroughly clean the fuel line unions.

Fig. 4-9. Typical Volkswagen fuel pump showing wiring harness connections (white arrows) and output fuel line with check valve (black arrow). Pump shown differs slightly from cars covered by this manual.

3. Clamp off the supply line from the fuel reservoir, then clamp off the output line leading to the fuel accumulator (CIS/CIS-E), or forward to the fuel filter (Digifant).
4. Place a large container beneath the pump to catch fuel, then disconnect the fuel lines from the fuel pump.

> ***WARNING —***
> *Fuel will be discharged. Do not smoke or work near heaters or other fire hazards. Keep a fire extinguisher handy.*

5. Remove the pump with the sound absorber from the mounting brackets. Separate the pump from the sound absorber and insert the new pump.
6. Mount the pump and the sound absorber in the brackets. Reattach the fuel lines and wiring connections. Use new sealing washers and hose clamps. Torque the fuel line union cap nut to 20 Nm (15 ft. lb.).
7. Reconnect the battery negative (–) cable. Start the engine and check for leaks.

Replacing Fuel Pump Check Valve

The check valve at the fuel pump outlet helps maintain fuel system pressure after the engine stops. This helps to prevent the fuel in the lines from overheating and causing vapor lock. It is also an aid to hot restarting. If there are no leaks in the fuel system, a faulty check valve is one possible cause of too little residual fuel pressure and difficult hot starting.

On 1985-1989 CIS and CIS-E cars, the check valve is available as a separate replacement part. On Digifant cars, the check valve is a part of the fuel pump and cannot be replaced separately.

To replace the check valve, first disconnect the battery. Clean and disconnect the fuel line, then remove the check valve from the pump. See **Replacing Fuel Pump** and Fig. 4-9. Always use new sealing washers. Torque the check valve to 20 Nm (15 ft. lb.), then hold it with a wrench and torque the fuel line cap nut to 20 Nm (15 ft. lb.)

4.4 Fuel Pressure Regulator

Each of the different fuel injection systems uses a different type of fuel pressure regulator to maintain a constant system pressure. They all recirculate excess fuel back to the fuel reservoir near the main fuel pump. For information about the Digifant pressure regulator, see **7. Digifant II Fuel injection**.

CIS Pressure Relief Valve

CIS system pressure is determined by the pressure relief valve mounted in the fuel distributor. See Fig. 4-10. System pressure can be adjusted by adding or subtracting shims. To measure and adjust system pressure, see **5.6 Fuel Pressure Tests and Specifications**.

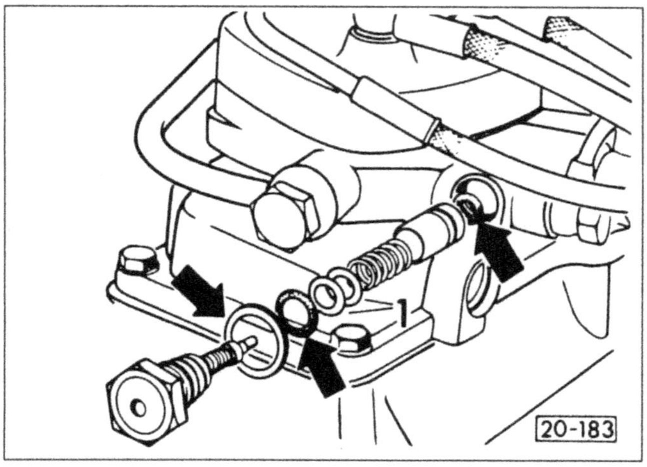

Fig. 4-10. Exploded view of CIS pressure relief valve showing location of adjusting shims (**1**). O-rings and sealing washer (arrows) should always be replaced.

CIS-E Diaphragm Pressure Regulator

The CIS-E diaphragm pressure regulator, shown in Fig. 4-11, is mounted externally on the fuel distributor. In addition to controlling system pressure, it acts as a one-way check valve to maintain residual pressure in the system after the engine is turned off. The diaphragm pressure regulator is not adjustable and should be replaced if system pressure is not within the specifications given in **6.5 Fuel Pressure Tests and Specifications**.

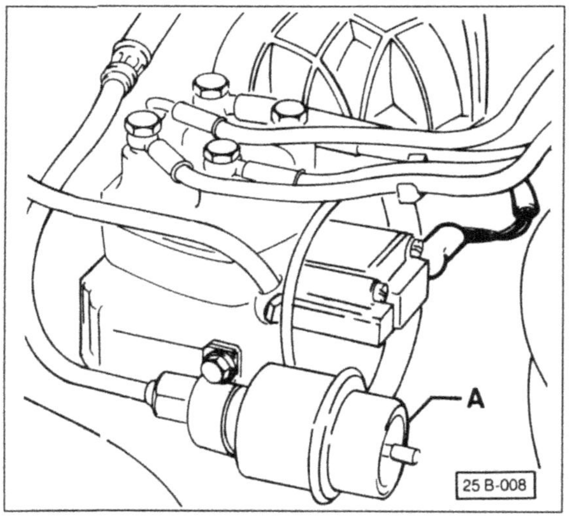

Fig. 4-11. CIS-E diaphragm pressure regulator (**A**). Mounting may differ on models covered by this manual.

4.5 Evaporative Emission Controls

Evaporative emission controls provide essential ventilation for the fuel tank, but trap the fuel vapors that would otherwise be vented into the atmosphere. The closed system collects evaporated fuel vapor and directs it into the main intake air steam.

The main components of the system are the charcoal canister (located in the left front wheelhouse), a valve that controls canister venting, the expansion tank (in the right front wheelhouse), and the gravity/vent valve shown in Fig. 4-12. System layout is shown in Fig. 4-13.

WARNING —
- *Disconnect the battery ground (-) cable before working on the fuel system. Do not smoke or work near heaters or other fire hazards.*
- *When working on the fuel system, always replace seals and fuel line hose clamps to ensure proper sealing and prevent leaks.*

Fig. 4-12. Gravity/vent valve.

Gravity/Vent Valve Testing

The gravity/vent valve in the fuel tank filler neck is normally open, venting the fuel tank. If the car is upset in an accident, the valve closes to prevent the liquid fuel from entering the vent system and flooding the charcoal canister, or escaping.

You can reach the valve working from inside the right rear wheelhouse. With the valve in an upright position, air should pass through ports freely. When it is tilted at a 45° angle, the valve should close and no air should be able to pass through. Refer to Fig. 4-12 above. A faulty gravity/vent valve would be hazardous in an accident and should be replaced.

Charcoal Canister Valve Testing

The charcoal canister valve (Fig. 4-14) is normally closed. To check it, disconnect the two vent hoses. You should not be able to pass any air through the vent ports.

Engine vacuum opens the valve. Use a hand-held vacuum pump to apply vacuum to the small hose connection. The valve should open and allow air to pass through the other ports.

On earlier cars, the canister is open to be purged by intake vacuum and the valve is a canister bypass valve. On 1990 and later cars, the valve opens to allow the canister to be purged.

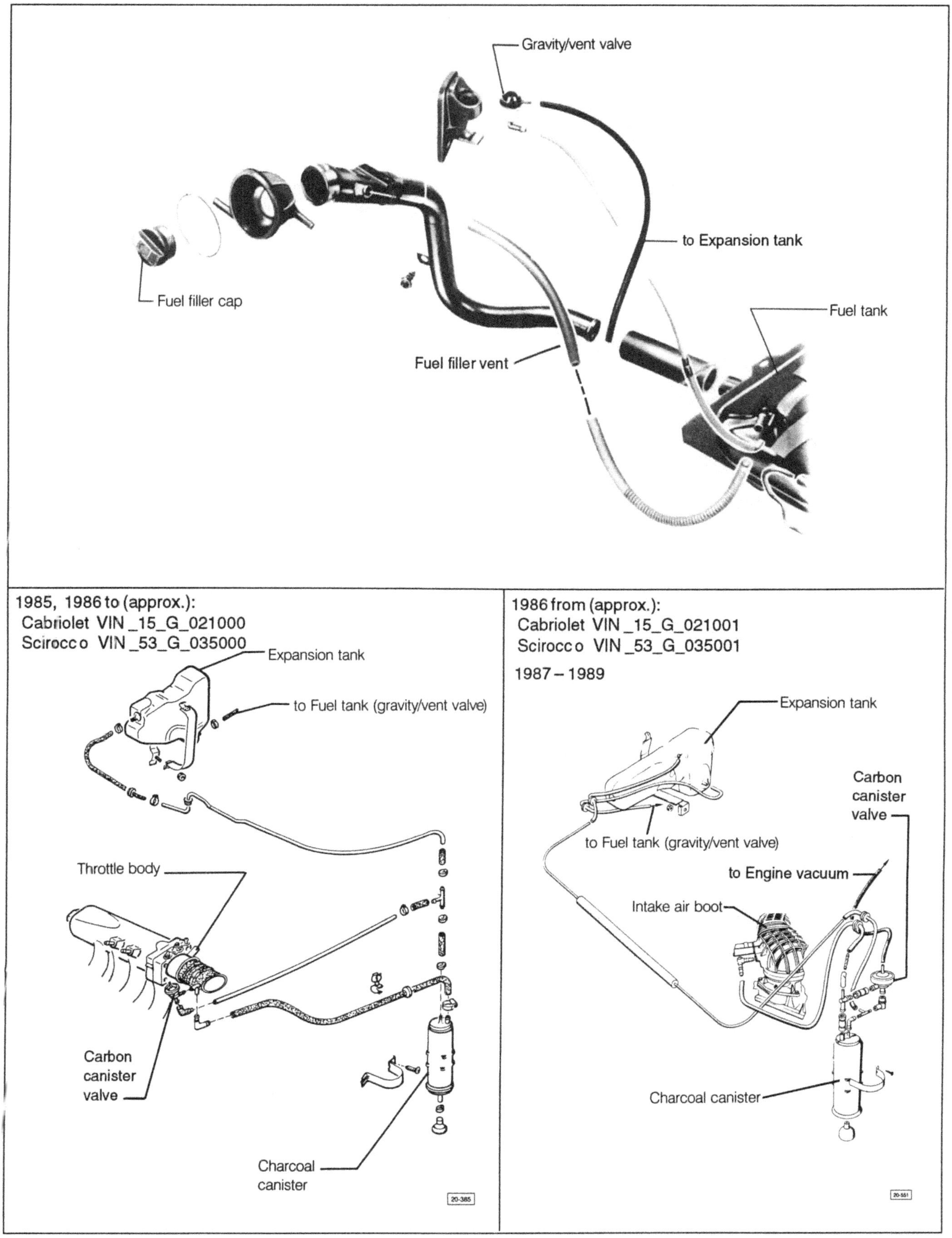

Fig. 4-13. Exploded views of evaporative emission controls. 1990 and later cars are similar.

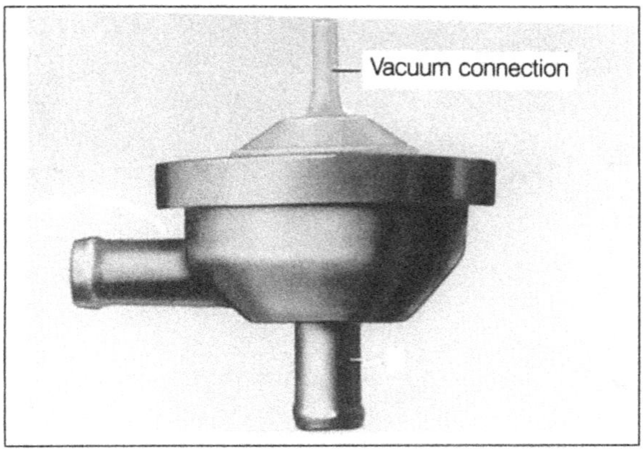

Fig. 4-14. Charcoal canister valve.

Leak Testing

To leak-test the system, make sure that the tank is at least 2/3 full and the fuel filler cap is tight. Disconnect the vent line to the charcoal canister from the gravity/vent valve. Pressurize the disconnected line to about 1.3 in. Hg (0.63 psi).

After five minutes, the system should still hold about 1 in. Hg (1/2 psi). If not, check for leaks at the fuel filler cap and in the vent system between the charcoal canister and the fuel tank.

4.6 Accelerator Cable

Cars with automatic transmission have two cables which connect the accelerator pedal to both the throttle valve and the transmission kickdown mechanism. Because the adjustment of the accelerator and throttle cables may affect the operation of the automatic transmission, adjustment and repair of these cables is covered in **AUTOMATIC TRANSMISSION**.

Cars with manual transmission have one accelerator cable linking the accelerator pedal to the throttle body. This cable is easily adjusted or replaced using ordinary hand tools.

> ***CAUTION —***
> *The cable adjustment removes slack and compensates for cable stretch. It is not for idle speed adjustment. To check and adjust idle speed, see the appropriate headings under* **5. Continuous Injection System (CIS), 6. Continuous Injection System-Electronic (CIS-E), 7. Digifant II Fuel Injection,** *or* **8. Digifant I Fuel Injection.**

To check the cable adjustment on manual transmission cars, have a helper push the accelerator pedal down to the full-throttle position.

> **NOTE —**
> With the pedal fully depressed, the throttle valve should barely reach the full-throttle stop on the housing, leaving a maximum clearance of 1 mm (.04 in.). If the throttle valve reaches the stop before full-throttle, the cable will be over-stressed and may fail.

On Scirocco 16V, remove the cable retaining clip from the bracket near the throttle valve. See Fig. 4-15. Adjust cable tension by repositioning the cable in the grommet and installing the retaining clip in a different groove. Use a groove closer to the throttle valve to tighten the cable and reduce slack at full-throttle. Use a groove farther from the throttle valve to loosen the cable and restore the necessary small gap at full-throttle.

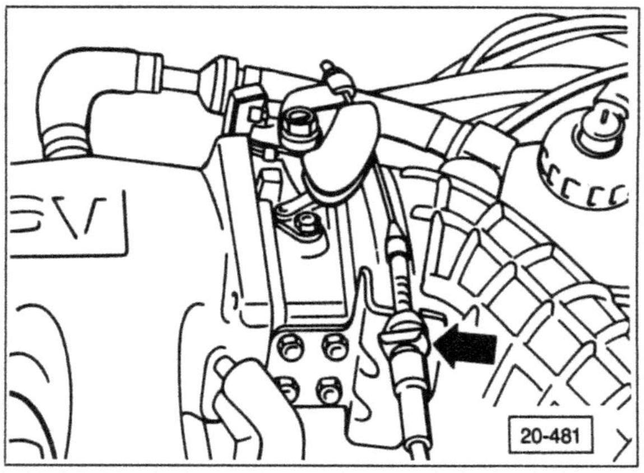

Fig. 4-15. Retaining clip and grommet (arrow) used to adjust accelerator cable tension.

On cars with 8-valve engines, loosen the locknuts at the cable bracket atop the cylinder head cover. See Fig. 4-16. Adjust the cable conduit position by turning the nuts. Move it farther from the throttle valve to tighten the cable and reduce slack at full-throttle. When the cable is properly adjusted, recheck it at the full-throttle position, then tighten the locknuts.

Fig. 4-16. Locknuts (arrows) that hold cable conduit to bracket on cylinder head cover. (Model shown is not one of those covered by this manual. Scirocco and Cabriolet models are similar).

To remove and install the cable, remove the cable conduit from its retaining bracket to slacken the cable, and disconnect the cable end from the throttle valve lever. Working from inside the car, pull the cable end from its ball socket in the accelerator pedal assembly. Then, working from the engine compartment, pull the cable through the firewall. Installation is the reverse of removal. Adjust the new cable as described above. Finally, depress the accelerator pedal several times to check for binding before driving.

4.7 Fuel Injectors (CIS and CIS-E)

This section covers the mechanical fuel injectors that are part of CIS and CIS-E fuel injection. For information on Digifant solenoid fuel injectors see **7. Digifant II Fuel Injection**.

The injectors are actually precision nozzles. Above a certain minimum opening pressure, they continuously discharge atomized fuel into the engine's intake ports. Injector inserts are threaded into the cylinder head or, on 16-valve engines, the intake manifold. Thick O-rings seal the injectors to prevent air and fuel leakage, and hold them in place in the inserts. One injector, inserts and sealing parts are shown in Fig. 4-17.

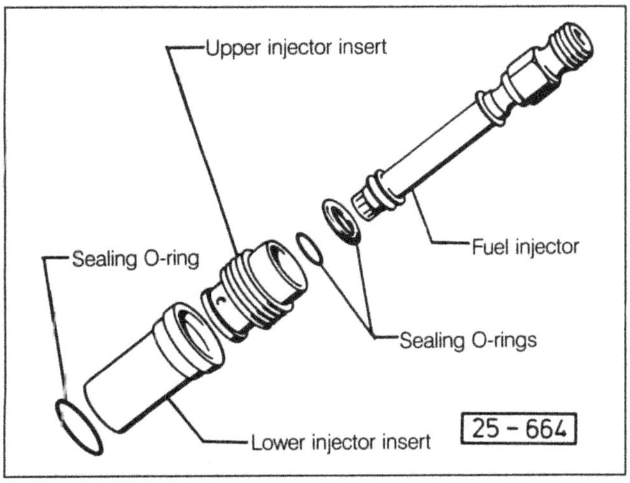

Fig. 4-17. Components of CIS and CIS-E fuel injector.

The most common fuel injector problems are low fuel flow or an uneven spray pattern, both caused by contaminated fuel or carbon deposits. Some signs of faulty injectors are rough idle, poor starting, and detonation (pinging) at full throttle. Keep injectors functioning properly by keeping them clean. See **LUBRICATION AND MAINTENANCE** for information on fuel filter maintenance, and on a fuel additive which Volkswagen recommends for reducing and preventing carbon deposits.

Removing and Installing Fuel Injectors

The fuel injectors must be removed for inspection or testing. As described above, the injectors are held in the inserts by thick rubber O-rings. To remove the injectors simply pull them straight out, forcing them out of the O-rings, as indicated in Fig. 4-18. Leave the fuel lines attached. Pull straight and pull only on the injectors themselves, not on the fuel lines.

It is not normally necessary to remove the injector inserts, except to replace the O-rings and cure air leaks which can cause rough idling. The inserts have an integral hex socket and are removed using a 12 mm hex wrench. Replace all the O-rings. See Fig. 4-17 above. Moisten the O-rings with fuel for installation. Clean the threads of the upper insert and reinstall using Loctite® T primer and Loctite® 620 thread locking compound, or equivalent. Torque the upper injector insert to 20 Nm (15 ft. lb.).

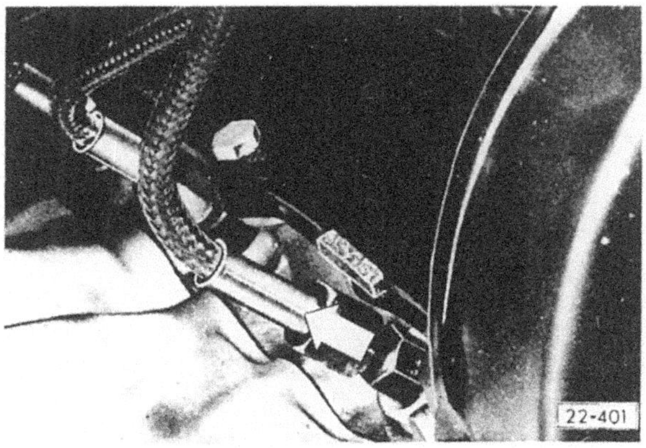

Fig. 4-18. Fuel injectors. Remove by pulling straight out (arrow). Do not pull on fuel line.

Injector installation is the reverse of removal. Replace the thick rubber O-rings, soaking the new ones in fuel for several minutes before installing them. Then press the injector into the insert until the O-ring is fully seated in its groove.

Quick-checking Injectors

Each individual injector can be easily quick-checked to evaluate its spray pattern and to check for leakage. A more involved test, using special equipment, compares the injection quantity from all four injectors to check that both the fuel distributor and the injectors are supplying the same amount of fuel to each cylinder. Finally, a special injector tester can be used to clean injectors, measure opening pressure, or check and flush new injectors prior to installation.

If the quick-check suggests that more testing is necessary and the special equipment is not available, we suggest having the injectors tested by an authorized Volkswagen dealer or other qualified and properly equipped repair shop.

To quick-check injectors:

1. Remove one injector, leaving the fuel hose connected, as shown in Fig. 4-18 above. Aim the removed injector into a fuel-resistant container.

2. Remove the large rubber duct from the top of the air flow sensor.

3. Remove the fuel pump relay from the fuse/relay panel and install a fused jumper wire to run the fuel pump, as described in **4.1 Fuses and Relays**.

4. Using a magnetic tool such as the one shown in Fig. 4-19, lift the air flow sensor plate and observe the injector's spray pattern. The injector should discharge a fine mist of fuel in an even, cone-shaped pattern.

WARNING —

Fuel will be discharged from all four injectors during this check. Keep the test time short and discharge as little fuel as possible. Do not smoke or work near heaters or other fire hazards. Keep a fire extinguisher handy.

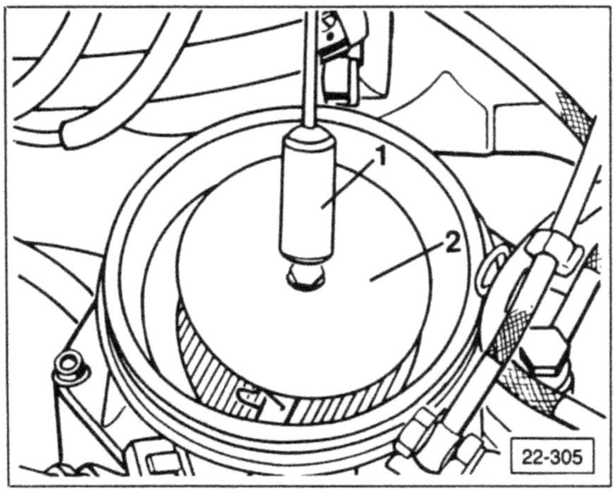

Fig. 4-19. Magnetic tool (**1**) being used to lift air flow sensor plate (**2**) for fuel injector quick-check.

5. Release the sensor plate and hold the injector in a horizontal position. It should not drip fuel.

6. Disconnect the fuel pump jumper wire to stop the pump.

NOTE —

An injector that has an irregular spray pattern or drips when the sensor plate is at rest with the fuel pump running should be replaced.

Testing Injector Fuel Delivery

A similar but more complicated test is used to compare the rate of fuel delivery from all four injectors as a means of isolating a bad injector or fuel distributor. Remove all four injectors from the engine, with fuel lines attached, and connect them to four identical measuring containers as shown in Fig. 4-20.

Use a jumper wire in place of the fuel pump relay to run the fuel pump during the test, as described in **4.1 Fuses and Relays**. Use Volkswagen special tool no. VW 1348/1 (order no. TV1 348 001 25 ZEL), to raise and hold the sensor plate in the various test positions.

Install the Volkswagen special tool as shown in Fig. 4-21. Carefully push the center sleeve all the way down, then turn the adjusting knob clockwise until the tool's magnetic tip just contacts the sensor plate center bolt. Turn the adjusting screw counterclockwise to raise the sensor plate to a point just before any of the injectors start spraying.

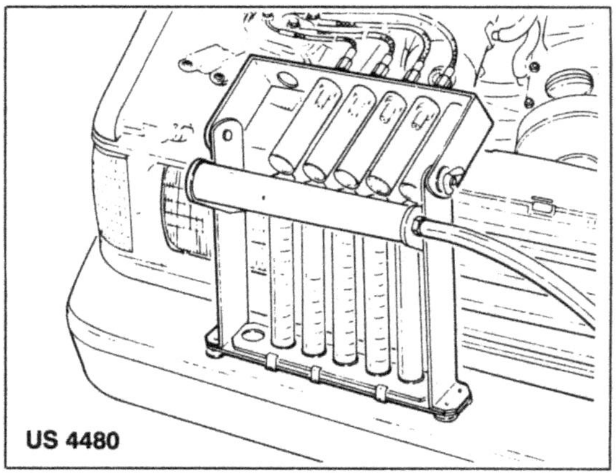

Fig. 4-20. Volkswagen special tool US 4480 being used to collect and measure fuel injectors' output. Injectors are inserted at top.

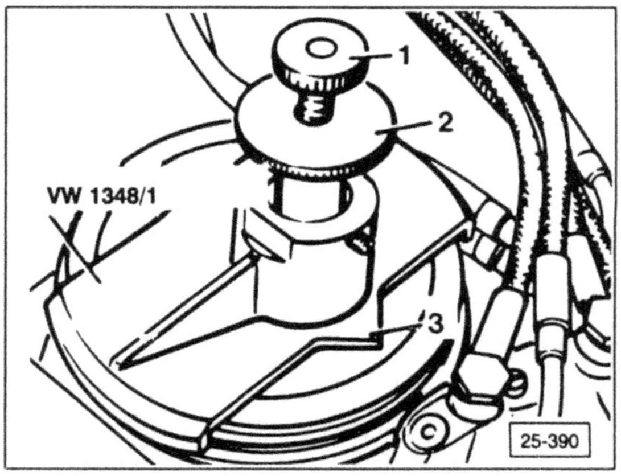

Fig. 4-21. Volkswagen special tool VW 1348/1 with adjusting knob (**1**) and sleeve (**2**) being used to simulate sensor plate position at idle, full throttle, or any point between. Pointer (**3**) faces center of fuel distributor.

WARNING —

Do not smoke or work near heaters or other fire hazards. Keep a fire extinguisher handy.

Run the fuel pump and raise the sensor plate to conduct the test. The tool's center sleeve has two detents—one raises the plate to a position simulating idle and the second raises it higher to simulate full throttle. At the idle setting, with 20 ml collected from each injector (approximately 2 minutes) there should be no more than 3 ml difference between any two. At the full throttle setting, with 80 ml collected from each injector (approximately 40 seconds) there should be no more than 8 ml difference.

If the difference in fuel quantity is greater than allowed, select the two injectors with the lowest and highest values. Loosen the unions, remove them from their fuel lines, exchange them, and repeat the test. If the same injector delivers a low amount of fuel, it is faulty and should be replaced. If the same container collects a low amount from a different injector, then that fuel line is blocked or kinked, or the fuel distributor is faulty. When reinstalling fuel line unions to injectors, torque them to 25 Nm (18 ft. lb.).

Cleaning Injectors

A special tool for cleaning injectors and testing their opening pressure is shown in Fig. 4-22. Remove the injector from its fuel line by loosening the union, and install the injector on the tester line as shown. Pump several times to purge the air from the line. When spraying normally, the injector will make a high-pitched squeal. Pump slowly and read the injector opening pressure on the gauge as the injector starts to spray. It should open at 3.0 to 4.1 bar (44 to 59 psi). Any injector which opens or leaks at a lower pressure or has an uneven spray pattern should be replaced.

WARNING —

Use only Shell mineral spirit 135 or equivalent when testing or cleaning injectors using pressure tester. Shell mineral spirit 135 is flammable. Do not smoke or work near heaters or other fire hazards. Keep a fire extinguisher handy.

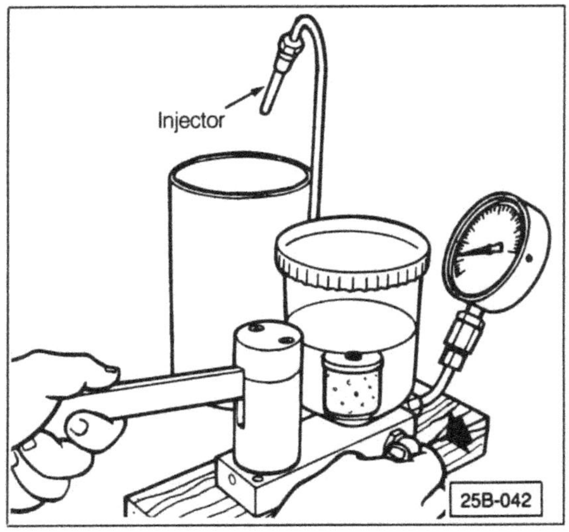

Fig. 4-22. CIS and CIS-E fuel injector pressure tester (Volkswagen special tool no. US 8034) being used to clean and test fuel injector.

5. CONTINUOUS INJECTION SYSTEM (CIS)

Fig. 5-1 is a schematic view of CIS fuel injection. The system is almost entirely mechanical, utilizing the influence and control of fuel pressure to accurately meter fuel. Other than the main fuel pump, electrically operated components are only used in an auxiliary role, to help adapt the fuel mixture to such extreme conditions as cold weather starting and warm-up.

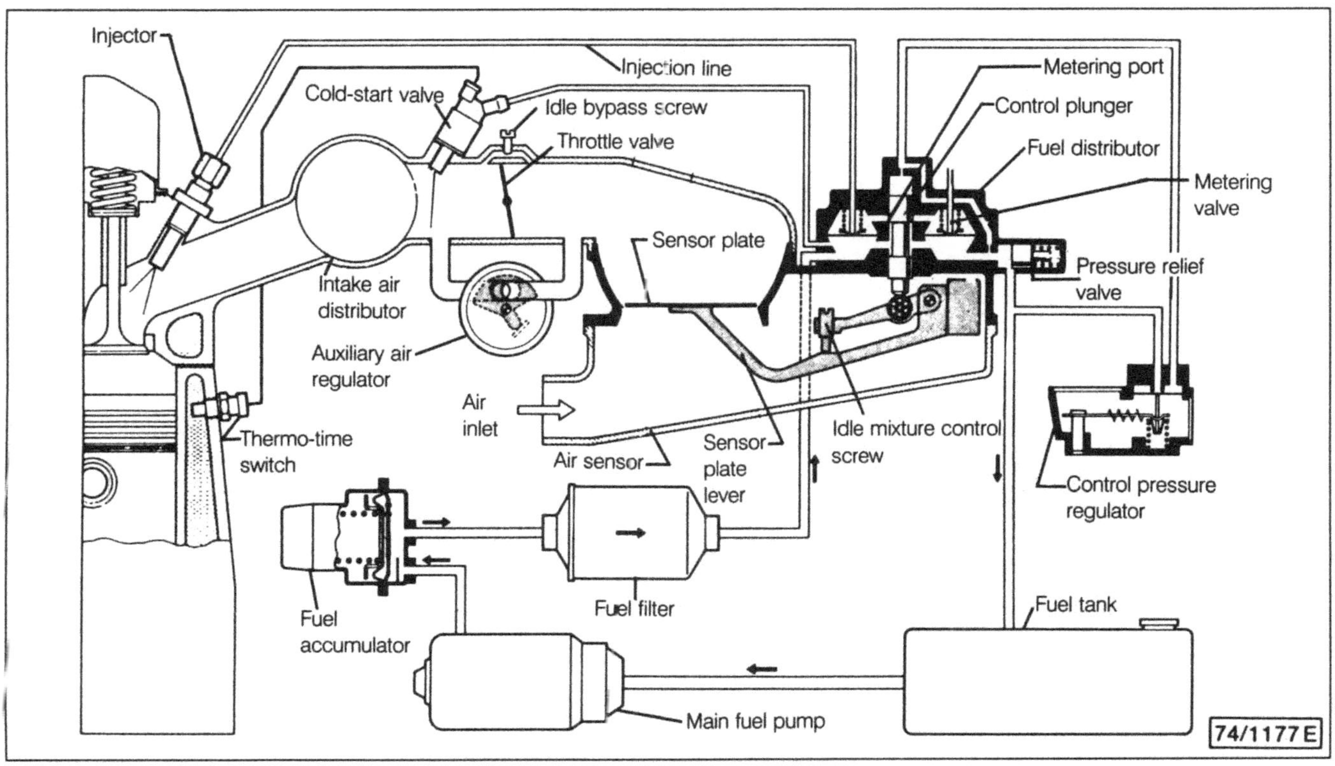

Fig. 5-1. Schematic view of CIS fuel injection. Arrows indicate direction of fuel flow.

The continuous injection system (CIS) is so-called because, unlike many other fuel injection systems, the injectors do not open and close – they deliver fuel continuously. The main function of the fuel injectors is to atomize the continuous flow of fuel being supplied under pressure from the fuel distributor.

The heart of CIS is the mixture control unit, consisting of the air flow sensor and the fuel distributor. Air entering the engine raises the air flow sensor plate. A lever connected to the sensor plate raises the control plunger in the fuel distributor, increasing the quantity of fuel flowing to the injectors in proportion to the air flow. Fuel pressure working against the control plunger, called control pressure, helps regulate the plunger's movement and pushes it back down for less fuel flow when there is less air flow.

The cold start valve, the auxiliary air regulator, and the control pressure regulator are used to adapt the system to different operating conditions. The cold-start valve is like a fifth injector, and performs a choke-like function. It injects additional fuel into the intake manifold during starting to provide a richer fuel mixture when the engine is cold. The cold-start valve is operated electrically.

The auxiliary air regulator opens when the engine is cold to allow a small amount of additional air to bypass the throttle plate and boost the idle speed. The control pressure regulator improves performance when the engine is cold by reducing control pressure. This not only creates a greater pressure differential across the metering valves, causing more fuel flow for a given air flow (a richer mixture), but also improves throttle response by allowing the control plunger in the fuel distributor to rise more easily.

CIS is designed so that all the auxiliary electrical components can be tested using only a test light and an ohmmeter, but the basic fuel metering system is highly dependent on fuel pressure. Evaluation of the system's basic function requires the use of a fuel pressure gauge. See **5.6 Fuel Pressure Tests and Specifications** for more information.

5.1 Air Flow Measurement and Fuel Metering

The throttle valve, which is operated by the accelerator cable, regulates the amount of air drawn in by the engine. The most basic function of the fuel injection system is to measure that incoming air and precisely meter a proportional amount of fuel. The components responsible for air flow measurement and fuel metering are the air flow sensor and the fuel distributor.

The throttle valve is adjusted at the factory. It does not require routine adjustment and should not be used to adjust idle speed. The procedure below for the throttle valve's basic adjustment is provided in the event that the factory adjustment has been altered. The air flow sensor plate can be adjusted if it is found to be out of alignment. After the adjustment, idle speed and idle mixture (% CO) will need to be adjusted also. The fuel distributor is precisely calibrated when it is manufactured and, except for adjusting system pressure, it cannot be adjusted or repaired.

Throttle Valve Basic Adjustment

The purpose of the throttle valve adjusting screw is to provide a positive stop and prevent the valve from closing so far that it becomes worn or damaged.

To correct a faulty throttle valve adjustment, use a screwdriver to back off the throttle valve adjusting screw until there is clearance between its tip and the throttle valve lever. The adjusting screw is shown in Fig. 5-2. Place a thin piece of paper between the throttle valve adjusting screw and the throttle valve lever. With the throttle valve closed, turn the screw in until it lightly contacts the paper. From this position, remove the paper and turn the screw in an additional ½ turn.

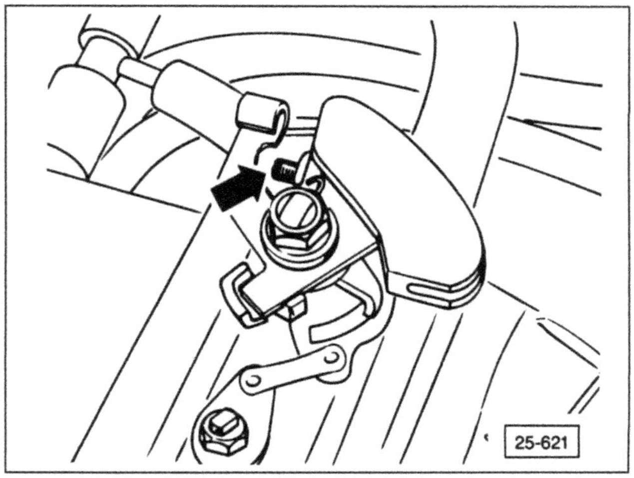

Fig. 5-2. Location of adjusting screw (arrow) for throttle valve basic adjustment (not used to adjust idle speed).

Air Flow Sensor

The air flow sensor measures all the air being drawn in by the engine. Air flows past the air flow sensor plate and the plate is lifted, in turn lifting the control plunger in the fuel distributor to meter the fuel. The sensor plate and its cone-shaped venturi are carefully designed to achieve the proper balance of air flow and fuel metering for all flow rates from idle to full throttle. If the sensor plate is off-center or binding, or the lever has too much resistance, the fuel distributor cannot respond properly to the air flow sensor.

To check and adjust:

1. Remove the rubber intake air boot from the top of the air flow sensor.

2. Disconnect the coil wire from the center tower of the distributor cap and connect it to ground, using a jumper wire. Actuate the starter for approximately 10 seconds to pressurize the fuel system.

3. Using a magnetic tool, lift the sensor plate slowly to check that the resistance, due to fuel pressure resisting movement of the control plunger, is uniform throughout the entire range of travel. Move the sensor plate quickly back to its rest position. There should be no resistance.

4. If the sensor plate binds as it is moved through the air cone, loosen the center bolt and center the sensor plate. Push down on the plate lightly to check that it is correctly centered in the air cone.

NOTE —
When properly centered, a 0.10 mm (.004 in.) feeler gauge should just fit between the sensor plate and the air cone at all points.

5. If the sensor plate is hard to move or moves with uneven resistance, remove the air flow sensor from the air filter housing. Clean and lubricate the sensor plate lever pivot so that it moves smoothly.

6. If the resistance encountered in moving the sensor plate is caused by a binding control plunger, remove and clean the control plunger as described below in **Fuel Distributor**. If it still binds, replace the fuel distributor.

NOTE —
Fuel distributors cannot be serviced. Due to the precision fit required, faulty control plungers cannot be replaced separately.

7. Reconnect the coil wire and run the engine until it reaches normal operating temperature. Then ground the coil wire and pressurize the fuel system as described in step 2.

8. Check the height of the air flow sensor plate. At rest, the front edge of the plate should be aligned with the narrowest point in the venturi as shown in Fig. 5-3. Adjust the height by bending the spring clip below the plate.

9. Check idle speed and the idle speed adjustment, and adjust if necessary. See **5.5 Idle Speed** and **5.7 Idle Specifications**.

Fuel Distributor

The fuel distributor's control plunger, moving up or down in its cylinder, uncovers or covers the metering ports leading to the four metering valves—one for each injector. The metering valves maintain a constant pressure differential for precise control at all fuel flow rates. A cross-section of one metering valve is shown in Fig. 5-4.

The fuel distributor is not normally a source of problems. It is, however, a precision device that is easily damaged by plugging or corrosion from contaminated fuel. For fuel filter maintenance information, see **LUBRICATION AND MAINTENANCE**. On some cars, a mini-filter is installed during manufacture in the fuel inlet bolt at the fuel distributor. This filter should be removed and a normal bolt installed in its place at a specified interval. For more information on identifying and replacing the mini-filter, see **LUBRICATION AND MAINTENANCE**. Check fuel distributor function by measuring injection quantity, as described in **4.7 Fuel Injectors**.

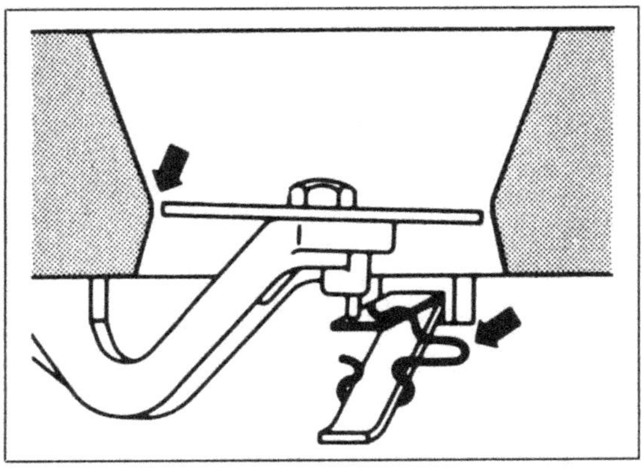

Fig. 5-3. Correct air flow sensor plate rest position. Front edge of plate (left arrow) should be neither above nor more than 0.5 mm (.020 in.) below narrowest point of venturi cone. Adjust height by bending wire spring stop (right arrow).

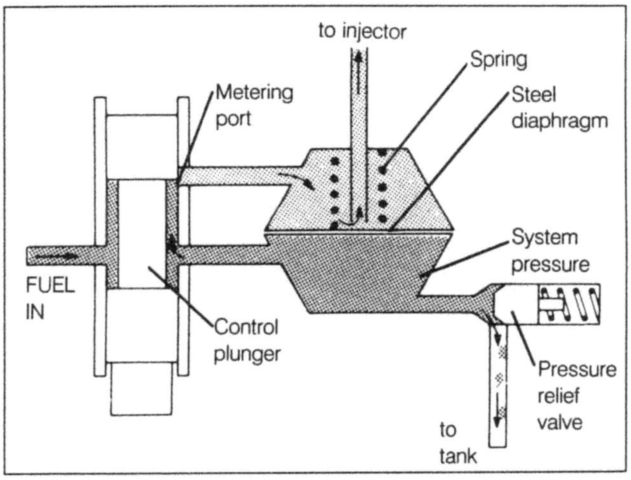

Fig. 5-4. Cross-section of one fuel distributor metering valve. Pressure in upper chamber, combined with spring pressure, deflects diaphragm and increases opening to injector.

Fig. 5-5 illustrates the fuel distributor's fuel line connections and mounting detail. If either the fuel distributor or the air flow sensor is repaired or replaced, idle speed and the idle mixture adjustment should be corrected. See **5.5 Idle Speed** and **5.7 Idle Specifications**. Before making the adjustments, make sure the air flow sensor plate position is correctly adjusted as described earlier in **Air Flow Sensor**.

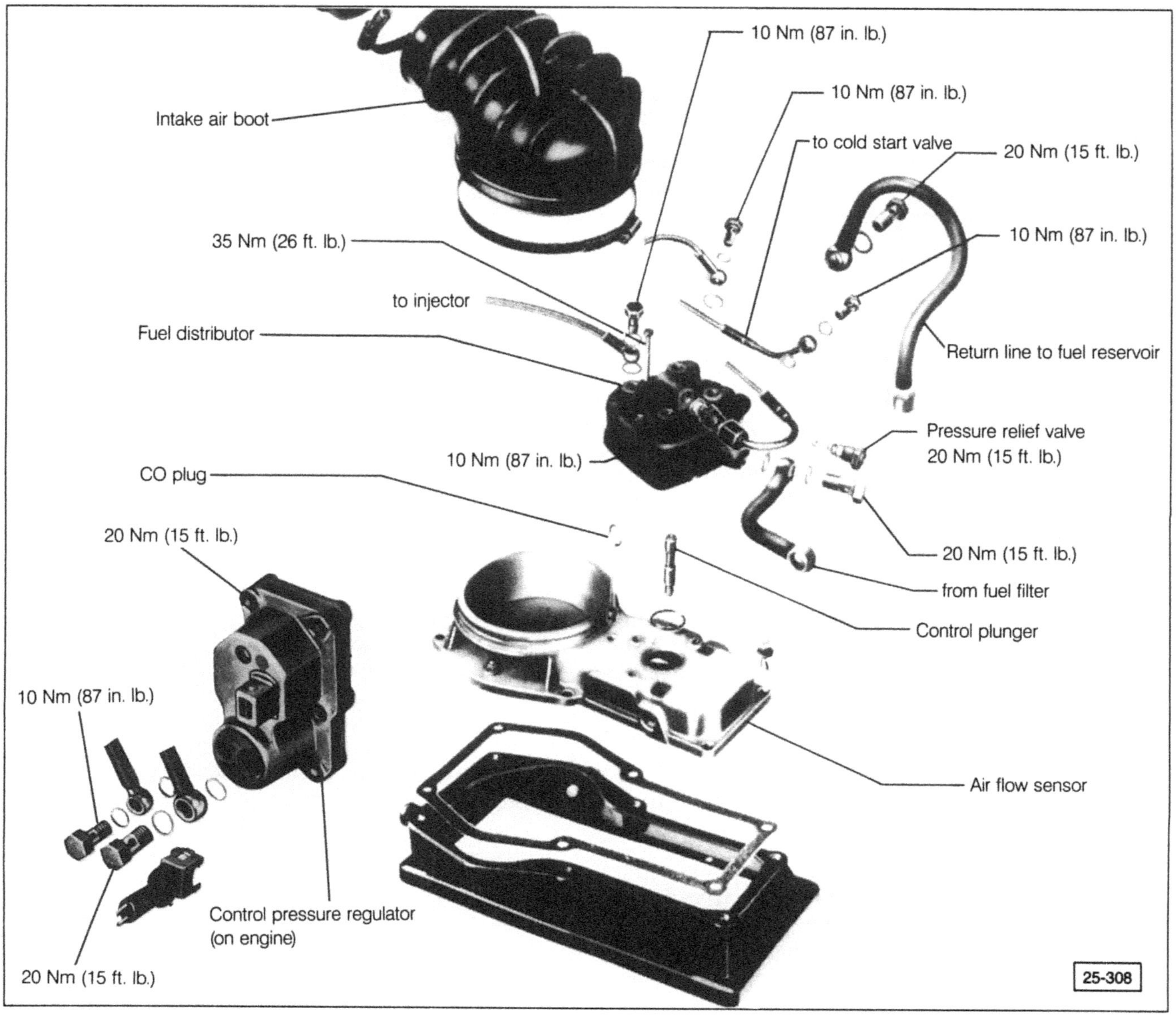

Fig. 5-5. CIS fuel distributor mounting and fuel line connections. Torque values shown are for installation.

To remove and clean the control plunger, remove the two mounting screws from the top of the fuel distributor and pull the fuel distributor from the air flow sensor assembly. Remove the control plunger and clean it with carburetor cleaner or equivalent. Lubricate the control plunger with a light coat of fuel and reinstall it with the small end facing down toward the air flow sensor plate lever.

WARNING —

Fuel will be expelled when disconnecting fuel unions. Disconnect the battery negative (–) terminal. Do not smoke or work near heaters or other fire hazards. Keep a fire extinguisher handy.

CAUTION —

When removing the fuel distributor from the air flow sensor housing, be careful not to let the control plunger fall. If the control plunger is allowed to fall, it may be damaged—rendering the entire fuel distributor unserviceable. The fuel distributor is precision calibrated at the factory and should not be disassembled.

CAUTION —

There may be a small spring on top of the control plunger. Remove the plunger carefully to avoid losing the spring.

NOTE —

Fuel distributors cannot be serviced. Due to the precision fit required, faulty control plungers cannot be replaced separately.

5.2 Cold Start Enrichment

The electrically operated cold-start valve sprays extra fuel into the intake manifold for several seconds when the starter is actuated and the engine is cold. The valve is controlled by the thermo-time switch, located in the cylinder head coolant outlet. When the engine is cold the switch is closed, allowing power to reach the valve and open it. When the engine is warm, the switch is open and the valve does not operate. To limit valve operation and prevent flooding, the electric current also warms the switch and opens the circuit after a few seconds. The cold-start valve and the thermo-time switch are shown in Fig. 5-6.

If the cold-start valve fails to inject fuel during cold starting, it will be difficult or impossible to start the engine. If the cold-start valve leaks, the engine may receive extra fuel at the wrong time and become flooded, especially if the engine is hot.

To test cold start valve:

1. Make sure engine coolant temperature is below 86°F (30°C). Preferably the engine should sit for several hours.
2. Remove the two screws holding the cold start valve to the intake manifold. Without disconnecting the fuel line or the harness connector, remove the cold start valve from the intake manifold.
3. Disconnect the coil wire from the center tower of the distributor cap and connect it to ground, using a jumper wire.
4. Wipe dry the nozzle of the cold start valve. Point the valve into a transparent container, and have a helper actuate the starter. See Fig. 5-7.

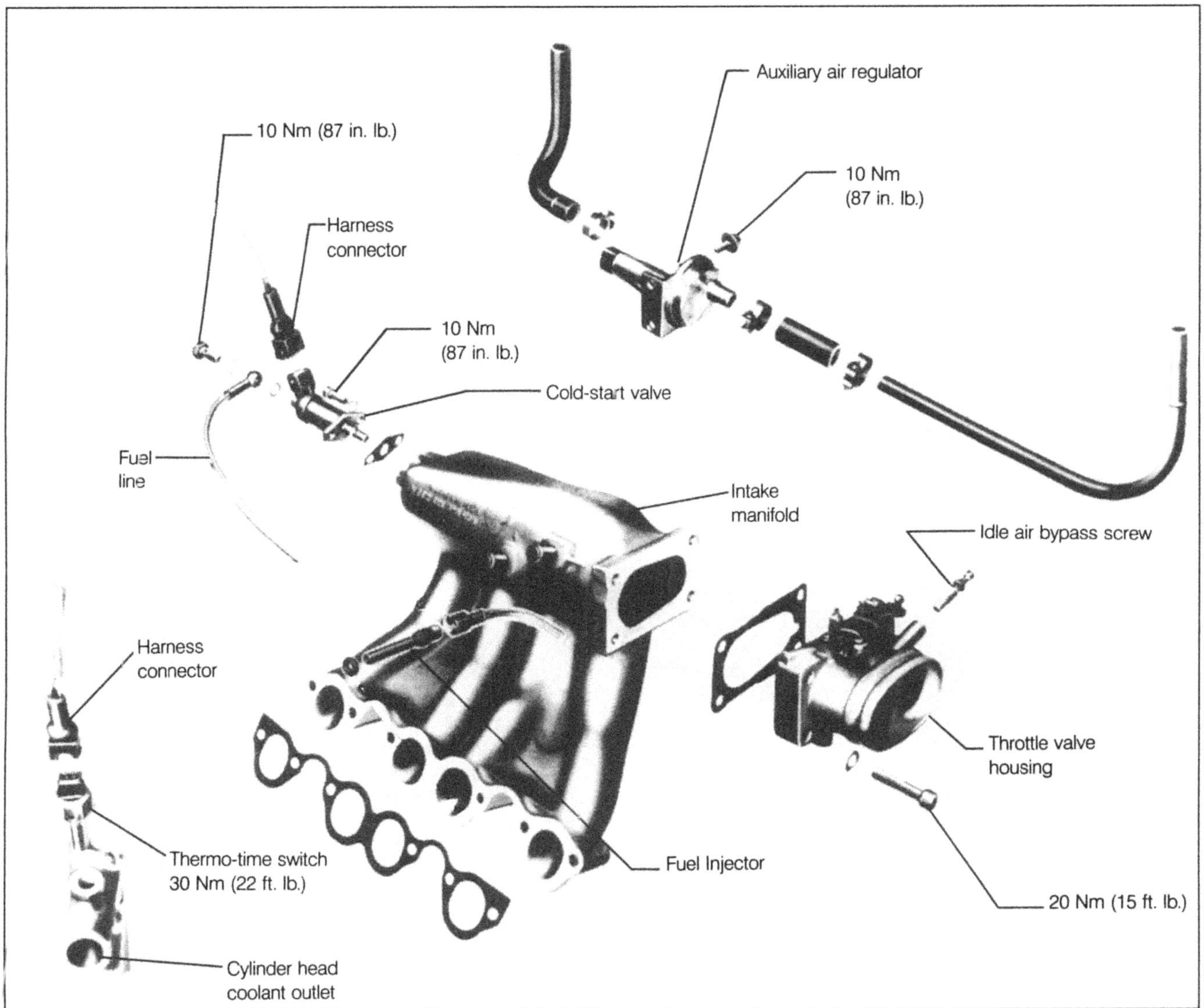

Fig. 5-6. Cold-start valve and thermo-time switch used to provide extra fuel for cold starting.

> ***WARNING —***
> *Fuel will be expelled. Do not smoke or work near heaters or other fire hazards. Keep a fire extinguisher handy.*

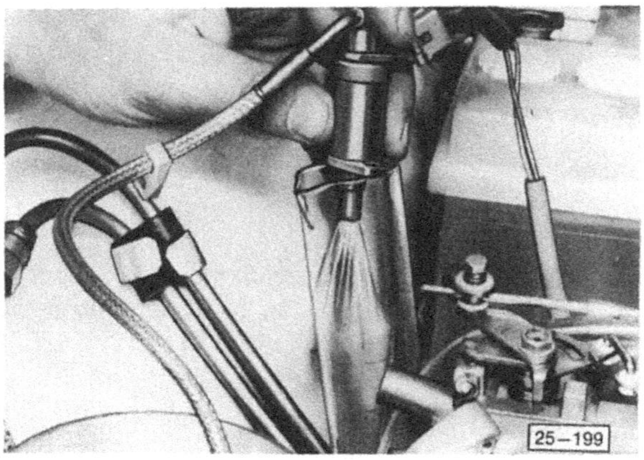

Fig. 5-7. Cold-start valve operation being checked.

> **NOTE —**
> The valve should spray in an even, cone-shaped pattern until the thermo-time switch interrupts the circuit. An irregular spray pattern indicates a dirty or faulty cold start valve. If the valve does not spray at all, test the thermo-time switch as described below.

5. Wipe the nozzle dry. The valve should not drip for at least a minute. If it does, it is faulty and should be replaced.
6. Reconnect coil wire when finished.

To test thermo-time switch:

1. Make sure engine coolant temperature is below 86°F (30°C). Preferably the engine should sit for several hours.
2. Disconnect the coil wire from the center tower of the distributor cap and connect it to ground, using a jumper wire.
3. Disconnect the cold-start valve harness connector and connect a test light across the connector terminals, as shown in Fig. 5-8.
4. Actuate the starter while observing the test light. The test light should light for several seconds, then go out. The amount of time that the light should stay on depends on coolant temperature. See the graph in Fig. 5-9.

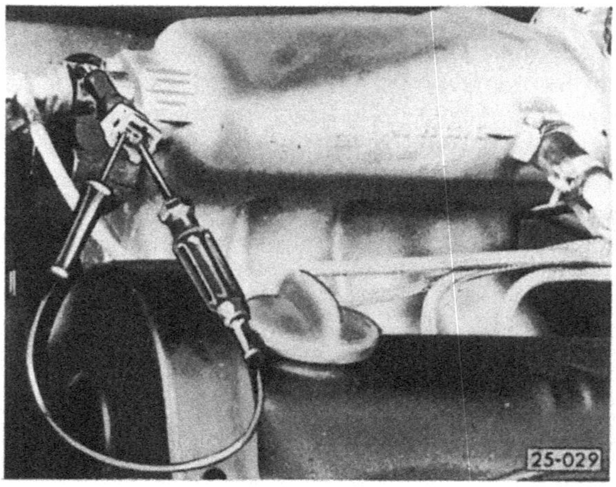

Fig. 5-8. Test light connected across cold-start valve harness connector.

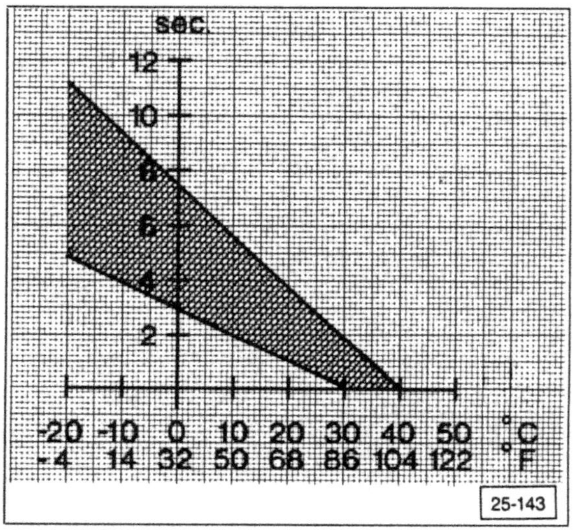

Fig. 5-9. Graph of thermo-time switch response to coolant temperature. Switch completes circuit to open cold-start valve for number of seconds shown. Example: at coolant temperature of 68°F (20°C), switch is closed for 1 to 3.7 seconds.

> **NOTE —**
> If the test light does not light, it is either because voltage is not reaching the connector due to faulty wiring, or because the path to ground (through the thermo-time switch) is interrupted by a faulty switch or wire. Consult the current flow diagrams in **ELECTRICAL SYSTEM** for help in identifying wires and terminals.

To replace the thermo-time switch, first drain the coolant from the engine as described in **COOLING SYSTEM**. Use a thread sealant and torque the new switch to 30 Nm (22 ft. lb.). Then refill the cooling system with coolant.

5.3 Hot Start Pulse Relay System

The hot start pulse relay is used as an aid to starting the engine when it is hot. The relay operates the cold-start valve in pulses by providing an intermittent ground whenever the starter is operated. The hot start pulse relay is located in position 14 on the fuse/relay panel; the cold-start valve is shown above in Fig. 5-6.

If the hot start pulse relay fails to operate during hot starting, it will be difficult or impossible to start the engine. A leaking cold-start valve may also cause hot starting problems, and should be tested as described above in **5.2 Cold Start Enrichment**.

To test hot start pulse relay:

1. Run the engine until the coolant temperature is at least 104°F (40°C).
2. Disconnect the coil wire from the center tower of the distributor cap and ground it to the engine, using a ground wire.
3. Remove the two screws holding the cold-start valve to the intake manifold. Without disconnecting the fuel line or the harness connector, remove the cold start valve from the intake manifold and point the valve into a transparent container as shown above in Fig. 5-7.
4. Observe the cold start valve while a helper actuates the starter. If the system is operating correctly, after approximately two seconds the cold-start valve will spray fuel in regular pulses.
5. If the valve does not pulse, remove the harness connector, connect a test light as shown above in Fig. 5-8, and actuate the starter. If the test light pulses, then the cold-start valve is faulty and should be replaced. If the test light does not pulse, it is either because the pulse relay is faulty, or because voltage or ground for the circuit is faulty. For more information on relay and circuit testing, see **ELECTRICAL SYSTEM**.
6. Reconnect coil wire when finished.

5.4 Cold Running Enrichment

To compensate for cold running conditions, the control pressure regulator provides a richer mixture (higher proportion of fuel) when the engine is cold. The function of the control pressure regulator is similar to that of the choke on a carburetor. The control pressure regulator is mounted to the cylinder block near the oil filter flange. See Fig. 5-5 above.

Fig. 5-10 is a schematic diagram that shows the relationship of the control pressure regulator to the fuel distributor. When the engine is cold, the control pressure regulator allows more fuel to be delivered to the injectors by reducing the control pressure above the control plunger. Above the desired pressure, the control pressure regulator bypasses excess fuel to the fuel return line.

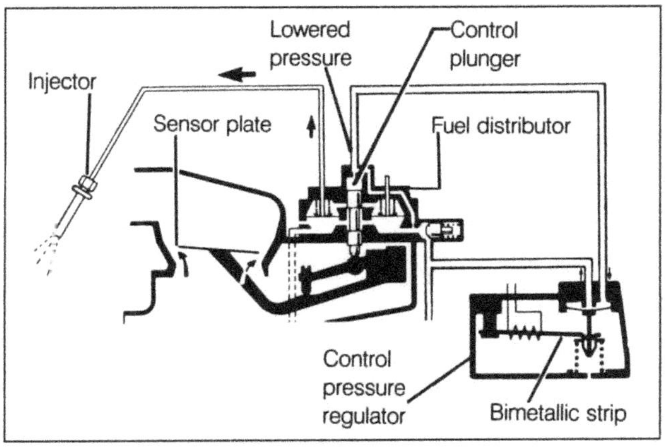

Fig. 5-10. Schematic diagram of control pressure regulator and fuel distributor. When powered with ignition on and engine running, heating element reduces valve opening to increase control pressure to normal warm-engine level.

When the engine is cold, a bimetallic strip holds the regulator wide open to minimize control pressure and provide the richest mixture. When the engine is started and running, electric current warms a heating element that surrounds the bimetallic strip. This in turn slowly closes the regulator's valve, which increases control pressure and leans the fuel mixture. When the regulator is fully warmed, the valve is fully closed, and the control pressure reaches a fixed value.

Testing and Replacing Control Pressure Regulator

To test the heating element of the control pressure regulator, disconnect the harness connector and measure the resistance across the regulator terminals, as shown in Fig. 5-11. If there is an open circuit (no continuity), the control pressure regulator is faulty and should be replaced.

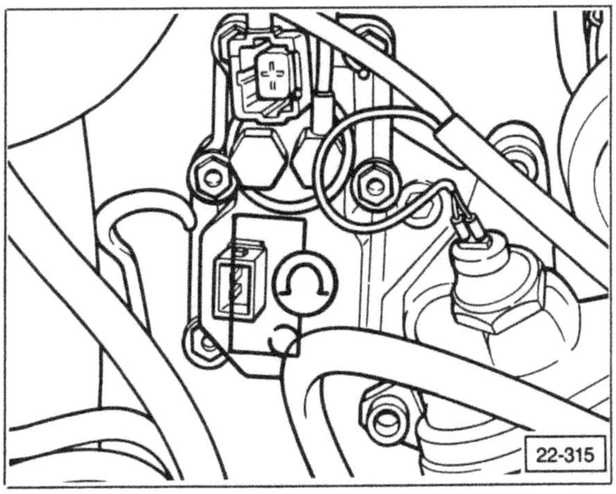

Fig. 5-11. Resistance of control pressure regulator heating element being checked with ohmmeter (shown schematically).

Also check to see that the control pressure regulator heating element is receiving voltage. Disconnect the harness connector and ground the coil wire as described above to prevent the engine from starting. When the starter is actuated, there should be a minimum of 8 VDC across the harness connector, as shown in Fig. 5-12.

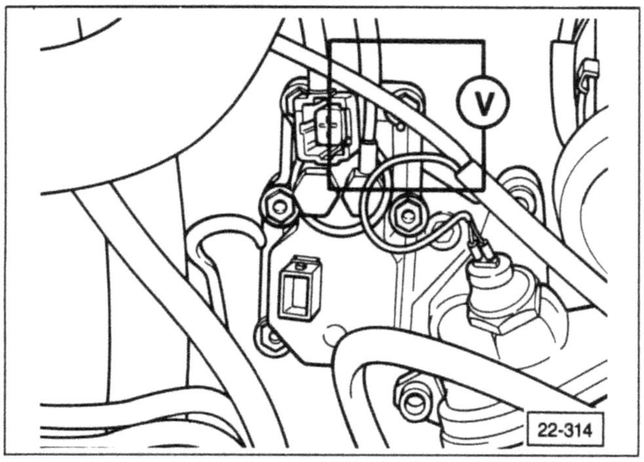

Fig. 5-12. Voltmeter (shown schematically) being used to check for voltage at control pressure regulator harness connector.

To replace the control pressure regulator, disconnect the fuel union bolts from the regulator, disconnect the electrical plug, and remove the mounting bolts that hold the control pressure regulator to the engine block. Use new sealing washers. Torque the larger fuel line union bolt and the two mounting bolts to 20 Nm (15 ft. lb.). Torque the smaller fuel line union bolt to 10 Nm (87 in. lb.).

> ***WARNING* —**
> *Fuel will be expelled as fuel unions are disconnected. Do not smoke or work near heaters or other fire hazards. Keep a fire extinguisher handy.*

5.5 Idle Speed

Idle speed is determined by the small amount of air that enters the engine intake when the throttle is closed. At idle, the throttle valve is fully closed. Idle air enters through a small port that allows air to bypass the closed throttle valve. Idle speed is adjusted by making this idle air bypass more open or more closed. Neither the accelerator cable nor the throttle valve regulate idle speed. The idle air bypass is shown in Fig. 5-13.

Additional inlet air for cold running is supplied by the auxiliary air regulator, which allows more air to bypass the throttle valve. Fig. 5-6 earlier in this section shows both the idle air bypass adjusting screw and the auxiliary air regulator.

A separate idle speed boost valve helps maintain the correct idle speed under all conditions. It boosts idle air to compensate for higher engine loads at idle due to, for example, high power demand from the electrical charging system. Cars with air conditioning have two idle speed boost valves.

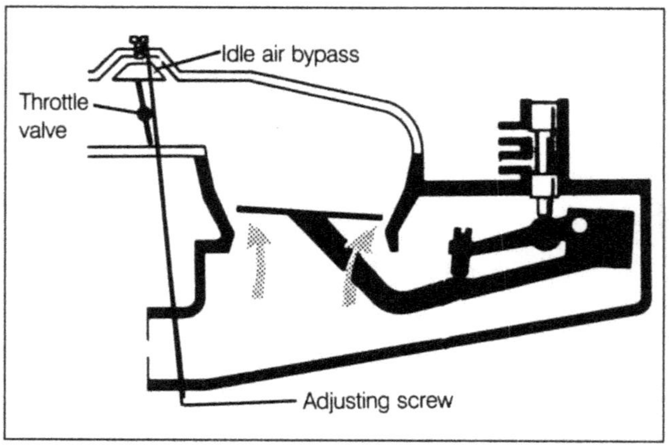

Fig. 5-13. Schematic view of air flow sensor and throttle valve showing idle air bypass and adjusting screw.

Idle Air Bypass Adjustment

Idle speed should always be checked and adjusted along with ignition timing and idle mixture (frequency valve duty cycle and CO content in the exhaust). Adjusting any of these in isolation will affect the others. It may cause driveability problems and increase exhaust emissions. First, make sure that the ignition timing is properly adjusted. See **IGNITION**.

Accurate adjustment of the idle mixture requires a CO meter to measure carbon monoxide in the exhaust. If you do not have access to the equipment necessary to accurately make these checks and adjustments, we suggest turning the job over to an authorized Volkswagen dealer or other qualified repair shop. In a properly equipped shop, these checks and adjustments can be made quickly and accurately.

For idle mixture adjustment specifications, see **5.7 Idle Specifications**. For more information on checking the oxygen sensor system and measuring the frequency valve duty cycle, see **EXHAUST SYSTEM AND EMISSION CONTROLS**.

The following test conditions apply:

1. The engine must be at normal operating temperature (oil temperature at least 176°F (80°C)). The radiator fan should have cycled on and off at least once.
2. All electrical accessories should be switched off. This includes air conditioning and the radiator cooling fan. Check values and make adjustments only when the fan is not running.
3. Fuel pressure measuring equipment should not be connected.
4. The exhaust system must be free of leaks.
5. The oxygen sensor system and frequency valve must be functioning properly (oxygen sensor connected).

NOTE —

If repairs have been made to the fuel system, raise the engine speed over 3000 rpm a few times, then let idle for approximately 2 minutes. This will bleed any trapped air from the fuel system.

Make sure that the auxiliary air regulator is fully closed. The engine rpm should not change when the hose to the auxiliary air regulator is pinched shut. If the rpm drops when the hose is pinched, check the auxiliary air regulator as described below.

Fig. 5-14 shows additional preparations for idle adjustment. Pinch off the hose leading from the idle speed boost valve(s). Disconnect the crankcase ventilation hoses from the intake manifold and the air flow sensor. Remove the charcoal canister hose from the "T" fitting on the air flow sensor boot.

Connect a tachometer according to the instrument manufacturer's instructions. The ignition signal lead from the tachometer should be connected to terminal 1 of the coil.

CAUTION —

Connect and disconnect the test leads only with the ignition off.

Fig. 5-14. Preparations for idle speed adjustment. Pinch shut idle boost valve hose (**1**), disconnect crankcase ventilation hose (**2** and **3**), remove hose (**4**) from "T" fitting (**5**).

Adjust the idle speed to 925 ± 75 rpm by turning the idle air bypass screw—clockwise to decrease idle speed, and counterclockwise to increase it. When all adjustments are complete, restore the hose and fitting connections described in Fig. 5-14 to their original configurations.

Auxiliary Air Regulator

Fig. 5-15 is a schematic diagram of the auxiliary air regulator. When open, the auxiliary air regulator lets additional air bypass the throttle valve to raise idle speed. Electrical current warms the heating element and causes the bimetallic strip to deflect gradually, closing the rotary valve and cutting off the additional air. The auxiliary air regulator is mounted on the intake manifold. See Fig. 5-6 above.

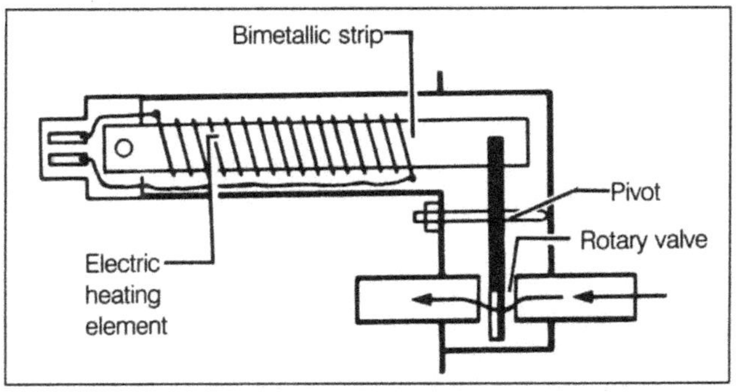

Fig. 5-15. Schematic view of auxiliary air regulator. Arrows indicate direction of air flow through valve.

Check the function of the auxiliary air regulator by checking its effect on engine idle speed under various conditions. When the engine is cold (engine coolant temperature below 86°F (30°C)), disconnect the auxiliary air regulator harness connector. Start the engine and pinch shut the hose from the auxiliary air regulator to the intake manifold as shown in Fig. 5-16. The valve should be open and the engine should lose rpm and perhaps stall.

When the engine is fully warm, with the harness connector attached, pinch shut the hose as shown in Fig. 5-16. The valve should be closed and engine rpm should not change.

Fig. 5-16. Auxiliary air regulator hose being pinched shut with wire-loop pliers.

If in doubt about the results, the valve can be inspected in place using a light and a small mirror. When the valve is open (cold engine), it should be possible to see light through the valve port. When the valve is closed (engine warm), it should not be possible to see through the valve port. If the valve remains open when the engine warms, check to see that the auxiliary air regulator is receiving voltage at the harness connector. There should be battery voltage (approximately 12VDC) at the connector any time the ignition is on.

To remove the auxiliary air regulator, disconnect the harness connector. Loosen the hose clamps and remove the hoses, then remove the socket-head mounting bolts. Installation is reverse of removal. Torque the bolts to 10 Nm (87 in. lb.).

Idle Speed Boost Valve

The idle speed boost valve maintains the idle speed within a 300 rpm range, compensating for varying engine loads at idle. A second valve is used on air-conditioned cars to boost idle speed when the air conditioning is on. The electrically operated valves, shown in Fig. 5-17, allow additional air to bypass the throttle plate and increase idle speed whenever the idle speed drops below 750 rpm. Above 1050 rpm, the valves close and idle speed is reduced.

Fig. 5-17. Idle speed boost valve (**1**) and connecting hoses. Cars with air conditioning are equipped with additional boost valve (**2**). (Shown with connecting hose (**3**) pinched off for test).

The idle speed boost valve is a solenoid valve controlled by an electronic control unit on the fuse/relay panel (position 16) that turns the valve on and off according to an engine speed signal from the coil. Below 750 rpm, the control unit sends a voltage signal which opens the valve. The valve should open with an audible click. When engine speed exceeds 1050 rpm, the control unit cuts the voltage signal to the valve and it closes. The second boost valve on cars with air conditioning is switched on and off at the same time as the air conditioning compressor.

If the valve does not respond as specified, check the voltage signal to the valve at the harness connector, using a test light or a voltmeter. If the valve is receiving a voltage signal below 750 rpm and not opening, the valve is faulty and should be replaced. If the valve is not receiving voltage below 750 rpm, check the signals to and from the control unit as described below.

Using the terminal numbers on the back of the control unit as a guide, test at the fuse/relay panel sockets. Test for continuity to ground at the socket for terminal 31. With the ignition on, check for battery voltage (approximately 12 VDC) at the socket for terminal 15. With the engine running, test for voltage between the socket for terminal 1 and ground. Also check the continuity of the wiring between the panel and the valve's harness connector. If no wiring faults are discovered, and the system does not function properly, then the control unit is faulty and should be replaced. See the current flow diagrams in **ELECTRICAL SYSTEM** for help in troubleshooting the wiring.

5.6 Fuel Pressure Tests and Specifications

As described earlier in this section, much of the function of CIS is dependent on precise fuel pressure. Fuel pressure influences all engine operating characteristics, such as idle, partial-throttle and full-throttle response, starting and warm-up, engine power, and emission levels. Any significant troubleshooting or repair of the system requires the use of a pressure gauge to measure fuel pressure in different parts of the system under different operating conditions.

There are three significant CIS fuel pressure values: 1) System pressure—the basic fuel pressure created by the main fuel pump and maintained by the pressure relief valve in the fuel distributor, 2) Control pressure—determined by the control pressure regulator and used to counter system pressure and regulate the movement of the control plunger, and 3) Residual pressure—the amount of pressure which remains in the closed system after the engine (and fuel pump) are shut off. Testing of each of these pressures is described below.

The pressure gauge shown in Fig. 5-18 (Volkswagen special tool no. VW 1318) has a range of 0 to 10 bar (150 psi) and a valve in one line. It is connected between the fuel distributor and the control pressure regulator. At the fuel distributor, disconnect the line that runs from the center of the fuel distributor to the control pressure regulator. Connect the gauge line without the valve to the fuel distributor. Connect the gauge line with the valve to the end of the control pressure regulator line. Bleed any air from the gauge by hanging it upside down while opening and closing the valve several times. Leave the gauge connected this way for all three pressure tests.

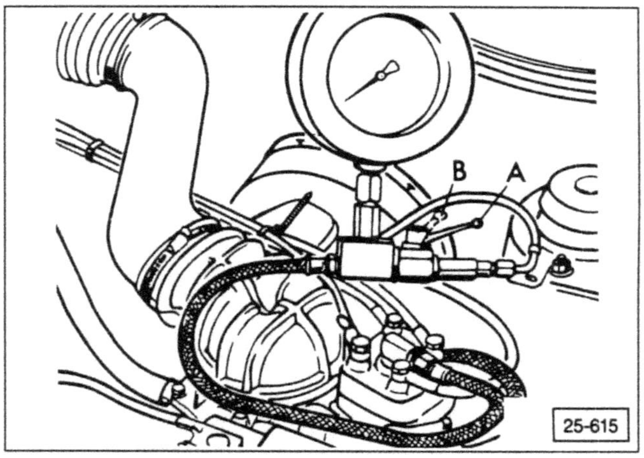

Fig. 5-18. Pressure gauge used to test system pressure, control pressure, and residual pressure. Gauge is open in position **A**, closed in position **B**.

WARNING —
Fuel will be expelled when fuel lines are disconnected. Do not disconnect wires that could cause sparks. Do not smoke or work near heaters or other fire hazards. Keep a fire extinguisher handy.

CAUTION —
Absolute cleanliness is essential when working with fuel circuit components of the CIS. Even a minute particle of dirt can cause trouble if it reaches an injector. Before disconnecting any of the fuel line connections, thoroughly clean the unions. Use clean tools.

System Pressure

To measure system pressure, install a pressure gauge as described above and close the valve in the line to the control pressure regulator. Start the engine and let it idle. System pressure should be 4.7 to 5.4 bar (68 to 77 psi). System pressure is preset during fuel distributor manufacture and should not require routine adjustment.

If system pressure is too low, look for fuel leaks, a clogged fuel filter, or a damaged fuel line blocking fuel flow. Check fuel pump delivery rate as described in **4.3 Fuel Pump**. If no other cause for low system pressure can be found, the pressure can be adjusted by adding shims to the pressure relief valve. See Fig. 5-19. An additional 0.50 mm (.020 in.) shim will increase system pressure by about 0.30 bar (4 psi). An additional 1.00 mm (.040 in.) shim will increase it by about 0.60 bar (8 psi).

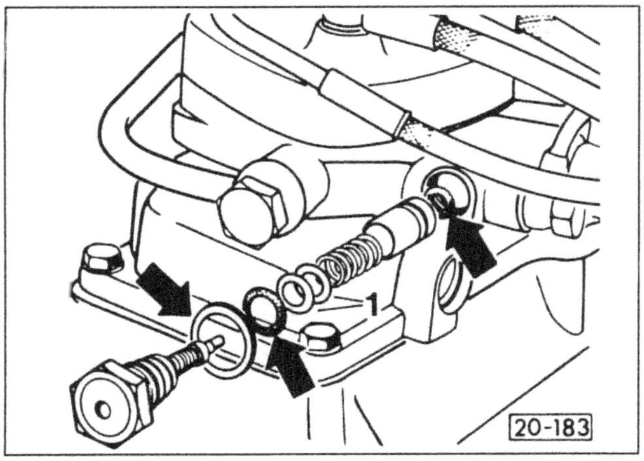

Fig. 5-19. Exploded view of pressure relief valve showing location of adjusting shims at (**1**). Always replace the sealing washer and the O-rings (arrows).

If system pressure is too high, check for a blocked or damaged fuel return line. If the fuel return line is in good condition, the pressure can be adjusted by reducing the thickness of the shims on the pressure relief valve. See Fig. 5-19 above. A change of 0.50 mm (.020 in.) total shim thickness will decrease system pressure by about 0.30 bar (4 psi). A change of 1.00 mm (.040 in.) total shim thickness will decrease it by about 0.60 bar (8 psi).

If the fuel pressure cannot be accurately adjusted, the fuel distributor is faulty and should be replaced. See **5.1 Air Flow Measurement and Fuel Metering**.

Control Pressure

Control pressure is that which is determined by the control pressure regulator. It helps regulate the response of the control plunger and, therefore, influences the fuel mixture. Measurement of control pressure is the primary way of evaluating the control pressure regulator. Control pressure is not adjustable, and if the pressure is not within specifications the control pressure regulator should be replaced.

System pressure influences control pressure, and should be checked, as described above, prior to checking control pressure. Check control pressure with the pressure gauge connected as described above at the beginning of **5.6 Fuel Pressure Tests and Specifications**, with the gauge valve open. A major function of the control pressure regulator is to adapt the fuel mixture for cold starting and cold running, so control pressure should first be checked with the engine completely cold.

To check cold control pressure, disconnect the harness connectors from control pressure regulator and the auxiliary air regulator, then start the engine. Cold control pressure varies with temperature, and should correspond to the chart in Fig. 5-20.

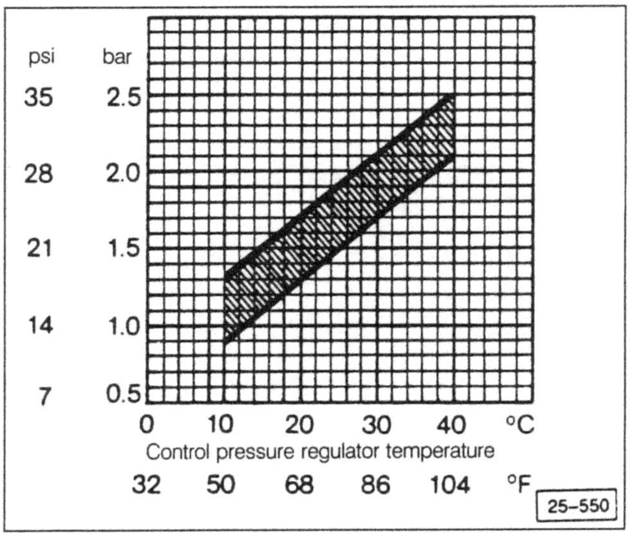

Fig. 5-20. Graph of cold control pressure vs. temperature. Example: at 86°F (30°C), control pressure should be between 1.7 and 2.1 bar (24 and 30 psi).

If cold control pressure is too high, check for a blocked or kinked fuel line. Also, remove the fuel union at the control pressure regulator and check for a clogged filter screen. If no such faults are found, the control pressure regulator is faulty and should be replaced.

WARNING —

Fuel will be expelled as the unions are disconnected. Do not smoke or work near heaters or other fire hazards. Keep a fire extinguisher handy.

To check warm control pressure, reconnect the harness connectors to the control pressure regulator and the auxiliary air regulator. Run the engine until the control pressure is no longer increasing (about 2 minutes). The warm control pressure should be between 3.4 and 3.8 bar (49 and 55 psi).

If the warm control pressure is too high, check for a blocked or kinked fuel line. Also, remove the fuel union at the control pressure regulator and check for a clogged filter screen. If no such faults are found, the control pressure regulator is faulty and should be replaced.

If warm control pressure is too low, or takes more than about 2 minutes to reach its peak, the fuel return line may be blocked. Also, test the resistance of the heating element and test for voltage reaching the harness connector, as described in **5.4 Cold Running Enrichment**.

Residual Pressure

The closed fuel system is designed to maintain pressure after the engine is shut off to help prevent the fuel in the injector lines from overheating, causing vapor lock and hard starting. The fuel pump's one-way check valve and the fuel accumulator help maintain this residual pressure. See **4. Fuel Supply**.

Check residual pressure with the pressure gauge connected as described above for measuring control pressure. When the engine is fully warm (control pressure between 3.4 and 3.8 bar (49 and 55 psi), shut off the engine. Leave the gauge connected.

After ten minutes, the pressure should not have dropped below 2.6 bar (38 psi). If the pressure drops excessively, check for leaks in the fuel lines, the fuel distributor including the pressure relief valve O-rings, the injectors, the cold-start valve, and the oxygen sensor frequency valve. To further isolate the cause, check residual pressure at the fuel supply line from the pump. See Fig. 5-5, shown earlier. Disconnect the gauge from the fuel distributor and the control pressure regulator line, and reconnect that line to the fuel distributor. Connect the pressure gauge as shown in Fig. 5-21.

WARNING —

The gauge line with the valve is left disconnected. Be sure to close the valve.

Run the fuel pump using a fused jumper wire at the fuse/relay panel, as described in **4.1 Fuses and Relays**, until fuel pressure indicated by the gauge reaches 5 bar (72.5 psi). Then turn the fuel pump off. Once again, after ten minutes, the pressure should not have dropped below 2.6 bar (38 psi).

If the pressure drops excessively and no leaks are found between the fuel pump and the gauge, repeat the test. This time shut off the pump and immediately pinch closed the fuel hose between the tank and the pump. If residual pressure now stays within specifications, the fuel pump check valve is faulty and should be replaced. See **4.3 Fuel Pump**.

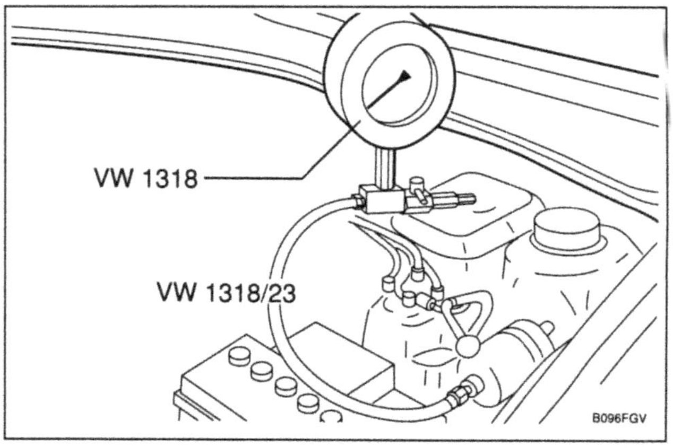

Fig. 5-21. Pressure gauge connected to fuel supply line from fuel pump for residual pressure test of fuel pump, check valve, fuel accumulator, and fuel line. Gauge valve must be closed. Gauge shown (Volkswagen special tool no. VW 1318) is connected using adapter (Volkswagen special tool no. VW 1318/23).

If the residual pressure still drops off too quickly, the fuel accumulator may be faulty. For a quick check, remove the plug from the end of the accumulator. If fuel drips out, then the diaphragm is leaking and the accumulator should be replaced.

5.7 Idle Specifications (rpm and % CO)

Idle mixture (frequency valve duty cycle and CO content) should only be adjusted with idle speed and ignition timing. The conditions and set-up required for these adjustments are described in **5.5 Idle Speed**. For information on checking the oxygen sensor and measuring frequency valve duty cycle, see **EXHAUST SYSTEM AND EMISSION CONTROLS**.

If you do not have access to the equipment necessary to accurately make these checks and adjustments, we suggest turning the job over to an authorized Volkswagen dealer or other qualified repair shop. In a properly equipped shop, these checks and adjustments can be made quickly and accurately.

The idle mixture adjustment changes the relationship between the air flow sensor plate and the control plunger in the fuel distributor. For any particular air flow or sensor plate position, the adjustment raises or lowers the control plunger. This increases or decreases the fuel delivered to the injectors for a particular intake air flow volume. The adjusting screw is inside the air flow sensor, as shown in Fig. 5-22.

The idle mixture adjusting screw can only be adjusted after removing the anti-tampering plug in the top of the air flow sensor housing. This small plug has a soft aluminum top and a hard steel bottom. Remove the plug by drilling a 2.5 mm ($^3/_{32}$ in.) hole in the aluminum only. Use a sheet metal screw threaded into the aluminum to extract the plug.

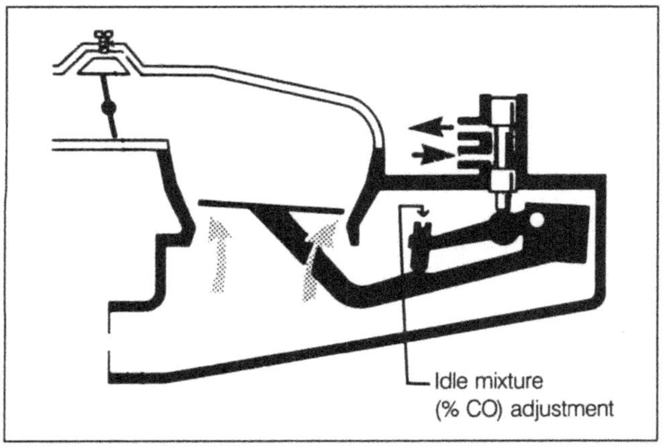

Fig. 5-22. Cut-away view of air flow sensor showing idle mixture (% CO) adjusting screw.

CAUTION —
Do not drill through the steel bottom. Do not allow metal shavings to fall into the air flow sensor.

With the plug removed, the adjustment is made using a long 3 mm hex wrench as shown in Fig. 5-23. Turning the screw clockwise makes the mixture richer (higher % CO). Turning it counterclockwise makes the mixture leaner (lower % CO). **Table c** lists idle speed and % CO specifications. A new plug should be installed when the adjustment is complete.

CAUTION —
Do not accelerate the engine or push on the adjusting wrench while it is in place. Doing so may damage the air flow sensor. Remove the wrench and briefly accelerate the engine after each adjustment.

NOTE —
The U.S. government and some states consider it "illegal tampering" for the car to leave a commercial repair facility without the anti-tampering plug correctly installed.

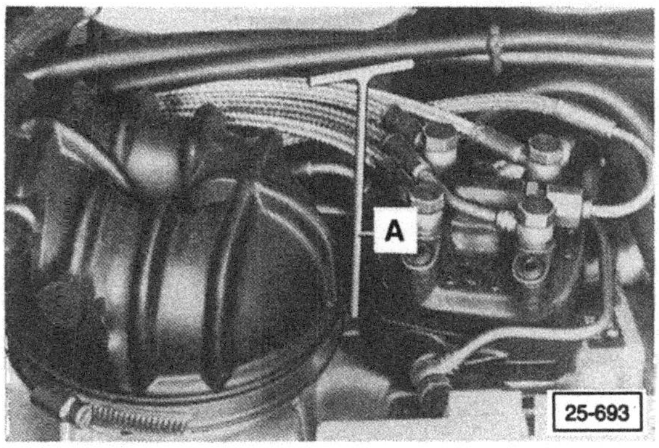

Fig. 5-23. Engine idle mixture (% CO) being adjusted with 3 mm hex wrench (**A**).

Table c. Idle Specifications (CIS)

Engine code JH	Checking value	Adjusting value
Idle speed	800 to 1000 rpm	925 ± 75 rpm
% CO (measured upstream of catalytic converter at exhaust manifold CO tap tube)	0.3% to 1.2%	–
Frequency valve duty cycle	25% to 65% (fluctuating)	50% ± 8% (average)
duty cycle (measured with dwell meter)	23° to 59° (fluctuating)	45° ± 7° (average)

6. CONTINUOUS INJECTION SYSTEM – ELECTRONIC (CIS-E)

Fig. 6-1 is a schematic view of CIS-E fuel injection. Much like the simpler CIS fuel injection described earlier, the basic part of CIS-E is mechanical, utilizing the influence and control of fuel pressure to accurately meter fuel. As the system's name implies, however, electronic control also plays a role in precisely adapting the fuel mixture to the running conditions.

The continuous injection system - electronic (CIS-E) is called continuous because, unlike many other fuel injection systems, the injectors do not open and close—they deliver fuel continuously. The main function of the fuel injectors is to atomize the flow of fuel being supplied under pressure from the fuel distributor. CIS-E differs from the earlier CIS system in that it also uses electronic controls to regulate fuel pressures and therefore control fuel mixture more precisely.

The heart of CIS-E is the mixture control unit, consisting of the air flow sensor and the fuel distributor. Air entering the engine raises the air flow sensor plate. A lever connected to the sensor plate raises the control plunger in the fuel distributor, increasing the quantity of fuel flowing to the injectors in proportion to the air flow. The differential pressure regulator is controlled by a fluctuating signal from the oxygen sensor control unit, based on inputs from various sensors. By regulating electric current flow in the differential pressure regulator circuit, fuel pressure in the lower chamber of the fuel distributor is used to control the volume of fuel delivered to the injectors by the metering valves.

In addition to the differential pressure regulator's constant inputs, the cold start valve, the idle air stabilizer, and the air flow sensor plate potentiometer (not shown in Fig. 6-1) help adapt the system to different operating conditions. The cold-start valve is like a fifth injector, and performs a choke-like function. It injects additional fuel into the intake manifold during starting to provide a richer fuel mixture when the engine is cold. The cold-start valve is operated electrically.

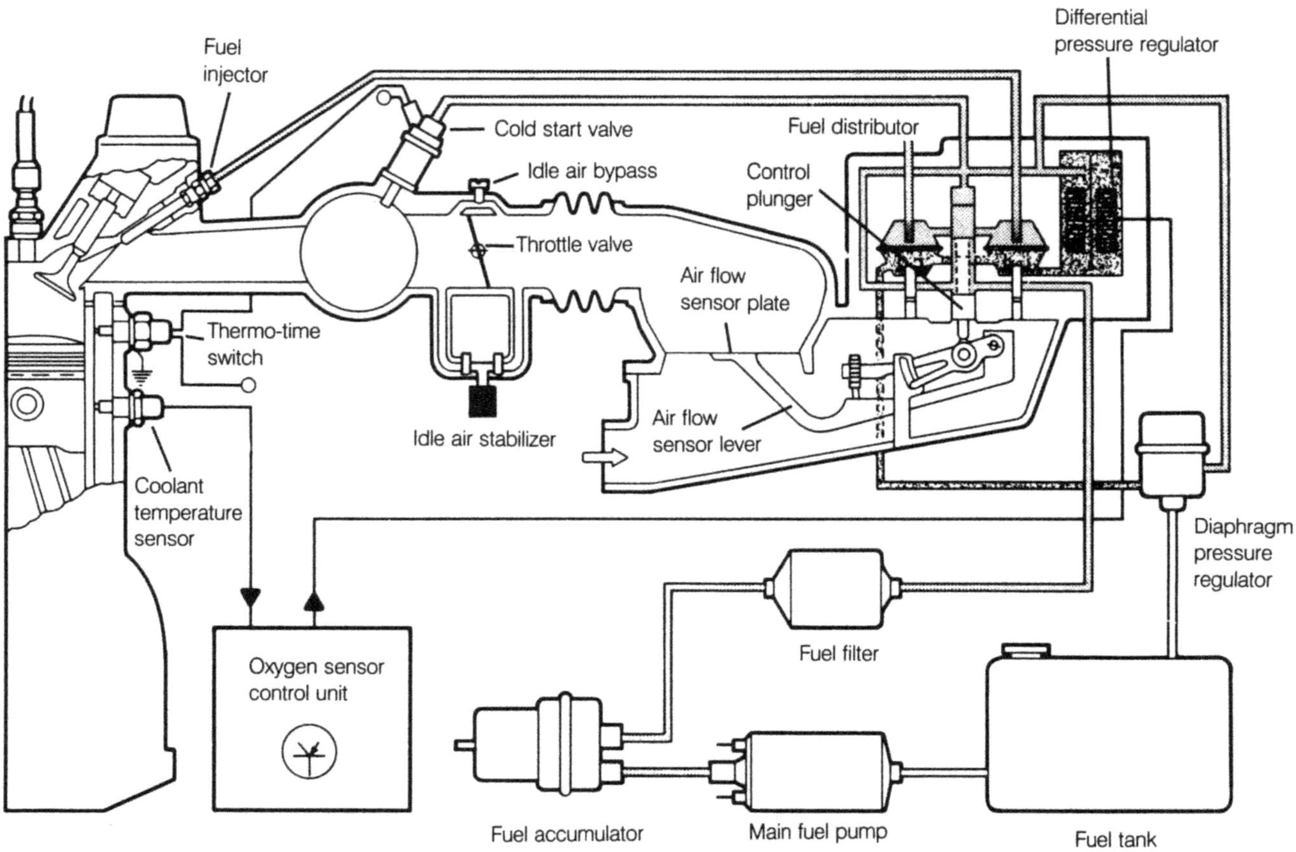

Fig. 6-1 Schematic view of CIS-E fuel injection.

The idle air stabilizer maintains a constant idle speed under all conditions by controlling the amount of additional air that bypasses the throttle plate.

The air flow sensor plate potentiometer is a sensor which provides the control unit with a signal indicating sensor plate position. The signal is primarily used for cold acceleration enrichment.

CIS-E is designed so that all the electrical components can be tested using only a low-current LED test light and a multimeter, but the basic fuel metering system is highly dependent on fuel pressure. Evaluation of the system's basic function requires the use of a fuel pressure gauge. See **6.5 Fuel Pressure Tests and Specifications** for more information.

6.1 Air Flow Measurement and Fuel Metering

The throttle valve, which is operated by the accelerator cable, regulates the amount of air drawn in by the engine. The most basic function of the fuel injection system is to measure that incoming air and precisely meter a proportional amount of fuel. The components responsible for air flow measurement and fuel metering are the air flow sensor and the fuel distributor.

The throttle valve stop screw is adjusted at the factory. It does not require routine adjustment and should not be used to adjust idle speed. The procedure below for the throttle valve basic adjustment should be used only in the event that the original factory adjustment has been altered. The air flow sensor plate can be adjusted if it is found to be out of alignment. After the adjustment, idle speed and the idle mixture adjustment will need to be corrected. The fuel distributor is precisely calibrated at the factory. Except for adjusting system pressure, it cannot be adjusted or repaired.

Throttle Valve Basic Adjustment

The purpose of the throttle valve adjusting screw is to provide a positive stop and prevent the valve from closing so far that it becomes worn or damaged.

To correct a faulty throttle valve adjustment, use a screwdriver to back off the throttle valve adjusting screw until there is clearance between its tip and the throttle valve lever. The adjusting screw is shown in Fig. 6-2.

Place a thin piece of paper between the throttle valve adjusting screw and the throttle valve lever. With the throttle valve closed, turn the screw in until it lightly contacts the paper. From this position, remove the paper and turn the screw in an additional ½ turn.

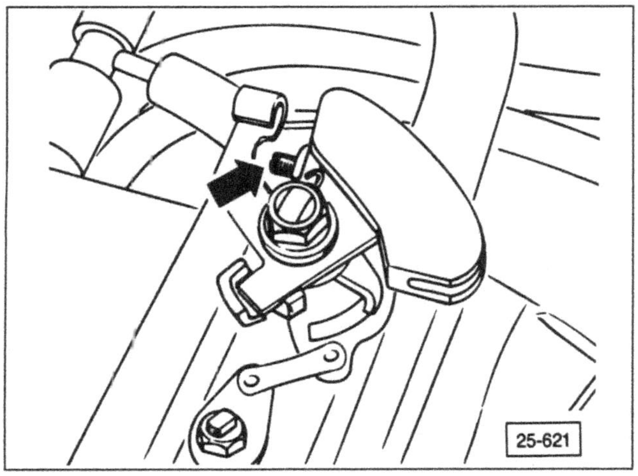

Fig. 6-2. Location of adjusting screw (arrow) for throttle valve basic adjustment (not used to adjust idle speed).

Throttle Switches

Two switches, an idle switch and a full throttle switch, are mounted on the throttle valve housing. The idle switch supplies the control unit with a signal when the throttle is fully closed, for idle air stabilizer control and deceleration fuel shut-off. The full-throttle switch supplies the control unit with a signal when the throttle valve is fully open, for full-throttle enrichment.

Fig. 6-3 shows Volkswagen special tool no. VW 1501 (order no. TV1 501 000 24 ZEL) attached to the throttle switch harness connectors. The harness, if available, is a useful test aid, but the switches can be tested without it.

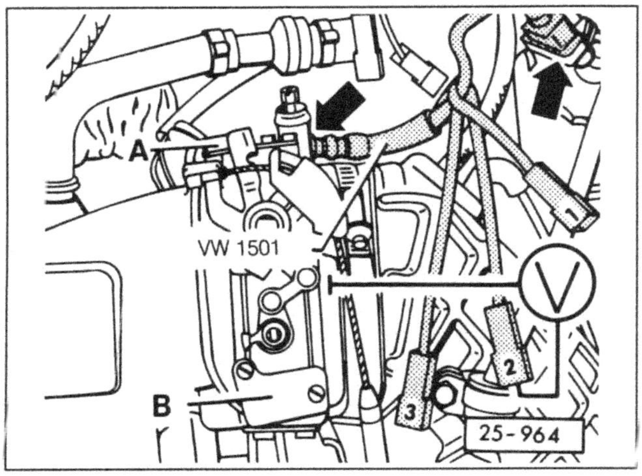

Fig. 6-3. Test harness (Volkswagen special tool no. VW 1501) connected to both sides of harness connector (arrows) for idle switch (**A**) and full throttle switch (**B**). Voltmeter shown here schematically.

First, check to see that voltage is reaching the switches when the ignition is on. Using the test harness, check for voltage between test lead no. 2 and ground. Without the test harness, check for voltage between the center terminal of the harness connector and ground. With the ignition on, there should be battery voltage (approximately 12 VDC). If not, check for a failed fuse (no. 17).

Next, check the switches. Using the test harness, check for voltage between each of the other two test leads and ground to indicate when each switch is open or closed. Test lead no. 1 is for the idle switch. Test lead no. 3 is for the full throttle switch.

Without the test harness, check at the switch connector for continuity between the center terminal and each of the outer terminals to indicate when each switch is open or closed. Terminal 1 is for the idle switch. Terminal 3 is for the full-throttle switch. Use the wire colors indicated on the current flow diagrams in **ELECTRICAL SYSTEM** to help identify the proper terminals.

To check the idle switch, open the throttle valve about half-way and slowly let it close. The idle switch should close, completing the circuit (and indicating voltage or continuity), when the throttle valve lever gets to within 0.15 to 0.5 mm (.006 to .020 in.) of its stop. Check the gap at which the switch closes with a feeler gauge as shown in Fig. 6-4. If necessary, the switch can be repositioned to adjust the switch-on point by loosening the switch mounting screws.

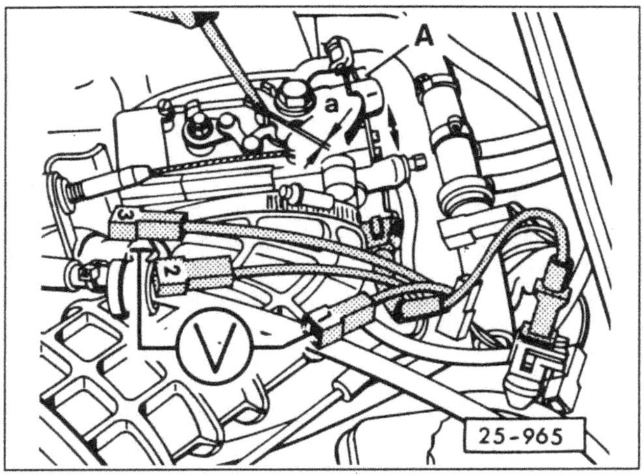

Fig. 6-4. Idle switch **A** being checked and adjusted using feeler gauge to measure gap (**a**) between throttle lever and stop.

To check the full-throttle switch, slowly open the throttle valve to the maximum, full-throttle position. The full-throttle switch should close, completing the circuit (and indicating voltage or continuity), when the throttle is 10°±2° from full open. Check the angle using a protractor as shown in Fig. 6-5. If necessary, the switch can be repositioned to adjust the angle by loosening the switch mounting screws.

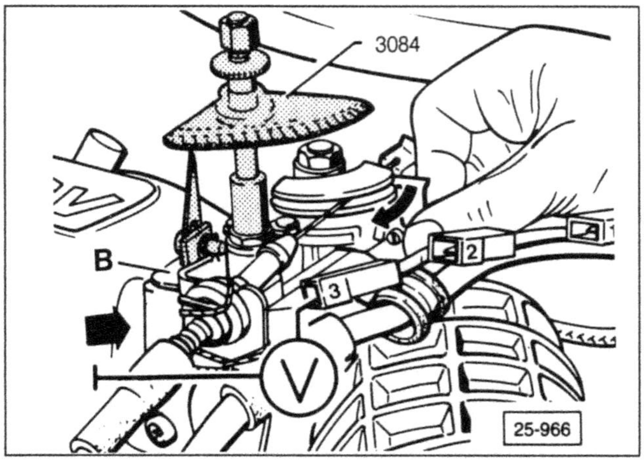

Fig. 6-5. Full throttle switch **B** being checked and adjusted using protractor (left arrow, Volkswagen special tool no. 3084, order no. T03 084 000 25 ZEL).

Air Flow Sensor

The air flow sensor measures all the air being drawn in by the engine. Air flows past the air flow sensor plate and the plate is lifted, in turn lifting the control plunger in the fuel distributor to meter the fuel. The sensor plate and its cone-shaped venturi are carefully designed to achieve the proper balance of air flow and fuel metering for all flow rates from idle to full throttle. Although it is seldom necessary, the sensor plate is easily adjusted. If the sensor plate is off-center or binding, or the lever has too much resistance, the fuel distributor cannot respond properly to the air flow sensor. If the sensor plate is adjusted, the idle speed and idle mixture adjustment should be checked and corrected, as described in **6.3 Idle Speed** and **6.6 Idle Specifications**.

The sensor plate potentiometer provides the control unit with information on the position of the sensor plate. This signal is primarily used for cold acceleration enrichment. For information on repair and adjustment of the air flow sensor plate potentiometer, see **Cold Acceleration Enrichment** under **6.4 Electrical Tests**.

To check the air flow sensor plate with the engine at normal operating temperature, remove the rubber intake air boot from the top of the air flow sensor. Using a magnetic tool, lift the sensor plate slowly and check to see that there is uniform resistance throughout the plate's entire range of travel. Move the sensor plate quickly from its raised position to its rest position. There should be no resistance.

If the sensor plate binds in the air cone, remove the bolt in the center of the sensor plate. Apply thread locking compound to the bolt and reinstall it, centering the sensor plate before tightening the bolt. If the plate cannot be centered, separate the air flow sensor housing from the air filter housing by releasing the wire clamps. Remove the clamp bolt from the end of the air flow sensor lever as indicated in Fig. 6-6. Coat the bolt threads with thread locking compound and reinstall the bolt hand-tight. Center the air flow sensor lever on the shaft and tighten the clamping bolt.

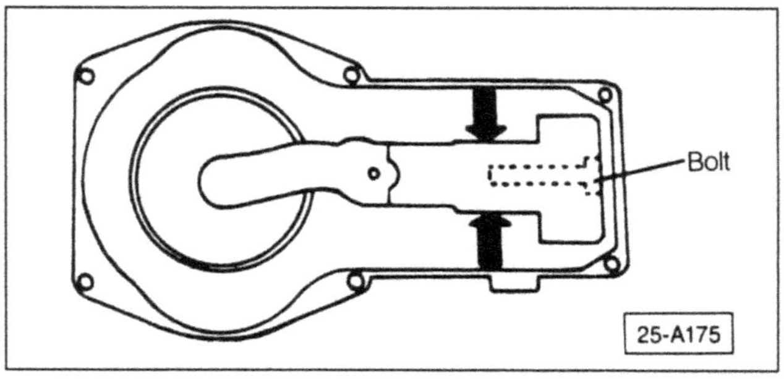

Fig. 6-6. Bottom view of air flow sensor showing location of clamping bolt. Center sensor plate lever (arrows) and tighten clamping bolt.

If the sensor plate is hard to move or moves with uneven resistance, clean and lubricate the sensor plate lever pivot so that it moves smoothly. If the resistance encountered in moving the sensor plate is caused by a binding control plunger, remove and clean the control plunger as described below in **Fuel Distributor**.

Check the sensor plate rest position with the engine at normal operating temperature. Check the position of the plate in the air cone as shown in Fig. 6-7. When correctly adjusted, the sensor plate should be 1.9 mm (.075 in.) below the narrowest point of the air cone. Up to 2.1 mm (.083 in.) is permissible. Adjust the height by bending the wire spring stop indicated in Fig. 6-8.

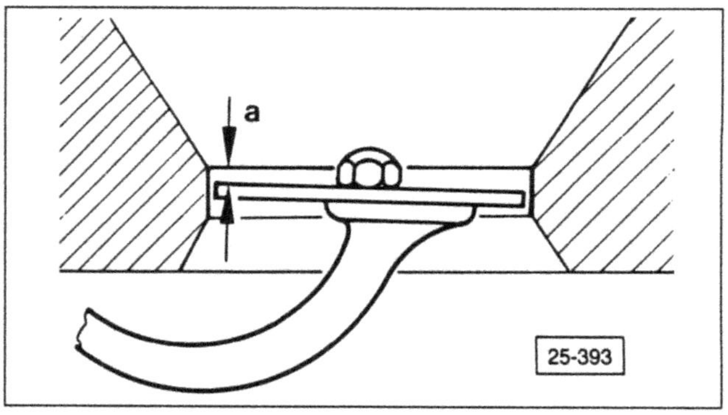

Fig. 6-7. Correct sensor plate position (**a**) is 1.9 mm (.075 in.) below narrowest point of cone. As much as 2.1 mm (.083 in.) is permissible.

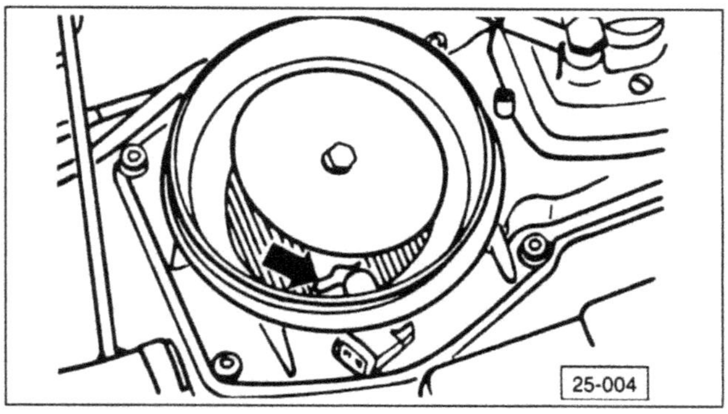

Fig. 6-8. Height of sensor plate can be adjusted by bending wire spring stop (arrow).

NOTE —
An ordinary U.S. nickel (approximately 2 mm thick) can be used to gauge the CIS-E sensor plate height adjustment.

When the engine is turned off, there should be a small gap between the sensor lever and the fuel distributor's control plunger, as shown in Fig. 6-9. This gap ensures that the control plunger seats against the O-ring at the bottom of its bore. The plunger seats on the O-ring to help maintain fuel pressure at the injectors, thus helping to eliminate fuel vaporization and hard starting.

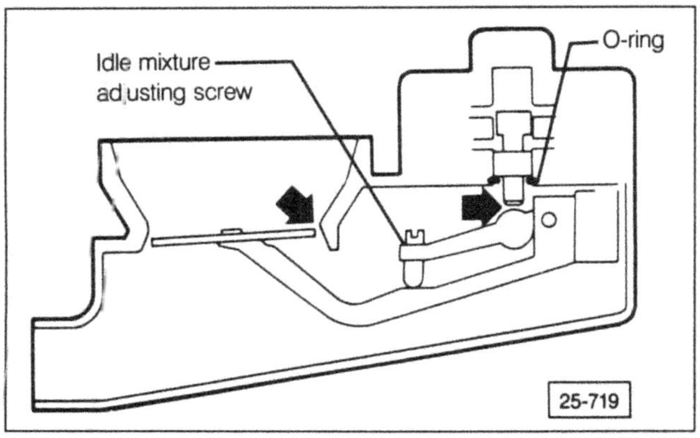

Fig. 6-9. Cut-away view of sensor plate housing showing gap between lever and control plunger (right arrow). Adjust gap by altering position of idle mixture screw. Use sensor plate position (left arrow) to check gap.

Check the gap with the engine at normal operating temperature. Slowly lift the air flow sensor plate until there is some resistance. There should be a small amount of freeplay, but no more than 2.1 mm (.083 in.) at the edge of the sensor plate. In other words, the freeplay should be just enough for the sensor plate to rise and be even with the narrowest point in the air cone. If the freeplay is incorrect, the fuel distributor must be removed from the air flow sensor assembly to make further adjustments, which are described below in **Fuel Distributor**.

Fuel Distributor

The fuel distributor's control plunger, moving up or down in its cylinder, uncovers or covers the metering ports leading to the four metering valves—one for each injector. The differential pressure regulator controls pressure in the lower chambers to control the metering valves which meter fuel to the injectors. A cross-section of the fuel distributor showing the differential pressure regulator and one metering valve is shown in Fig. 6-10.

The fuel distributor is not normally a source of problems. It is, however, a precision device that is easily damaged by plugging or corrosion from contaminated fuel. For fuel filter maintenance information, see **LUBRICATION AND MAINTENANCE**. On some cars, a mini-filter is installed during manufacture in the fuel inlet bolt at the fuel distributor. This filter should be removed and a normal bolt installed in its place at a specified interval. For more information on identifying and replacing the mini-filter, see **LUBRICATION AND MAINTENANCE**. Check fuel distributor function by measuring injection quantity, as described in **4.7 Fuel Injectors**.

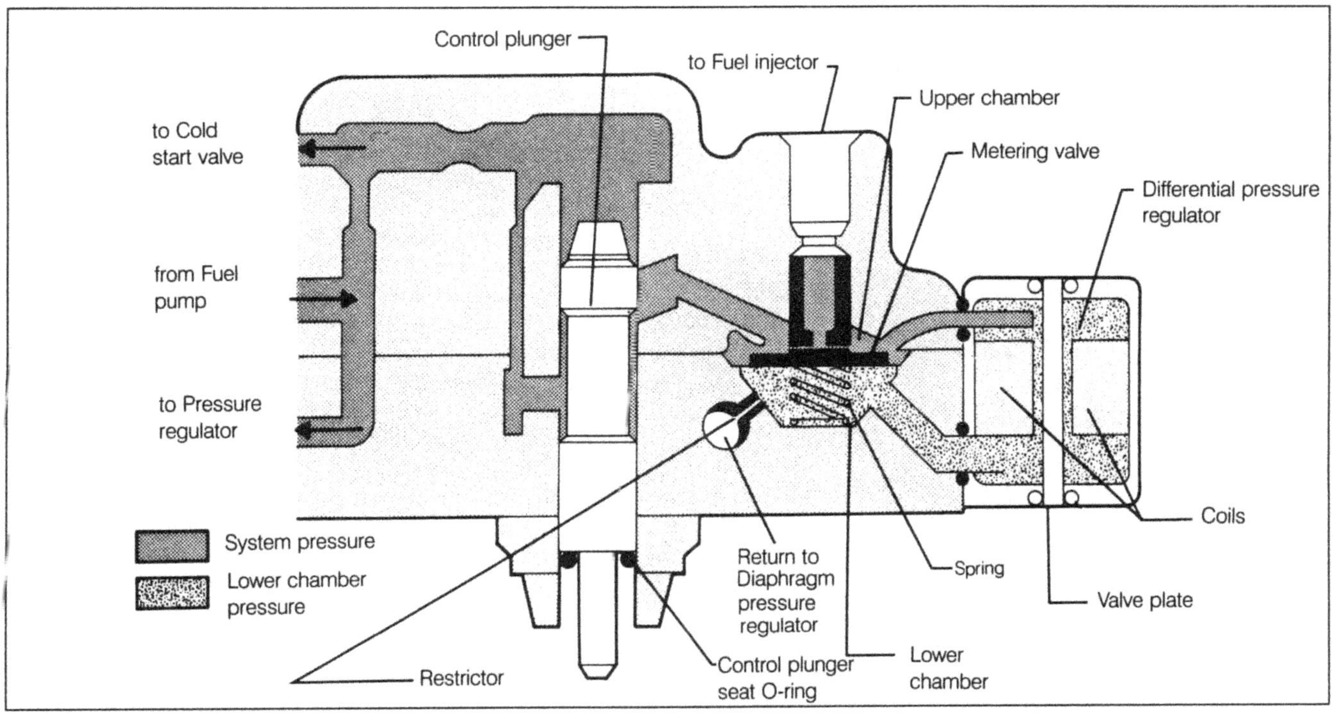

Fig. 6-10. Cross-section of fuel distributor showing differential pressure regulator and one metering valve. Varying pressure in lower chamber regulates metering valve opening and therefore fuel flow to injectors.

Fig. 6-11 illustrates the fuel distributor's fuel line connections and mounting detail. If either the fuel distributor or the air flow sensor is repaired or replaced, the idle speed and idle mixture adjustment must be checked and corrected. See **6.3 Idle Speed** and **6.6 Idle Specifications**. Before making the adjustments, make sure the air flow sensor plate position is correctly adjusted as described above in **Air Flow Sensor**.

If the fuel distributor is removed or replaced, or the air flow sensor plate free play is incorrect as described earlier in **Air Flow Sensor**, basic adjustments to the air flow sensor lever and control plunger will be required. See the checking and adjusting procedure below.

To remove and clean the control plunger, remove the two mounting screws from the top of the fuel distributor and pull the fuel distributor from the air flow sensor assembly. Remove the stop screw and the O-ring. Remove the control plunger. Inspect the O-ring and replace it if it is deformed or otherwise damaged. Clean the plunger with carburetor cleaner or equivalent, and then lubricate the plunger with a light coat of fuel and install it. Install the O-ring with the smaller diameter side facing the control plunger. Install the stop screw and make the control plunger basic adjustment as described below.

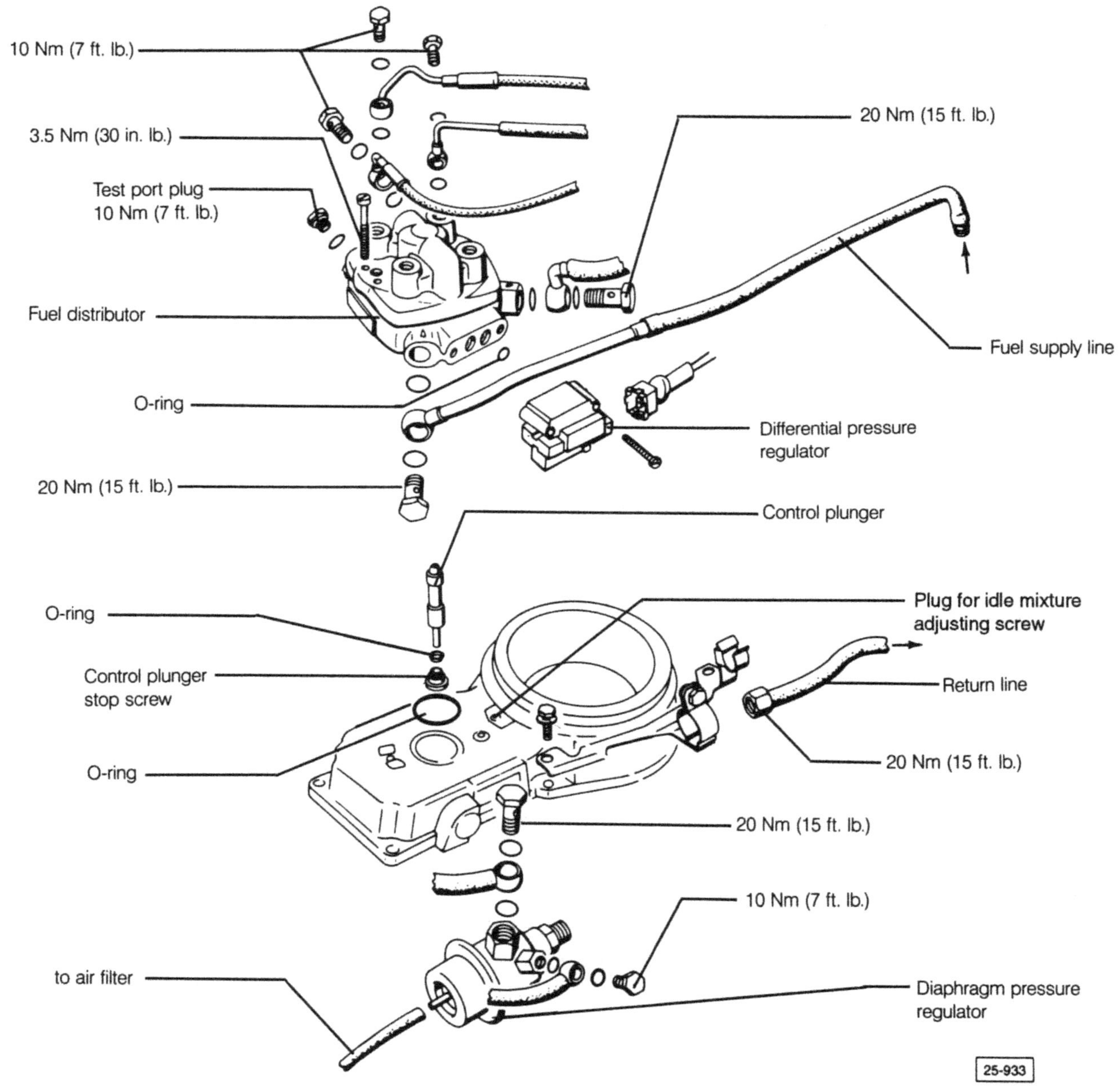

Fig. 6-11. Exploded view of CIS-E fuel distributor and related components. Tightening torques are given for removal and installation of fuel distributor and related components.

WARNING —
Fuel will be expelled when disconnecting fuel unions. Disconnect the battery negative (−) terminal. Do not smoke or work near heaters or other fire hazards. Keep a fire extinguisher handy.

CAUTION —
When removing the fuel distributor from the air flow sensor housing, be careful not to let the control plunger fall. If the control plunger is allowed to fall, it may be damaged—rendering the entire fuel distributor unserviceable. The fuel distributor is precision calibrated at the factory and should not be disassembled.

To check the air flow sensor lever basic adjustment, measure the distance from the fuel distributor mounting surface to the sensor plate lever roller when the sensor plate is at rest, as shown in Fig. 6-12. The height should be 19.0 ± 0.1 mm (0.75 ± .004 in.). If the height measurement is incorrect, remove the tamper-proof plug in the top of the air flow sensor housing and use a 3 mm hex wrench to turn the idle mixture adjusting screw until the measurement is correct. For more information on the mixture adjusting screw, see **6.6 Idle Specifications**.

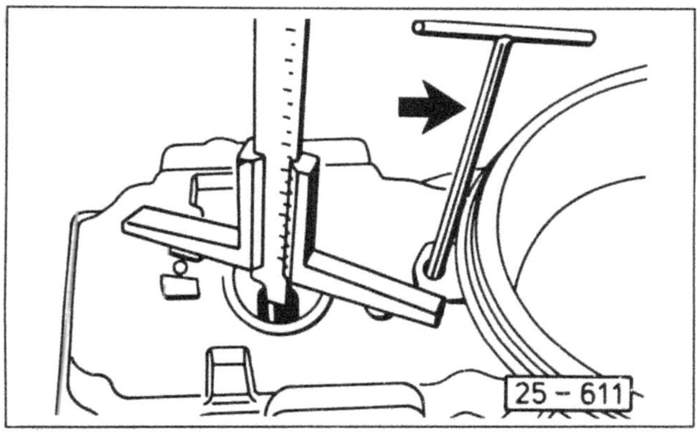

Fig. 6-12. Air flow sensor lever basic adjustment being measured. Adjust by turning idle mixture screw using 3 mm hex wrench (arrow).

To check the control plunger basic adjustment, measure the distance from the gland nut to the stop screw on the bottom of the fuel distributor, as shown in Fig. 6-13. It should be approximately 0.6 mm (.024 in.). If not, adjust it to this height by turning the stop screw. Turning the screw in (clockwise) will increase free play; turning the screw out will reduce free play.

NOTE —
If any adjustments or repairs are made to the fuel distributor or the air flow sensor lever, the idle must be checked as described in **6.3 Idle Speed** and **6.6 Idle Specifications**.

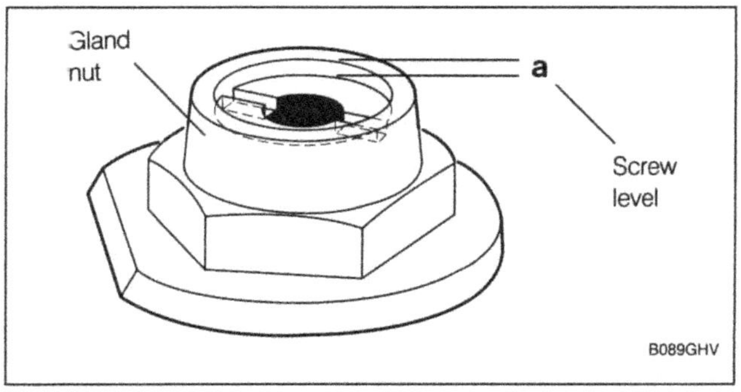

Fig. 6-13. Control plunger basic measurement measured at dimension **a**. (Gland nut and stop screw shown with control plunger omitted for clarity).

Differential Pressure Regulator

The differential pressure regulator controls pressure in the lower chamber of the fuel distributor, thereby controlling the pressure differential across the metering valves and the volume of fuel that is delivered to the injectors. It is operated by the control unit which continuously makes adjustments based on information received from the oxygen sensor and other inputs. Fig. 6-14 is a cut-away view of the fuel distributor showing the differential pressure regulator. Information on testing the actual differential pressure with a pressure gauge is found in **6.5 Fuel Pressure Tests and Specifications**.

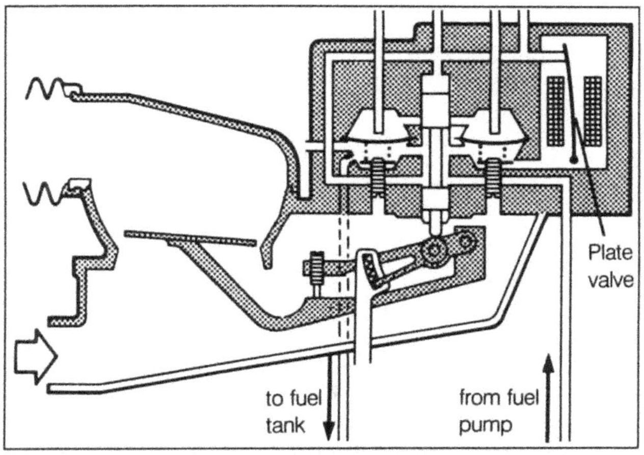

Fig. 6-14. Cut-away view of fuel distributor showing differential pressure regulator. Position of plate valve is controlled by varying electrical current flow.

The differential pressure regulator is an electro-mechanical valve which responds to changes in electrical current flow. The control unit controls it by varying the amount of current flowing in the circuit. In the event that the differential pressure regulator does not receive a signal from the control unit, due to an electrical failure for example, the differential pressure regulator mechanically reverts back to a constant pressure value which will still allow the car to be driven, but perhaps with the engine running roughly.

The action of the differential pressure regulator is the single most important factor controlling and adapting fuel mixture. The measurement of differential pressure regulator current is therefore the most important and significant means of checking the operation of CIS-E fuel injection.

Fig. 6-15 shows the connection of a test harness, Volkswagen special tool no. VW 1315A/1 (order no. TV1 315 0A1 25 ZEL), used to measure differential pressure regulator current. The harness and a schematic representation of the test circuit are shown in Fig. 6-16. Alternatively, a harness can be fabricated to allow hookup of an ammeter as shown.

CAUTION —

The small female terminals in the differential pressure regulator harness connector are easily damaged. If fabricating a test harness, use the appropriate male terminals (Volkswagen part no. N 17 457 2) to make connections with the harness connector.

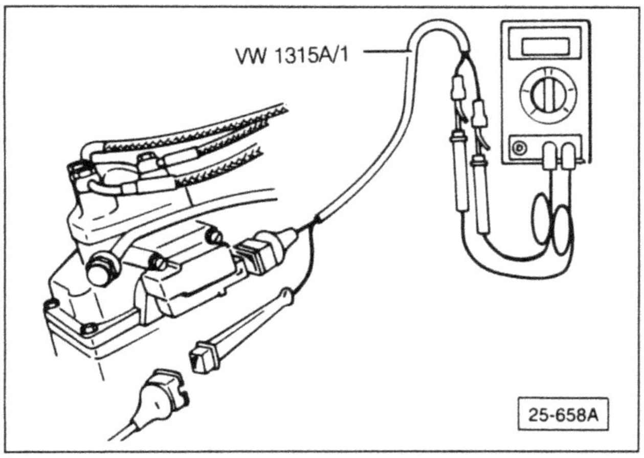

Fig. 6-15. Test harness (Volkswagen special tool no. VW 1315A/1) and multimeter shown connected for measuring differential pressure regulator current.

Complete electrical tests and current specifications are found in **6.4 Electrical Tests**. The differential pressure regulator cannot be repaired and if found to be faulty should be replaced. To replace the regulator, remove the harness connector and the two mounting screws. Remove the regulator and the O-rings from the fuel distributor. Installation is the reverse of removal. Replace the two O-rings.

NOTE —

The differential pressure regulator mounting screws are special anti-magnetic screws. If a screw must be replaced, obtain a correct replacement part from an authorized Volkswagen dealer.

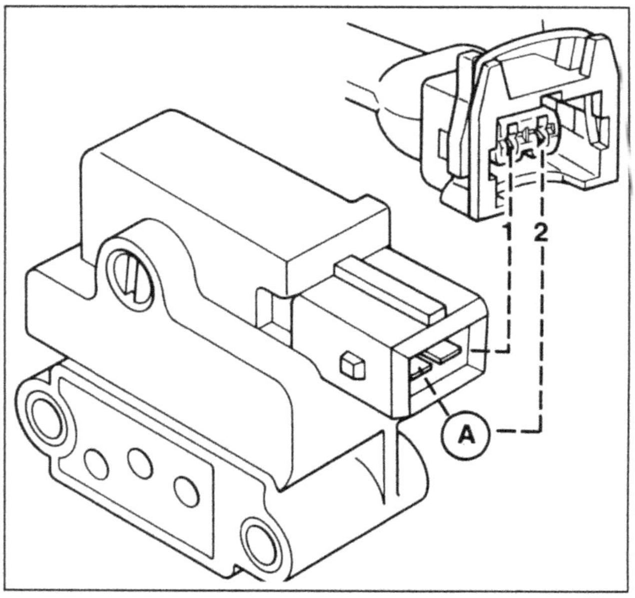

Fig. 6-16. Differential pressure regulator with harness connector removed for connection of ammeter (shown schematically) for current measurement. Numbers shown (**1** and **2**) identify terminals and are found on regulator housing.

6.2 Cold Start Enrichment

The electrically operated cold-start valve sprays extra fuel into the intake manifold for several seconds when the starter is actuated and the engine is cold. The valve is controlled by the thermo-time switch which senses engine coolant temperature. When the engine is cold the switch is closed, allowing power to reach the valve and open it. When the engine is warm, the switch is open and the valve does not operate. To limit valve operation and prevent flooding, the electric current also warms the switch and opens the circuit after a few seconds. On PL engines, the cold-start valve is located in the intake manifold and the thermo-time switch is mounted in the end of the cylinder head, as shown in Fig. 6-17.

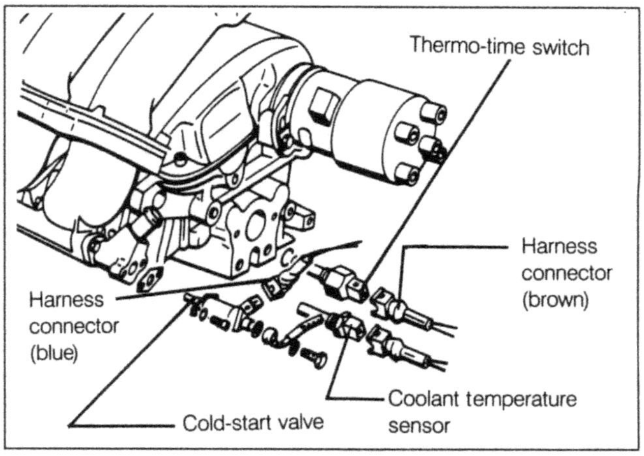

Fig. 6-17. Locations of cold-start valve and thermo-time switch on PL (16-valve) engines.

If the cold-start valve fails to inject fuel during cold starting, it will be difficult or impossible to start the engine. If the cold start valve leaks, the engine may receive extra fuel at the wrong time and become flooded, especially if the engine is hot.

To test cold-start valve:

1. Make sure engine coolant temperature is below 86°F (30°C). Preferably the engine should sit for several hours.
2. Remove the two screws holding the cold start valve to the intake manifold. Without disconnecting the fuel line or the harness connector, remove the cold start valve.
3. Disconnect the coil wire from the center tower of the distributor cap and connect it to ground, using a jumper wire.
4. Use a jumper wire to connect the metal cold-start valve housing to ground.
5. Wipe dry the nozzle of the cold start valve. Point the valve into a transparent container, and have a helper actuate the starter. See Fig. 6-18.

WARNING —
Fuel will be expelled. Do not smoke or work near heaters or other fire hazards. Keep a fire extinguisher handy.

NOTE —
The valve should spray in an even, cone-shaped pattern until the thermo-time switch interrupts the circuit. An irregular spray pattern indicates a dirty or faulty cold start valve. If the valve does not spray, test the thermo-time switch as described below.

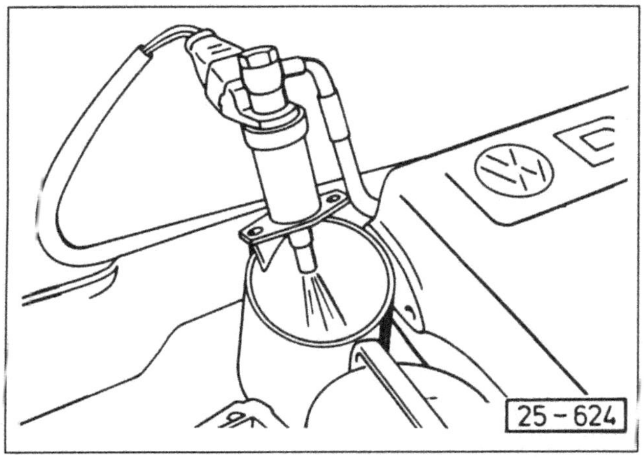

Fig. 6-18. Cold-start valve being checked by observing as fuel sprays into transparent container.

6. Wipe the nozzle dry. The valve should not drip for at least a minute. If it does, it is faulty and should be replaced.

When replacing the cold-start valve, the sealing washers on either side of the fuel line fitting should be replaced. Torque the fuel line bolt and the valve's mounting bolts to 10 Nm (87 in. lb.).

To test thermo-time switch:

1. Make sure engine coolant temperature is below 86°F (30°C). Preferably the engine should sit for several hours.
2. Disconnect the coil wire from the center tower of the distributor cap and connect it to ground, using a jumper wire.
3. Disconnect the cold-start valve harness connector and connect a test light across the connector terminals, as shown in Fig. 6-19.

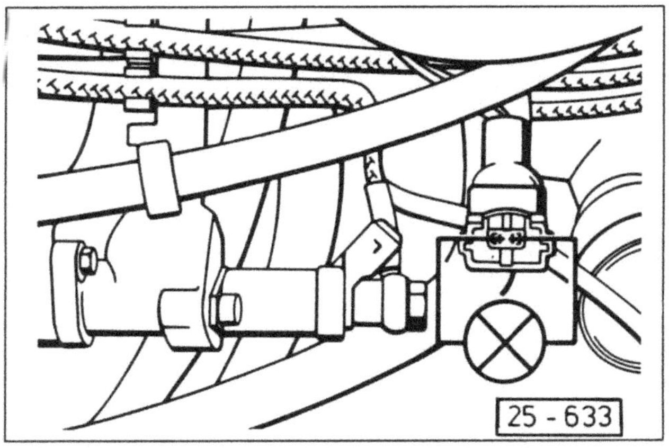

Fig. 6-19. Test light (shown schematically) connected to cold start valve connector.

4. Actuate the starter while observing the test light. The test light should light for several seconds, then go out. The amount of time that the light should stay on depends on coolant temperature. See the graph in Fig. 6-20.

NOTE —
If the test light does not light, it is either because voltage is not reaching the connector due to faulty wiring, or because the path to ground (through the thermo-time switch) is interrupted by a faulty switch or wire. Consult the current flow diagrams in **ELECTRICAL SYSTEM** for help in identifying wires and terminals.

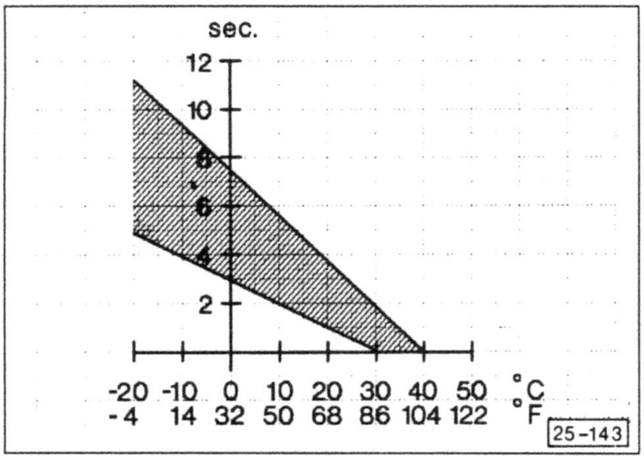

Fig. 6-20. Graph of thermo-time switch response to coolant temperature. Switch completes circuit to open cold-start valve for number of seconds shown. Example: at coolant temperature of 68°F (20°C), switch is closed for 1 to 3.7 seconds.

To replace the thermo-time switch, first drain the coolant from the engine as described in **COOLING SYSTEM**. Use a thread sealant and torque the new switch to 30 Nm (22 ft. lb.). Reconnect the cold-start valve harness connector and the coil wire, and refill the cooling system with coolant.

6.3 Idle Speed

Engine idle speed is controlled not by a fine adjustment of the throttle valve, but by the position of the idle air bypass adjusting screw. By controlling a small amount of air that is allowed to bypass the throttle, idle speed is more reliable, as it is independent of accelerator cable stretch or wear in the throttle valve mechanism. The idle air bypass is shown schematically in Fig. 6-21.

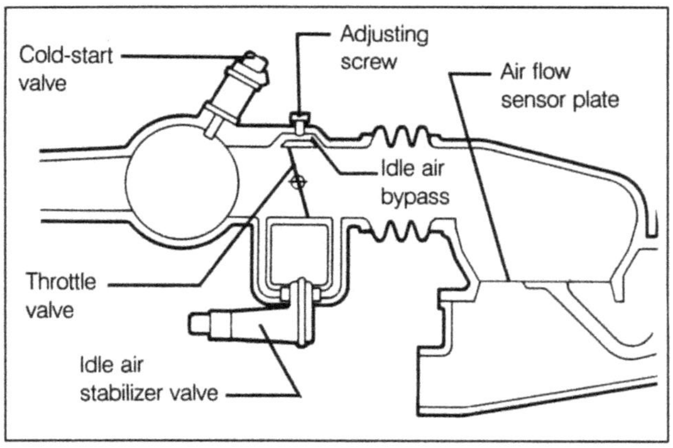

Fig. 6-21. Schematic view of air flow sensor and throttle valve showing idle air bypass and adjusting screw.

The need for additional inlet air during cold running is handled in much the same way, by allowing additional air to bypass the throttle valve. For this function, an electronically controlled idle air stabilizer valve is used to maintain idle speed within a prescribed range. Idle speed itself is subject to electronic control and is not adjustable. See **Idle Air Stabilizer Valve** below for information on checks and adjustments to engines equipped with the idle air stabilizer. Fig. 6-22 shows one example of the idle air stabilizer.

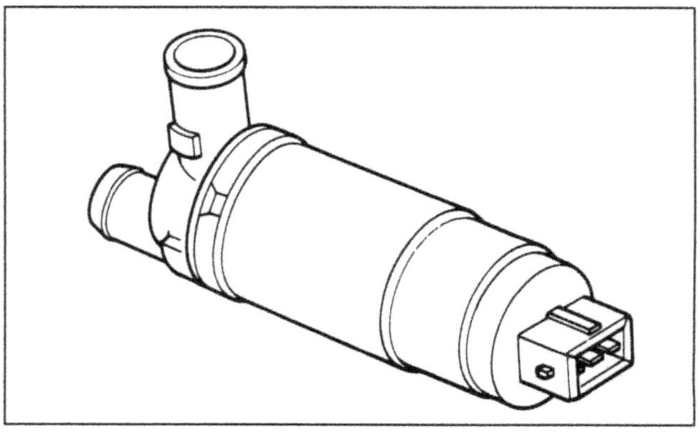

Fig. 6-22. Electronically controlled idle air stabilizer valve used with CIS-E fuel injection.

For best results, idle speed should always be checked and adjusted along with ignition timing and the idle mixture adjustment (differential pressure regulator current). Isolated adjustment of any of these will tend to affect the others. It may cause driveability problems and increase exhaust emissions. The first step is to make sure ignition timing is properly adjusted. See **IGNITION**. Then check and adjust idle speed and idle mixture (differential pressure regulator current), repeating the adjustments until all are simultaneously within specifications. For information on measuring differential pressure regulator current, see **6.1 Air Flow Measurement and Fuel Metering**. Also see **6.6 Idle Specifications.**

Making these checks and adjustments together is very important to both performance and emissions control. If the equipment necessary to accurately perform this work is not available, we suggest turning the job over to an authorized Volkswagen dealer or other qualified repair shop. In a properly equipped shop, these checks and adjustments can be made quickly, accurately, and at reasonable cost.

When adjusting idle, the following requirements apply:

1. The engine must be at normal operating temperature (oil temperature at least 176°F (80°C)). The radiator fan should have cycled on and off at least once.
2. All electrical accessories should be off (including the radiator cooling fan—make adjustments only when the fan is not on).
3. Fuel pressure measuring equipment should not be connected.
4. The exhaust system must be free of leaks.
5. The oxygen sensor must be functioning correctly.

NOTE —

If repairs have been made to the fuel system, raise the engine speed over 3000 rpm a few times, then let idle for approximately 2 minutes. This will bleed any trapped air from the fuel system.

Connect a tachometer according to the instrument manufacturer's instructions in order to accurately measure rpm. The ignition signal lead from the tachometer should be connected to terminal 1 of the coil.

CAUTION —

Connect and disconnect the test leads only with the ignition off.

Idle Air Stabilizer Valve

The electronic idle air stabilizer valve controls the amount of air allowed to bypass the throttle valve and increase engine idle speed. Whenever the throttle valve idle switch is closed, the valve receives a cycled voltage signal from the oxygen sensor control unit based on engine rpm and other inputs, as shown in Fig. 6-23. This voltage signal cycles on and off to incrementally open or close the valve to adjust idle speed. This on/off signal is referred to as the valve's duty cycle, and is measured with a duty cycle meter or a dwell meter.

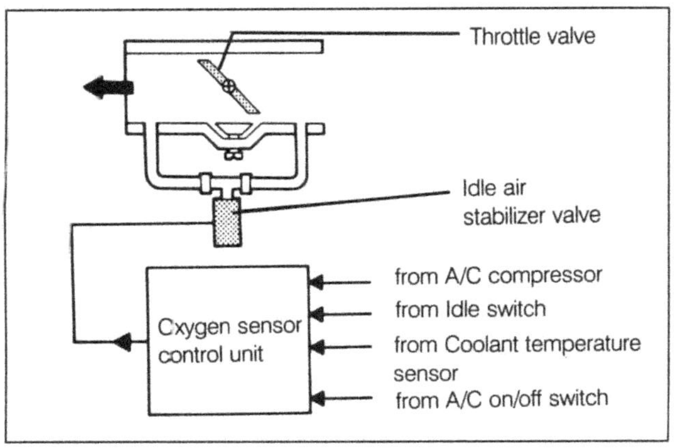

Fig. 6-23. Schematic representation of idle air stabilizer valve and electronic control inputs.

When the idle air stabilizer valve is functioning correctly, it should vibrate and hum slightly whenever the engine is running at idle. If not, turn the engine off, check for a voltage signal reaching the valve, and check the valve's resistance.

Check for a voltage signal at the valve's harness connector as shown in Fig. 6-24. With the ignition on, there should be battery voltage (approximately 12 VDC) between the center terminal and ground, and approximately 10 VDC between the center terminal and each of the two outer terminals. If not, see **6.4 Electrical Tests**.

Check resistance at the connector terminals on the valve as shown in Fig. 6-24. There should be continuity between the center terminal and each of the outer terminals. If not, the valve is faulty and should be replaced.

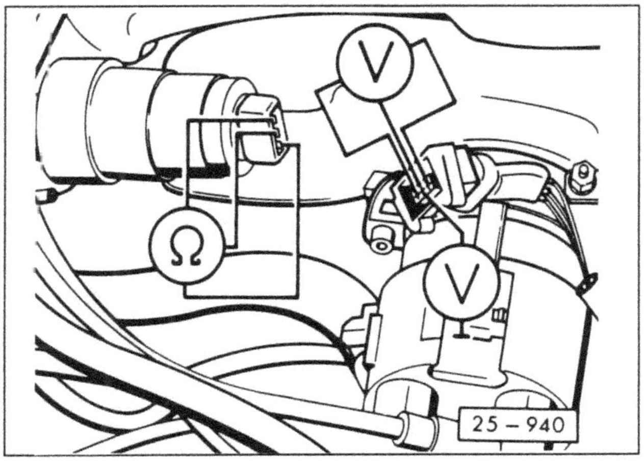

Fig. 6-24. Ohmmeter and voltmeter (shown schematically) being used to test idle stabilizer valve resistance and voltage signal.

Adjustment of the idle air stabilizer system is limited to measuring the duty cycle and adjusting it to specification by adjusting the idle air bypass. Duty cycle measures the on time of the cycling on/off voltage signal, and is expressed as a percentage. It is measured by connecting a duty cycle meter, such as the Siemans 451, to a test connector in the wiring harness near the ignition coil. Connect the duty cycle meter according to the instrument manufacturer's instructions.

A common dwell meter can also be used to measure duty cycle. The 90° scale of a dwell meter corresponds to the 100% scale of a duty cycle meter. For example, a 50% duty cycle would be read as 45° on a dwell meter. The tests below express test values for both instruments. Connect a dwell meter by connecting the meter's positive (+) lead to the test connector with the blue/gray wire, and the negative (−) lead to ground.

Fig. 6-25 shows the temporary set-ups necessary to make accurate idle checks. Disconnect the crankcase ventilation hoses as shown. Remove the charcoal canister line "T" fitting from the air flow sensor boot and turn it to insert the side with the 1.5 mm (.059 in.) restrictor hole, as shown in Fig. 6-26.

With the engine running, the duty cycle should be 26 to 30% (23° to 27° on a dwell meter). If the duty cycle is incorrect, adjust it by turning the idle air bypass adjusting screw as shown in Fig. 6-27.

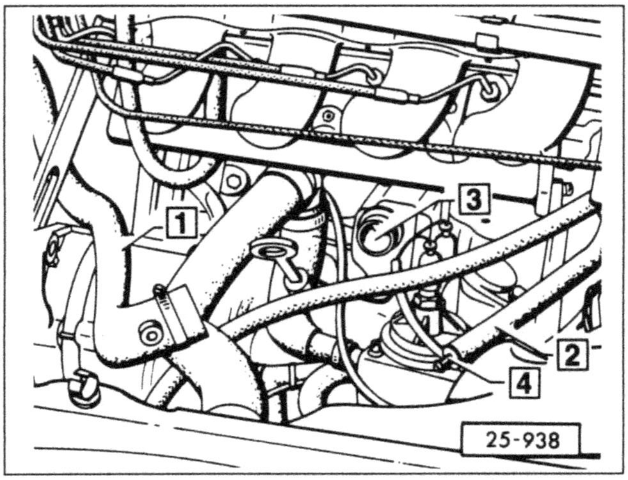

Fig. 6-25. Temporary set-up for idle speed adjustment. Disconnect crankcase ventilation hose (**2**) and connector (**4**) from hose (**1**). Disconnect hose (**1**) from ventilation housing (**3**).

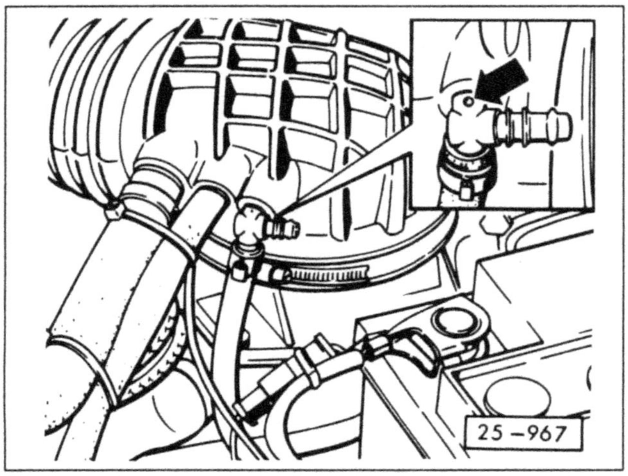

Fig. 6-26. Charcoal canister "T" fitting shown turned with 1.5 mm (.059 in.) restrictor in place in boot.

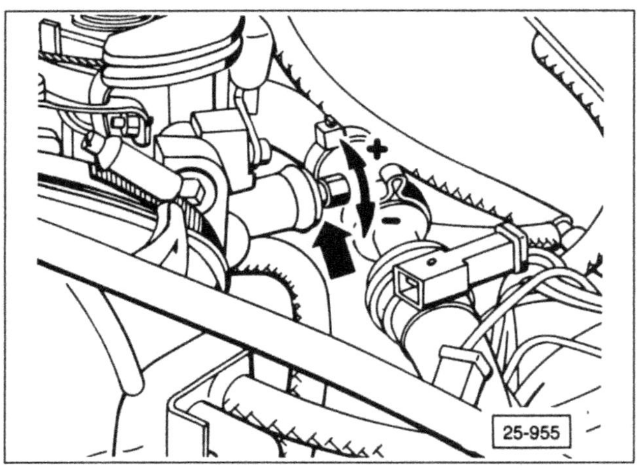

Fig. 6-27. Idle air bypass adjusting screw on throttle valve.

6.4 Electrical Tests

Two types of electrical tests are described here. First, several fuel injection electronic control functions can be checked and evaluated by measuring differential pressure regulator current under different running conditions. Next, the voltage and continuity tests listed under **Control Unit Inputs** can be used to help determine whether there are faults in the wiring or components which provide information to the control unit. If all control unit inputs are found to be correct and the system does not perform as specified, the control unit itself may be faulty.

For purposes of general troubleshooting, the electrical tests below are arranged in a specific order designed to logically isolate the cause of a problem. For best results, the tests should be carried out in the order in which they appear.

All tests of differential pressure regulator current require connecting an ammeter in series with the differential pressure regulator as described in **Differential Pressure Regulator** under **6.1 Air Flow Measurement and Fuel Metering**.

After-start and Cold Running Enrichment

To ensure smooth running, the system supplies additional fuel to a cold engine by increasing differential pressure regulator current for a short period after starting and during warm-up. A coolant temperature sensor provides the control unit with temperature information. The starter provides an additional signal for after-start enrichment. After-start enrichment lasts approximately 40 seconds. Cold running enrichment decreases gradually as the engine warms up, finally reaching the normal warm running value.

To test either function, remove the harness connector from the coolant temperature sensor to simulate a cold engine. The connector is shown earlier in Fig. 6-17. To check cold running enrichment, turn the ignition on. Differential pressure regulator current should be 80 to 100 mA.

After-start enrichment is checked after operating the starter. To prevent the engine from starting during the test, disconnect the ignition coil wire from the center of the distributor cap and use a jumper wire to connect it to ground. Actuate the starter for 2 to 3 seconds, then leave the ignition on. The differential pressure regulator current should increase to more than 120 mA for 30 to 60 seconds and then return to the cold running value given above. When the test is complete, turn the ignition off. Reconnect the coil wire and the coolant temperature sensor harness connector.

Test the coolant temperature sensor by detaching its harness connector and checking resistance across the sensor's terminals. The correct resistance varies with engine coolant temperature. A sensor with resistance not corresponding to the values given in the chart in Fig. 6-28 is faulty and should be replaced. When replacing the sensor, drain and refill the engine coolant as described in **COOLING SYSTEM**.

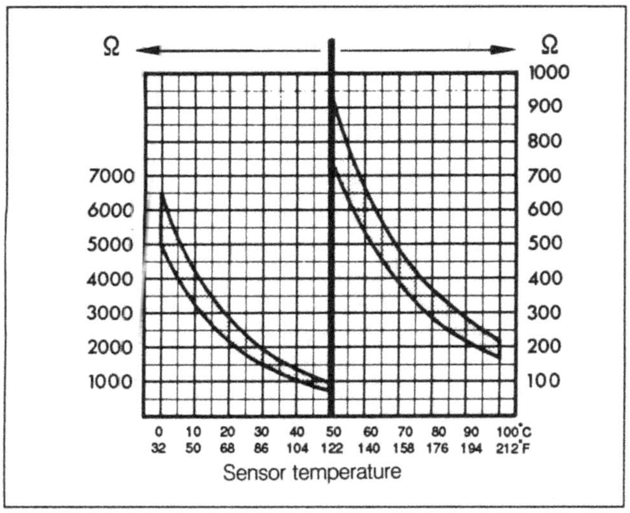

Fig. 6-28. Graph showing coolant temperature sensor resistance (ohms) vs. temperature (°C). Example: at 122°F (50°C), resistance is 725 to 925 ohms.

Cold Acceleration Enrichment

To ensure good throttle response when the engine is cold, the system briefly delivers additional fuel when the throttle valve is opened suddenly. The air flow sensor plate potentiometer, shown in Fig. 6-29, sends a signal to the control unit based on sensor plate travel.

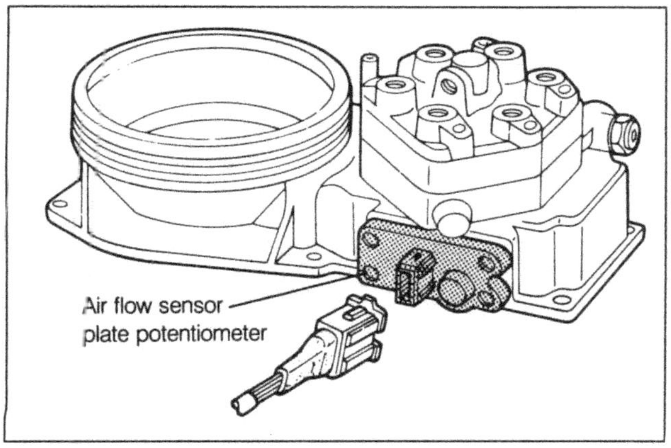

Fig. 6-29. Air flow sensor plate potentiometer mounted to air flow sensor housing. No adjustment is necessary unless it is replaced.

The potentiometer is adjusted at the factory and the mounting screws are sealed. Its operation can be checked by making resistance measurements with an ohmmeter. No adjustment is required unless the potentiometer is replaced. Adjusting a new potentiometer requires the use of an accurate voltmeter. First make certain that the air flow sensor plate is correctly adjusted, as described in **6.1 Air Flow Measurement and Fuel Metering**.

CAUTION —

Do not attempt to adjust the potentiometer unless it is necessary because of replacement. The mechanism is very sensitive to position and precise adjustment may take several tries.

To test cold acceleration enrichment, disconnect the coolant temperature sensor harness connector and remove the black rubber intake air boot from above the air flow sensor plate. Open the throttle so that the idle switch is not contacted by the throttle lever. Turn the ignition on and quickly raise the air flow sensor plate to its stop. The differential pressure regulator current should briefly exceed and then return to the cold running enrichment value given above. Turn the ignition off and reconnect the temperature sensor harness connector.

To check the potentiometer's operation, detach the harness connector and use an ohmmeter to measure resistance between terminals, as shown in Fig. 6-30. The resistance between terminals 2 and 3 should be greater than 4000 ohms. The resistance between terminals 1 and 2 should be less than 1000 ohms. Check the resistance between terminal 1 and terminal 2 while slowly and evenly lifting the sensor plate. The resistance should steadily increase, without any flat spots, to approximately 4000 ohms. A potentiometer which does not perform as specified is faulty and should be replaced.

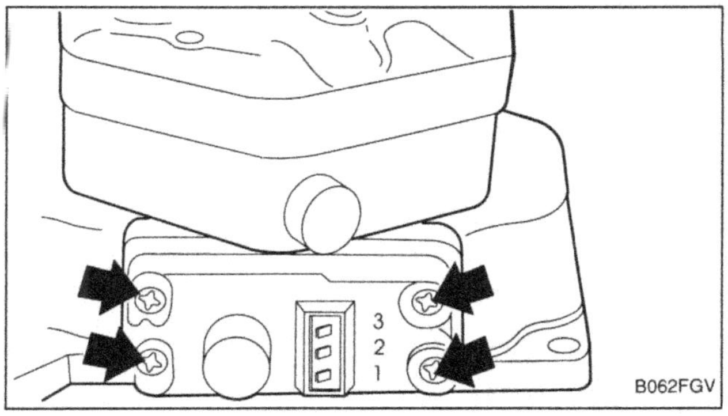

Fig. 6-30. Air flow sensor plate potentiometer terminal identification. When replacing potentiometer, leave mounting screws (arrows) slightly loose to allow adjustment.

To replace the potentiometer, remove its mounting screws and pull it from the fuel distributor. Install the new potentiometer with the mounting screws, leaving them just loose enough to allow some position adjustment. Lift the air flow sensor plate until it is flush with the narrowest point in the air cone, as shown in Fig. 6-31, and hold the plate steady in this position.

Adjust the position of the potentiometer until the voltage between the center and bottom terminals is 0.02 to 0.2 volts. Carefully tighten the mounting screws and recheck the voltage reading.

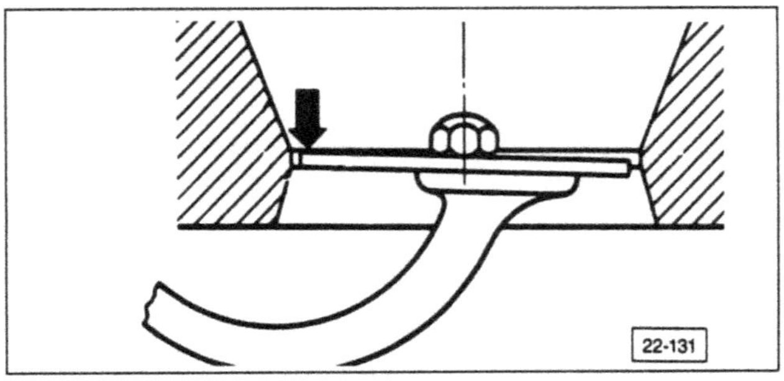

Fig. 6-31. Air flow sensor plate correctly positioned prior to adjusting potentiometer.

Raise the sensor plate fully to its stop. The voltage should increase to approximately 7.0 volts. If not, check the rest position of the sensor plate and recheck the initial position setting of the potentiometer.

Full-throttle Enrichment

The full-throttle enrichment function operates only when the engine is warm, when engine speed is above 2500 rpm, and when the throttle is fully open. The full-throttle switch supplies a signal to the control unit when the throttle is fully open.

To check full-throttle enrichment, disconnect the harness connector from the coolant temperature sensor and, using a jumper wire, bridge the connector terminals. The sensor is shown earlier in Fig. 6-17. Start the engine. With the engine speed raised above 2500 rpm, actuate the full-throttle switch on top of the throttle valve by hand, as shown in Fig. 6-32. The differential pressure regulator current should increase to approximately 16 mA.

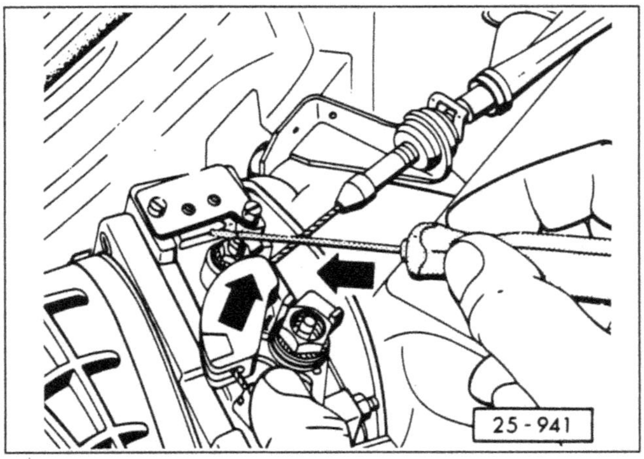

Fig. 6-32. Small screwdriver being used to operate full-throttle switch on throttle valve and test full-throttle enrichment. (Model shown is not one of those covered by this manual).

If the differential pressure regulator current reads excessively high, recheck the installation of the jumper wire at the temperature sensor connector. If the current does not increase when the full-throttle switch is closed, check for a faulty full throttle switch or a faulty rpm signal from the ignition control unit as described below under **Control Unit Inputs**.

Deceleration Fuel Shutoff

When decelerating, fuel flow to the injectors is shut off to improve fuel economy. At normal operating temperature, fuel flow is shut off when the throttle is closed at engine speeds above 1600 rpm, and fuel flow resumes when rpm drops to below 1300 rpm. Below normal operating temperature, these limits are higher. The closed throttle signal is provided to the control unit by the idle switch mounted underneath the throttle body.

To check the deceleration fuel shutoff function, disconnect the harness connector from the coolant temperature sensor and bridge the connector terminals with a jumper wire as described above under **Full-throttle Enrichment**. Start the engine and increase engine speed to about 3000 rpm, then release the throttle. The positive differential pressure regulator current should briefly change to a negative value (approximately −45 mA) and then return to its normal positive value.

If the current does not temporarily change to a negative value, check for a faulty or incorrectly adjusted idle switch or a faulty rpm signal from the ignition control unit as described below in **Control Unit Inputs**.

Control Unit Inputs

The circuit information in the table below can help diagnose a faulty CIS-E oxygen sensor control unit by checking the components and the wiring harness which supply information to the control unit. The terminals of the control unit connector are identified in Fig. 6-33.

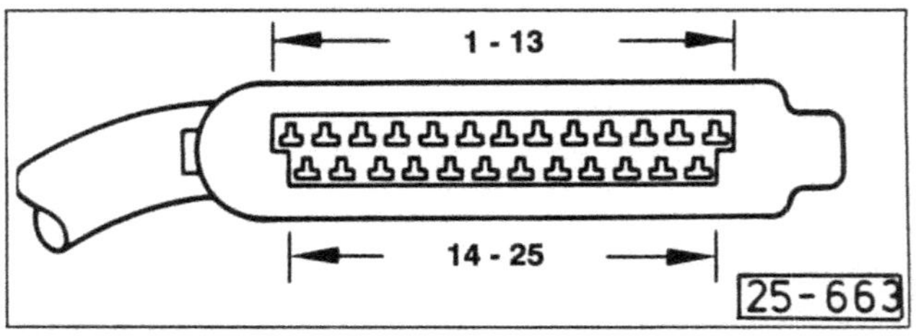

Fig. 6-33. Identification of terminals on CIS-E oxygen sensor control unit.

As a general rule, a complete absence of the voltage or continuity specified in the table suggests an open circuit in the wiring harness. For more specific information, consult the current flow diagrams in **ELECTRICAL SYSTEM**.

Test results which differ from the values specified in the table do not necessarily mean that a component is faulty. Check for loose connections or connections which are inadequate due to contamination or corrosion.

CAUTION —

Use only a high input impedance voltmeter, ohmmeter, multimeter, or LED test light. The electrical characteristics of other types of test equipment may cause inaccurate results or damage to the electronic components. For more information, see **FUNDAMENTALS**.

Use care to avoid damaging the delicate connector terminals with meter probes during testing. For best and safest results, fabricate a set of test leads as shown in Fig. 6-34. Use the flat connectors (Volkswagen part no. N 17 457 2 or equivalent) to make contact with female connector terminals, then connect the meter leads to the insulated alligator clips.

Before performing the tests indicated by the table, disconnect the coil wire from the center tower of the distributor cap and connect it to ground on the engine block with a jumper wire. With the ignition turned off, disconnect the 25-point connector from the oxygen sensor control unit and test as indicated in **Table d**. The control unit is located in the upper left side of the engine compartment, beneath the drip tray. The drip tray is clipped into place. Pull it up carefully to remove it for access to the control units.

CAUTION —

Always connect or disconnect the control unit connector and the meter probes with the ignition off, to avoid damage to the electronic components.

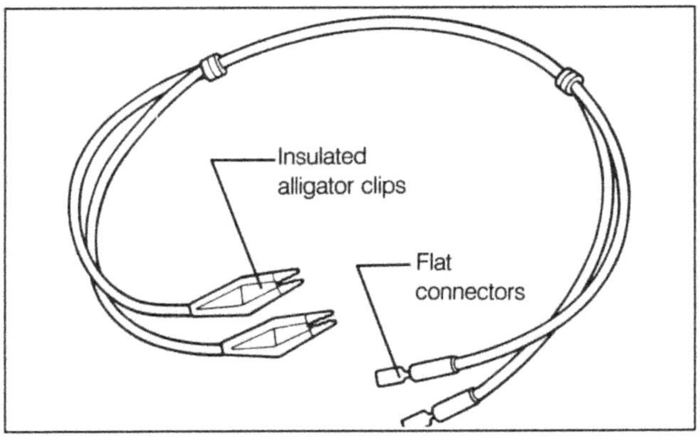

Fig. 6-34. Test leads made with flat connectors (Volkswagen part no. N 17 457 2 or equivalent) and insulated alligator clips, used to make contact without connector damage.

Table d. CIS-E Electrical Tests

Component or circuit	Test terminals	Test conditions	Correct test value
Voltage supply and control unit ground	1 and 2	Ignition on	Battery voltage (approximately 12 VDC)
	1 and 2	Starter operated	8 volts minimum
Signal from starter to control unit	2 and 24	Starter operated	8 volts minimum
Differential pressure regulator	10 and 12	Ignition off	17-22 ohms
Air flow sensor plate potentiometer	17 and 18	Air flow sensor plate at rest	more than 4000 ohms
	14 and 17	Air flow sensor plate at rest	less that 1000 ohms
	14 and 17	Air flow sensor plate lifted to stop	more than 4000 ohms
Coolant temperature sensor	2 and 21	Ignition off	Resistance per Fig. 6-28
Wire from oxygen sensor to control unit	2 and 8	Green oxygen sensor wire disconnected and touched to ground	Continuity
	2 and 8	Green oxygen sensor wire connected to black wire	No continuity
Ground connection to intake manifold for ground cable bridge	2 and 15	Ignition off	Continuity

continued on next page

Table d. CIS-E Electrical Tests (continued)

Component or circuit	Test terminals	Test conditions	Correct test value
Ground connection to intake manifold for oxygen sensor cable shield	2 and 7	Ignition off	Continuity
Ground connection to intake manifold (manual transmission only)	2 and 9	Ignition off	Continuity
Idle air stabilizer valve	2 and 3 2 and 4	Ignition on	Battery voltage (approximately 12 VDC)
Full throttle switch	2 and 5	Ignition on; Throttle fully open (full-throttle switch closed)	Battery voltage (approximately 12 VDC)
Idle switch	2 and 13	Ignition on; Throttle closed	Battery voltage (approximately 12 VDC)
RPM signal from knock sensor control unit	2 and 25	LED test light connected between terminals 2 and 25; Operate starter	LED must flicker
Voltage signal from air conditioning system	2 and 6	Ignition on	Battery voltage (approximately 12 VDC)
	2 and 16 2 and 19	Ignition on A/C switch on	Battery voltage (approximately 12 VDC)
Battery ground	2 and 20	Ignition off	Continuity

6.5 Fuel Pressure Tests and Specifications

As described earlier in this section, proper function of the CIS-E fuel injection system depends to a great extent on precise fuel pressure. Fuel pressure influences engine operating characteristics such as idle, partial throttle and full-throttle response, starting and warm-up, engine power, and emission levels. Any significant troubleshooting or repair of the system requires the use of a pressure gauge to measure fuel pressure in different parts of the system and under different operating conditions.

The three significant CIS-E fuel pressure values are system pressure, differential pressure and residual pressure. System pressure is the basic fuel pressure created by the main fuel pump and maintained by the diaphragm pressure regulator. Differential pressure is the pressure ratio in the fuel distributor, controlled by the differential pressure regulator, which regulates the flow of fuel to the injectors. Residual pressure is the pressure that remains in the closed system after the engine (and fuel pump) are shut off. Procedures for measuring each of these pressure values are described below.

The pressure gauge shown in Fig. 6-35 (Volkswagen special tool no. VW 1318) has a range of 0 to 10 bar (150 psi) and a valve in one line. It is connected between the fuel distributor test port and the end of the fuel line which supplies the cold-start valve. Bleed any air from the gauge by hanging it upside down while opening and closing the valve several times. Leave the gauge connected this way for all three pressure tests.

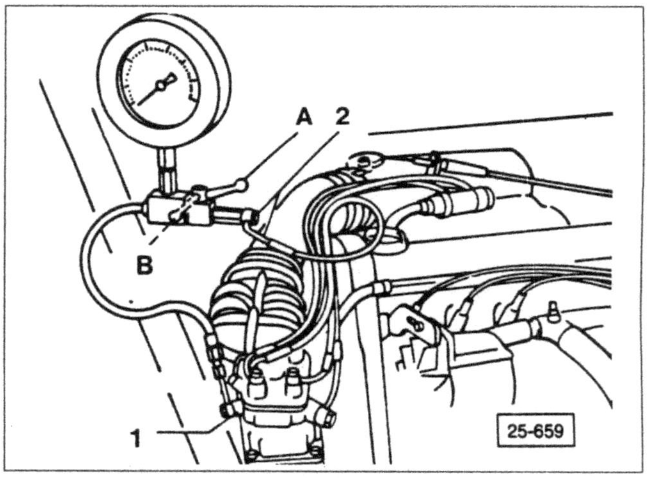

Fig. 6-35. Fuel pressure gauge and valve installed between fuel distributor test port (**1**) and cold-start valve supply line (**2**, removed from cold-start valve). Gauge valve is open in position **A** and closed in position **B**. (Model shown is not one of those covered by this manual).

CAUTION —

Fuel will be expelled when fuel lines are disconnected. Do not disconnect wires that could cause sparks. Do not smoke or work near heaters or other fire hazards. Keep a fire extinguisher handy.

CAUTION —

Absolute cleanliness is essential when working with fuel circuit components of CIS-E. Even a minute particle of dirt can cause trouble if it reaches an injector. Before disconnecting any of the fuel line connections, thoroughly clean the unions. Use clean tools.

System Pressure

System pressure is the pressure created by the fuel pump and regulated by the diaphragm pressure regulator shown in Fig. 6-36. Above the predetermined pressure limit of the regulator, the regulator opens and routes fuel back to the fuel reservoir. System pressure is not adjustable and a faulty pressure regulator must be replaced as described in **4.4 Fuel Pressure Regulator**.

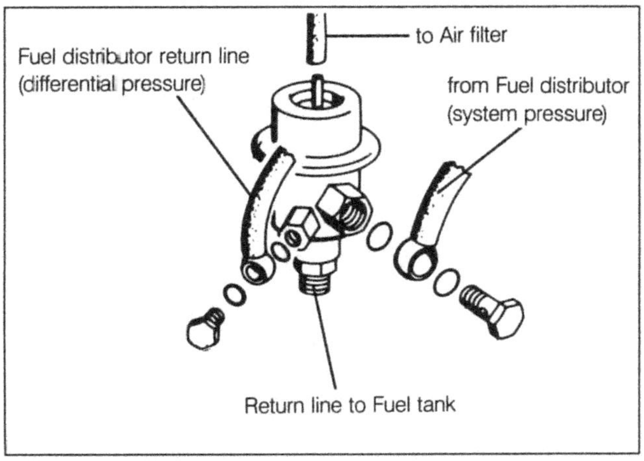

Fig. 6-36. CIS-E diaphragm pressure regulator, and fuel line connections. Always use new sealing washers. Regulator is mounted next to fuel filter in engine compartment.

To check system pressure, connect the fuel pressure gauge as described above and disconnect the harness connector from the differential pressure regulator. See **6.1 Air Flow Measurement and Fuel Metering**. Run the fuel pump, to pressurize the fuel system, using a jumper wire at the fuse/relay panel as described in **4.1 Fuses and Relays**. System pressure should be 5.2 to 5.6 bar (75 to 82 psi). After reading system pressure on the gauge, disconnect the fuel pump jumper.

If the system pressure is too high, check for a blocked or damaged fuel return line. If the system pressure is too low, look for fuel leaks, a clogged fuel filter, or a damaged fuel line blocking fuel flow. Check fuel pump delivery rate as described in **4.3 Fuel Pump**. If no faults are found with the return line or the fuel pump, the pressure regulator is faulty and should be replaced.

Differential Pressure

Differential pressure is the pressure difference between the upper and lower chambers of the fuel distributor, regulated by the differential pressure regulator, which controls the air/fuel mixture by regulating fuel flow to the injectors.

Check differential pressure with the gauge connected as described above and the gauge valve closed. Also, connect an ammeter to measure differential pressure regulator current as described in **Differential Pressure Regulator** under **6.1 Air Flow Measurement and Fuel Metering.** To test, run the fuel pump using a fuse/relay panel jumper wire as described in **4.1 Fuses and Relays**. The differential pressure should be 0.2 to 0.5 bar (2.9 to 7.0 psi) less than the system pressure measured above. Disconnect the fuel pump jumper wire to stop the fuel pump.

If the differential pressure is too low, measure the volume of fuel coming from the fuel distributor return line. Disconnect the fuel distributor return line from the diaphragm pressure regulator (see Fig. 6-36 above) and place the disconnected line in a measuring container suitable for catching fuel. Plug the open port on the diaphragm pressure regulator and run the fuel pump. After sixty seconds, there should be 130 to 150 cc (4.4 to 5.1 oz.) of fuel in the container. When reconnecting the line, use new sealing washers and torque the union to 10 Nm (87 in. lb.).

If the fuel quantity is correct, the differential pressure regulator is faulty and should be replaced. If system pressure is within specifications but the fuel quantity is incorrect, the fuel distributor is faulty and should be replaced. See **6.1 Air Flow Measurement and Fuel Metering**.

The differential pressure is lower at lower temperatures. Using Volkswagen special tool no. VW 1490 (order no. TV1 490 000 25 ZEL) as shown in Fig. 6-37, with the 15,000 ohm side plugged into the coolant temperature sensor harness connector and testing as described above, the differential pressure should be 0.7 to 1.2 bar (10 to 17.5 psi.) below system pressure and the differential pressure regulator current should be 50 to 70 mA.

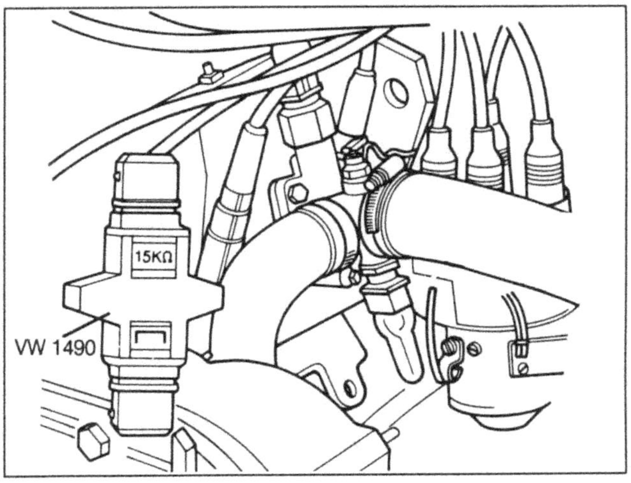

Fig. 6-37. Volkswagen special tool no. VW 1490 used to simulate cold temperature condition with 15,000 ohm resistor side inserted into coolant temperature sensor harness connector. On 16-valve engines (not shown), coolant temperature sensor is located on end of cylinder head.

If the differential pressure regulator current is within specifications but the differential pressure is not, the differential pressure regulator is faulty and should be replaced. If both the pressure and the regulator current are out of specifications, look for an electrical problem. Check that the resistance across the terminals of the differential pressure regulator is between 17.5 and 21.5 ohms. Check all wires and connections. Check for a good ground connection at the cold-start valve. If no other faults can be found, the oxygen sensor control unit is probably faulty.

Residual Pressure

The closed fuel system is designed to maintain pressure after the engine is shut off to help prevent the fuel in the injector lines from overheating, causing vapor lock and hard starting. The fuel pump's one-way check valve and the fuel accumulator help maintain this residual pressure. See **4. Fuel Supply**.

Check residual pressure with the pressure gauge connected as described above for measuring system pressure. Pressurize the fuel system by operating the fuel pump, using a jumper wire at the fuse/relay panel as described in **4.1 Fuses and Relays**, then disconnect the jumper wire. After ten minutes, the pressure should not have dropped below 2.6 bar (38 psi). If the pressure drops excessively, check the fuel pump check valve and the fuel accumulator as described below. Also check for leaks in the fuel lines, the fuel distributor O-rings, and the injectors. If no leaks are found, check the air flow sensor plate clearance as described in **6.1 Air Flow Measurement and Fuel Metering.**

To check the fuel pump check valve and the fuel accumulator, check residual pressure at the fuel filter outlet, as shown in Fig. 6-38. Connect the pressure gauge to the fuel supply line coming from the pump.

WARNING —
One gauge line (the one with the valve) is left disconnected. Be sure to close the valve.

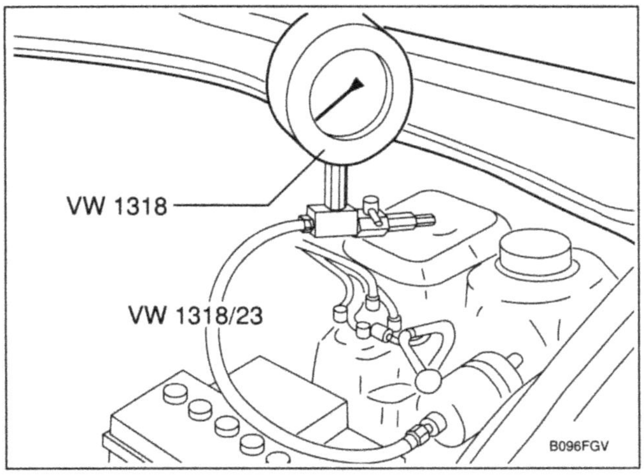

Fig. 6-38. Gauge (VW 1318) connected to fuel filter outlet with adapter (VW 1318/23) for testing fuel pump check valve and accumulator.

Run the fuel pump using a jumper wire at the fuse/relay panel, as described in **4.1 Fuses and Relays**, until fuel pressure indicated by the gauge reaches 5 bar (73 psi). Then turn the fuel pump off. Once again, after ten minutes, the pressure should not have dropped below 2.6 bar (38 psi).

If the pressure drops excessively and no leaks are found between the fuel pump and the gauge, repeat the test. This time shut off the pump and immediately pinch closed the fuel hose between the tank and the pump. If residual pressure now stays within specifications, the fuel pump check valve is faulty and should be replaced. See **4.3 Fuel Pump**.

If the residual pressure still drops off too quickly, the fuel accumulator may be faulty. For a quick check, remove the plug from the end of the accumulator. If fuel drips out, then the diaphragm is leaking and the accumulator should be replaced.

6.6 Idle Specifications

Idle speed, ignition timing, the idle mixture adjustment and idle mixture checks (differential pressure regulator current and CO content) must be checked and adjusted together. See **6.3 Idle Speed**.

If the equipment necessary to accurately perform this work is not available, we suggest turning the job over to an authorized Volkswagen dealer or other qualified repair shop. In a properly equipped shop, these checks and adjustments can be made quickly, accurately, and at reasonable cost.

Fuel mixture at idle is adjusted by turning an adjusting screw inside the air flow sensor as shown in Fig. 6-39. Turning the screw raises or lowers the control plunger to adjust the amount of fuel delivered to the injectors for a given air flow. A small CO test pipe connection to the exhaust manifold provides a place to connect a CO meter and measure CO upstream of the catalytic converter. The pipe is about 8 inches long with a blue rubber plug on the end.

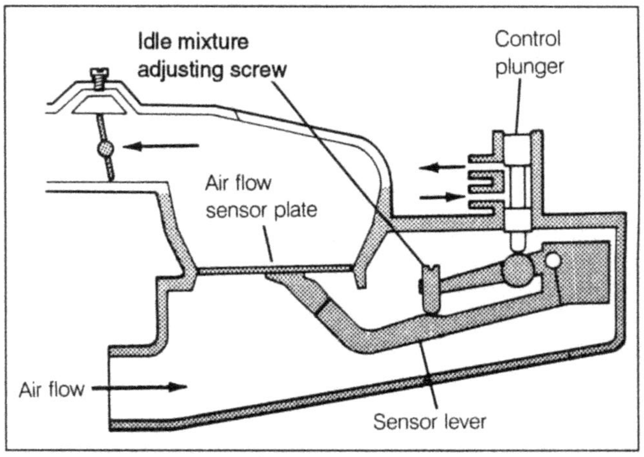

Fig. 6-39. Cut-away view of sensor plate, lever, fuel distributor control plunger and mixture adjusting screw.

The idle mixture screw can only be reached and adjusted after removing the anti-tampering plug in the top of the air flow sensor housing. This is a small plug with a soft aluminum top and a hard steel bottom, removed by drilling a 2.5 mm (3/32 in.) hole in the aluminum and using a sheet metal screw to extract the plug.

CAUTION —
Do not drill through the steel bottom. Drill only about 3.5 to 4.0 mm (9/64 to 5/32 in.) deep. Do not allow metal shavings to fall into the air flow sensor.

With the plug removed, the adjustment is made using a long 3 mm hex wrench as shown in Fig. 6-40. Turning the screw clockwise makes the mixture more rich (increases % CO). Turning it counterclockwise leans the mixture (lower % CO). **Table e** lists idle speed and mixture specifications.

CAUTION —
Do not accelerate the engine or push on the adjusting wrench while it is in place. Doing so may damage the air flow sensor. Remove the wrench and briefly accelerate the engine after each adjustment.

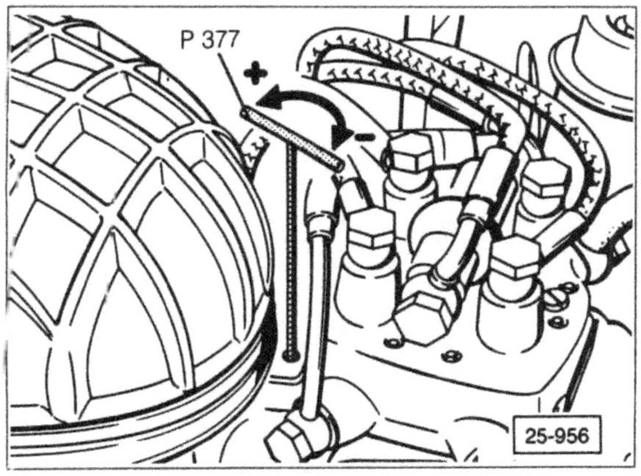

Fig. 6-40. Engine idle mixture adjustment being made with 3 mm hex wrench. Volkswagen special tool P377 shown (order no. TPO 377 000 25 ZEL).

Table e. Idle Specifications (CIS-E)

	Checking value	Adjusting value
Idle speed	800 to 900 rpm	850 ± 50 rpm
Differential pressure regulator current — milliamps (mA)	4 to 16 mA (fluctuating)	10 ± 1 mA (average)
CO content	0.3 to 1.2%	–
Idle air stabilizer valve duty cycle (0 to 100%)	26% to 30%	29% ± 1%
duty cycle measured with dwell meter (0 to 90°)	23° to 27°	26° ± 1°

7. DIGIFANT II FUEL INJECTION

Fig. 7-1 shows the main components of the Digifant engine management system. Digifant combines fuel injection, ignition and other engine management functions in one integrated system, managed by a single electronic control unit (ECU).

The ECU determines basic fuel metering according to intake air flow and engine speed. The intake air volume and temperature signals are provided by the air flow sensor. An engine rpm signal is taken from the ignition distributor. Fuel is delivered to the engine through electrically-operated solenoid-valve fuel injectors. The ECU makes and breaks the injectors' connection to ground to regulate injector open time.

The ECU makes additional fuel metering adjustments based on input from an array of sensors. Switches at the throttle body indicate throttle position and initiate electronic idle stabilization. A coolant temperature sensor helps the ECU determine fuel metering adjustments for starting and warm-up.

A signal from the oxygen sensor gives an indication of combustion efficiency that the ECU uses to fine-tune the air-to-fuel mixture. For more information about the oxygen sensor, see **EXHAUST SYSTEM AND EMISSION CONTROLS**.

7.1 Digifant Troubleshooting

The general fuel supply and fuel system troubleshooting information elsewhere in this manual also applies to Digifant systems. See **3. Troubleshooting** and **3.1 Basic Troubleshooting Principles** earlier in this section before beginning any specific Digifant troubleshooting that is described here.

Beginning with the 1991 model year, Cabriolet models built for sale in California are equipped with a version of the Digifant engine management system called Digifant I. One important difference is the Digifant I ECU's internal fault memory and on-board diagnostic (OBD) capability. Before going ahead with any other troubleshooting on a California car, see **8. Digifant I Fuel Injection (California)** and **8.1 Digifant I On-board Fault Diagnosis**.

> ***CAUTION —***
> *Volkswagen specifies that the first step in any Digifant I system troubleshooting must be to check the OBD fault memory, and to find and correct any faults that may be indicated. See* **8.1 Digifant On-board Fault Diagnosis**.

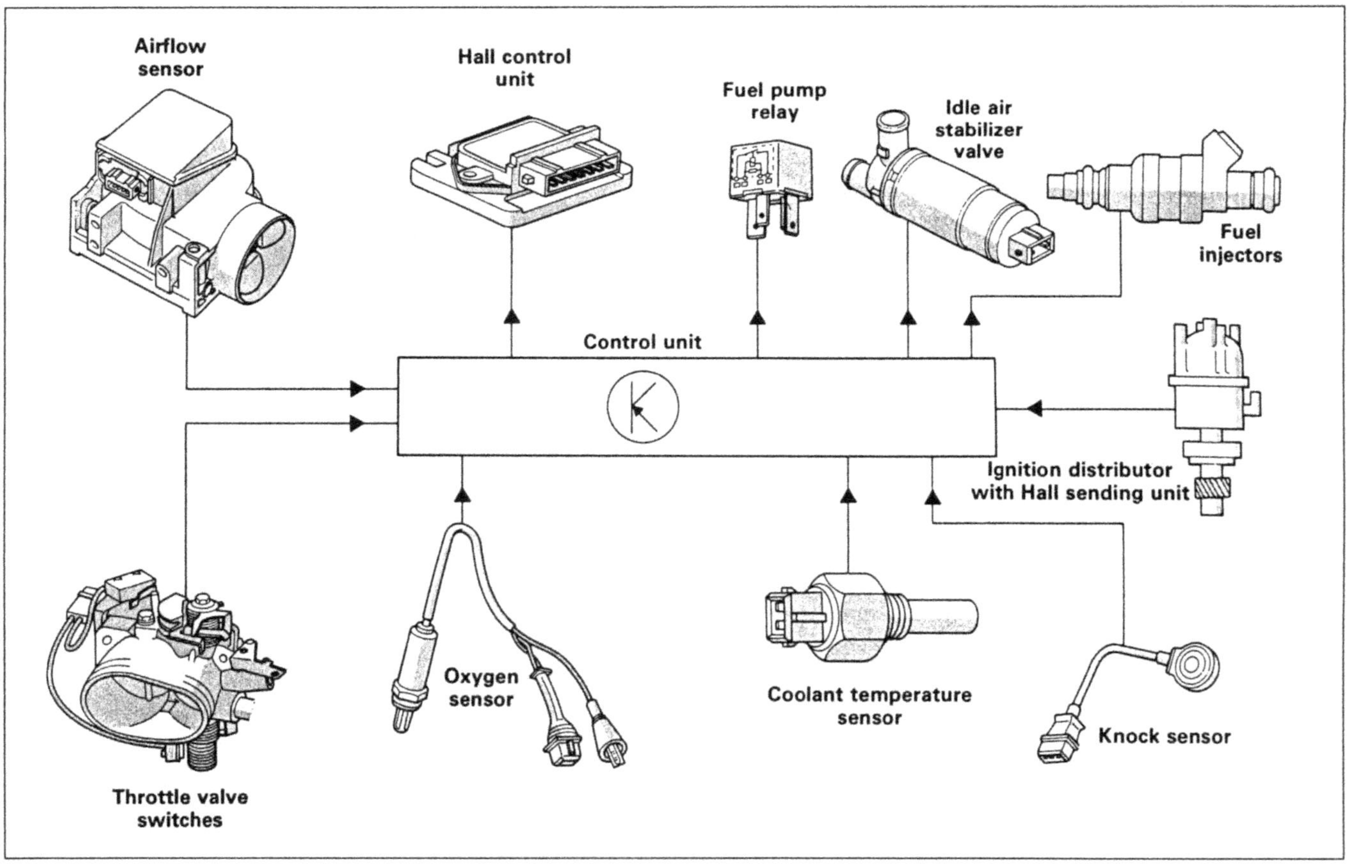

Fig. 7-1. Schematic illustration of Volkswagen Digifant engine management system showing electronic control unit and components that either provide input signals or respond to output signals.

The basic adjustments to ignition timing, idle speed and idle mixture (% CO) are critical because they influence all other functions of the Digifant system. Troubleshooting without the assurance that basic settings are correct can be misleading.

Ignition timing, idle speed and idle mixture (% CO) should always be checked and adjusted together. See **7.4 Digifant II Basic Adjustments—Idle Speed and Idle Mixture**.

The Digifant systems have been designed so that most electrical components can be tested with a low-current LED test light and a multimeter. Most of these tests can be performed right at the ECU harness connector.

CAUTION —
Always switch the ignition off before disconnecting or connecting the Digifant II control unit connector or any test equipment.

Table f lists many of the most likely symptoms of Digifant fuel system problems and their probable causes. The third column suggests appropriate corrective action. Numbers in bold type in the corrective action column refer to numbered headings in this **FUEL SYSTEM** section where you can find test and repair information.

Table f. Digifant Troubleshooting

Symptom	Probable cause	Corrective action
1. Engine will not start (warm or cold)	a. No ignition spark	a. See **IGNITION**
	b. Fuel pump not running	b. Check fuel pump relay. **4.1**
	c. Inadequate fuel pump output	c. Test fuel pump delivery rate. **4.3**
	d. Fuel pressure too low	d. Check for leaks. Test fuel pressure. **7.6**
	e. Fuel filter restricted	e. Check fuel pump delivery rate downstream of filter outlet. **4.3** Replace fuel filter. See **LUBRICATION AND MAINTENANCE**
	f. Fuel injector power supply or electrical connections faulty	f. Test fuel injectors. **7.5**, **8.5**
	g. Air flow sensor flap binding or faulty	g. Inspect air flow sensor. Test resistance values at harness connector. Clean, repair or replace as necessary. **7.2**
	h. No power to electronic control unit.	h. Check fuse #17. Test for power at control unit connector. **7.7**, **8.7** Check Digifant control unit relay. See **ELECTRICAL SYSTEM**
	i. Electronic control unit faulty	i. Make electrical tests at control unit connector. Replace faulty control unit. **7.7**, **8.7**
2. Hard starting (warm or cold)	a. Coolant temperature sensor faulty (incorrect cold-start enrichment)	a. Test coolant temperature sensor and replace as necessary. **7.3**
	b. Idle stabilizer valve faulty.	b. Inspect idle stabilizer valve, clean or replace as necessary. **7.4**
	c. Intake air (vacuum) leak(s)	c. Check for loose oil fill cap or dip stick. Inspect intake air components for leaking hoses, hose connections, and cracks or other leaks. Repair as necessary.
	d. Inadequate fuel pump output	d. Test fuel pump delivery rate. **4.3**
	e. Inadequate residual fuel pressure or fuel pressure too low	e. Check for leaks. Test residual fuel pressure. Test fuel pressure regulator. **7.6**
	f. Fuel filter restricted	f. Check fuel pump delivery rate downstream of filter outlet. **4.3** Replace fuel filter. See **LUBRICATION AND MAINTENANCE**
	g. Evaporative emissions system faulty	g. Inspect evaporative emissions control system. **4.5**
	h. Weak ignition spark	h. See **IGNITION**
	i. Electronic control unit faulty	i. Make electrical tests at control unit connector. Replace faulty control unit. **7.7**, **8.7**
3. Rough idle, stalling (engine cold)	a. Coolant temperature sensor faulty	a. Test coolant temperature sensor and replace as necessary. **7.3**
	b. Idle stabilizer valve faulty.	b. Inspect idle stabilizer valve, clean or replace as necessary. **7.4**
	c. Intake air preheating system faulty	c. Test intake air preheating system and replace faulty components as necessary. **7.2**
	d. Intake air (vacuum) leak	d. Check for loose oil fill cap or dip stick. Inspect intake air components for leaking hoses, hose connections, and cracks or other leaks. Repair as necessary.

(continued on next page)

Table f. Digifant Troubleshooting (continued)

Symptom	Probable cause	Corrective action
3. Rough idle, stalling (engine cold) (continued)	**e.** Oxygen sensor heating too slowly	**e.** Test oxygen sensor heater circuit. **7.7**, **8.7** Also see **EXHAUST SYSTEM AND EMISSION CONTROLS**
	f. Electronic control unit faulty	**f.** Make electrical tests at control unit connector. Replace faulty control unit. **7.7**, **8.7**
4. Rough idle, stalling (engine warm)	**a.** Intake air (vacuum) leak(s)	**a.** Check for loose oil fill cap or dip stick. Inspect intake air components for leaking hoses, hose connections, and cracks or other leaks. Repair as necessary
	b. Idle switch faulty (Digifant II only)	**b.** Test idle switch and replace as necessary. **7.2**
	c. Throttle valve potentiometer faulty (Digifant I—California only)	**c.** Test throttle valve potentiometer and adjust as necessary. Replace faulty parts. **8.2**
	d. Air flow sensor flap binding or faulty	**d.** Inspect air flow sensor. Test resistance values at air flow sensor harness connector. **7.7**, **8.7**
	e. Idle stabilizer valve faulty.	**e.** Inspect idle stabilizer valve, clean or replace as necessary. **7.4**
	f. Electronic idle stabilization faulty	**f.** Make electrical tests at control unit connector. Replace faulty control unit. **7.7**, **8.7**
	g. Evaporative emissions system faulty	**g.** Inspect evaporative emissions control system. **4.5**
	h. Intake air preheating system faulty	**h.** Test intake air preheating system and replace faulty components as necessary. **7.3**
5. Engine stalls at highway speed	**a.** Inadequate fuel pump output	**a.** Test fuel pump delivery rate. **4.3**
	b. Fuel pressure too low	**b.** Check for leaks. Test fuel pressure. **7.6**
	c. Fuel filter restricted	**c.** Check fuel pump delivery rate downstream of filter outlet. **4.3** Replace fuel filter. See **LUBRICATION AND MAINTENANCE**
	d. Evaporative emissions system faulty	**d.** Inspect evaporative emissions control system. **4.5**
	e. Incorrect ignition timing, idle mixture and/or idle speed adjustments	**e.** Check and correct Digifant basic adjustments. **7.4** (Digifant II), **8.4** (Digifant I—California only). Also see **IGNITION**
	f. Faulty electrical (power and ground) connections	**f.** Make electrical tests at control unit connector. Repair wiring harness or replace faulty components as necessary. **7.7**, **8.7**
6. Engine idles too fast	**a.** Accelerator pedal, cable, or throttle valve binding	**a.** Inspect for worn or broken parts, kinked cable, or other damage. Replace faulty parts. **4.6** Also see **BODY AND INTERIOR**
	b. Coolant temperature sensor or sensor wiring faulty	**b.** Test sensor. **7.3** Check wiring harness for damage
	c. Idle stabilizer valve faulty.	**c.** Inspect idle stabilizer valve, clean or replace as necessary. **7.4**
	d. Idle switch faulty (Digifant II only)	**d.** Test idle switch and replace as necessary. **7.2**
	e. Throttle valve potentiometer faulty (Digifant I—California only)	**e.** Test throttle valve potentiometer and adjust as necessary. Replace faulty parts. **8.2**
	f. Electronic idle stabilization faulty	**f.** Make electrical tests. Replace faulty control unit. **7.7**, **8.7**
7. Poor performance (during warm-up)	**a.** Incorrect ignition timing, idle mixture and/or idle speed adjustments	**a.** Check and correct Digifant basic adjustments. **7.4** (Digifant II), **8.4** (Digifant I—California only). Also see **IGNITION**
	b. Coolant temperature sensor faulty	**b.** Test coolant temperature sensor and replace as necessary. **7.2**
	c. Idle switch faulty (Digifant II only)	**c.** Test idle switch and replace as necessary. **7.2**
	d. Throttle valve potentiometer faulty (Digifant I—California only)	**d.** Test throttle valve potentiometer and adjust as necessary. Replace faulty parts. **8.2**
	e. Electronic idle stabilization faulty	**e.** Make electrical tests at control unit connector. Replace faulty control unit. **7.7**, **8.7**
	f. Intake air preheating system faulty	**f.** Test intake air preheating system and replace faulty components as necessary. **7.3**

Table f. Digifant Troubleshooting (continued)

Symptom	Probable cause	Corrective action
8. Poor performance (engine warm)	a. Incorrect ignition timing, idle mixture and/or idle speed adjustments	a. Check and correct Digifant basic adjustments. **7.4** (Digifant II), **8.4** (Digifant I—California only). Also see **IGNITION**
	b. Idle switch faulty (Digifant II only)	b. Test idle switch and replace as necessary. **7.2**
	c. Throttle valve potentiometer faulty (Digifant I—California only)	c. Test throttle valve potentiometer and adjust as necessary. Replace faulty parts. **8.2**
	d. Oxygen sensor system faulty	d. Check oxygen sensor and replace as necessary. See **EXHAUST SYSTEM AND EMISSION CONTROLS**. Also see **LUBRICATION AND MAINTENANCE**.
	e. Inadequate fuel pump output	e. Test fuel pump delivery rate. **4.3**
	f. Fuel pressure too low	f. Check for leaks. Test fuel pressure. **7.6**
	g. Fuel filter restricted	g. Check fuel pump delivery rate downstream of filter outlet. **4.3** Replace fuel filter. See **LUBRICATION AND MAINTENANCE**
	h. Electronic idle stabilization faulty	h. Make electrical tests at control unit connector. Replace faulty control unit. **7.4**, **8.7**
	i. Intake air preheating system faulty	i. Test intake air preheating system and replace faulty components as necessary. **7.3**
	j. Intake air flow restriction(s)	j. Check for loose oil fill cap or dip stick. Inspect intake air components for leaking hoses, hose connections, and cracks or other leaks. Repair as necessary
9. Poor performance under load (full throttle)	a. Weak ignition spark	a. See **IGNITION**
	b. Intake air flow restriction(s)	b. Inspect air filter element and replace as necessary. See **LUBRICATION AND MAINTENANCE** Check air filter housing, preheat system and intake air ducting. Clean, repair or replace as necessary. **7.2**, **7.3**
	c. Air flow sensor flap binding or faulty	c. Inspect air flow sensor. Test resistance values at air flow sensor harness connector. Clean, repair or replace air flow sensor as necessary. **7.2**
	d. Inadequate fuel supply to injectors	d. Test fuel pump delivery rate. **4.3** Test fuel pressure. **7.6**
	e. Fuel filter restricted	e. Check fuel pump delivery rate downstream of filter outlet. **4.3** Replace fuel filter. See **LUBRICATION AND MAINTENANCE**
10. Poor fuel economy	a. Oxygen sensor system faulty	a. Check oxygen sensor and replace as necessary. See **EXHAUST SYSTEM**. Also see **LUBRICATION AND MAINTENANCE**
	b. Inadequate fuel pump output	b. Test fuel pump delivery rate. **4.3**
	c. Fuel pressure too low	c. Check for leaks. Test fuel pressure. **7.6**
	d. Fuel filter restricted	d. Check fuel pump delivery rate downstream of filter outlet. **4.3** Replace fuel filter. See **LUBRICATION AND MAINTENANCE**
	e. Fuel injectors leaking or fouled	e. Test fuel injectors. **7.5**, **8.5**
	f. Evaporative emissions system faulty	f. Inspect evaporative emissions control system. **4.5**
	g. Intake air flow restriction(s)	g. Inspect air filter element and replace as necessary. See **LUBRICATION AND MAINTENANCE** Check air filter housing, preheat system and intake air ducting. Clean, repair or replace as necessary. **7.2**
	h. Air flow sensor flap not fully opening	h. Inspect air flow sensor. Check movement of sensor plate. Clean, repair or replace air flow sensor as necessary. **7.2**
	i. Electronic control unit faulty	i. Make electrical tests at control unit connector. **7.7**, **8.7**
11. Engine "runs-on" or "diesels" with ignition switched off	a. Incorrect ignition timing, idle mixture and/or idle speed adjustments	a. Check and correct Digifant basic adjustments. **7.4** (Digifant II), **8.4** (Digifant I—California only). Also see **IGNITION**
	b. Fuel injectors leaking	b. Inspect fuel injectors. Test injector fuel delivery and check for leaks. **7.5**, **8.5**
	c. Engine overheated	c. See **COOLING SYSTEM**

7.2 Intake Air Flow

Intake air drawn through the air flow sensor pushes against a spring-loaded mechanical flap. Changing air flow changes the position of the flap, which in turn changes the electrical signal from the air flow sensor potentiometer. The Digifant ECU interprets the changing electrical signal to determine the volume of air entering the engine. The air flow sensor assembly is shown in Fig. 7-2.

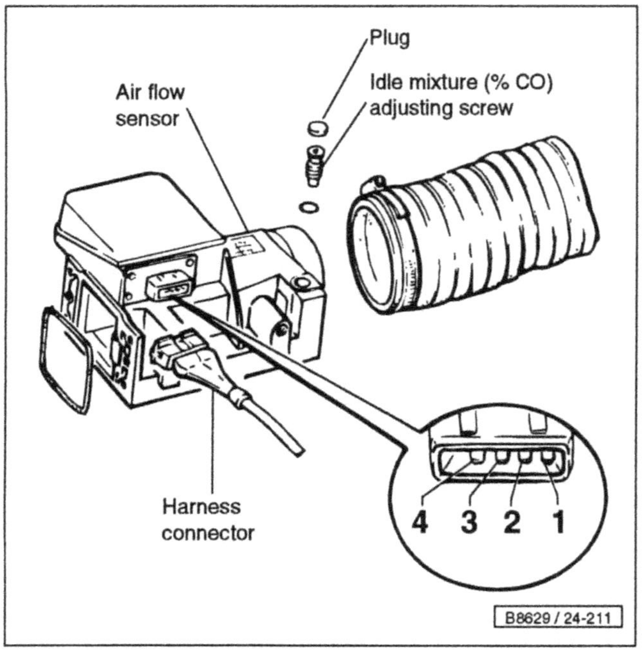

Fig. 7-2. Digifant air flow sensor, mounted at side of intake air filter housing. Inset identifies connector terminals used for electrical testing.

Air Flow Sensor

The spring-loaded flap inside the main air passage must move freely in order to accurately measure intake air. To inspect the air flow sensor, check for smooth movement of the flap through it's entire range of travel. If the flap binds or sticks and this cannot be corrected by cleaning, the air flow sensor is faulty and should be replaced.

To remove the air flow sensor, separate the upper and lower parts of the air filter housing. Remove the air flow sensor mounting bolts from inside. Installation is the reverse of removal. Replace the sealing gasket if it is deformed or otherwise damaged. Use care not to over-tighten the mounting bolts. The correct tightening torque is just 5 Nm (44 in. lb.).

Check the air flow sensor potentiometer by measuring resistance across its electrical terminals. Disconnect the harness connector from the sensor and test with an ohmmeter. Terminals are identified earlier in Fig. 7-2. If the measured resistance values are incorrect, the air flow sensor is faulty and should be replaced.

Air flow sensor potentiometer resistance:

- terminals 3 and 4 — 500 to 1000 ohms
- terminals 2 and 3 — resistance varies with sensor flap position

Intake air temperature sensor resistance:

- terminals 1 and 4 — see Fig. 7-3

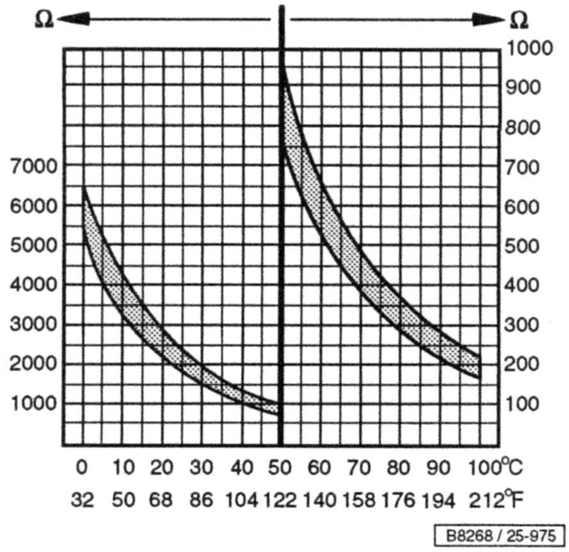

Fig. 7-3. Temperature sensor resistance vs. temperature. Example: At 68°F (20°C), resistance should be 2200 to 2800 ohms. Resistance values shown apply to both intake air temperature sensor and engine coolant temperature sensor.

Throttle Valve

The throttle valve (Fig. 7-4) regulates intake air flow to the engine. It is opened and closed by the accelerator cable. An idle switch mounted on the throttle body signals the Digifant ECU when the throttle valve is closed (idle). The throttle valve is adjusted at the factory and should not be tampered with unless the factory settings have been altered.

Throttle Valve Basic Adjustment

The factory-set throttle valve stop screw is not used to adjust idle speed. Its only function is to provide a mechanical stop for the throttle mechanism, to prevent contact (and wear) between the throttle valve and the housing. It should only be adjusted, using the procedure that follows, if the factory setting has been changed.

Loosen the locknut and back out the throttle valve stop screw until there is a gap between the lever and the tip of the stop screw. See Fig. 7-5. Place a thin piece of cellophane or paper between the screw and the lever to gauge the gap.

To adjust, turn the stop screw in until it is just gripping the paper, until the paper is binding slightly. Remove the paper and turn the screw in an additional ½ turn. Tighten the locknut.

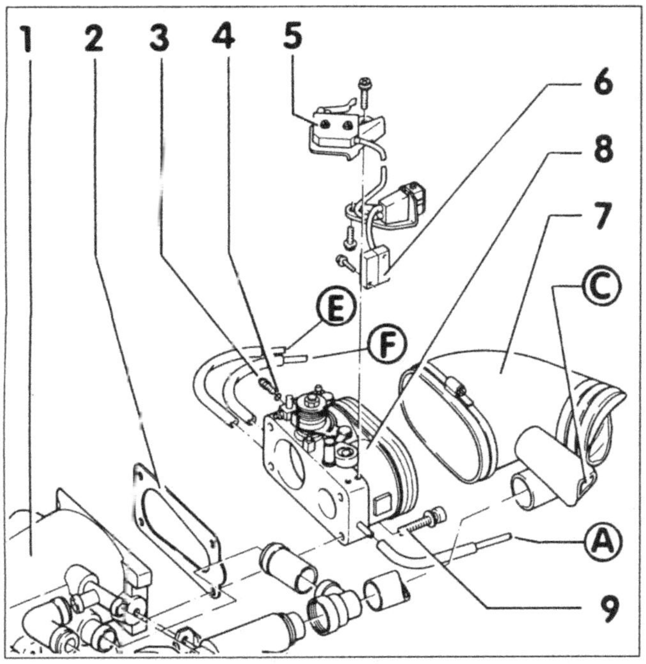

Fig. 7-4. Throttle body and throttle valve detail.
1 – Intake manifold
2 – Gasket
3 – Idle speed adjusting screw
4 – O-ring
5 – Full-throttle switch
6 – Idle switch
7 – Intake air duct
8 – Throttle body
9 – 20 Nm (15 ft. lb.)
A – to Air filter housing
E – to Carbon canister shut-off valve, small dia.
F – to Carbon canister shut-off valve, larger dia.

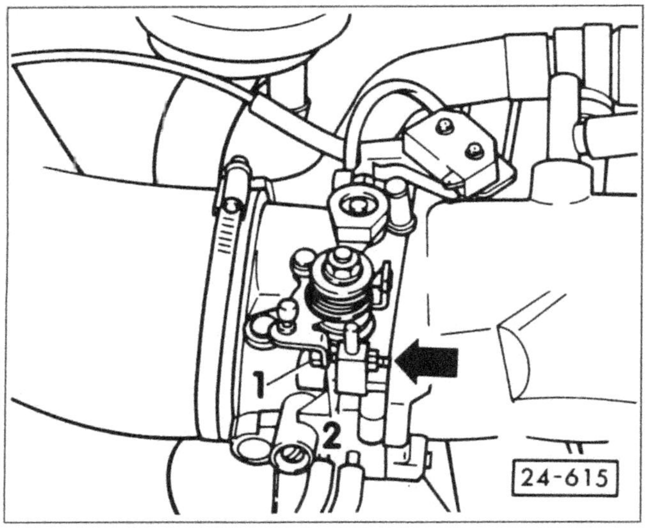

Fig. 7-5. Throttle valve stop screw with locknut. Closed throttle adjustment (arrows) is factory set. Do not adjust unless screw has been tampered with.

Idle and Full-throttle Switches

The idle and full-throttle switches are mounted on the throttle body as shown earlier in Fig. 7-4. The idle switch informs the Digifant control unit when the throttle valve is fully closed (at idle). The full-throttle switch provides a signal to the Digifant control unit that is used for full-throttle enrichment.

To be sure power is reaching the switch harness connector, disconnect it (beneath the throttle body) and measure DC voltage across the bottom two connector sockets. With the ignition switched on, there should be at least 5 volts.

To check switch functions, operate the throttle by hand and check continuity at the switch connector. Check the idle switch while approaching idle from part-throttle. Check the full-throttle switch by slowly approaching full-throttle. Adjust by loosening and repositioning the appropriate switch. Tighten the mounting screws, then recheck the adjustment.

Idle switch adjustment:

- throttle closed (idle position) — continuity
- part throttle — no continuity
- idle switch actuates (continuity) — 0.20 mm (0.008 in.) to 0.60 mm (0.024 in.) before throttle closed

Full-throttle switch adjustment:

- full-throttle — continuity
- part throttle — no continuity
- throttle switch actuates (continuity) — 10° ± 2° before full-throttle

7.3 Cold Start and Cold Running Enrichment

A cold engine requires a richer mixture—more fuel for any given amount of air. The intake air preheat system helps to improve cold-running driveability and reduce the need for extra fuel by using exhaust system heat to warm the cold intake air. A coolant temperature sensor provides an engine temperature input to the Digifant control unit that is used to boost fuel delivery increasing fuel injector open time.

Intake Air Preheat System

Fig. 7-6 shows the air filter housing, the vacuum-operated servo for the regulator flap, and the thermostatic vacuum control valve in the upper air filter housing. Intake manifold vacuum operates the flap when the vacuum control valve is open.

Fig. 7-7 shows the warm air inlet and the regulator flap inside the air filter housing. Below 68°F (20°C) the flap should seal the fresh air intake. Above 86°F (30°C), the vacuum control valve should be closed and the flap should seal the warm air intake. In between, restricted engine vacuum should keep the flap partially actuated, with intake air blending from both intakes.

Test the system by applying vacuum to the control valve or to the servo using a hand-held vacuum pump. The flap should operate smoothly throughout its range of travel.

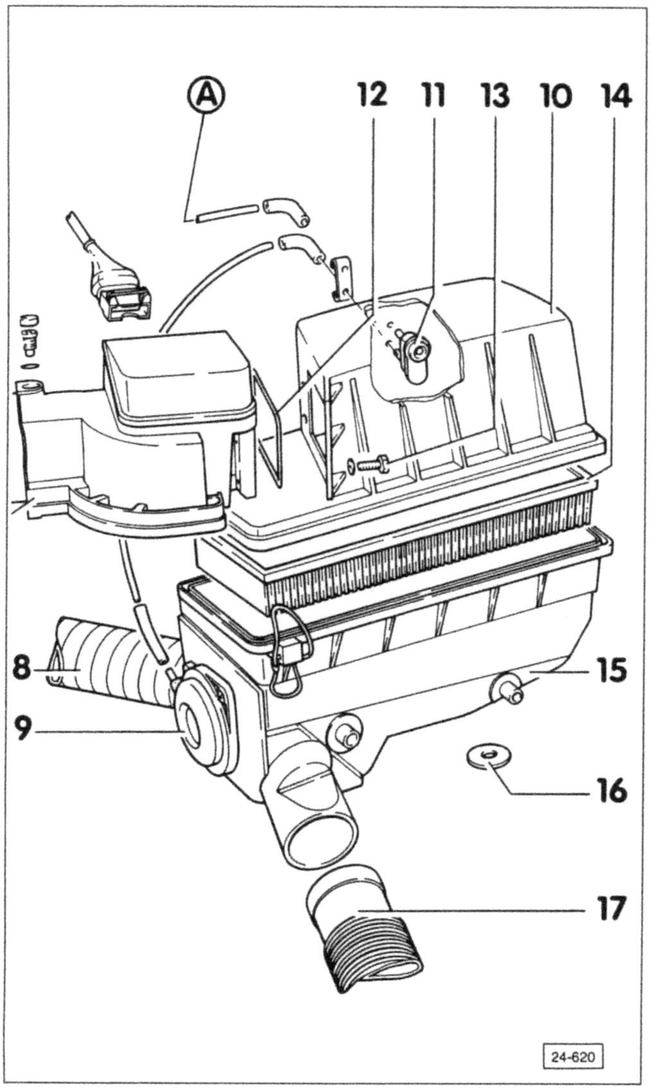

Fig. 7-6. Intake air preheat system in air filter housing. Cutaway shows thermostatic vacuum control valve that regulates engine vacuum.
- **A** – Vacuum hose from throttle body
- **8** – Warm air inlet
- **9** – Temperature control
- **10** – Air filter housing, upper
- **11** – Vacuum control valve
- **12** – Seal
- **13** – 5 Nm (44 in. lb.)
- **14** – Air filter element
- **15** – Air filter housing, lower
- **16** – Rubber washer
- **17** – Fresh air inlet

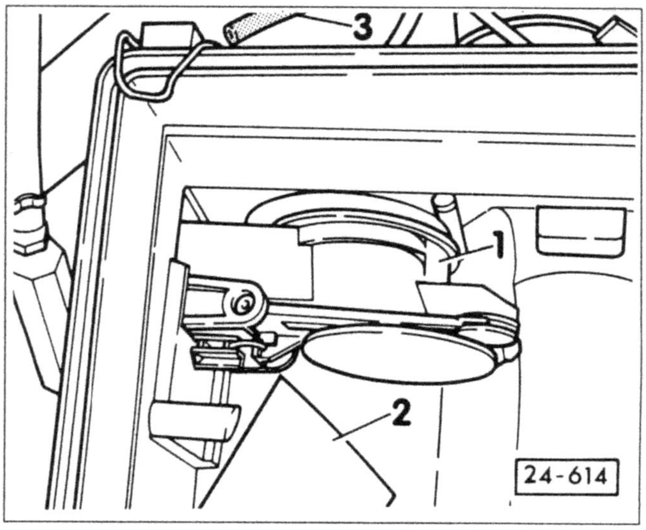

Fig. 7-7. Intake air preheat system inside air filter housing. Warm air inlet (**1**) is normally blocked by regulator flap (**2**). Cold temperature opens vacuum control valve. Engine vacuum actuates flap to open warm air inlet and seal fresh air inlet.

Coolant Temperature Sensor

The coolant temperature sensor for Digifant has a blue harness connector and is mounted in the coolant outlet on the cylinder head. See Fig. 7-6. Resistance measured across the sensor terminals should match the graph shown earlier in Fig. 7-3. If not, the sensor is faulty and should be replaced. This requires draining and replacing some of the engine coolant. See **COOLING SYSTEM.**

A large clip holds the sensor in place, pressed against a sealing O-ring. When installing, make sure that the sensor is fully seated before installing the clip. Always use a new O-ring.

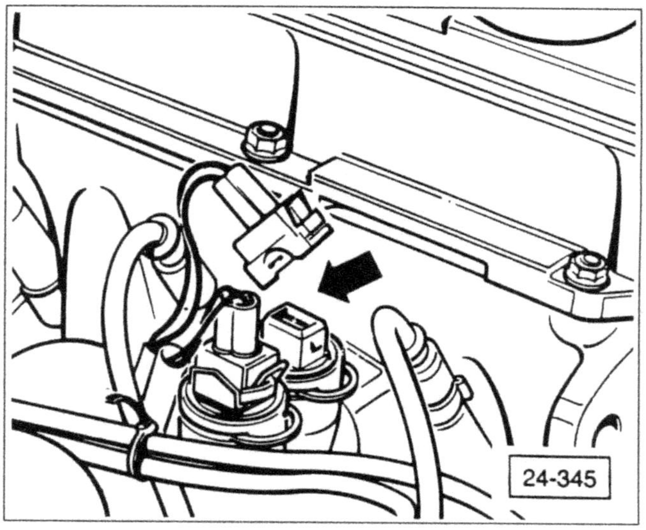

Fig. 7-8. Digifant coolant temperature sensor (arrow) on cylinder head coolant outlet has blue harness connector (shown disconnected).

7.4 Digifant II Basic Adjustments—Idle Speed and Idle Mixture

The basic Digifant engine settings are idle speed, idle mixture (% CO) and ignition timing. They establish baseline values on which all other system adjustments are based. They must always be checked and adjusted together.

This section covers the checking and adjusting procedure for cars with Digifant II. A different procedure applies for 1991 and later California cars with Digifant I. See **8.4 Digifant I Basic Adjustments—Idle Speed and Idle Mixture**.

Requirements:
- engine warm—oil temperature at least 176°F (80°C)
- radiator cooling fan must come on at least once
- idle switch functioning and correctly adjusted
- exhaust system free of leaks
- air conditioning switched off
- tachometer connected according to the instrument manufacturer's instructions to accurately measure engine rpm
- timing light or equivalent connected to measure ignition timing, see **IGNITION**
- CO tester (Sun 105 or an EPA-approved equivalent) connected to the CO tap tube (tube with blue cap, right side of exhaust manifold)

CAUTION —
- *Connect and disconnect the test leads only with the ignition off.*
- *Do NOT allow terminal 1 of the ignition coil to be shorted to ground when connecting test equipment. This will damage the Digifant II control unit.*
- *Do not make checks or adjustments while the radiator cooling fan is running. Wait for it to stop.*
- *The connection to the CO tap tube must be secure. Even small leaks will result in inaccurate measurements.*

To check basic adjustments:

1. Start the engine. Raise engine speed above 2100 rpm four times. Let the throttle close completely each time.

   NOTE —
   This step erases the control unit fault memory and cancels hot-start idle speed boost. It must be repeated any time the engine is stopped and restarted during this procedure.

2. Disconnect the blue harness connector from the coolant temperature sensor. See Fig. 7-6.

3. Raise engine speed to 2300±50 rpm and check ignition timing. If necessary, adjust ignition timing before going farther. See **IGNITION**.

   Ignition timing specification:
   - checking: 4° to 8° BTDC
   - adjusting: 6°±1° BTDC
   - engine speed: 2300±50 rpm

4. Reconnect the coolant temperature sensor connector and switch the ignition off.

5. Disconnect the crankcase ventilation hose from the crankcase emission control valve on top of the cylinder head cover. Plug the hose as shown in Fig. 7-9.

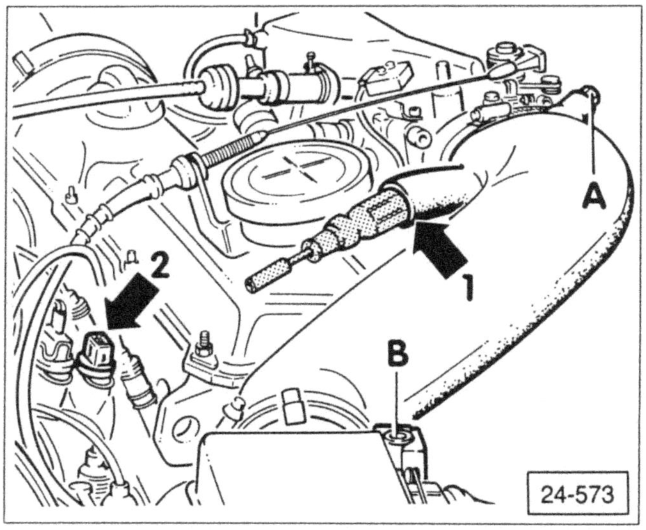

Fig. 7-9. Hose (**1**) disconnected from crankcase emission control valve and plugged, and coolant temperature sensor (**2**) disconnected for checking idle. Also shown are idle air bypass screw (**A**) and idle mixture adjusting screw (**B**).

6. Start the engine. Repeat step 1, to clear the fault memory and fast idle.

7. Disconnect the blue harness connector from the coolant temperature sensor.

8. Check idle speed. If necessary, idle speed should be corrected before going on. See **Idle Air Bypass**.

   Idle speed specifications:
   - 1990 (to July, 1990): 900 ± 50 rpm
   - from 1991 (August, 1990): 850 ± 50 rpm

9. With the engine running at idle, read the exhaust CO content and compare the measured value to the specified range shown Fig. 7-10.

10. If necessary, correct the idle mixture adjustment as described in **Idle Mixture Adjustment** later in this section. Repeat idle speed (step 9 above) and idle mixture adjustments until both are within specifications.

11. Reconnect the coolant temperature sensor connector.

12. Raise engine speed briefly to at least 3000 rpm, then let it return to idle. Repeat this step at least three times.
    - idle speed should drop to 800 ± 50 rpm
    - CO should be 0.7± 0.4% (regardless of altitude)

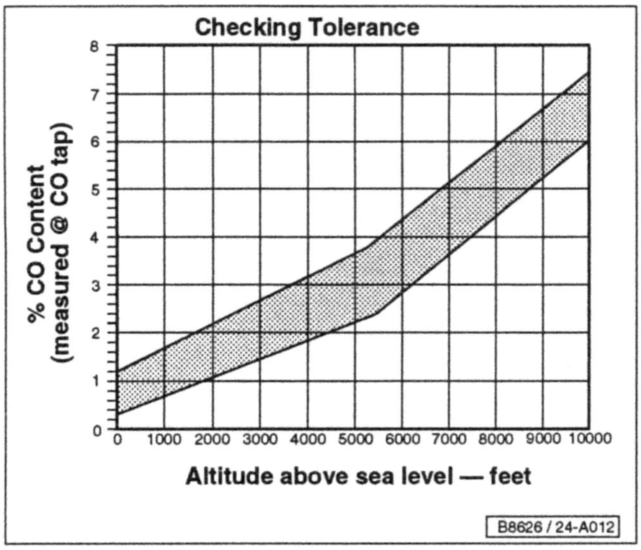

Fig. 7-10. CO content (% CO) vs. altitude specification for checking idle mixture adjustment. Measured CO value, corrected for altitude, must fall within shaded area or mixture adjustment is needed.

NOTE —
- If CO content is OK but idle speed is not, check the idle switch and the idle stabilization system.
- If idle speed is OK but CO content is not, check the oxygen sensor system. See **EXHAUST SYSTEM AND EMISISON CONTROLS**.
- If both idle speed and CO are wrong, check the coolant temperature sensor and wiring.

13. Reconnect the crankcase ventilation hose and recheck the CO content.

NOTE —
If the % CO changes when the crankcase ventilation hose is reconnected, it is not because an incorrect idle mixture adjustment. Repeated short trips that do not allow the engine to fully warm up can cause fuel vapor to contaminate the engine oil. Longer trips will return CO to normal levels. Changing oil is a temporary solution.

14. Stop the engine and disconnect the test equipment. If the CO anti-tampering plug has been removed, install a new plug.

Idle Air Bypass

Baseline idle speed is determined by the small amount of air that is allowed to bypass the throttle valve. Engine idle speed should never be adjusted by changing the throttle valve stop position. If the throttle valve has been tampered with, correct it as described in **Throttle Valve Basic Adjustment** elsewhere in this section.

The idle air bypass adjusting screw is on the back side of the throttle body as shown in Fig. 7-11. Turn the screw clockwise to restrict bypass air and reduce idle speed. Turn it counterclockwise to allow more bypass air and increase idle speed.

CAUTION —
- *Idle speed checks and adjustments are only accurate under certain precise test conditions. Make changes only when also checking ignition timing. See **7.4 Digifant II Basic Adjustments**.*

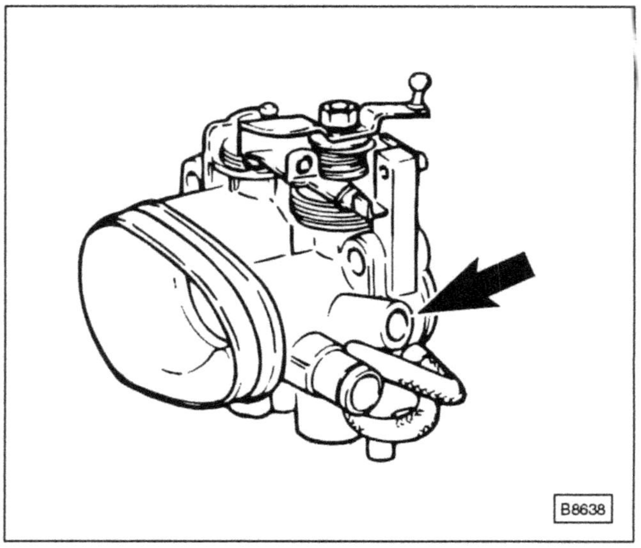

Fig. 7-11. Idle air bypass adjusting screw on throttle body. Adjust idle speed only in combination with the other basic adjustments. See text.

Idle Mixture Adjustment

The idle mixture adjustment screw is on the air flow sensor housing, as shown earlier in Fig. 7-2. Also see Fig. 7-9.

For access to the adjusting screw, remove the anti-tampering plug. Lightly center-punch and drill the plug using a 2.5 mm ($^3/_{32}$ in.) bit. Thread a sheet metal screw into the hole and grip it with a pliers to remove the plug.

CAUTION —
*CO checks and idle mixture adjustments are only accurate under certain precise test conditions. Make changes only when also checking ignition timing and idle speed. See **7.4 Digifant II Basic Adjustments**.*

The adjustment requires a 5 mm hex wrench to turn the adjusting screw. The precise adjusting specification depends on location and the altitude above sea level. Use the Fig. 7-12 graph to determine the proper adjusting value. After final adjustments a new anti-tampering plug should be installed.

CAUTION —
The radiator cooling fan must not be running while checking CO content.

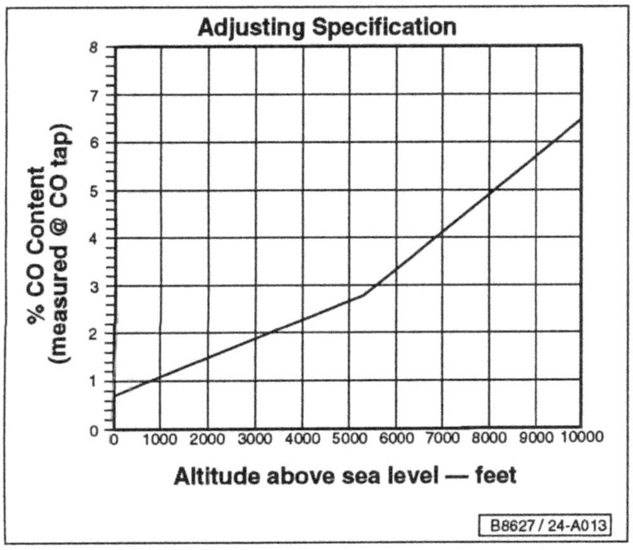

Fig. 7-12. Idle mixture adjustment (% CO) specification vs. altitude above sea level.

Idle Stabilization

The electro-mechanical idle stabilizer valve controls extra air bypassing the closed throttle valve. For cold-start and cold running, it passes extra air to boost idle speed. Under normal warm running conditions it continuously adjusts bypass air to maintain a stable idle speed.

With the ignition switched on, the valve should always hum and vibrate slightly. If not, check resistance across the valve terminals. If not OK, the valve is faulty and should be replaced.

Idle stabilizer valve:
- resistance: 2 to 10 Ω (ohms)

If resistance is OK but the valve does not buzz and vibrate with the ignition switched on, there is a fault in the wiring harness or in the Digifant control unit. See **7.7 Digifant II Electrical Tests**.

The idle stabilization function can be checked by measuring control current in the idle stabilizer valve circuit.

Test conditions:
- engine at normal operating temperature—176°F (80°C); the radiator cooling fan should have come on at least once
- coolant temperature sensor OK
- idle stabilizer valve resistance OK
- system free of intake air leaks

To check idle stabilization:

1. Disconnect the harness connector from the idle stabilizer valve (engine not running).

CAUTION —
Disconnecting the stabilizer valve with the engine running will damage the electronic control unit.

2. Connect a test harness and a multimeter (0 to 1000 mA range) to measure control current. See Fig. 7-13.

CAUTION —
- *Connecting the test harness with the meter set to measure voltage will damage the electronic control unit.*
- *The small connector terminals are easily damaged. Use the Volkswagen V.A.G. 1594 adapter kit or equivalent to make connections and prevent terminal damage.*

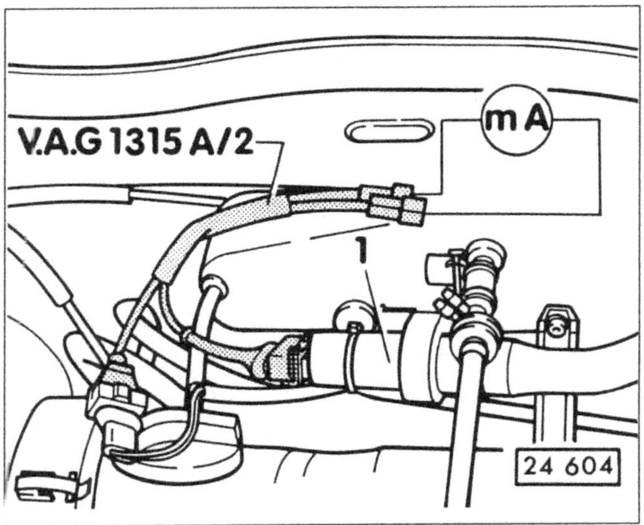

Fig. 7-13. Volkswagen special tool V.A.G. 1315A/2 test harness (order no. TV1 315 0A2 025 ZEL) being used to connect multimeter and measure idle stabilizer valve (**1**) control current.

3. Start the engine and let it idle for at least one minute.
4. Briefly raise the engine speed over 3000 rpm and let it return to idle. Repeat this step at least two more times.
5. Check idle stabilizer current.
   - approximately 390 to 450 mA and fluctuating

NOTE —
Control current normally varies between 400 and 1000 mA according to engine load at idle. It will be higher if the engine is cold, if the basic idle speed adjustment is incorrect, if there are intake air leaks, if the air conditioning or other accessories are switched on, or if the coolant temperature sensor is faulty.

6. Disconnect the blue coolant temperature sensor harness connector. See Fig. 7-8 earlier in this section.
   - control current should be a steady 420±30 mA (not fluctuating)
   - if not, the Digifant control unit is faulty and should be replaced

7.5 Fuel Injectors

The Digifant fuel injectors are solenoid valves, operated electrically by the Digifant control unit. As the control unit triggers the injectors, each solenoid opens a needle valve and discharges fuel into the intake port.

Fig 7-14 is a cut-away view of one injector and the O-rings that seal the injectors in place. The four injectors are connected to a common fuel rail, as shown in Fig. 7-15.

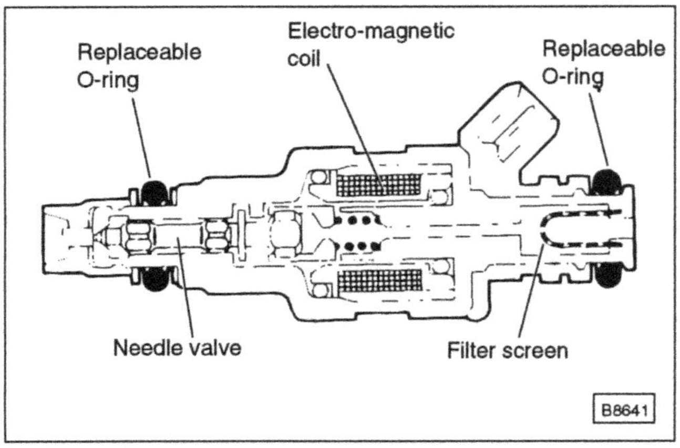

Fig. 7-14. Cut-away view of Digifant fuel injector.

The rate at which the injectors open and close varies with engine rpm, but all four injectors open and close at the same time. Fuel injection is not synchronized to crankshaft rotation.

The control unit regulates fuel delivery by controlling the amount of time that the injectors are open. The injectors are not repairable, and faulty ones must be replaced.

Testing Injectors

Fuel injector electrical faults can be detected by checking electrical resistance at the main harness connector near the end of the fuel rail, as shown in Fig. 7-16. More than 5.0 ohms resistance indicates that at least one injector is faulty.

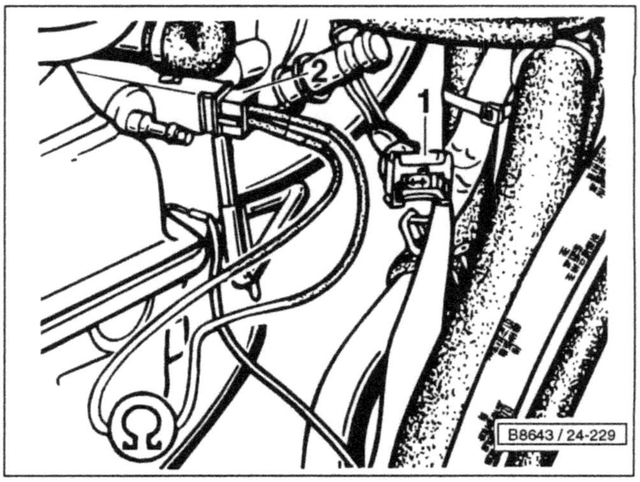

Fig. 7-16. Checking resistance of Digifant II fuel injectors. Disconnect harness connector (**1**) and connect ohmmeter (shown schematically) across injector main harness connector (**2**).

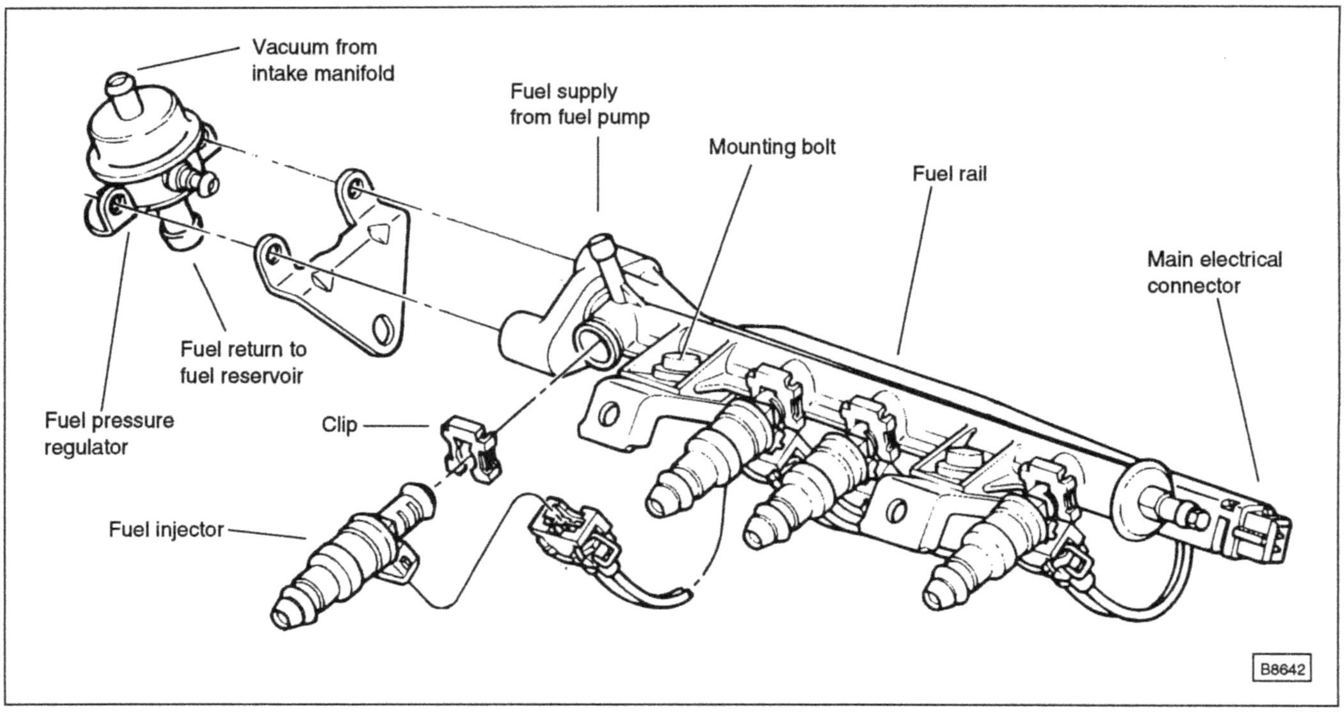

Fig. 7-15. Digifant fuel rail is common fuel supply for all injectors. On Digifant II, one two-point connector powers injectors wired in parallel.

To isolate a faulty injector, disconnect the harness connectors from individual injectors one at a time, rechecking the resistance as you go. The resistance should increase each time another injector is disconnected. Approximate test values are listed in **table g**.

Table g. Digifant Fuel Injector Resistance

Resistance — ohms	
At main injector harness connector:	
total for all 4 injectors	3.7 to 5.0 ohms
with 1 injector disconnected	5.0 to 6.7 ohms
with 2 injectors disconnected	7.5 to 10.0 ohms
with 3 injectors disconnected	15.0 to 20.0 ohms
Individual injector	15 to 20 ohms

If resistance does not increase when one injector is disconnected, check the resistance of that injector. If it is between 15 and 20 ohms, the injector is OK and the wiring or the connector for that injector is faulty. If not, the injector is faulty and should be replaced.

If the engine does not start and run, check to see whether the injectors are powered and receiving control signals from the ECU. Connect an LED test light to the wiring harness connector as shown in Fig. 7-17.

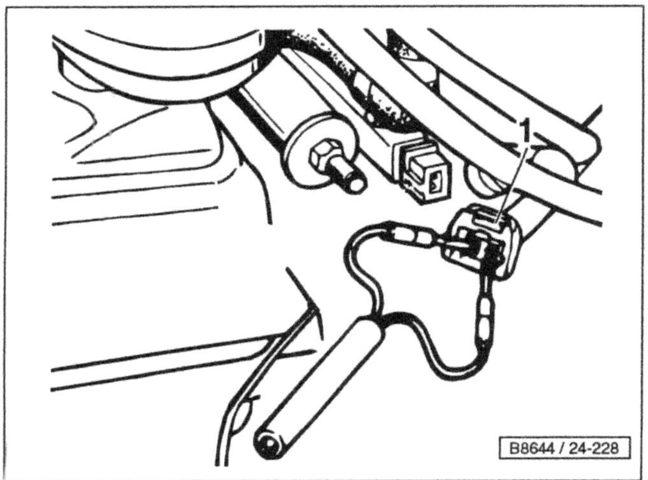

Fig. 7-17. LED test light connected to injector wiring harness (**1**) to check for voltage signal to injectors.

With the transmission in neutral and the parking brake set, actuate the starter. The LED should flicker. If it does not light at all, check for a fault in the wiring harness. If it stays lighted all the time and does not flicker, check the battery-to-transaxle ground strap and the ground connection for the electronic control unit. Also see **7.7 Digifant II Electrical Tests**.

Removing and Installing Injectors

The fuel injectors are removed and installed by removing the fuel rail. Although the entire assembly of fuel rail and injectors can be removed as a unit, it may be easier to loosen the injectors and their O-rings from the inserts if you first remove the fuel rail from the injectors. Gently pry off the injector mounting clips using a small, flat-bladed screwdriver.

> ***WARNING —***
> *Fuel will be expelled. Do not smoke. Do not work near heaters or other fire hazards. Keep a fire extinguisher handy.*

Remove each injector from its insert, and make sure that the inserts are tight. Remove loose inserts using a 10 mm hex wrench. Clean the threads and apply thread locking compound (Volkswagen D6 or equivalent) to the threads before installing.

Inspect the O-rings on the injectors (both ends) and at the pressure regulator connection to the fuel rail. Replace any that are brittle, cracked or otherwise damaged. Always use new hose clamps.

Tightening torques:	
• injector inserts	20 Nm (15 ft. lb.)
• fuel rail mounts	10 Nm (7 ft. lb.)

7.6 Fuel Pressure Tests and Specifications

Fuel delivery is mainly determined by the amount of time that the injectors are open, but correct fuel pressure is also important. The fuel pressure generated by the main fuel pump is regulated by a diaphragm-type pressure regulator mounted on the end of the fuel rail (Fig. 7-18).

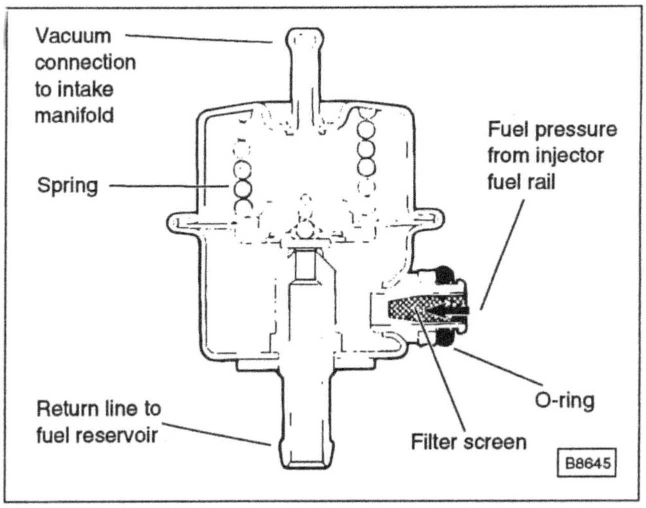

Fig. 7-18. Cut-away view of Digifant pressure regulator. Normal fuel pressure is boosted by high engine load (low intake manifold vacuum).

The regulator, with a vacuum connection to the intake manifold, controls fuel pressure as a function of engine vacuum. When the throttle is opened and there is less vacuum, fuel pressure increases slightly to richen the air/fuel mixture and improve throttle response.

CAUTION —
Cleanliness is essential when working with Digifant II fuel circuit components. Before disconnecting any fuel lines, thoroughly clean the unions. Use clean tools.

System Pressure

System pressure is the normal operating pressure maintained by the pressure regulator. When fuel pressure from the pump is too high, the regulator opens and routes fuel back to the fuel reservoir near the main pump. System pressure is not adjustable. The pressure regulator should be replaced if pressure measurements do not match the specification.

Residual Pressure

Residual pressure is fuel pressure is retained in the system after the engine has stopped running. Its purpose is to help prevent fuel vaporization as an aid to hot restarting. Residual pressure is maintained by check valves in the pressure regulator and in the fuel pump.

To test fuel pressures:

1. Wrap the fuel rail with a clean rag, then disconnect the fuel supply hose from the fuel rail. Connect a 0 to 3.5 bar (0 to 50 psi) pressure gauge as shown in Fig. 7-19.

   WARNING —
   - *Fuel will be expelled. Use a clean rag to soak up fuel. Do not disconnect any wires that could cause electrical sparks. Do not smoke or work near heaters or other fire hazards. Keep a fire extinguisher handy.*
   - *The gauge must be securely connected to prevent it coming loose under pressure.*

2. Start the engine and let it idle. Check system fuel pressure. If it is too high, the pressure regulator is faulty and should be replaced.
   - fuel pressure: approx. 2.5 bar (36 psi)
3. Disconnect the fuel pressure regulator vacuum hose from the intake manifold (Fig. 7-20) and plug the hose nipple. Fuel pressure should increase.
   - fuel pressure: approx. 3.0 bar (44 psi)
4. Switch the ignition off to stop the engine. Wait ten minutes, then check residual pressure.
   - residual pressure: min. 2.0 bar (29 psi) after 10 minutes
5. If residual pressure is OK, this test is over. Proceed to the last step.
6. If residual pressure is not OK, briefly run the engine again to restore full system pressure.

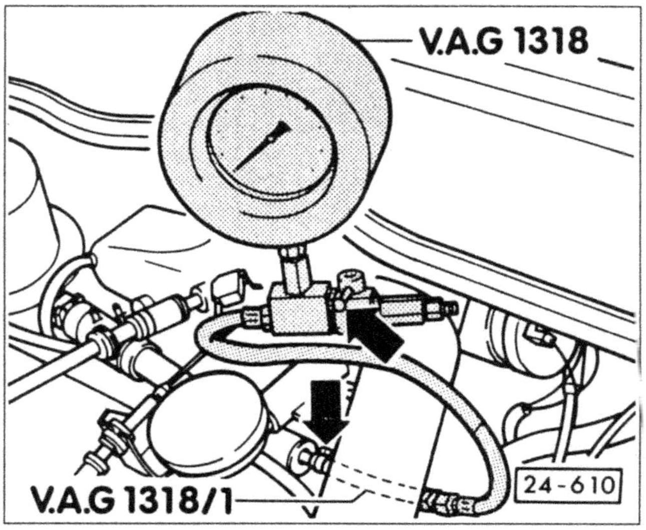

Fig. 7-19. Fuel pressure gauge (Volkswagen special tool V.A.G. 1318) connected to measure fuel pressures. V.A.G. 1318/1 Adapter connects gauge to fuel rail (bottom arrow). Gauge shut-off valve (top right arrow) should be closed.

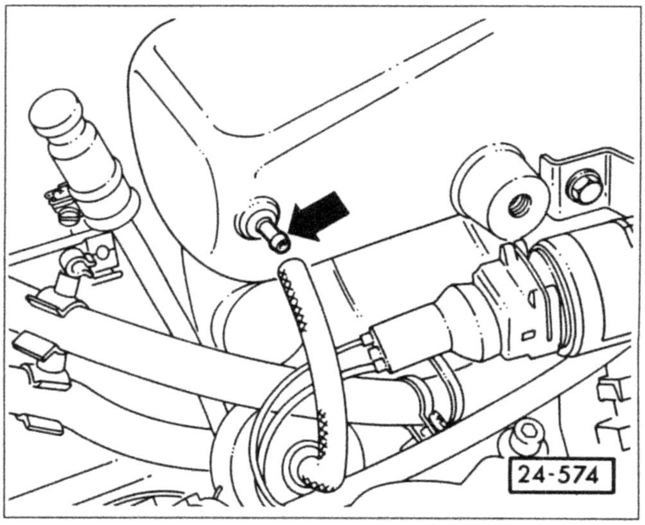

Fig. 7-20. Fuel pressure regulator vacuum hose disconnected from intake manifold connection (arrow).

7. Stop the engine and quickly clamp shut the fuel return line from the pressure regulator. If residual pressure is OK after 10 minutes with the return line blocked, the pressure regulator is faulty and should be replaced.
8. Reconnect the vacuum line at the intake manifold. Disconnect the gauge from the fuel rail, and reconnect the fuel supply line. Be prepared to catch any excess fuel. Always use new hose clamps.

If residual pressure still cannot be maintained, check for leaks in the fuel system and at the gauge connections. Other possible faults are a leaky fuel injector, leaky injector connections, damaged O-rings, or a faulty fuel pump check valve.

7.7 Digifant II Electrical Tests

This section lists a series of Digifant II electrical tests used to check component circuits and the wiring harness at the control unit harness connector. First some general advice and safety rules that apply to all Digifant electrical testing. For more information about tools and test equipment, see **FUNDAMENTALS**.

Test Equipment

Most of the electrical tests described here call for measuring voltage, current or resistance with a digital volt-ohm meter (DVOM) or digital multi-meter (DMM). It is most important that the meter have high input impedance—at least 10 megohm (10,000,000 Ω). This means that the meter itself places little or no current load on the circuit being tested. Most digital meters will meet this specification. Always check to be sure.

An LED test light is an inexpensive tool that can be used to perform many simple electrical tests that would otherwise require a multimeter. The LED indicates when there is voltage potential between any two test-points in a circuit.

CAUTION —
Use an LED test light, or a meter with at least 10 megohm input impedance. High current flow of a conventional test light or an analog meter with too little input impedance will damage sensitive components in the Digifant control unit.

Testing Precautions

Test probes and jumper wires must be used carefully to avoid damaging connectors. Some Volkswagen special tools are used to make test connections easily and without damage. Use of this or equivalent equipment is highly recommended.

CAUTION —
- *The harness connector terminals and sockets are small and easily damaged. Volkswagen special tool VW 1594 Adapter Kit is highly recommended to make safe connections that will not damage the small terminals.*
- *Home-made jumper wire ends should be flat-blade connectors that are the same size as the connector or relay terminals.*
- *Floating ground connections will damage the control unit. Make sure that system ground connections are clean and tight, especially at the engine cylinder head, transmission, body and negative (–) battery cable connections.*
- *Always switch meter functions and range as necessary before making test connections with the meter probes.*
- *Connect and disconnect test equipment only with the ignition switched off to avoid damaging electronic components.*

Test conditions:
- battery **OK**
- fuel pumps and fuel pump relay **OK**
- ground connections and ground straps **OK**
- fuses #5 (S5) and #17 (S17) **OK**
- switch ignition **OFF**
- coil wire from the center of the distributor cap disconnected and grounded with a jumper wire

The Digifant control unit is mounted under the hood, in the heater intake plenum in front of the windshield. Remove the control unit from its mounting bracket to gain better access to the control unit connector. Fig. 7-21 shows the control unit connector and identifies the terminals. Fig. 7-22 shows the Volkswagen special tools used to connect to the Digifant control unit for the electrical tests listed in **table h**.

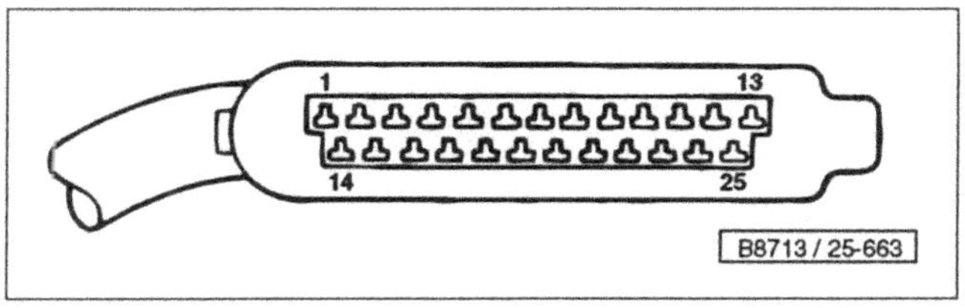

Fig. 7-21. Terminal identification for 25-pt. Digifant control unit connector. Use care to avoid damaging contacts.

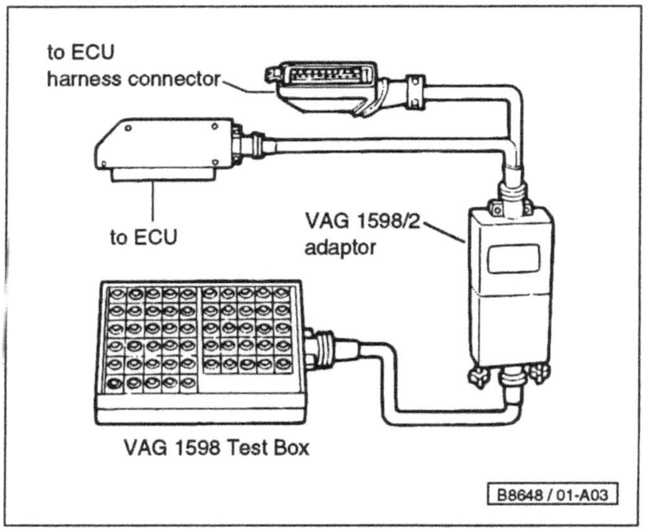

Fig. 7-22. Volkswagen special tool V.A.G. 1598 Test Box and adapter installed between ECU and ECU connector allows access to test points without risk of damage to delicate connector pins. Numbered test points on VAG 1598 correspond to ECU connector terminal numbers.

Table h. Digifant II Electrical Tests

Test step	Digifant control unit connector terminals	Checking . . .	Test conditions, additional steps	Specified value or result
DC voltage measurements (20-volt scale, control unit **NOT** connected)				
1	**13** + **14** **14** + **19**	voltage supply to Digifant control unit	– switch ignition **ON**	approximate battery voltage (about 12 volts DC)
2	bridge **3** + **13** with jumper wire	fuel pumps / fuel pump relay	– switch ignition **ON**	fuel pumps must run (audibly)
3	**1** + **13**	wiring from starter terminal 50	– operate starter	8 volts DC (minimum)
Resistance measurements (ohms scale, control unit **NOT** connected)				
4	**6** + **10**	coolant temperature sender		see graph — Fig. 7-3
5	**6** + **9**	intake air temperature sensor		see graph — Fig. 7-3
6	**22** + **23**	idle stabilizer valve		2 to 10 ohms
7	**6** + **8** **6** + **18**	wiring to Hall sender	– disconnect Hall sender connector from distributor – bridge all three connector terminals	continuity (0.5 ohms maximum)
8	**4** + **5** **4** + **7**	wiring to knock sensor	– disconnect knock sensor – bridge all three connector terminals	continuity (0.5 ohms maximum)
9	**25** to terminal **6** of Hall ignition control unit connector	wiring to Hall ignition control unit	– disconnect Hall ignition control unit connector	continuity (0.5 ohms maximum)
10	**6** + **11**	idle switch	– throttle valve closed (idle)	continuity (0.5 ohms max.)
		idle and full-throttle switches	– throttle valve between idle and full-throttle	open (infinite ohms)
		full-throttle switch	– throttle valve open (full-throttle)	continuity (0.5 ohms max.)
11	**12** to pin **5** of 5-pt. connector near coolant reservoir	fuel injectors	– disconnect 5-pt. connector near coolant reservoir	3.7 to 5.0 ohms (for all 4 injectors)
12	For cars with A/C only: **13** + **16**	wiring to A/C compressor	– disconnect connector (on brake booster) and jumper green wire to ground	continuity (0.5 ohms maximum)
13	**2** + **13**	oxygen sensor wiring	– disconnect 4-pt. oxygen sensor harness connector – jumper terminal **4** to ground	continuity (0.5 ohms maximum)
			– reconnect oxygen sensor	open (infinite ohms)
14	**6** + **17**	air flow sensor potentiometer		500 to 1000 ohms
	17 + **21**		– move air flow sensor flap or actuate starter	resistance must change
DC voltage measurement (20-volt scale, with control unit connected) Perform these tests only if the engine will not start				
15	**6** + **8**	voltage supply to Hall sender	– switch ignition **ON**	10 volts DC (minimum)
With LED test light (with control unit connected) Perform these tests only if the engine will not start				
16	**6** + **18** (with LED test light)	signal from Hall sender	– disconnect main harness connector for fuel injectors – operate starter	LED test light must flicker
17	**13** + **25** (with LED test light)	ignition switching function of Digifant II control unit	– disconnect main harness connector for fuel injectors) – operate starter	LED test light must flicker

CAUTION —
Connect and disconnect test equipment only with the ignition switched off to avoid damaging electronic components.

NOTE —
The voltage and resistance values specified in **table h** are accurate within the range of ambient temperature from 32° to 104°F (0° to 40°C).

8. DIGIFANT I FUEL INJECTION (CALIFORNIA)

The Digifant I engine management system is a version of the Digifant II system developed for use on California cars. Beginning with the 1991 model year, all Cabriolet models built for sale in California are equipped with Digifant I.

Digifant I is much like Digifant II in its basic design and overall operation. The engine's ignition and fuel injection functions are combined into one system, informed by a single array of sensors and managed by a single electronic control unit. The significant differences included in Digifant I are:

- New 38-pin Digifant I control unit with permanent fault memory and expanded on-board diagnostic capability (rapid data transfer)
- Adaptive learning capability for idle stabilization and fuel mixture control
- Throttle valve potentiometer to signal throttle position (replacing idle and full-throttle switches)
- Simplified idle mixture adjustment procedure
- New ignition coil with separate power stage

This section is a Digifant I supplement. It mainly describes the Digifant I components, functions and procedures that are different from Digifant II. See **7. Digifant II Fuel Injection** for more general information about the design and operation of both Digifant engine management systems, and for those procedures that are not covered in this section. For information about the ignition functions of Digifant I, see **IGNITION**.

The most important feature affecting service of Digifant I is the control unit fault memory and its on-board diagnosis (OBD) capability. Volkswagen specifies reading the fault memory and correcting the indicated faults before starting any other work on the Digifant I system. If you lack the skills and equipment to do the job properly, Volkswagen service technicians are uniquely equipped and trained to test and service the system for you.

CAUTION —

- *Volkswagen specifies that the first step in any Digifant I system troubleshooting must be to check the OBD fault memory, and to find and correct any faults that may be indicated.*

- *Always check the Digifant I OBD fault memory and correct any indicated faults before checking or adjusting any of the basic engine settings such as ignition timing or the idle mixture adjustment. This especially applies to a car that may still be protected by any of Volkswagen's extensive warranties.*

- *Always switch the ignition off before disconnecting or connecting the Digifant I control unit harness connector, and before disconnecting or connecting test equipment. Doing so with the ignition on risks permanent control unit damage.*

- *Do NOT allow terminal 1 of the ignition coil to be shorted to ground when connecting test equipment. Doing so will damage the Digifant control unit.*

8.1 Digifant I On-Board Fault Diagnosis

Built-in on-board diagnosis (OBD) circuitry monitors the range of system values in normal operation, detects unusual values and other faults, and stores fault information for later recall. If the system detects a fault, the "CHECK" warning light on the instrument panel will be illuminated.

The warning light is illuminated every time the ignition is switched on, and should go out shortly after the engine starts. If the light stays on it means that the system has detected a fault. The light may go out some time after the engine starts if the fault is temporary, but the fault will remain stored in the control unit's permanent memory until it is deliberately erased.

Volkswagen dealer service technicians use a special test instrument, Volkswagen special tool VAG 1551, connected to the ECU through a set of dedicated diagnostic connectors in the center console (Fig. 8-1). This powerful tool is the only way to use the full diagnostic and troubleshooting capability of the Digifant I control unit—to test system components and check ECU output functions in addition to accessing the control unit fault memory and displaying the coded fault information.

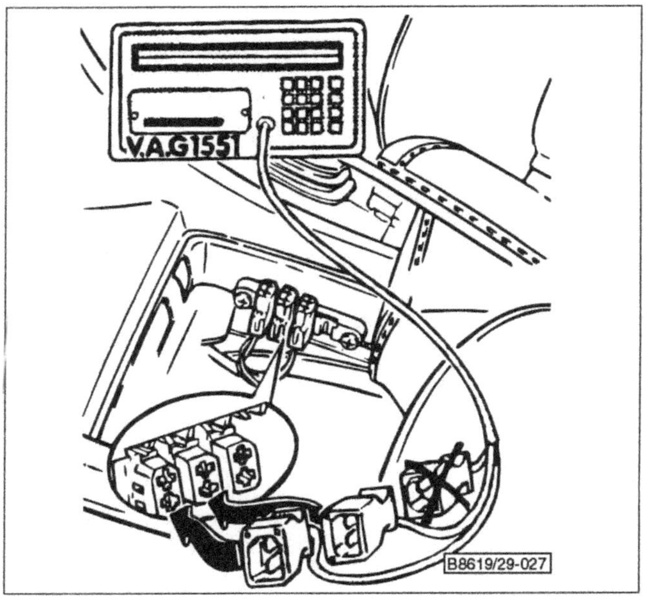

Fig. 8-1. Volkswagen special tool VAG 1551 Tester connected to diagnostic connectors to read Digifant I OBD fault codes.

The control unit fault memory can be accessed without the VAG 1551 tester by connecting a special jumper cable to the test connectors, Volkswagen special tool no. 357 971 514E. With the jumper connected, the "CHECK" light will flash codes to indicate faults stored in memory. For example, a sequence of two flashes, three flashes, four flashes and two flashes indicates the code 2-3-4-2.

Each flash code sequence corresponds to a particular fault listed in **table i**. The numbers in bold type in the probable cause column refer to pertinent information in other numbered headings within this section.

Table i. Digifant I On-Board Diagnosis (OBD) Fault Codes

Flash code	Fault location	Probable cause
4444	No faults recorded	—
1111	Control unit	Defective control unit. **8.7**
2141 or **2142**	Knock control system	Defective knock sensor, open or short circuit in wiring, or faulty Digifant control unit. See **IGNITION**.
2212	Throttle valve potentiometer	Defective potentiometer or wiring. **8.2**, **8.7**
2312	Coolant temperature sensor	Defective coolant temperature sensor or sensor wiring. **7.3**, **8.7**
2322	Intake air temperature sensor	Defective temperature sensor or wiring. **7.2**, **8.7**
2323	Air flow sensor potentiometer	Defective air flow sensor or wiring. **7.2**, **8.7**
2341	Oxygen sensor control limit exceeded	Basic adjustments faulty. **8.4** Oxygen sensor wiring short circuit to positive (+) or to ground (–). **8.7**; Intake air or exhaust leaks; Evaporative emission controls faulty. **4.5**
2342	Oxygen sensor	Defective oxygen sensor or sensor wiring. **8.7** Also see **EXHAUST SYSTEM**.
4411	Fuel injector	Open circuit in injector wiring, or faulty injector. **8.7**
0000	End of fault code sequence	—

To access fault memory using jumper cable:
(Volkswagen Part No. 357 971 514E)

1. Switch the ignition on.
2. Connect the black end of the jumper cable to the black diagnostic connector. Connect the white end of the jumper to the white diagnostic connector.

   CAUTION —
   Part numbers are listed for reference only. Always check with an authorized Volkswagen dealer Parts department for up-to-date information.

3. After five seconds the OBD CHECK light should begin to flash. Remove the jumper, but do not switch off the ignition. Record the number of flashes in sequence.
4. When code **4444** or **0000** appears, the fault memory is clear and no more faults are stored. Switch the ignition off or reconnect the jumper to end the code display.

   NOTE —
   0000 is indicated by 2 1/2 second flashes at 2 1/2 second intervals.

If wiring or component faults are indicated, check further as described in **8.7 Digifant I Electrical Tests** before replacing a suspected component. Once the faults are corrected, the permanent fault memory should be erased.

To erase fault memory using jumper cable:

1. Make sure the ignition is switched off.
2. Connect the jumper cable as in the previous fault code access procedure (black to black, white to white).
3. Switch the ignition on.
4. After 5 seconds remove the jumper cable. The OBD light should flash the code **4444**.
5. Switch the ignition off or reinstall the jumper cable. If no new faults exist, the fault memory will be cleared.

Resetting Digifant I Control Unit

Because of its adaptive capability, the Digifant I control unit must be returned to its reference settings:

- if the coolant temperature sensor (blue connector) has been disconnected while the engine is running
- if any of these components have been replaced:
  - Digifant control unit
  - air flow sensor
  - throttle valve potentiometer
  - throttle body

Requirements:

- engine at normal operating temperature—176°F (80°C); the radiator cooling fan should have come on at least once
- exhaust system must be free of leaks
- idle stabilization system OK
- all electrical accessories must be switched off
- engine not running

To reset the Digifant I control unit:

1. Disconnect the crankcase ventilation hose from the emission control valve atop the cylinder head cover, then plug the hose. See Fig. 7-9 earlier in this section.
2. Start the engine and let it idle.
3. Disconnect the coolant temperature sensor (blue connector. See earlier Fig. 7-8 earlier in this section.
4. After one minute, reconnect the harness connector.

   CAUTION —
   Do not reconnect the coolant temperature sensor while the radiator cooling fan is running.

5. Stop the engine. Unplug and reconnect the crankcase ventilation hose.

6. Check and clear the OBD fault memory as described earlier.

> **NOTE —**
> The fault memory needs to be erased because disconnecting the coolant temperature sensor with the engine running registers as a fault.

8.2 Digifant I Intake Air Flow

On Digifant I engines, the idle and full-throttle switches are replaced by a throttle valve potentiometer. Otherwise the air flow sensor, intake air temperature sensor, and all applicable tests and adjustments are the same as for Digifant II. See **7.2 Intake Air Flow** for procedures and specifications.

> ***CAUTION —***
> • *Volkswagen specifies that the first step in any Digifant I system troubleshooting must be to check the OBD fault memory, and to find and correct any faults that may be indicated. See* **8.1 Digifant I On-Board Fault Diagnosis**.
>
> • *If the air flow sensor or throttle valve is replaced, the control unit must be returned to its reference settings. See* **Resetting Digifant I Control Unit.**

Throttle Valve Potentiometer

The throttle valve potentiometer is a throttle position sensor, used by the Digifant control unit as an input for acceleration enrichment. At closed-throttle, the potentiometer doubles as an idle switch, activating idle speed stabilization and deceleration fuel shut-off. It also provides a substitute engine-load signal in the event of an air flow sensor failure. The potentiometer is mounted beneath the throttle housing. See Fig. 8-2.

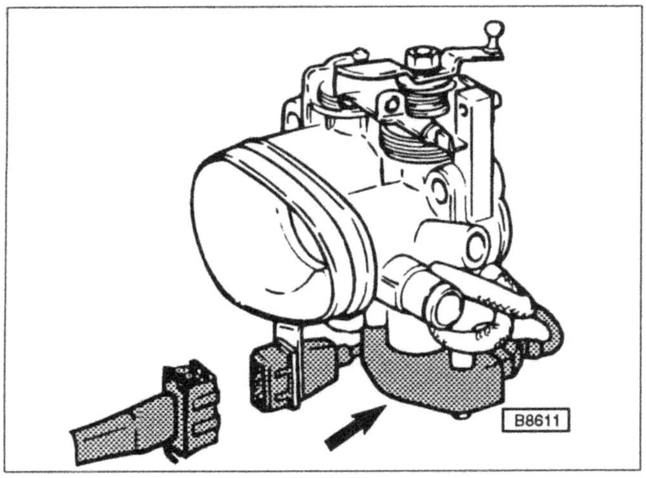

Fig. 8-2. Digifant I throttle valve potentiometer (arrow).

The control unit responds to varying resistance across the potentiometer. You can check this at the harness connector with a digital multimeter. As you open and close the throttle by hand, the resistance between potentiometer terminals 1 and 2 should increase and decrease smoothly with no dead spots. Also see **8.7 Digifant I Electrical Tests.**

> ***CAUTION —***
> • *Always switch the ignition off before changing meter ranges, and before disconnecting or connecting the meter probes. Doing so with the ignition on risks permanent control unit damage.*
>
> • *If the throttle valve potentiometer is replaced, the control unit must be returned to its reference settings. See* **Resetting Digifant I Control Unit.**

8.3 Cold Start and Cold Running Enrichment

The Digifant I intake air temperature sender, intake air preheat system, and all applicable tests and adjustments are the same as for Digifant II. See **7.3 Cold Start and Cold Running Enrichment** for procedures and specifications.

> ***CAUTION —***
> • *Volkswagen specifies that the first step in any Digifant I system troubleshooting must be to check the OBD fault memory, and to find and correct any faults that may be indicated. See* **8.1 Digifant I On-Board Fault Diagnosis**.
>
> • *If the coolant temperature sensor is replaced, the control unit must be returned to its reference settings. See* **Resetting Digifant I Control Unit.**

8.4 Digifant I Basic Adjustments— Idle Speed and Idle Mixture

On Digifant I engines, idle mixture control is adaptive. The control unit "learns" the normal range of values used to maintain proper control of the fuel mixture, then alters its operating range to compensate for changing conditions such as altitude changes, intake air leaks, engine wear and other variations.

Because of its adaptive capability, Digifant I basic adjustments are slightly different. The complete procedure is detailed here. Otherwise the basic idle speed adjustment, idle speed boost valves and idle speed stabilization are the same as for Digifant II. See **7.4 Digifant II Basic Adjustments.**

> ***CAUTION —***
> *Volkswagen specifies that the first step in any Digifant I system troubleshooting must be to check the OBD fault memory, and to find and correct any faults that may be indicated. See* **8.1 Digifant I On-Board Fault Diagnosis**.

Requirements:

- engine warm—oil temperature at least 176°F (80°C)
- radiator cooling fan must come on at least once
- exhaust system free of leaks
- idle stabilization system OK
- air conditioning switched off

Test equipment:

- tachometer to accurately measure engine rpm—connect according to instrument manufacturer's instructions
- timing light or equivalent to measure ignition timing, see **IGNITION**
- CO tester (Sun 105 or an EPA-approved equivalent) connected to the CO tap tube (tube with blue cap, right side of exhaust manifold)

CAUTION —

- *Connect and disconnect test equipment only with the ignition switched off.*
- *Do NOT allow terminal 1 of the ignition coil to be shorted to ground when connecting test equipment,. This will damage the Digifant control unit.*
- *Do not make any checks or adjustments while the radiator cooling fan is running. Wait for it to stop.*
- *The connection to the CO tap tube must be secure. Even small leaks will result in inaccurate measurements.*

To check basic adjustments:

1. Start the engine and let it idle.
2. Disconnect the blue harness connector from the coolant temperature sensor. See Fig. 7-8.
3. Raise engine speed to between 2000 and 2500 rpm and check ignition timing. If necessary, adjust ignition timing before going farther. See **IGNITION**.

   Ignition timing specification:

   - checking: 4° to 8° BTDC @ 2000 to 2500 rpm

4. When ignition timing is correct, reconnect the blue coolant temperature sensor connector.

   CAUTION —
   Do not reconnect the coolant temperature sensor while the radiator cooling fan is running.

5. Check idle speed and CO content (% CO).

   Specifications:

   - idle speed (checking) 750 to 850 rpm
   - CO content (tap tube) 0.3 to 1.2% (volume)

   CAUTION —

   - *Do not attempt to adjust idle mixture (% CO) using the sealed screw in the air flow sensor.*
   - *Do not attempt to adjust idle speed using the sealed bypass adjusting screw at the throttle body.*

6. Switch the ignition off. If idle speed and CO are OK, no further adjustments are necessary. Disconnect the test equipment. Then erase the control unit fault memory to clear the coolant temperature sensor fault. See **8.1 Digifant I On-board Fault Diagnosis**.
7. To proceed with adjustments, disconnect the crankcase ventilation hose from the crankcase emission control valve on top of the cylinder head cover. Plug the hose as shown in Fig. 8-3.

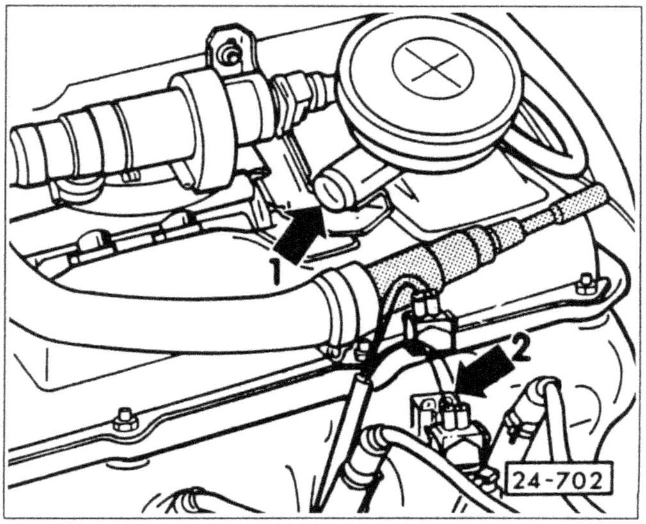

Fig. 8-3. Hose (**1**) disconnected from crankcase emission control valve and plugged for checking idle speed and CO content (% CO).

8. Start the engine and let it idle until the radiator cooling fan cycles on and off at least once.
9. With the engine running, disconnect the blue coolant temperature sensor connector.
10. After one full minute, reconnect the coolant temperature sensor connector.

    CAUTION —
    The radiator cooling fan must not be running.

11. Reconnect the crankcase ventilation hoses.
12. Check and erase the OBD fault memory as described in **8.1 Digifant I On-board Fault Diagnosis**.

    NOTE —
    The fault memory needs to be erased because disconnecting the coolant temperature sensor with the engine running registers a fault.

8.5 Digifant I Fuel Injectors

All four Digifant I fuel injectors open and close at the same time just like Digifant II, but the injector ground circuit connections to the control unit are separate for each injector. Separate wiring allows the on-board diagnosis (OBD) system to monitor individual injectors. Remember that injectors are opened and closed by the control unit making and breaking the injector ground circuit.

The procedure that follows covers basic Digifant I injector tests. Also see **8.7 Digifant I Electrical Tests**. Otherwise, the basic layout of the fuel rail and fuel injectors and all other tests and procedures are the same as for Digifant II, as covered in **7.5 Fuel Injectors**.

> ***CAUTION —***
> *Volkswagen specifies that the first step in any Digifant I system troubleshooting must be to check the OBD fault memory, and to find and correct any faults that may be indicated. See* **8.1 Digifant I On-Board Fault Diagnosis.**

Testing Injectors

Fuel injector electrical faults can be detected by testing at the main harness connector, as shown in Fig. 8-4.

Requirements:
- Hall sender signal from ignition system OK.

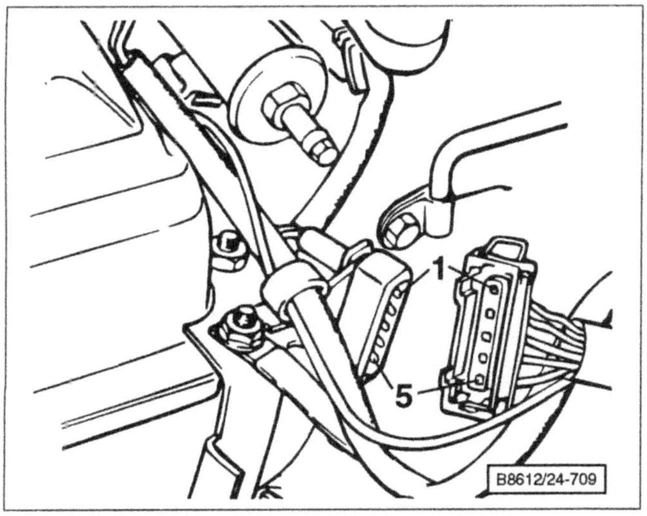

Fig. 8-4. Main connector for Digifant I fuel injectors.

To check power and ground:

1. Disconnect the fuel injector main harness connector.
2. Connect an LED test light between connector terminals 1 and 5.
3. Briefly operate the starter. The test light should flicker, indicating that voltage is reaching the connector and that the control unit is switching the injector ground.
4. Repeat the LED test for terminals 2 and 5, terminals 3 and 5, and terminals 4 and 5.

> ***CAUTION —***
> *Connect and disconnect the test light only with the ignition switched off.*

If the LED stays on continuously, there is a faulty ground connection. For problems with one or more injector circuits, check the wiring between the injector connector and the Digifant control unit.

If the LED stays on for tests of all four circuits, especially check the ground strap and connections between the engine/-transmission assembly and the battery.

If the LED does not light at all, the problem may be a faulty ground connection or a lack of supply voltage. See **8.7 Digifant I Electrical Tests.**

To check injector resistance:

1. Connect a digital multimeter to measure resistance (ohms) between terminals 1 and 5 of the open connector on the fuel rail. Repeat the resistance check for terminals 2 and 5, terminals 3 and 5, and terminals 4 and 5.

   Digifant I fuel injector resistance (ohms):

• terminals **1** and **5**	15 to 20 Ω
• terminals **2** and **5**	15 to 20 Ω
• terminals **3** and **5**	15 to 20 Ω
• terminals **4** and **5**	15 to 20 Ω

   > ***CAUTION —***
   > *Connect and disconnect test equipment only with the ignition switched off.*

2. If any of the injector resistance readings is incorrect, check the wiring between the main connector and the individual injector connectors. If the wiring is OK, then the injector is probably faulty and should be replaced.
3. Reconnect the fuel injector harness connector.
4. Return the Digifant control unit to its basic settings. See **Resetting Digifant I Control Unit.**

8.6 Fuel Pressure Tests and Specifications

The Digifant I fuel pumps, fuel pressure regulator and all applicable tests and procedures are the same as for Digifant II. See **7.6 Fuel Pressure Tests and Specifications.**

8.7 Digifant I Electrical Tests

This section includes a series of Digifant I electrical tests used to check components and the wiring harness at the control unit harness connector. First, please read the advice on test equipment and testing precautions that apply to all Digifant electrical testing. See **7.7 Electrical Tests.**

CAUTION —

- *Use an LED test light, or a meter with at least 10 megohm input impedance. High current flow from an analog meter or a conventional test light may damage sensitive electronic components in the Digifant control unit.*
- *Floating ground connections will damage the control unit. Check for good clean system ground connections, especially at the engine cylinder head, transmission, body and negative (–) battery cable connections.*
- *Always switch meter functions and range as necessary before making test connections with the meter probes.*
- *Connect and disconnect test equipment only with the ignition switched off to avoid damaging electronic components.*
- *Volkswagen specifies that the first step in any Digifant I system troubleshooting must be to check the OBD fault memory, and to find and correct any faults that may be indicated. See* **8.1 Digifant I On-board Fault Diagnosis.**

Test conditions:
- battery **OK**
- fuses **5, 17** and **18 OK**
- fuel pumps and fuel pump relay **OK**
- ground connections and ground straps **OK**
- switch ignition **OFF**
- disconnect the coil wire from the center of the distributor cap and connect it to ground with a jumper wire

The Digifant control unit is mounted under the hood, in the heater intake plenum in front of the windshield. Remove the control unit from its mounting bracket to gain better access to the control unit connector. Fig. 8-5 shows the control unit connector and identifies the terminals. Fig. 8-6 shows the Volkswagen special tools used to connect to the Digifant I control unit for the electrical tests listed in **table j.**

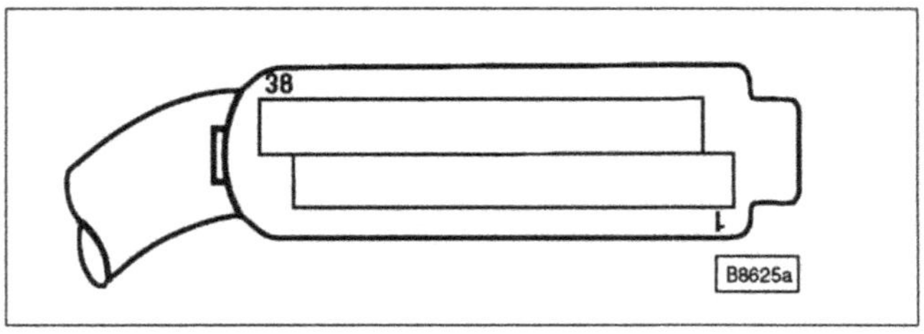

Fig. 8-5. Terminal orientation for 38-pin Digifant I control unit connector. Use care to not damage connector terminals with test probes.

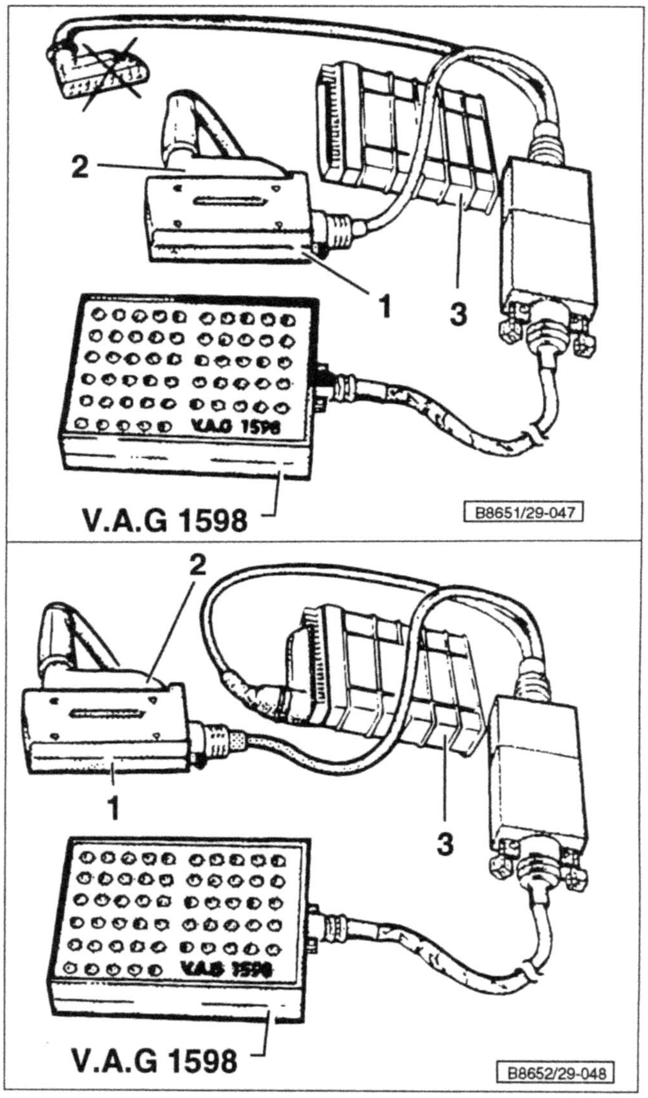

Fig. 8-6. Volkswagen special tool VAG 1598 Test Box with VAG 1598/9 adapter (**1**) allows access to ECU connector (**2**) and ECU (**3**) for testing without risk of damage to delicate connector pins. Test point numbers on VAG 1598 correspond to ECU connector terminal numbers. Connect as shown to test with control unit disconnected (top) or connected (bottom). See **table j.**

Table j. Digifant I Electrical Tests

Test step	Digifant control unit connector terminals	Checking . . .	Test conditions, additional steps	Specified value or result
DC voltage measurements (20-volt scale, control unit **NOT** connected)				
1	**1** + **19** **19** + **35**	voltage supply to Digifant control unit	– switch ignition **ON**	approximate battery voltage (about 12 volts DC)
2	**1** + **6**	wiring to ignition coil	– switch ignition **ON**	approximate battery voltage (about 12 volts DC)
3	**1** + **4**	fault indicator (OBD) light	– switch ignition **ON**	approximate battery voltage (about 12 volts DC)
4	**1** + **23**	idle stabilizer valve	– switch ignition **ON**	approximate battery voltage (about 12 volts DC)
5	bridge **1** + **24** with jumper wire	wiring to fuel pump relay	– switch ignition **ON**	fuel pump must run (audibly)
6	**1** + **20** **1** + **21** **1** + **3** **1** + **2**	wiring to fuel injectors* • cylinder #1 • cylinder #2 • cylinder #3 • cylinder #4	– remove fuse #5 – bridge terminals **1** + **24** – switch ignition **ON** * check injector resistance. **8.5 Digifant I Fuel Injectors**	approximate battery voltage (about 12 volts DC)
7	For automatic trans. only: **19** + **25**	wiring to starter lockout	– switch ignition **ON** – transmission selector lever in Park (**P**) or Neutral (**N**)	approximate battery voltage (about 12 volts DC)
8	For manual trans. only: **1** + **25**	wiring from starter terminal 50	– operate starter – switch ignition **OFF**	8 volts DC (minimum)
Resistance measurements (ohms scale, control unit **NOT** connected, ignition switched **OFF**)				
9	**11** + **28**	coolant temperature sender		see graph — Fig. 7-3
10	**28** + **30**	intake air temperature sensor		see graph — Fig. 7-3
11	**17** + **36**	diagnostic connector	– bridge terminals of white diagnostic connector	1.5 ohms maximum
12	**15** + **28**	air flow sensor potentiometer		500 to 1000 ohms (depends on flap position)
	13 + **15**		– move air flow sensor flap or actuate starter	resistance must change
13	**34** (of ECU) + terminal **1** (of potentiometer connector) **33** (of ECU) + terminal **2** (of potentiometer connector) **28** (of ECU) + terminal **3** (of potentiometer connector)	throttle valve potentiometer wiring	– switch ignition **OFF** – disconnect 3-point connector from throttle valve potentiometer	1.5 ohms maximum
14	**28** (of ECU) + terminal **1** (of Hall sender connector) **37** (of ECU) + terminal **2** (of Hall sender connector) **38** (of ECU) + terminal **3** (of Hall sender connector)	Hall sender wiring	– disconnect harness connector from Hall sender at distributor	1.5 ohms maximum
15	**29** (of ECU) + terminal **1** (of Hall sender connector) **27** (of ECU) + terminal **2** (of Hall sender connector) **8** (of ECU) + terminal **3** (of Hall sender connector)	knock sensor wiring	– disconnect knock sensor from harness connector	1.5 ohms maximum

(continued on next page)

CAUTION —
Connect and disconnect test equipment only with the ignition switched off to avoid damaging electronic components.

NOTE —
The voltage and resistance values specified in **table j** are accurate within the range of ambient temperature from 32° to 104°F (0° to 40°C).

Table j. Digifant I Electrical Tests (continued)

Test step	Digifant control unit connector terminals	Checking . . .	Test conditions, additional steps	Specified value or result
16	**1** (of ECU) + terminal **2** (of oxygen sensor connector) **16** (of ECU) + terminal **3** (of oxygen sensor connector)	oxygen sensor wiring	– disconnect oxygen sensor harness connector	1.5 ohms maximum
	9 (of ECU) + terminal **3** (of oxygen sensor connector)			infinite ohms (open circuit)
	1 + 16		– reconnect harness connector to oxygen sensor	
17	For cars with A/C only: **32** (of ECU) + terminal for green wire of A/C connector	wiring to A/C compressor	– disconnect single pin connector (near coolant reservoir)	1.5 ohms maximum
DC voltage measurement (20-volt scale, with control unit connected)				
18			– check control unit fault memory, then erase, see **8.1 Digifant I On-board Fault Diagnosis**	
19	**34** + **28**	voltage supply for throttle valve potentiometer	– disconnect throttle valve potentiometer – switch ignition **ON**	approx. 4.5 to 5.5 volts DC
20			– check control unit fault memory, then erase	
21	**33** + **28**	throttle valve potentiometer	– switch ignition **ON** – reconnect throttle valve potentiometer harness connector throttle valve: • closed (idle) • full-throttle	 0.3 to 1.7 volts DC 3 to 5 volts DC
22	**28** + **38** (perform this step only if the engine will not start)	voltage supply to Hall sender	– switch ignition **ON**	10 volts DC (minimum)
With LED test light (with control unit connected) Perform these tests (#23 and #24) only if the engine will not start				
23	**37** + **38** (with LED test light)	signal from Hall sender	– disconnect main harness connector for fuel injectors (on/near fuel rail) – operate starter	LED test light must flicker
24	**1** + **6** (with LED test light)	ignition switching function of Digifant I control unit	– disconnect main harness connector for fuel injectors – operate starter	LED test light must flicker

CAUTION —
Connect and disconnect test equipment only with the ignition switched off to avoid damaging electronic components.

NOTE —
The voltage and resistance values specified in **table j** are accurate within the range of ambient temperature from 32° to 104°F (0° to 40°C).

9. FUEL SYSTEM TECHNICAL DATA

I. CIS Fuel Injection Specifications

Fuel Pump Delivery Rate

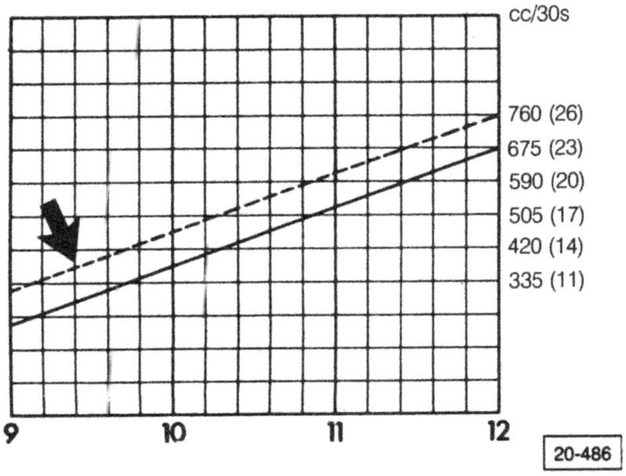

Voltage at fuel pump — DC volts

Fuel pressures
- system pressure................4.7 to 5.4 bar (68 to 78.3 psi)
- control pressure – cold
  (harness connectors for control pressure regulator and auxiliary air regulator disconnected)..... see graph below

Control Pressure

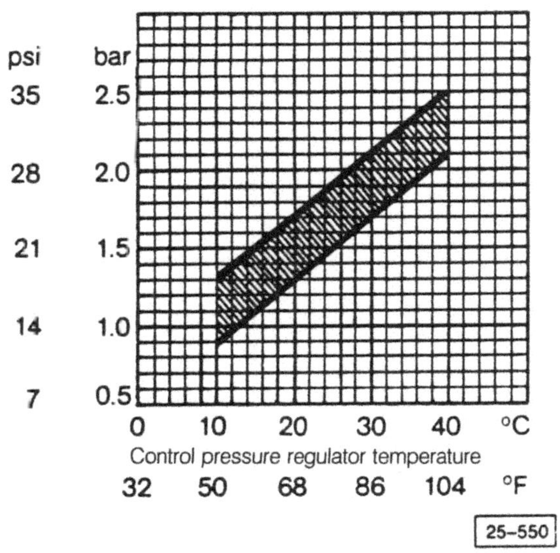

NOTE —
Control pressure variations of less than 0.05 bar (1 psi) are not significant.

Fuel pressure (continued)
- control pressure – warm.......... 3.4 to 3.8 bar (49 to 55 psi)
- residual pressure (minimum after 10 min.) ..... 2.4 bar (35 psi)

Idle speed .... 925±75 rpm

Idle mixture (CO content)
- checking .... 0.3 to 3.0 %
- adjusting (only if not within checking range)....... 1.0 ± 0.2 %

II. CIS-E Fuel Injection Specifications

Fuel Pump Delivery Rate

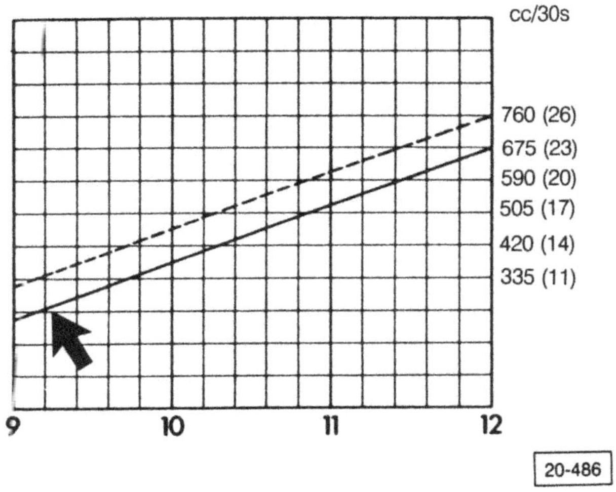

Voltage at fuel pump — DC volts

Fuel pressure
- system pressure .... 5.2 to 5.6 bar (75 to 82 psi)
- differential pressure
  - (with differential pressure regulator disconnected) .... 0.2 to 0.5 bar (2.9 to 7.0 psi) **less than** system pressure
  - (with differential pressure regulator connected, coolant temperature sensor disconnected, and special tool (fixed resistance) connected at temp. sensor harness connector ..... 0.7 to 1.2 bar (10 to 17.5 psi) **less than** system pressure
- residual pressure (minimum after 10 min. ......2.6 bar (38 psi)

Idle speed (duty cycle)
- PL (16-valve)...................... 850±50 rpm (26 to 30%)

Idle mixture adjustment (differential pressure regulator current)
- checking.......................... 4 to 16 mA (fluctuating)
  - corresponding CO content .... 0.3 to 1.2% CO
- adjusting............................. 10±1 mA (average)

III. Digifant I and Digifant II Fuel Injection Specifications

Fuel injector resistance
- single injector ....3.7 to 5.0 Ω
- total for four injectors (at main injector harness connector, Digifant II only) .... 15 to 20 Ω

Fuel pressure
- system pressure
  - normal running ....2.5 bar (36 psi)
  - with vacuum hose to regulator disconnected ..3.0 bar (44 psi)
- residual pressure
  - after 10 minutes .... 2 bar (29 psi)

Idle speed
- checking.................................. 800±50 rpm
- adjusting... see text

Section 6

EXHAUST SYSTEM AND EMISSION CONTROLS

Contents

TABLES

Exhaust System and Emission Controls

Introduction

The exhaust system has two main functions: first, to provide a conduit for the rapidly expanding gasses of combustion, and second, to quiet the noise and cool the temperature of the exhaust. On cars equipped with a catalytic converter, the exhaust system and the emission control system remove most of the pollutants resulting from combustion.

Exhaust system function depends on each component being free from holes, with airtight seals at all joints. Emission control system function depends on proper adjustment of the fuel injection system and the ignition system, as well as on the action of the catalytic converter. Any misadjustment of the fuel injection or ignition systems can adversely affect the emission control system.

During operation, all exhaust system components are subjected to extremes in temperature and pressure, vibration, and damaging contact with all manner of road hazards. Even though the exhaust system is designed for relatively maintenance-free operation, the environment to which it is exposed means that regular inspection of the system is warranted. The only scheduled recommended emission control maintenance is the replacement of the oxygen sensor.

This section covers the maintenance, troubleshooting, and repair of the exhaust system and the emission controls. Special tools are required to remove and install the front pipe spring clamps on certain models, but otherwise only basic hand tools are required for complete servicing. Testing of the emission control system requires special equipment and skills. If you lack the necessary skills or equipment, we suggest you leave these tests or repairs to an authorized Volkswagen dealer or other qualified and properly equipped repair shop. We especially urge you to consult your authorized Volkswagen dealer service department before attempting any repairs on a car that may be subject to Volkswagen's warranty coverage.

1. GENERAL DESCRIPTION

1.1 Exhaust System

The basic exhaust system components are the exhaust manifold mounted to the cylinder head, the main exhaust pipe or "front pipe", an intermediate exhaust pipe or front muffler, and the rear muffler with integral tailpipe. All U.S. models and some Canada models also have a catalytic converter.

The manifold is cast iron to withstand the extreme heat, vibration, and pressure. The corrosion-resistant, alloy steel front pipe mounts to the exhaust manifold at the top, and to the catalytic converter at the rear. The rest of the exhaust system is also mild steel treated to resist corrosion. The converter housing is stainless steel. On Cabriolet from 1990, and all Scirocco 16V, the catalytic converter and the front pipe are one piece.

The front end of the system is bolted to the exhaust manifold, while the middle and rear portions of the system are suspended from the underbody by bonded rubber mounts or hangers. The positive but non-rigid mounting allows expansion and contraction with changes in temperature and helps to isolate noise and vibration.

1.2 Emission Controls

Emission controls are necessary to reduce harmful exhaust emissions. For all U.S. models and most Canada models, emission controls in the exhaust system include the catalytic converter and an oxygen sensor.

Catalytic Converter

The catalytic converter is similar in appearance to a small muffler. Its honeycombed core contains many small passages, the surfaces of which are coated with precious metal catalysts. The catalysts promote chemical reactions with the exhaust gasses. Most of the polluting chemical compounds in the exhaust are transformed into relatively harmless nitrogen, carbon dioxide and water vapor.

The catalytic converter is the final step in exhaust emission control. It works best and lasts longest when the fuel injection's air/fuel mixture is carefully controlled and combustion is as complete as possible. Higher levels of pollutants, created by the wrong air/fuel mixture and inefficient combustion, will overload the converter and impair its ability to control emissions.

Oxygen Sensor System

The oxygen sensor is part of a closed-loop feedback system used to control fuel delivery and emissions. The oxygen left in the exhaust is an indicator of combustion efficiency. The sensor reacts to the difference in oxygen content between the exhaust gas and the outside air. A low-voltage signal generated by the reaction becomes an input to the fuel injection system.

The oxygen sensor is mounted in the exhaust manifold or at the inlet to the catalytic converter depending on the model. Getting the sensor quickly up to operating temperature is important. On later models, the sensor is electrically heated.

1.3 Exhaust System and Emission Control Identification

Exhaust system details depend on the model and the engine. Procedures, specifications, and component illustrations are often identified by the engine code for the engine to which they apply. Where no engine code is indicated, it can be assumed that the information applies to all cars covered by this manual. For information on engine codes, see **ENGINE**.

Emission controls are usually identified by the fuel injection or engine management system. CIS refers to engine code JH, most 1985-1989 Cabriolets and Sciroccos. CIS-E refers to the Scirocco 16V. Since 1990, all Cabriolets have the Digifant engine management system. Also see **FUEL SYSTEM**.

2. MAINTENANCE

Volkswagen specifies the maintenance steps listed below, to be carried out at particular time and mileage intervals for proper maintenance of the exhaust system.

A number in bold type indicates that the procedure is covered in this section, under that numbered heading. See **LUBRICATION AND MAINTENANCE** for information about other exhaust system maintenance and for the prescribed maintenance intervals.

1. Inspecting exhaust system
2. Replacing oxygen sensor. **5.**

> ***WARNING —***
> - *A leaking exhaust system may allow deadly exhaust gas to enter the passenger compartment.*
> - *The exhaust system should be inspected immediately if it becomes unusually noisy or if the smell of exhaust is noticeable inside the car.*

> **NOTE —**
> Volkswagen recommends inspecting the exhaust system every 15,000 miles (24,000 km), but it is also a good idea to inspect it whenever the underside of the car is accessible.

3. TROUBLESHOOTING

Exhaust leakage or excessive noise are pretty obvious indications of a faulty exhaust system, but poor performance or increased exhaust emissions may have more complex causes. This troubleshooting section addresses problems with the exhaust and emission control systems, including the catalytic converter and the oxygen sensor. Also see **IGNITION**, **ENGINE**, and **FUEL SYSTEM** for additional troubleshooting information on driveability and performance problems.

Table a lists common exhaust and emission control problems, probable causes, and corrective actions. The numbers in bold type in the corrective action column refer to numbered headings in this **EXHAUST SYSTEM AND EMISSION CONTROLS** section where tests and repairs are covered.

3.1 Basic Troubleshooting Principles

Sudden changes in the exhaust noise level are probably caused by damage or by failure of one component. A gradual noise increase probably indicates more general deterioration throughout the system. Noise may increase suddenly when one part of the system, weakened by corrosion, finally fails.

Exhaust restriction or backpressure helps the engine produce power smoothly over a wide range of engine speeds, but damage that increases backpressure may decrease performance or cause rough idling and stalling. Too much unburned fuel, or any leaded fuel, will cause harmful chemical reactions that may plug the catalytic converter.

The oxygen sensor plays an important role in engine management—control of fuel injection, ignition and exhaust emissions. Performance, driveability or emissions problems may be caused by a fault or misadjustment in these other systems.

Digifant I (Cabriolet from 1991, California only)

Beginning in 1991, cars built for sale in California have a special version of the Digifant engine management system called Digifant I. Built-in on-board diagnostic (OBD) circuitry monitors the range of system values in normal operation, detects unusual values and other faults, and stores fault information for later recall. See **FUEL SYSTEM**.

NOTE —
Volkswagen specifies that the first step in any Digifant I system troubleshooting must be to check the control unit's OBD fault memory and to find and correct any faults that are indicated. See **FUEL SYSTEM**.

4. EXHAUST SYSTEM

The exhaust components bolt together at welded flanges or clamp together at slip-joints. Gaskets seal the system at the cylinder head, and at the joint between the exhaust manifold and the front pipe. The pipe is fastened by spring clamps or by a bolted flange, depending on the engine. Exhaust manifolds and front pipes are shown in Fig. 4-1 and Fig. 4-3. Behind the catalytic converter, the pipes slip together and are secured with clamps, as shown in Fig. 4-2 and Fig 4-4.

The front pipe and catalytic converter are made of special corrosion-resistant materials for long life. The rest of the system is more vulnerable because moisture in the exhaust stream condenses more readily on the cooler parts and eventually causes corrosion. If one part of the system is perforated by rust, it is likely that other parts are too.

4.1 Removing and Installing Exhaust System

If replacing only part of the system, try to make sure that the new parts will fit properly before removing the old ones. Parts from different sources may not mate properly. For partial replacement of an original system, genuine Volkswagen parts from an authorized Volkswagen dealer are more certain to fit.

Table a. Exhaust and Emission Control Troubleshooting

Symptom	Probable cause	Corrective action
1. Hissing, rumbling, loud noise during acceleration	**a.** Exhaust system leaks **b.** Internally damaged muffler or catalytic converter	**a.** Examine all system joints (see **LUBRICATION AND MAINTENANCE**). Replace faulty components. **4.1** **b.** Replace faulty components. **4.1**
2. Exhaust system rattles	**a.** Exhaust system out of alignment **b.** Missing or broken rubber exhaust system retaining ring	**a.** Re-align exhaust system. **4.1** **b.** Replace retaining ring(s). **4.1**
3. Reduced power, poor mileage, hesitation on initial acceleration, rough idle	**a.** Oxygen sensor system faulty/out of adjustment **b.** Excessive backpressure due to plugged catalytic converter **c.** Excessive backpressure due to damaged muffler or pipes	**a.** Test oxygen sensor system function, replace faulty components. **5.** **b.** Replace catalytic converter. **4.2** **c.** Inspect components and replace as required. **4.1**
4. Cold running hesitation, stalling at idle	**a.** Faulty oxygen sensor thermoswitch (CIS)	**a.** Test and replace if necessary. **5.1**
5. Failed emissions test	**a.** Oxygen sensor system faulty/out of adjustment **b.** Catalytic converter failed	**a.** Test system and replace faulty components. **5.** **b.** Replace catalytic converter. **4.2**

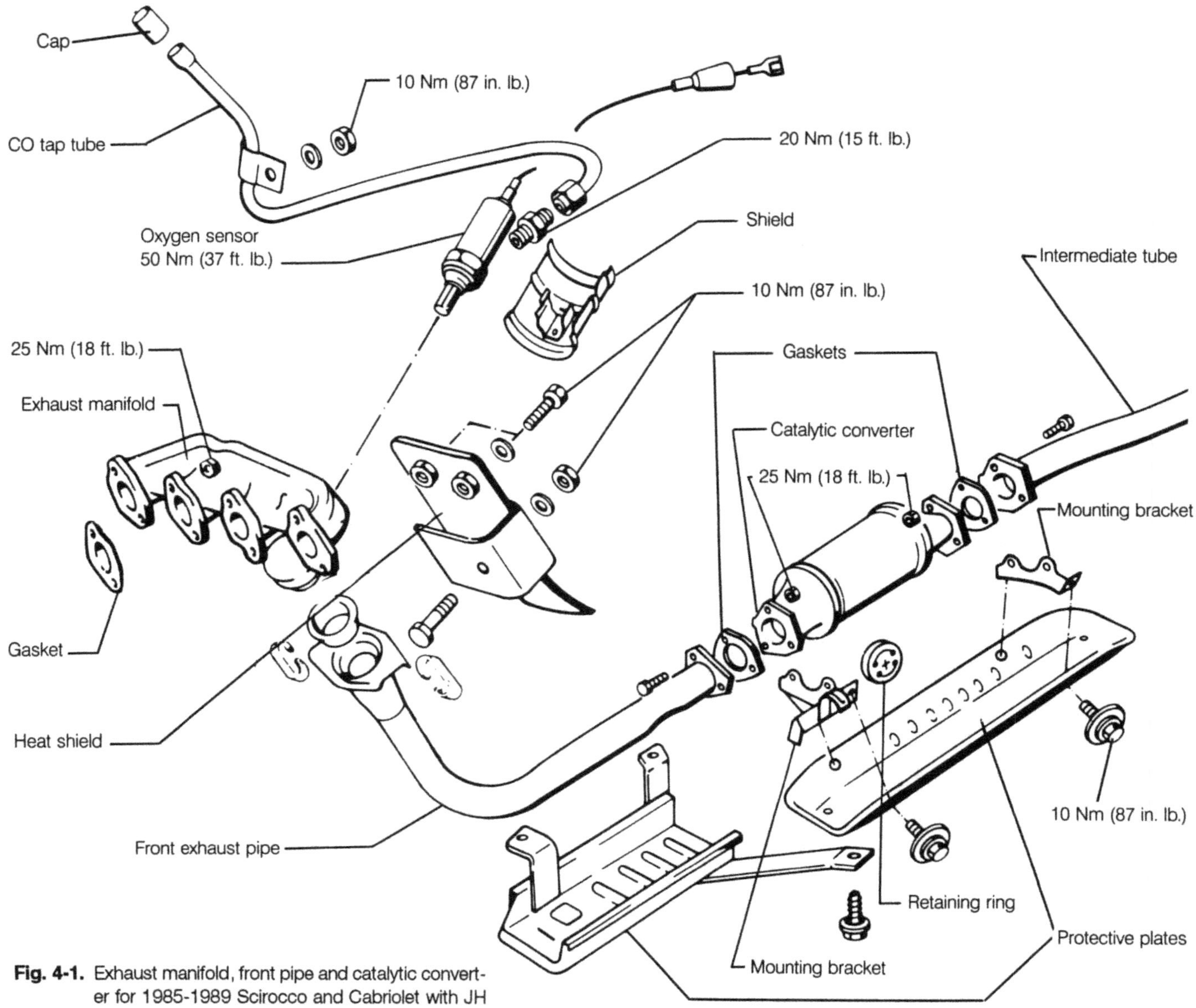

Fig. 4-1. Exhaust manifold, front pipe and catalytic converter for 1985-1989 Scirocco and Cabriolet with JH engine.

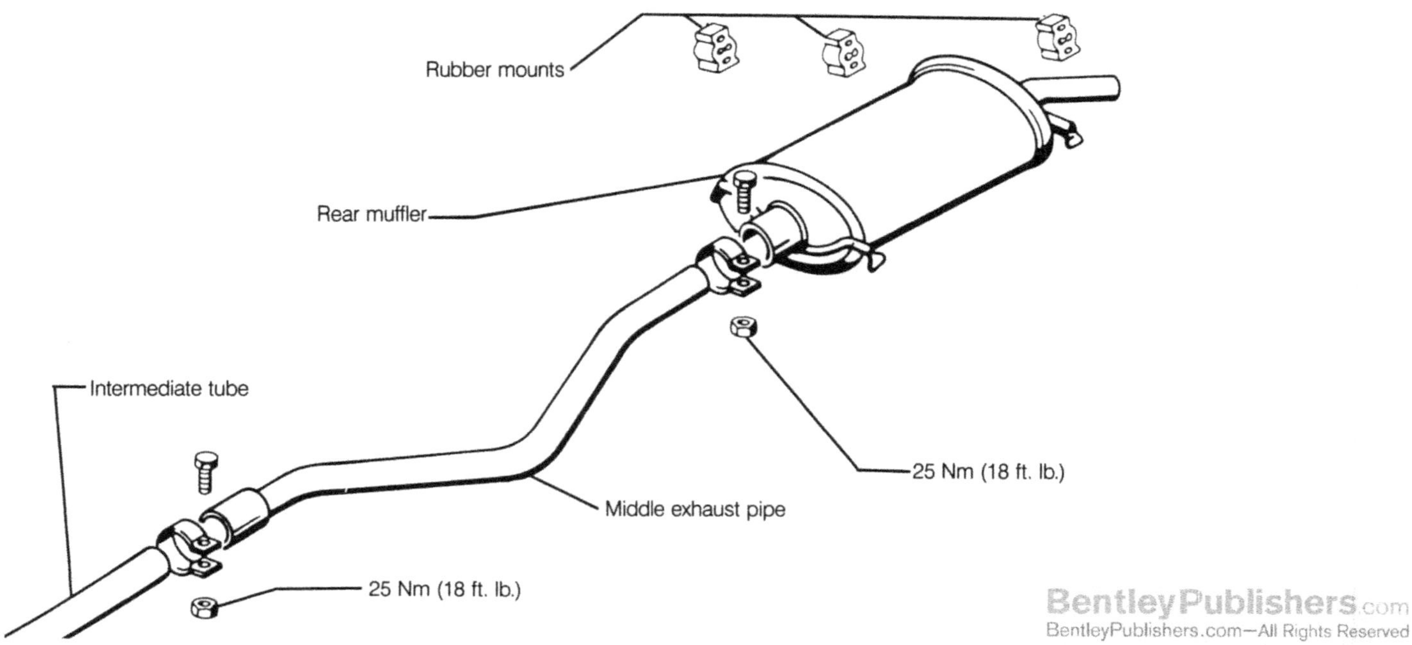

Fig. 4-2. Rear section of exhaust system for 1985-1989 Scirocco and Cabriolet with JH engine.

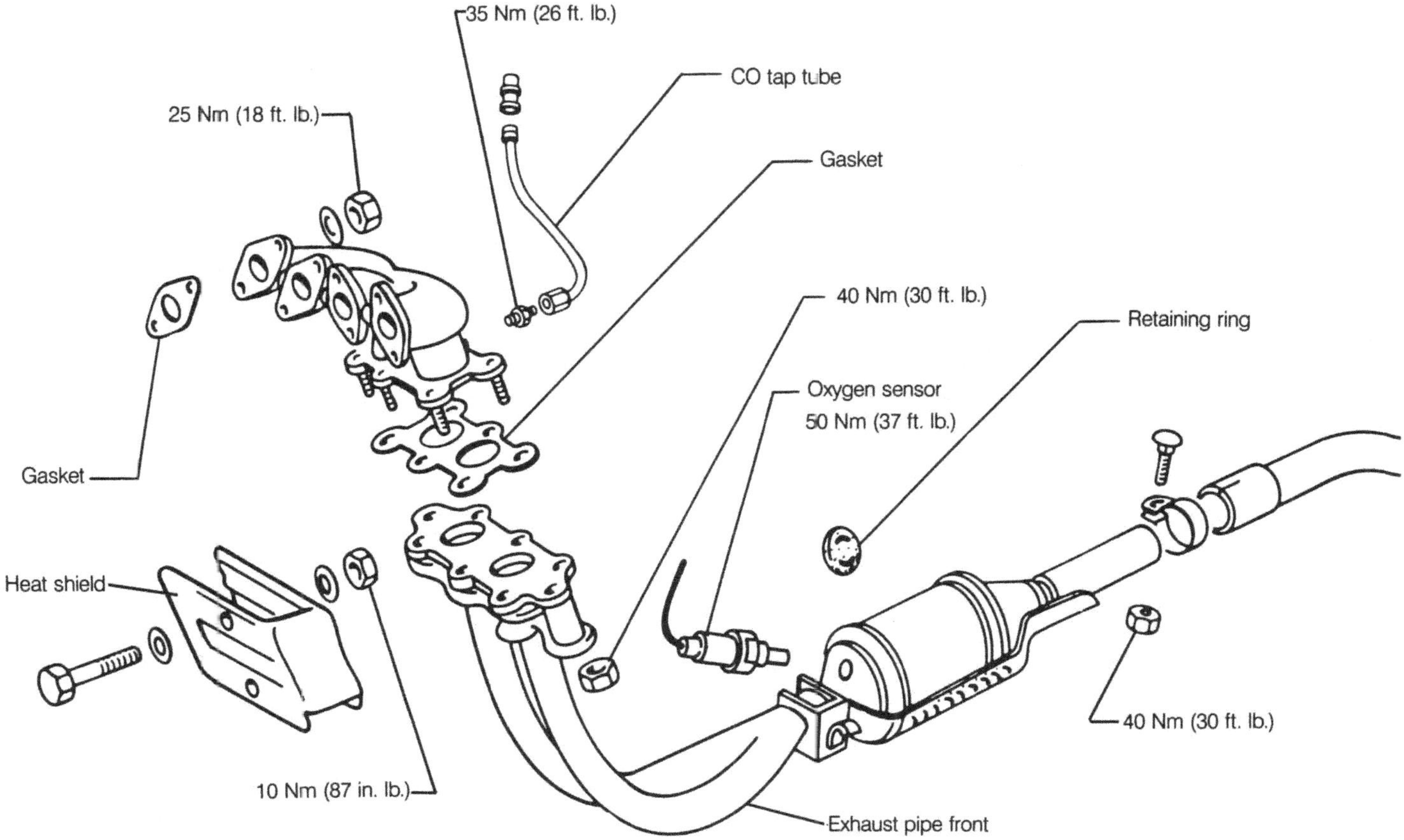

Fig. 4-3. Exhaust manifold, front pipe and catalytic converter for Scirocco 16V. Cabriolet from 1990 is similar, except for single front pipe and oxygen sensor mounted in exhaust manifold.

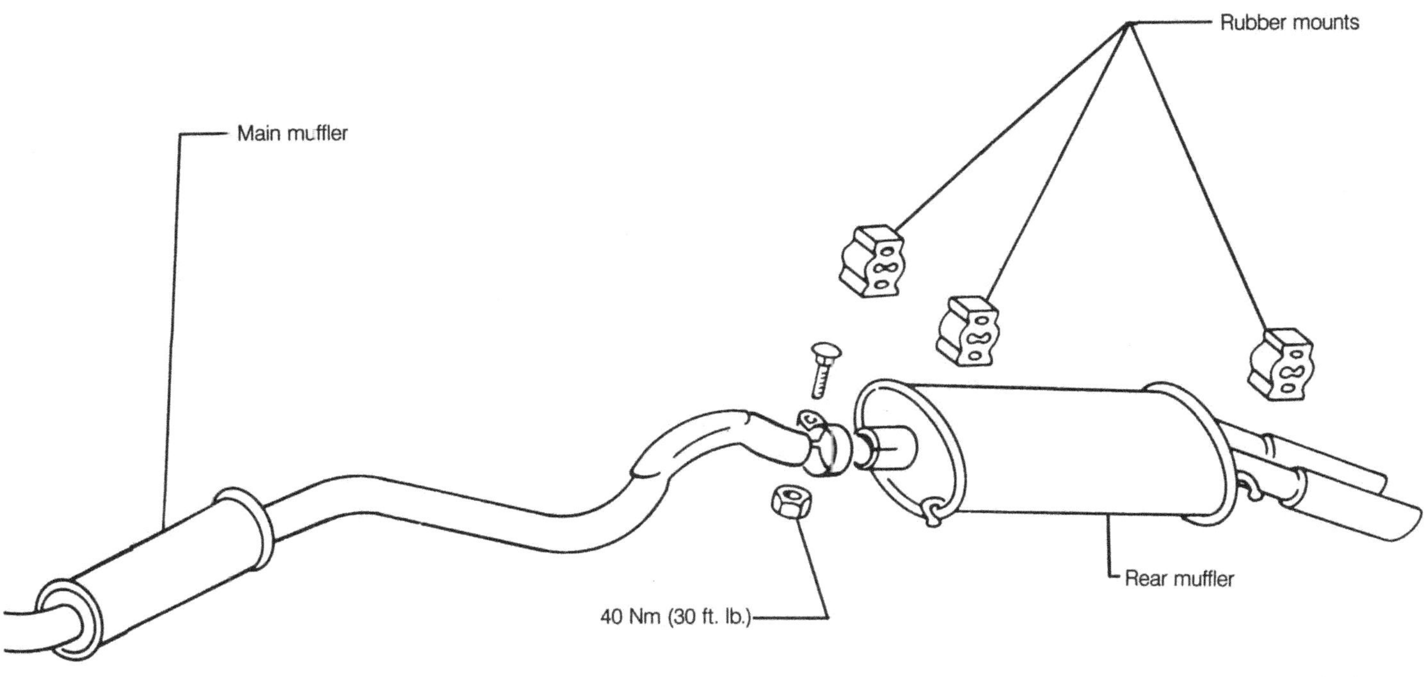

Fig. 4-4. Rear section of exhaust system used on Scirocco 16V. Cabriolet from 1990 is similar.

New clamps and fasteners are always recommended. The chance of getting the old ones off in reusable condition is slim anyway. On cars so equipped, the nuts that mount the front pipe to the exhaust manifold are self-locking nuts. They are designed to be used only once and should always be replaced. If the front pipe is disconnected from the exhaust manifold, the gasket should also be replaced.

WARNING —
Use eye protection and heavy gloves when working on the rusty parts of the old exhaust system.

Removing the front pipe on cars with the JH engine (1985-1989 Scirocco and Cabriolet) requires a special tool set to safely remove the front pipe spring clamps. The procedure for using these tools is described below.

The rubber retaining rings allow some movement—perhaps enough to remove front components without completely removing the rear for example. Penetrating oil applied to all fasteners and slip joints several hours ahead of time will make disassembly easier.

Removing and Installing Front Pipe Spring Clamps

On JH engines, spring clamps fasten the front exhaust pipe to the exhaust manifold. A set of clamp-expanding wedges is required to remove and install the clamps safely. The tool set shown here is Volkswagen special tool 3140A (order no. T03 140 A00 19 ZEL). See Fig. 4-5.

CAUTION —
Volkswagen part numbers and special tool order numbers are listed in this manual for reference only. Always rely on an authorized Volkswagen dealer Parts department for the most accurate and up-to-date information.

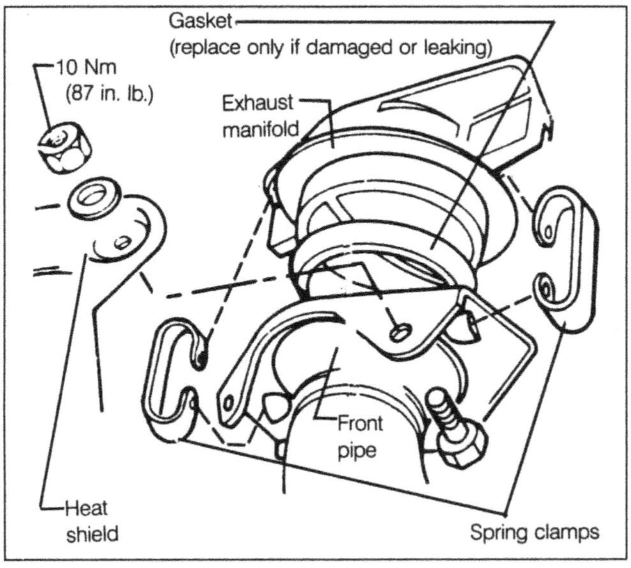

Fig. 4-5. Exploded view of front pipe connection with spring clamps.

WARNING —
Due to the spring tension in the clamps, serious injury may result from improper handling or from attempts to remove or install the clamps without proper tools.

To remove:

1. Push the front pipe to one side to expand the opposite clamp. Insert one short wedge in the expanded clamp as shown in Fig. 4-6. Push the pipe the opposite way to expand the other clamp and insert the other short wedge.

CAUTION —
When inserting wedge into spring clamp be sure it is completely inserted. The shoulders of the wedge should butt against the edge of the clamp.

WARNING —
Use extra caution when handling the expanded spring clamps. Spring tension can expel the wedges with considerable force and may cause serious injury. Wear safety goggles or safety glasses.

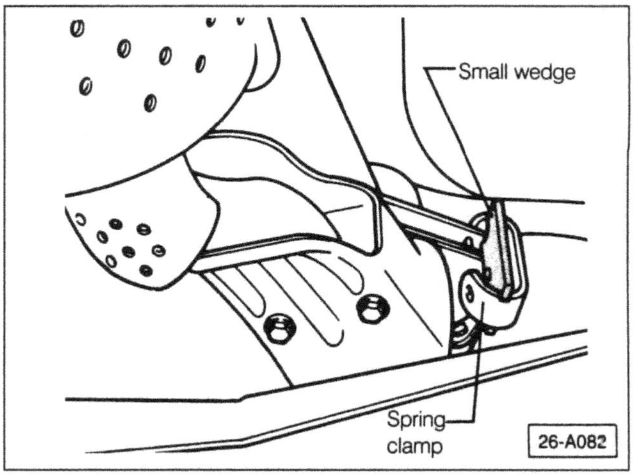

Fig. 4-6. Short wedge being installed in expanded spring clamp.

2. Push the front pipe to one side, and use a pair of locking pliers to grasp the clamp and wedge on the opposite side and remove them as a unit.

3. Repeat the operation to remove second clamp and wedge.

4. If the old clamps are to be reused, the wedges should remain in the clamps until they are reinstalled. If it is necessary to replace the clamps, place the old clamp with the installed wedge between the jaws of a vise, as shown in Fig. 4-7. Open the jaws of the vise approximately 15 mm (9/16 in.).

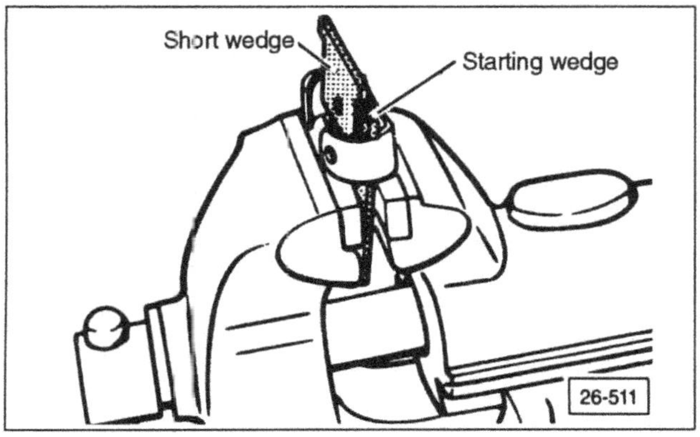

Fig. 4-7. Longer starting wedge being used to remove short wedge from clamp.

5. Use a soft-faced mallet to drive the longer starting wedge into the clamp next to the short wedge, to further expand the clamp.

6. Turn clamp over, knock out the short wedge, then knock out longer starting wedge. Repeat this process to remove the wedge from the other clamp.

WARNING —
The wedge will be expelled with considerable force. Provision should be made to catch the wedge in a safe container or trap to avoid injury.

To install:

1. If the spring clamps are being replaced, place a new clamp on its side in vise jaws and, using a soft-faced mallet, drive the longer starting wedge into clamp until it is spread enough to allow insertion of the short wedge next to the starting wedge. Drive in the short wedge until its shoulder butts against the edge of the spring clamp.

WARNING —
Do not attempt to insert the short wedge until first spreading the clamp with the starting wedge. Serious injury and tool and clamp damage may result.

2. Turn the clamp over and knock out the starting wedge, leaving the short wedge in place. Repeat the operation for the second clamp.

3. To install the clamps, position the front pipe against the exhaust manifold. If the gasket has been removed, it should be placed in position at this time. Push the pipe to one side and install an expanded clamp and wedge on that side.

4. Push the pipe to the other side and install the remaining clamp and wedge on that side. With the pipe in this position, remove the first wedge from its clamp. Use the mallet if necessary.

5. Push the pipe to the other side and remove the second wedge.

Installing Exhaust System

Loosely install and align the complete system before tightening any clamps or mounting bolts. Anti-seize compound used on all joints and threaded fasteners will extend service life and make any future replacement easier. Make sure that no part of the exhaust system contacts any part of the car body, and that the rubber mounts are straight and unstressed as shown in Fig. 4-8. If necessary, adjust the muffler or exhaust pipes by twisting or by pushing in a longitudinal direction.

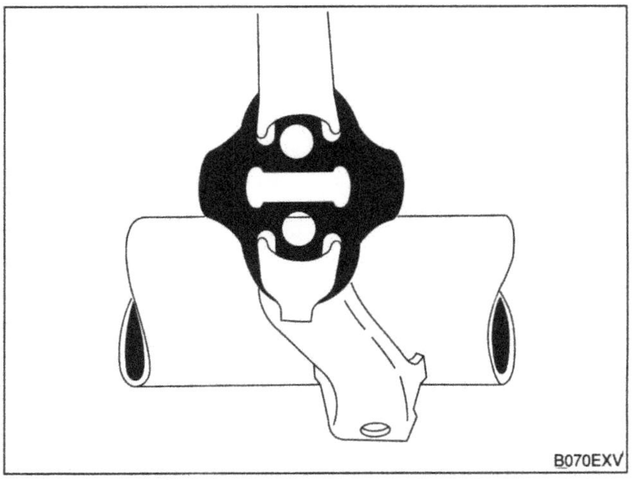

Fig. 4-8. Alignment of rubber mounts.

After the system is loosely installed and aligned, tighten the bolts and clamps. Torque values are given in Fig. 4-1 through Fig. 4-4 above. After completing the installation, start the engine and check for any exhaust leakage. Some slight smoking and odor as the new parts become hot for the first time are normal. System alignment can be checked by driving the car over a rough road and listening for sounds of the exhaust system striking the body.

4.2 Catalytic Converter

All of the cars covered by this manual which have an oxygen sensor system also have a three-way catalytic converter in the exhaust system. Fig. 4-9 shows the construction of the catalytic converter and illustrates the chemical reaction that turns harmful compounds into nitrogen, carbon dioxide, and water.

The catalytic converter is designed to be maintenance free, and under normal operating conditions it should last at least 60,000 miles (96,000 km). However, conditions which result in elevated temperature in the converter can cause damage and leave the catalytic converter partially plugged or inoperative. Some of those conditions are: incorrect ignition timing, incorrect fuel injection CO adjustment, engine misfiring, prolonged idling, prolonged high load such as towing, and the use of leaded gasoline.

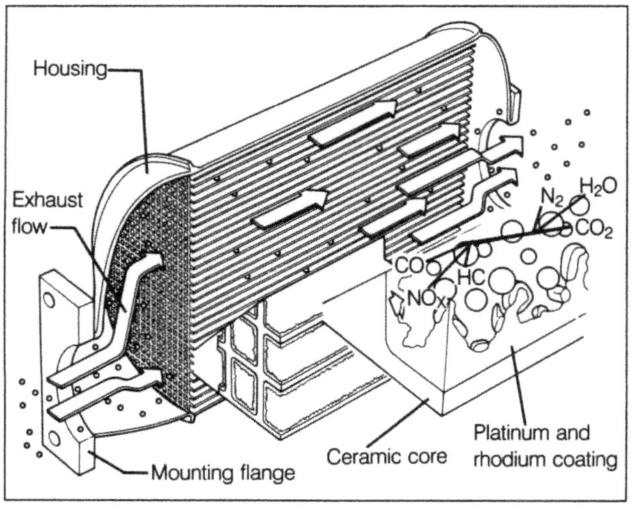

Fig. 4-9. Cutaway view of catalytic converter illustrating reaction of harmful compounds to form nitrogen, carbon dioxide, and water.

Reduced power, stalling at idle, rattles in the exhaust system, and measured excessive emissions are all possible indications of a faulty catalytic converter. The converter can be removed and visually inspected if converter failure is suspected. See **Checking Catalytic Converter** below.

Removing and Installing Catalytic Converter

The catalytic converter is one of the components of the exhaust system and can be removed and installed with the rest of the system as described in **4.1 Removing and Installing Exhaust System**. On 16-valve models, where the converter is an integral part of the front pipe, removal of the front pipe is required to thoroughly inspect the converter.

Checking Catalytic Converter

To check a one-piece front pipe and converter, remove the pipe and hold it vertically with the outlet down. Firmly tap the end on a block of wood, then turn the pipe over and tap the other end in the same manner. A knocking sound inside the converter housing indicates that the ceramic core has become dislodged, and the converter should be replaced.

In addition to the check above, a separate catalytic converter can be visually inspected by holding it up to a strong light and looking through both ends. If the core appears broken or melted down as shown in Fig. 4-10, the converter should be replaced.

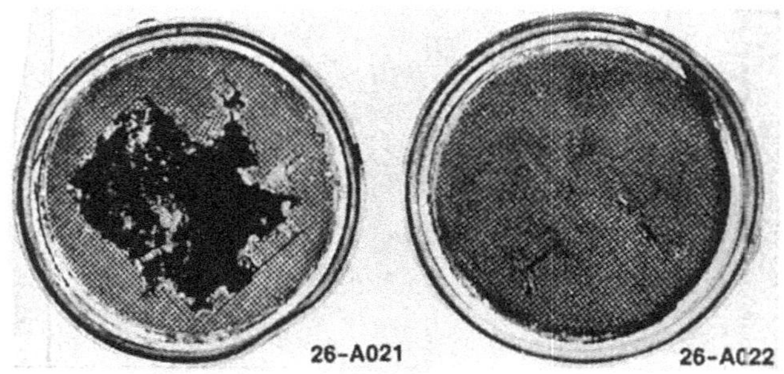

Fig. 4-10. Examples of catalytic converter cores showing severe meltdown (left) and minor meltdown (right).

5. OXYGEN SENSOR

The oxygen sensor provides information about combustion efficiency by measuring the oxygen content in the exhaust. It is constructed of ceramic material coated with platinum.

Inside the sensor, one surface is exposed to the exhaust gas, while the other is exposed to atmosphere. The difference in oxygen content between the two surfaces results in a chemical reaction that generates a low-voltage signal (100 to 1000 millivolts). This signal is monitored by an electronic control unit which, in turn, signals for changes in the air/fuel ratio in the fuel injection system.

Since precise control of fuel delivery relies on the oxygen sensor's low-voltage signal, the system is very sensitive to corrosion and poor connections. Making sure that the electrical connections are clean, dry and free of corrosion may solve driveability and performance problems easily without the need for more complicated testing.

The different fuel injection and engine management systems covered in this manual process the oxygen sensor signal differently, so the repair information in this section is organized according to fuel injection type. For help in identifying the fuel injection system, see **FUEL SYSTEM**.

Oxygen Sensor Maintenance

Oxygen sensor replacement is scheduled maintenance. See **LUBRICATION AND MAINTENANCE** for more information on servicing the oxygen sensor.

For 1985-1987 cars, a mileage counter signals the replacement interval by illuminating a light on the instrument panel. Whenever the oxygen sensor is replaced, the light must be reset. The resetting procedure is also covered in **LUBRICATION AND MAINTENANCE**.

> **NOTE —**
> Emission controls, including the oxygen sensor system, are covered by an extended warranty. See **LUBRICATION AND MAINTENANCE** for maintenance requirements. Consult an authorized Volkswagen dealer about warranty coverage before beginning any repairs.

5.1 Testing Oxygen Sensor System (OXS)—CIS
(1985-1990 except Scirocco 16V)

CIS fuel metering is determined mainly by regulating and adjusting fuel pressure in various parts of the system. The oxygen sensor and oxygen sensor control unit also manipulate fuel pressure. The frequency valve, mounted in a fuel line near the fuel distributor, opens and closes at a variable rate to finely adjust fuel pressure, and therefore fuel delivery.

The oxygen sensor provides an accurate signal only when fully warm. When the engine is cold a thermo-time switch interrupts the circuit. The system reverts to a fixed or "open-loop" mode, bypassing the oxygen sensor until it has had a chance to reach operating temperature. For more information on CIS fuel injection, see **FUEL SYSTEM**.

Component Locations

The oxygen sensor control unit on Sciroccos is mounted in the fresh air plenum, in front of the windshield under the hood. The Cabriolet control unit, shown in Fig. 5-1, is inside behind the glovebox trim. The oxygen sensor is at the back side of the engine, mounted in the exhaust manifold. The thermo-time switch is mounted in the cylinder head coolant outlet at the front of the engine.

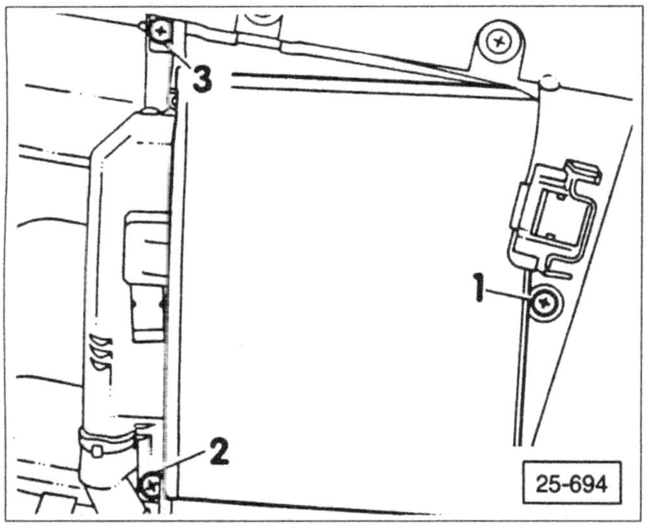

Fig. 5-1. Fasteners securing Cabriolet OXS control unit. To remove, loosen screw **1** about two turns, then remove screws **2** and **3**.

OXS Quick-check

The frequency valve gets its name from the pulsed control unit signal that cycles it rapidly on and off (open and closed). You can quickly check basic system function with the engine running, or in the few seconds that the fuel pump runs after actuating the starter. Listen at the valve for a clicking or buzzing sound, and feel to make sure it is vibrating slightly. If the valve does not buzz and vibrate, check the frequency valve itself before testing any other components or functions.

Checking OXS Frequency Valve

Requirements:
- fuel pump fuse OK

Checking the frequency valve also requires bypassing the fuel pump relay to operate the fuel pump without running the engine. See **FUEL SYSTEM**.

> ***CAUTION —***
> *Avoid running the fuel pump any longer than necessary for each test.*

With the fuel pump running, the valve should click or buzz. If not, disconnect the frequency valve harness connector and measure resistance across the valve terminals with an ohmmeter, shown in Fig. 5-2.

Frequency valve:
- resistance 2 to 3 ohms

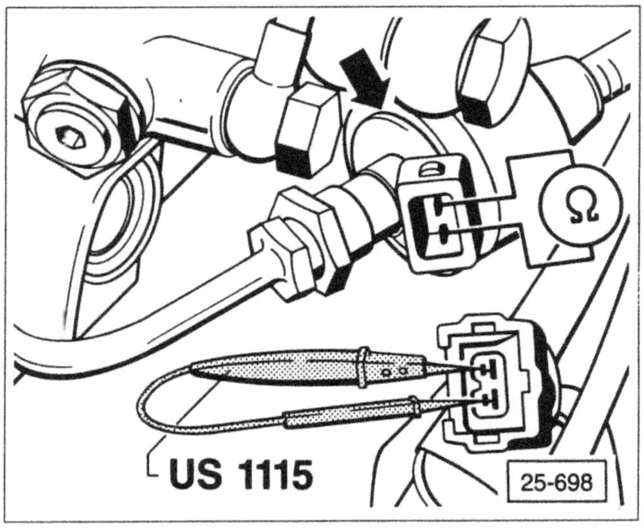

Fig. 5-2. Frequency valve resistance being checked at valve. Voltage from OXS control unit being checked at harness connector with LED test light.

If resistance is OK, use an LED test light to check for a voltage signal at the connector. With the fuel pump running, the LED should light. If not, check the fuel pump fuse, the fuel injection power supply relay, the wiring harness between these components, and the control unit ground connections. See **ELECTRICAL SYSTEM** for wiring diagrams.

To replace the frequency valve, clean and disconnect the fuel line unions. Loosen the mounting clamp bolt and press the valve out of the rubber bushing in the clamp. Installation is the reverse of removal. Always use new sealing washers.

> ***WARNING —***
> *Fuel will be expelled as the lines are disconnected. Do not smoke or work near heaters or other fire hazards. Have a fire extinguisher handy.*

Measuring Frequency Valve Duty Cycle

The pulsed signal from the OXS control unit determines the amount of time the valve is open. This is called its duty cycle. Duty cycle is used to check the signals from the OXS control unit. It is measured with a duty cycle meter or a dwell meter.

A duty cycle meter such as the Siemans 451 measure duty cycle as a percentage of time, from 0 to 100%. With a simple conversion, duty cycle can also be measured with an ordinary dwell meter, such as the Sun TDT-11. The 0 to 90° (4-cylinder) scale of a dwell meter corresponds to the 0 to 100% scale of a duty cycle meter. Multiply duty cycle by 0.9 to get an equivalent "dwell" reading.

Either meter gets its signal from the duty cycle test connector, in the engine compartment near the ignition coil, with a blue/white wire. Volkswagen special tool US1112 (order no. TU1 112 000 00 HRN) is an adapter that will help make an easy, positive connection.

Checking OXS Control Unit Regulating Functions

Requirements:
- frequency valve OK
- fuel pump relay bypassed to operate fuel pump
- duty cycle meter or dwell meter connected

CAUTION —
Avoid running the fuel pump any longer than necessary for each test.

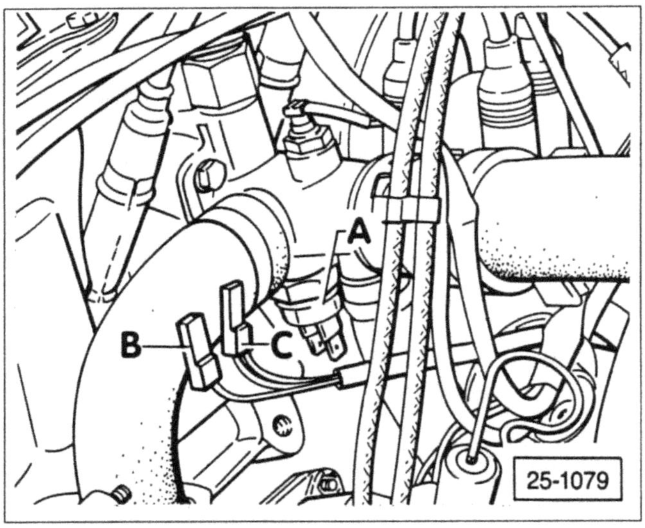

Fig. 5-3. Thermoswitch **A** set up for OXS control unit tests with harness connectors **B** and **C** disconnected. Later system (from July 1987) shown, with two wires at connector **C**.

Check OXS control unit regulating functions as described in **table b**. If the duty cycle is not as specified, check for faults in the wiring harness to the control unit and check control unit ground connections. See **ELECTRICAL SYSTEM** for wiring diagrams. If no other faults are found, the control unit is faulty and should be replaced.

Table b. OXS Control Unit Tests (CIS)

Test	Conditions/Steps	Duty cycle
Emergency running or "limp-home" function (open-loop)	– thermoswitch disconnected (Fig. 5-3) – operate fuel pump	50% ± 2% (45° on a dwell meter)
Full-throttle enrichment	– thermoswitch disconnected (Fig. 5-3) – operate fuel pump and actuate throttle switch (see **FUEL SYSTEM**)	65% ± 2% (58.5°)
Normal warm running – oxygen sensor regulating function (closed-loop)	– thermoswitch disconnected (Fig. 5-3) – disconnect oxygen sensor – touch harness connector (green wire) to ground, to simulate max. lean signal	> 87% (increasing)
	– touch harness connector to "+" side of 1.5 volt battery ("–" side to ground), to simulate rich signal	< 20% (decreasing)
Cold-running enrichment (up to July 1987)	– thermoswitch disconnected (Fig. 5-3) – bridge thermoswitch connectors with jumper wire – operate fuel pump	80% ± 2% (72°)
Cold-running enrichment (from July 1987)	– thermoswitch disconnected (Fig. 5-3) – connect single connector (brown/red wire) to ground – operate fuel pump	65% ± 2% (58.5°)
	– bridge thermoswitch connectors with jumper wire (not connected to ground) – disconnect oxygen sensor – start engine (run at idle) • if duty cycle reverts to open-loop (~50%), check cold-running enrichment vacuum switch	65% ± 2% (58.5°)
	– disconnect cold-running enrichment vacuum switch harness connector (Fig. 5-4)	50% ± 2% (45°)
	– reconnect harness connector – disconnect vacuum hose (Fig. 5-4)	50% ± 2% (45°)
	– switch ignition off – reconnect vacuum hose – reconnect oxygen sensor – disconnect coil wire from distributor cap and connect to ground with jumper wire – thermoswitch connectors bridged – actuate starter • if duty cycle reverts to open-loop (~50%), check cold-start enrichment relay (and relay wiring) See **FUEL SYSTEM**; See wiring diagrams in **ELECTRICAL SYSTEM**	65% ± 2% (58 ½°) while cranking

Checking Cold-running Enrichment Vacuum Switch
(from July 1987 only)

Engine manifold vacuum normally closes the switch at idle to complete a thermoswitch ground circuit for cold-running enrichment. Off idle, when manifold vacuum is low, the switch opens. Its purpose is to interrupt cold-running enrichment above idle. The switch is shown in Fig. 5-4.

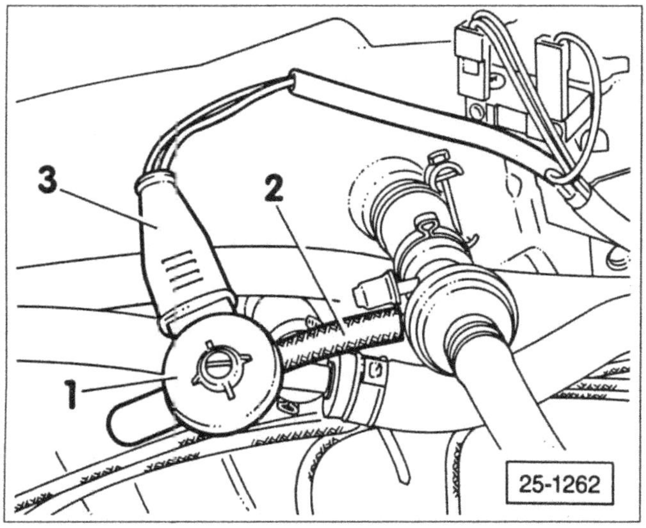

Fig. 5-4. Cold-running enrichment vacuum switch (**1**), vacuum hose (**2**) and harness connector (**3**) at front of intake manifold.

Requirements:
- oxygen sensor disconnected
- thermoswitch connectors disconnected and bridged with jumper wire
- duty cycle meter or dwell meter connected
- engine running at idle

If results are not as specified in **table b**, disconnect the harness connector and bridge the connector terminals with a jumper wire. If the duty cycle is 50% ± 2% with the connector disconnected and 65% ± 2% with the terminals bridged, the vacuum switch or vacuum hose are faulty and should be replaced. If not, check for poor ground connections or faults in the wiring harness.

Checking Oxygen Sensor

Requirements:
- engine at normal operating temperature – radiator cooling fan cycled on and off at least once
- idle speed OK
- fuel pump relay bypass or other remote switch NOT connected
- oxygen sensor connected
- thermoswitch connected
- OXS control unit regulating functions OK
- duty cycle meter or dwell meter connected

To check oxygen sensor (CIS):

1. Start the engine and let it idle for at least two minutes.
2. Clamp off the hose from the idle speed boost valve and disconnect the crankcase ventilation hose. See Fig. 5-5.

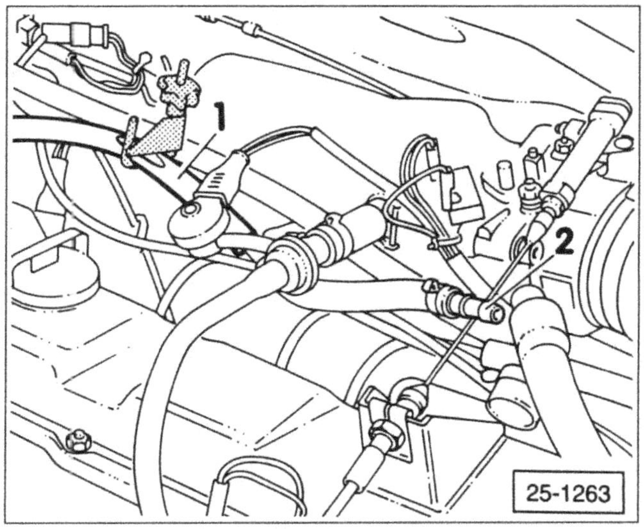

Fig. 5-5. Idle speed boost valve hose (**1**) clamped shut and crankcase ventilation hose (**2**) disconnected and plugged in preparation for checking oxygen

3. Note the duty cycle reading, then plug the crankcase ventilation hose.
   - duty cycle should drop, then rise and fluctuate
   - if not, oxygen sensor is faulty and should be replaced

Installing:
- apply anti-seize compound (threads only)
- tightening torque 50 Nm (37 ft. lb.)

CAUTION —
- *Do not let anti-seize compound come into contact with the slit portion of the sensor body.*
- *Grip only the hex-shaped part of the sensor.*
- *Use tools that will not damage the sensor wire.*

Checking Thermoswitch

The thermoswitch is shown earlier in Fig. 5-3. Use a multimeter (ohms scale) to measure resistance across the switch terminals. See **table c**. When installing, always use a new sealing washer. Check the coolant level and add coolant as necessary. See **LUBRICATION AND MAINTENANCE**.

Table c. Thermoswitch Specifications

Part No.	Temperature	Specification
035 919 369 D	above 77°F (25°C) below 59°F (15°C)	open 0.5 Ω max.
027 919 369 B	above 177°F (47°C) below 59°F (15°C)	open 0.5 Ω max.
035 919 369 B	above 104°F (40°C) below 86°F (30°C)	open 0.5 Ω max.

5.2 Testing Oxygen Sensor—CIS-Electronic
(Scirocco 16V)

The oxygen sensor signal is just one of many inputs that the CIS-Electronic (CIS-E) fuel injection control unit uses to adjust fuel metering and control the air/fuel mixture. For more information on the operation of the control unit and the rest of the CIS-E system, see **FUEL SYSTEM**.

CIS-E controls fuel metering by manipulating control current through the differential pressure regulator. Measurement of differential pressure regulator current is also used to evaluate oxygen sensor function. See **FUEL SYSTEM** for more information on measuring differential pressure regulator current.

To test oxygen sensor:

Requirements:
- engine warm—oil temp. at least 176°F (80°C)
- radiator cooling fan must have cycled on and off at least once
- all electrical consumers switched off
- A/C off
- exhaust system free of leaks
- fuel pressure gauge not connected—if fuel lines were disconnected, run the engine to about 3000 rpm several times before testing

1. Disconnect the differential pressure regulator harness connector. Connect the test harness meter to measure differential pressure regulator control current.
2. Disconnect the crankcase ventilation hoses from the ventilation housing as shown in Fig. 5-6.

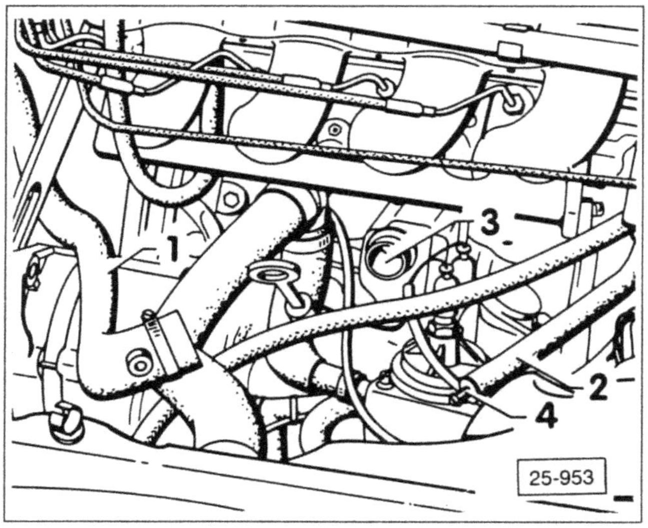

Fig. 5-6. Crankcase ventilation system configured to check oxygen sensor. Disconnect hoses **1** and **2** from ventilation housing **3**. Connector piece **4** with calibrated restrictor hole remains in hose **2**. Position hoses to draw in only fresh air.

3. Remove the "T" connector from the intake air boot and reinsert it as shown in Fig. 5-7.

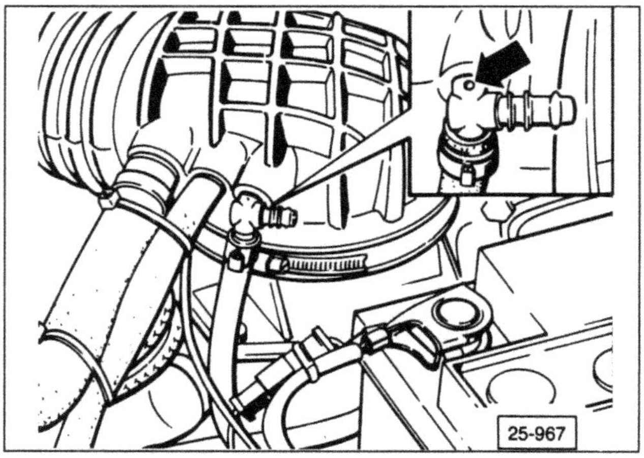

Fig. 5-7. "T" connector at intake air boot repositioned for test. Position calibrated restrictor hole (arrow) to draw in fresh air.

4. Start the engine and let it idle. Check differential pressure regulator current as the engine and oxygen sensor warm up. After about two minutes, the current reading should be between 1 and 10 mA and fluctuating.
5. If the current reading does not fluctuate at idle, raise the engine speed to about 3000 rpm and check again. Fluctuating current means the system is responding to oxygen sensor input, and the sensor is OK.

CAUTION —
Do not make checks or adjustments while the radiator cooling fan is running. Wait for it to stop.

NOTE —
- If differential pressure regulator current fluctuates at 3000 rpm but does not fluctuate at idle, the idle mixture adjustment should be checked as described in **FUEL SYSTEM**.
- If differential pressure regulator current is a constant 9 to 11 mA, check control unit response as described below.

6. Stop the engine. Reinstall the hoses in their original configurations. If the results of this test were OK, disconnect the test harness and reconnect the harness connector.

To test control unit response:

1. Connect a meter to measure differential pressure regulator current as described for the previous test. Make sure the engine is completely warmed up.
2. Disconnect the oxygen sensor connector. Switch the ignition on. The control unit should revert to "open-loop" or "limp-home" mode.
   - regulator current 9 to 11 milliamps (mA)

3. Touch the male end of the oxygen sensor connector (at the end of the green wire) to ground on some metal part of the engine or body. After about 20 seconds, differential pressure regulator current should increase.
   - regulator current 19 to 22 mA

4. Switch the ignition off. Disconnect the test harness and reconnect the differential pressure regulator harness connector.

If the system does not respond as specified, check the wiring and harness connections between the oxygen sensor and the CIS-E control unit. Also see **FUEL SYSTEM** for detailed CIS-E system electrical tests.

Replacing Oxygen Sensor

On Scirocco 16V (engine code PL) the oxygen sensor is in the exhaust system, mounted in the front of the catalytic converter housing.

Installing:
- apply anti-seize compound (to threads only)
- tightening torque 50 Nm (37 ft. lb.)

CAUTION —
- *Do not let anti-seize compound come into contact with the slit portion of the sensor body.*
- *Use a wrench or socket only on the hex-shaped part of the sensor.*
- *Use tools that will not pinch or otherwise damage the sensor wire.*

5.3 Testing Oxygen Sensor—Digifant
(Cabriolet from 1990)

Requirements:
- engine warm—oil temperature at least 176°F (80°C)
- radiator cooling fan must have cycled on and off at least once
- idle speed OK
- air conditioning and all other electrical consumers switched off
- exhaust system between cylinder head and catalytic converter must be free of leaks
- oxygen sensor heater and voltage supply to heater **OK**
- engine coolant temperature sensor remains connected
- CO tester (Sun 105 or EPA-approved equivalent) connected to the CO tap tube (right side of exhaust manifold)

CAUTION —
The connection to the CO tap tube must be secure. Even small leaks will result in inaccurate measurements.

CAUTION —
- *Connect and disconnect the test leads only with the ignition switched off.*
- *Do NOT allow terminal 1 of the ignition coil to be shorted to ground when connecting test equipment,. This will permanently damage the Digifant control unit.*
- *Do not check CO values while the radiator cooling fan is running. Wait for it to stop.*
- *Volkswagen specifies that the first step in any Digifant I system troubleshooting must be to check the control unit's OBD fault memory and to find and correct any faults that are indicated. See* **FUEL SYSTEM**.

To check oxygen sensor:

1. Start the engine and let it idle for at least two minutes.
2. Check and record the CO content measured on the CO meter.
3. Disconnect the fuel pressure regulator vacuum hose from the intake manifold, then plug the opening at the nipple. See Fig. 5-8.
   - CO content should increase briefly, then drop back to normal
   - if so, the oxygen sensor is functioning normally and the test is over
   - if not, continue testing control unit response

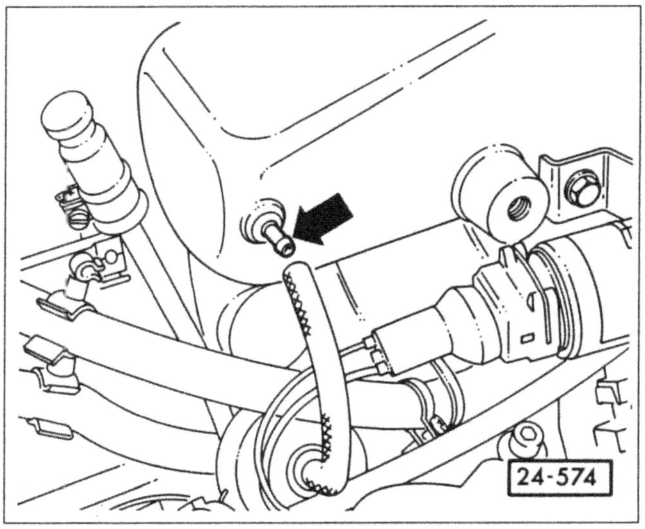

Fig. 5-8. Fuel pressure regulator vacuum hose disconnected from intake manifold nipple (arrow).

4. Disconnect the 4-pin oxygen sensor harness connector at the bracket on the cylinder head cover.

5. Use a jumper wire as shown in Fig. 5-9 to make a connection between the oxygen sensor signal terminal (with the violet-colored wire) and the battery, alternating between positive (+) and negative (–) terminals.
   - CO content should increase and decrease, depending on which battery terminal is connected

NOTE —
On California cars, with Digifant I, the connector has only three wires.

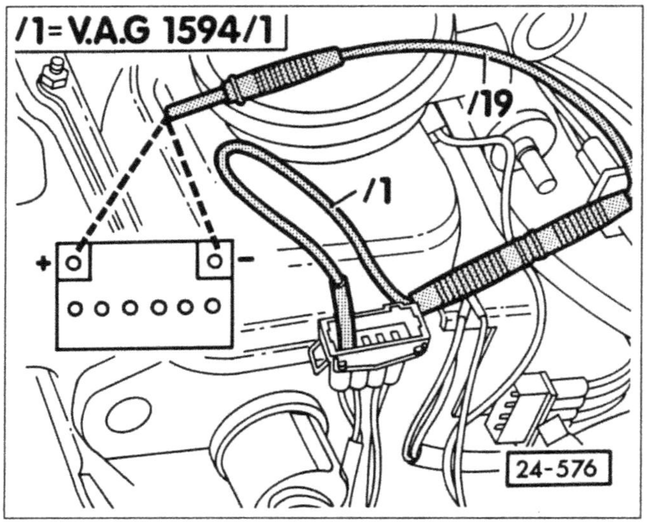

Fig. 5-9. Jumper wire (Volkswagen special tool V.A.G. 1594/1 shown) used to make temporary contact between oxygen sensor signal wire and battery, to simulate very lean and very rich input signals to Digifant control unit.

If the CO content still does not change, there is a fault in the wiring harness between the oxygen sensor harness connector and the Digifant control unit. If the CO content changed in step #5, but did not change in step #3, the oxygen sensor is faulty and should be replaced.

Replacing Oxygen Sensor

The oxygen sensor is in the exhaust system, mounted in the exhaust manifold.

Installing:
- apply anti-seize compound (to threads only)
- tightening torque 50 Nm (37 ft. lb.)

CAUTION —
- *Do not let anti-seize compound come into contact with the slit portion of the sensor body.*
- *Use a wrench or socket only on the hex-shaped part of the sensor.*
- *Use tools that will not pinch or otherwise damage the sensor wires.*

6. TECHNICAL DATA

I. Tightening Torques

CO tap tube clamp (bolt)	10 Nm (87 in. lb.)
CO tap tube union to exhaust manifold (JH)	20 Nm (15 ft. lb.)
CO tap tube union to exhaust manifold (PL)	35 Nm (26 ft. lb.)
Exhaust manifold to cylinder head (nut)	25 Nm (18 ft. lb.)
Exhaust pipe to exhaust manifold (engine codes PI, 2H)	40 Nm (30 ft. lb.)
Front pipe or intermediate pipe flange to catalytic converter (engine code JH)	25 Nm (18 ft. lb.)
Heatshield mounting (nut or bolt)	10 Nm (87 in. lb.)
Slip-joint clamps (engine codes PL, 2H)	40 Nm (30 ft. lb.)
Slip-joint clamps (engine code JH)	25 Nm (18 ft. lb.)
Oxygen sensor	50 Nm (37 ft. lb.)

II. Oxygen Sensor System (CIS)

Thermoswitch resistance (see text for temperature specifications)	
warm	open
cold	continuity (0.5 Ω maximum)
Frequency valve resistance – ohms	2–3 Ω
Control unit duty cycle (dwell meter equivalent)	
open-loop	50%±2% (45±2°)
closed-loop	fluctuating
cold-running enrichment	
up to July 1987	80% ± 2%
from July 1987	65% ± 2%
rich-mixture limit	87% (78°)
lean-mixture limit	20% (18°)
CO content	0.3 to 1.2%

Section 7

IGNITION

Contents

TABLES

Ignition

Introduction

The ignition systems used on the cars covered by this manual incorporate many technologically advanced components to maximize engine performance and reliability in the face of increasing restrictions on engine exhaust emissions. The ignition system, in cooperation with the fuel system, precisely controls engine operation to ensure maximum power and efficiency from the engine under all operating conditions.

Due to the use of sophisticated electronics to operate and control these modern ignition systems, the need for routine maintenance is reduced and troubleshooting can be more difficult. Any service or repair work must be carried out carefully, with special attention to the cautions and warnings and proper working procedures. Most necessary service and repair information is provided here, although some of the operations that are described may be of practical value only to professional mechanics. If you lack the skills, or the special tools and equipment, we suggest you leave the service and repair of the ignition system to an authorized Volkswagen dealer or other qualified repair shop. We especially urge you to consult your authorized Volkswagen dealer before attempting any repairs on a vehicle still covered by any warranties.

1. GENERAL DESCRIPTION

The spark-ignition system provides each cylinder with a precisely timed high-voltage spark to ignite the air and fuel mixture in the combustion chamber. The system adjusts the timing of that spark to compensate for changing engine speed and load.

1.1 Hall-sender Electronic Ignition

The basic ignition system for cars covered by this manual is a transistorized coil ignition with Hall sender, represented in Fig. 1-1. The Hall sender in the distributor is the electronic equivalent of breaker points. It provides a timing signal to the ignition control unit, which in turn switches the 12-volt coil primary circuit to discharge the high-voltage spark. The basic system is equipped with mechanical advance mechanisms in the distributor that adjust the ignition timing for load and speed.

1.2 Electronic Ignition With Knock Sensor

For the Scirocco 16V, the basic electronic ignition has the added capability to sense pre-ignition or detonation (commonly called knock or ping), and adjust ignition timing electronically to eliminate it. The system, represented in Fig. 1-2, includes a knock sensor (Fig. 1-3) and knock sensor control unit. With this system, the distributor has no mechanical advance mechanisms. Ignition timing advance is controlled electronically.

1.3 Digifant Engine Management Systems

Cabriolet models from 1990 are equipped with a version of the Digifant engine management system. Digifant controls all aspects of fuel injection, ignition, knock control and emission control. The coil primary circuit is still switched by a separate ignition control unit, but all timing advance and knock regulation are controlled electronically by the Digifant control unit.

From 1991 (California only)

Beginning with the 1991 model year, cars built for sale in California are equipped with a special version of Digifant called Digifant I. Significant differences are:

- special ignition coil with power output stage
- no separate ignition control unit (primary circuit)
- Digifant I control unit with fault memory and on-board diagnosis (OBD) capability

CAUTION —
Volkswagen specifies that the first step in any Digifant I system troubleshooting must be to check the OBD fault memory, and to find and correct any faults that may be indicated. See **FUEL SYSTEM**.

1.4 System Identification and Applications

Table a lists engine codes and the corresponding ignition system for cars covered by this manual. For more information on identifying a particular engine, see **ENGINE**.

Table a. Ignition System Applications

Engine code	Ignition system	Application
JH	Hall-sender electronic	1985-1989 exc. 16V
PL	Electronic w/knock sensor	Scirocco 16V only
2H	Digifant	from 1990

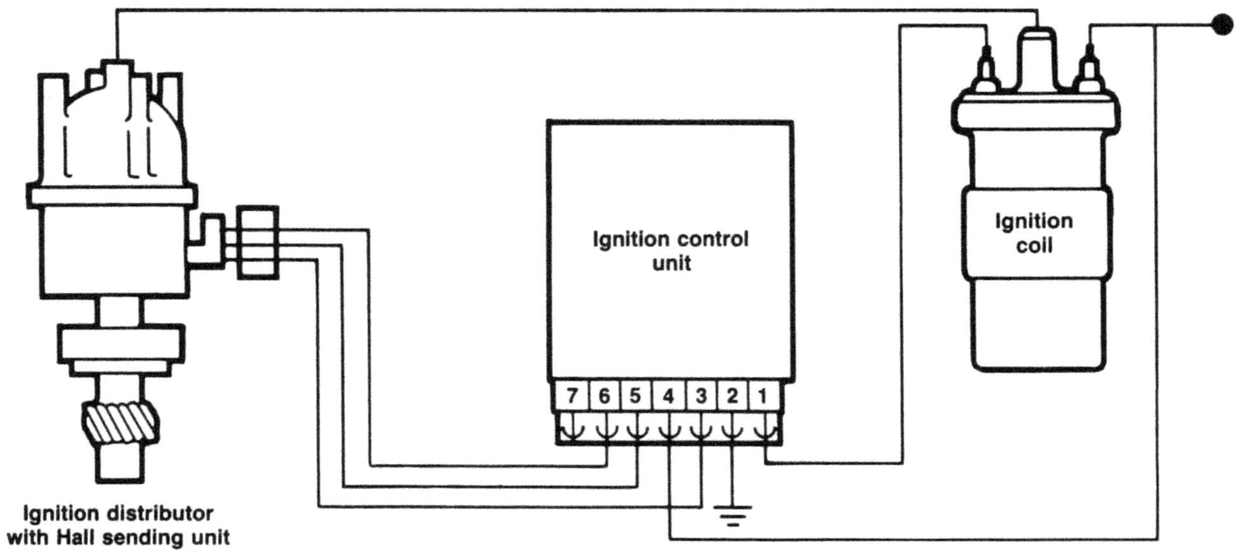

Fig. 1-1. Schematic diagram showing basic components of Hall-sender electronic ignition for all 1985-1989 models (engine code JH) except Scirocco 16V.

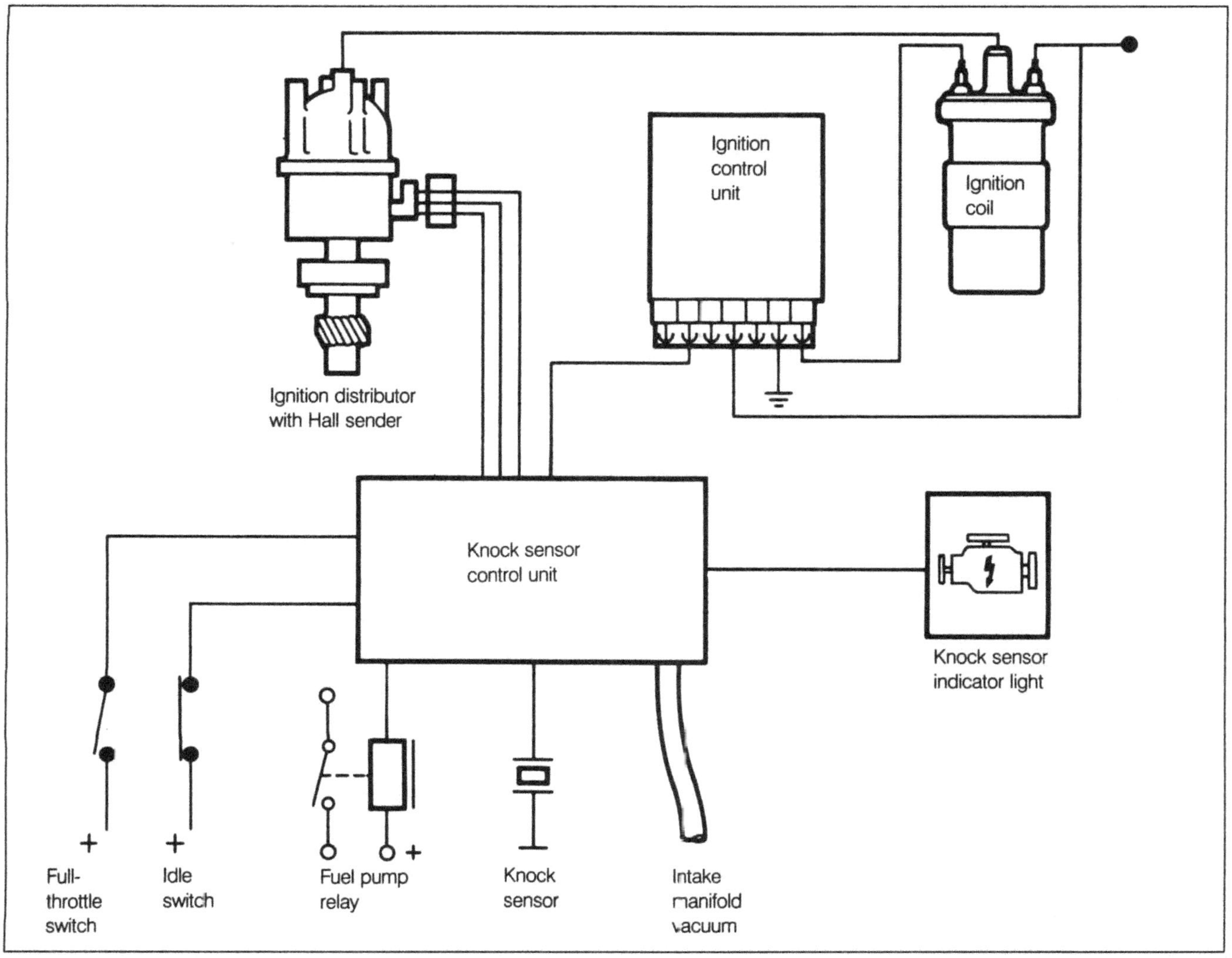

Fig. 1-2. Schematic view of Hall-sender electronic ignition with knock sensor.

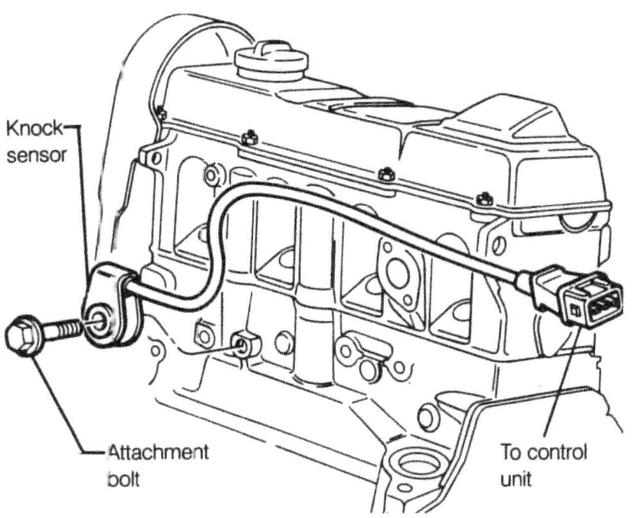

Fig. 1-3. Typical knock sensor and mounting location on side of cylinder block.

2. MAINTENANCE

Volkswagen specifies the maintenance steps below to be carried out at particular time or mileage intervals for proper maintenance of the ignition system.

A number in bold type indicates maintenance that is covered in this section, under the heading with that number. See **LUBRICATION AND MAINTENANCE** for information about other listings, and for the prescribed maintenance intervals.

1. Replacing spark plugs
2. Inspecting and replacing distributor cap and rotor **3.2**
3. Inspecting and replacing spark plug wires **3.3**

3. TROUBLESHOOTING

Complete failure to produce spark at the spark plugs is almost certainly an ignition system problem. For other problems such as poor starting, rough idle, or misfiring, the cause may not be as clear. Start by checking the ignition system, but if no ignition problems are discovered, proceed elsewhere. Low engine compression or fuel system faults will also cause such problems. Also see **ENGINE** and **FUEL SYSTEM**.

3.1 Basic Troubleshooting Principles

The ignition system's main function is to provide a properly timed high-voltage spark to the spark plugs. If cranking speed is normal, not starting may be caused by a complete failure of the system to produce spark. A basic check to see if spark is being produced will be the most important first troubleshooting step. If a strong spark is observed, then the failure to start has another cause, perhaps incorrect ignition timing or a mechanical engine fault.

If the engine starts and runs, the ignition system is basically sound—it is delivering voltage to the spark plugs. An engine that is hard to start or runs badly may have problems affecting how well the spark is delivered.

A faulty coil, poor spark plug wires, a worn or cracked distributor cap or rotor, and worn or fouled spark plugs are also possible causes of reduced spark intensity and inefficient combustion. For cars with catalytic converters, this is an especially serious problem.

> ***WARNING —***
> *A catalytic converter overheated by rough running or misfire may be a fire hazard. Avoid running the engine for very long if it is running badly.*

The ignition system contains delicate electronic components. It is sensitive to corrosion and loose connections, especially faulty ground connections. For your safety, and to protect the system, read these Cautions and follow them during any troubleshooting or other work on the ignition system.

> ***CAUTION —***
>
> - *Never touch or disconnect any of the high tension cables from the coil, distributor, or spark plugs while the engine is running or being cranked by the starter.*
> - *Connect or disconnect ignition system wires, connectors, and test equipment only while the ignition is switched off.*
> - *Before operating the starter without starting the engine (as when making a compression test), disconnect the wire from the center tower of the distributor cap and positively connect it to ground on the engine, or (on Digifant I only) disconnect the coil power output stage.*
> - *Do not connect test instruments with a 12-volt current supply to terminal 15 (+) of the ignition coil. Reverse voltage flow will damage the ignition control unit. Make connections only as specified by Volkswagen, as described in this manual or by the instrument manufacturer.*
> - *Do not connect a condenser to negative (–) coil terminal 1. This may damage the ignition system or cause misfiring.*

> ***CAUTION —***
>
> - *Your Volkswagen uses a special resistance-type distributor rotor. Take care to install only the correct part. Do not substitute any other Volkswagen or Bosch part.*
> - *Do not disconnect the battery while the engine is running.*
> - *Do not quick-charge or boost the battery for longer than one minute, and do not exceed 16.5 volts at the battery with the booster cables connected. Wait at least one minute before boosting the battery a second time.*
> - *Do not tow a car suspected of having a defective ignition system without first disconnecting the ignition control unit or (on Digifant I only) disconnecting ignition coil terminal 1 (green wire).*
> - *Do not wash the engine while it is running, or anytime the ignition is switched on.*
> - *Disconnect both battery terminals when doing any electric arc or spot welding on the car.*
> - *Do not try to start the engine of a car which has been heated above 176°F (80°C), (for example, in a paint drying booth) until allowing it to cool to normal temperature.*
> - *Do not test the ignition system with a normal incandescent bulb test light. The high current load may damage the electronic components.*

Test Equipment

Many of the tests in this manual call for measuring voltage, current or resistance using a volt-ohm meter (VOM) or a multimeter. A digital volt-ohm meter (DVOM) or digital multimeter is highly recommended to prevent circuit damage.

An LED (light emitting diode) test light such as Volkswagen special tool VAG 1527B (order no. TAG 152 7B0 28 ZEL) is an inexpensive tool that is used for many simple electrical tests. The LED lights to indicate when there is voltage potential between any two test points in a circuit. The LED's low current draw makes it safe for testing sensitive electronic circuits. For more information on test equipment and safe electrical testing, see **FUNDAMENTALS**.

> ***CAUTION —***
>
> - *Choose test equipment carefully. Use an LED test light or a meter with at least 10 megohm input impedance. Other tools may damage sensitive electronic components in the control unit.*
> - *Volkswagen part numbers and special tool order numbers listed in this manual are provided for reference only. Check with an authorized Volkswagen dealer Parts department for the most accurate and up-to-date information.*

Quick-check of Ignition System

If the engine does not start, the most fundamental troubleshooting step is to determine whether or not the system is creating spark at the spark plug. If a spark is present, then it is the ignition system's efficiency and accuracy or another system's function that is in question. If no spark is present, then more detailed testing of the ignition system is necessary.

To check for spark, switch the ignition off and disconnect one of the spark plug wire connectors from the spark plug. Temporarily connect it to a known good spark plug, preferably a new one. Hold the spark plug with insulated pliers so that the outer shell or electrode are grounded on the engine.

> ***CAUTION —***
> *Any test set-up other than the one described above may give inconclusive results or damage the ignition system.*

While a helper actuates the starter, look for spark in the gap between the two spark plug electrodes. A bright blue spark indicates a healthy ignition system. A yellow-orange spark is weaker and indicates that, while spark is present and the system is functioning, it is not operating at peak efficiency. Check the condition of the ignition system components as described in **3.2 Ignition System Visual Inspection** and replace any faulty components.

> ***WARNING —***
> - *If the ignition system is not faulty, the engine may start during this test. Be prepared to switch off the ignition immediately.*
> - *Running the engine too long with a spark plug wire disconnected will contaminate and damage the catalytic converter.*

> **NOTE —**
> Before checking the ignition system when there is no spark or a weak spark, make sure that the battery is fully charged. See **ELECTRICAL SYSTEM**.

If there is no spark, test for primary voltage at the ignition coil. Connect a test light between terminal 1 (–) of the ignition coil and ground. While the starter is actuated, the test light should flash, indicating that the coil is receiving primary voltage.

If the coil is not receiving primary circuit voltage, there is a fault in the wiring harness or the ignition switch. If the coil is receiving voltage, or if a strong spark is observed but the engine still will not start, refer to **table b** for more troubleshooting information. Bold numbers in the corrective action column refer to information under numbered headings in this section.

> ***CAUTION —***
> *For Digifant cars (from 1990), also see the engine management system electrical tests in* **FUEL SYSTEM**.

Table b. Ignition System Troubleshooting

Symptom	Probable cause	Corrective action
1. Engine will not start—cranking speed too low	**a.** Starting system fault	**a.** See **ELECTRICAL SYSTEM**
2. Engine will not start—cranking speed OK	**a.** No spark **b.** Wet distributor cap and/or spark plug wires **c.** Weak coil **d.** Defective spark plug wires **e.** Faulty Hall sender **f.** Incorrect ignition timing **g.** Engine mechanical failure **h.** Fuel system fault	**a.** Spark test ignition. **3.1** **b.** Remove cap and wires. Dry and reinstall **c.** Test and replace as needed. **3.3** **d.** Test and replace as needed. **3.3** **e.** Test and replace as needed. **3.4** **f.** Check and adjust timing. **4.1** **g.** See **ENGINE** **h.** See **FUEL SYSTEM**
3. Engine starts but runs badly, misfires, loses power, backfires	**a.** Incorrect ignition timing **b.** Wet distributor cap and/or spark plug wires **c.** Fouled spark plugs **d.** Weak coil	**a.** Check and adjust timing. **4.** **b.** Remove cap and wires. Dry and reinstall **c.** Replace spark plugs. See **LUBRICATION AND MAINTENANCE** **d.** Test and replace as needed. **3.3**

continued on next page

Table b. Ignition System Troubleshooting (continued)

Symptom	Probable cause	Corrective action
3. Engine starts but runs badly, misfires, loses power, backfires (cont'd)	**e.** Defective spark plug wires **f.** Worn distributor cap/rotor **g.** Faulty knock sensor control unit (TCI-h with knock sensor only) **h.** Faulty centrifugal advance mechanism (TCI-h only) **i.** Faulty control unit **j.** Faulty throttle switches **k.** Engine mechanical damage	**e.** Test and replace as needed. **3.3** **f.** Inspect and replace as needed. **3.2** **g.** Test knock sensor system. **4.3, 5.2** **h.** Test and replace as needed. **4.2** **i.** Test and replace as needed. **4.** **j.** Test switches and replace as needed. See **FUEL SYSTEM** **k.** See **ENGINE**
4. Knocking or pinging noise while accelerating	**a.** Incorrect ignition timing **b.** Faulty centrifugal advance mechanism (TCI-h only) **c.** Faulty ignition control unit **d.** Faulty knock sensor and/or control unit **e.** Faulty control unit **f.** Pre-ignition due to carbon buildup or burned valves in engine combustion chambers	**a.** Check and adjust timing. **4.1** **b.** Test and replace as needed. **4.2** **c.** Test and replace as needed. **3.5** **d.** Test knock sensor system. **4.3, 5.2** **e.** Test and replace as needed. **f.** Overhaul or replace cylinder head. See **ENGINE**

3.2 Ignition System Visual Inspection

Accumulated dirt and moisture, wear and electrical breakdown may keep the distributor cap, rotor and spark plug wires from delivering a precise and powerful spark. Thorough visual inspection is the way to check on these conditions. Some more detailed assembly information can be found in **5. Distributor**.

Remove the cap and inspect the contacts inside the distributor cap and at the tip of the rotor for corrosion, wear or pitting. You can stall for time by cleaning and reusing parts with slightly corroded contacts but replacement is highly recommended, especially if there is noticeable wear and pitting. The center black carbon brush inside the cap is spring-loaded to ensure contact, and should spring back when depressed.

Replace a distributor cap that shows any sign of cracks, white burn spots or carbon tracking. Carbon tracks are faint black lines, usually between two wire terminals or from one terminal to ground. The cracks may be fine and difficult to see. Check carefully, especially around the contacts. Be sure to check under the radio-suppression shield.

Gently bend the spark plug wires to expose cracks in the insulation that cause voltage "leaks". The rubber boots should be soft and pliable to seal out dirt and moisture. Replace wires that are cracked, oil-soaked, or hardened and brittle.

Listen for the sound of arcing while the engine runs. At night, high-voltage arcing to ground because of a crack in the cap or a poorly insulated wire may be visible as a small blue spark. For another quick check of cap and wire condition, use a spray bottle to spray a fine mist of water around them while the engine runs. If the cap and wires are in good condition and insulated properly, the added moisture should have no effect. If their condition is marginal, the added moisture may promote arcing and cause the engine to run rough.

3.3 Testing Coil and Spark Plug Wires

Use an ohmmeter to test coil primary and secondary circuits as shown in Fig. 3-1. Replace the coil if primary or secondary resistance measurements exceed the values listed in **table c**.

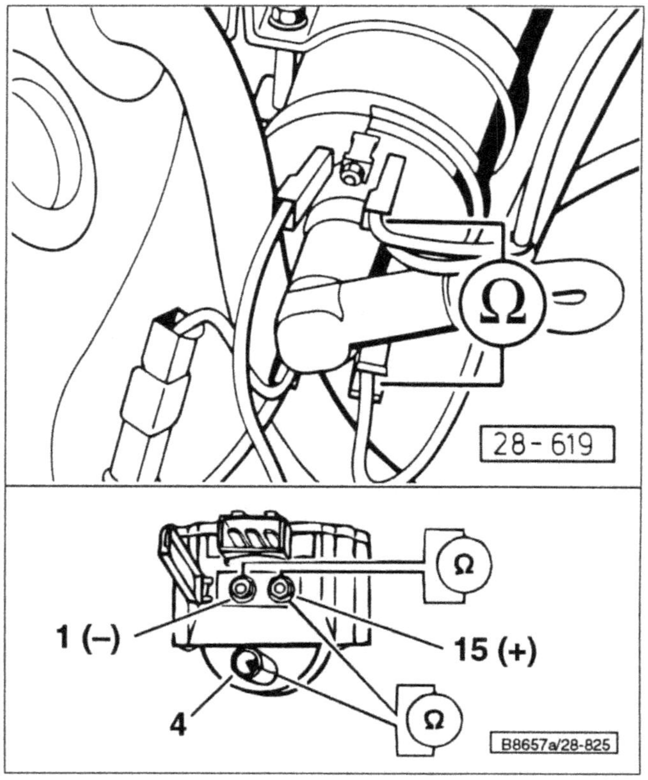

Fig. 3-1. Measuring coil resistance on conventional coil (top) and Digifant I coil (bottom) with power output stage disconnected.

Table c. Ignition Coil Resistance

Test	Engine	Resistance – ohms
Primary resistance terminals **1** (–) and **15** (+)	JH (1985-1989 except 16-valve)	0.52 to 0.76 Ω
	2H (except Digifant I California)	
	PL (Scirocco 16V)	0.6 to 0.8 Ω
	2H (Digifant I only, California from 1991)	0.5 to 0.7 Ω
Secondary resistance terminals **15** (+) and **4** (coil center tower)	JH (1985-1989 except 16-valve)	2400 to 3500 Ω
	2H (except Digifant I California)	
	PL (Scirocco 16V)	6500 to 8500 Ω
	2H (Digifant I only, California from 1991)	3000 to 4000 Ω

Disconnect the spark plug wires from the spark plugs and from the distributor cap. Use an ohmmeter to check resistance as shown in Fig. 3-2. If resistance is too high, check the wire and the wire end connectors separately, as shown in Fig. 3-3.

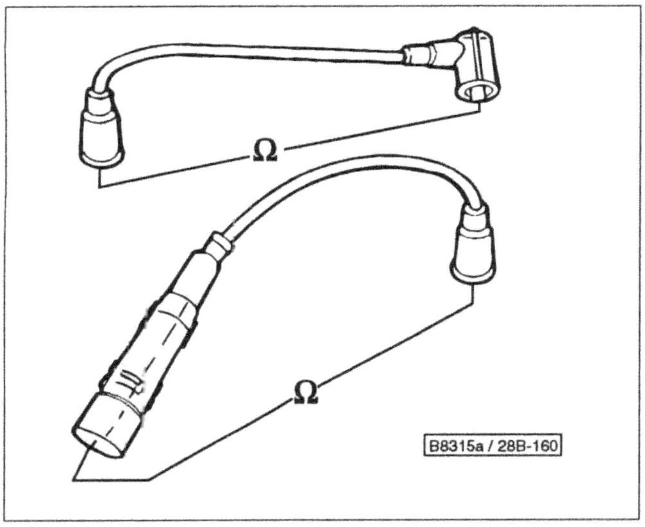

Fig. 3-2. Resistance of spark plug wire (bottom) and coil wire (top) being measured with an ohmmeter (shown schematically).

CAUTION —
To avoid damaging the distributor cap, do not wiggle the connectors. If necessary, twist to loosen them. Then, pull straight out from the cap.

Resistance specifications (with connectors):
- spark plug wire 4600 to 7400 Ω
- coil wire 1200 to 2800 Ω

Remove the connectors carefully. Twist, then pull. If the metal wire end stays in the connector, the connection is ruined and both the connector and the wire must be replaced. Check for corrosion at the connections. Wires or connectors with too

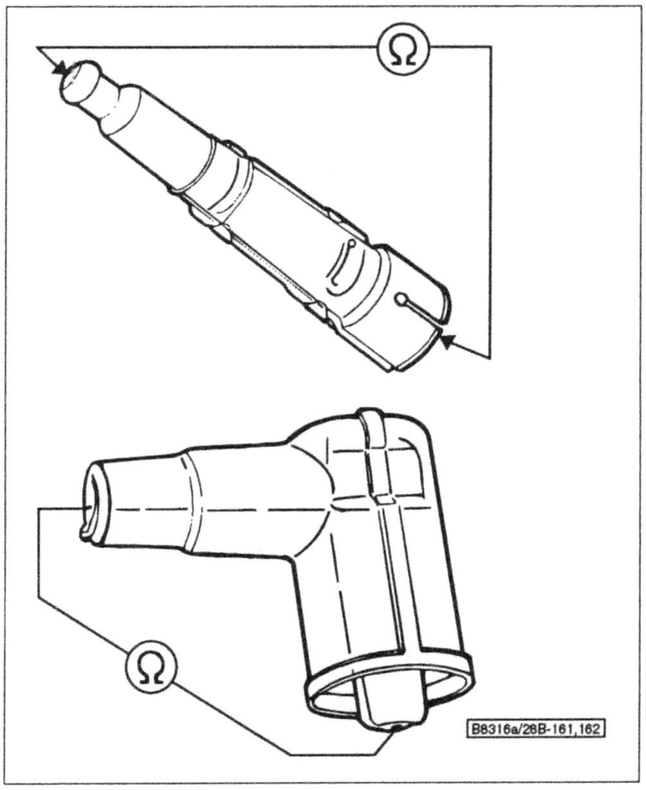

Fig. 3-3. Resistance of spark plug wire (bottom) and coil wire (top) being measured with an ohmmeter (shown schematically).

much resistance should be replaced. The distributor rotor also has a specific resistance value. Test between the center terminal and the end. Do not interchange rotors.

Resistance specifications:
- wire (alone) continuity (approx. 0 Ω)
- spark plug connector 4000 to 6000 Ω

CIS and Digifant (all except Scirocco 16V):
- coil/distributor cap connectors 600 to 1400 Ω
- distributor rotor 600 to 1400 Ω

CIS-E (Scirocco 16V):
- coil/distributor cap connectors 800 to 1200 Ω
- distributor rotor 800 to 1200 Ω

3.4 Testing Hall Sender and Ignition Control Unit

The Hall sender and ignition control unit only need to be tested if there is no spark at the spark plugs. To proceed, it must first be confirmed that the ignition coil is OK. Check the spark plug wires and coil as described in **3.3 Testing Coil and Spark Plug Wires**. The ignition control unit is in the fresh air plenum beneath the hood in front of the windshield. The Hall sender connector is on the side of the distributor.

The tests that follow require the use of a high-impedance voltmeter or low-current LED test light. For more information about test equipment and testing precautions, see **3.1 Basic Troubleshooting Principles**.

CAUTION —
Always switch multimeter function or range before connecting the test probes. Connect or disconnect probes only with the ignition switched off.

Ignition Control Unit Power and Ground
(except Digifant I, California from 1991)

With the ignition switched off, disconnect the harness connector from the ignition control unit. Connect a multimeter to measure voltage (20 volt scale) between connector terminals 2 (–) and 4 (+) as shown in Fig. 3-4. With the ignition switched on, the meter should indicate battery voltage (about 12 volts DC). If not, check for faults in the wiring harness between terminal 2 and ground, and between terminal 4 and the coil. Wiring diagrams are found in **ELECTRICAL SYSTEM**.

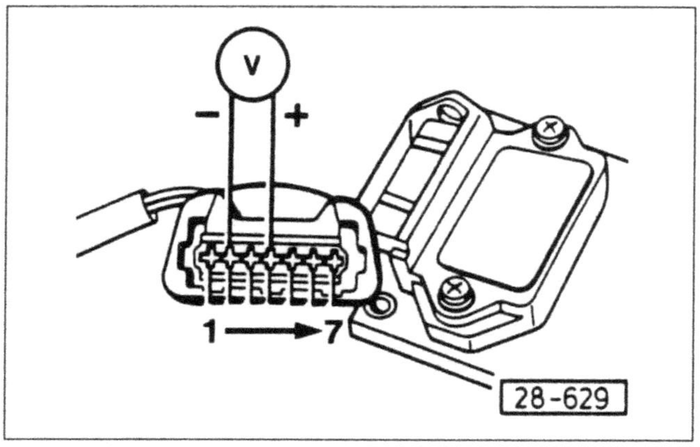

Fig. 3-4. Ignition control unit voltage supply being checked at connector terminals **2** and **4**.

Coil Power Output Stage Power and Ground
(Digifant I only, California from 1991)

With the ignition switched off, disconnect the 3-pin harness connector from the power output stage. Connect a multimeter to measure voltage (20 volt scale) between connector terminals 1 and 3 as shown in Fig. 3-5. With the ignition switched on, the meter should indicate battery voltage (about 12 volts DC). If not, check for faults in the wiring harness.

Power to Coil Primary Circuit

With the ignition switched off, connect a multimeter to measure voltage between ignition coil terminals 1 (–) and 15 (+). See Fig. 3-6 or, for Digifant I (California from 1991), Fig. 3-1 earlier in this section. Do not disconnect the existing coil wiring.

CIS:
- disconnect Hall sender connector at distributor

CIS-E (Scirocco 16V):
- disconnect knock sensor control unit connector

Digifant II (from 1990):
- disconnect Digifant control unit connector

Digifant I (California from 1991):
- reconnect coil power output stage connector

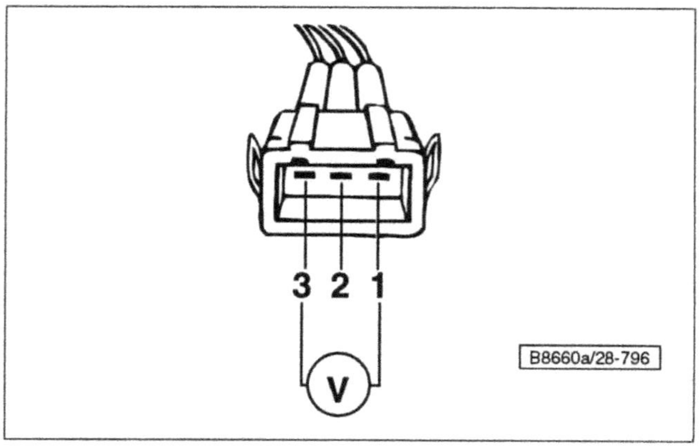

Fig. 3-5. Voltage being measured at Digifant I power output stage harness connector terminals **1** and **3**.

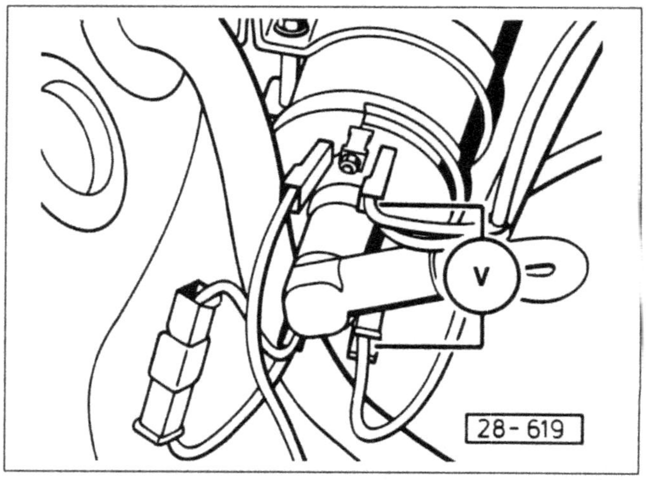

Fig. 3-6. Voltage being measured at ignition coil terminals **1** (–) and **15** (+). (Digifant I see Fig. 3-1.)

With the ignition switched on, the meter should indicate at least 2 volts, then drop to 0 volts after 1 to 2 seconds. If not, check the ignition coil. It should be replaced if there is any sign of fluid leakage. If the coil is OK, replace the ignition control unit or, on Digifant I cars, the coil power output stage.

Checking Hall Sender

With the ignition switched off, disconnect the Hall sender harness connector at the distributor. Connect a multimeter to measure voltage between connector terminals 1 (–) and 3 (+), as shown in Fig. 3-7. With the ignition is switched on, the meter should indicate battery voltage (about 12 volts DC). If not, check for faults in the wiring harness.

Disconnect the coil wire from the center of the distributor cap (terminal 4) and connect it to ground with a jumper wire to disable the secondary ignition (high voltage) circuit. On Digifant cars, disconnect the main fuel injector harness connector to disable the solenoid-type fuel injectors.

Carefully push back the rubber boot to expose the back of the Hall sender harness connector. Connect an LED test light

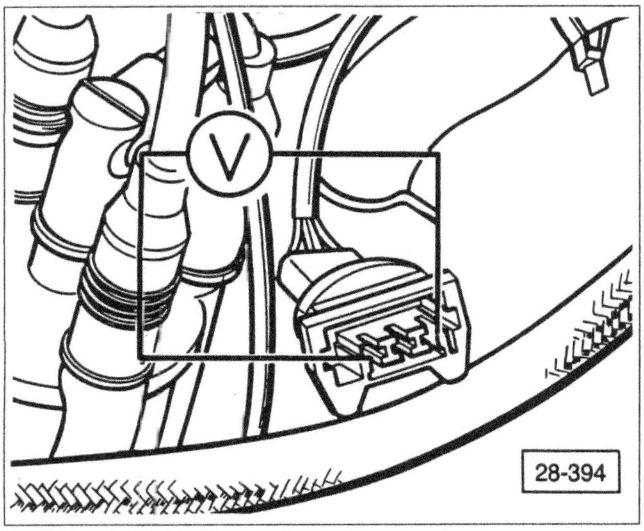

Fig. 3-7. Power and ground being checked at Hall sender harness connector at distributor.

between the center terminal and one of the other terminals, as shown in Fig. 3-8. As you briefly actuate the starter, the LED should flicker on and off. If not, the Hall sender is defective and should be replaced. See **6. Distributor**.

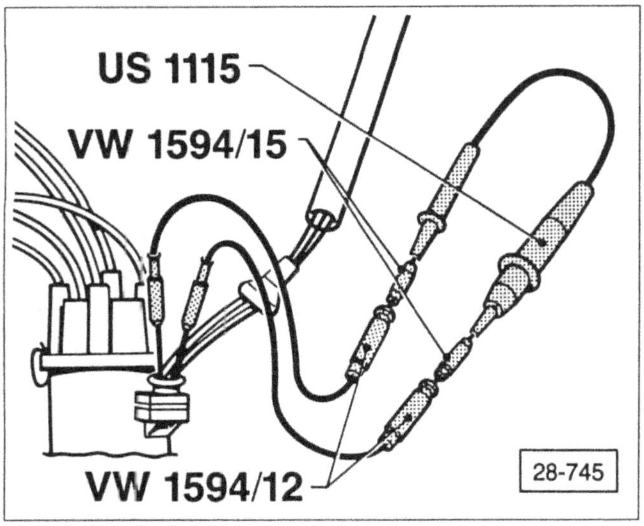

Fig. 3-8. Hall sender function being checked with LED test light. Volkswagen special tools VAG 1527B test light and VAG 1527B adapter kit shown.

On Digifant cars, connect the LED test light at the ignition coil and repeat the test. If the LED does not flicker on and off, there are faults in the wiring harness, or the Digifant control unit is faulty and should be replaced.

- Digifant II: coil terminals **1** and **15**
- Digifant I (California from 1991): power output stage connector terminals **1** and **2** (Fig. 3-5)

Checking Ignition Control Unit

The primary switching function can be checked by simulating a signal from the Hall sender. Connect a multimeter to measure voltage between ignition coil terminals 1 (–) and 15 (+).

CIS:
- disconnect Hall sender connector at distributor
- test with connector terminal **2** (center)

CIS-E (Scirocco 16V):
- disconnect knock sensor control unit connector
- test with connector terminal **12** (Fig. 3-9)

Digifant II (from 1990):
- disconnect Digifant control unit connector
- test with connector terminal **25** (Fig. 3-10)

Switch the ignition on, then briefly connect the designated connector terminal to ground. The meter should indicate a brief rise to at least 2 volts at the coil. If not, the ignition control unit is defective and should be replaced.

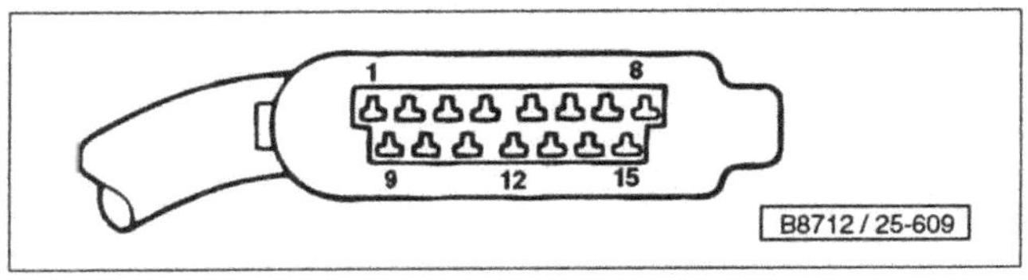

Fig. 3-9. CIS-E knock sensor control unit connector.

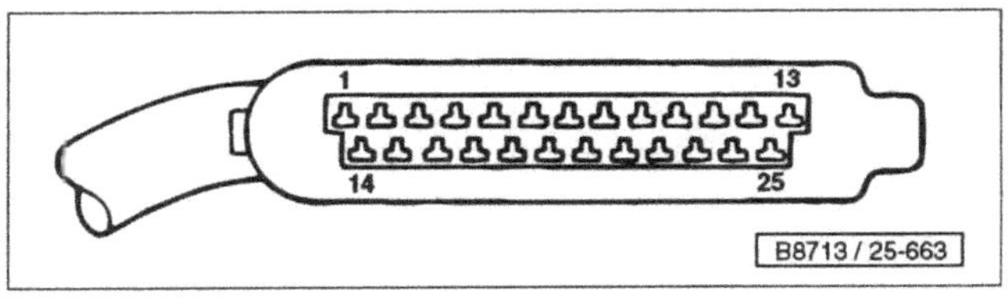

Fig. 3-10. Digifant II control unit connector.

4. HALL-SENDER ELECTRONIC IGNITION

The main components of the Hall-sender electronic ignition system are the distributor with Hall sender, the coil, an electronic control, and the spark plugs (Fig. 4-1). The knock sensor system for 16-valve engines (engine code PL) adds the knock sensor and a knock sensor control unit, for electronic control of timing advance (Fig. 4-2).

On Digifant cars (from 1990), the Digifant control unit is responsible for all timing advance and knock control. Digifant I (California from 1991) also takes over control of the ignition primary circuit, and eliminates the separate ignition control unit.

This section covers ignition timing and timing advance, with procedures for adjusting basic timing adjustment and checking timing advance functions. Ignition system specifications are listed in **table d** and **table e**. Troubleshooting for the 16-valve Scirocco's knock sensor system is covered later in **5. Knock Sensor System.**

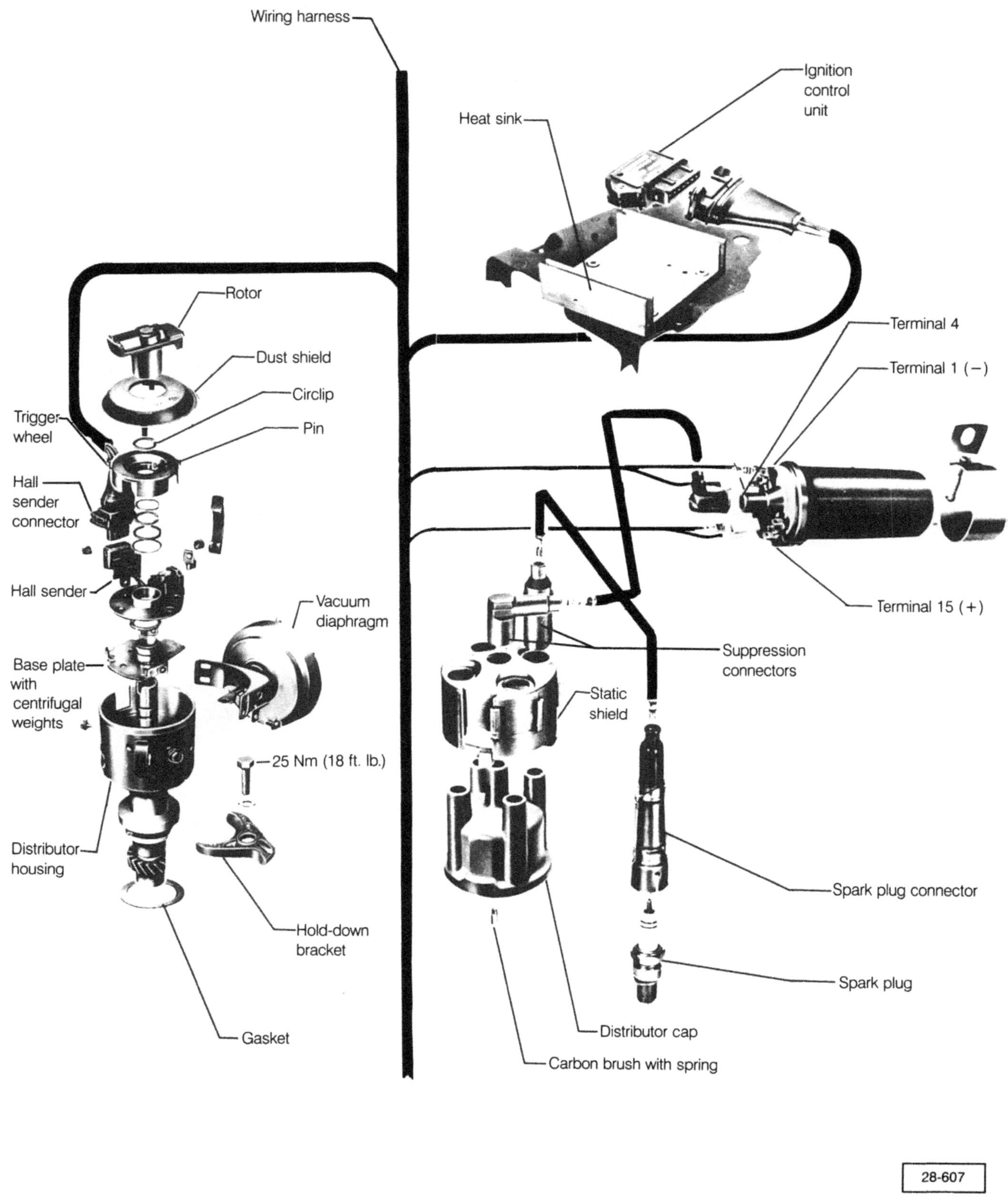

Fig. 4-1. Basic Hall-sender electronic ignition for JH engines, 1985-1989 except Scirocco 16V. Distributor assembly includes internal centrifugal advance mechanism and externally mounted vacuum advance diaphragm.

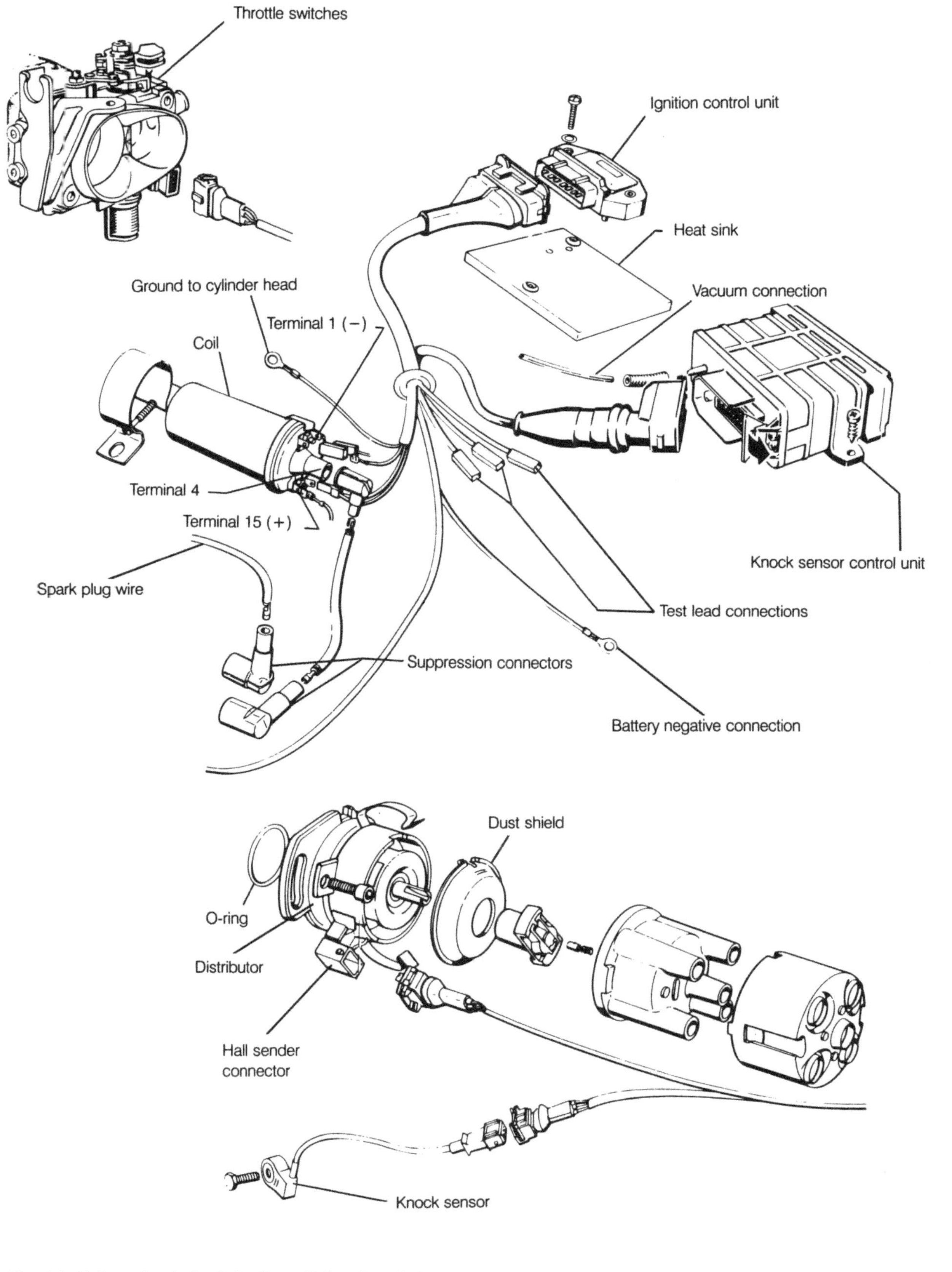

Fig. 4-2. Hall-sender electronic ignition with knock control for PL (Scirocco 16V) engine. Distributor has no mechanical timing advance mechanisms.

Table d. Electronic Ignition System Specifications
(1985-1989, CIS and CIS-E)

Engine codes	JH, PL
Firing order	1–3–4–2
Ignition timing	
at idle (see **FUEL SYSTEM**)	
vacuum hoses	
to distributor (engine code JH)	connected
to knock sensor control unit (PL)	connected
checking	4 to 8° BTDC
adjusting	6 ± 1° BTDC
Centrifugal advance (engine code JH only)	
starts	1050 – 1300 rpm
degrees advance @ rpm	15 – 19° @ 2600 RPM
total advance @ rpm	22 – 26° @ 4500 RPM
Vacuum advance (engine code JH only)	
start	4 – 8° BTDC (basic timing) @ 120 – 160 mbar (90 - 120 mm Hg)
total	10 – 14° BTDC @ 240 – 260 mbar (180 - 195 mm Hg)
Spark plug tightening torque	20 Nm (15 ft. lb.)

Table e. Electronic Ignition System Specifications
(Digifant, engine code 2H)

Engine code	2H
Firing order	1–3–4–2
Ignition timing	
coolant temperature sensor	disconnected
engine speed	2000 to 2500 rpm
checking	4 to 8° BTDC
adjusting	6 ±1° BTDC
Timing advance (Digifant II)	
coolant temperature sensor	connected
engine speed	2000 to 2500 rpm
total advance	basic timing plus 30° ± 3°
Timing advance (Digifant I)	
coolant temperature sensor	connected
engine speed	4500 rpm
total advance	basic timing plus 30° (minimum)
Spark plug tightening torque	20 Nm (15 ft. lb.)

4.1 Ignition Timing

Correct ignition timing is essential if the ignition and fuel systems are to deliver top performance and maintain control of harmful exhaust emissions. Checking and adjusting ignition timing is an important part of any fuel system adjustments such as idle speed and idle mixture. See **FUEL SYSTEM**.

Ignition timing also needs to be adjusted whenever the distributor has been removed or taken apart, and whenever the camshaft drive belt has been adjusted or replaced.

Requirements:
- Engine warm—minimum oil temperature 176°F (80°C)
- Idle speed OK (see **FUEL SYSTEM**)
- tachometer and timing light connected
- throttle switch(es) OK (see **FUEL SYSTEM**)

To check basic ignition timing:

1. On manual transmission cars, remove the plastic plug from the timing check hole at the top of the transaxle bellhousing.

   NOTE —
   Remove the entire plug. The smaller hole is for a special timing indicator used by Volkswagen dealer Service technicians.

2. Start the engine and let it idle. Briefly raise the engine speed above 3000 rpm three or four times.
3. On Digifant cars only (from 1990), disconnect the blue harness connector from the coolant temperature sensor. See Fig. 4-3.

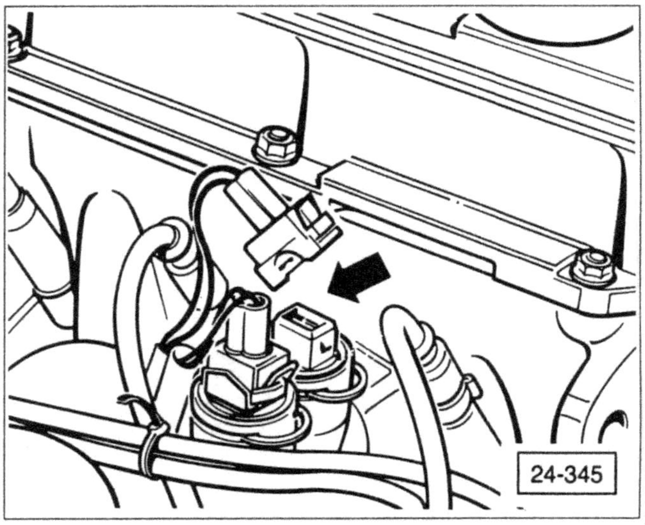

Fig. 4-3. Coolant temperature sensor harness connector (blue) disconnected for ignition timing check on Digifant I and Digifant II cars (from 1990).

4. Check the timing. With a timing light, the timing mark on the flywheel should appear at the pointer in the timing check hole in the transaxle, as shown in Fig. 4-4.

   Ignition timing specifications:
   - checking — 4° to 8° before TDC
   - CIS, CIS-E — check at idle
   - Digifant (I and II) — at 2000–2500 rpm

   WARNING —
   Keep hands and other objects clear of the radiator cooling fan. The fan may start at any time, even when the ignition is switched off.

   CAUTION —
   - *The timing indication is accurate only when the engine idle speed is within specifications.*
   - *Check the timing marks only when the radiator cooling fan is not running.*

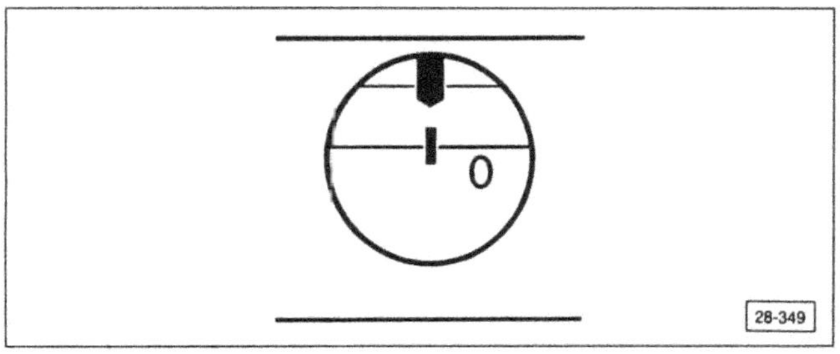

Fig. 4-4. Timing mark on flywheel aligned with pointer in timing check hole (top of bellhousing). Timing mark indicates 6° before top dead center (BTDC).

5. Switch the ignition off. If the ignition timing is OK, disconnect the test equipment and, on Digifant cars, reconnect the coolant temperature sensor. If timing is outside the specified checking specification, it should be adjusted.

NOTE —
For Digifant I only, (California cars from 1991), disconnecting the coolant temperature sensor registers a fault that must be cleared from the Digifant OBD fault memory. See **FUEL SYSTEM**.

To adjust basic ignition timing:

1. With the ignition switched off, loosen the distributor hold-down bolt. On PL (16-valve) engines, there are two.

CAUTION —
The distributor should be just loose enough to be moved by hand, but not so loose that it engine vibration alone can turn it.

2. Start the engine and check the timing as described in the previous procedure.
3. To adjust, gradually turn the distributor housing until the timing mark on the flywheel lines up with the pointer in the timing check hole in the transaxle, as shown above in Fig. 4-4.

Ignition timing specifications:
- adjusting — 6°±1° before TDC
- CIS, CIS-E — check at idle
- Digifant (I and II) — at 2000–2500 rpm

4. Stop the engine and tighten the distributor hold-down bolt(s) without moving the distributor.

Tightening torque:
- JH, 2H (all except 16V) — 25 Nm (18 ft. lb.)
- PL (16V) — 10 Nm (87 in. lb.)

5. Start the engine and recheck the ignition timing. Repeat the adjustment if necessary.
6. Turn the ignition off and remove the test equipment. On cars with manual transmission, replace the plastic plug in the timing check hole.

NOTE —
- For Digifant I only, (California cars from 1991), the Digifant OBD fault memory must be cleared.
- If the timing has been changed, the Digifant I control unit must be returned to its basic setting. See **FUEL SYSTEM**.

Unsteady ignition timing may be caused by worn distributor shaft bearings. If the shaft can be wiggled in the housing, the distributor housing assembly should be replaced.

4.2 Centrifugal and Vacuum Spark Advance
(1985-1989 except Scirocco 16V)

Checking timing advance requires an adjustable timing light or an engine analyzer such as Volkswagen special tool VW 1367 to accurately read ignition timing throughout the range of timing advance. Follow the instrument manufacturer's operating instructions. See **table d** earlier in this section for centrifugal and vacuum advance specifications.

The centrifugal advance mechanism inside the distributor advances ignition timing with engine rpm. When checking centrifugal advance, first disable the vacuum advance. Disconnect the vacuum hose from the vacuum diaphragm on the distributor, and plug it.

Vacuum advance helps establish the basic ignition timing at low load when intake manifold vacuum is high. It gives less advance at higher load when intake manifold vacuum is low. To check vacuum advance, disconnect the vacuum hose from the vacuum diaphragm on the distributor and substitute a hand-held vacuum pump.

The vacuum advance diaphragm must not leak. It should hold about 500 mbar vacuum, losing no more than 50 mbar in one minute. Also make sure that the hose does not leak.

Also check the plate on which the Hall sender is mounted, It should move back and forth smoothly without sticking, and spring back to its original position when released. If not, then the plate is binding or the vacuum advance diaphragm is faulty.

4.3 Electronic Timing Advance—Scirocco 16V

Checking timing advance requires an adjustable timing light or an engine analyzer such as Volkswagen special tool VW 1367 to accurately read ignition timing throughout the range of timing advance. Follow the instrument manufacturer's instructions.

RPM-dependent Timing Advance

Start the engine and let it idle for a few minutes to establish a base timing value in the knock sensor control unit. Check and record this base reading. Disconnect the vacuum hose from the knock sensor control unit. Raise engine rpm as indicated, then compare maximum timing advance with the base value.

CIS-E rpm-dependent timing advance:
- total timing @ 2500 rpm = measured base timing value (4° to 8° BTDC) plus approximately 13°

After this check, stop the engine and reconnect the vacuum hose to the knock sensor control unit. If timing does not advance at increased rpm as specified, the knock sensor or the control unit is at fault. See **5. Knock Sensor System**.

Vacuum Advance

Check vacuum advance only after determining that the rpm-dependent timing advance is correct. Start the engine and increase engine speed to about 3400 rpm. Record the total timing. Disconnect the vacuum hose from the knock sensor control unit and again raise the engine speed to about 3400 rpm. Timing should advance approximately 6° from the previous value. If not, check the vacuum hose for kinks or leaks. If the hose is OK, the knock sensor control unit may be faulty.

5. KNOCK SENSOR

Details on removing and installing the knock sensor apply to all cars that have a knock sensor. The rest of this section covers tests of the knock sensor and knock sensor control unit that are part of the Scirocco 16V ignition system. For information on system electrical checks for later cars with the integrated Digifant engine management system, see **FUEL SYSTEM**.

5.1 Removing and Installing Knock Sensor

The two types of knock sensors used on engines covered by this manual are shown in Fig. 5-1. The Scirocco 16V may be equipped with either type. All other (Digifant) cars are equipped with type II. Correct mounting bolt tightening torque is very important. The knock sensor will not operate properly if the mounting bolt is too loose or too tight. **Table f** lists knock sensor resistance values and tightening torques.

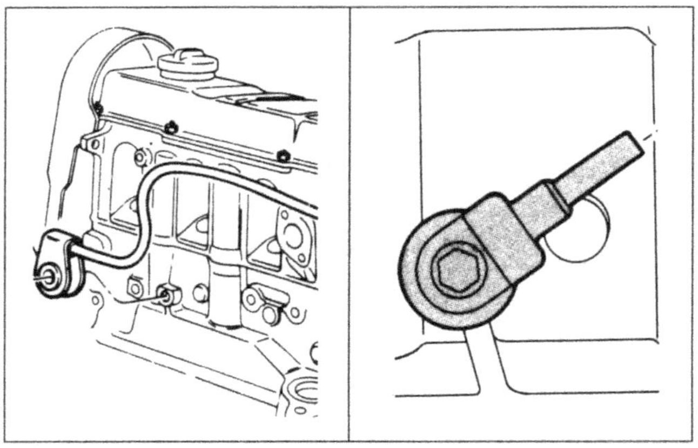

Fig. 5-1. Knock sensors type I (left) with wiring offset from center and type II (right) with wiring centered. Tightening torque is critical. See text.

CAUTION —
- *Disconnect battery negative (–) cable before loosening or tightening the knock sensor.*
- *The knock sensor must be installed using the precise tightening torque as specified.*
- *Do not use washers with the mounting bolt.*

CAUTION —
- *Volkswagen replacement knock sensors have gold-plated connector terminals. When replacing an older sensor, a new harness connector (Volkswagen Part No. 443 906 233) and gold-plated terminals (Volkswagen Part No. N 903 352 07) must also be installed to prevent corrosion.*
- *Volkswagen part numbers are provided for reference only. Check with an authorized Volkswagen dealer Parts department for the most accurate and up-to-date information.*

Table f. Knock Sensor Specifications

Type	Application	Resistance	Tightening torque
I	Scirocco 16V	approx. 300 kΩ	10 to 12 Nm (7 to 9 ft. lb.)
II	Scirocco 16V Digifant I & II	∞ ohms	15 to 25 Nm (11 to 18 ft. lb.)

5.2 Knock Sensor System Diagnosis
(Scirocco 16V only)

The knock sensor control unit stores certain system faults in electronic fault memory. These can later be recalled using a test connector in the wiring harness. The test connector is near the ignition coil. It has a single wire, usually blue/brown in color, leading to terminal 4 at the knock sensor control unit harness connector.

To display faults stored in memory:

1. Connect an LED test light between the test connector and the battery positive (+) terminal. See Fig. 5-2.
2. Switch the ignition on.
   - the LED should light
3. Start the engine.
   - the LED should go out (a slight glow is OK)
4. Briefly raise the engine speed to at least 3000 rpm. If the LED test light comes on again, proceed with the next step. If not, there are no faults stored in memory.
5. With the engine idling and the LED test light connected, use a jumper wire to ground the test connector for at least three seconds, as shown in Fig. 5-2.
   - the LED test light should begin to flash

   NOTE —
   Do not switch the ignition off during this step. Doing so will permanently erase the fault memory.

Series of two flashes indicate a fault in the wiring harness, the knock sensor, or the knock sensor control unit. Loosen and retighten the knock sensor using the proper tightening torque. Check the wiring harness using the appropriate wiring diagrams in **ELECTRICAL SYSTEM** as a guide. Fig. 5-3 and Fig. 5-4 identify the control unit connector terminals. Repair wiring defects as necessary and reconnect the harness connectors.

Series of three flashes indicate a problem with the vacuum connection to the knock sensor control unit. Check the vacuum

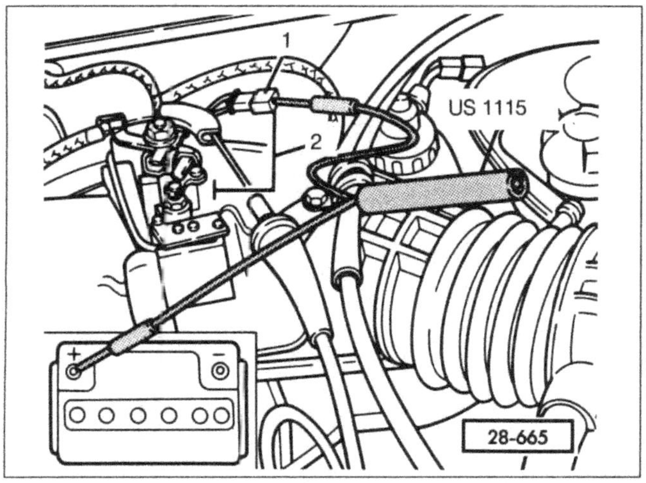

Fig. 5-2. LED test light connected to knock sensor control unit test connection (**1**), and test connection connected by jumper wire (**2**) to ground. LED test light shown is Volkswagen special tool US 1115.

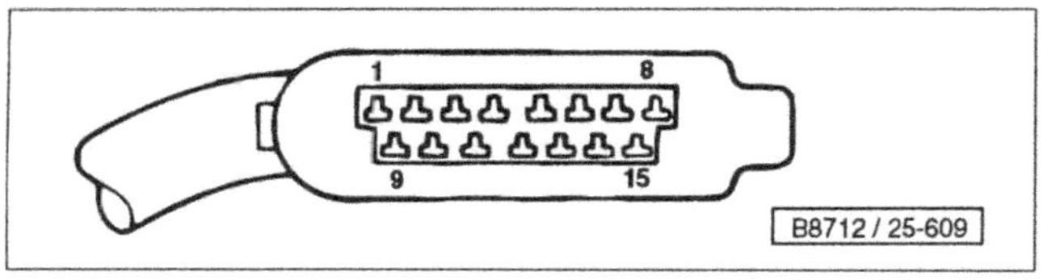

Fig. 5-3. CIS-E knock sensor control unit connector.

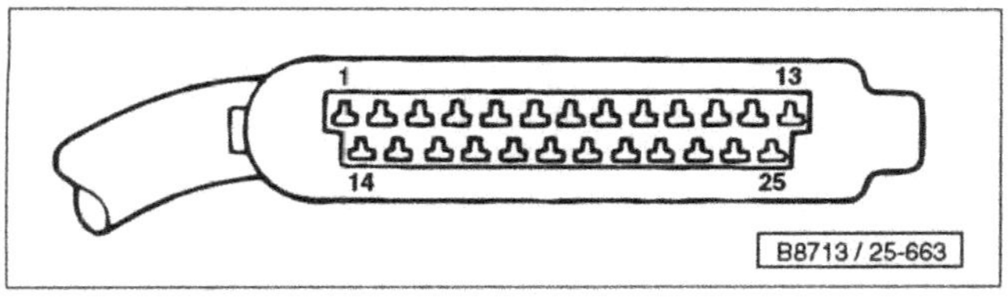

Fig. 5-4. CIS-E oxygen sensor control unit connector.

hose for kinks, damage or leaks and replace it if necessary. If the vacuum hose and hose connections are OK, the knock sensor control unit is faulty and should be replaced.

6. DISTRIBUTOR

This heading covers removing and installing the distributor, the distributor cap and rotor, the spark plug wires, and the Hall sender. Inspection and testing procedures for these components are covered in **3. Troubleshooting**. Disassembly or removal of the distributor will require checking and adjusting the ignition timing. See **4.1 Ignition Timing**. Circlips, gaskets and O-rings should always be replaced.

Fig. 6-1 is an exploded view of the distributor with centrifugal and vacuum advance mechanisms used on JH engines (1985-1989 except Scirocco 16V).

Fig. 6-2 and Fig. 6-3 show the distributor versions used on Scirocco 16V and Digifant cars, with ignition systems featuring

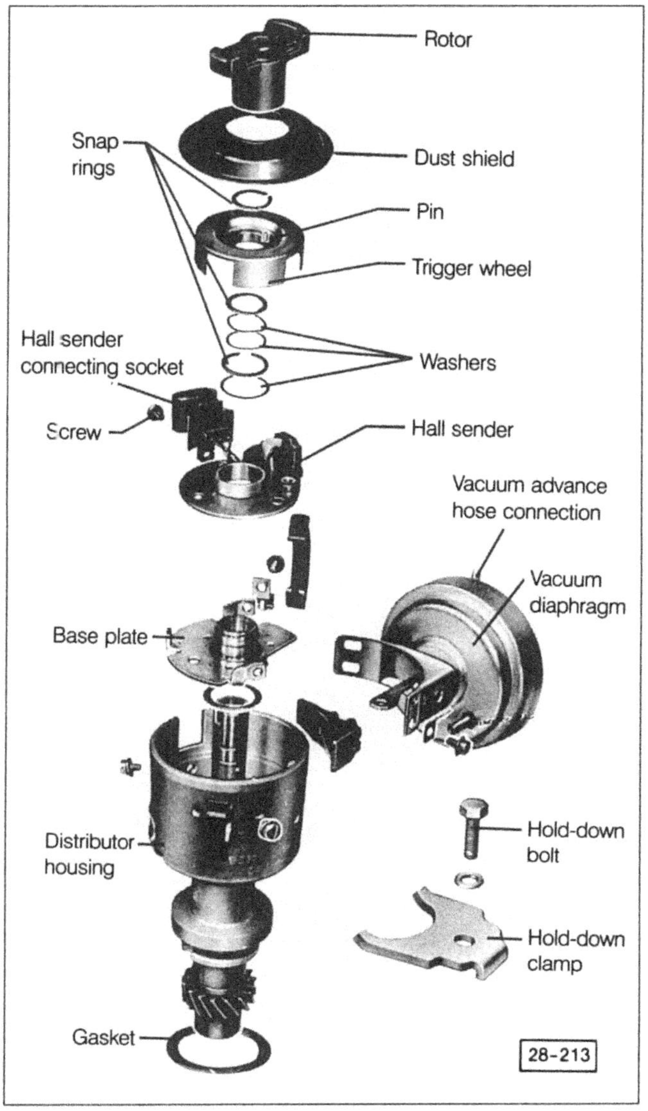

Fig. 6-1. Distributor assembly for basic Hall-sender ignition with centrifugal and vacuum advance (engine code JH, 1985-1989 except Scirocco 16V).

electronic control of ignition timing advance. The Hall sender is rigidly mounted and there are no mechanical advance mechanisms. The Scirocco 16V distributor, because of its cylinder head mounting, has a shorter housing with a different drive.

The Hall sender, the trigger wheel and the vacuum advance diaphragm are available as replacement parts, but other internal parts of the distributor are not. If the centrifugal advance mechanism or the distributor shaft bushings are worn, the complete distributor housing must be replaced as a unit.

6.1 Removing and Installing Distributor Cap, Rotor and Spark Plug Wires

The cap, rotor and wires all carry high voltage to the spark plugs, so proper engine performance depends on getting the best possible spark at the spark plug. These components should be replaced if they have worn or corroded contacts, or

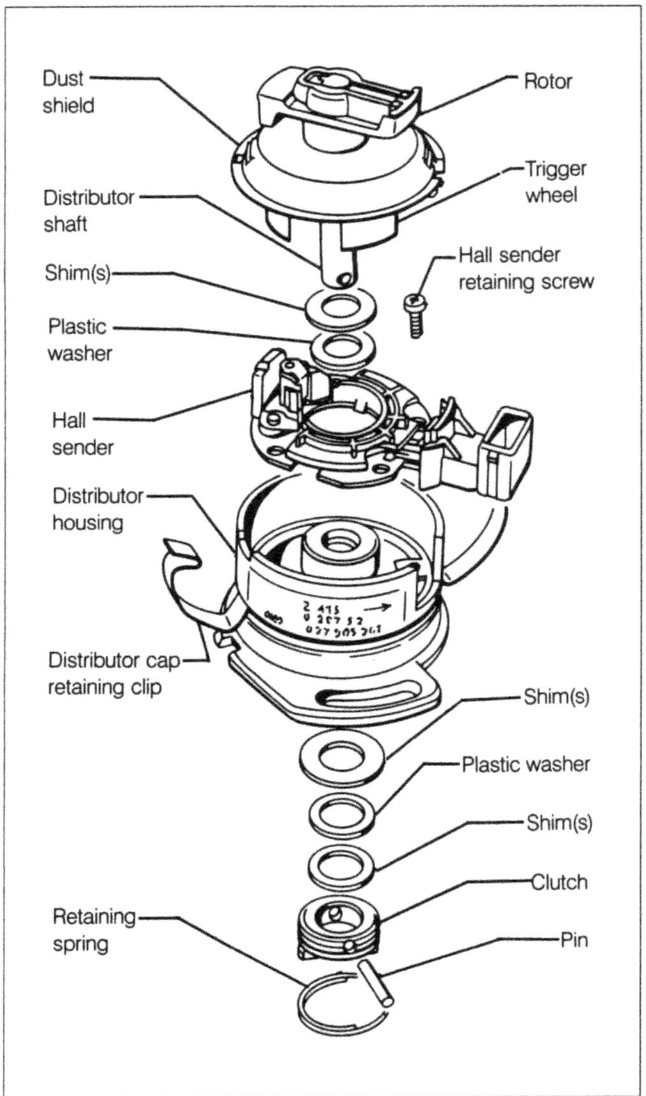

Fig. 6-2. Distributor assembly for Scirocco 16V (engine code PL).

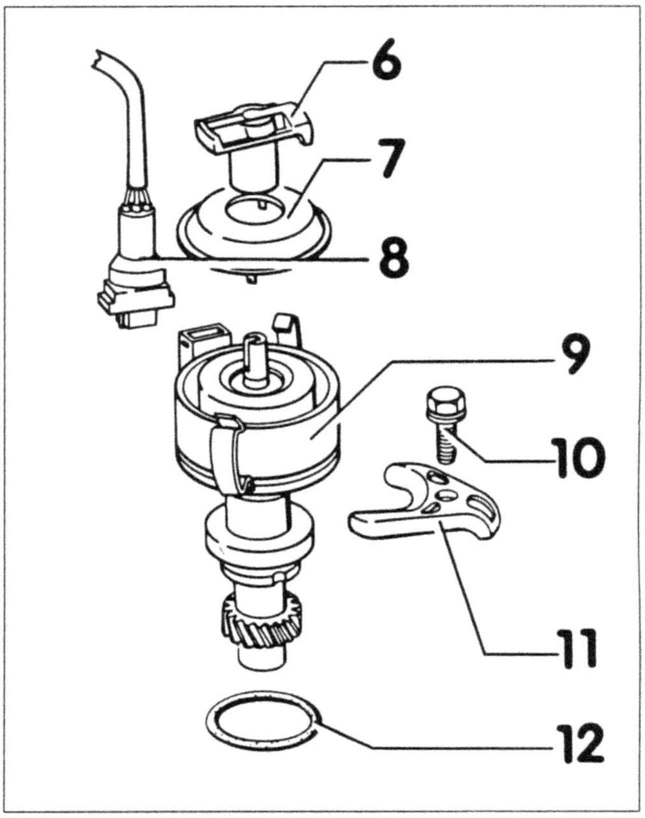

Fig. 6-3. Distributor assembly for Digifant I and Digifant II engine management systems (engine code 2H).
6 – Rotor
7 – Dust shield
8 – Hall-sender harness connector
9 – Distributor
10 – 25 Nm (18 ft. lb.)
11 – Base clamp
12 – O-ring

broken down insulation that can allow the high-voltage spark to short to ground.

Replacing Spark Plug Wires

Each spark plug wire leads from a specific terminal on the distributor cap to a specific spark plug. When removing the wires, label their positions so that they can be reinstalled in the proper places. Also see **Spark Plug Firing Order**.

The spark plug wires are removed from the distributor cap and spark plugs by pulling the boots straight up. Pull only on the connector, not on the wire itself. Twisting the connector first may make it easier to remove the wire without damage.

CAUTION —
Wiggling the connectors from side to side when removing the spark plug wires from the distributor cap may damage the cap.

Spark Plug Firing Order

Spark plug wires must be installed so that the spark plugs have the proper firing order. For all engines covered by this manual, the firing order is 1–3–4–2. Looking at the top of the distributor cap, the spark plug wires are connected in clockwise order for cylinders #1, #3, #4 and #2 on non-16-valve engines. On 16-valve engines, the wires connect to the distributor cap terminals according to the firing order (1–3–4–2), in counterclockwise direction.

NOTE —
- Cylinders are numbered 1 through 4. Cylinder #1 is the one on the right (passenger) side, closest to the camshaft drive belt.
- The distributor rotor points to the #1 terminal inside the cap when #1 cylinder is near or at top dead center (TDC). See Fig. 6-4.

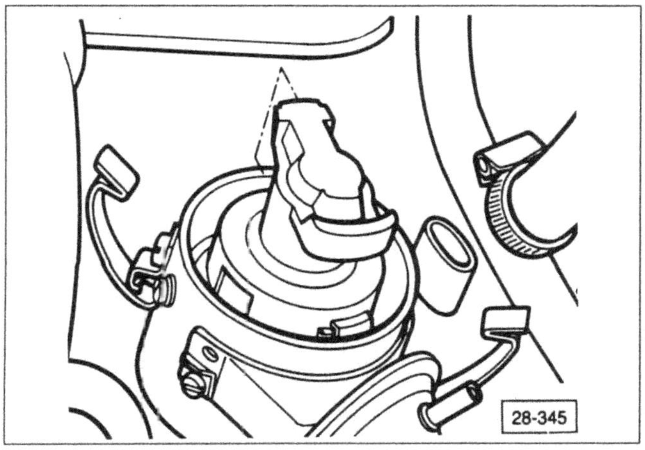

Fig. 6-4. Distributor rotor aligned with #1 cylinder mark on distributor housing.

Replacing Distributor Cap and Rotor

The distributor cap is held in place with two spring clips which can be pried loose using the blade of a screwdriver. Avoid bending the clips. To replace the cap remove the spark plug wires, then remove the suppression shield by releasing the spring clips and sliding it off the cap. Fig. 6-5 shows the distributor cap and suppression shield used on distributors for all engines covered by this manual.

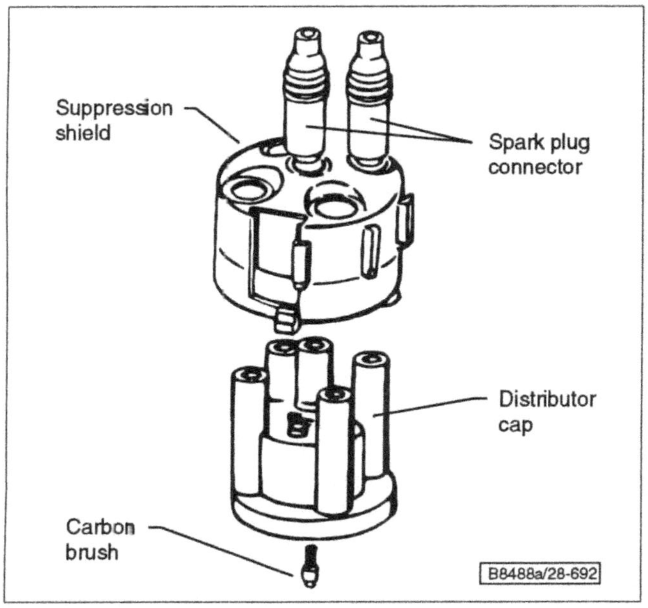

Fig. 6-5. Distributor cap with suppression shield. Shield is grounded by contact with metal distributor cap attachment clips.

On JH and 2H engines (all except Scirocco 16V), simply pull straight up on the rotor to remove it. When installing a rotor, make sure that the molded key inside fits into the notch on the distributor shaft.

On PL engines (Scirocco 16V), the rotor is attached to the distributor shaft with an adhesive. To replace the rotor, the part of the old rotor that fits over the distributor shaft must be carefully crushed using a pair of pliers or other suitable tool.

> ***CAUTION —***
> *Do not hit the rotor or the distributor shaft with any tool in an effort to remove the rotor.*

> **NOTE —**
> Volkswagen part numbers listed in this manual are for reference only. Check with an authorized Volkswagen dealer Parts department for the most accurate and up-to-date information.

Clean off all of the old adhesive. When installing a new rotor, attach it to the shaft using Volkswagen Part No. AMV 185 100 01 or an equivalent instant adhesive such as Loctite® 325 Speed Bonder. Make sure that the molded key inside the rotor fits into the notch on the distributor shaft.

6.2 Removing and Installing Distributor

Ignition timing must be checked and adjusted any time the distributor is loosened or removed. See **4.1 Ignition Timing**.

To remove:

1. Rotate the engine crankshaft by hand until the tip of the distributor rotor is aligned with the #1 cylinder top dead center (TDC) mark on the distributor housing, as shown earlier in Fig. 6-4.
2. Disconnect the Hall sender harness connector from the distributor. On JH engines, also disconnect the vacuum line from the vacuum diaphragm.
3. Remove the distributor hold-down, and remove the distributor by pulling it straight out. Retrieve the sealing washer or O-ring from the bottom of the distributor. Cover the opening.

> **NOTE —**
> On JH and 2H engines (all except Scirocco 16V), the distributor will want to rotate counter-clockwise as the beveled drive gears disengage. Hold the rotor and shaft steady and let the whole assembly turn. Keep track of how much it turns and start at that angle to reinstall it.

To install:

1. If the crankshaft has been turned since the distributor was removed, rotate the crankshaft so that the cylinder #1 is at TDC. Use the flywheel, front pulley and camshaft timing marks as described in **ENGINE**.
2. On JH and 2H engines (all except Scirocco 16V), align the lug on the oil pump drive shaft so that it is parallel to the crankshaft as shown in Fig. 6-6.

3. Using a new sealing washer or O-ring, install the distributor. Align the rotor with the distributor housing, as shown in Fig. 6-4 above. Be sure that the distributor housing is seated firmly and that the distributor shaft engages properly.

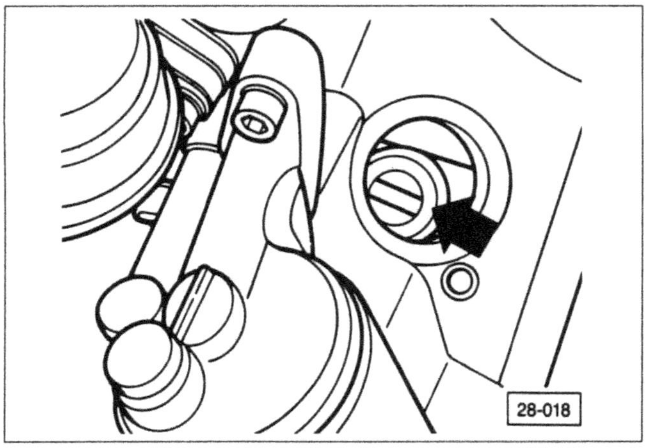

Fig. 6-6. Lug (arrow) on oil pump drive shaft should be aligned parallel to engine crankshaft.

NOTE —
- The rotor will rotate as the beveled drive gears engage, so check again for proper rotor alignment after installation.
- On PL (16-valve) engines, make sure that the distributor drive clutch is properly engaged with the notches in the camshaft.

4. Install the distributor hold-down and the distributor cap. Reconnect the Hall sender harness connector. On JH engines, reconnect the vacuum line.
5. Adjust the ignition timing as described earlier in this section under **4.1 Ignition Timing**.

6.3 Disassembling and Assembling Distributor

Some disassembly is required to replace the Hall sender. It will require a snap-ring pliers and, for PL (16-valve) engines, a thin drift or punch. The base plate, the distributor shaft and the centrifugal advance mechanism on JH engines can be removed for cleaning and inspection, but these are not available as individual replacement parts. Fig. 6-1, Fig. 6-2 and Fig. 6-3 earlier in this section are exploded views showing details of the distributor assemblies.

First remove the distributor cap and suppression shield, the rotor, and the dust shield as described in **6.1 Removing and Installing Distributor Cap, Rotor and Spark Plug Wires**.

JH, 2H Engines
(all except Scirocco 16V)

Use a snap-ring pliers to remove the snap-ring that secures the Hall sender trigger wheel. To remove the wheel, use two screwdrivers positioned at opposite sides of the trigger wheel to pry it up as shown in Fig. 6-7.

CAUTION —
Push the screwdrivers in as far as possible, and pry up only under the strongest, center portion of the trigger wheel. Take care not to lose the small pin that keys the trigger wheel to the shaft.

NOTE —
A bent trigger wheel must be replaced.

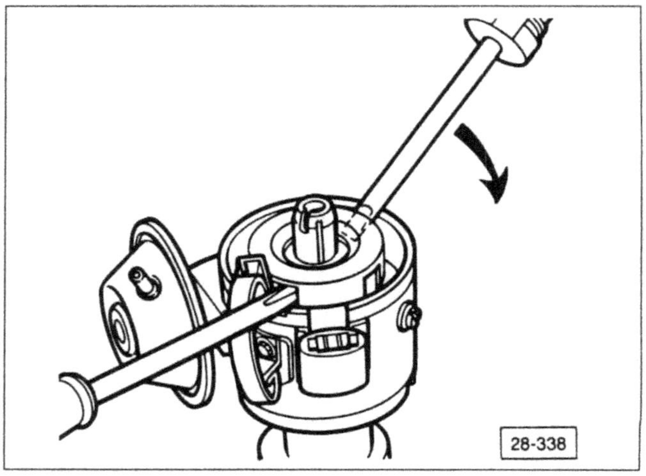

Fig. 6-7. Trigger wheel being pried off of distributor shaft. Pry evenly under center of trigger wheel, from both sides with two flat-bladed screwdrivers.

Remove the Hall sender by removing the screws holding it to the base plate.

PL (16-valve) Engines

Use a thin drift or punch to drive out the retaining pin in the distributor drive clutch, and pull the distributor shaft, the rotor, the dust shield, and the trigger wheel out from the top as a unit. The Hall sender is removed from the base of the distributor housing by removing the mounting screws.

7. TECHNICAL DATA

I. Electronic Ignition System Specifications— 1985 to 1989, CIS and CIS-E

Engine codes	JH, PL
Firing order	1–3–4–2
Ignition timing	
at idle (see **FUEL SYSTEM**)	
vacuum hoses	
to distributor (engine code JH)	connected
to knock sensor control unit (PL)	connected
checking	4 to 8° BTDC
adjusting	6 ± 1° BTDC
Centrifugal advance (engine code JH only)	
starts	1050 – 1300 rpm
degrees advance @ rpm	15 – 19° @ 2600 RPM
total centrifugal advance @ rpm	22 – 26° @ 4500 RPM
Vacuum advance (engine code JH only)	
start	4 – 8° BTDC (basic timing) @ 120 – 160 mbar (90 - 120 mm Hg)
total vacuum advance	10 – 14° BTDC @ 240 – 260 mbar (180 - 195 mm Hg)
Spark plug tightening torque	20 Nm (15 ft. lb.)

II. Electronic Ignition System Specifications— Digifant I and Digifant II from 1990

Engine code	2H
Firing order	1–3–4–2
Ignition timing	
coolant temperature sensor	disconnected
engine speed	2000 to 2500 rpm
checking	4 to 8° BTDC
adjusting	6 ± 1° BTDC
Timing advance (Digifant II)	
coolant temperature sensor	connected
engine speed	2000 to 2500 rpm
timing	basic timing plus 30° ± 3°
Timing advance (Digifant I)	
coolant temperature sensor	connected
engine speed	4500 rpm
timing	basic timing plus 30° (minimum)
Spark plug tightening torque	20 Nm (15 ft. lb.)

MANUAL TRANSMISSION AND CLUTCH

Contents

TABLES

Manual Transmission and Clutch

Introduction

The five-speed manual transmission, standard on the vehicles covered by this manual, is fully synchronized in all forward gears. A two-piece lightweight alloy case houses both the gear train and the final drive. As a unit, the transmission, the final drive, and the case are called the transaxle. From the transaxle, power is transmitted to the drive wheels by the drive axles and constant velocity joints. The service and repair of drive axles and constant velocity joints is covered in **SUSPENSION AND STEERING**.

Because of the transverse engine mounting, the transmission of power from the transmission to the differential is by spur gears, rather than by bevel gears—the kind used in other Volkswagen transaxles. The transaxle is also unique in that the mainshaft is hollow, allowing the clutch pushrod to pass through its center.

The transaxle can be removed and installed without removing the engine, or the engine and transaxle can be removed as a unit, using the procedure given in **ENGINE**, and separated later.

Though internal repair of the transaxle requires specialized knowledge and equipment, significant expense may be saved by removing and installing the transaxle, using the procedures described in this section. Once removed and cleaned, the transaxle can be taken to an authorized Volkswagen dealer or other qualified shop for repairs. A partially disassembled transaxle in a box is a mechanic's nightmare, and partial disassembly is not a way to save money. Do not attempt to begin a job which cannot be properly finished.

If performing work on the transaxle, make sure that the correct tools are available before beginning the job—particularly for procedures given only with metric specifications. Volkswagen vehicles, and especially their drivetrains, are covered by an extensive new-car warranty. We urge you to consult an authorized Volkswagen dealer before beginning any repairs on a car which may still be covered by the new-car warranty.

1. GENERAL DESCRIPTION

Fig. 1-1 is cross-section of the transaxle with five-speed manual transmission. The gears on the mainshaft and the drive pinion shaft are called the transmission gear train. The drive pinion, the ring gear, and the differential gears—all of which carry driving torque from the transmission gear train to the front wheel driveshafts—are called the final drive.

1.1 Clutch

The clutch is cable-operated. Two types of clutch-actuating cable are used on cars covered by this manual. On some later models, the conventional clutch cable has been upgraded with a self-adjusting mechanism. The clutch release mechanism, on the driver's side of the transaxle, activates a push rod through a hollow mainshaft. This pushrod moves the release plate of the pressure plate assembly to engage or disengage the clutch.

The pressure plate assembly consists of a diaphragm-spring type pressure plate, a release plate, and a retaining ring to hold them together. The friction clutch disc is splined to accept the mainshaft of the transmission and has integral cushion springs and dampening springs. The cushion springs—which are between the friction surfaces of the disc—help to reduce the shock when the clutch is engaged. The dampening springs—which are visible in the center hub—help to absorb the rotating power pulses of the engine. Unlike most other cars, the clutch's pressure plate assembly is bolted to the crankshaft with the flywheel bolted to the periphery of the pressure plate assembly. The ring gear, used to engage the starter, is integral with the flywheel and cannot be separated.

1.2 Transaxle

The transaxle case, which contains the transmission gear train and the final drive gears, is a lightweight alloy die casting with an integral flywheel bellhousing. There is no partition between the transmission gear train and the final drive, as both are lubricated by the common hypoid gear oil supply in the transaxle case.

Five-Speed Manual Transmission

The transmission gears are of the constant-mesh type with balk ring synchronizers. Each gear (1st through 5th) is actually a mating pair of gears, and constant-mesh simply means the mating gears are always meshing. When shifting gears, synchronizers cause the moving parts to rotate at the same speed before engaging. This prevents damage to the gears, minimizes wear, and eases shifting. The 3rd, 4th, and 5th gear synchronizers are on the mainshaft; the 1st and 2nd gear synchronizers are on the pinion shaft. Gear selection is made through the gearshift lever in the passenger compartment, which is linked to the transmission by the gearshift linkage.

Final Drive

The ring and pinion gearset consists of two helical-cut gears (curved teeth), designed for quiet running. The driven gear, or ring gear, is mounted on the differential housing. The driving gear, or pinion, is an integral part of the pinion shaft. The differential gearset, which consists of two differential sidegears

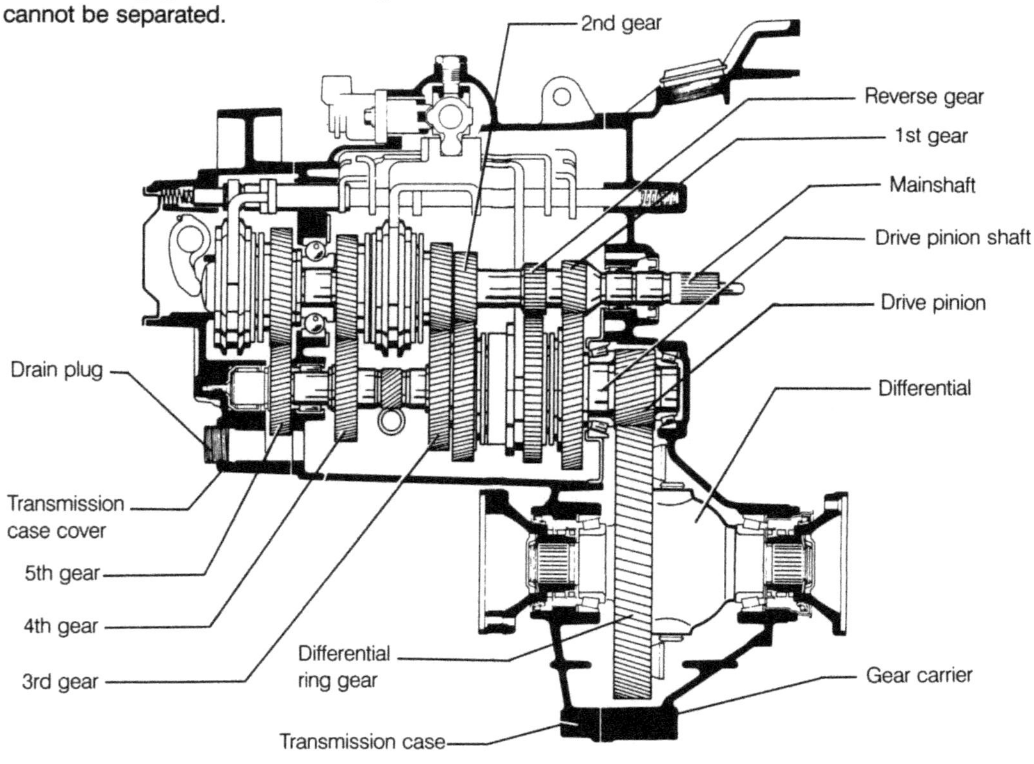

Fig. 1-1. Cross-section of transaxle with manual transmission.

and two differential pinions, allows the front wheels to turn at different speeds, as is necessary when making turns (the outside wheel must travel farther than the inside wheel in the same amount of time).

1.3 Identification Codes and Specifications

Transaxles are identified by a series of characters stamped on the bottom of the bellhousing, as shown in Fig. 1-2. A two or three character code identifies the transaxle type, followed by series of numbers identifying the date of manufacture. A manual transmission type number is located on the driver's side of the transmission case near the drive flange as shown in Fig. 1-3. All of the manual transmissions used in cars covered by this manual are type 020. **Table a** lists manual transmission specifications.

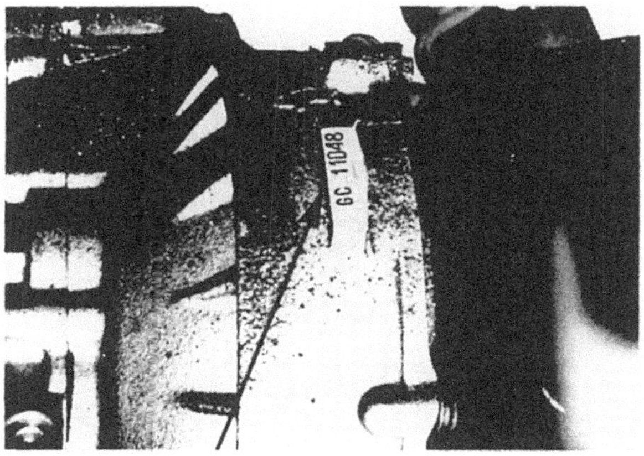

Fig. 1-2. Transaxle code letters and date of manufacture on five speed manual transmissions. Example shows code letters GC and date 11 (day), 04 (month, April), 8 (year 1978).

Fig. 1-3. Transaxle type number (020) on five-speed manual transmission.

Table a. Manual Transaxle Specifications

Transmission Type no.	020	020	020
Transmission Codes	9A, ACD	AUG	AGB, 2Y
Gear Ratios			
1st Gear	3.45	3.45	3.45
2nd Gear	2.12	1.94	2.12
3rd Gear	1.44	1.44	1.44
4th Gear	1.13	1.13	1.13
5th Gear	0.89	0.89	0.91
Final Drive Ratio	3.67	3.67	3.67
Engine code	JH (to 9/89)	JH, 2H (from 9/89)	PL (16-valve)

2. MAINTENANCE

Checking clutch pedal freeplay at a prescribed mileage, except on models with self-adjusting clutch cable, is the only required maintenance procedure. The transaxle is designed to be lubricated for its operational life, so checking and correcting the transaxle oil level is not necessary unless the transaxle is leaking or has been repaired. Checking pedal freeplay and checking transaxle oil level are described in **LUBRICATION AND MAINTENANCE**.

3. TROUBLESHOOTING

This troubleshooting section applies to problems affecting the correct functioning of the transaxle—including the gearshift mechanism, the gear train, and the final drive—as well as the clutch.

The source of most transaxle problems is apparent from the symptoms. For example, difficulty in engaging a gear, shuddering when releasing the clutch, or a vague shifting mechanism are all transaxle problems. Other symptoms, such as power loss or noises from the front end, are less specific. Power loss may be due to an engine problem, or to a slipping clutch. Noises may be the result of a failure in the gear train, worn constant velocity joints, or a bad wheel bearing. What appears to be a transmission oil leak may be engine oil leaking from a faulty rear crankshaft oil seal, especially if the leak is near the bottom of the transaxle bellhousing. See **ENGINE** for information on crankshaft oil seals. See **SUSPENSION AND STEERING** for information on constant velocity joints and wheel bearings.

3.1 Basic Troubleshooting Principles

Transaxle problems fall into two categories: those that can be fixed by external adjustments, and those which require removal and disassembly of the transaxle. Problems which at first appear to be caused by internal faults, such as gear shifting difficulty or noisy operation, may be correctable with external adjustments.

Begin any transaxle troubleshooting with a thorough visual inspection, both in the engine compartment and from beneath. Check all parts of the gearshift mechanism for wear which might cause misalignment and shifting difficulty. Look for wet spots which may indicate oil leaks. Low lubricant level may be the cause of hard shifting or noise. Accurate pinpointing of leaks may require that the suspected area be cleaned and reinspected.

To avoid removing the transaxle unnecessarily, check and adjust the gearshift mechanism and correct leaks and oil level before acting on suspected internal problems. The gearshift mechanism is covered in **4. Shift Mechanism**. Replacement of oil seals is covered in **6.2 Transaxle Oil Seals**. Checking and correcting oil level is covered in **LUBRICATION AND MAINTENANCE**.

NOTE —

It is not necessary to routinely check the transaxle oil level unless a leak is suspected.

As with most other troubleshooting, logical analysis of the observed symptoms is the key to isolating and identifying transaxle problems. A road test is an important step when troubleshooting the transaxle. Determining whether the problem is present in all gears, only during acceleration, when the clutch is engaged, or in some other special conditions may help isolate the source of the problem.

Table b lists manual transmission, final drive, and clutch symptoms, their probable causes, and recommended corrective actions. The numbers in bold type in the corrective action column refer to numbered headings in this section where the suggested repairs are described.

Table b. Manual Transaxle Troubleshooting

Symptom	Probable cause	Corrective action
1. Difficult or noisy shifting	**a.** Clutch not fully releasing **b.** Clutch disc binding on transmission mainshaft **c.** Worn or misadjusted gearshift linkage **d.** Insufficient gear oil **e.** Worn or damaged internal transmission components	**a.** Adjust clutch pedal freeplay. **5.1** **b.** Remove transaxle and inspect splines of clutch disc and transmission mainshaft. Lubricate splines. If necessary, replace clutch disc. **5.** **c.** Adjust shift mechanism and, if necessary, replace worn parts. **4.1** and **6.1** **d.** Check for transmission oil leaks. Check and correct gear oil level if needed. See **LUBRICATION AND MAINTENANCE** **e.** Remove and repair transmission. **6.3**
2. Transaxle noisy	**a.** Insufficient gear oil **b.** Worn or damaged internal transaxle components	**a.** Check for transmission oil leaks. Check and correct gear oil level if needed. See **LUBRICATION AND MAINTENANCE** **b.** Remove and repair transaxle. Especially checking the differential gears. **6.3**
3. Grinding noise when shifting (1st or reverse gears)	**a.** Excessive idle speed **b.** Worn or damaged internal transmission components	**a.** Adjust idle speed. See **FUEL SYSTEM** **b.** Remove and repair transmission. Especially checking the synchronizers of the noisy gear. **6.3**
4. Transmission fails to engage a gear or jumps out of gear	**a.** Worn or misadjusted gearshift linkage **b.** Worn or damaged internal transmission components	**a.** Adjust shift mechanism and, if necessary, replace worn parts. **4.1** and **6.1** **b.** Remove and repair transmission. **6.3**
5. Poor acceleration, clutch slipping on hills or when accelerating	**a.** Clutch friction surfaces worn or burnt **b.** Clutch not fully engaging **c.** Clutch disc, pressure plate or flywheel are oil soaked	**a.** Inspect clutch components and replace faulty parts. **5.2** **b.** Adjust clutch pedal freeplay. **5.1** **c.** Inspect clutch components. If necessary, clean pressure plate and flywheel; replace clutch disc. **5.2**
6. Clutch grabs or chatters when the pedal is released	**a.** Binding clutch cable **b.** Clutch disc binding on transmission mainshaft **c.** Contaminated or glazed (overheated) clutch lining **d.** Faulty engine mount	**a.** Inspect the clutch cable. Replace the cable if it is severely bent or otherwise faulty. **5.1** **b.** Remove transaxle and inspect splines of clutch disc and transmission mainshaft. Lubricate splines. If necessary, replace clutch disc. **5.** **c.** Inspect clutch components and replace faulty parts. **5.2** **d.** Inspect and, if necessary, replace broken engine mounts. See **ENGINE**

3.2 Diagnostic Tests

Most internal transaxle problems cannot be accurately diagnosed unless the unit is removed and disassembled; however, a quick test can be used to determine whether the clutch is performing satisfactorily or slipping.

To quick-check clutch performance, start the engine and set the parking brake. Depress the clutch pedal and place the gearshift lever in 3rd or 4th gear, then slightly accelerate the engine and slowly release the clutch pedal. The engine should immediately stall, indicating that the clutch is engaging properly and will not slip enough to allow the engine to continue to run.

If the engine stalls slowly, or does not stall at all, the clutch is probably faulty. Adjust the clutch pedal freeplay, as described in **5.1 Clutch Cable**, and repeat the test. If the clutch is still slipping, the transaxle will have to be removed to make further repairs to the clutch.

4. SHIFT MECHANISM

The transverse mounting of the transaxle in vehicles covered by this manual requires a relatively complex mechanism of shafts and levers to transmit the movements of the gearshift lever to the transmission for gear changes. For any shifting problems, the gearshift mechanism should be inspected for wear and misalignment which might affect smooth or accurate shifting. Fig. 4-1 shows the arrangement of the shift mechanism and its directions of movement.

4.1 Disassembling, Assembling, and Adjusting Shift Mechanism

Fig. 4-2 is an exploded view of the shift mechanism used with five-speed manual transmissions, which can be used as a guide for disassembling and assembling the linkage, to replace any worn bushings, or when removing the transaxle. When removing the selector rods, use a screwdriver to press back the clips on the plastic rod ends before pulling them off of the ball pins on the levers. When assembling, remove any white grease from the linkage and lubricate all joints and pivot points with molybdenum disulfide grease (Volkswagen part no. G 000 602). Also, make sure that the connecting link is installed to the selector shaft lever and relay lever as shown in Fig. 4-3.

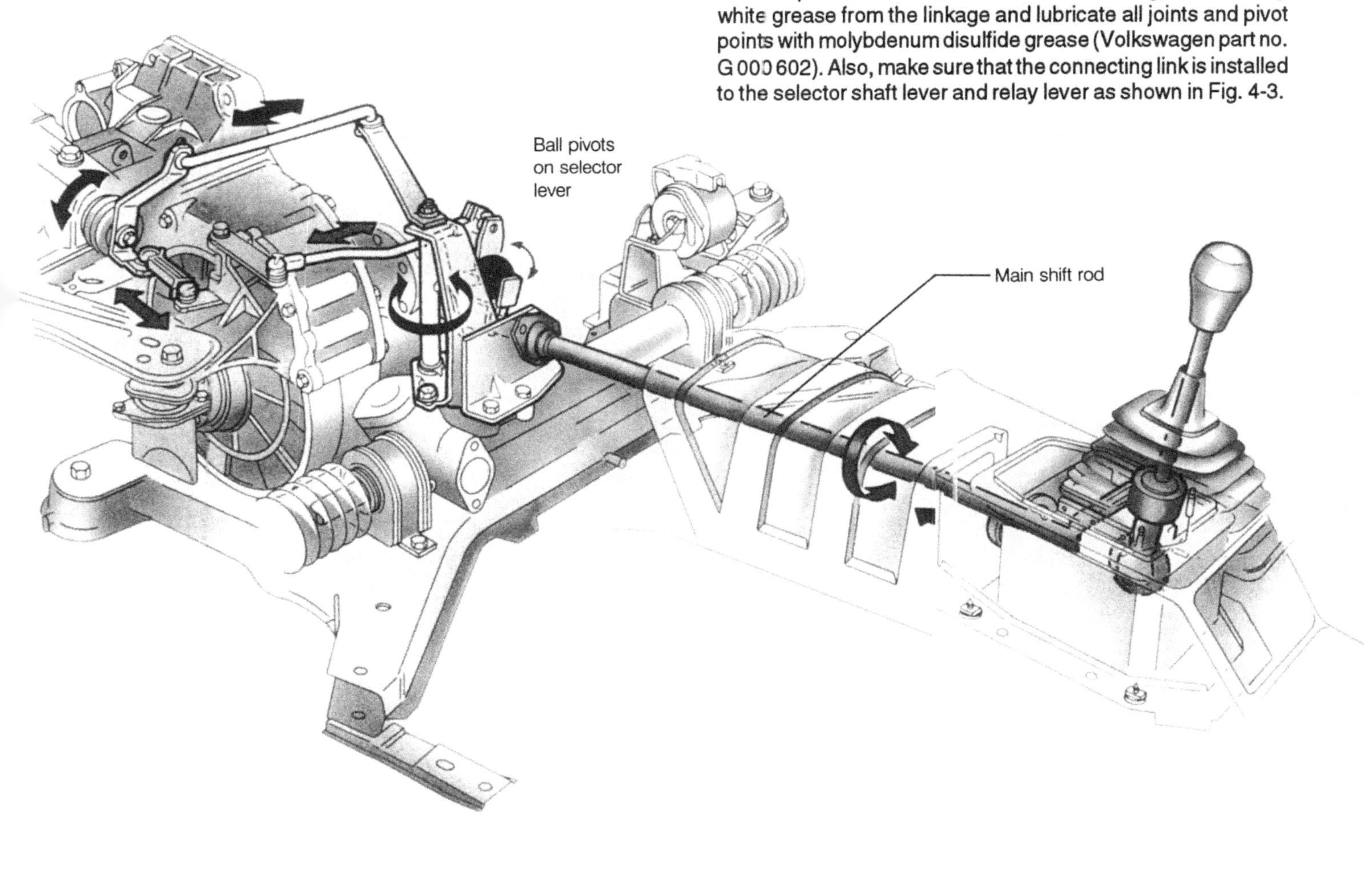

Fig. 4-1. Typical Volkswagen shift mechanism with directions of movement (arrows) shown. Details of Cabriolet and Scirocco models may differ slightly from the assembly shown.

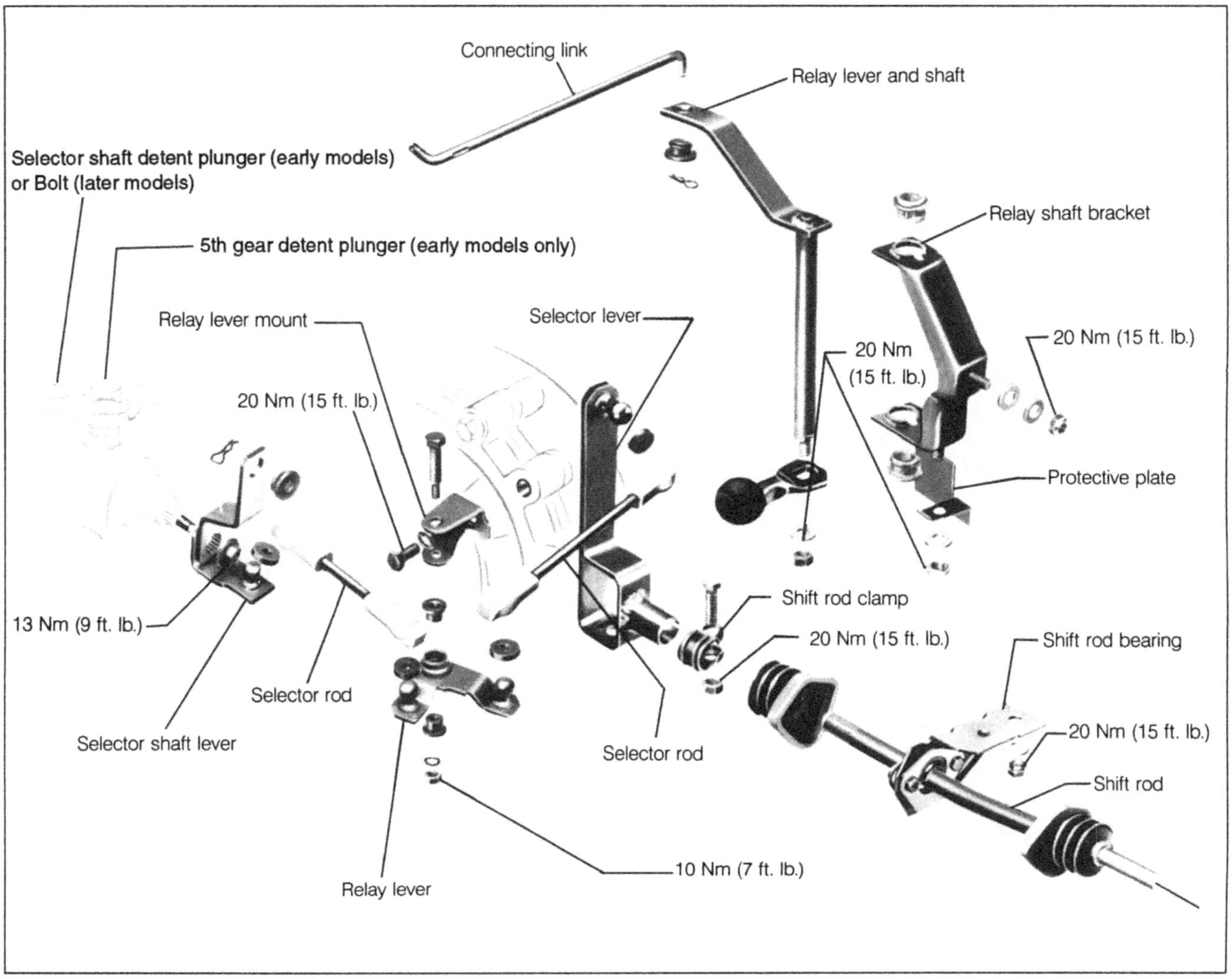

Fig. 4-2. Exploded view of shift mechanism.

Fig. 4-3. Connecting link showing orientation of ends. End with 95° bend and notch (arrow) is connected to selector shaft lever.

Shift Mechanism Adjustment

The position of the selector lever on the shift rod is the only adjustment that can be made to the shift mechanism. Before adjusting the shift rod to correct shifting difficulties, first check for any worn or broken parts of the shift mechanism.

To adjust:

1. Park the car on a level surface. Set the parking brake and place the transmission in neutral.

2. Loosen the lock nut on the shift rod clamp bolt, shown in Fig. 4-4, so that the shift rod moves freely in the selector lever.

3. Working under the car, remove the cover from the weatherseal boot at the base of the gearshift lever, or, if there is no cover, remove the boot and slide it forward on the shift rod.

4. Move the gearshift lever so that it is positioned in relation to the lockout plate as shown in Fig. 4-5. Then tighten the clamp locknut to 20 Nm (15 ft. lb.).

5. Check that all the gears engage smoothly and easily. Reinstall the gearshift weatherseal boot and cover.

Fig. 4-4. Shift rod clamp bolt (arrow).

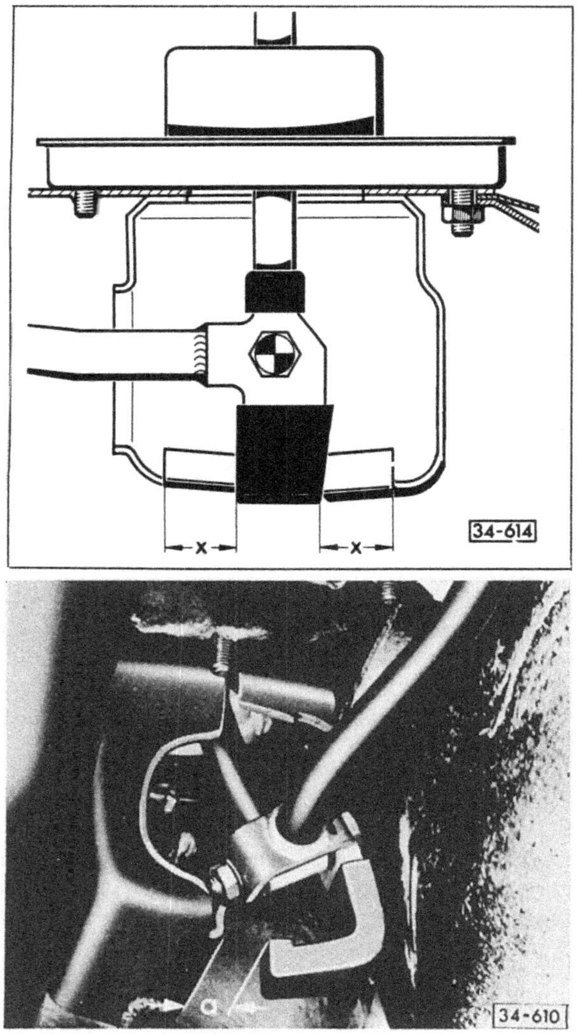

Fig. 4-5. Gearshift lever being adjusted. Shift finger must be positioned so that dimensions **x** (top) are equal, and so that distance **a** (bottom) between shift finger and lockout plate is 15 mm (9/16 in.).

NOTE —

If the shifting action is still spongy or if it jams after adjustment, reduce the distance between the shift finger and the lockout plate to 13 mm (1/2 in.). If the adjustment cannot be maintained, a new bolt (part no. N 010 247 B) and locknut (part no. N 023 003 3) should be installed and the nut torqued to 13-17 Nm (9-12 ft. lb.).

NOTE —

Part numbers are subject to change. Always rely on an authorized Volkswagen dealer parts department for the latest information.

Removing and Installing Shift Lever and Shift Rod

Fig. 4-6 is an exploded view of the shift lever and related parts. Removal is necessary to replace the bearing assembly or the ball if either is determined to be cracked or worn. After installing the shift lever, the mechanism should be adjusted as described in **Shift Mechanism Adjustment**.

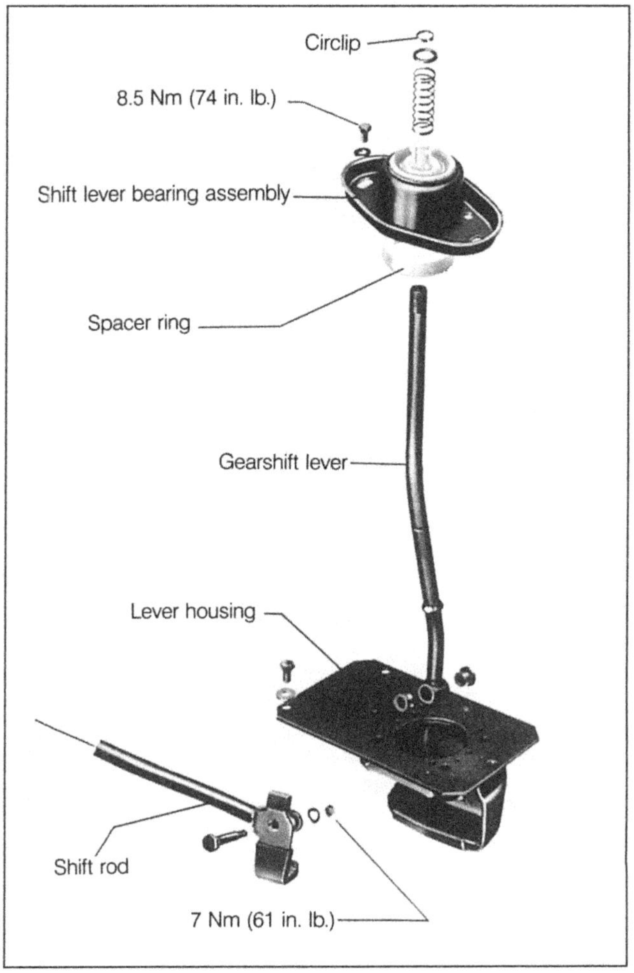

Fig. 4-6. Exploded view of gearshift lever and related parts.

To remove and install:

1. Unscrew the shift knob. Remove the boot by detaching it at the bottom and pulling it off over the gearshift lever.
2. Working beneath the car, remove the cover from the weatherseal boot at the bottom of the shift lever, or, if there is no cover, detach the boot and slide it forward on the shift rod.
3. Remove the bolt that holds the gearshift lever to the shift rod.
4. Working in the passenger compartment, remove the two mounting bolts for the shift lever bearing assembly and withdraw the shift lever and bearing.
5. Installation is the reverse of removal. The bearing is only available as a complete replacement part, but it can be disassembled as shown in Fig. 4-7 for cleaning and greasing. Remove any old grease from the joints and pivot points, and lubricate them with molybdenum disulfide grease (Volkswagen part no. G 000 602). Install the bearing assembly as shown in Fig. 4-8, and torque the bolts to 8.5 Nm (74 in. lb.). Torque the bolt that joints the shift rod and shift lever to 7 Nm (61 in. lb.).

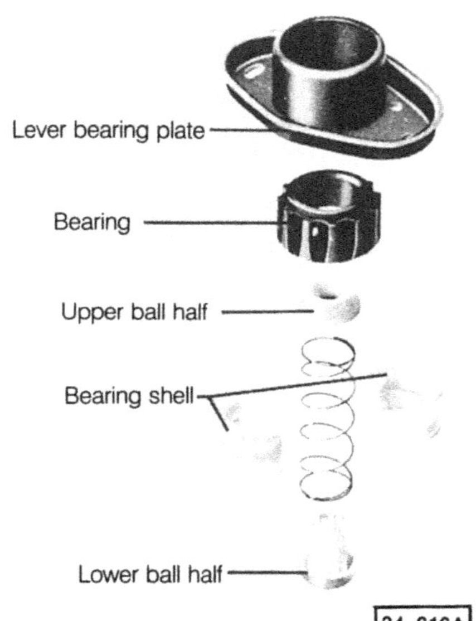

Fig. 4-7. Exploded view of shift lever bearing assembly. When assembling bearing, first place bearing shells into bearing, then install remaining parts.

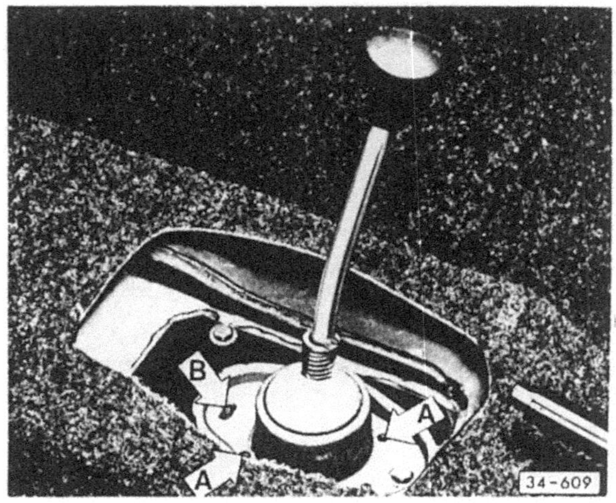

Fig. 4-8. Gearshift lever being installed. Round holes (**A**) must align with holes in lever plate. Center bolts (**B**) in elongated holes. If bolts cannot be centered, turn bearing 180° and install bolts.

5. CLUTCH

Except for replacement or adjustment of the clutch cable, any repairs to the clutch require that the transaxle be removed from the car. The flywheel and the clutch assembly remain bolted to the crankshaft when the transaxle is separated from the engine. The clutch release bearing and its related parts are located at the opposite end of the transaxle, and operate the clutch by means of a pushrod running through the center of the mainshaft. Fig. 5-1 is an exploded view of the clutch components.

5.1 Clutch Cable

Two types of clutch cables are used on the vehicles covered by this manual. The more conventional, manually-adjusted cable needs periodic adjustment to compensate for the wear of the clutch friction material. For mileage intervals, see **LUBRICATION AND MAINTENANCE**.

Some models are equipped with a self-adjusting clutch cable. It has a built-in mechanism which automatically compensates for wear, adjusting every time the clutch pedal is actuated. The self-adjusting cable can be identified by the rubber boot atop the clutch cable sleeve on the transaxle end of the cable, as shown in Fig. 5-2. The two cables are not interchangeable.

Removal and installation of the clutch cable can be accomplished without any special tools. Replacement self-adjusting cables come with a holding strap that relieves the spring tension and allows easy installation. If the cable does not have the strap attached, a helper may be needed during installation.

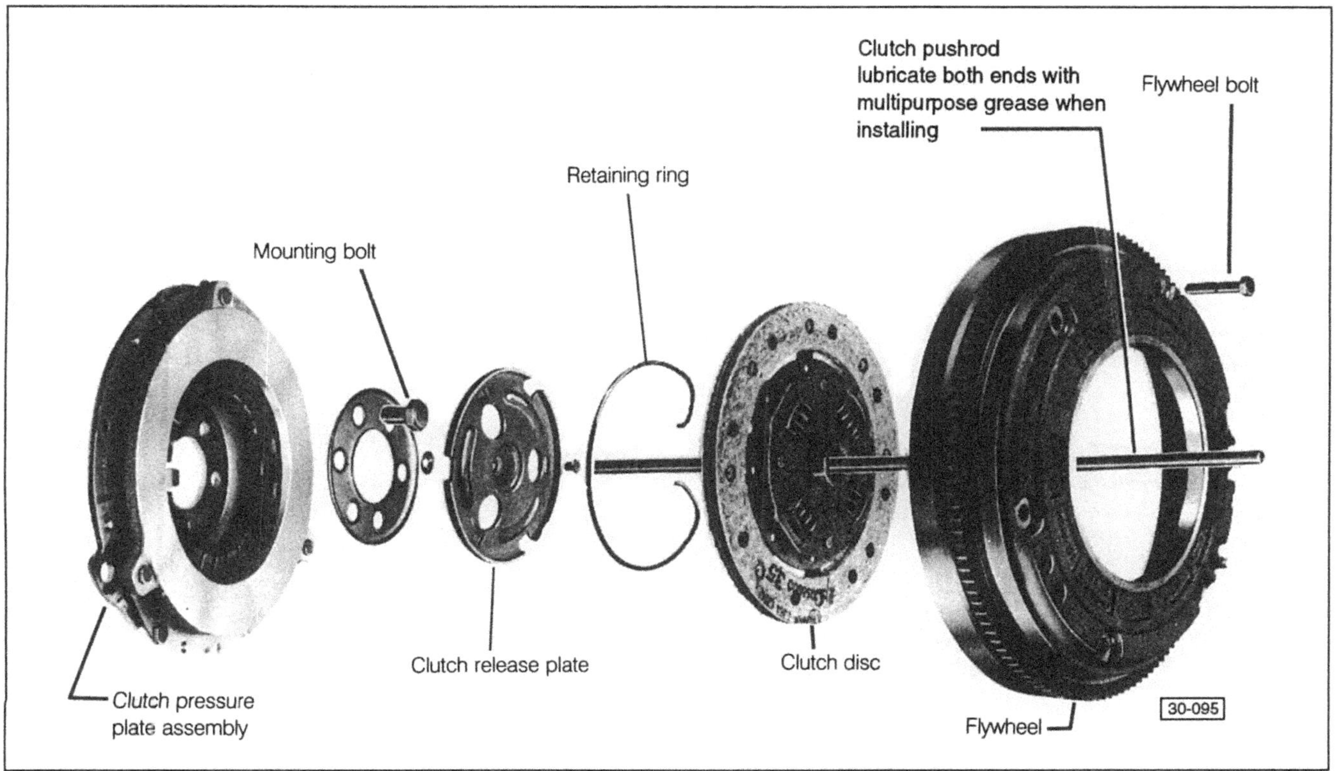

Fig. 5-1. Exploded view of flywheel and clutch.

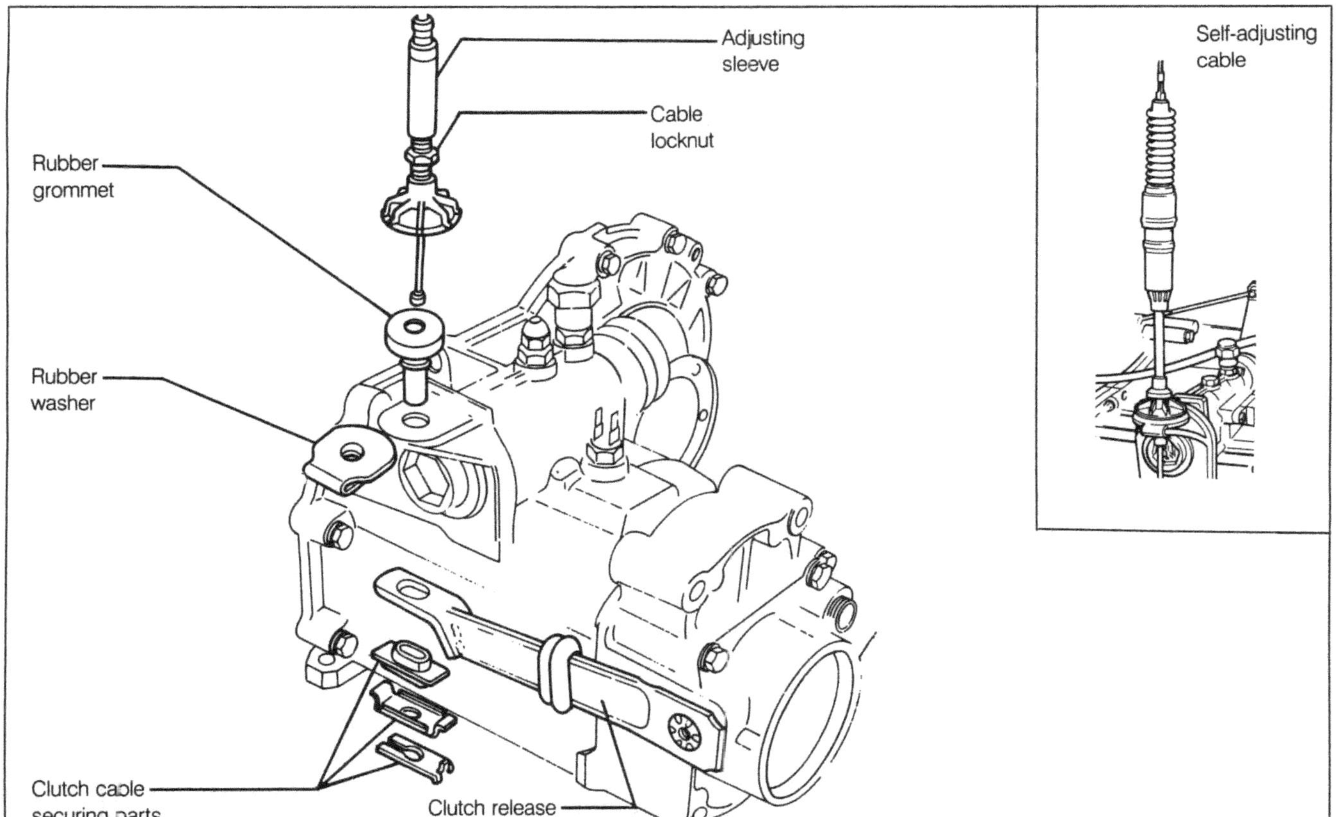

Fig. 5-2 Exploded view of clutch cable and related parts. Inset shows self-adjusting cable.

To replace clutch cable:

1. On manually-adjusted cables, loosen the adjusting sleeve locknut and turn the sleeve clockwise until the cable's tension is relieved. Remove the securing parts from the cable end beneath the clutch release lever and pull the cable end from the lever.

2. On self-adjusting clutch cables, depress the clutch pedal several times. Compress the spring (under the rubber boot) and have a helper remove the securing parts from beneath the clutch release lever. Pull the cable end from the lever.

CAUTION

Do not open the automatic adjustment mechanism. If the adjustment mechanism is opened, the cable assembly will need to be replaced.

3. Unhook the cable end from the top of the clutch pedal and pull the cable through the firewall from the engine compartment side.

4. Install the clutch cable. Lubricate both ends with multi-purpose grease, then insert the clutch cable end through the firewall grommet and hook the cable onto the clutch pedal.

5. Position the rubber washer and the rubber grommet as shown in Fig. 5-3.

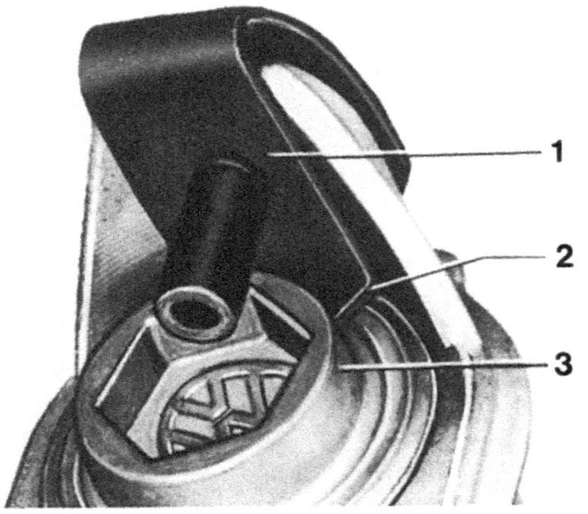

Fig. 5-3. Rubber washer and rubber grommet correctly installed. Rubber washer **1** is fitted with lip **2** parallel to selector shaft cover **3** to protect vent.

CAUTION

On 1985 and 1986 models it is especially important that the rubber parts are installed correctly. If not, the transaxle vent which is protected by the washer may become clogged, and cause oil leakage past the seals. For more information see **6.2 Transaxle Oil Seals**.

6. If reinstalling a used self-adjusting cable, compress the spring (beneath the rubber boot).

7. Insert the cable end through the rubber grommet and the release lever. Attach the securing parts to the end of the cable.

8. On self-adjusting cables, adjust clutch pedal freeplay by removing the spring retaining strap or relieving the spring tension, and then operating the clutch pedal several times.

To adjust (except self-adjusting clutch cable):

1. Operate the clutch pedal several times. Loosen the cable locknut.

2. Raise the clutch release lever by hand until resistance is felt.

3. Adjust freeplay by turning the adjusting disc until the clearance between the adjusting disc and the transaxle housing is 12.0±0.3 mm (.472±.012 in.). See Fig. 5-4

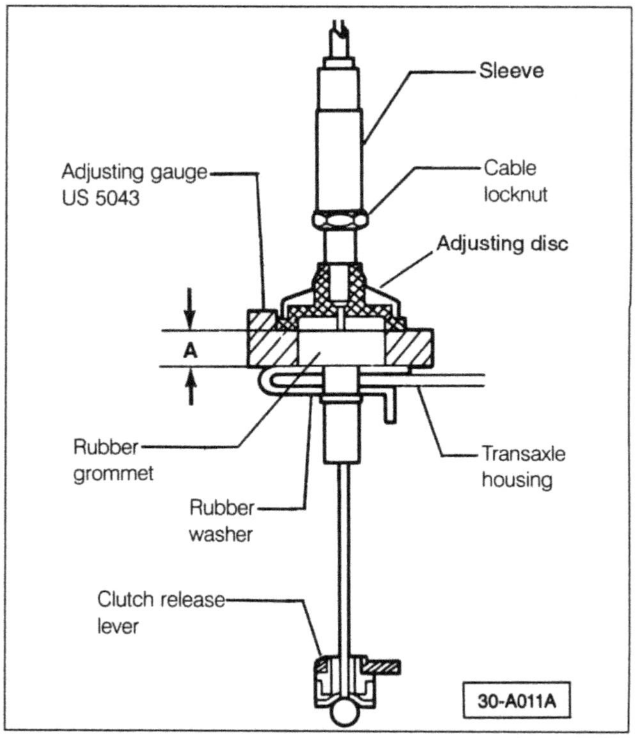

Fig. 5-4. Adjusting clutch freeplay. Dimension **A** should be 12 mm ($\frac{15}{32}$ in.) while raising clutch release lever by hand. Adjusting gauge shown is Volkswagen special tool no. US 5043 (order no. TU5 043 000 30 ZEL).

4. Tighten the cable locknut. Freeplay at the clutch release lever should now be 6.0±0.3 mm (.236±.012 in.). Torque the locknut to 4 to 6 Nm (36 to 48 in. lb.).

5.2 Removing and Installing Clutch

The transaxle must be removed from the engine to gain access to the clutch. See **6.3 Removing and Installing Transaxle**. It is normally recommended that the pressure plate be replaced when a new clutch disc is installed. If the clutch pressure plate is removed from the crankshaft, new mounting bolts should be used to reinstall it.

The flywheel should be carefully inspected when replacing the clutch. Replacing the clutch disc without replacing other worn components may accelerate clutch disc wear.

An inexpensive alignment tool is used to install and center the clutch disc. Use of this tool or its equivalent will greatly aid the installation of the transaxle. For JH engines, use Volkswagen special tool no. VW 547 (order no. TV0 547 000 30 ZEL). For PL (16-valve) engines, use Volkswagen special tool no. 3178 (order no. T03 178 000 30 ZEL).

Some lubrication is recommended when reassembling the clutch. To avoid accelerated failure of the new parts, use only the Volkswagen recommended lubricants. If signs of oil are present at the bottom of the transaxle, carefully inspect the rear crankshaft oil seal while the clutch is removed. A faulty oil seal should be replaced.

To replace clutch:

1. Remove the transaxle from the engine as described in **6.3 Removing and Installing Transaxle**.

2. Remove the flywheel, clutch disc, and release plate. A holding fixture, Volkswagen special tool number VW 558 (order no. TV0 558 000 13 ZEL), is shown in Fig. 5-5. Loosen the flywheel mounting bolts evenly, one-quarter turn at a time, until the clutch pressure is relieved. Remove the flywheel and the clutch disc.

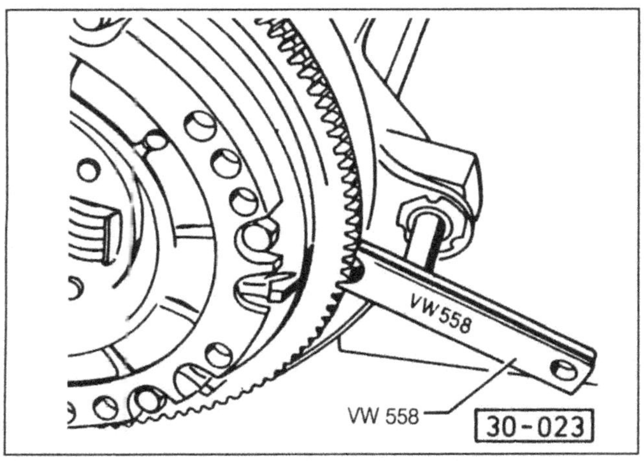

Fig. 5-5. Holding fixture being used while loosening flywheel bolts.

3. Use a screwdriver to pry out the retaining ring for the clutch release plate.

WARNING —

The retaining ring is a spring which will be released with some force. To avoid injury, hold on to it firmly while prying it loose.

4. Remove the six center mounting bolts and remove the clutch pressure plate from the crankshaft. Use of the holding fixture is illustrated in Fig. 5-6.

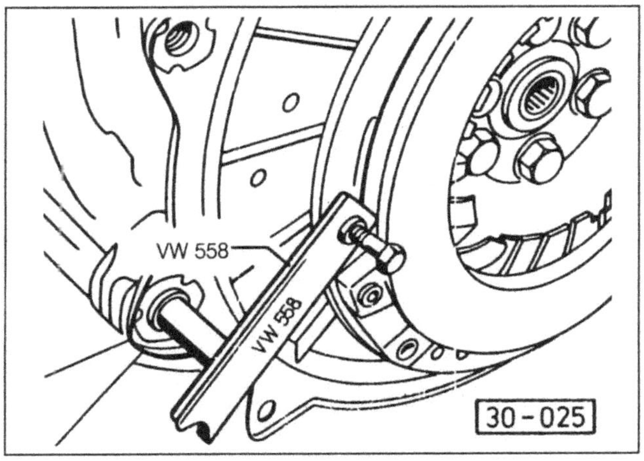

Fig. 5-6. Holding fixture used to remove clutch pressure plate assembly.

To inspect and install:

1. Inspect the clutch pressure plate. Check the release levers for bending or misalignment. Using a straightedge as shown in Fig. 5-7, check that the mating surface does not taper inward more than 0.3 mm (.012 in.). Check to see that no rivets are loose. Carefully inspect the friction surface for cracks, scoring, discoloration due to heat or oil contamination, or other damage. A clutch pressure plate showing any of these signs of damage should be replaced.

Fig. 5-7. Pressure plate being checked for taper. If feeler gauge over 0.30 mm (.012 in.) can be inserted between pressure plate and straightedge on inside edge, pressure plate should be replaced.

2. Clean the flywheel friction surface and inspect it for wear, cracks, and grooves. Check for loose or worn dowel pins. Replace a flywheel with any of these faults.

3. Check the thickness, runout, and general condition of the clutch disc. See Fig. 5-8. If runout exceeds 0.40 mm (.016 in.), or if there is any evidence of contamination by oil, the disc should be replaced.

Fig. 5-8. Dial indicator being used to measure clutch disc runout while disc is rotated.

NOTE —
The clutch disc is a relatively low-cost part which commonly wears out and requires replacement. Many experienced mechanics routinely install a new clutch disc anytime the transaxle is removed.

4. Inspect the splines of the clutch disc and the transmission input shaft. Check that the clutch disc is free to slide on the shaft. If the clutch disc is in any way unserviceable, it should be replaced. Clean all traces of grease from the shaft and clutch splines.

5. Position the clutch pressure plate on the crankshaft. Using new bolts, position the washer and thread the bolts into place. The holding fixture shown earlier in Fig. 5-6 can be used to hold the pressure plate. Torque the bolts to 30 Nm (22 ft. lb.) plus 1/4 turn.

CAUTION —
Always use new pressure plate mounting bolts with locking compound. Once torqued, they are deformed and should not be reused.

6. Place the clutch release plate in position and install the retaining ring. Lightly lubricate the contact surface of the plate and the center seat for the clutch pushrod with lithium grease. Align the ends of the retaining ring as shown in Fig. 5-9.

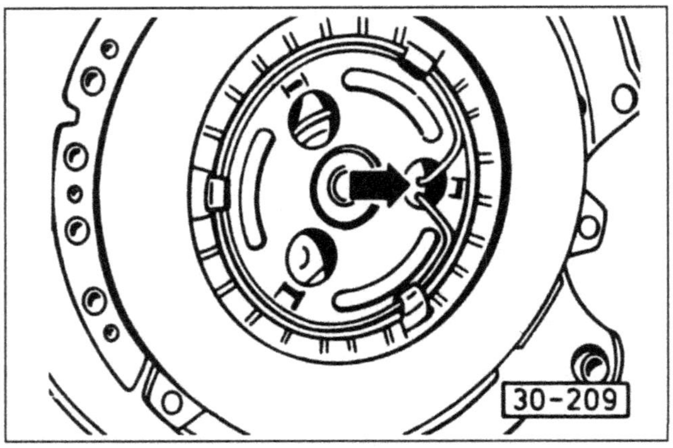

Fig. 5-9. Installation of clutch release plate retaining ring. Ends of ring rest in hole of release plate (arrow).

7. Lightly lubricate the transmission mainshaft splines and the clutch splines using molybdenum disulfide grease, Volkswagen part no. G 000 602, then place the clutch disc inside the flywheel.

CAUTION —
Use only the recommended molybdenum disulfide grease on the splines. Otherwise the clutch disc may bind on the input shaft and cause hard shifting. Other lubricants may contaminate the clutch disc.

NOTE —
The projecting part of the clutch hub faces the flywheel, away from the engine.

8. Install the flywheel. Position the flywheel on the dowel pins and start the mounting bolts in their holes. Leave the bolts loose enough so that the clutch disc can still move.

CAUTION —
The two guide pins in the flywheel must align with the corresponding holes in the clutch pressure plate, or the pressure plate will be damaged during installation. Do not mistake balancing holes for guide pin holes.

9. Center the clutch disc. The Volkswagen special tool used to accurately center the disc is shown in Fig. 5-10. Tighten the flywheel mounting bolts evenly until the clutch disc is firmly held in place.

10. Remove the centering tool, if used, and torque the bolts to 20 Nm (15 ft. lb.) in the sequence shown in Fig. 5-11.

NOTE —
New replacement flywheels for gasoline engines have no timing mark. They are marked only with a 0° TDC mark. Before installing, make a timing mark as shown in Fig. 5-12.

Fig. 5-10. Clutch disc being centered in flywheel with centering tool, Volkswagen special tool no. VW 547 (order no. TV0 547 000 30 ZEL). For PL (16-valve) engines, use Volkswagen special tool no. 3178 (order no. T03 178 000 30 ZEL).

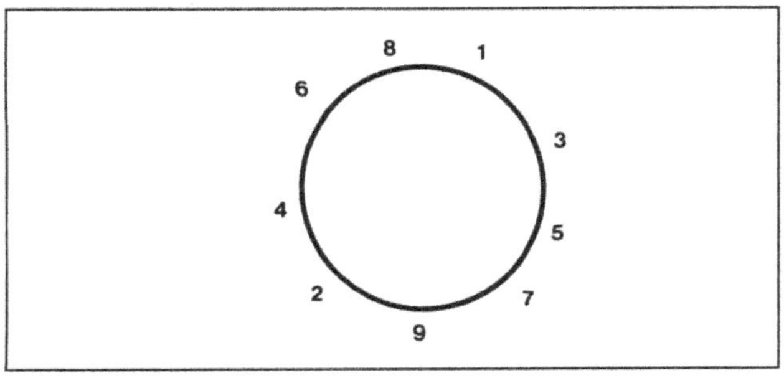

Fig. 5-11. Flywheel bolt torque sequence.

Fig. 5-12. Location of timing mark to be made on new replacement flywheels. Dimension **a** (arc length) is 12 mm (.472 in.).

11. Reinstall the transaxle as described in **6.3 Removing and Installing Transaxle.**

6. TRANSMISSION AND FINAL DRIVE

This section covers external adjustments, replacing external oil seals, and removal and installation. The reverse/upshift light switch atop the transaxle housing is covered in **ELECTRICAL SYSTEM**. For information covering internal repair, see **7. Transmission** and **8. Final Drive**.

> ***CAUTION —***
> *Disassembly and internal repair requires special tools and some prior knowledge of transmission rebuilding practices. Do not begin disassembly of the transmission if you do not have the equipment or the experience to complete the job.*

6.1 External Adjustments

On early models, adjustment of the selector shaft detent plunger and the fifth gear lockout plunger may correct shifting problems that are not fully remedied by adjusting the shifting mechanism. In addition, these adjustments will be necessary if the transaxle has been partially disassembled. The selector shaft adjusters are located on the transaxle housing as illustrated in Fig. 6-1. The adjustments should be made with the transaxle removed, or with the shift linkage removed from the selector shaft. The selector shaft adjusting mechanisms have been eliminated on later models

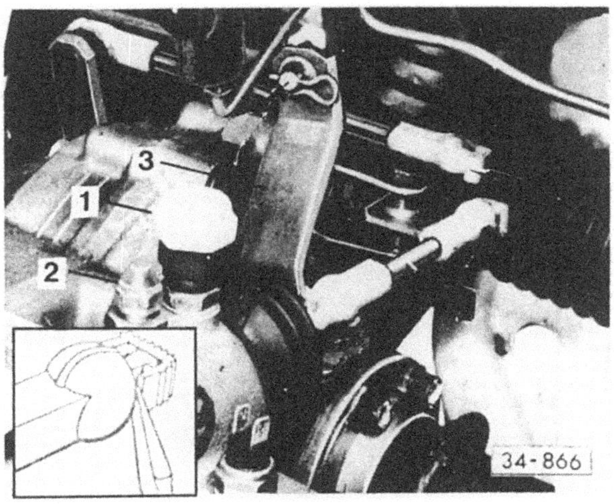

Fig. 6-1. Location of selector shaft adjusters on early 020 transmissions. Selector shaft detent plunger **1** with plastic cap, 5th gear lockout plunger **2**, and selector shaft lever **3**. Inset shows screwdriver being used to remove selector rod.

The selector shaft detent plunger is shown in Fig. 6-2. To adjust, place the transmission in neutral, remove the plastic cap, and loosen the locknut. Turn the adjusting bolt clockwise until the lockring lifts away from the adjusting bolt, then turn the adjusting bolt counterclockwise until the lockring just contacts the adjusting bolt. While holding the adjusting bolt stationary, torque the locknut to 20 Nm (15 ft. lb.). To check the adjustment, hand turn the selector shaft slightly. If the lockring does not lift as soon as the selector shaft is turned, repeat the adjustment. When satisfied with the adjustment, reinstall the plastic cap.

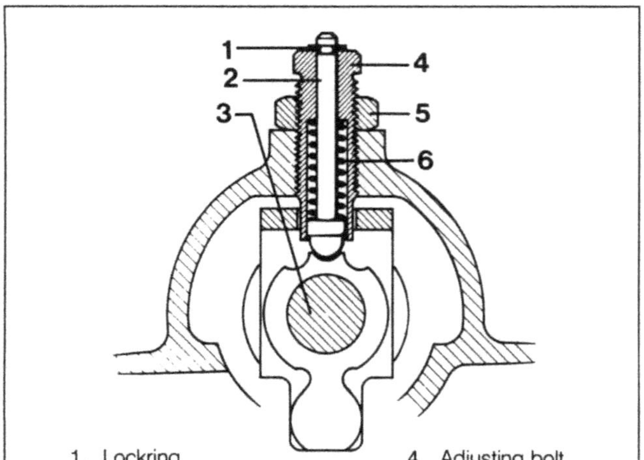

Fig. 6-2. Cross-section of main selector shaft detent plunger.

The fifth gear lockout plunger is shown in Fig. 6-3. To adjust, place the transmission in neutral, remove the plastic cap, and loosen the locknut. Tighten the adjusting sleeve until the detent plunger just begins to move up. From this position, loosen the adjusting sleeve 1/3 turn, then hold the sleeve stationary and tighten the locknut.

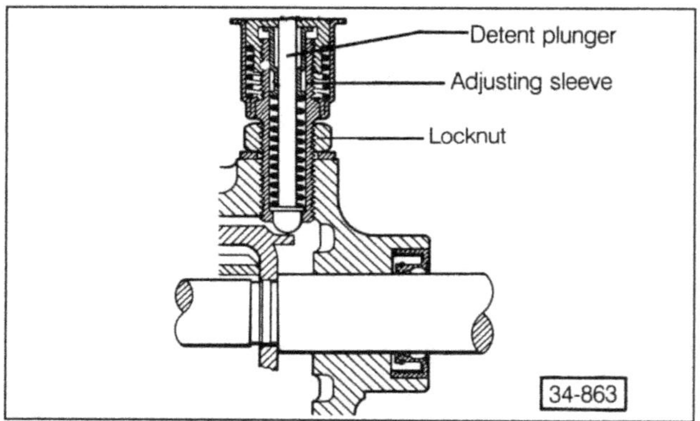

Fig. 6-3. Cross-section of 5th gear lockout plunger.

6.2 Transaxle Oil Seals

Oil leakage due to faulty transaxle oil seals may be the cause of problems such as hard shifting, jumping out of gear, and transaxle noise. The final drive oil seals, the selector shaft oil seal, and the clutch release shaft oil seal can all be replaced while the transaxle is installed in the car. Read the procedure to determine what other new parts are required before beginning repairs. When replacing the clutch release shaft oil seal, some disassembly of the transaxle will make the repair easier. The procedure for replacing the final drive oil seals includes the use of Volkswagen special tools which are helpful, but not essential, to the repair.

To replace final drive oil seals:

1. Detach the drive axle from the transaxle as described in **SUSPENSION AND STEERING**.

> ***CAUTION—***
> *Suspend the detached drive axle from the car body with a stiff wire hook to avoid damage to the outer constant velocity joint.*

2. Pry the dust cap from the center of the drive flange. Remove the circlip and the spring washer.

3. Remove the drive flange from the transaxle using a puller as shown in Fig. 6-4. Inspect the flange and plan to replace it if there is a groove worn where it contacts the oil seal.

> ***CAUTION—***
> *The internal threads of the shaft are used to reinstall the drive flange. Avoid thread damage during removal.*

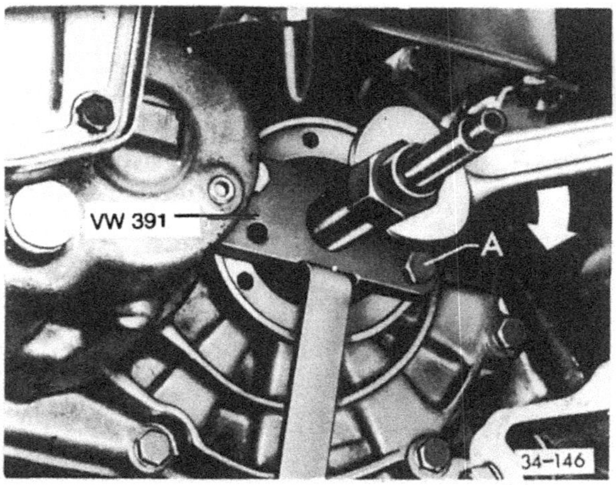

Fig. 6-4. Drive flange being removed. Puller VW 391 shown. Puller mounts to drive flange with M8 x 30 bolts (**A**).

4. Pry the faulty seal from its recess using a hooked seal removal tool, as shown in Fig. 6-5, or a large screwdriver.

5. Make sure the new seal has a spring behind its lip. Then pack the open side of the new seal with multipurpose grease. Drive the new seal into place until it is fully seated.

6. Install the drive flange as shown in Fig. 6-6.

7. Install the spring washer, a new circlip, and a new dust cap. Attach the drive axle as described in **SUSPENSION AND STEERING**.

Fig. 6-5. Drive flange oil seal being removed. Volkswagen special tool VW 681 shown.

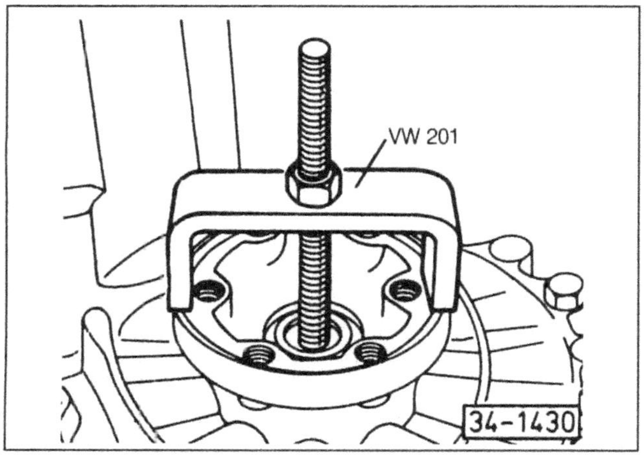

Fig. 6-6. Drive flange being installed. Installation tool shown (VW 201) threads into shaft.

To replace selector shaft oil seal:

1. Shift the transmission into neutral.
2. Remove the selector shaft lever. Unscrew the nut holding the lever to the selector shaft, then disconnect the connecting link from the top of the selector shaft lever. See the earlier Fig. 4-2.
3. Pull the lever off the selector shaft, noting its position on the shaft. Remove the flexible boot and clean the exposed end of the selector shaft.
4. On early models, remove the detent plunger and the 5th gear lockout plunger from the top of the transaxle housing. See Figs. 6-1, 6-2, and 6-3 above. On later models, remove the bolt from the top of the transaxle housing. Remove the reverse/upshift light switch by removing the two mounting bolts.
5. Remove the clutch cable and rubber stop/guide from the transaxle as described in **5.1 Clutch Cable.**
6. To gain access to the selector shaft, remove the selector shaft cover as shown in Fig. 6-7. Use care to avoid losing the spring as the cover is removed. Remove the selector shaft by hand-pressing it out through the cover opening.

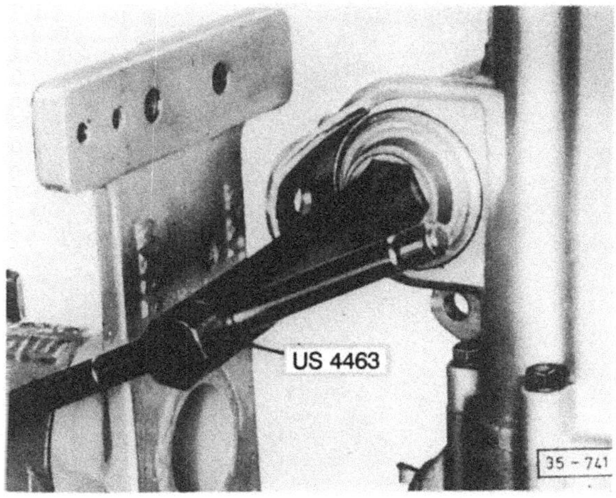

Fig. 6-7. Selector shaft cover being removed. Hex socket size is 27mm. Use short bolt (27mm head) and locknuts with wrench in place of special tool shown.

7. Pry out the old seal, as shown in Fig. 6-8, and install a new seal.

NOTE —
Beginning with transmission number 07 02 9 (2/7/89 production), a modified selector shaft oil seal is installed. The new seal can be identified by a collar on the seal's inner diameter.

Fig. 6-8. Selector shaft oil seal being removed.

8. Lubricate the new oil seal and all surfaces of the selector shaft with multipurpose grease, then carefully install the selector shaft in the transaxle case. See Fig. 6-9.
9. The rest of the installation is the reverse of removal. Torque the selector shaft cover to 50 Nm (37 ft. lb.). On early models, adjust the selector shaft adjusters as described in **6.1 External Adjustments.** Torque the selector shaft lever mounting nut to 13 Nm (10 ft. lb.).

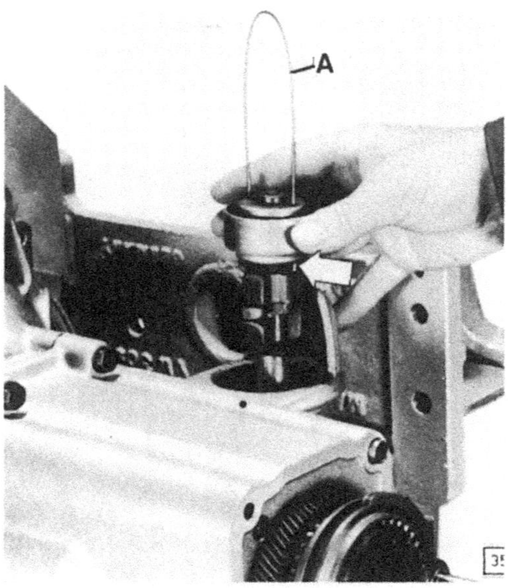

Fig. 6-9. Selector shaft (arrow) being installed. Bent coat hanger wire or welding rod (**A**) can be inserted as shown to keep shaft components from moving while engaging shift forks.

To replace clutch release shaft oil seal:

1. Remove the clutch cable from the clutch release shaft as described in **5.1 Clutch Cable**, then remove the stop clip from side of the release lever. Drain the transaxle as described in **LUBRICATION AND MAINTENANCE**.

2. Pierce the end cap and pry the cap out of its recess in the 5th gear/clutch release bearing housing. See Fig. 6-10.

3. Remove the circlip from each side of the clutch release shaft, then pull the shaft from the transaxle case. Remove clutch lever and the return spring.

4. Remove the oil seal as indicated in Fig. 6-11. Drive in a new oil seal as shown in Fig. 6-12.

5. Position the clutch lever and the return spring in the housing. The central hook of the return spring should lie atop the lug on the clutch lever, and the spring's end hooks should point downward so that the clutch lever is held away from the clutch release bearing.

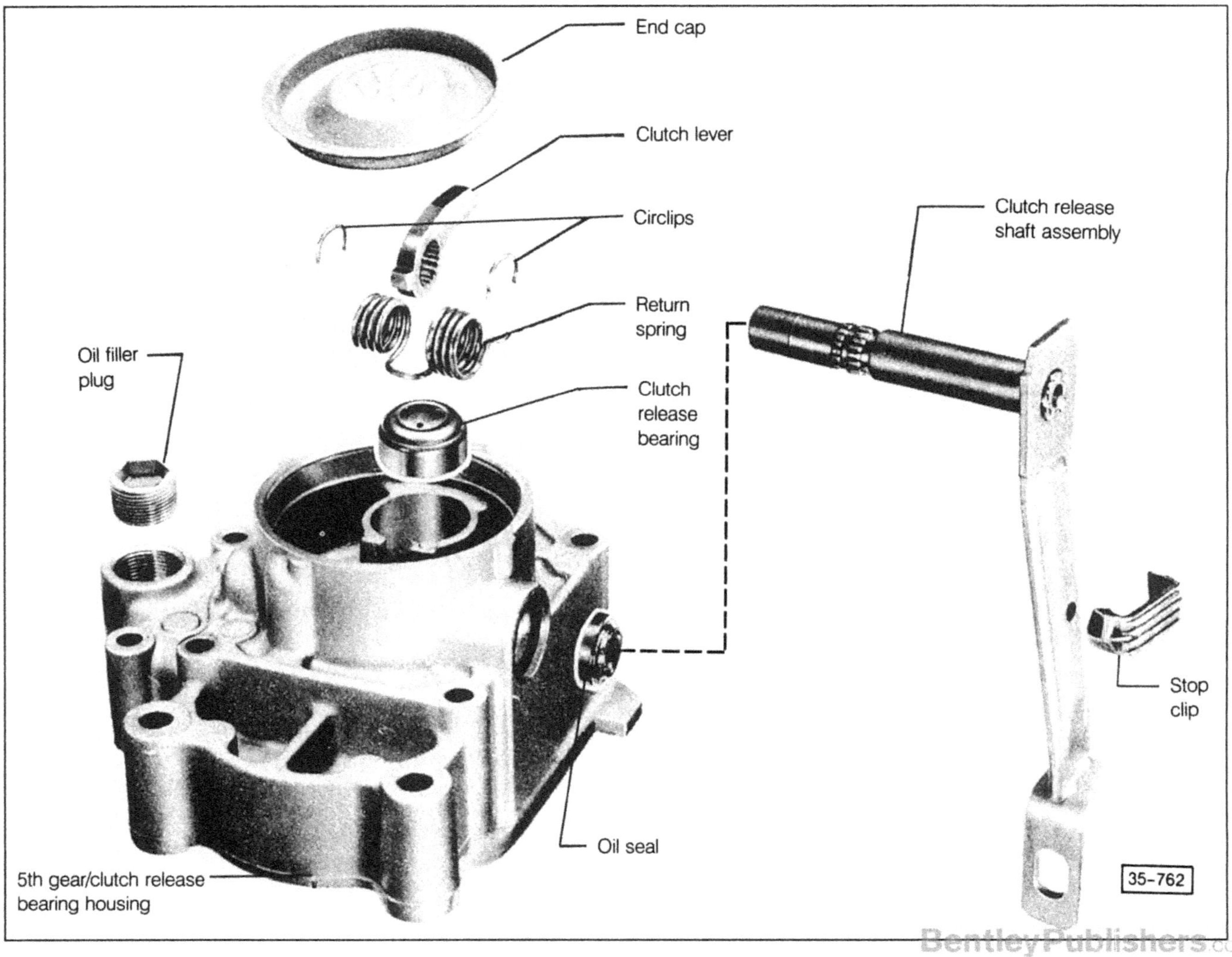

Fig. 6-10. Exploded view of clutch release mechanism.

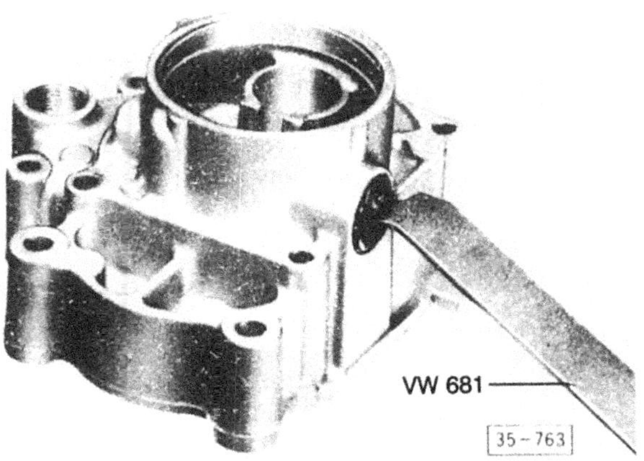

Fig. 6-11. Clutch release shaft oil seal being removed.

Fig. 6-12. Clutch release shaft oil seal being installed.

6. Lightly coat the release bearing with multipurpose grease. Then lubricate the seal lip and pack the lubricant into the seal's open side.

7. Insert the clutch release shaft, being careful that the splines do not damage the oil seal, and engage its splines with the clutch lever. The splines will mate only one way.

8. Install the retaining circlips and check that the return spring is correctly tensioned when the operating lever is lifted to its normal operating position.

NOTE —

Check the release bearing by hand-turning it. If the bearing feels rough or makes noise, it should be replaced at this time.

9. Install a new end cap as shown in Fig. 6-13. Use Volkswagen special tool no. VW 792/2 or equivalent to install the cap evenly.

10. Install the stop clip and the clutch cable to the clutch release lever. If necessary, adjust the clutch pedal freeplay as described in **LUBRICATION AND MAINTENANCE**.

Fig. 6-13. New end cap being installed.

Transaxle Vent

To compensate for pressure changes within the transaxle due to changes in temperature, the transaxle is vented. If the vent becomes plugged, the buildup of pressure inside the transaxle can cause lubricant leaks past the oil seals. If replacement of an oil seal does not correct a leak, the transaxle vent should be inspected.

On 1985 and 1986 models, the transaxle vent is located above the selector shaft cover, as shown in Fig. 6-14. The clutch cable must be removed from the transmission as described in **5.1 Clutch Cable** in order to inspect the vent.

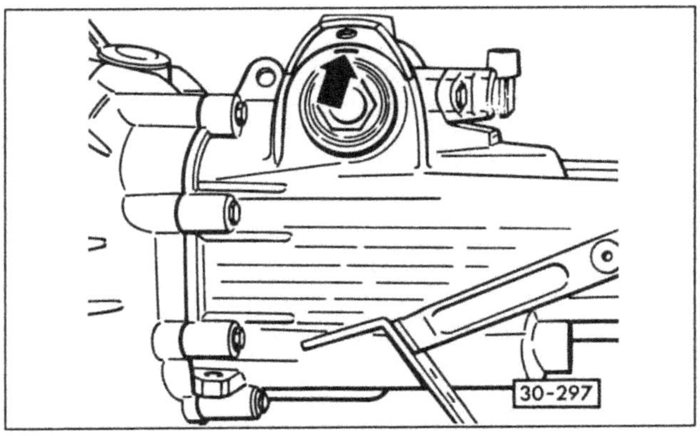

Fig. 6-14. Location of transaxle vent (arrow) on 1985 and 1986 models.

On 1987 and later models, the transaxle vent is located atop the transaxle, as shown in Fig. 6-15, and is inspected by removing the plastic cap. Replacement transmission housings for all years have the top vent hole open, and the plastic vent sleeve and cap must be installed separately. Both parts are available from an authorized Volkswagen dealer.

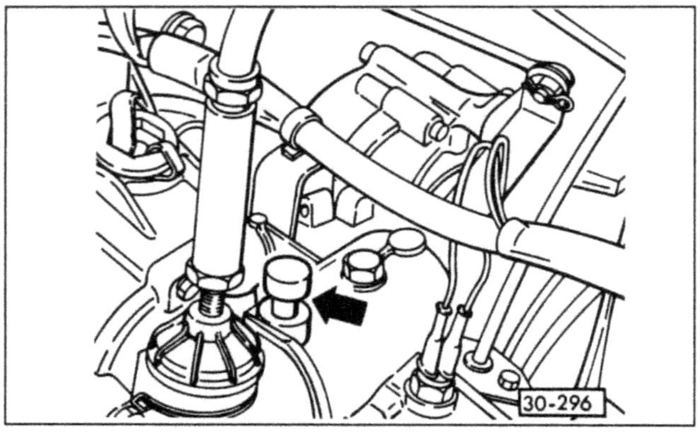

Fig. 6-15. Location of transaxle vent plastic sleeve and cap on 1987 and later models.

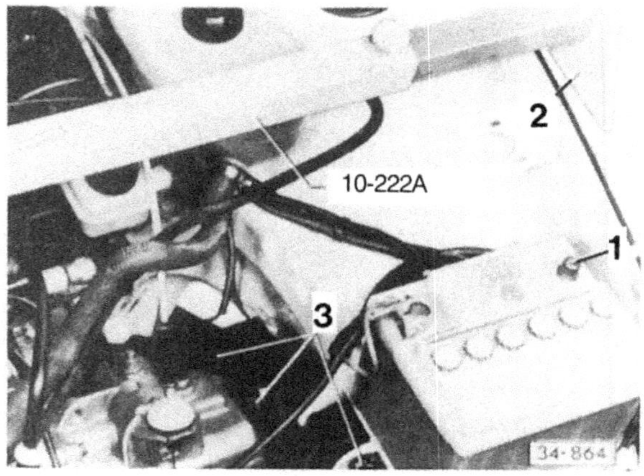

Fig. 6-16. Engine being supported by support tool. Volkswagen special tool no. 10-222A shown.

8. Remove the plastic timing plug on the transmission bellhousing, then, using a wrench on the crankshaft pulley, turn the engine clockwise so that the lug on the flywheel aligns with the boss on the bellhousing as shown in Fig. 6-17.

Fig. 6-17. Lug on flywheel aligned with boss on bellhousing.

9. Remove the rear transaxle/engine mount from the transaxle and the body.
10. Remove the left (driver's side) transaxle mount from the transaxle and the body.
11. Detach the drive axles from the transaxle drive flanges as described in **SUSPENSION AND STEERING**.

CAUTION —
Suspend the detached drive axles from the car body with stiff wire hooks to avoid damaging the outer constant velocity joint.

6.3 Removing and Installing Transaxle

The transaxle must be removed from the car for most internal repairs or for replacement. In the procedure given below, the transaxle is separated from the engine, supported on a jack, and removed from below. Alternatively, the engine and transaxle assembly can be removed, as described in **ENGINE**, and the transaxle then separated from the engine.

Removal of the transaxle requires a means of lifting the engine, a transmission jack or a floor jack with transmission adaptor, and jack stands to support the car. The engine lift is necessary because the engine mounts are removed. Use extreme caution when working beneath the car and lowering the transaxle.

To remove:

1. Disconnect the battery negative (−) terminal. Disconnect the ground strap from the transaxle. Disconnect the electrical connector from the reverse/upshift light switch atop the transaxle.
2. Remove the speedometer cable and insert a plug in the hole to prevent loss or contamination of the transaxle lubricant.
3. Disconnect the clutch cable from the clutch release lever as described in **5.1 Clutch Cable**.
4. Remove the starter as described in **ELECTRICAL SYSTEM**.
5. Remove the short selector rod and the long connecting link from the two relay levers of the shift linkage. If necessary, see **4. Shift Mechanism**.
6. Install an engine support, as shown in Fig. 6-16, and increase tension until it supports the weight of the engine.
7. Remove the upper bolts that hold the transaxle to the engine.

12. Remove the bolts from the clutch cover plate as indicated in Fig. 6-18 (the cover plate will remain on the engine when the transaxle is removed). Remove the bolts from the small cover plate (located behind the left drive flange).

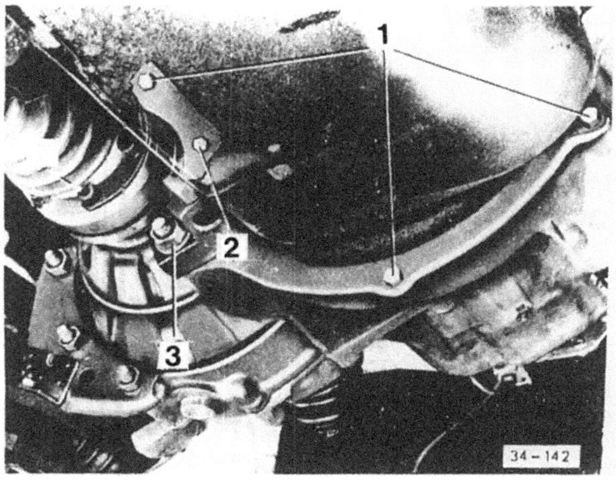

Fig. 6-18. Clutch cover plate bolts at **1**, small cover plate bolts at **2**, and left mounting bolt at **3**.

13. Place a transmission jack or a floor jack with transmission adaptor beneath the transaxle. Raise the jack just enough to support the transaxle. Do not raise the engine.

14. Slightly lower the transaxle and remove the left mounting bolt below the small cover plate, shown above in Fig. 6-18.

> ***CAUTION —***
> *At no time should the weight of the transaxle be supported by the transmission mainshaft. Such a load will damage the clutch and transaxle components.*

15. Push the engine and transaxle assembly as far as possible toward the passenger side of the car (approximately 4 mm or 5/32 in.). Remove the remaining lower mounting bolts holding the transaxle to the engine.

16. Separate the transaxle from the engine, taking care not to place any strain on the transmission mainshaft, and either lower the jack or raise the car until there is enough clearance to remove the transaxle. Pull the transaxle out from under the car.

> ***WARNING —***
> *Make sure the car is stable and well supported at all times during the removal procedure. Use jack stands which are designed for the purpose. A floor jack alone is not adequate support.*

To install:

1. Lightly lubricate the input shaft splines with molybdenum disulfide grease, then position the transaxle on the engine, being careful not to damage any of the clutch components. Check that the large clutch plate is correctly positioned. If necessary, use a wrench on the crankshaft pulley to align the flywheel.

2. Install the lower transaxle to engine mounting bolts and then install the upper bolts. Torque the bolts to 80 Nm (59 ft. lb.).

3. Install the left transaxle mount, but do not fully tighten the bolts and nuts.

4. Reconnect the drive axles to the transaxle drive flanges. Torque the 8mm 12-point socket-head bolts to 45 Nm (33 ft. lb.).

5. Install and torque the large and small cover plate bolts to 9 Nm (78 in. lb.).

6. Install the rear transaxle/engine mount, but do not fully tighten the bolts and nuts.

7. Install the starter as described in **ELECTRICAL SYSTEM**. Install the reverse gear/upshift light switch and attach its connector. Torque the bolts to 10 Nm (87 in. lb.). Install the plastic plug in the timing check hole.

8. Reconnect the shift mechanism and loosely install the clutch cable to the release lever. Connect the speedometer cable to the transaxle housing.

9. Remove the engine lift or hoist. Reconnect the battery negative (–) terminal and the ground strap.

10. Align the engine and transaxle mounts as described in **ENGINE**. With the mounts aligned, and the fasteners properly torqued, adjust clutch pedal freeplay as described in **5.1 Clutch Cable**, and adjust the shifting mechanism as described in **4. Shift Mechanism**.

11. Refill the transaxle with the appropriate lubricant before starting or towing the car. See **LUBRICATION AND MAINTENANCE** for information on the proper lubricant and how to check the level or fill the transaxle.

7. TRANSMISSION

All the work described under in this section requires that the transaxle first be removed from the car. See **6.3 Removing and Installing Transaxle.** This section covers the repair and replacement of internal transmission components including gears, synchronizers and bearings. The work is broken down into two main sub-sections.

The first sub-section, **7.1 Transmission Disassembly and Assembly**, covers disassembling the transmission for access to the main internal component assemblies. The second sub-section, **7.2 Transmission Gear Train Disassembly and Assembly**, covers repair and rebuilding procedures for those internal assemblies.

See **6.3 Removing and Installing Transaxle**. Repair procedures for the final drive portion of the transaxle are covered in **8. Final Drive**.

CAUTION —
For the most part, the internal transmission work covered here requires special tools and prior knowledge of proper transmission rebuilding practices. Do not begin disassembly of the transmission if you do not have the tools or the experience to complete the job.

Many of the procedures described in this section of the manual suggest or require the use of special tools. Unless otherwise specified, the numbers which appear in the illustrations identify Volkswagen special tools by their Volkswagen tool numbers.

7.1 Transmission Disassembly and Assembly

Fig. 1-1, near the beginning of this transmission section, is a cross-section view of the transaxle assembly, identifying the location of the main components.

Fig. 7-1 on the page that follows is an exploded view of the type 020 manual transmission, the type fitted to all Cabriolet and Scirocco models covered by this manual.

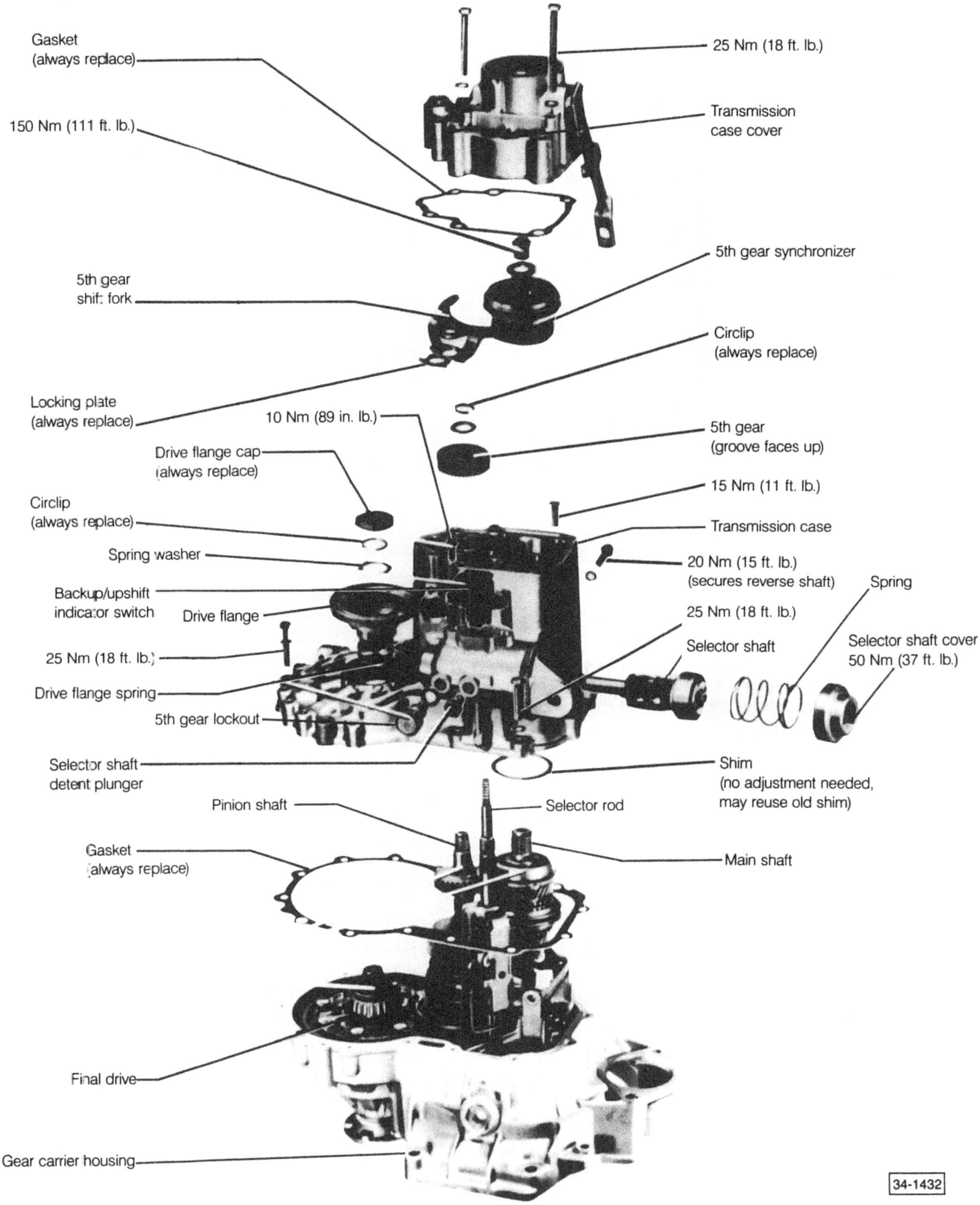

Fig. 7-1. Exploded view of 020 manual transmission and final drive

Before beginning work on the transmission, thoroughly clean the outside of the transmission case to prevent dirt from contaminating the interior components as you remove them. Drain the transaxle oil and mount the transmission vertically in a secure stand, supporting the main shaft as shown in Fig. 7-2 to prevent damage to the gear train. Remove the clutch pushrod.

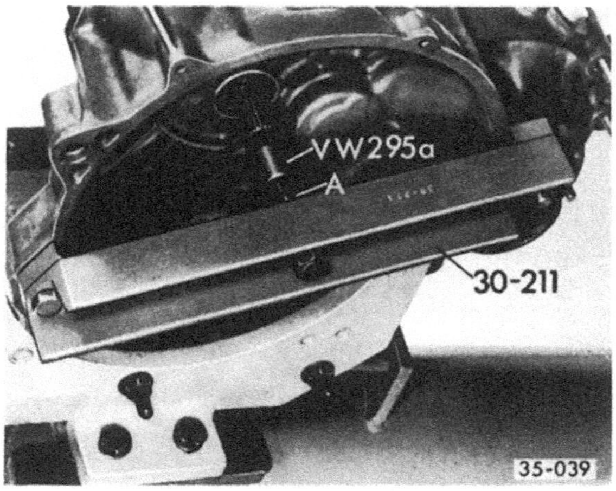

Fig. 7-2. Transmission main shaft supported for disassembly of transmission case. Use Volkswagen special tool no. 30-211 and thrust pad VW 295a. Thrust pad locknut is at **A**.

Removing and Installing 5th Gear

The 5th gear must be removed for complete disassembly of the transmission. 5th gear and its synchronizer can be replaced without further transmission disassembly.

Many of the procedures described here suggest or require the use of special tools. Unless otherwise specified, the numbers which appear in the illustrations identify Volkswagen special tools by their Volkswagen tool numbers.

To remove 5th gear:

1. Remove the bolts that secure the transmission case cover and remove the cover and clutch release bearing. Remove the clutch pushrod. Check the pinion shaft needle bearing in the transmission case cover. It is held in place by a screw. Replace the bearing if it shows signs of wear or pitting.

2. Remove the backup/upshift indicator light switch, the selector shaft detent plunger, and the 5th gear lockout.

3. Remove the selector shaft cover, then set the shift forks in Neutral and remove the selector shaft and spring.

CAUTION —

If the shift forks cannot be set in neutral, the selector shaft must be driven out. Refer to Fig. 7-3. Replace damaged parts (selector shaft, shift forks) before reassembly.

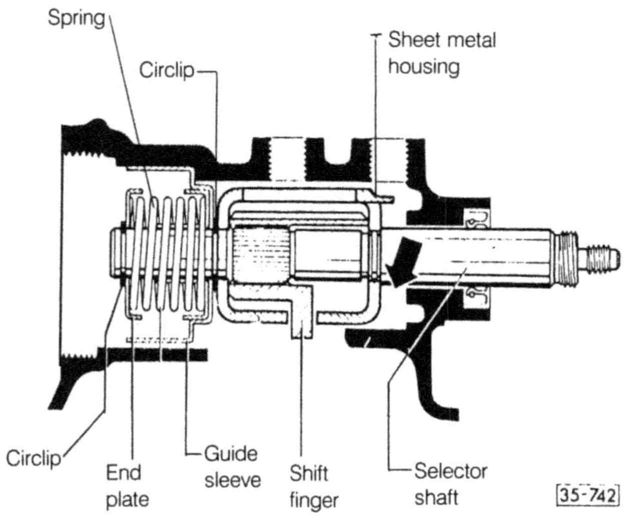

Fig. 7-3. Cutaway view of components to be removed if shift forks cannot be set in neutral. Remove circlip, end plate, spring, guide sleeve, circlip. For support, place nut at arrow and drive out selector shaft. Sheet metal housing and shift finger will remain in transmission case.

4. Using a 12 mm socket head wrench, remove the bolt that secures the 5th gear synchronizer to the main shaft. It is necessary to engage both 5th and reverse gears, as shown in Fig. 7-4, so that the nut can be removed without the transmission gears turning.

Fig. 7-4. 5th and reverse gears are engaged by pushing down on front shift fork (arrow) so that transmission will be locked to remove synchronizer mounting bolt.

5. Remove the 5th gear synchronizer and shift fork. First remove the selector tube by prying up the locking plate as shown in Fig. 7-5 so that the selector tube can rotate. Then, using the special tool as shown in Fig. 7-6, screw

the tube counterclockwise out of the shift fork and remove the 5th gear synchronizer and shift fork together. Check the shift fork for damage and replace if necessary.

CAUTION —

Do not pull the selector rod out of the tube with the tool. This may dislodge the shift forks, and the transmission will have to be disassembled to reinstall them. If necessary, hold the shift rod down with a screwdriver when taking off the special tool.

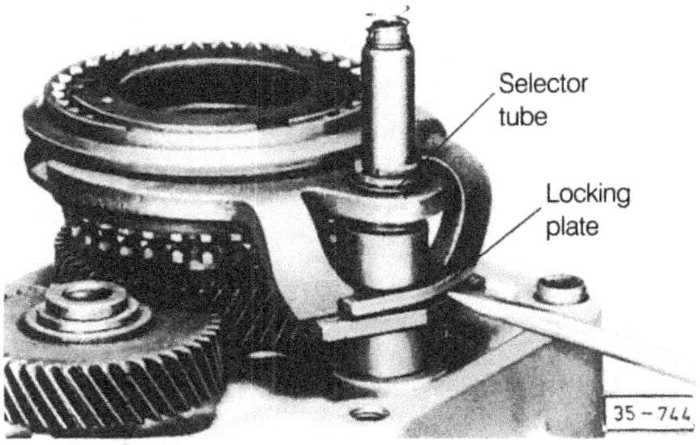

Fig. 7-5. Using screwdriver to pry up locking plate so that selector tube can be removed.

Fig. 7-6. Volkswagen special tool no. 3059 being used to remove selector tube.

6. Remove the 5th gear circlip and thrust washer, then use a puller to remove 5th gear from the pinion shaft.

To install and adjust 5th gear:

1. Heat 5th gear to about 212°F (100°C) and install it on the pinion shaft. The groove on the gear should face upwards. Install the thrust washer and new circlip.

2. Install the 5th gear synchronizer and shift fork on the main shaft, along with a new locking plate or adjusting sleeve.

NOTE —

A grooved selector tube uses the adjusting sleeve. A selector tube with a shoulder uses a locking plate.

3. Using Volkswagen special tool no. 3059 as shown above in Fig. 7-6, turn the selector tube clockwise to install it. Then adjust the tube by screwing it out until it projects 5.0 mm (.20 in.), as shown in Fig. 7-7.

CAUTION —

Do not pull the selector rod out of the tube with the tool. This may dislodge the shift forks, and the transmission will have to be disassembled to reinstall them. If necessary, hold the shift rod down with a screwdriver when taking off the special tool.

Fig. 7-7. Selector tube adjustment. Distance **x** is 5.0 mm (.20 in.).

4. Coat a new synchronizer mounting bolt with thread locking compound and tighten it to 150 Nm (111 ft. lb.). Make sure the 5th and reverse gears are engaged as shown earlier in Fig. 7-4 to prevent the main shaft from rotating.

5. Place the shift forks in Neutral, then install the selector shaft spring and its cover as shown earlier in Fig. 6-9. Reinstall and adjust the selector shaft detent plunger and the 5th gear lockout as described in **6.1 External Adjustments**.

6. Adjust the 5th gear shift fork by first putting the transmission in 5th gear. To do this, install the shift lever on the selector shaft, and then pull the lever out fully and turn it to the left.

7. Lightly lift the shift fork to eliminate any play, then check the engagement of the synchronizer sleeve with the 5th gear teeth as shown in Fig. 7-8. The overlap should be 1.0 mm (.039 in.). If the overlap is incorrect, adjust it by turning the selector tube.

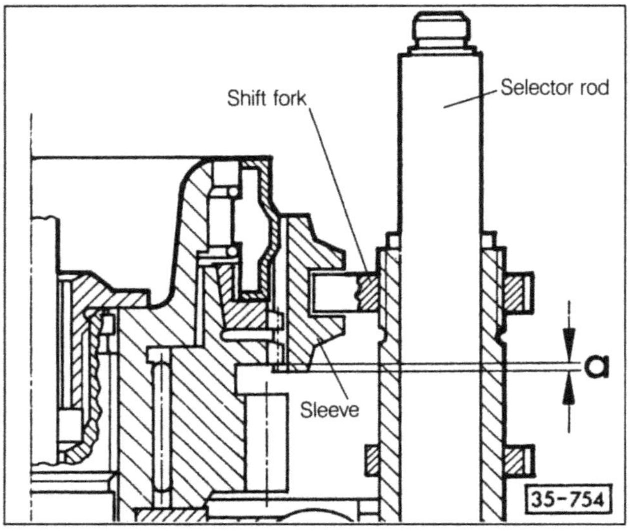

Fig. 7-8. Engagement of synchronizer sleeve with 5th gear, showing overlap to be measured (**a**).

8. Lock the selector tube in position by securing the locking plate as shown in Fig. 7-9. Support the shift fork with a tool at least 12 mm (.5 in.) thick, then press the locking plate down using Volkswagen special tool no. 3097.

CAUTION —
Protect the sealing surface of the transmission case from damage.

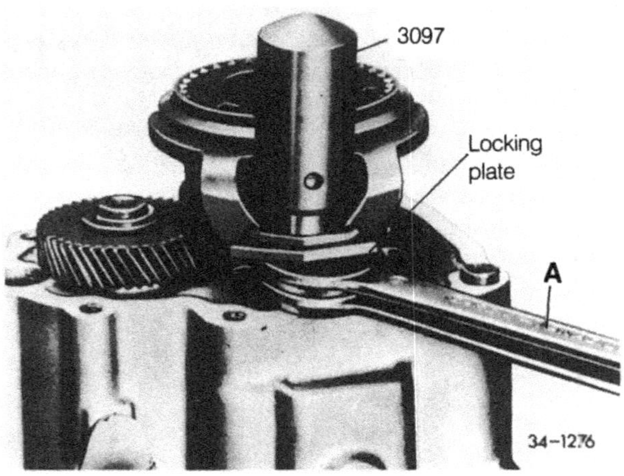

Fig. 7-9. Shift fork being supported with wrenches (**A**) so that locking plate can be locked using Volkswagen special tool no. 3097.

9. Reinstall the clutch pushrod and release bearing, and the light switch. Using a new gasket, install the transmission case cover and torque the bolts to 25 Nm (18 ft. lb.).

Removing and Installing Main Shaft and Pinion Shaft

This is a continuation of transmission disassembly. The 5th gear should be removed as described above before beginning this procedure. Fig. 7-10 shows an exploded view of the transmission main shaft and pinion shaft.

Many of the procedures described here suggest or require the use of special tools. Unless otherwise specified, the numbers which appear in the illustrations identify Volkswagen special tools by their Volkswagen tool numbers.

To remove main shaft and pinion shaft:

1. Remove the bolt that secures the reverse gear shaft.

2. Remove the drive flange on the transmission case side, as described in **6.2 Transaxle Oil Seals**.

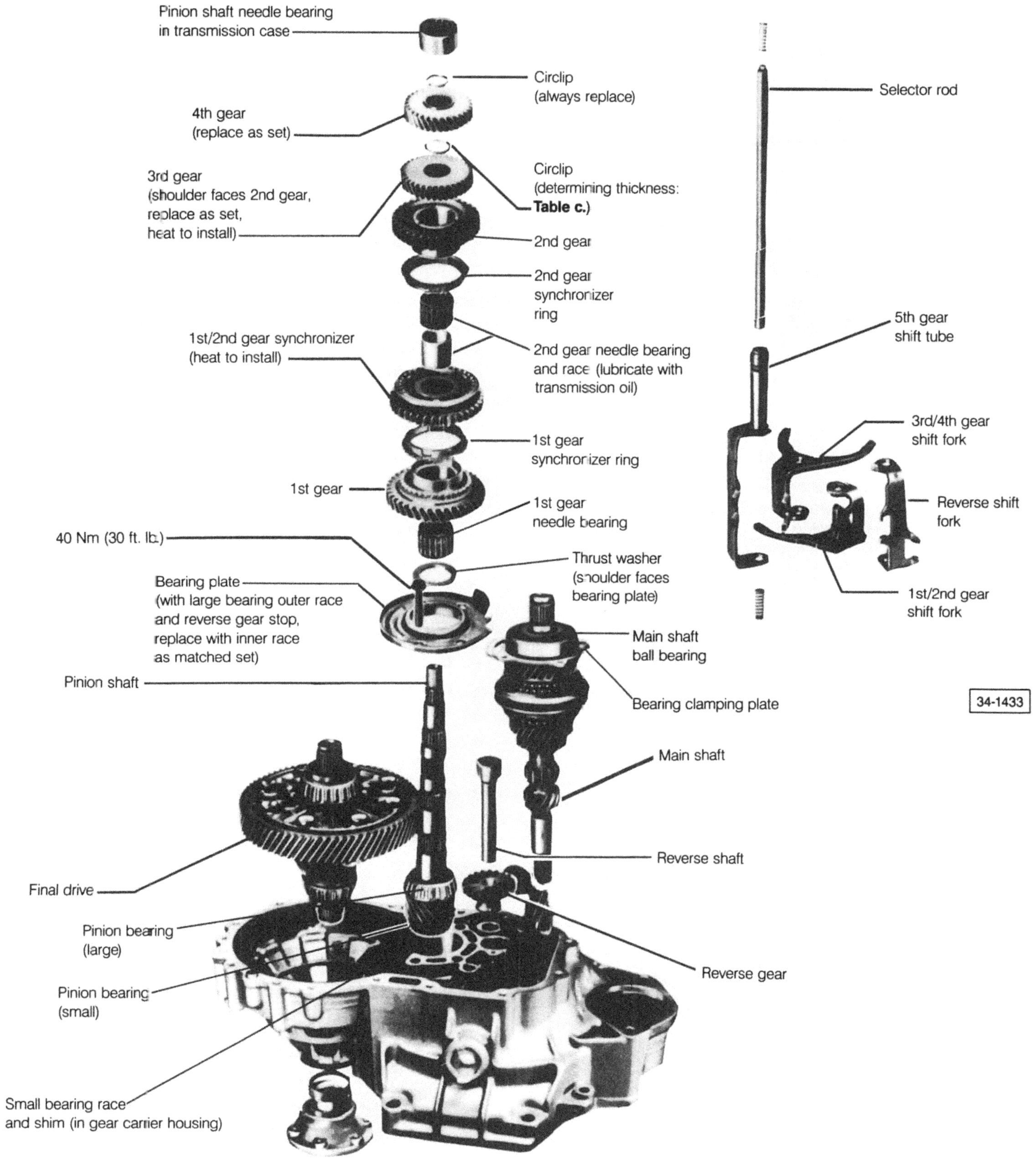

Fig. 7-10. Exploded view of main shaft and pinion shaft.

3. Using a 12-point M6 socket wrench, remove the four bolts that hold the bearing clamping plate for the main shaft bearing. See Fig. 7-11.

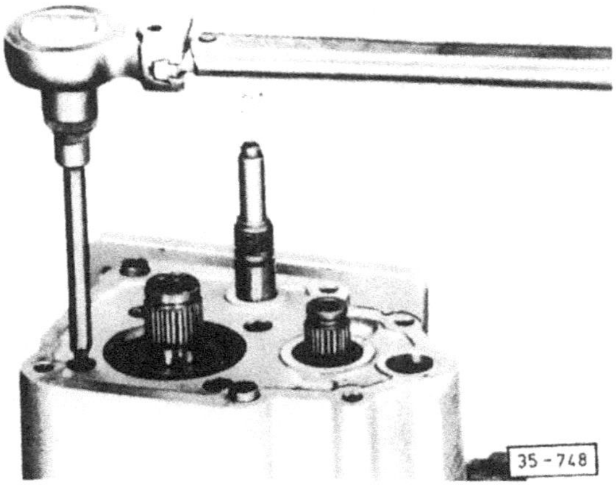

Fig. 7-11. Main shaft bearing clamping plate bolts being removed.

4. Remove the transmission-case-to-gear-carrier bolts and separate the case from the gear carrier housing using a puller as shown in Fig. 7-12. The main shaft ball bearing should remain on the main shaft. If there is a shim atop the ball bearing, save it for reinstallation of the bearing.

NOTE —
Do not lose the magnet that is held between the transmission case and gear carrier housing.

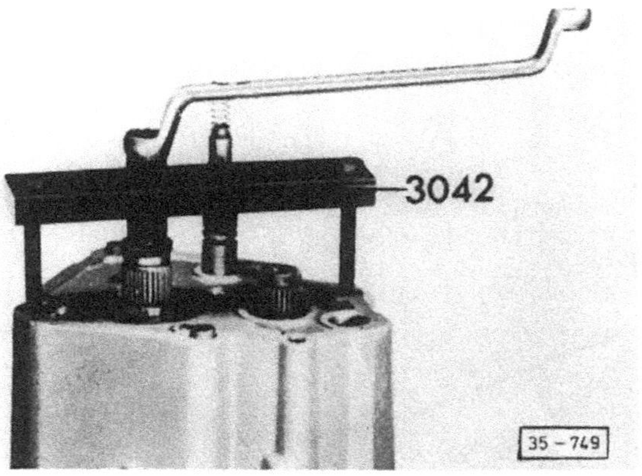

Fig. 7-12. Transmission case being separated from gear carrier housing. Puller presses against main shaft.

5. Lift the shift fork assembly up so that the selector rod is pulled from the gear carrier housing, and then remove the shift fork assembly.

6. Remove the 4th gear circlip from the pinion shaft, then remove the main shaft and 4th gear at the same time.

7. On the pinion shaft, remove the 3rd-gear circlip, and then remove 3rd gear, 2nd gear, the 2nd gear synchronizer, and the 2nd gear needle bearing. Remove reverse gear and the reverse gear shaft.

8. Using a puller, pull off the 1st/2nd gear synchronizer and 1st gear then remove the 1st gear needle bearing and thrust washer. Unbolt the pinion shaft bearing retaining plate and remove the pinion shaft.

To install pinion shaft and main shaft:

1. Install the pinion shaft and bearing retaining plate and torque the bearing plate bolts to 40 Nm (30 ft. lb.).

CAUTION —
If the pinion shaft, its bearings, or the transmission case are being replaced, it will be necessary to first adjust the pinion shaft. See **7.2 Transmission Gear Train Disassembly and Assembly**.

2. Install the thrust washer, 1st gear needle bearing, 1st gear, and synchronizer ring on pinion shaft.

3. Turn the synchronizer ring until the grooves align with the keys, then heat the synchronizer hub to 248°F (120°C) and press it on as shown in Fig. 7-13.

Fig. 7-13. 1st/2nd gear synchronizer being installed. Heat synchronizer to 248°F (120°C) before installing.

4. If the reverse gear relay lever was removed, install it and torque the bolts to 25 Nm (18 ft. lb.). Make sure the plastic insert is installed on the relay lever, then install the reverse gear shaft and reverse gear.

NOTE —
Beginning with transmission number 13 08 7 (8/13/87 production), a modified reverse gear relay lever is installed. The modified lever has a pressure spring, with a guide piece that fits into a new recess in the gear carrier housing. The relay levers were also modified. The old and new parts are not interchangeable. The newer parts can only be installed in the new style gear carrier housing.

5. Press on the 2nd gear needle bearing race, using the tools shown above in Fig. 7-13, then install the 2nd gear synchronizer ring, 2nd gear, and 3rd gear.

6. Install the 3rd gear circlip and use a feeler gauge to check the end play as shown in Fig. 7-14. Play should be between 0.00 to 0.20 mm, trying to keep as close as possible to the lower limit. If end play is incorrect, it can be adjusted using circlips of differing thickness. **Table c** lists the circlips that are available.

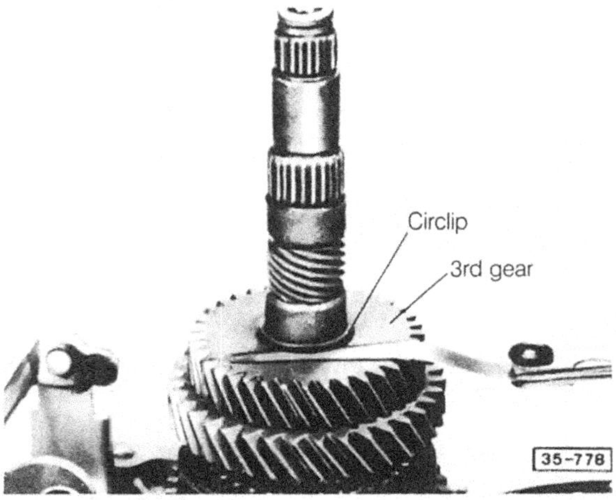

Fig. 7-14. 3rd gear endplay should be between 0.00 to 0.20 mm. Adjust by using a circlip of different thickness.

NOTE —
Part numbers are for reference only. Always rely on an authorized Volkswagen parts department for the most accurate and up-to-date parts information.

7. If the main shaft has not been disassembled, press the ball bearing off the main shaft.

Table c. Circlips for Adjusting 3rd Gear End Play

Thickness in mm	Color	Part No.
2.5	brown	020 311 381
2.6	black	020 311 381 A
2.7	silver	020 311 381 B
2.8	copper	020 311 381 C
2.9	brass	020 311 381 D
3.0	blue	020 311 381 E
End play specification: 0.00 to 0.20 mm		

8. Install the main shaft, without the ball bearing, into the gear carrier housing, at the same time installing 4th gear on the pinion shaft. Then press on the 4th gear circlip.

CAUTION —
To prevent damage to the gear carrier housing, the main shaft support must be installed as shown earlier in Fig. 7-1 before installing the main shaft.

9. Press the main shaft ball bearing into the transmission case as shown in Fig. 7-15. If the ball bearing was installed with a shim, the old shim or one of the same thickness should be used. If the ball bearing had no shim, do not use one.

NOTE —
The wide shoulder on the ball bearing inner race should face 4th gear.

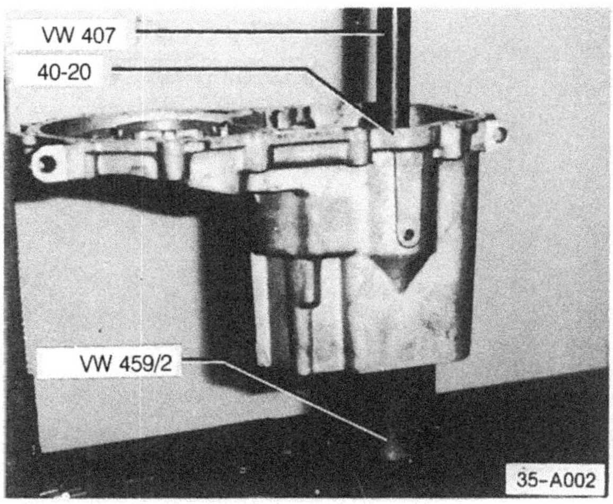

Fig. 7-15. Main shaft ball bearing being pressed into transmission case. Use Volkswagen special tool no. VW 459/2 (as shown) to support case.

10. Position the bearing clamping plate as shown in Fig. 7-16. Install but do not fully tighten the plate bolts.

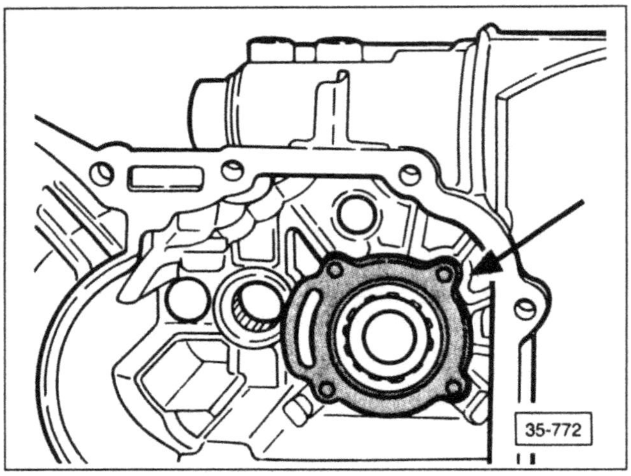

Fig. 7-16. Correct installation of main shaft ball bearing clamping plate (arrow).

11. Insert the lower shift fork spring into the gear carrier housing, then install the shift forks. First insert the 1st/2nd gear shift fork set into the synchronizer sleeve, then lift the selector rod slightly to swing the fork set around the pinion shaft. Next, guide the 3rd/4th gear fork into the synchronizer sleeve, and the reverse fork into the relay lever. Then slip the selector rod into the gear carrier housing.

12. Before installing the transmission case, check the alignment of the reverse gear shaft as shown in Fig. 7-17.

Fig. 7-17. Reverse gear shaft alignment. Distance **x** should be equal on both sides. Reverse gear shaft lock bolt shown temporarily installed to aid alignment. Remove bolt before installing transmission case.

13. Using a new gasket, install the transmission case as shown in Fig. 7-18, then tighten the case bolts to 25 Nm (18 ft. lb.).

14. Install and tighten the reverse gear shaft lock bolt to 30 Nm (22 ft. lb.). Tighten the main shaft ball bearing clamping plate to 15 Nm (11 ft. lb.).

15. Install the drive flange as shown earlier in Fig. 6-6.

16. Install 5th gear and the transmission case cover as described earlier in **Removing and Installing 5th Gear**.

CAUTION —
To prevent damage to the gear carrier housing, the main shaft support must be installed as shown above in Fig. 7-1 before installing transmission case.

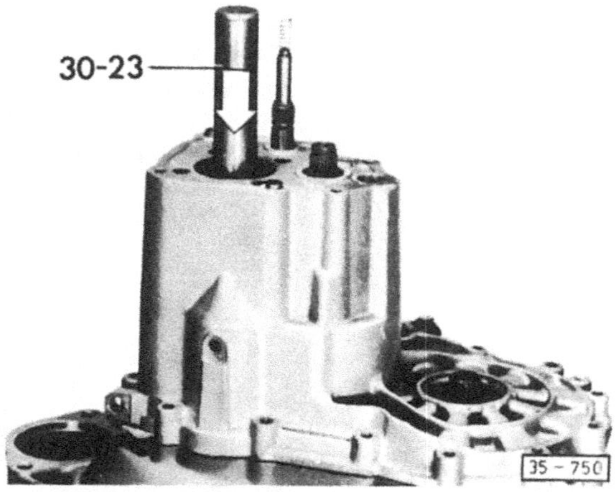

Fig. 7-18. Transmission case being installed to gear carrier housing. Use drift on main shaft ball bearing to seat bearing firmly on main shaft.

7.2 Transmission Gear Train Disassembly and Assembly

Fig. 7-19 is an exploded view of the transmission main shaft components. The pinion shaft components are shown earlier in Fig. 7-10. Many of the procedures described in this section of the manual suggest or require the use of special tools. Unless otherwise specified, the numbers which appear in the illustrations identify Volkswagen special tools by their Volkswagen tool numbers.

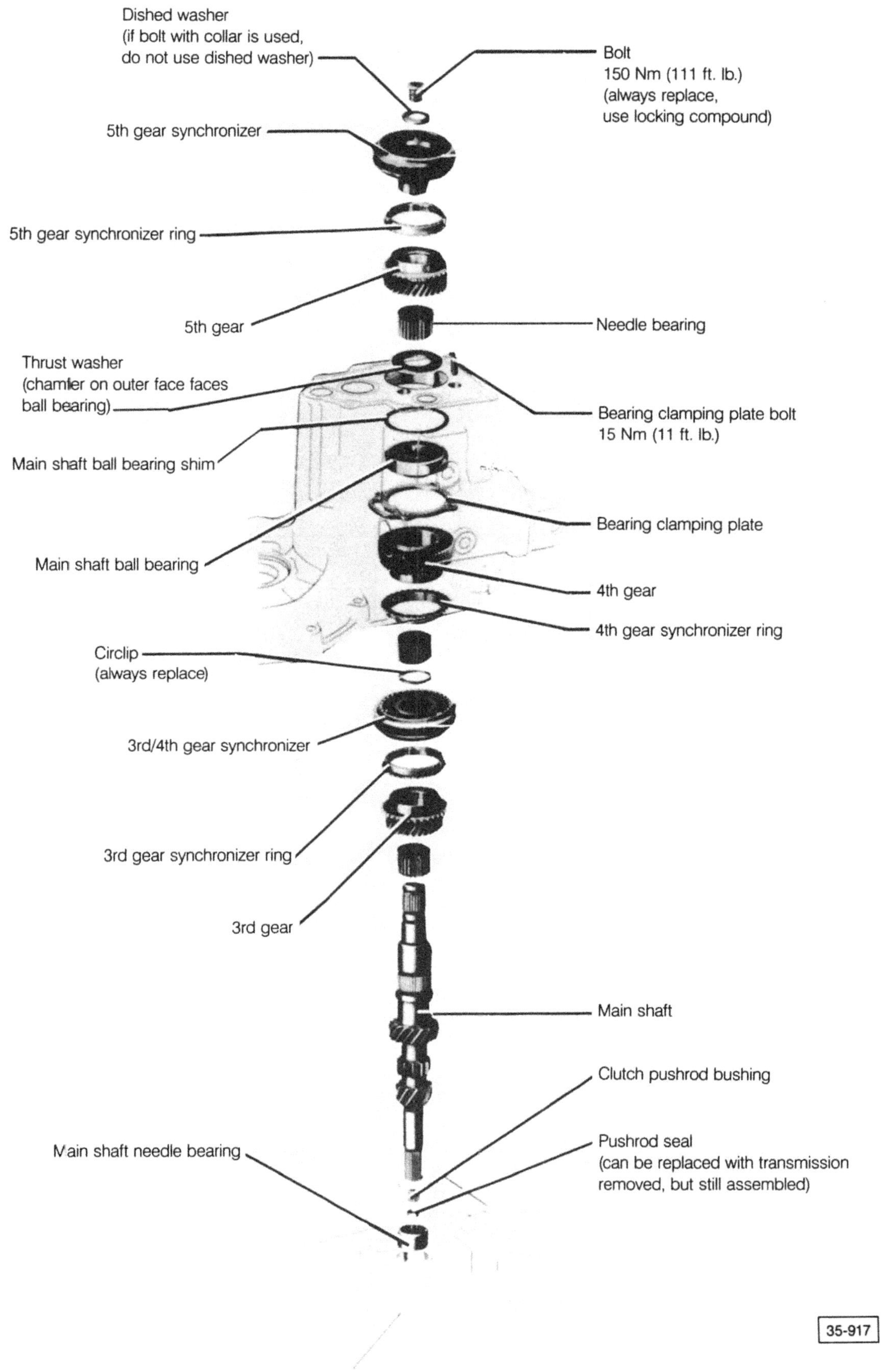

Fig. 7-19. Exploded view of transmission main shaft.

Synchronizers and Synchronizer Rings

All synchronizer components are available as replacement parts. All synchronizers except the 3rd/4th gear synchronizer are removed and installed as part of the transmission disassembly procedure. Remove and install 3rd gear and the 3rd/4th gear synchronizer as shown in Fig. 7-20 and Fig. 7-21.

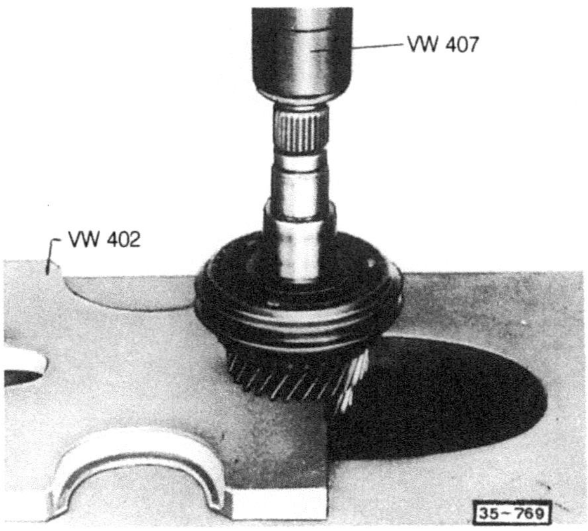

Fig. 7-20. 3rd gear and 3rd/4th gear synchronizer being pressed off of main shaft. Remove circlip and press off together.

Fig. 7-21. 3rd gear and 3rd/4th gear synchronizer being pressed onto main shaft. Turn synchronizer ring until grooves are in line with synchronizer keys.

Check the wear of the synchronizer rings and their gears as shown in Fig. 7-22. **Table d** gives wear specifications.

Fig. 7-22. Checking wear of synchronizer ring and gear. Measure gap **a** with feeler gauge. **Table d** gives specifications and wear limits.

Table d. Synchronizer Ring Wear Specifications

Ring and gear	New part mm (in.)	Wear limit mm (in.)
1st and 2nd gear	1.1 to 1.7 (.04 to .07)	0.5 (.02)
3rd gear	1.15 to 1.75 (.045 to .068)	0.5 (.02)
4th and 5th gear	1.3 to 1.9 (.05 to .07)	0.5 (.02)

NOTE —
The 1st gear synchronizer ring can only be installed on 1st gear. The original ring is not available as a spare part. If the ring is worn and must be replaced, use part no. 014 311 295 D.

When assembling synchronizers, install the retaining springs as shown in Fig. 7-23. Fig. 7-24, Fig. 7-25, and Fig. 7-26 show identification and assembly of the different synchronizers.

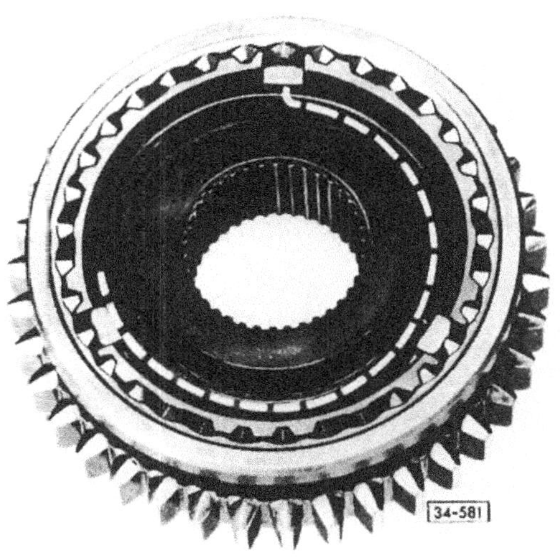

Fig. 7-23. Installed position of synchronizer retaining springs (dotted line). Springs are on both sides of synchronizer. Angled spring ends must engage keys. Install so springs are offset by 120°.

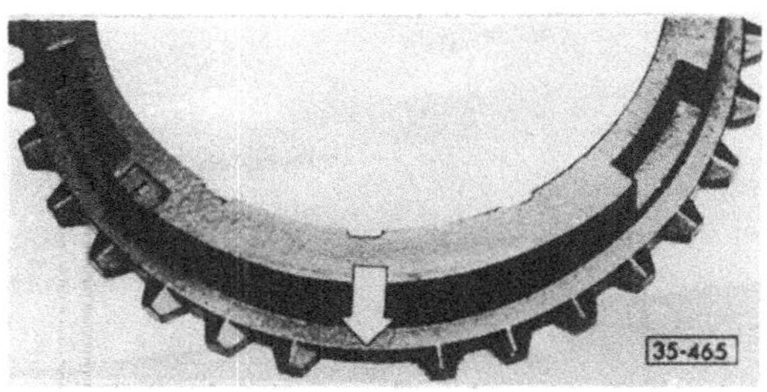

Fig. 7-24. 1st/2nd gear synchronizer. **Top:** 1st gear synchronizer ring identified by missing teeth (arrow). **Bottom:** When assembling, groove on hub should face 1st gear. Groove may be on front face (black arrow) or on splines (white arrow).

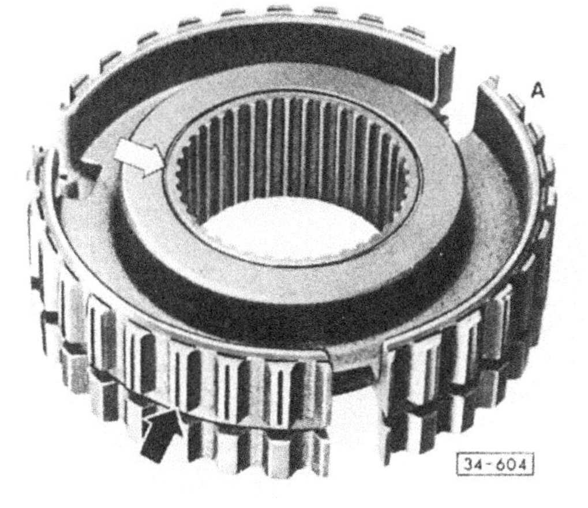

Fig. 7-25. 3rd/4th gear synchronizer. **Top:** Synchronizer hub identified by groove on 4th gear side (white arrow) or on splines (black arrow). **Bottom:** When assembling, chamfer on hub inner splines (black arrow) faces 3rd gear, groove at outer splines (white arrow) faces 4th gear.

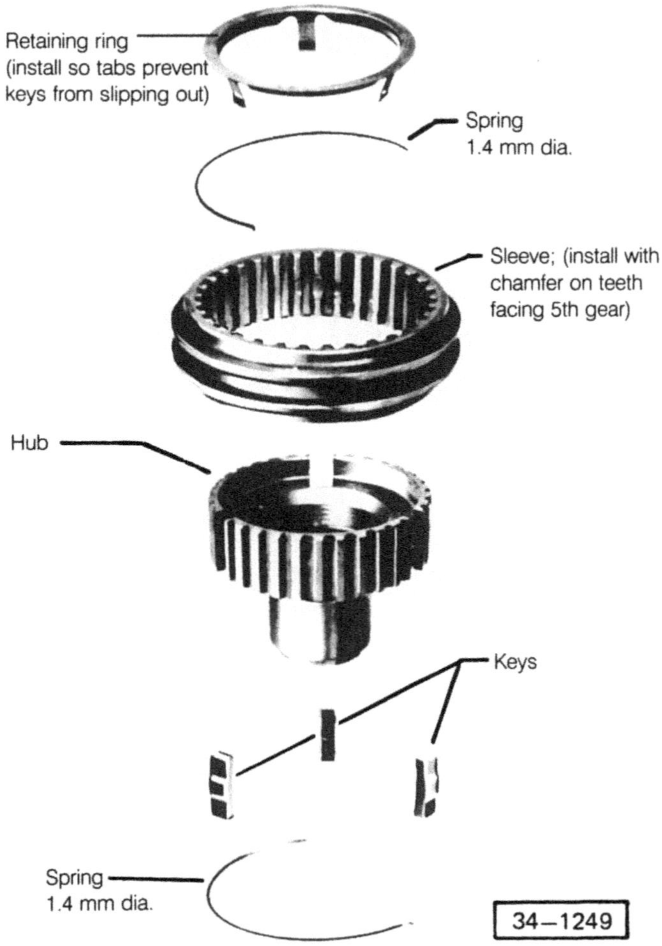

Fig. 7-26. 5th gear synchronizer.

Bearings and Races

In addition to the bearings pressed onto the main shaft and pinion shaft, there are needle bearings in the transmission case and gear carrier housing, as well as the bearing races and shims. While the main shaft ball bearing is removed with the main shaft when the transmission is disassembled, it must be pressed off the main shaft as shown in Fig. 7-27 before reassembly. It is then installed in the transmission case as shown earlier in Fig. 7-15.

Remove the main shaft needle bearing and press the new bearing into place as shown in Fig. 7-28.

The pinion shaft small bearing outer race is also pulled out as shown in Fig. 7-29. The bearing should be adjusted as described below before final installation of the race.

Drive the pinion shaft needle bearing out from the outside of the case as shown in Fig. 7-30, and drive the new bearing in from the inside.

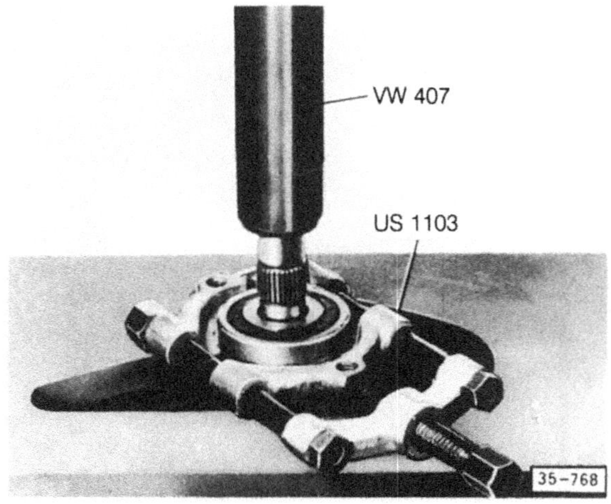

Fig. 7-27. Main shaft ball bearing being pressed off main shaft. For reassembly, ball bearing is first installed in transmission case, not on main shaft.

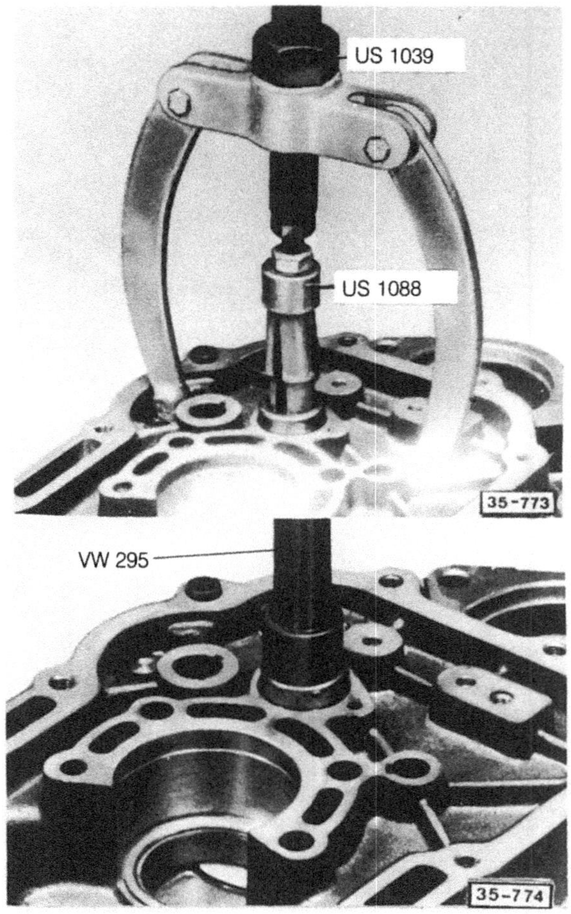

Fig. 7-28. Main shaft needle bearing being removed (**top**), and installed (**bottom**).

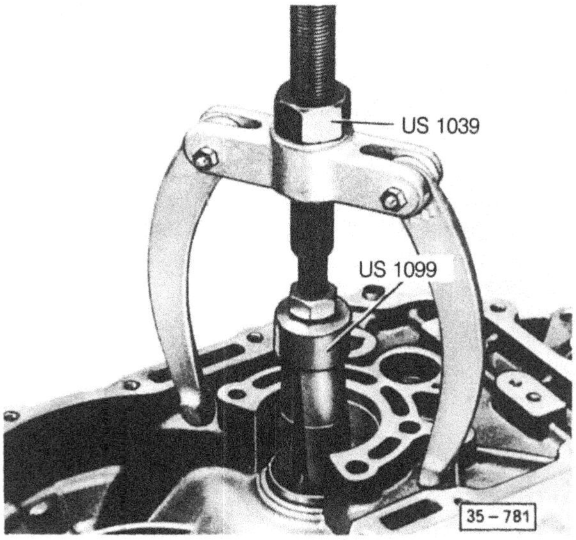

Fig. 7-29. Pinion shaft small-bearing outer race being pulled out. Check adjustment of bearing as described below before final installation of race.

Fig. 7-30. Pinion shaft needle bearing being removed from (**top**), and installed in (**bottom**) transmission case.

Remove and install the pinion shaft large bearing as shown in Fig. 7-31. Remove and install the pinion shaft small bearing as shown in Fig. 7-32. Both bearings must be heated to approximately 212°F (100°C) for installation.

CAUTION —

The small and large pinion bearings cannot be reused after being removed.

The clutch pushrod seal is pried out of the pinion shaft. The transmission must be removed from the car, but the seal can be replaced with the transmission still assembled. Install the seal as shown in Fig. 7-33. The pushrod bushing can be driven out using a 10 mm (⅜ in.) rod (such as an old automatic transmission pump rod) inserted into the pinion shaft from the opposite end. Drive the new bushing in until it is fully seated.

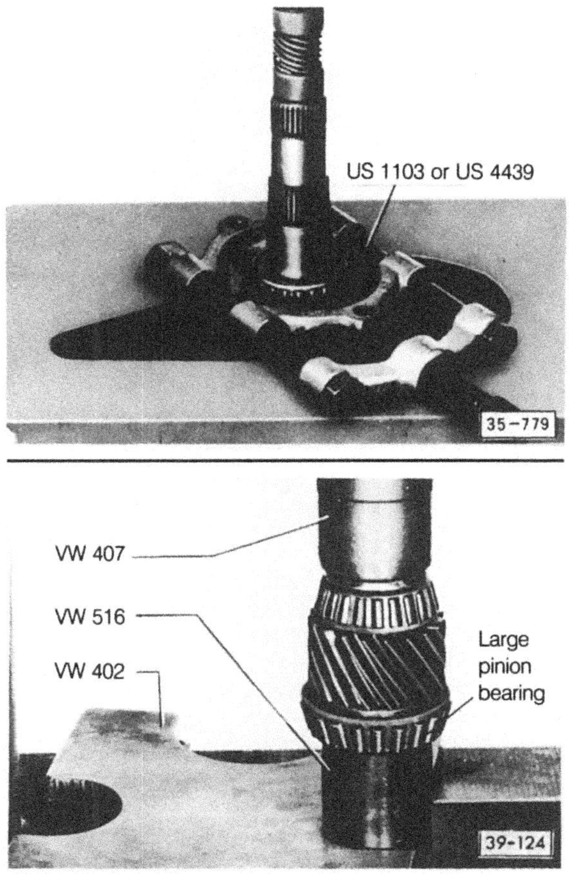

Fig. 7-31. Pinion shaft large bearing being removed (**top**) and installed (**bottom**). Heat new bearing to 212°F (100°C) before pressing it on.

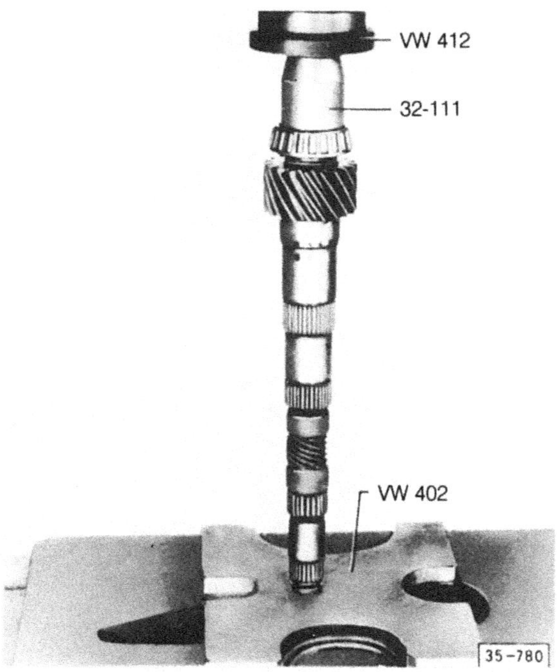

Fig. 7-32. Pinion shaft small bearing being removed (**top**) and installed (**bottom**). Heat new bearing to 212°F (100°C) before pressing it on.

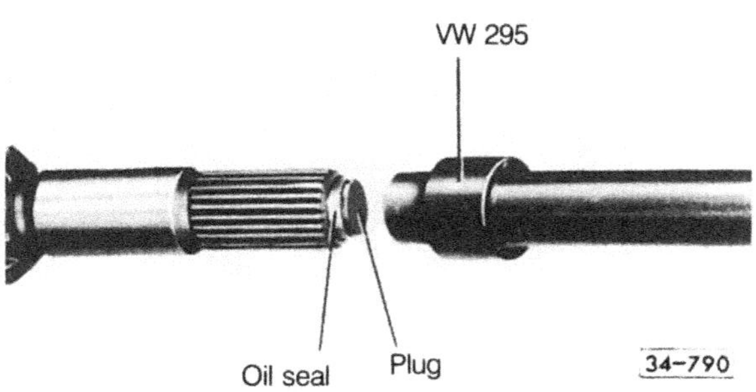

Fig. 7-33. Drive in clutch pushrod oil seal until plug supplied with new oil seal is flush, then remove plug. If plug is not on oil seal, install to depth of 0.8 to 1.3 mm (.03 to .05 in.).

Adjusting Pinion Shaft Bearings

The pinion shaft bearing preload must be adjusted if the gear carrier housing, the ring and pinion, or the pinion bearings are replaced. Adjust the pinion shaft with only the large and small pinion bearings installed on the shaft.

The procedures described here suggest or require the use of special tools. Unless otherwise specified, the numbers which appear in the illustrations identify Volkswagen special tools by their Volkswagen tool numbers.

To adjust:

1. With the small pinion bearing outer race removed, place a 0.65 mm shim in the gear carrier housing and then press in the outer race as shown in Fig. 7-34.

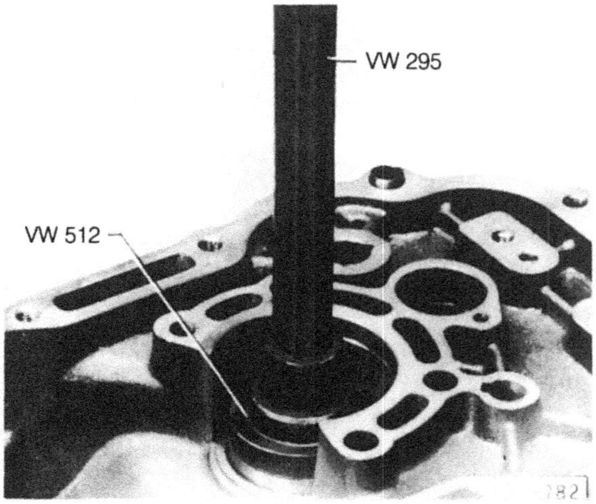

Fig. 7-34. Installing small pinion bearing outer race prior to adjustment. Temporary shim (0.65 mm) is installed between race and gear carrier housing to allow measurement of preload and calculation of correct shim thickness.

2. Temporarily install the pinion shaft and the large-bearing retaining plate, and torque the plate bolts to 40 Nm (30 ft. lb.).

3. Install a dial indicator as shown in Fig. 7-35 and set the dial to show 1 mm of preload.

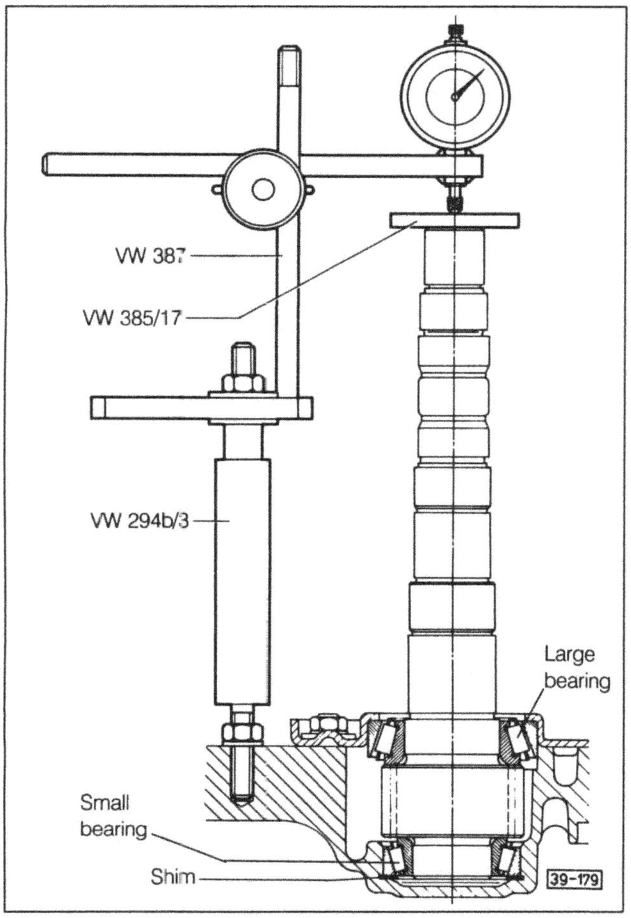

Fig. 7-35. Dial indicator set-up for testing pinion shaft bearing preload.

4. Move the pinion shaft up and down and note the amount of movement indicated by the dial indicator. The shim size needed to give the correct bearing preload is equal to the amount of movement, plus the thickness of the installed shim (0.65 mm), plus 0.20 mm. For example, if the pinion shaft moves up and down 0.30 mm, and the thickness of the installed shim is 0.65 mm, then the shim thickness needed for correct preload is 0.30 mm + 0.65 mm + 0.20 mm = 1.15 mm **Table e** lists the shims available to set the bearing clearance based on this calculation.

CAUTION —

Do not turn the pinion shaft when measuring its movement. This will cause the bearings to settle and give an incorrect reading.

NOTE —

Parts numbers are for reference only. Always rely on an authorized Volkswagen parts department for the most accurate and up-to-date parts information.

Table e. Shims For Adjusting Pinion Shaft Bearing Preload

Shim thickness (mm)	Part No.
0.65	020 311 391 P
0.70	020 311 391 Q
0.75	020 311 391
0.80	020 311 391 A
0.85	020 311 391 B
0.90	020 311 391 C
0.95	020 311 391 D
1.00	020 311 391 E
1.05	020 311 391 F
1.10	020 311 391 G
1.15	020 311 391 H
1.20	020 311 391 I
1.25	020 311 391 K
1.30	020 311 391 L
1.35	020 311 391 M
1.40	020 311 391 N

5. Remove the pinion shaft, and then remove the small bearing outer race as shown above in Fig. 7-29. Remove the 0.65 shim and install the shim of the correct thickness (calculated in step 4). Reinstall the pinion shaft.

6. As a final check of bearing preload, lubricate the bearings with transmission oil and check the pinion shaft turning torque as shown in Fig. 7-36. **Table f** lists turning torque specifications. Torque that is too high indicates inadequate clearance, while torque that is too low indicates too much clearance.

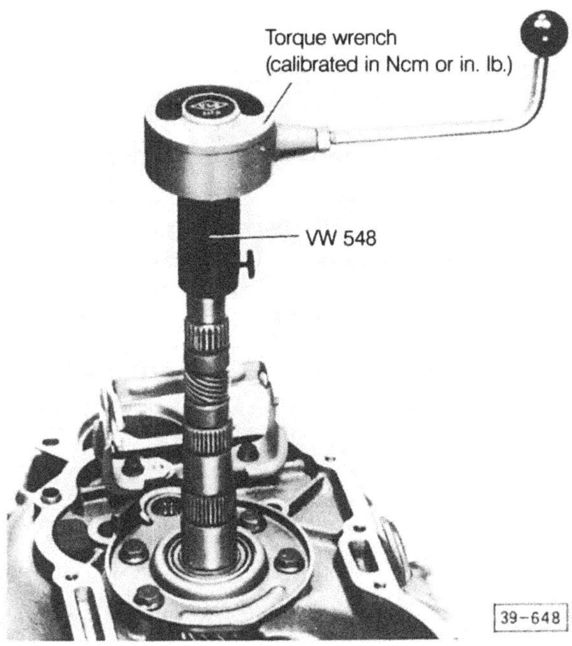

Fig. 7-36. Checking pinion shaft bearing preload by checking turning torque. Specifications are given in **Table f**.

Table f. Pinion Shaft Bearing Turning Torque

Bearings	Turning torque
New bearings	50 to 150 Ncm (4.4 to 13.3 in. lb.)
Used bearings	30 Ncm (2.7 in. lb.)

8. FINAL DRIVE

This heading covers repair and replacement of the final drive components of the transaxle, including the ring and pinion and the bearings. Repair procedures for the transmission portion of the transaxle are covered under **7. Transmission**. All the work described under this heading requires that the transaxle first be removed from the car. See **6.3 Removing and Installing Transaxle**.

CAUTION —
For the most part, the internal transaxle work covered here requires special tools and prior knowledge of proper rebuilding practices. Do not begin disassembly of the transmission or final drive if you do not have the tools or the experience to complete the job.

Many of the procedures described in this section of the manual suggest or require the use of special tools. Unless otherwise specified, the numbers which appear in the illustrations identify Volkswagen special tools by their Volkswagen tool numbers.

8.1 Final Drive Removal and Installation

Fig. 8-1 is an exploded view of the final drive portion of the transaxle. For access to the final drive components, the transmission must be fully disassembled as described in **7.1 Transmission Disassembly and Assembly**. Once the pinion shaft is removed, the final drive can be lifted out of the gear carrier housing. When installing the final drive, it may be necessary to adjust bearing preload as described in **8.2 Final Drive Disassembly and Assembly**.

8.2 Final Drive Disassembly and Assembly

The final drive gear set in the transaxle consists of the pinion gear, on the transmission pinion shaft, and the ring gear mounted to the differential. If the final drive gears are to be changed or replaced, the pinion shaft and ring gear must be replaced as a set. For information on removing, installing and adjusting the pinion shaft, see **7.2 Transmission Gear Train Disassembly and Assembly**.

Many of the procedures described in this section of the manual suggest or require the use of special tools. Unless otherwise specified, the numbers which appear in the illustrations identify Volkswagen special tools by their Volkswagen tool numbers.

Ring Gear

During production, the ring gear is mounted to the differential housing using rivets. Removal and installation of the ring gear requires replacing the original rivets with nuts, bolts, and washers. The repair kit containing these parts is Volkswagen part no. 171 498 088 A.

NOTE —
Part numbers are for reference only. always rely on an authorized Volkswagen parts department for the most accurate and up-to-date parts information.

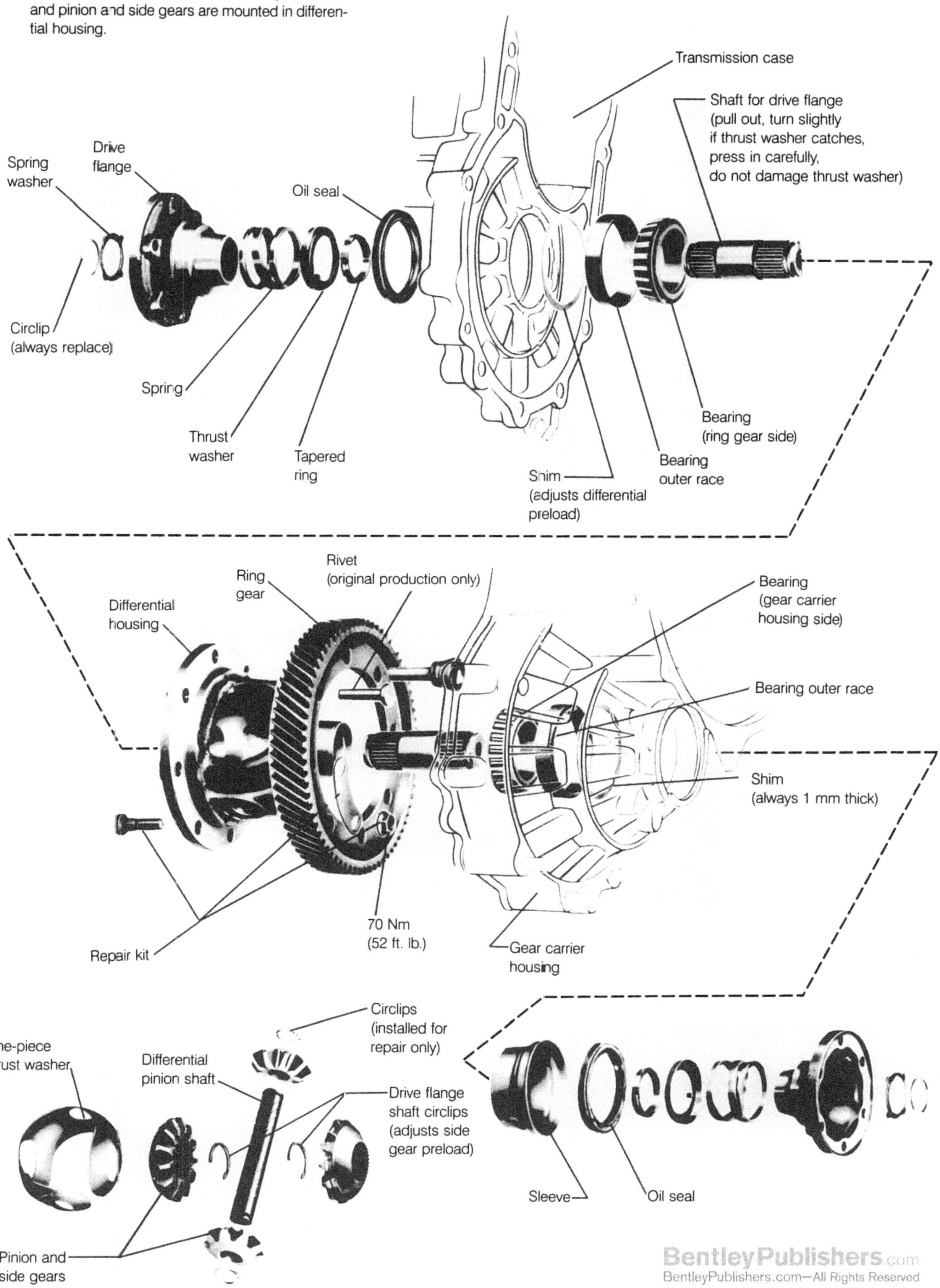

Fig. 8-1. Exploded view of final drive. Differential pinion shaft and pinion and side gears are mounted in differential housing.

To replace:

1. Remove the ring gear rivets by first using a 6 mm (1/4 in.) drill bit to drill halfway through the rivets as shown in Fig. 8-2. Next use a 12 mm (15/32 in.) bit to drill off the rivet heads, and drive out the rivets using a suitable drift.

> ***CAUTION —***
> *Protect the differential from metal chips. Clean the differential housing thoroughly after drilling.*

Fig. 8-2. Removing ring gear rivets. First use 6 mm (1/4 in.) bit to drill halfway, then use 12 mm (15/32 in.) bit to drill off rivet heads.

2. Press off the ring gear as shown in Fig. 8-3.

Fig. 8-3. Using press to remove ring gear.

3. Install two of the repair kit bolts in the differential housing as guide studs, then heat the ring gear to approximately 212°F (100°C) and press it onto the differential. Fig. 8-4 shows the correct orientation of the bolts with the differential housing. Tighten the bolts to 70 Nm (52 ft. lb.).

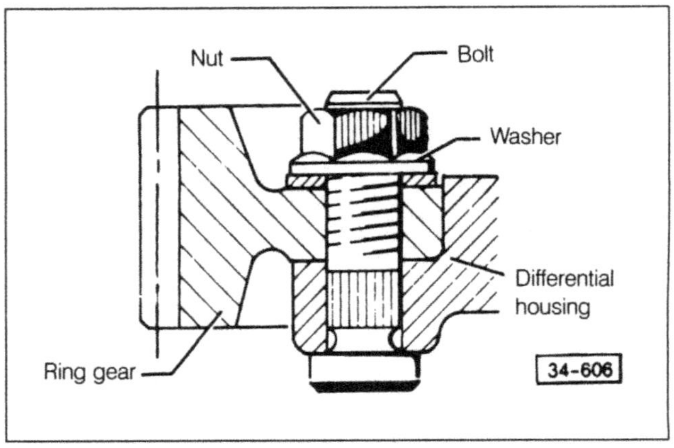

Fig. 8-4. Installation of repair kit bolts for mounting ring gear.

Differential Pinion Gears and Side Gears

In production, the differential pinion shaft is pressed in place during assembly. If the shaft or gears are removed and replaced, circlips must be installed to secure the shaft. The differential pinion shaft and the pinion and side gears are available as individual replacement parts. They are also available as a complete set, Volkswagen part no. 020 498 081 A.

> **NOTE —**
> Part numbers are for reference only. Always rely on an authorized Volkswagen parts department for the most accurate and up-to-date parts information.

To replace:

1. Remove the drive flange shaft circlips and remove the drive flange shafts.

2. Drive the differential pinion shaft out of the differential and remove the pinion and side gears, and the one-piece thrust washer.

3. Lubricate the thrust washer with transmission oil, then install the washer, the pinion gears, the side gears and the differential pinion shaft. Secure the shaft using new circlips at each end.

4. Install the drive flange shafts.

5. Adjust the shafts by pressing them against the differential pinion shaft while pressing the side gears against the differential housing, then install the circlips. See Fig. 8-5. Use the largest circlip that does not jam the side gears. Two different size circlips are available. One is 2.0 mm thick, part no. 020 409 299. The other is 2.3 mm thick, part no. 020 409 299 A.

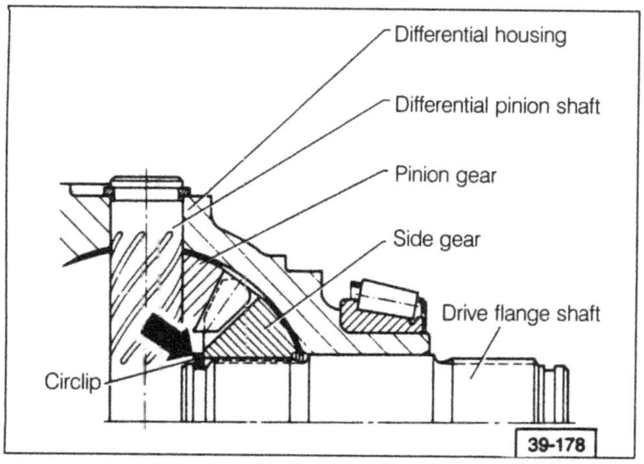

Fig. 8-5. Cutaway view of differential showing circlip (arrow) used to adjust drive flange shafts.

Bearings and Races

The final drive bearings and bearing races are matched sets and must not be interchanged. Remove the bearings as shown in Fig. 8-6, and install them as shown in Fig. 8-7. Remove the bearing races as shown in Fig. 8-8, and install them as shown in Fig. 8-9.

Fig. 8-6. Puller set up for removal of final drive bearings.

Fig. 8-7. Final drive bearings being pressed onto differential housing.

Fig. 8-8. Final drive bearing race being removed.

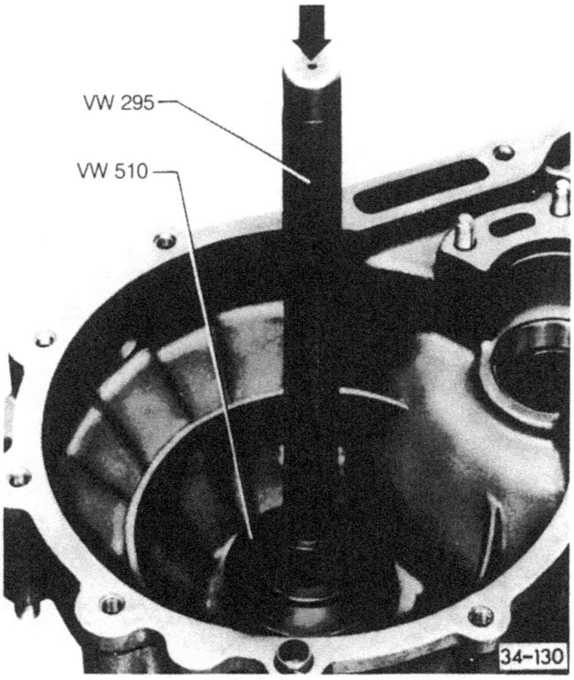

Fig. 8-9. Final drive bearing race being driven in.

Adjusting Final Drive Bearings

The final drive bearing preload must be adjusted whenever the transmission case, gear carrier housing, differential housing, or differential bearings are replaced.

To adjust:

1. If the bearing race is not installed in the gear carrier housing, install the shim and then drive in the bearing race as shown above in Fig. 8-9.

    CAUTION —
    Do not interchange the bearings and their races. They are matched sets.

    NOTE —
    The bearing race shim in the gear carrier housing is always 1 mm thick.

2. Without using a shim, install the bearing race into the transmission case.

3. Install the differential in the gear carrier housing and, using a new gasket, assemble the housing and the case. Install and torque five of the transmission case bolts to 25 Nm (18 ft. lb.).

4. Install a dial indicator as shown in Fig. 8-10 and set the indicator to show a preload of 1 mm.

Fig. 8-10. Dial indicator set up for testing final drive bearing preload.

5. Move the final drive up and down and note the amount of movement indicated by the dial indicator. The shim size needed to give the correct bearing preload is equal to the amount of movement plus 0.40 mm. For example, if the final drive moves up and down 0.90 mm, the shim thickness needed for correct preload is 0.90 mm + 0.40 mm = 1.30 mm. It may be necessary to use two shims to get the correct shim thickness and preload. **Table g** lists the shims available to set the preload based on this calculation.

    NOTE —
    Do not turn the differential when measuring its movement up and down. This will cause the bearings to settle and give an incorrect reading.

Table g. Shims For Adjusting Final Drive Bearing Preload

Shim thickness (mm)	Part No.
0.15	020 409 231
0.30	020 409 231 A
0.60	020 409 231 D
0.70	020 409 231 E
0.80	020 409 231 F
1.00	020 409 298

NOTE —

Part numbers are for reference only. Always rely on an authorized Volkswagen parts department for the most accurate and up-to-date parts information.

6. Disassemble the transmission case and the gear carrier, then drive out the bearing race in the transmission case. Install the shim of the correct thickness, as determined in step 5, and reinstall the race.

NOTE —

If two shims are being used, install the thicker one first into the transmission case.

7. As a final check of bearing preload, reassemble the transmission case and gear carrier as described above and check the final drive turning torque as shown in Fig. 8-11. **Table h** gives turning torque specifications. Torque that is too high indicates inadequate clearance. Torque that is too low indicates too much clearance.

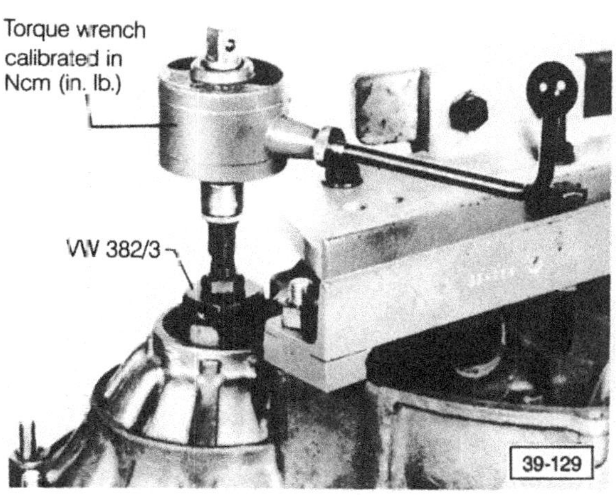

Fig. 8-11. Checking final drive bearing preload by checking turning torque. Specifications are given in **Table h**.

Table h. Final Drive Bearing Turning Torque

Bearings	Turning torque
New bearings	120 to 350 Ncm (11 to 31 in. lb.)
Used bearings	Minimum 30 Ncm (2.7 in. lb.)

9. TECHNICAL DATA

I. Tightening Torques

Shift rod to selector lever (clamp bolt and nut)	20 Nm (15 ft. lb.)
Shift lever retaining plate to shift lever housing (bolt)	8.5 Nm (74 in. lb.)
Gearshift lever housing to body (bolt)	23 Nm (17 ft. lb.)
Relay shaft bracket to body (nut)	20 Nm (15 ft. lb.)
Relay lever to relay shaft (nut)	20 Nm (15 ft. lb.)
Selector shaft lever to selector shaft (nut)	13 Nm (10 ft. lb.)
Pressure plate to crankshaft	30 Nm (20 ft. lb.) plus $\frac{1}{4}$ turn (90°)
Flywheel to pressure plate (bolt)	20 Nm (15 ft. lb.)
Clutch cable locknut to cable sleeve (nut)	4 to 6 Nm (36 to 48 in. lb.)
Selector shaft cover	50 Nm (37 ft. lb.)
Driveshaft to drive flange (socket-head bolt)	45 Nm (33 ft. lb.)
Transaxle to engine (M12 bolt)	80 Nm (59 ft. lb.)
Starter to transaxle (M10 bolt)	60 Nm (44 ft. lb.)
Selector shaft detent plunger to transaxle housing (locknut)	20 Nm (15 ft. lb.)
5th gear lockout plunger to transaxle housing (locknut)	20 Nm (15 ft. lb.)
Reverse gear/upshift light switch to transaxle (bolt)	10 Nm (89 in. lb.)
5th gear/release housing to transaxle case (bolt)	25 Nm (18 ft. lb.)
5th gear synchronizer to main shaft (bolt)	150 Nm (111 ft. lb.)
Transmission case cover (bolts)	25 Nm (18 ft. lb.)
Pinion shaft bearing retaining plate to transmission case (bolt)	40 Nm (30 ft. lb.)
Main shaft ball bearing clamping plate to transmission case (bolt)	15 Nm (11 ft. lb.)
Reverse gear relay lever supports to gear carrier housing (bolt)	25 Nm (18 ft. lb.)
Transmission case to gear carrier housing (bolt)	25 Nm (18 ft. lb.)
Reverse gear shaft lock bolt	30 Nm (22 ft. lb.)
Ring gear to differential housing (bolt) (repair kit only)	70 Nm (52 ft. lb.)

II. Transmission Codes and Gear Ratios

Application	Cabriolet, Scirocco 1985-1989 (except 16V)	Scirocco 16V	Cabriolet from 1990
Type no.	020	020	020
Code	9A, ACD	AGB, 2Y	AUG
Gear ratios			
1st	3.45	3.45	3.45
2nd	2.12	2.12	1.94
3rd	1.44	1.44	1.44
4th	1.13	1.13	1.13
5th	0.89	0.91	0.89
Final drive	3.67	3.67	3.67

III. Clutch Specifications

Clutch lever freeplay . . . . 12 mm ($^{15}/_{32}$ in.)
Clutch pedal freeplay . . . . 6 mm ($^{1}/_{4}$ in.)
Clutch disc runout (maximum) . . . . 0.5 mm (.02 in.)
Pressure plate taper (max. inward) . . . . 0.2 mm (.008 in.)

IV. Synchronizer Ring Wear Specifications

Ring and gear	Gap thickness –new	Wear limit
1st and 2nd gear	1.1 to 1.7 mm (0.04 to 0.07 in.)	0.5 mm (0.02 in.)
3rd gear	1.15 to 1.75 mm (0.045 to 0.068 in.)	
4th and 5th gear	1.3 to 1.9 mm (0.05 to 0.07 in.)	

V. Pinion Shaft Bearing Turning Torque

	Turning torque
New bearings	50 to 150 Ncm (4.4 to 13.3 in. lb.)
Used bearings	30 Ncm (2.7 in. lb.)

VI. Final Drive Bearing Turning Torque

	Turning torque
New bearings	120 to 350 Ncm (11 to 31 in. lb.)
Used bearings	minimum 30 Ncm (2.7 in. lb.)

Section 9

AUTOMATIC TRANSMISSION

Contents

Automatic Transmission

Introduction

The automatic transmission is housed in a cast aluminum alloy case that is attached by studs to a separate final drive housing. Taken as a unit, the transmission and final drive assembly are called the transaxle. Though the transaxle with automatic transmission operates on entirely different principles from those of the transaxle with manual transmission, the constant velocity joints and the front wheel drive axles of the two transaxles are identical. Please consult **SUSPENSION AND STEERING** for repairs related to the drive axles and constant velocity joints.

Cleanliness and a careful approach are very important when repairing the transaxle. Become familiar with the procedure before beginning a job and make sure that you have the necessary tools—particularly for procedures with specifications given in metric units only. Specifications that lack U.S. equivalents require that the related work be carried out using metric tools and instruments.

With normal use and regular maintenance, the transaxle with automatic transmission is very reliable and does not require internal repairs. Repairs to the internal sections of the transaxle require special knowledge and equipment, and as such are beyond the scope of this manual. In the event that internal repairs or overhaul are required, you may be able to save service time and expense by removing the transaxle for repair by a properly equipped and qualified repair shop. The procedures for transaxle removal and installation, as well as the necessary external adjustments are included in this manual. We recommend that the transaxle be thoroughly cleaned on the outside and taken to the shop as is—partial disassembly will not save money and may complicate repairs. We especially urge you to consult an authorized Volkswagen dealer before beginning any work on a car that may be subject to Volkswagen's warranty coverage.

1. GENERAL DESCRIPTION

The transaxle with automatic transmission is shown in Fig. 1-1. At the right end of the transaxle is the final drive housing, which has an integral bellhousing for the torque converter. Inside the final drive housing are the differential, the ring and pinion gearset, and the intermediate gear—which is between the drive pinion and the ring gear. The governor for the automatic transmission is located beneath the round cover atop the final drive housing.

The final drive housing, containing hypoid gear oil for lubricating the final drive gears, is completely sealed off from the transmission case, which contains automatic transmission fluid (ATF). The transmission case contains the ATF pump, the hydraulic controls, and the planetary gear system. The planetary gear system is lubricated solely by the ATF that is circulated through the transmission by the ATF pump. The ATF does not circulate unless the engine is running, so the transmission parts are only partially lubricated when the engine is turned off.

CAUTION —

Towing a Volkswagen with an automatic transmission while the front wheels are on the ground can cause damage due to lack of lubrication. Volkswagen recommends that cars with automatic transmission be flat-towed (four wheels on the ground) for no more than 30 miles (48 km), at no more than 30 mph (48 km/h). Do not flat-tow the car backward.

1.1 Automatic Transmission 010

All automatic transmissions used on the Volkswagens covered by this manual are type number 010 and are hydraulically controlled with three forward speeds. The automatic transmission can best be described and understood by dividing it into five subsystems. These are the torque converter, the ATF pump, the planetary gear system, the hydraulic controls, and the final drive. See Fig. 1-2.

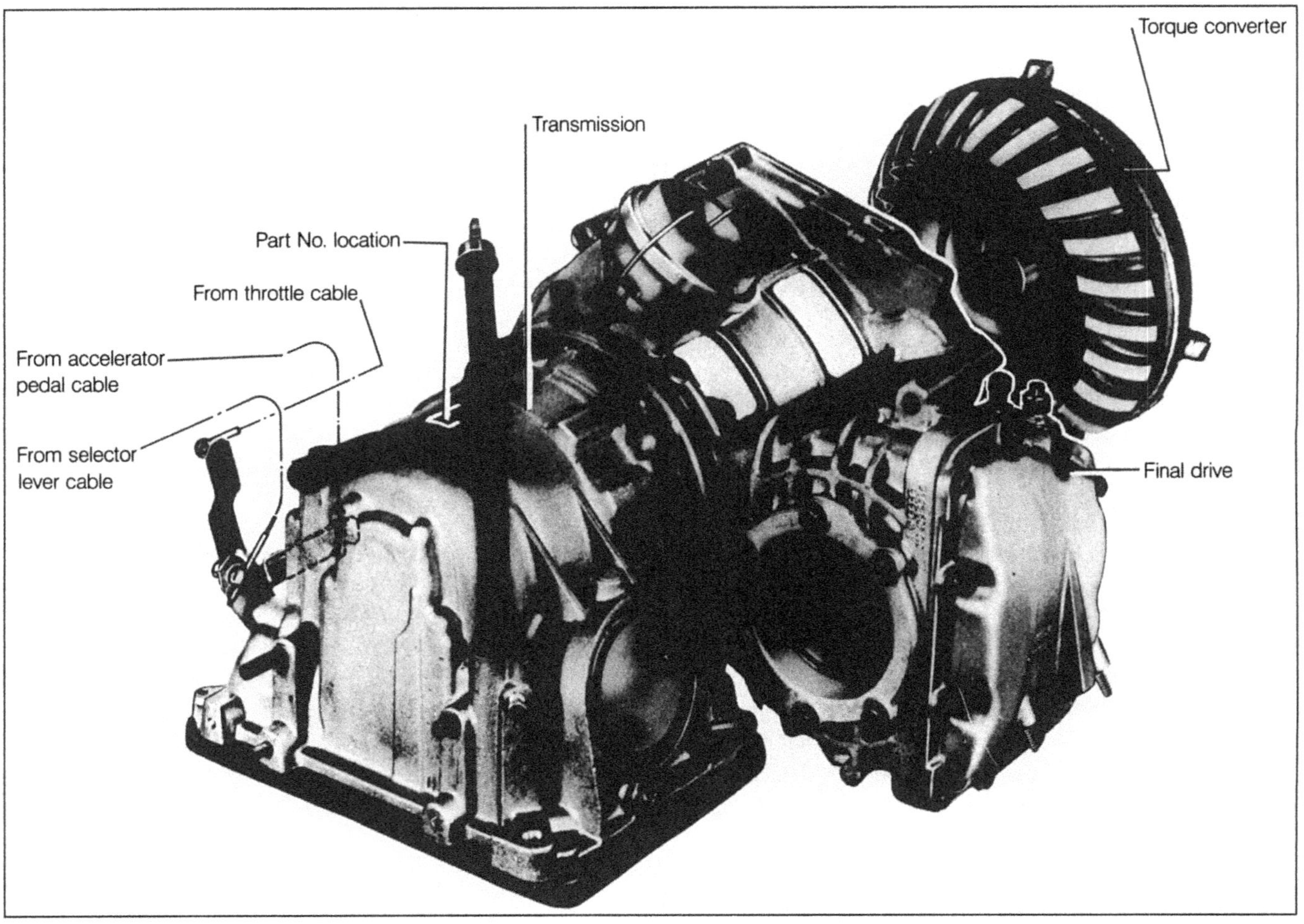

Fig. 1-1. Transaxle with automatic transmission.

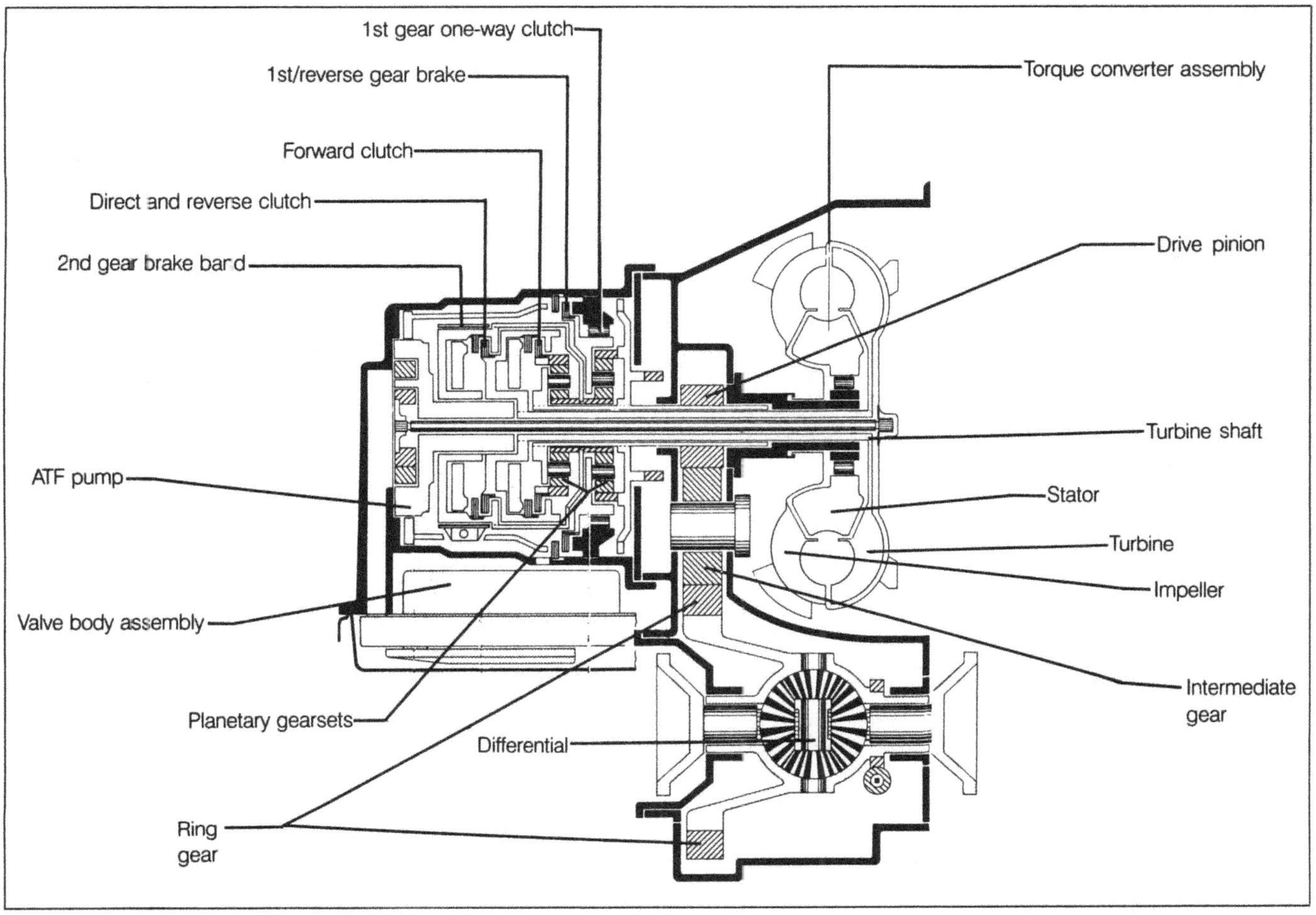

Fig. 1-2. Cross-section of transaxle with automatic transmission showing orientation of internal parts.

Torque Converter

The torque converter is a doughnut-shaped assembly located between the engine and the transaxle inside the bellhousing. The torque converter transmits engine output to the transmission, and also multiplies engine torque at low vehicle speeds. The torque converter is a fluid coupling. Curved impeller vanes inside the housing set up a flow of ATF that drives another vaned wheel opposite it, called the turbine. The turbine drives a hollow shaft that transmits power to the planetary gear system in the transmission. The central set of vanes in the torque converter is called the stator. At low speeds, a one-way clutch keeps the stator stationary so that its curved vanes will redirect the flow of ATF in such a way that torque is multiplied. At higher speeds, when there is no longer a speed differential between the impeller vanes and the turbine, the stator freewheels and torque multiplication ceases.

ATF Pump

The automatic transmission fluid (ATF) circulates through the transmission under pressure. The ATF pump that creates this pressure is located in the extreme left-hand end of the transaxle case, the end farthest from the engine. A long pump driveshaft passes through the center of the hollow turbine shaft to drive the ATF pump. The splined pump driveshaft is connected directly to the torque converter housing, so ATF circulates whenever the engine is running.

Planetary Gear System

A torque converter alone cannot supply the torque multiplication needed for all driving conditions. The torque converter therefore drives through a planetary gear system which can operate at different drive ratios. The planetary gear system consists of two clutches, two brakes, and two planetary gearsets—one gearset for forward operation of the vehicle and the other gearset for reverse.

The two hydraulically-operated multiple-disc clutches control the routing of power to the planetary gearsets. The clutch at the right-hand end of the transmission is called the forward clutch because it transfers power to the planetary gears in all forward gears. The clutch at the left-hand end of the transmission is called the direct and reverse clutch because it transfers power to the planetary gears only when the transmission is in direct drive (3rd or top gear) or reverse.

The first of the two brakes is the 2nd-gear brake band. It contracts around the direct and reverse clutch drum to provide 2nd gear. The second of the two brakes is the 1st/reverse gear brake. It is applied hydraulically to obtain reverse. This brake is also applied when the selector lever is at **1** in order to provide additional engine braking by locking the normally freewheeling 1st gear one-way clutch.

Hydraulic Controls

The hydraulic control system directs and regulates hydraulic pressure from the ATF pump to control shifting of the planetary gear system. Shifts are produced by the piston-type servo for the second gear brake band, by the ring-shaped piston for the multiple-disc 1st/reverse gear brake, and by the two ring-shaped clutch pistons.

Three primary control devices regulate the movement of the control valves in the automatic transmission's valve body assembly. The first is the manual valve, which is connected to the selector lever by a flexible cable. Moving the lever changes the setting of the valve and produces the necessary application of hydraulic pressure for the drive range selected. The second primary control device, the throttle pressure valve, is operated by the accelerator cable and makes the transmission responsive to engine speed and load. The third primary control device, the governor, controls ATF pressure relative to the output shaft rotational speed, making the transmission responsive to rear wheel speed.

Final Drive

The final drive consists of the drive pinion—which is mounted around the turbine shaft and the ATF pump driveshaft—the intermediate gear, the ring gear, and the differential. The teeth on the ring gear, the intermediate gear, and the drive pinion are helically-cut for quieter running. The differential allows the front wheels to turn at different speeds, as is necessary when turning.

1.2 Identification Codes and Specifications

There are two versions of the basic automatic transmission transaxle installed on the cars covered by this manual. The different versions are identified by the code letters TN or TNA. TN transaxles are installed on some early 1985 cars. All other cars with automatic transmissions covered by this manual are equipped with TNA transaxles. The two versions are nearly identical. The later TNA transaxle has larger drive flanges (which mate to larger CV joints) and some internal differences.

The transmission code letters and a transmission number corresponding to date of manufacture are located on the top of the bellhousing. The automatic transmission type number 010 is found atop the transmission case near the dip stick. See Fig. 1-3. Torque converter code letters are found on the converter as shown in Fig. 1-4. **Table a** lists transmission specifications for the automatic transmission models covered by this manual. If replacing the transaxle or the torque converter, the replacement part's code letters should correspond to those of the original part.

NOTE —
Replacement parts are available for installing the larger drive flanges on TN transaxles. Consult an authorized Volkswagen dealer parts department for more information.

Fig. 1-3. Automatic transmission type number (**A**) and code letters with date of manufacture (**B**). In example shown, code letters are EG. Numbers indicate date of manufacture, 15 is day, 08 is month, and 8 is last digit of year.

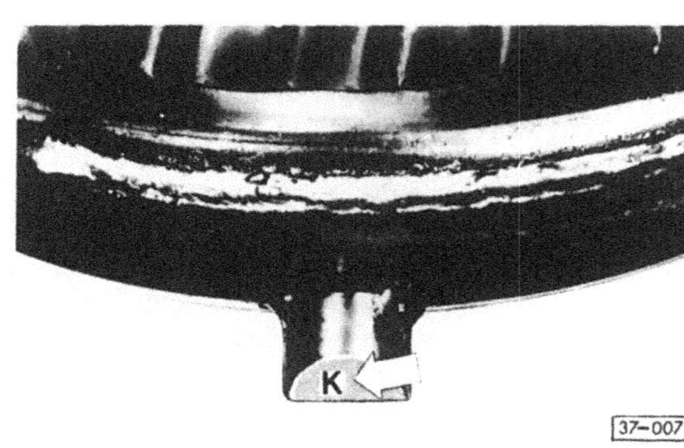

Fig. 1-4. Torque converter code letter location (arrow). Code letter "K" shown.

Table a. Automatic Transaxle Specifications

Engine type engine code (application)	**1.8 liter JH (early 1985)**	**1.8 liter JH, 2H (from late 1985)**
Transmission type number and code letter	010 TN	010 TNA
Torque converter code letter	K	K
Gear ratios		
1st gear	2.71	2.71
2nd gear	1.50	1.50
3rd gear	1.00	1.00
Final drive ratio	3.41	3.41

2. MAINTENANCE

The following routine maintenance should be performed at the time or mileage intervals prescribed in **LUBRICATION AND MAINTENANCE.** Maintenance procedures are provided there, as are the specifications and quantities for the required lubricants.

1. checking ATF level and adding ATF
2. checking kickdown operation
3. checking final drive hypoid oil level
4. changing ATF

3. TROUBLESHOOTING

This troubleshooting section applies to problems affecting the transaxle, including the torque converter, the automatic transmission, and the final drive. The basic functions of the automatic transmission are to circulate clean ATF under pressure, to provide the correct drive ratio in response to both the hydraulic and mechanical controls, and to transfer engine power to the front wheels.

The information in this section applies to the diagnosis of problems which are remedied either by external adjustment or by internal disassembly and repair. Most external adjustments to the transmission, as well as removal and installation of the transaxle, are covered in this manual; however, internal transmission repairs require specialized knowledge and equipment and should be left to an authorized Volkswagen dealer or other qualified automatic transmission repair shop.

3.1 Basic Troubleshooting Principles

In order for automatic transmission troubleshooting to provide meaningful results, the engine must be in good condition and properly tuned. See **ENGINE** for troubleshooting engine problems.

Inspect the transaxle for external damage, loose or missing fasteners, and for any obvious leaks. Check both the ATF and the final drive hypoid oil levels. The transmission and final drive use completely separate fluids. If the ATF level is too low and there is no sign of leakage, it may be because ATF is leaking into the final drive. Check for abnormally high final drive oil level. See **LUBRICATION AND MAINTENANCE**.

Many automatic transmission problems can be traced to an incorrect ATF level, to contaminated ATF, or to misadjusted transmission controls. Check for leaks, check the ATF level, and check to see if the fluid is dirty or has a burned odor. A burned odor indicates overheated fluid, which may be accompanied by burned clutches or brake bands. Friction material may be clogging the valve body passages.

Less severe automatic transmission problems may be solved merely by correcting the ATF level, by draining the ATF and refilling the transmission with new fluid as described in **LUBRICATION AND MAINTENANCE**, or by inspecting the adjustment and operation of the controls as described in **4.2 External Transmission Adjustments.**

Table b lists symptoms of automatic transmission and final drive problems, their probable causes, and recommended corrective action.

NOTE —
Probable causes and corrective actions may discuss specific internal transmission faults for reference only. The publisher recommends that internal transmission repair and overhaul be left to an authorized Volkswagen dealer or other qualified repair shop.

Table b. Automatic Transmission Troubleshooting

Symptom	Probable cause	Corrective action
1. No drive (car will not move)	**a.** ATF level low	**a.** Check and, if necessary, correct ATF level. See **LUBRICATION AND MAINTENANCE**
	b. Drive plate or torque converter not correctly installed	**b.** Check and correct drive plate and torque converter installation. See **ENGINE**
	c. Selector lever cable detached from manual valve	**c.** Connect and adjust cable. **4.2**
	d. Sticking main pressure valve, faulty ATF pump, or broken gear or shaft – no pressure	**d.** Test ATF pressure. **3.2** Remove, disassemble, and inspect transaxle. To remove and install, see **5.1**

continued on next page

Table b. Automatic Transmission Troubleshooting (continued)

Symptom	Probable cause	Corrective action
2. Irregular drive in all forward gears	**a.** ATF level low **b.** ATF pump pickup strainer partially clogged	**a.** Check and correct ATF level. See **LUBRICATION AND MAINTENANCE** **b.** Remove ATF pan. Clean or replace strainer. See **LUBRICATION AND MAINTENANCE**
3. No drive in 1st gear (shift position **D** or **2**)	**a.** Forward clutch or 1st gear one-way clutch faulty	**a.** Overhaul or replace transmission. To remove and install transaxle, see **5.1**
4. No drive in 1st gear (shift position **1**)	**a.** 1st/reverse gear brake defective	**a.** Overhaul or replace transmission. To remove and install transaxle, see **5.1**
5. No drive in 2nd gear (shift position **D** or **2**)	**a.** 2nd gear brake band incorrectly adjusted **b.** 2nd gear brake band damaged	**a.** Adjust band. **4.2** **b.** Overhaul or replace transmission. To remove and install transaxle, see **5.1**
6. No drive in 3rd gear (shift position **D**)	**a.** Direct/reverse clutch faulty	**a.** Overhaul or replace transmission. To remove and install transaxle, see **5.1**
7. No drive in reverse	**a.** 1st/reverse gear brake, direct/reverse clutch, or forward clutch faulty	**a.** Overhaul or replace transmission. To remove and install transaxle, see **5.1**
8. No upshift into 2nd or 3rd gear	**a.** Faulty governor or governor drive **b.** Governor dirty **c.** Shift valve(s) sticking **d.** Faulty valve body	**a.** Replace governor shaft **b.** Remove, clean, and reinstall governor **c.** Remove and clean valve body **d.** Repair or replace valve body
9. No downshift into 2nd or 1st gear	**a.** Governor dirty **b.** Shift valves sticking	**a.** Remove, clean, and reinstall governor **b.** Remove and clean valve body
10. Downshift to 1st gear delayed (jerky engagement)	**a.** 1st gear one-way clutch slipping	**a.** Overhaul or replace transmission. To remove and install transaxle, see **5.1**
11. Gear engagement delayed in upshift to all gears	**a.** ATF level low **b.** Valve body contaminated **c.** 2nd gear brake band incorrectly adjusted **d.** 2nd gear brake band faulty	**a.** Check and correct ATF level. See **LUBRICATION AND MAINTENANCE** **b.** Remove and clean valve body **c.** Adjust band. **4.2** **d.** Overhaul or replace transmission. To remove and install transaxle, see **5.1**
12. Gear engagement jerky when moving selector lever from **N** to **D**, or from **N** to **R**	**a.** ATF level low **b.** Engine idle speed too high	**a.** Check and correct ATF level. See **LUBRICATION AND MAINTENANCE** **b.** Adjust idle. See **FUEL SYSTEM**
13. Kickdown fails to operate (no full-throttle downshift)	**a.** Accelerator cable incorrectly adjusted	**a.** Adjust cable. **4.2**
14. Poor acceleration, poor high speed performance	**a.** Engine out of tune **b.** Throttle cable incorrectly adjusted **c.** Accelerator cable incorrectly adjusted	**a.** Tune up engine. See **LUBRICATION AND MAINTENANCE** **b.** Adjust cable. **4.2** **c.** Adjust cable. **4.2**
15. Transmission slips during hard cornering or after hard stops	**a.** ATF level low **b.** Clogged ATF pump strainer, incorrect strainer cover	**a.** Check and correct ATF level. See **LUBRICATION AND MAINTENANCE** **b.** Remove and clean strainer. Install new strainer cover as described in **LUBRICATION AND MAINTENANCE**
16. Chronic low ATF level with no visible leaks	**a.** ATF leaking into final drive housing	**a.** Check final drive hypoid oil level. See **LUBRICATION AND MAINTENANCE**. Remove transaxle, separate transmission and final drive, and replace oil seals, O-ring, or gaskets as required. **5.2**
17. ATF appears dirty, smells burnt	**a.** Contaminated ATF **b.** Damaged clutch and brake friction linings	**a.** Drain and replace ATF. Remove and clean ATF strainer. See **LUBRICATION AND MAINTENANCE** **b.** Overhaul or replace transmission. To remove and install transaxle, see **5.1**
18. Parking lock fails to engage	**a.** Parking lock mechanism damaged **b.** Selector lever cable misadjusted	**a.** Overhaul or replace transmission. To remove and install transaxle see **5.1** **b.** Adjust cable. **4.2**
19. Selector lever will not move	**a.** Selector lever cable binding **b.** Internal transmission failure	**a.** Remove and re-route cable. **4.1** **b.** Overhaul or replace transmission. To remove and install transaxle, see **5.1**

3.2 Diagnostic Tests

Although most internal transaxle problems cannot be fully diagnosed unless the unit is removed and disassembled, these tests may help further diagnose transaxle trouble prior to its removal. To obtain accurate test results, the ATF must be clean, in good condition (not previously overheated or burnt), and at the proper level.

Stall Speed Test

This test is used to check for faults in the torque converter and planetary gear system when there is no other apparent cause for poor performance and acceleration. The test results are meaningless if the engine is not running properly. A precise tachometer must be used for the rpm measurements, as dashboard instruments are not sufficiently accurate.

CAUTION —
The stall speed test should be as short as possible, and should never extend beyond 20 seconds maximum. Prolonging the test may overheat the transmission and damage the seals or internal components.

To test:

1. Drive the vehicle to warm the engine and transmission to normal operating temperature.
2. Connect a tachometer according to the instrument manufacturer's instructions, so that it can be read from the driver's seat, then start the engine.
3. Set the parking brake and depress the foot brake firmly to hold the vehicle stationary.
4. Place the selector lever in **D**.
5. While holding the car stationary with the brakes, floor the accelerator for no more than 20 seconds. Notice the tachometer readings. The rpm should increase, and then hold steady. For transmission code TN the rpm should be between 2200 and 2500. For transmission code TNA, the rpm should be between 2320 and 2570. Maximum rpm achieved under these conditions is the stall speed.

NOTE —
It is normal for the stall speed to be 125 rpm lower than specified for each 1000 meters (3200 ft.) above sea level. High ambient air temperature will also cause a slight drop in stall speed.

A stall speed that is a few hundred rpm below the specified range is probably due to reduced engine performance. A stall speed that is 400 rpm more below the specified range indicates a faulty torque converter.

If the stall speed is above the specified range, there is slippage in the forward clutch or in the 1st gear one-way clutch. To determine which is faulty, repeat the test, this time with the selector lever in **1**. If the stall speed is now within specifications, then the 1st gear one-way clutch is defective. If the stall speed rpm is still high, then the forward clutch is defective.

The test can also be made with the selector lever in **R**. If the stall speed is too high, it indicates slippage in either the direct/reverse clutch or in the 1st/reverse clutch.

All of these faults require that the transaxle be removed and disassembled for repair.

Pressure Test

A main pressure test will reveal internal leaks, sticking control valves, or other troubles in the hydraulic controls. Although the full-throttle part of this test can be performed while driving the car, whenever possible this test should be carried out on a dynamometer. The pressure gauge should have a range of at least 0 to 15 bar or 0 to 200 psi, and a hose long enough to allow it to be read from the passenger compartment. Engine idle speed must be correctly adjusted.

To test:

1. Connect the pressure gauge to the main pressure tap on the transmission, as shown in Fig. 3-1. Route the hose so that the gauge can be read from inside the car.

Fig. 3-1. Main pressure tap on transmission case, located on rear of transaxle near driver's side drive axle flange.

2. Drive the car to warm the engine and transmission to normal operating temperature. Stop the car and set the parking brake.

3. Place the selector lever at **D**. With car stationary and idling, the main pressure should be between 2.90 and 3.00 bar (42 and 43.5 psi).

4. Place the selector lever at **R**. With the car stationary and idling, the main pressure should be between 9.00 and 10.00 bar (130.5 and 145 psi).

5. Either on a dynamometer or during a road test, with the selector lever at **D**, and the speedometer indicating a speed over 25 mph (40 km/h), floor the accelerator. During full-throttle acceleration the main pressure should be between 5.85 and 5.95 bar (85 and 86 psi).

6. Shut off the engine. Disconnect the pressure gauge hose and reinstall the plug for the pressure tap. Torque the plug to 10 Nm (87 in. lb.).

Any pressure that is higher or lower than specification usually indicates a malfunctioning valve body and valves, probably due to contamination. The valve body can be removed, cleaned, and reinstalled with the transaxle in place by removing the ATF pan and strainer. However, it contains many precision parts which must be reassembled in their exact locations. Because of the complexity of the valve body assembly, the publisher recommends that these repairs be left to an authorized Volkswagen dealer or other qualified repair shop.

Low pressure may also indicate a worn ATF pump or internal ATF pump leaks past seals, gaskets, and metal mating surfaces. These repairs require that the transaxle be removed and disassembled.

4. CONTROLS

The automatic transmission controls include the selector lever cable, the accelerator cable, and the throttle cable. The selector lever cable allows the transmission to be manually shifted from inside the passenger compartment. The accelerator cable and the throttle cable make the transmission responsive to throttle position.

The selector lever cable activates the manual valve on the transmission. When the lever is moved out of Park, the parking mechanism is released and the manual valve is correctly positioned according to the gear selected.

The accelerator cable and the throttle cable are connected to an operating lever on the side of the transmission. This operating lever controls both the movement of the throttle and the engagement of the kickdown valve. The lever is connected to a shaft which runs through the center of the manual valve. On the opposite end of the shaft is the operating lever for the kickdown valve.

The accelerator and throttle cables are shown in Fig. 4-1. When the accelerator pedal is depressed the accelerator cable changes the position of the operating levers, which in turn change the position of the throttle valve via the throttle cable. The kickdown device allows for maximum acceleration when the accelerator pedal is fully depressed either by downshifting to a lower gear or by delaying the shift into a higher gear.

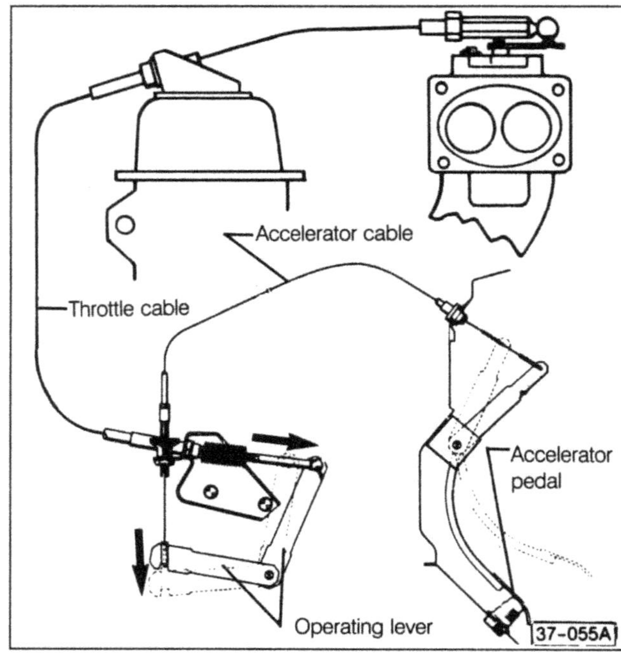

Fig. 4-1. Accelerator and throttle cable layout. Arrows show movement of operating levers as accelerator pedal is depressed.

4.1 Removing and Installing Control Cables

All three cables operate within a conduit or sheath. They are available for replacement only as a *complete assembly*, with conduit.

Selector Lever Cable

The selector lever cable and related parts are shown in Fig. 4-2. The neutral/park/reverse light switch prevents the car from starting when the selector lever is in gear. The switch also actuates the reverse lights when the shift selector is in reverse. For information on testing the switch, see **ELECTRICAL SYSTEM**.

To replace:

1. Disconnect the battery ground strap.

2. Place the selector lever in **P**. If the cable is broken, select Park by hand-moving the operating lever on the transaxle all the way to the left.

3. Remove the knob on the top of the selector lever by removing its set screw. Then pry off the indicator plate.

4. Remove the screws that fasten the console, and remove the console. Remove the cable from the selector lever by prying off the E-clip.

5. Remove the bolts that hold the selector lever mounting plate to the body, then loosen the locknuts that hold the cable conduit to the mounting plate and remove the cable from the plate.

6. Working in the engine compartment on the transmission end of the cable, loosen the nut on the clamping sleeve, then loosen and remove the cable from the transmission bracket. Pull the cable through to the engine compartment.

7. Installation of the cable is the reverse of removal. Adjust the cable as described below in **4.2 External Transmission Adjustments**.

Accelerator Cable and Throttle Cable

To remove the accelerator cable, place the selector lever in **P** and set the parking brake. Loosen the adjusting nut on the transmission operating lever to make the cable slack. Unhook the cable from the top of the accelerator pedal by prying off the clip on the cable end, and then pull the cable from the pedal. Unhook the cable from the transmission operating lever and pull it through the firewall into the passenger compartment.

To remove the throttle cable, place the selector lever in **P** and set the parking brake. Remove the cable end from the throttle valve by removing the retaining clip. Loosen the nuts mounting the cable to its bracket. Remove the retaining clip *from the transmission end* of the cable and pull the cable end from its ball socket.

Installation of either cable is the reverse of removal. Adjust the cables as described below in **4.2 External Transmission Adjustments**.

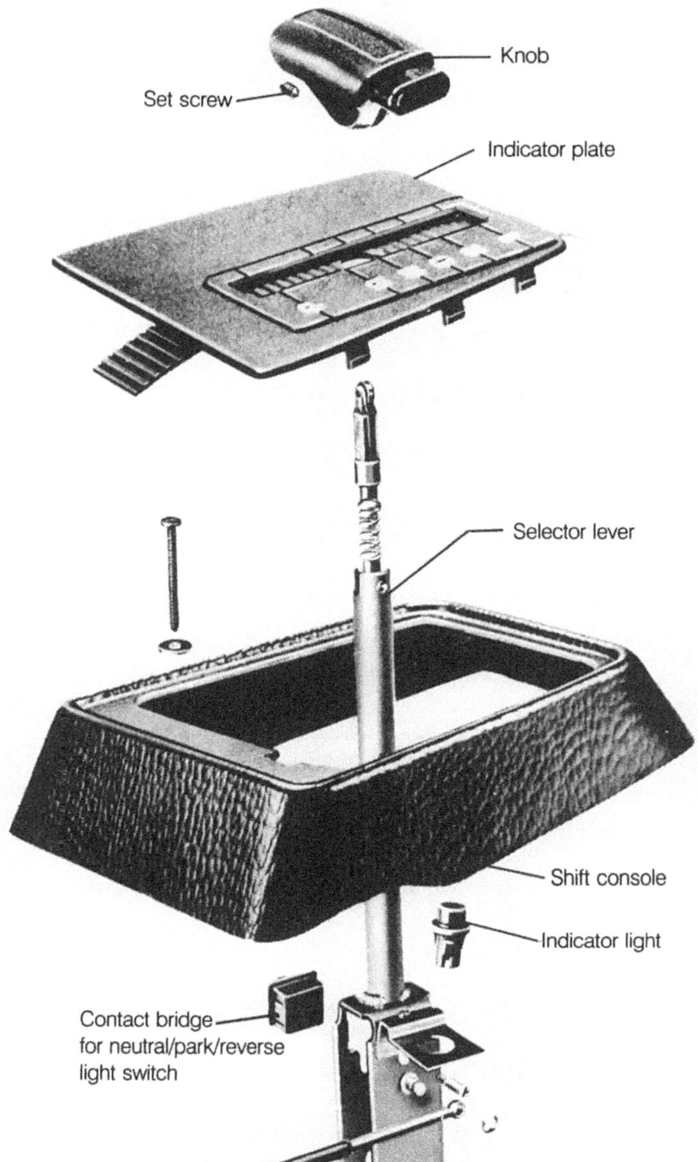

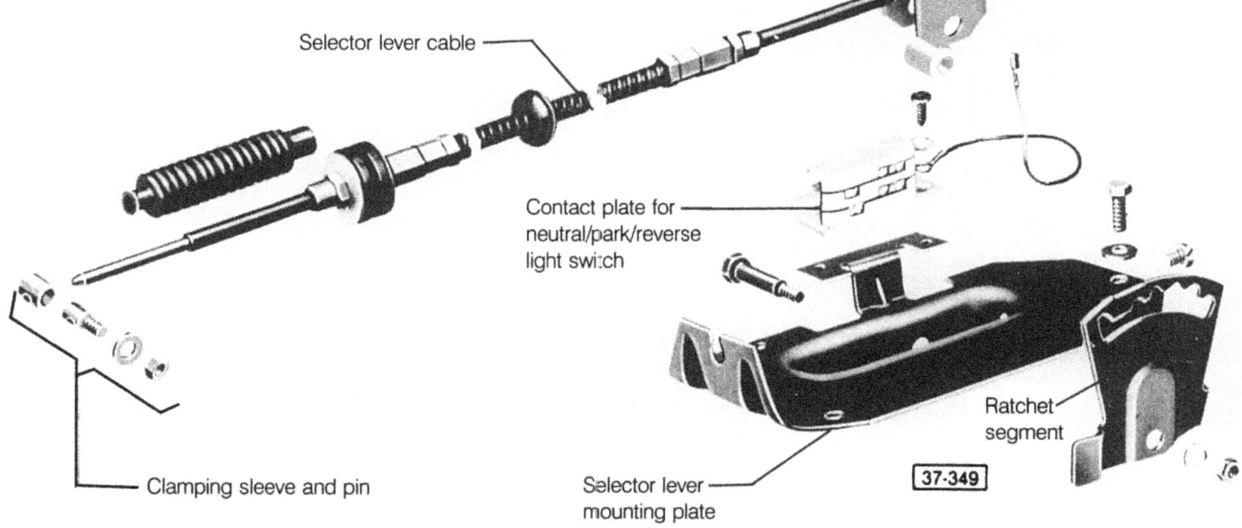

Fig. 4-2. Exploded view of automatic transmission selector lever and related parts.

4.2 External Transmission Adjustments

External adjustments to the automatic transmission include adjustments of the control cables and the 2nd gear brake band. The cables should be adjusted any time a cable is installed. A helper will be required to hold the accelerator pedal at full throttle while adjustments are made. These external adjustments may correct imprecise shifting without the need for removal of the transaxle or costly overhaul work.

Adjusting Selector Lever Cable

Adjust the selector lever cable with the shift lever in **P**. Remove the transmission protection plate. Loosen the nut clamping the cable to the actuating lever on the transmission. Make sure that the transmission lever is in the Park position by hand-pushing the actuating lever fully to the left. Tighten the nut to 8 Nm (71 in. lb.) and reinstall the protection plate. Check that the cable is correctly adjusted by starting the engine and moving the selector lever through all shift positions.

Adjusting Accelerator and Throttle Cables

Proper throttle control and transmission operation depend on accurate adjustment of the accelerator and throttle cables.

To adjust:

1. Place the selector lever in **P** and set the parking brake. Loosen the adjusting nut on the transmission's operating lever and disconnect the accelerator cable from the lever. See Fig. 4-3.

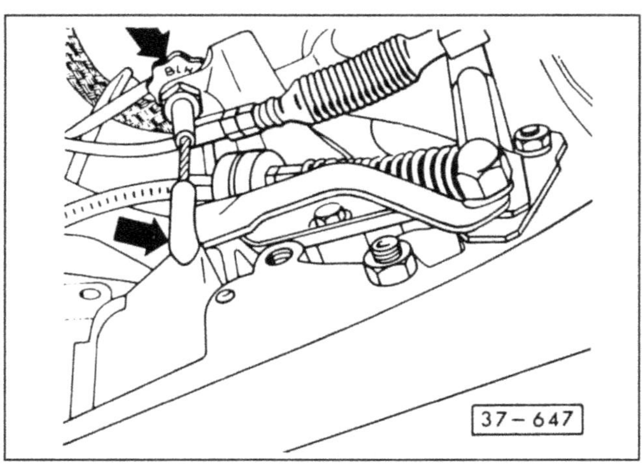

Fig. 4-3. Accelerator cable adjusting nut (top arrow) and accelerator cable end attachment to operating lever (bottom arrow).

2. Loosen the locknuts that hold the cable conduit to the bracket on the cylinder head cover. Make certain that the cable moves freely and that the throttle valve is fully closed.

3. Lightly tension the throttle cable as shown in Fig. 4-4 by adjusting the position of the locknuts until there is no freeplay in the cable. Make sure the throttle valve is still fully closed. Torque the locknuts to 10 Nm (87 in. lb.).

Fig. 4-4. Accelerator cable being adjusted. Move cable in direction of arrow before tightening locknuts **1** first and **2** last.

4. Connect the hooked end of the accelerator cable to the transmission operating lever. With the accelerator pedal depressed to its kickdown position (contacting the floor stop), push the transmission lever to its kickdown stop and turn the adjusting nut on the accelerator cable until there is no free play in the cable. See Fig. 4-5.

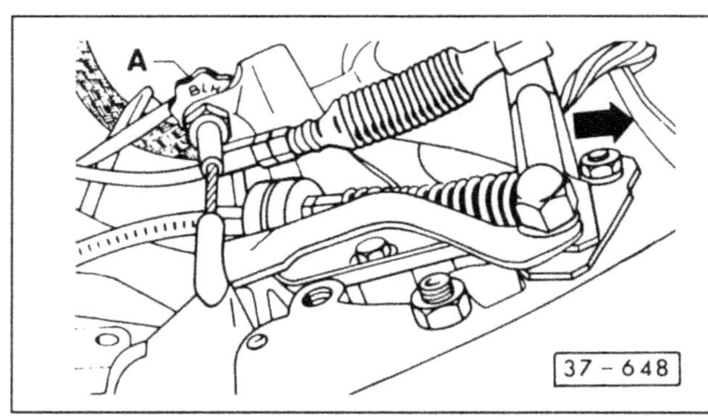

Fig. 4-5. Accelerator cable being adjusted. Move operating lever of automatic transmission to kickdown position (in direction of arrow until it contacts stop) and adjust cable to remove slack. Cable adjusting nut is at **A**.

5. Release and depress the accelerator pedal several times to check that the cable operates smoothly and that the operating lever rests against the kickdown stop when the pedal is fully depressed.

Adjusting 2nd Gear Brake Band

The 2nd gear brake band can be adjusted with the transaxle installed or removed. In either case, the transaxle must be in its normal horizontal position or the brake band may jam during adjustment. The 2nd gear brake band adjusting screw is located just to the left of the operating lever, as viewed from the front of the car.

To adjust:

1. Loosen the locknut and temporarily torque the adjusting screw to 10 Nm (87 in. lb.), as shown in Fig. 4-6, to center the brake band.

Fig. 4-6. 2nd gear brake band being adjusted.

2. Loosen the screw and retorque it to 5 Nm (44 in. lb.).
3. Back the screw off (counterclockwise) exactly 2 ½ turns. Hold the adjusting screw stationary and tighten the locknut, torquing it to 20 Nm (15 ft. lb.).

5. TRANSAXLE ASSEMBLY

Repair of internal transmission or final drive components requires that the transaxle be removed for disassembly. Once removed, the final drive and transmission can be separated and replaced or repaired individually. Remanufactured transaxles are available through authorized Volkswagen dealers, and offer some savings over the outright purchase of a new unit.

To correct fluid leaks, the drive flange oil seals and the governor O-ring are replaceable with the transaxle installed in *the vehicle.* The transaxle must be removed to replace any other faulty oil seals.

5.1 Removing and Installing Transaxle

In the procedure given here, the transaxle assembly is separated from the engine, supported on a floor jack, and taken out from below. Alternately, the engine/transaxle assembly may be removed from above and the transaxle then separated from the engine as described in **ENGINE**.

The procedure which follows requires a means of externally supporting the weight of the engine, a transmission jack or floor jack with a transmission adaptor, and jack stands designed for the purpose of solidly supporting the raised car. The engine support device is necessary because, in this procedure, three of the four engine mounts are removed.

CAUTION —
The car must be raised and supported off its front wheels during the removal procedure. Use extreme caution when working beneath the car.

To remove:

1. Disconnect the negative (−) and positive (+) battery cables from the battery, in that order. Disconnect the speedometer cable from the transaxle and plug the hole.
2. Set the parking brake and loosen the left and right front wheel lug bolts.
3. Raise the car and support it on jack stands, and remove the wheel lug bolts and the front wheels.

CAUTION —
Moving the car even a short distance with the drive axle nuts loosened or removed may cause wheel bearing damage and lead to premature wheel bearing failure.

4. Install a device to support the engine, as shown in Fig. 5-1.

Fig. 5-1. Engine being supported using Volkswagen special tool no. 10-222A.

5. Remove the nuts from the rear transaxle mount and remove the mount as shown in Fig. 5-2.
6. Place a transmission jack or a floor jack with a transmission adaptor beneath the transaxle. Raise the jack just enough to support the transaxle, but do not raise the engine.
7. Remove the upper transaxle to engine mounting bolts.

Fig. 5-2. Rear transaxle mount nuts (arrows).

8. Remove the bolts that hold the transmission protection plate and the torque converter cover plate, then remove the plates.

NOTE —
The transmission protection plate is at the lower left (driver's side) front of the transaxle; the torque converter cover plate is at the bottom of the bellhousing.

9. With the selector lever at **P**, remove the selector lever cable and the accelerator and throttle cables from the transmission as described in **4.1 Removing and Installing Control Cables**.
10. Remove the bolts from the left engine/transaxle mount and remove the mount.
11. Remove the starter as described in **ELECTRICAL SYSTEM**.
12. Detach the drive axles from the transmission drive flanges and suspend them out of the way using a stiff piece of wire.
13. Working through the lower opening in the bellhousing, remove the three bolts that hold the torque converter to the drive plate. See Fig. 5-3.

NOTE —
Use a socket wrench on the crankshaft sprocket bolt and turn the crankshaft clockwise to move the bolts into position for removal.

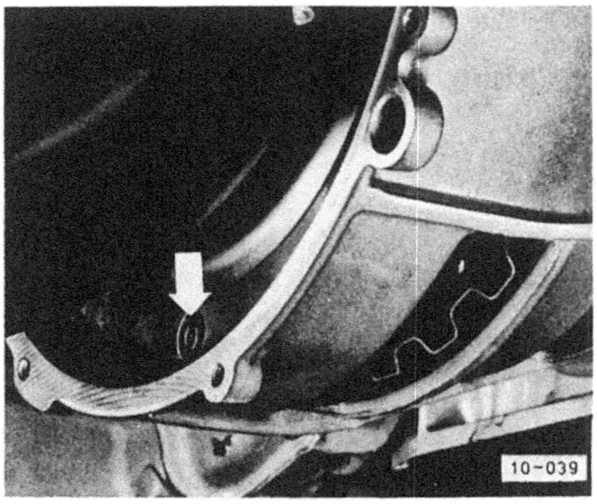

Fig. 5-3. Bolt (arrow) fastening torque converter to drive plate. Remove three bolts.

14. Remove the front transaxle mount.
15. Remove the lower engine-to-transaxle mounting bolt. Slide the transaxle off its dowels and lower it out of the car.

To install:

1. To avoid serious damage when installing the transaxle, make certain the torque converter has not slipped off of its support. Turn the converter back and forth to check that it has engaged its splines.
2. Position the transaxle until it can be pushed onto its locating dowels. Install and evenly tighten all the engine-to-transaxle mounting bolts until the transaxle is firmly in place. Torque the bolts and nuts to 80 Nm (59 ft. lb.).
3. Install all of the transaxle and engine mounts. Snug the nuts and bolts, but do not tighten them.
4. Install the driveshafts onto the drive flanges. Install the socket-head inner bolts and torque them to 42 Nm (30 ft. lb.).
5. Install the mounting bolts which attach the torque converter to the drive plate. Using a wrench on the front crankshaft bolt, turn the crankshaft (clockwise only) to align the bolt holes. Torque the bolts to 30 Nm (22 ft. lb.).
6. Remove the transmission jack and install the transmission protection plate and the torque converter cover plate, torquing the bolts to 24 Nm (18 ft. lb.).
7. Install the starter as described in **ELECTRICAL SYSTEM**.

8. Install the road wheels and lower the car to the ground. Torque the wheel lug bolts to 110 Nm (81 ft. lb.).

9. Reconnect the battery cables. Reconnect the speedometer cable to the transaxle and torque it to 10 Nm (87 in. lb.).

10. Align the engine/transaxle mounts, as described in **ENGINE**. With the mounts aligned and the nuts and bolts torqued, install and adjust the selector lever cable and the accelerator and throttle cables as described earlier in **4.2 External Transmission Adjustments**.

CAUTION —
If the ATF or the final drive hypoid gear oil have been drained, refill as described in **LUBRICATION AND MAINTENANCE** *before starting the engine or moving the car.*

5.2 Oil Seals

The torque converter oil seal and the drive flange oils seals can be replaced without extensive disassembly of the transaxle. The drive flange oil seals can be replaced without removing the transaxle from the car; torque converter oil seal replacement requires that the transaxle be removed. See **5.1 Removing and Installing Transaxle**.

Torque Converter Oil Seal

A leaking torque converter oil seal is often caused by a worn *bushing in the torque converter hub*. The bushing should always be checked if the seal is replaced. A worn bushing will promote rapid wear of the new seal.

To replace oil seal and check bushing:

1. Pry the seal from the final drive housing using a chisel as shown in Fig. 5-4.

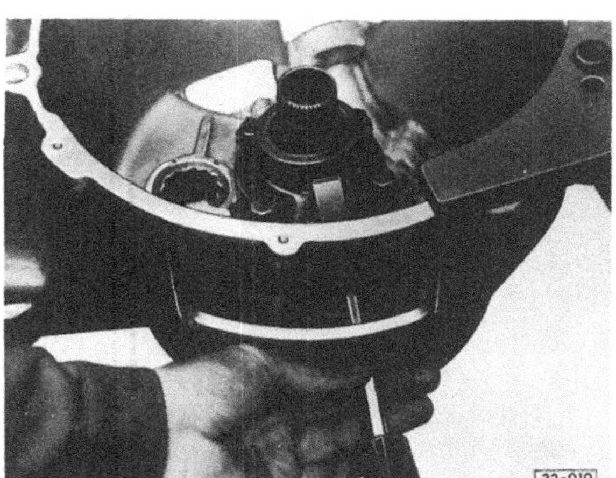

Fig. 5-4. Faulty torque converter seal being removed with chisel.

2. Coat the new seal in ATF and drive the seal into place as far as it will go. See Fig. 5-5.

CAUTION —
The seal is soft silicone rubber and easily damaged. Silicone seals must not contact gasoline or other solvents. Replace seals that are torn or have been exposed to solvents.

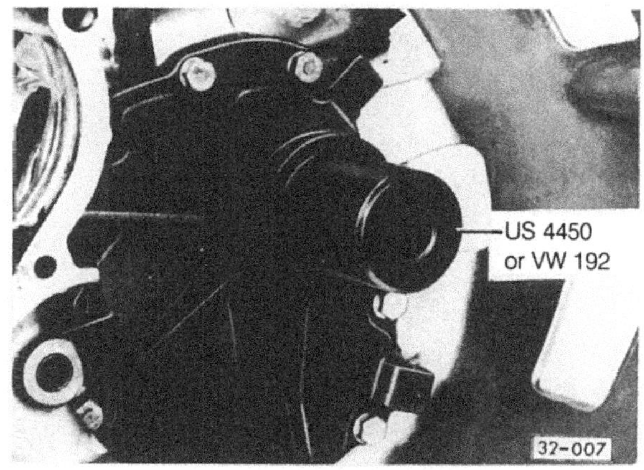

Fig. 5-5. Seal driver being used to install new torque converter seal onto one-way clutch support on final drive housing. Volkswagen special tool no. US 4450 or VW 192 shown. Use tool of approximate seal diameter to apply even installation pressure.

3. Check the torque converter oil seal hub for wear. See Fig. 5-6. Remove sharp edges and burrs with emery cloth. If the hub is deeply scored, the torque converter should be replaced.

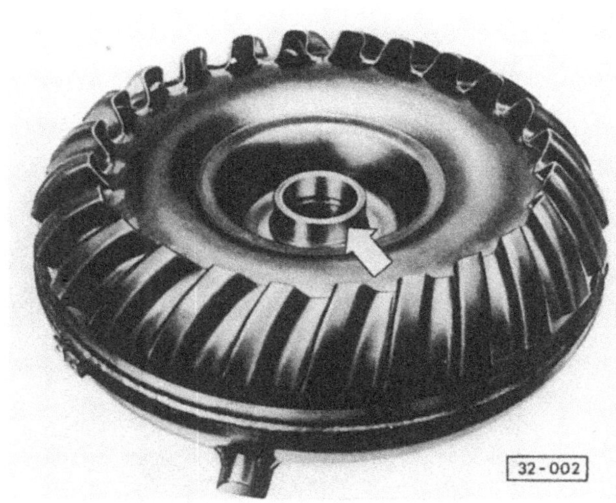

Fig. 5-6. Torque converter oil seal hub (arrow).

4. Measure the bore of the bushing inside the torque converter hub using an inside micrometer. If the diameter exceeds 34.25 mm (1.348 in.) or if the bore is out of round by more than 0.03 mm (.001 in.), the bushing should be replaced.

5. To replace the bushing, remove it using the Volkswagen special tools shown in Fig. 5-7. Press in the new bushing as shown in Fig. 5-8.

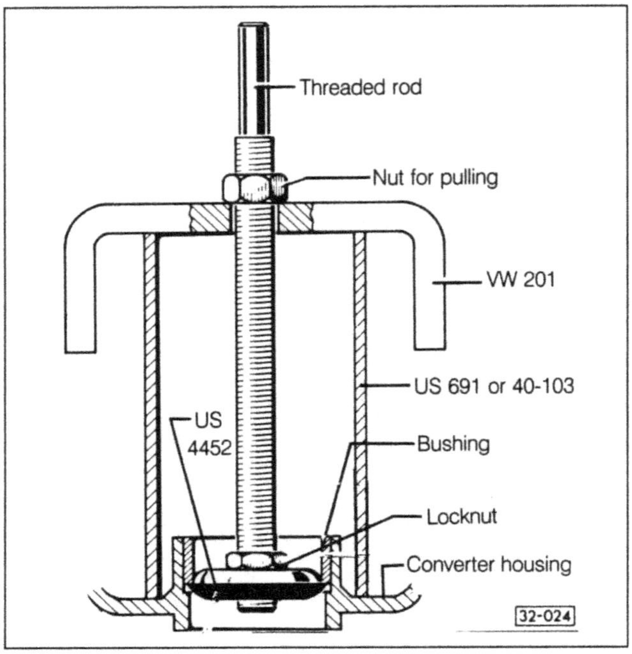

Fig. 5-7. Torque converter bushing being removed. Numbers shown identify Volkswagen special tools.

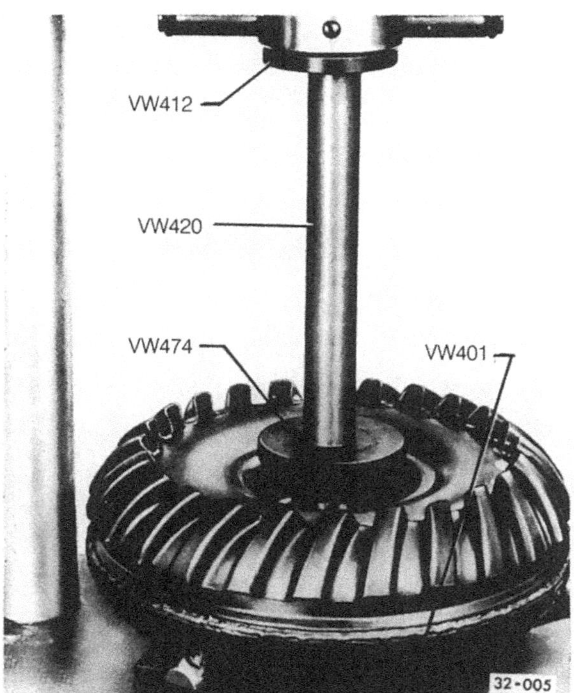

Fig. 5-8. Torque converter bushing being installed. Numbers shown identify Volkswagen special tools used with hydraulic press.

NOTE —

Replacement bushings are manufactured to the correct size and do not require additional reaming or honing.

Drive Flange Oil Seals

The drive flange oil seals can be replaced without removing the transaxle from the car. Check the new seals before beginning the job to make sure that each is equipped with a tension spring behind the seal lip.

To replace:

1. Remove the socket-head bolts to separate the front drive axle and constant velocity joint from the transaxle drive flange. Suspend the detached end of the drive axle with a stiff wire hook to avoid outer constant velocity joint damage.

   CAUTION —

   Do not let the driveshaft hang freely. This may damage the outer constant velocity joints, causing them to fail soon after the vehicle is returned to service.

   NOTE —

   It may be necessary or at least helpful to separate the ball joint from the lower control arm to allow moving the drive axle farther out of the way. For additional information, see **SUSPENSION AND STEERING**.

2. Pry the dust cap from the center of the drive flange. Remove the circlip which retains the flange, and remove the spring washer. Using a puller such as that shown in Fig. 5-9, pull the drive flange from the drive flange shaft.

   NOTE —

   Notice the way that the spring washer is installed so that it can be reinstalled the same way.

3. Pry the oil seal from its recess using a hooked tool or a screwdriver, using care not to damage the sealing surface.

4. Pack the open side of the new seal with multipurpose grease, and then drive the seal into place, as shown in Fig. 5-10, until it is fully seated.

5. Inspect the drive flange for wear at the point where it is contacted by the oil seal. Replace the drive flange if the seal contact area is grooved.

6. Install the drive flange as shown in Fig. 5-11.

7. Install the spring washer. Fit a new circlip over the end of the drive flange shaft.

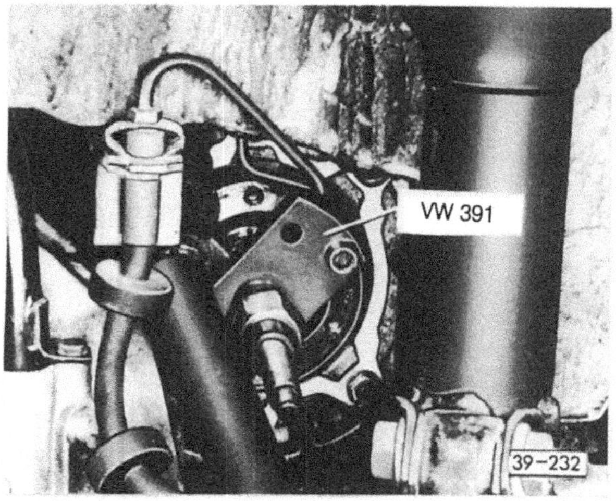

Fig. 5-9. Drive flange being pulled from drive flange shaft. Volkswagen special tool no. VW 391 shown. If another puller is used, be careful not to damage threaded hole in drive flange shaft.

Fig. 5-10. Seal driver (Volkswagen special tool no. US 4450 or VW 194) being used to install new drive flange oil seal. Alternately, use tool of approximate seal diameter to apply even installation pressure.

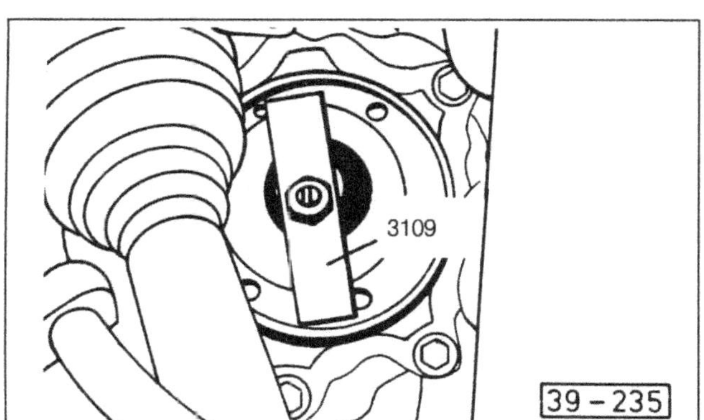

Fig. 5-11. Drive flange being installed. Volkswagen special tool no. 3109 (shown) threads into center of drive flange shaft. Nut is tightened to pull flange into place evenly and avoid seal damage.

8. Lightly drive the circlip into place until it is firmly seated in its groove in the drive flange shaft.
9. Reattach the drive axle to the drive flange. Torque the socket-head bolts to 45 Nm (33 ft. lb.).
10. If the ball joint was disconnected, raise the lower suspension control arm and insert the ball joint stud into the bearing housing. Install the clamp bolt and nut, and torque to 50 Nm (37 ft. lb.).

6. TECHNICAL DATA

I. Automatic Transmission Test Data

Stall speed (brakes locked, wide-open throttle)	
transmission code TN	2200 to 2500 rpm
transmission code TNA	2320 to 2570 rpm
Main pressure	
at idle (car stationary), selector lever at **D**	2.90 to 3.00 bar (42 to 43.5 psi)
at idle (car stationary), selector lever at **R**	9.00 to 10.00 bar (131 to 145 psi)
at full throttle, indicated speed above 25 mph (40 km/h), selector lever at **D**	5.85 to 5.95 bar (85 to 86 psi)

II. Tightening Torques

Driveshaft to drive flange (socket-head bolt)	45 Nm (33 ft. lb.)
Engine to transaxle (bolt and/or nut)	80 Nm (59 ft. lb.)
Locknut for 2nd gear brake band adjusting screw (nut)	20 Nm (15 ft. lb.)
Lower ball joint to wheel bearing housing (clamp bolt)	50 Nm (37 ft. lb.)
Main pressure tap plug in transmission case (threaded plug)	10 Nm (87 in. lb.)
Protection plate to transaxle (bolt)	24 Nm (17 ft. lb.)
Selector support to floor board (bolt)	9.2 Nm (80 in. lb.)
Selector lever cable to transmission operating lever (nut)	8 Nm (71 in. lb.)
Starter to bellhousing (bolt)	23 Nm (17 ft. lb.)
Throttle cable to valve cover	10 Nm (87 in. lb.)
Torque converter to drive plate (bolt)	30 Nm (22 ft. lb.)
Torque converter cover plate to bellhousing (bolt)	15 Nm (11 ft. lb.)
Transmission case to final drive housing (nut)	30 Nm (22 ft. lb.)

III. Automatic Transaxle Specifications

Engine type engine code (application)	**1.8 liter JH (early 1985)**	**1.8 liter JH, 2H (from late 1985)**
Transmission type number and code letter	010 TN	010 TNA
Torque converter code letter	K	K
Gear ratios		
1st gear	2.71	2.71
2nd gear	1.50	1.50
3rd gear	1.00	1.00
Final drive ratio	3.41	3.41

Section 10

SUSPENSION AND STEERING

Contents

TABLES

Suspension and Steering

Introduction

Both Cabriolet and Scirocco are equipped with a strut-type independent front suspension and a torsion-beam rear axle and trailing arm assembly. The front suspension struts are integrated spring and shock absorber units. The rear suspension has separate shock absorbers mounted inside coil springs. This type of suspension system is one of the most commonly used systems in contemporary automotive design for the specific reason that it provides excellent steering and handling in a compact size with comparatively light weight.

The front suspension struts have been designed to provide a negative steering roll radius. That is, the steering axis of each strut intersects the road surface at a point outboard of the wheel's vertical centerline. This suspension geometry tends to steer the car automatically in the direction of a skid caused by unequal front wheel traction. Conventional front suspension geometry, which places the steering axes inboard of the tire centerlines, tends to steer the car in the opposite direction, thereby increasing the severity of the skid.

The rear suspension has been designed for low unsprung weight, easy replacement of springs and shock absorbers, and good ride and handling qualities. The main axle beam is somewhat flexible. Uneven rear suspension loads cause the axle beam to twist slightly, thus acting as a stabilizer bar. This suspension design yields many desired performance benefits which are otherwise only available from more complicated and expensive kinds of independent suspension.

The steering utilizes a rack and pinion steering gear to minimize weight and space requirements, with tie rods connecting the rack to the front suspension. The steering column is connected to the steering gear by a universal joint shaft. The universal joint shaft is angled in order to prevent collision impacts from forcing the steering wheel toward the driver. Beginning with 1990 models, the steering column assembly also houses an airbag to protect the driver in a frontal collision. Special procedures and Warnings apply to any repairs on airbag-equipped cars.

The information in this section of the manual is intended as a resource for professional automotive technicians as well as a guide to Cabriolet and Scirocco owners. Some of the repairs or procedures may require special equipment and prior knowledge of typical suspension and steering repair practices. If you lack the skills, tools, or a suitable workplace for a particular repair, we suggest that you leave the job to an authorized Volkswagen dealer or other qualified shop. We especially urge you to check with your authorized Volkswagen dealer Service department before beginning any repairs on a car which may be covered wholly or in part by any of the extensive warranties issued by Volkswagen United States, Inc.

1. GENERAL DESCRIPTION

Cabriolet and Scirocco suspension and steering systems are designed to provide excellent handling and comfort while minimizing weight and space requirements. Although the front and rear suspension systems are independent sub-systems, and are of completely different designs, they work together to achieve Volkswagen's outstanding combination of precise handling and ride comfort. Troubleshooting, maintenance, and repair should always consider the condition of both front and rear suspension systems, as well as the steering system.

1.1 Front Suspension

Fig. 1-1 shows the components of the front suspension system. The lower control arms, sometimes called wishbones or A-arms, connect to mounting points on the unitized body.

Each front suspension strut is a shock absorber and coil spring in one unit. The hydraulic shock absorber is inside the tubular strut housing. The front wheel bearing housing with its integral steering arm is a separate component that is bolted to the lower end of the strut and mounted to the control arm with a ball joint. Suspension travel is limited by rubber bump stops that are part of each strut assembly.

The two Y-shaped control arms position the lower ends of the suspension struts. The ball joints are riveted to the control arms during manufacture—replacements are installed with bolts. There are no adjustments for front wheel alignment on the control arms. Toe is adjusted by altering the length of the right-hand steering tie rod. Camber is adjusted by means of an eccentric bolt at the lower strut-mounting point. There is no provision for caster adjustment. A stabilizer bar mounted in rubber bushings is connected to both control arms and reduces body roll during cornering. Scirocco 16V models feature an additional lower front suspension support to help stiffen the suspension mounting points.

1.2 Rear Suspension

The rear suspension system is shown in Fig. 1-2. The beam-type rear axle is a one-piece welded assembly of axle beam and trailing arms which yields constant toe and camber throughout the range of suspension travel. The axle beam itself acts as a stabilizer bar, its torsional stiffness helping to reduce body roll during cornering. Rear wheel stub axles are bolted to the trailing arms. The rear suspension "struts" are actually separate coil springs mounted concentrically with shock absorbers. They are easily removed and installed and, unlike the front struts, do not require a special spring compressor for assembly and disassembly.

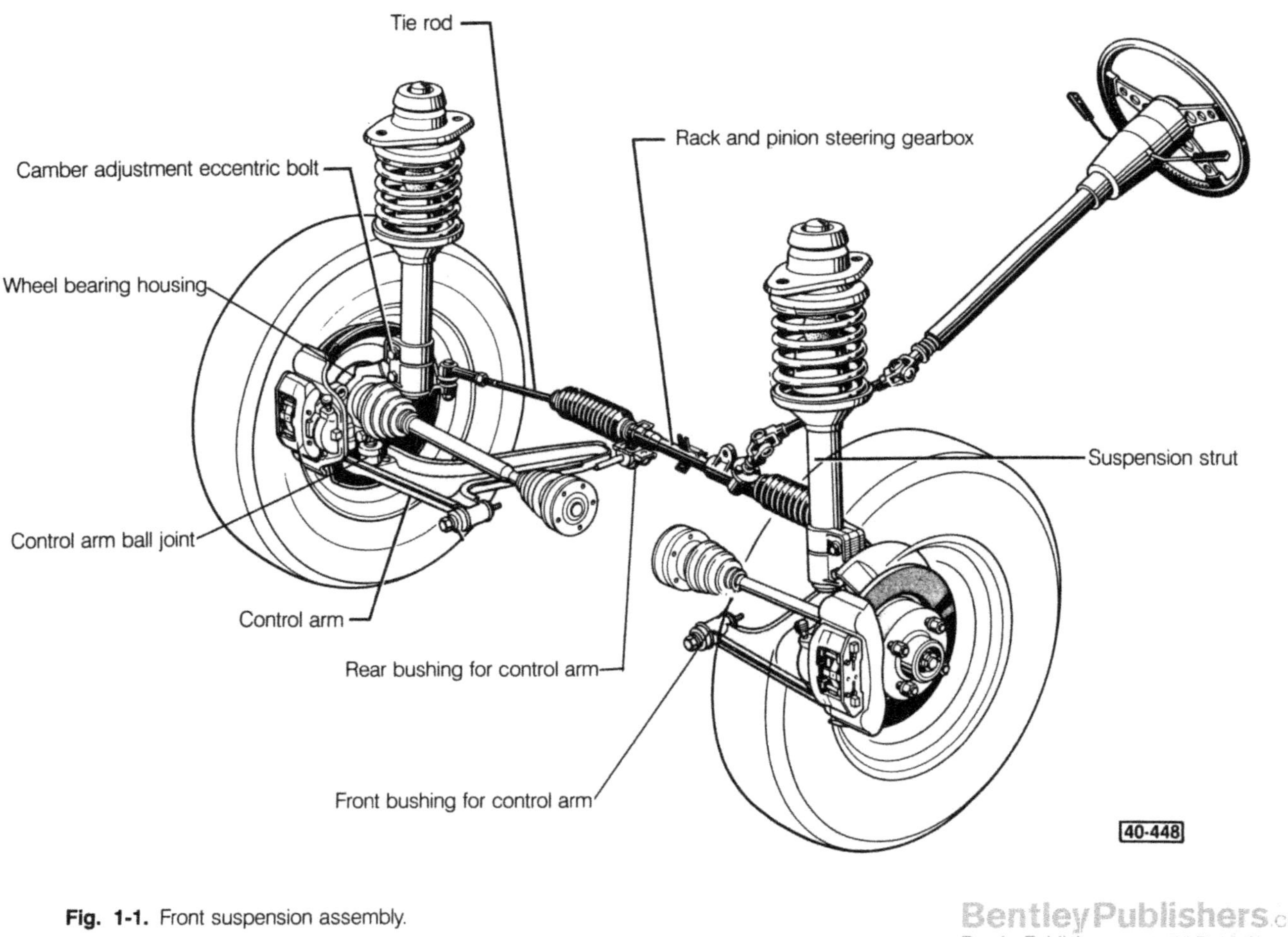

Fig. 1-1. Front suspension assembly.

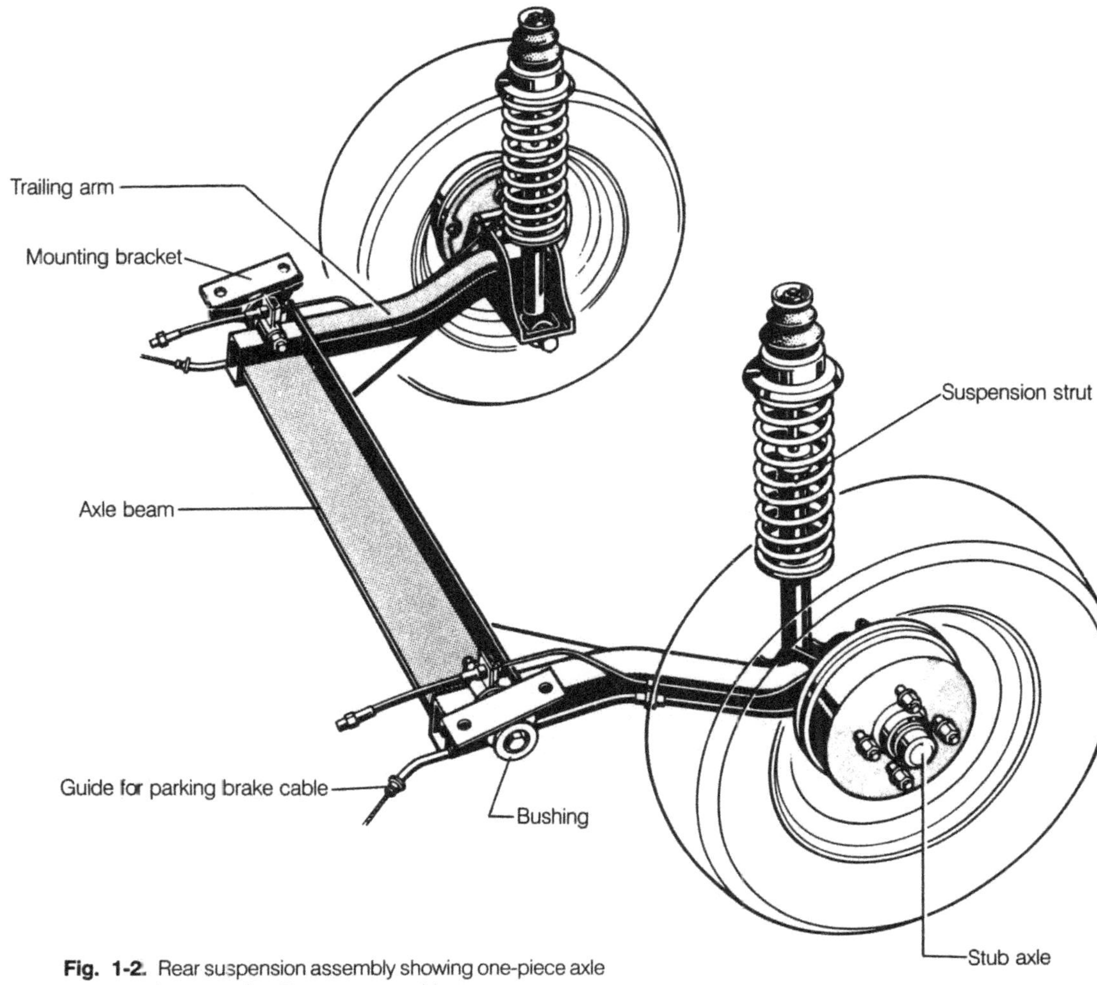

Fig. 1-2. Rear suspension assembly showing one-piece axle beam and trailing arm assembly.

1.3 Steering

The rack and pinion steering gear is mounted behind the engine and the transaxle. This location protects the steering gear from weather and from damage in all but the most severe collisions. The steering gear and the tie rod ends are permanently lubricated and do not require any routine maintenance in normal service. An adjusting screw on the steering gear provides means of adjustment, but no routine adjustment is required.

Beginning with 1990 models, the Cabriolet is equipped with a driver-side airbag. Special procedures and Warnings apply to any repair work on airbag-equipped cars.

> ***WARNING —***
> - *The airbag is operated by an explosive device. Handled improperly or without adequate safeguards, it can be accidently activated and cause serious personal injury.*
> - *Thoroughly read all Cautions and Warnings at the front of this manual and in **8. Airbag** before beginning any work on an airbag-equipped car.*

2. MAINTENANCE

Volkswagen specifies the maintenance steps listed below, to be carried out at particular time or mileage intervals for proper maintenance of the suspension and steering. A number in bold type refers to information in this section, under that numbered heading. See **LUBRICATION AND MAINTENANCE** for information about the other listed items, and for the prescribed maintenance intervals.

1. Checking power steering fluid level
2. Inspecting constant velocity (CV) joints and boots. **5.2**
3. Inspecting ball joints. **5.4**
4. Inspecting tie rods and tie rod boots. **7.2**

3. TROUBLESHOOTING

This troubleshooting section applies to problems affecting ride comfort, steering, stability and handling. Such problems are likely to be caused by worn or damaged suspension or steering parts, but a problem such as a consistent pull to the left or right may also be caused by faulty brakes. For brake system troubleshooting, see **BRAKES**.

3.1 Basic Troubleshooting Principles

Ride comfort and stability both depend on the suspension and steering systems controlling the positions and movements of the wheels. These systems must allow controlled movement so that the wheels can steer and react to bumps, but they also limit and precisely position the wheels so that the car is stable, controllable, and predictable. Any symptom of instability or imprecise road feel may be caused by worn or damaged suspension and steering components.

Aside from inspection for worn parts, troubleshooting must also consider tires, wheels, and their alignment. Tire wear and inflation pressures can dramatically affect handling. Subtle irregularities in wheel alignment angles also affect stability. Tires of mixed brands or sizes affect alignment and may unbalance a car's handling.

Table a lists symptoms of suspension and steering problems and their probable causes, and suggests corrective actions. The boldface numbers in the Corrective action column refer to headings in this section where the repairs are described.

Table a. Suspension and Steering Troubleshooting

Symptom	Probable cause	Corrective action
1. Pull to one side, wandering	**a.** Incorrect tire pressure	**a.** Check and correct tire pressures. See **LUBRICATION AND MAINTENANCE**
	b. Defective/unevenly worn tire	**b.** Inspect tires and replace as needed. **4.1**
	c. Incorrect wheel alignment	**c.** Check and adjust wheel alignment. **4.2**
	d. Worn rear axle beam pivot bushings	**d.** Replace axle beam bushings. **6.3**
	e. Faulty brakes	**e.** See **BRAKES**
2. Steering heavy, poor self-centering	**a.** Incorrect tire pressure	**a.** Check and correct tire pressures. See **LUBRICATION AND MAINTENANCE**
	b. Worn front strut bearing(s)	**b.** Remove and disassemble front struts. Replace strut bearings. **5.1**
	c. Faulty universal joint shaft	**c.** Replace shaft assembly. **7.1**
	d. Binding steering gear	**d.** Adjust steering gear. **7.2**
	e. Low power steering fluid	**e.** Check power steering fluid and add as required. See **LUBRICATION AND MAINTENANCE**
	f. Loose or broken power steering pump V-belt	**f.** Inspect V-belt. Tighten or replace as necessary. See **LUBRICATION AND MAINTENANCE**
	g. Faulty power steering pump	**g.** Test and, if necessary, replace pump. **7.3**
	h. Air in power steering fluid	**h.** Repair air leak and, if necessary, add fluid. **7.3**
3. Front-end vibration or shimmy	**a.** Incorrect tire pressure	**a.** Check and correct tire pressures. See **LUBRICATION AND MAINTENANCE**
	b. Loose wheel lug bolts	**b.** Tighten lug bolts to proper torque. **4.**
	c. Bent wheel rim	**c.** Inspect wheels and replace as necessary. **4.**
	d. Unbalanced wheels/tires	**d.** Balance tires. Check for uneven wear patterns. **4.1**
4. Shudder or vibration when accelerating	**a.** Faulty inner constant velocity (CV) joint(s)	**a.** Replace inner CV joint(s). **5.2**
5. Steering loose, imprecise	**a.** Incorrect tire pressure	**a.** Check and correct tire pressures. See **LUBRICATION AND MAINTENANCE**
	b. Worn tie rod end(s)	**b.** Replace tie rod(s) and align. **7.2**
	c. Worn control arm ball joint(s)	**c.** Replace ball joint(s) and align. **5.4**
	d. Faulty front wheel bearing(s)	**d.** Replace wheel bearing(s). **5.5**
	e. Worn steering gear	**e.** Check and adjust steering gear. **7.2**
	f. Worn universal joint shaft	**f.** Replace shaft assembly. **7.1**

continued on next page

Table a. Suspension and Steering Troubleshooting (continued)

Symptom	Probable cause	Corrective action
8. Front wheel noise, continuous growling, may be more noticeable when turning	**a.** Worn front wheel bearing(s) **b.** Faulty transaxle	**a.** Replace front wheel bearing(s). **5.5** **b.** See **MANUAL TRANSMISSION AND CLUTCH** or **AUTOMATIC TRANSMISSION**
9. Front wheel noise, ratcheting, clicking, most severe when turning and accelerating	**a.** Failed outer constant velocity (CV) joint(s)	**a.** Replace outer CV joint(s). **5.2**
10. Rear wheel noise, scraping, dragging	**a.** Worn/improperly adjusted rear wheel bearings **b.** Faulty rear brakes	**a.** Adjust or replace bearings. **6.2** **b.** See **BRAKES**
11. Steering surges or makes noise when turning (power-assisted steering only)	**a.** Low power steering fluid level **b.** Loose power steering pump V-belt **c.** Faulty power steering pump **d.** Air in power steering fluid	**a.** Check power steering fluid and add as required. See **LUBRICATION AND MAINTENANCE** **b.** Inspect V-belt. Tighten or replace as necessary. See **LUBRICATION AND MAINTENANCE** **c.** Test and, if necessary, replace pump. **7.3** **d.** Repair air leak and, if necessary, add fluid. **7.3**

3.2 Diagnostic Inspection and Testing

Suspension and steering problems can usually be isolated and at least partially diagnosed by careful consideration of the symptoms and inspection of the components which are most likely to cause such a symptom.

Tire Wear

Tire tread wear, while not a source of immediate feedback, can be an indicator of suspension and steering problems. Proper tire tread wear is difficult to notice, so tires are made with wear indicator bars which visually indicate when the tire is nearly worn-out. These wear indicator bars show up as evenly spaced bald "stripes" running across the tread surface, as shown in Fig. 3-1.

Fig. 3-1. Tread wear indicator bars showing on worn-out tire.

Uneven tire wear usually indicates improper tire inflation pressures or a misalignment condition. Fig. 3-2 illustrates how tire inflation pressures influence tire wear. Tire wear that is uneven across the tread—one side worn more than the other or unusual wear of individual tread blocks—probably indicates an alignment problem. Cupping or scalloping—wear that is uneven around the circumference of the tire—is a telltale sign of a tire balance problem or a worn-out shock absorber.

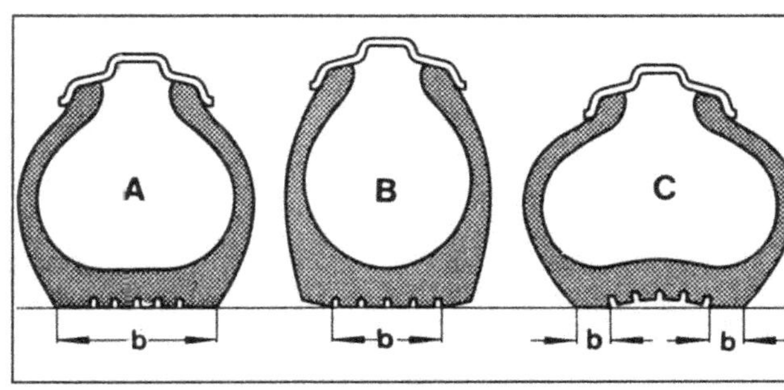

Fig. 3-2. Effect of tire inflation pressure on wear pattern. Condition **A** is normal. Condition **B** is overinflation which increases wear in center of tire tread. Condition **C** is underinflation which increases wear on outer edges of tire tread. Dimension **b** is tire contact area.

Isolating Pulling Symptoms

Consistent pulling to one side in a car driven straight ahead on a level road may be caused by either suspension misalignment or a faulty tire. In more unusual cases, a brake problem may be the cause.

To help decide whether alignment or tires are at fault, temporarily swap the front tires and then road test the car. If a tire problem is the cause of the pulling symptom, the problem should switch to the other side of the car when the tires are switched, and it should now pull to the other side. If the symptom is unchanged, then the problem is probably alignment, which is unaffected by swapping tires.

CAUTION —
Tire manufacturers generally recommend leaving radial tires, once mounted, on one side of the car so that they always turn in the same direction. Temporary side-to-side swaps are acceptable for diagnostic purposes. Volkswagen also specifies that diagonal tire rotation (right front to left rear, etc.) is acceptable as a remedy for uneven tire wear.

Vibration

Abnormal vibration, if not caused by a mechanical problem, will most often be the result of one or more unbalanced wheels. Before spending the money for wheel balancing, look for an obvious cause of wheel imbalance. Caked-on mud or ice and snow can dramatically affect wheel balance. Clean the wheels and tires and road test the car before investigating more serious causes.

4. WHEELS, TIRES, AND ALIGNMENT

For stability and control, wheels and tires must be in good condition and be properly aligned. Tire inflation pressures, tire wear, and wheel alignment will all influence how the car feels and responds on the road.

Precise wheel alignment can only be accomplished when the tires, the suspension, and the steering are in good condition. Uneven tire wear, different size tires, or worn suspension and steering parts all affect wheel alignment. Reputable wheel alignment technicians will always inspect the suspension and steering for worn parts before an alignment, and will recommend that any necessary repairs be made before proceeding. A wheel alignment on a car with excessively worn suspension and steering parts is a complete waste of time and money.

4.1 Wheels and Tires

Wheels and tires are subject to many stresses. They will perform as intended only if undamaged, properly inflated, and correctly balanced. Properly maintained, the factory-installed tires will provide long service with comfort and safety. See **LUBRICATION AND MAINTENANCE** for information on wheel size, tire size, and recommended inflation pressures.

Wheels

The wheels on the Scirocco and Cabriolet are designed for use with tubeless, radial tires of a specific size. Replacement tires of non-standard size or construction should be used only if the tire manufacturer specifies them for your make and model. See **LUBRICATION AND MAINTENANCE** for information on the original equipment wheel sizes.

When installing accessory wheels, be sure to install wheels with the proper offset. Measured across the width of the wheel rim, the offset is how far the center of the wheel is offset from the wheel mounting surface. The wheels on Volkswagens covered by this manual have negative offset. In other words, from the wheel mounting surface the wheel rim is offset to the inside. The use of other than original equipment wheels requires special care since many aftermarket wheels do not have the correct offset. Incorrect offset may cause the tires to interfere with the body, overload the wheel bearings, or affect the steering.

Tires

Radial-ply tires are factory-installed on the Scirocco and Cabriolet. To retain the car's excellent handling characteristics, it is recommended that the tires only be replaced with those of radial-ply construction and of the original size.

Winter Tires

Though inferior to regular tires for dry-road wear and handling, winter (mud and snow) tires offer a big improvement on snowy winter roads. Studded winter tires improve traction on icy surfaces, but may be more dangerous on dry roads. They should be used only if icy conditions predominate. Check local laws. The use of studded tires may be restricted or completely prohibited in your area.

WARNING —
When winter tires are installed on only two of the four wheels, they should be installed on the front driving wheels. To prevent dangerous handling characteristics, they must be of the same type of (radial-ply) construction as those on the rear of the car.

4.2 Alignment

Wheel alignment is the precise adjustment of wheel angles, used to achieve the best compromise between comfort, control, and tire wear. Alignment specifications differ from model to model, depending on suspension design. The alignment can also be fine-tuned for specific driving conditions. For example, if the car consistently carries only one person, or is consistently heavily loaded, this information will be useful to the alignment technician. Like tire pressure, wheel alignment also has some influence on fuel economy, because of the effect on rolling resistance.

The important alignment angles are camber, caster, and toe. Camber is the angle that the wheels tilt from vertical when viewed from front or rear, as illustrated in Fig. 4-1. Wheels which tilt out at the top have positive (+) camber. Wheels which tilt in at the top have negative (−) camber. Camber influences cornering, directional stability, and tire wear. Different camber on the two front wheels may cause the car to pull to one side. Misadjusted camber will cause uneven tire wear.

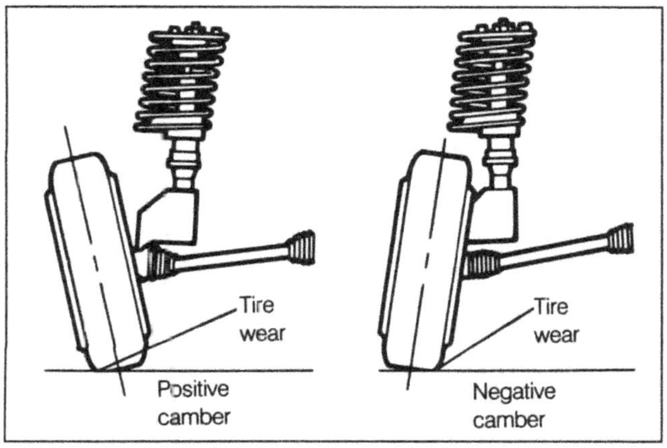

Fig. 4-1. Camber angle viewed from front or rear.

Caster, the angle by which the steering axis deviates from vertical, is illustrated in Fig. 4-2. The steering axis is an imaginary line about which the front wheels turn. Most cars are designed with a steering axis which is inclined toward the rear at the top (positive caster), giving them directional stability and self-centering steering. Caster should be checked as part of any alignment. It is, however, not an adjustable angle on cars covered by this manual. Caster which is out of specification suggests worn or damaged components. Uneven front wheel caster will cause the car to pull to one side. Too little caster reduces directional stability. Too much caster increases turning effort.

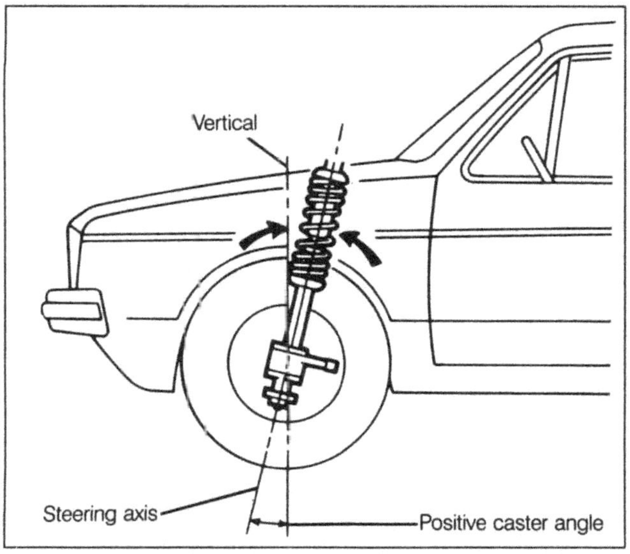

Fig. 4-2. Caster is angle of steering axis inclination from vertical. Caster on Scirocco and Cabriolet is non-adjustable.

Toe is a measurement of the amount that two wheels point toward each other (toe-in) or away from each other (toe-out), as viewed from above or below. Toe-in and toe-out are illustrated in Fig. 4-3. Toe affects directional stability and tire wear, and at the front wheels also has some effect on how the car responds to steering changes. Toe is the primary alignment adjustment for the Scirocco and Cabriolet. See **Toe Adjustment** below for information on adjusting toe. Too much toe will cause the tires to "scrub" and to wear unevenly and more quickly. Too little toe—too near zero—may cause the car to be less stable and wander from side to side.

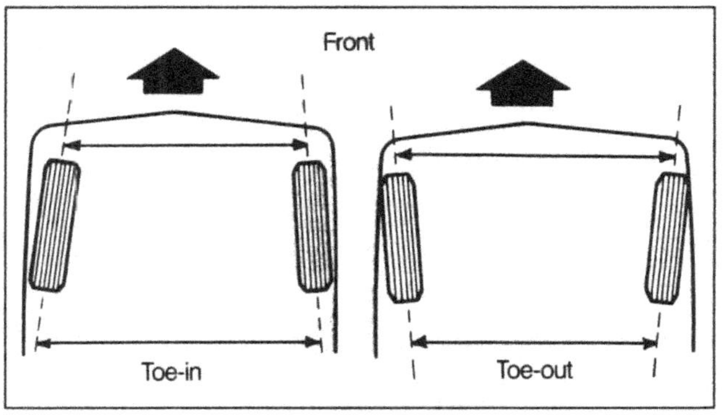

Fig. 4-3. Front wheel toe.

Because of the emphasis on precision, wheel alignment has become very sophisticated. Modern computerized equipment, using special optics or laser beams for measurement, help the qualified technician do the job more quickly and more accurately than is possible for the do-it-yourselfer. For this reason, this manual covers only the specific adjustment methods, and not the basic procedures for measuring wheel alignment angles. The reasonable cost of a professional wheel alignment is money well spent.

Four-wheel Alignment

A four-wheel alignment considers not just the individual wheels or pairs of wheels, but also the alignment of the wheels relative to the car's centerline and the relationships between front and rear wheels. Four-wheel alignment is more sophisticated and slightly more expensive, but is also the best, most precise alignment for any car. The Volkswagen Cabriolet and Scirocco, with their trailing beam axle rear suspension, can particularly benefit from this type of alignment.

Camber Adjustment

Camber is adjusted by means of an eccentric bolt, which is the upper bolt that holds the bottom of the suspension strut to the wheel bearing housing. If measurements indicate that front wheel camber needs to be adjusted, slightly loosen the two bolts and nuts shown in Fig. 4-4, then turn the (upper) eccentric bolt until the camber is within specification. Both front wheels must be on the ground to accurately check the adjustment. When the adjustment is complete, torque the bolts to 80 Nm (59 ft. lb.).

Rear wheel camber is not adjustable. Rear wheel camber which does not meet specifications suggests a damaged rear axle beam or rear stub axle. See **6. Rear Suspension** for more information.

Fig. 4-4. Bolts (arrows) to be loosened for camber adjustment. Eccentric bolt (top bolt) is turned to make adjustment.

Toe Adjustment

Toe at the front wheels is adjusted by changing the length of the right (passenger-side) steering tie rod. Because camber also affects toe, toe adjustments should be made only after camber adjustments are final. Loosen the locknut, and turn the tie rod to thread it into or out of the tie rod end, increasing or decreasing the overall tie-rod length. Large toe adjustments may also require centering the steering rack or the steering wheel. See **7. Steering** for more information.

Rear wheel toe is not adjustable. Rear wheel toe which does not meet specifications suggests a damaged rear axle beam or rear stub axle. See **6. Rear Suspension** for more information.

Centerline Adjustment
(four-wheel alignment only)

Although individual rear wheel toe is not adjustable, loosening the mounting bolts which hold the beam axle mounting brackets to the car body will allow a small amount of movement which can be used to adjust the rear wheel centerline.

4.3 Alignment Specifications

Alignment specifications for the Scirocco and Cabriolet are listed in **Table b**. See **4.2 Alignment** for more information.

Table b. Alignment Specifications

	Scirocco, Cabriolet	
Front wheels	specification	tolerance
Caster (not adjustable)		
wheels straight ahead	+ 1° 50'	± 30'
corresponding camber difference from 20° left lock to 20° right lock	1° 15'	± 20'
maximum permissible difference between sides	1°	–
Camber		
wheels straight ahead	+ 20'	± 30'
maximum permissible difference between sides	1°	–
Toe		
total toe, wheels not pressed	–15'	± 10'
toe angle difference at 20° lock (each side, wheels not pressed)	– 1° 30'	± 30'
Rear Axle (not adjustable)		
Camber		
each side	– 1° 15'	± 35'
maximum permissible difference between sides	40'	–
Toe		
total	+ 20'	± 30'
maximum permissible deviation from driving direction	30'	–

5. FRONT SUSPENSION

Although the Scirocco and Cabriolet have front wheel drive, the front suspension is no more complex than that of a rear-drive car. All of the front suspension parts that are subject to wear can be easily replaced. Some of these repairs require special tools and equipment. To avoid starting a job that may be difficult to complete, please read the entire procedure before beginning any repairs.

When doing any front suspension maintenance or repair, please observe the following general cautions.

> ***WARNING —***
> - *Do not re-use fasteners that are worn or deformed in normal use. Many fasteners are designed to be used only once and become unreliable and may fail when used a second time. This includes, but is not limited to, nuts, bolts, washers, self-locking nuts or bolts, circlips and cotter pins. Always replace these fasteners with new parts.*
> - *Do not install bolts and nuts coated with undercoating wax as correct tightening torque cannot be assured on installation. Always clean the threads of removed bolts and nuts with solvent before installation, or replace them with new parts.*
> - *Do not attempt to weld or straighten the suspension strut, wheel bearing housing, control arm, or subframe. Damaged parts should be replaced.*

5.1 Front Suspension Struts

Fig. 5-1 is an exploded view of the front strut assembly. The spring and the upper strut bearing are replaceable separately. If the shock absorber is worn-out or the strut itself is damaged, then the entire strut must be replaced. To maintain even side-to-side shock absorber performance, the struts should always be replaced in pairs.

The entire strut assembly is to be removed from the car as a unit before being disassembled. To disassemble the strut, special tools are required. Most importantly, the coil spring must be compressed with a spring compressor before removing the strut's top nut.

> ***WARNING —***
> *If you lack the tools or experience necessary to safely disassemble and reassemble the front suspension struts, we strongly urge that this work be left to an authorized Volkswagen dealer or other qualified repair shop. Attempts to disassemble the struts without proper tools are likely to cause serious injury.*

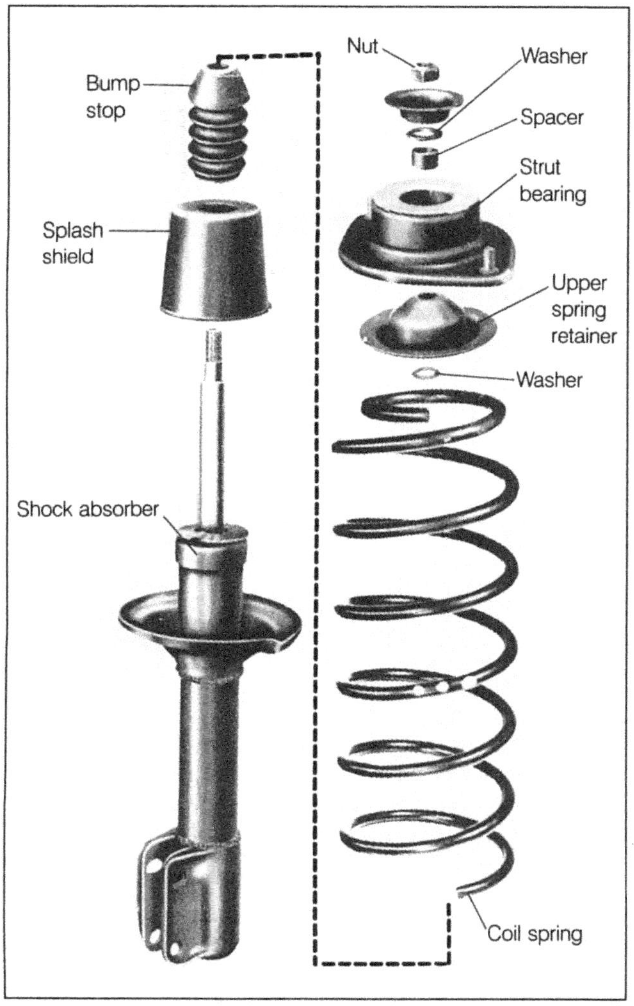

Fig. 5-1. Front strut assembly showing coil spring, strut bearing, and related parts removed.

Checking Shock Absorbers

Springs are what support the car and allow the suspension to bounce smoothly over bumps and other road irregularities. The shock absorber's function is to moderate the spring action—to quickly slow the bounce and help the spring return to its normal position. Shock absorbers do not require maintenance. An adequate supply of fluid is placed in them during manufacturing to compensate for small leaks. Minor leakage is acceptable if the shock absorber still functions correctly. Shock absorbers cannot be serviced, and need to be replaced when they lose their ability to control the suspension.

To the careful observer, a decline in shock absorber performance can be detected in normal driving. Worn shock absorbers will cause extra skittishness over bumps, and a less-controlled and wallowing feel in corners and after bumps. When seriously worn, the shock absorbers present little resistance to spring oscillations. Because they easily reach the limits of their travel, they may knock when going over bumps.

For more deliberate evaluation, the most common—though not entirely accurate—test of shock absorber function involves vigorously bouncing each end or corner of the car, and then releasing and observing how quickly the bouncing stops. More than one bounce usually indicates shock absorbers that need replacing.

Some evaluation of a shock absorber's condition can be made by removing and disassembling the strut and checking the shock absorber's resistance to movement. When held vertically, the shock absorber shaft should move smoothly and very firmly over its entire range of travel. If possible, compare the movement of a used shock absorber to a new one.

Removing and Installing Front Suspension Struts

The struts can be easily removed for replacement or for access to other components. Use a stiff wire, or equivalent, to support the free-hanging lower control arm and avoid damage to the flexible brake lines. Also, the strut mountings affect wheel alignment. To ensure correct reinstallation, the position of the eccentric camber adjustment bolt should be marked before removal. See **4.2 Alignment** for more information. Any time the struts have been removed and reinstalled, a wheel alignment is highly recommended.

To remove front struts:

1. Loosen the wheel lug bolts, then raise the car and support it with jack stands which are designed for the purpose. The front wheels should be off the ground so that the suspension is not bearing any weight.

2. Remove the wheel.

3. Remove the nuts from the two bolts that hold the bottom of the strut to the wheel bearing housing, leaving the bolts in place. Remove the brake line from its guide at the bottom of the strut and secure it out of the way.

4. Support the suspension and brake assembly so that it will not fall when the strut is removed.

5. Working in the engine compartment, remove the two nuts that hold the upper strut bearing to the body.

6. Remove the bolts from the bottom of the strut, and remove the strut.

> ***WARNING —***
> *Further disassembly of the strut requires special tools and great care to avoid serious personal injury. Do not proceed without thoroughly reading the procedure below.*

Installation is the reverse of removal. Torque the nuts that hold the strut to the body to 20 Nm (15 ft. lb.). Make sure that the eccentric bolt is installed correctly, and then torque the lower strut mounting bolts and nuts to 80 Nm (59 ft. lb.).

Disassembling and Assembling Front Suspension Strut

In the assembled strut, the coil spring is under some compression, even when the strut is removed from the car. Attempting to disassemble the strut without first compressing the spring is extremely dangerous. Read the procedure carefully. Do not attempt to disassemble the struts without a proper spring compressor.

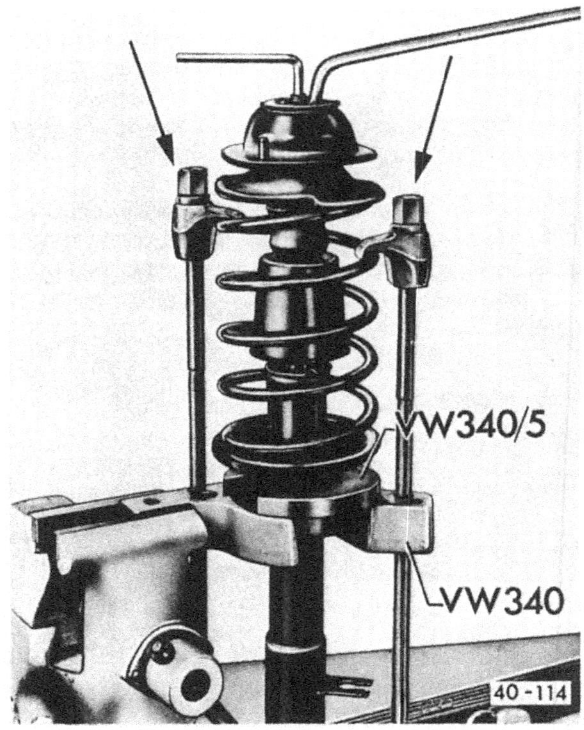

Fig. 5-2. Spring compressor being used to safely compress spring while removing strut nut using box wrench and hex wrench. On compressor shown (Volkswagen special tool no. VW 340), turn nuts (arrows) on threaded shafts to compress spring.

To disassemble front strut:

1. Install a spring compressor that is designed for the purpose. Relieve the pressure on the upper spring retainer by compressing the coil spring, as shown in Fig. 5-2.

> ***WARNING —***
> *Work slowly. Keep watch to make sure the spring compressor stays firmly in place on the spring.*

2. Remove the strut nut on the shock absorber rod using a 22 mm offset box wrench. Use a 7 mm hex wrench to hold the shock absorber rod.

> ***WARNING —***
> *Do not remove the nut from the shock absorber rod until the coil spring is compressed enough that there is no force pushing on the upper spring retainer.*

3. Slide the strut bearing, the upper spring retainer, the washer, the rubber bump stop, and the splash shield off the shock absorber rod.

4. Carefully release and remove the spring compressor. Remove the coil spring.

Assembly is the reverse of disassembly. Always use new self-locking nuts. Replace any other fasteners that are worn or damaged. Replace the upper strut bearing if it is damaged or does not turn smoothly. Fully extend the shock absorber rod by hand and compress the spring. Tighten the strut nut to 70 Nm (52 ft. lb.). Align the ends of the coil spring and the spring retainers. If the springs are to be replaced, note the paint color code on the coils. Replace springs only in matched pairs.

5.2 Drive Axles and Constant Velocity Joints

On cars with front wheel drive, the drive wheels must also be able to turn for steering. The drive axles are equipped at each end with constant velocity (CV) joints which allow power to be continuously delivered to the drive wheels—even when all the connections are not precisely straight. The CV joint is similar in function to older-type universal joints, but it operates much more smoothly at more radical drive axle angles.

The inner CV joints attach to the transaxle drive flanges, while the outer CV joints are equipped with splined stub axles which engage the splines of the wheel hubs and are secured by a large nut. Each joint is packed with a special molybdenum disulfide grease and completely sealed by a rubber boot. Although some inspection of the CV joints—and even replacement of the inner joints—is possible while the drive axles are installed, complete inspection and thorough lubrication of the joints requires that the drive axles be removed from the car.

If the drive axles themselves are damaged, complete remanufactured drive axle assemblies are available from an authorized Volkswagen dealer at lower cost than a new unit.

Beginning with 1988 models, a revised CV joint was introduced, and all replacement joints are now of this new type. The new joint has a larger shoulder diameter. See Fig. 5-3. When installing these replacement CV joints on 1985 to 1987 cars, it is necessary also to replace the wheel bearing circlip with the corresponding new part.

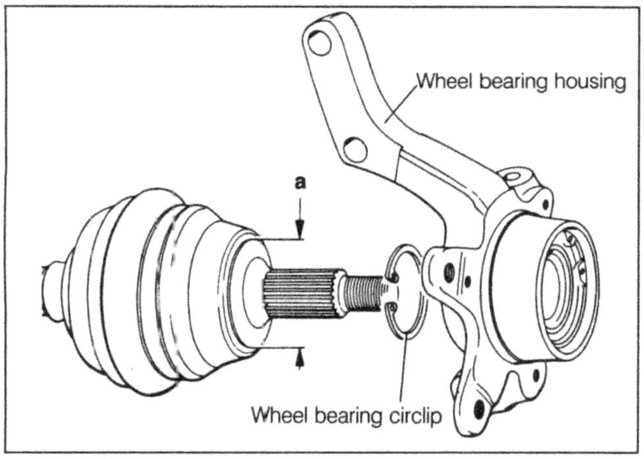

Fig. 5-3. Revised late model CV joint. On 1985-1987 models, dimension **a** is 50 mm. On 1988 and later models and on replacement CV joints, dimension **a** is 53 mm.

CAUTION —
Moving a car with drive axles removed will quickly damage the wheel bearings. If moving the car is necessary, always secure the bearing races by temporarily installing an old outer CV joint in the hub, or hold the bearing races in place with two large washers and a suitable bolt and nut.

Removing and Installing Drive Axles

The large nut securing the stub axle and outer CV joint to the hub is very tight. Removing it and torquing it properly on installation may require special tools. The inner CV joints are mounted with 8 mm, 12-point socket-head bolts which require a special wrench. Both inner and outer CV joints are held onto the drive axle shafts with circlips which should be replaced any time the joints are removed and installed. Use a new gasket when installing the inner CV joints to the transaxle drive flanges.

To remove:

1. With the vehicle on the ground, pry off the dust cap in the center of the front wheel hub and loosen, but do not remove, the axle shaft nut.

WARNING —
Loosen axle shaft nuts only while the car is on the ground. The leverage required to do this could topple the car from a lift or jack stands.

2. Raise the front end of the vehicle and support it securely on jack stands. Loosen the clamping bolt which holds the control arm ball joint to the wheel bearing housing, and remove the ball joint from the housing.
3. Clean around the inner CV joint and remove the six 8 mm, 12-point socket-head bolts that hold the inner CV joint to the transaxle drive flange. Support the inner end of the drive axle to prevent damage to the outer CV joint.
4. Remove the axle shaft nut and pull the drive axle out of the wheel hub. Cover the exposed inner CV joints to prevent contamination.

NOTE —
A light tap with a soft-faced mallet will help dislodge a stubborn axle.

Installation is the reverse of removal. Always use a new self-locking axle nut, and a new self-locking nut on the ball-joint clamping bolt. Replace any other fasteners that are worn or damaged. Apply a new adhesive-backed gasket to the clean inner CV joint.

Tightening torques:

• 12-pt. socket-head bolts	45 Nm (33 ft. lb.)
• ball joint clamping bolt	50 Nm (37 ft. lb.)
• axle shaft self-locking nut	240 Nm (177 ft. lb.)

WARNING —
- *Torque the axle shaft nut only while the car is on the ground. The leverage required to do this could topple the car from a lift or jack stands.*
- *Always replace self-locking nuts.*

Removing and Installing Constant Velocity Joints

Fig. 5-4 is an exploded view of a drive axle and its two constant velocity joints. Notice the ridges that are forged into the shaft for the purpose of holding the rubber boots in place. New circlips should be used when reinstalling the joints, and new clamps should be used to fasten the boots.

NOTE —
Because the CV joints must be removed to replace the boots, it is good preventive maintenance to replace the boots whenever the CV joints are removed.

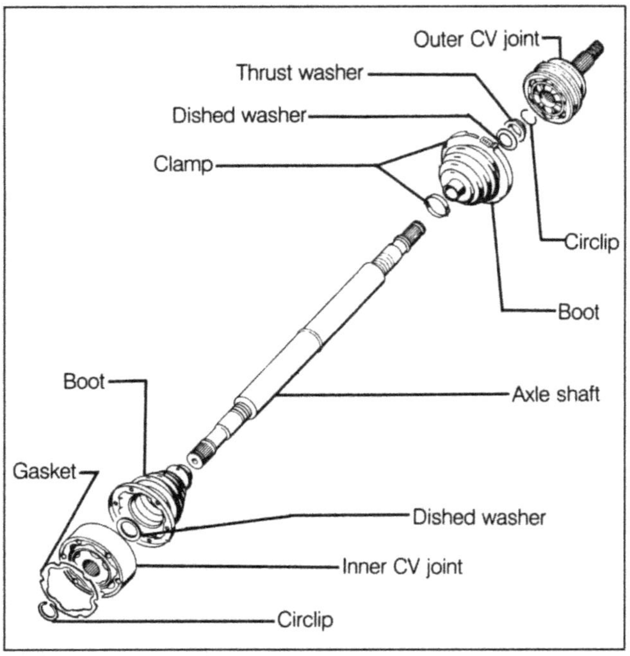

Fig. 5-4. Drive axle assembly showing inner and outer constant velocity joints and attachment parts.

CV joints are not rebuildable, and must be replaced as complete units. Volkswagen specifies a special molybdenum disulfide grease for packing CV joints. See **LUBRICATION AND MAINTENANCE** for more information on CV joint lubricant. Some special tools and equipment are required to remove and install CV joints. Read the procedure carefully before starting work. If you lack the necessary tools and equipment, or a suitable workplace, we suggest leaving these repairs to an authorized Volkswagen dealer or other qualified repair shop.

To remove and install outer CV joint:

1. Remove the drive axle as described above in **Removing and Installing Drive Axles**.

2. Cut and remove the large clamp holding the rubber boot to the CV joint. Fold the boot back from the joint, inside out over the shaft.

3. Clamp the shaft in a vise, and drive the CV joint off the shaft, as shown in Fig. 5-5, using a soft-faced mallet. Remove the thrust washer, the dished washer, and the circlip from the axle and inspect the shaft for wear.

NOTE —
The CV joint is secured to the shaft with a circlip that is inaccessible. The force used to drive the joint off the shaft must be sufficient to override the clip, but not enough to cause damage to the joint.

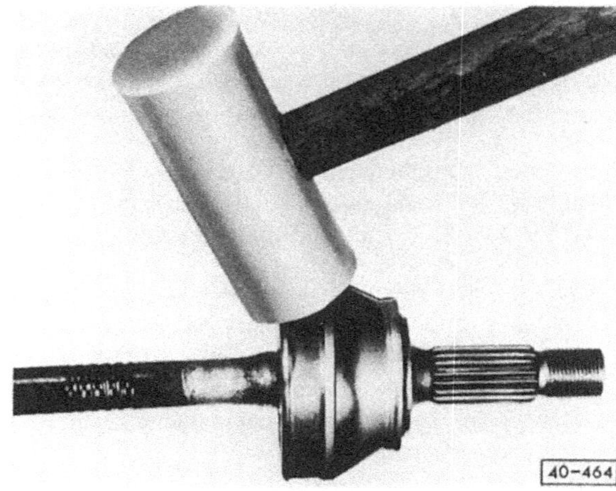

Fig. 5-5. Outer CV joint being driven off axle shaft with soft-faced mallet (shown with boot removed for clarity).

4. When installing the CV joint, use a new circlip in the groove at the end of the axle shaft. See Fig. 5-6. Drive the joint onto the shaft until the circlip is seated. Fold the boot back over the joint, using a new clamp to hold it.

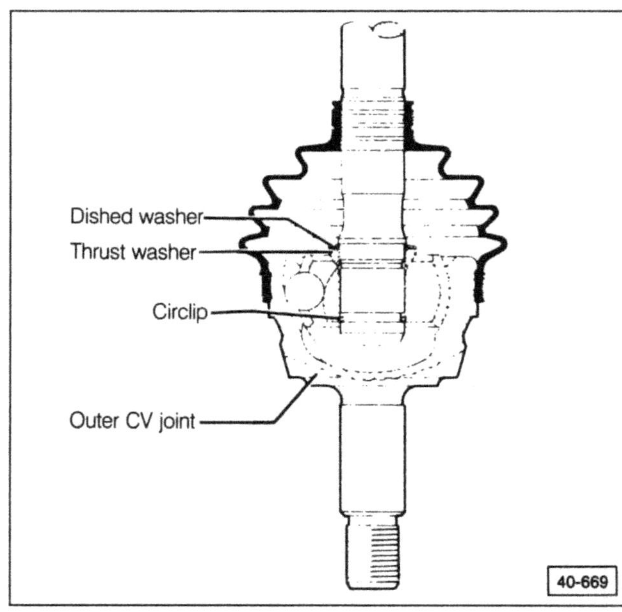

Fig. 5-6. Cutaway view of outer CV joint, showing positions of circlip and washers.

To remove and install inner CV joint:

1. Remove the drive axle as described above in **Removing and Installing Drive Axles**.

2. Remove the circlip from the end of the axle shaft, as shown in Fig. 5-7.

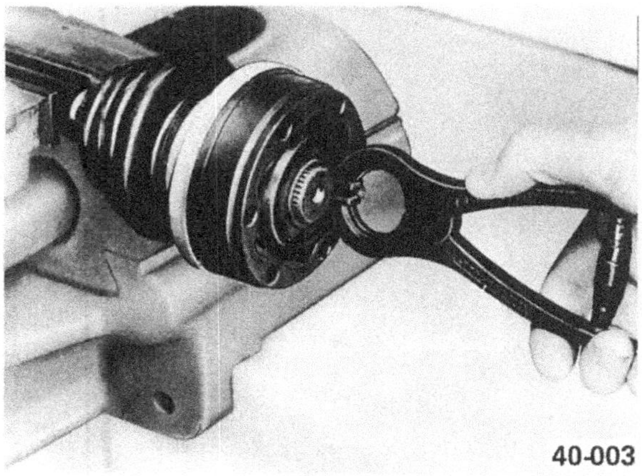

Fig. 5-7. Removing circlip which holds inner CV joint to axle shaft.

3. Drive the protective cap off of the outer ring of the CV joint, as shown in Fig. 5-8, and fold the boot away from the joint, inside out over the axle shaft.

CAUTION —
Do not pivot the ball hub more than 20° in the outer ring of the CV joint. The balls will fall out if the hub is pivoted too far.

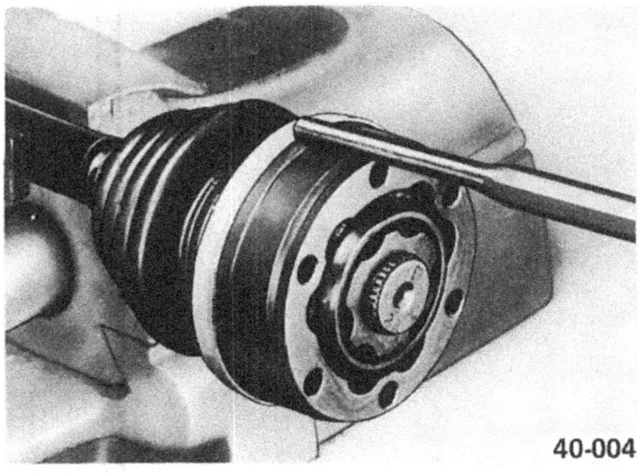

Fig. 5-8. Driving off CV joint protective cap with drift and hammer.

4. Support the hub of the CV joint and press the axle shaft out of the joint, as shown in Fig. 5-9.
5. Remove the dished washer and inspect the shaft and the splines for wear and damage.
6. When installing the CV joint, use a new dished washer and install it with the concave surface facing the end of the shaft, as shown in Fig. 5-10.
7. Press the CV joint onto the shaft as shown in Fig. 5-11. While using the press to hold the hub of the CV joint against the dished washer, install the circlip in the groove on the end of the shaft.

Fig. 5-9. Driving axle shaft out of inner CV joint with punch (drift) and thrust plate (support).

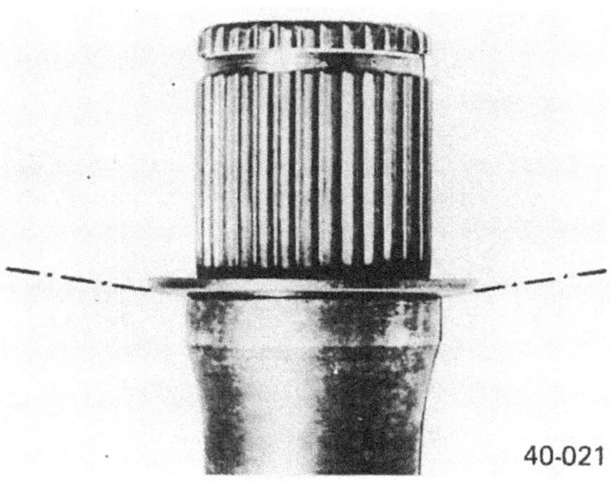

Fig. 5-10. Correct installation of dished washer used with inner CV joint on end of drive axle.

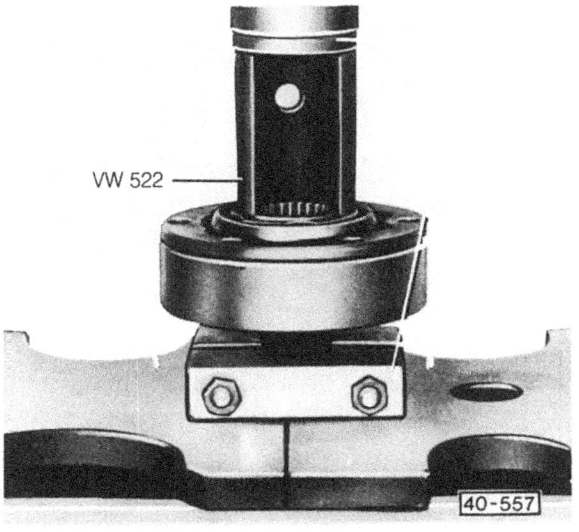

Fig. 5-11. Pressing inner CV joint onto drive axle shaft. Press tool shown, Volkswagen special tool no. VW 522 (order no. TV0 522 000 00 ZEL) allows installation of circlip while CV joint and washer are being pressed.

8. Release the press and make sure that the circlip is completely seated in the groove. Install the protective cap around the edge of the CV joint.

Disassembling and Inspecting Constant Velocity Joints

The components of each CV joint are precisely matched during manufacture and they cannot be serviced individually. The joints can, however, be disassembled for cleaning, inspection, and repacking with lubricant.

Begin the inspection of a CV joint before it is disassembled. The balls and grooves allow the inner hub to move relative to the outer housing, but the parts should fit snugly and move with some effort. A joint with loose fitting parts—such that there is obvious freeplay—should be replaced. Polished interior surfaces or visible ball tracks alone are not necessarily cause for replacement.

To disassemble outer CV joint:

1. With the CV joint removed from the drive axle as described above, mark the relative positions of the hub, the cage, and the outer housing.
2. Swivel the cage and hub as shown in Fig. 5-12 to remove the balls one at a time. Keep track of the removed balls so that they can be reinstalled in the same relationship to each other and to the marked hub and cage.

Fig. 5-12. Hub and cage of outer CV joint swivelled for removal of balls. Notice mark on outer housing. Similar marks should be made on hub and cage.

3. Remove the hub and cage as a unit, turning them so that the two large rectangular openings in the cage are positioned as shown in Fig. 5-13, and remove the hub and cage.

Fig. 5-13. Large rectangular openings in cage (arrow) in position for cage and hub removal.

4. Remove the hub from the cage. Turn the hub so that one of the splines is in line with one of the large rectangular openings in the cage. See Fig. 5-14. The hub now has sufficient clearance to be tipped out of the ball cage.

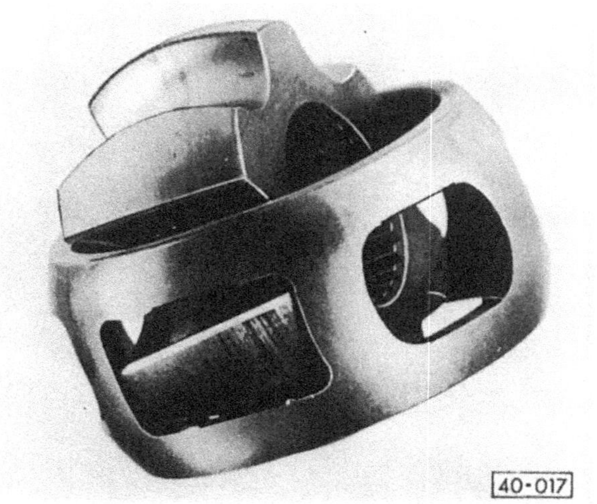

Fig. 5-14. Hub in position for removal or installation. Rectangular opening just accommodates spline of hub.

5. Clean and inspect the joint for galling, pitting, and other wear or damage. Replace any joint that is worn or damaged in these ways.

Assembly of the outer CV joint is the reverse of disassembly. Thoroughly coat all parts with molybdenum disulfide grease, and pack the joint with approximately 45 grams (approx. 1 ½ oz.) of the grease. This is half the total amount required for one CV joint. Make sure that the markings made on the housing, hub, and cage during disassembly line up. Install the joint on the drive axle as described in **Removing and Installing Constant Velocity Joints**. Finish packing the joint with another 45 grams of molybdenum disulfide grease.

To disassemble inner CV joint:

1. With the CV joint removed from the drive axle as described above, mark the relative positions of the hub, the cage, and the outer housing.

2. Pivot the hub and cage 90° to the housing, as shown in Fig. 5-15, and then push it out of the housing.

Fig. 5-15. Inner CV joint hub and cage with balls being separated from housing. Pivot hub and cage until perpendicular to housing, then push in direction indicated by arrow.

3. Remove the balls one at a time. Keep track of the removed balls so that they can be reinstalled in the same relationship to each other and to the marked hub and cage.

CAUTION —
The cage, housing, and balls are precisely matched during manufacture. When disassembling more than one joint, do not intermix components between joints.

4. Rotate the ball hub into the position shown in Fig. 5-16. The ball grooves on the hub must be in line with the outer edge of the cage. The hub now has sufficient clearance to be tipped out of the ball cage.

To assemble inner CV joint:

1. Clean and inspect the joint for galling, pitting, and other wear or damage. Replace any joint that is worn or damaged in these ways.

2. Thoroughly coat all components with molybdenum disulfide grease, and install the hub in the cage as illustrated in Fig. 5-16.

3. Install the balls in the cage, as shown in Fig. 5-17, using the adhesion of the molybdenum disulfide grease to hold the balls in place.

Fig. 5-16. Hub being removed or installed. Arrows indicate alignment of hub ball groove with edge of cage.

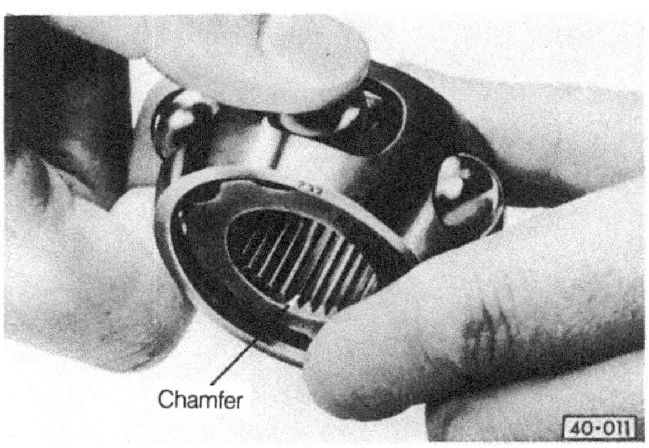

Fig. 5-17. Balls being installed in inner CV joint cage and hub.

4. Insert the hub and cage with the balls into the outer housing.

CAUTION —
The chamfer on the inner splines of the hub (visible in Fig. 5-17) must face the drive axle and the larger diameter side of the outer housing.

NOTE —
Align the components so that when the hub and cage are rotated into place, the widely-spaced grooves in the hub will line up with the widely-spaced grooves in the outer housing (**a**), and the closely-spaced grooves in the hub (**b**) will line up with the closely-spaced grooves in the outer housing, as shown in Fig. 5-18.

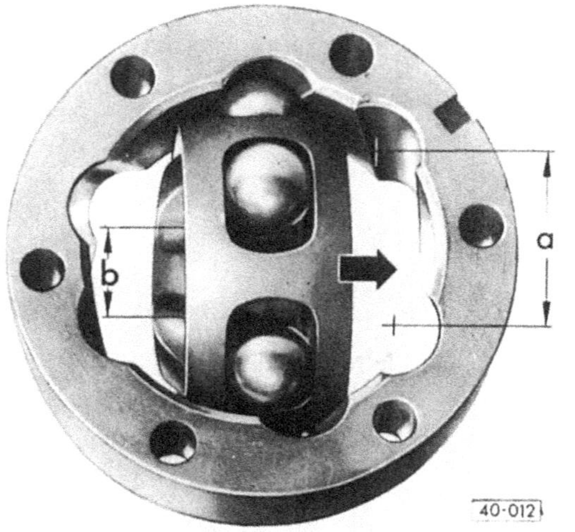

Fig. 5-18. Inner CV joint hub and cage positioned for assembly in outer housing. Wide grooves in housing (**a**) mate with wide grooves in hub. Narrow grooves in hub (**b**) mate with narrow grooves in housing. Arrow indicates direction in which hub and cage assembly is turned to install.

5. Pivot the hub and cage, as shown in Fig. 5-19, until the balls align with the grooves in the outer housing.

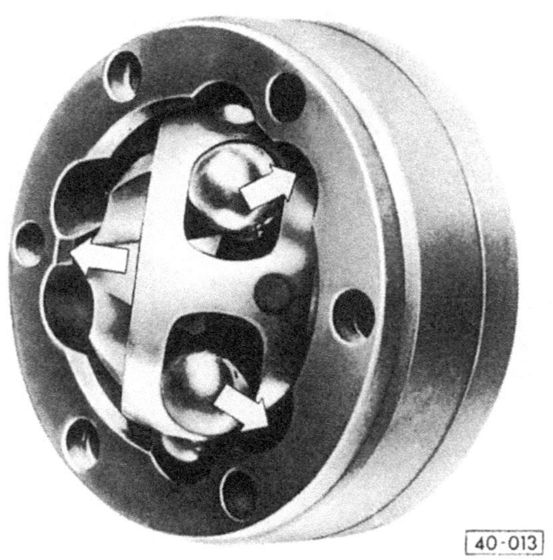

Fig. 5-19. Hub being rotated within cage to align balls with grooves of outer housing (arrows).

6. When the alignment is correct, press on the cage as indicated in Fig. 5-20 until it swings into place.

CAUTION —

The CV joint should go together firmly but smoothly. Heavy force should not be required. If in doubt, start over and recheck the alignment. A joint which is forced together incorrectly may not come apart again.

Fig. 5-20. Balls being engaged in grooves of outer housing. Apply only hand pressure at point indicated by arrow.

7. Check the operation of the joint. The hub should be able to move smoothly throughout its entire range of travel.

8. Pack the joint with 90 grams (approx. 3 oz.) of molybdenum disulfide grease, using 2/3 of the grease on the outer side of the joint, and the remaining 1/3 on the other side.

5.3 Control Arms

A control arm and its bushings are shown in Fig. 5-21. The control arms precisely position the wheel bearing housing between the suspension strut and the body. The control arm is also a mounting point for the stabilizer bar.

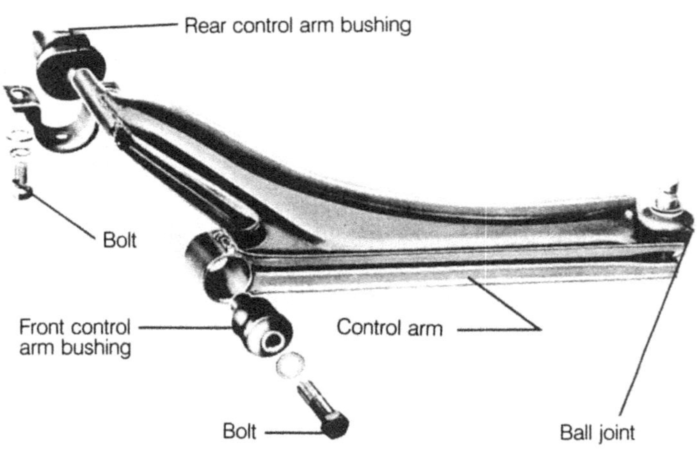

Fig. 5-21. Exploded view of control arm showing control arm bushings.

Control arms damaged in an accident or otherwise distorted can result in alignment that cannot be brought into specification, or in a noticeable and possibly dangerous deterioration in handling. The control arms do not require routine maintenance, but they should be checked for distortion whenever the rest of the suspension is inspected. Using a straightedge, check the control arms as shown in Fig. 5-22 and Fig. 5-23. Replace a control arm with any faults, as described below.

> ***WARNING —***
> *Do not attempt to straighten a bent control arm. Doing this can seriously weaken it. Damaged control arms should be replaced.*

Fig. 5-22. Control arm being checked with straightedge for rearward distortion. Replace control arm if area in circle is kinked or creased.

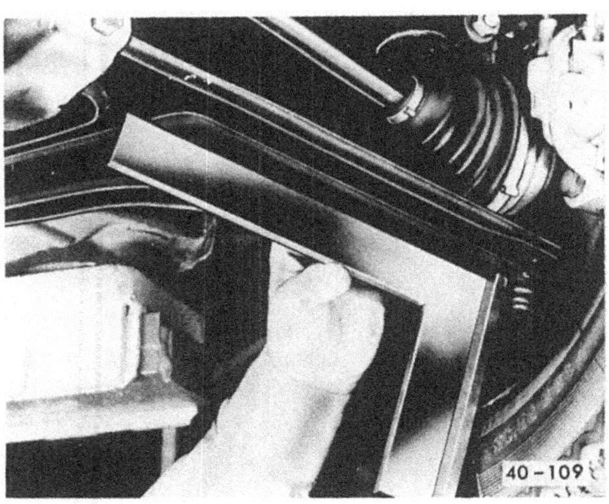

Fig. 5-23. Control arm being checked for horizontal distortion. Replace if not straight and smooth.

Removing and Installing Control Arms

To remove the control arms, raise the car and support it on jack stands. Remove the control arm mounting bolts and nuts. On models with automatic transmission, it is necessary to raise the engine to remove the left (driver's) side front mounting bolt. Install an engine lifting device, remove the front, rear, and left engine/transaxle mounts, and then raise the engine until the bolt can be removed. On all models, remove the nuts that hold the stabilizer bar to the control arm, then remove the ball joint clamping bolt and separate the ball joint from the wheel bearing housing. See **5.4 Ball Joints**.

Installation is the reverse of removal. Use a new self-locking nut on the ball joint clamping bolt and torque it to 50 Nm (37 ft. lb.). Torque the stabilizer bar nuts to 25 Nm (18 ft. lb.). Torque the front control arm mounting bolts to 70 Nm (52 ft. lb.), and the rear mounting nuts to 45 Nm (33 ft. lb.). On cars with automatic transmission, torque the engine mount bolts and nuts to 35 Nm (26 ft. lb.).

Control Arm Bushings

The control arm bushings are replaced with the control arms removed from the car as described above. The rubber bushings are somewhat flexible, but they are a tight fit, and a press is recommended for removal and installation. If a press is not available, arrangements should be made to have this press work done at an authorized Volkswagen dealer or other automotive machine shop.

Press out the old control arm bushings, as shown in Fig. 5-24 and Fig. 5-25. On installation, the front bushing must be oriented as shown in Fig. 5-26. Use soft soap to lubricate new control arm bushings for installation.

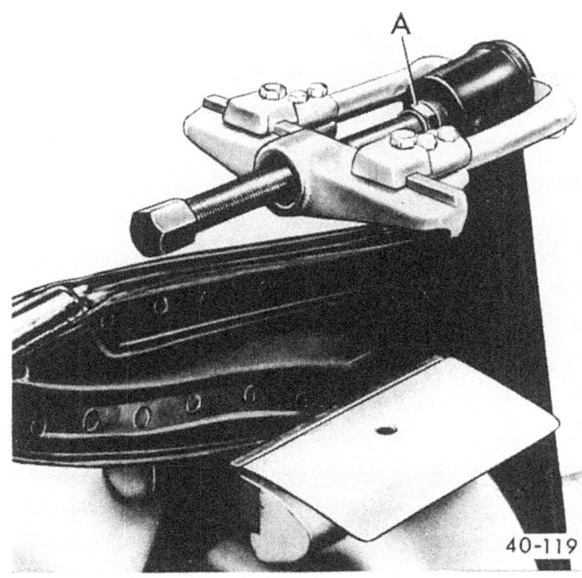

Fig. 5-24. Front control arm bushing being pressed out of control arm using hand puller. Use bolt (**A**) with depression in head to keep puller centered.

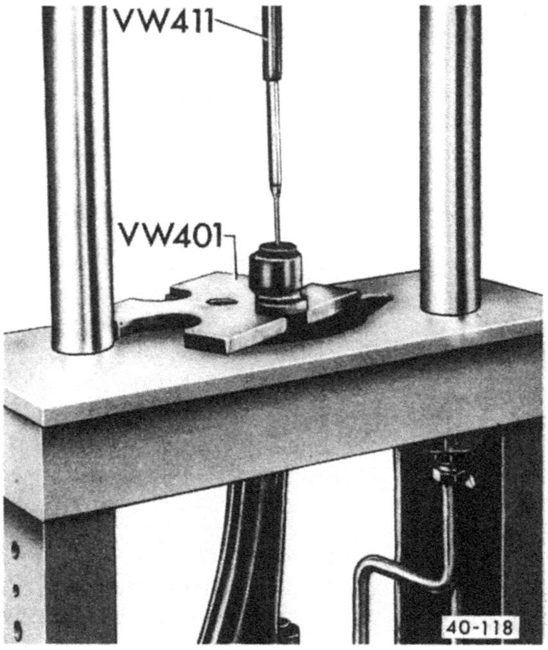

Fig. 5-25. Punch being used to force rear control arm bushing off of control arm pivot.

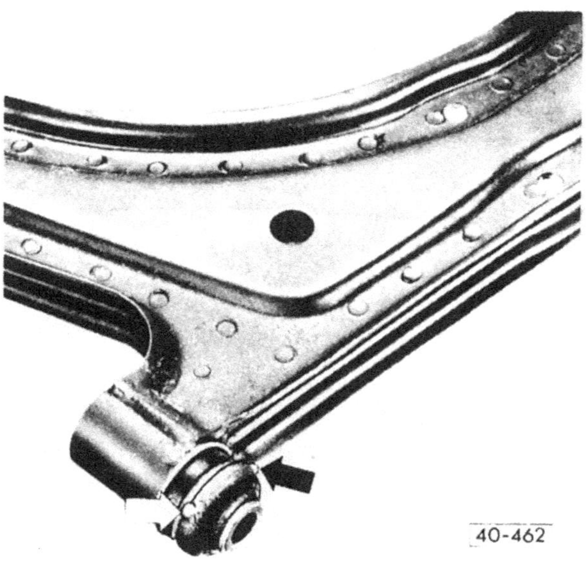

Fig. 5-26. Correct orientation of front bushing in control arm. Lugs (arrows) must point to front. Bushing may differ in appearance, depending on manufacturer.

5.4 Ball Joints

The ball joints provide a precise but moveable connection between the wheel bearing housing and the lower suspension control arm. The ball joints are permanently lubricated and surrounded by a protective rubber boot. No routine maintenance is required, except for periodic inspection to check for wear.

Inspecting and Replacing Control Arm Ball Joints

The ball joints should be checked regularly for freeplay. A simple inspection of the ball joints involves using a pry bar to check freeplay and a caliper or ruler to measure it. Factory installed ball joints are riveted to the control arm. Replacement ball joints are bolted in place. Remove the old joint by drilling out the rivets. Although a wheel alignment is recommended after any suspension repair, it is not strictly necessary if only the ball joint is replaced.

To inspect:

1. Lift the front end of the car and support it securely on jack stands designed for the purpose.
2. Place the tip of the pry bar directly under the socket of the ball joint, and place the metal spacer between the shank of the pry bar and the wheel rim as shown in Fig. 5-27.

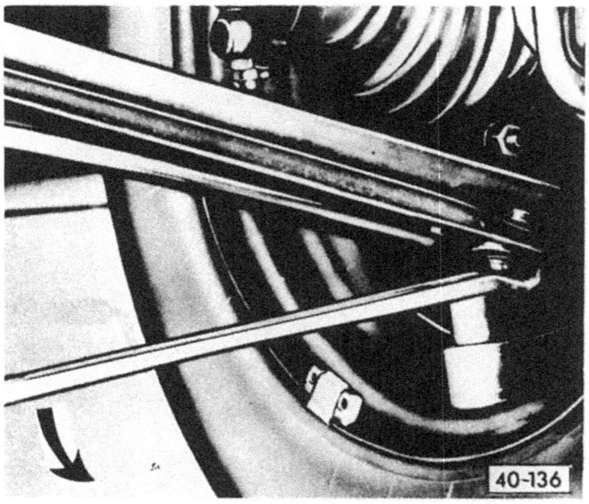

Fig. 5-27. Checking ball joint wear with pry bar on ball joint, and spacer on wheel rim for use as support. Moving lever (arrow) will compress joint.

3. Check for movement of the ball joint socket relative to the wheel bearing housing. Replace any ball joints that can be compressed 2.5 mm (.10 in.) or more. For reference, the freeplay in new ball joints is about 1.0 mm (.04 in.).

To replace:

1. With the front of the car raised and supported securely on jack stands, remove the wheel.
2. Disconnect the stabilizer bar from the control arm.
3. Loosen and remove the clamping bolt that holds the ball joint stud in the wheel bearing housing, and then pull or pry the stud out of the housing.

4. Remove the worn ball joint. Use a 6 mm (1/4 in.) drill bit to drill out the original rivets. If necessary, remove the rivet heads with a cold-chisel after drilling.

Installation is the reverse of removal. Install the new ball joint with bolts and self-locking nuts as shown in Fig. 5-28. Torque to 25 Nm (18 ft. lb.). Torque the ball joint clamping bolt and nut to 50 Nm (37 ft. lb.) and the stabilizer bar nuts to 25 Nm (18 ft. lb.). Always replace self-locking nuts with new parts.

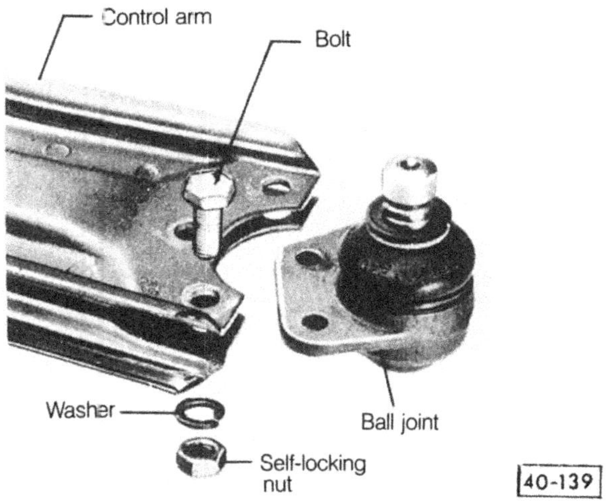

Fig. 5-28. Replacement ball joint being installed on control arm. Center replacement ball joint in slotted mounting holes.

5.5 Bearings

The front suspension includes wheel bearings and strut bearings. Both are subject to wear and deterioration, and both are replaceable.

The strut bearings are replaced by disassembling the struts as described in **5.1 Front Suspension Struts**. The wheel bearings are press-fit into the wheel bearing housing and the wheel hub, and can be replaced after removing the wheel bearing housing.

Removing and Installing Wheel Bearing Housings and Wheel Bearings

The wheel bearings are permanently sealed and maintenance-free. The bearing inner race is press-fit on the wheel hub and the outer race is press-fit into the wheel bearing housing. The bearing is therefore normally destroyed as it is removed.

Removal of the wheel bearing and installation of the new one both require a press. If a press is not available, arrangements should be made to have this press work done at an authorized Volkswagen dealer or other automotive machine shop.

Plan to use new self-locking nuts and cotter pins on the tie rod end and ball joint clamping bolt, as well as new circlips to install the new wheel bearing.

To remove wheel bearing housing:

1. With the vehicle on the ground, pry off the dust cap in the center of the front wheel hub and loosen, but do not remove, the axle shaft nut.

> ***WARNING —***
> *Loosen axle shaft nuts only while the car is on the ground. The leverage required to do this could topple the car from a lift or jack stands.*

2. Raise the front end of the vehicle and support it securely on jack stands.

3. Remove the two bolts holding the brake caliper to the bearing housing, and suspend it from the coil spring using a stiff wire. Remove the countersunk screw that holds the brake rotor to the wheel hub and remove the brake rotor. Also see **BRAKES** for more information on disassembly and assembling the brake components.

> ***CAUTION —***
> *Do not allow the caliper to hang by the brake line. Doing this could damage the line and cause brake failure.*

> **NOTE —**
> By leaving the brake line attached to the caliper, there will be no need to bleed the brake system.

4. Remove the cotter pin and nut holding the tie rod end to the steering arm, and use a puller to remove the tie rod end, as shown in Fig. 5-29.

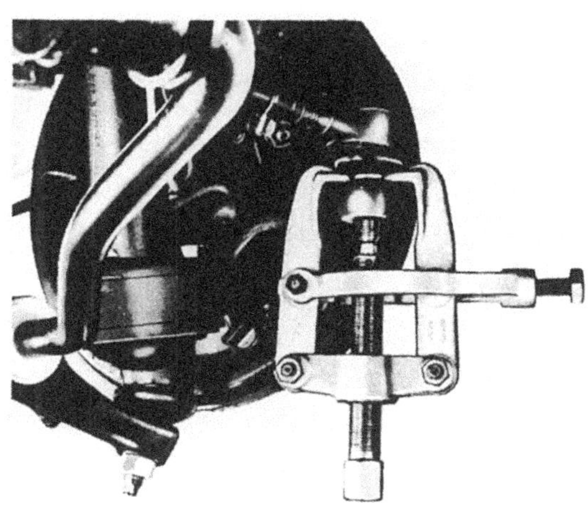

Fig. 5-29. Puller being used to disconnect tie rod end (car shown is not one of those covered by this manual.

CAUTION —
The puller is used to prevent thread damage. Do not use a hammer.

5. Disconnect the stabilizer bar from the control arm.

6. Remove the nut and clamping bolt holding the ball joint stud in the bottom of the wheel bearing housing, and remove the ball joint from the housing so that the control arm hangs free.

7. While supporting the bearing housing and drive axle, remove the bolts mounting the strut to the bearing housing.

NOTE —
Mark the position of the eccentric camber adjustment bolt (upper bolt) before removal. If the bolt is reinstalled in a different position, the front wheels will need to be aligned.

8. Remove the axle shaft nut and pull the bearing housing off the stub axle.

NOTE —
A light tap with a soft-faced mallet will help dislodge a stubborn splined stub axle.

Installation is the reverse of removal. Torque the bolts holding the strut to the bearing housing to 80 Nm (59 ft. lb.). Install the ball joint with a new self-locking nut and torque the clamping bolt to 50 Nm (37 ft. lb.). Torque the stabilizer bar nuts to 25 Nm (18 ft. lb.). Install the tie rod end in the steering arm and torque the nut to 30 Nm (22 ft. lb.). Install a new cotter pin. Remount the brake caliper to the wheel bearing housing and torque the mounting bolts to 40 Nm (29 ft. lb.). Install a new axle nut and torque it to 240 (177 ft. lb.). Torque the wheel lug bolts to 110 Nm (81 ft. lb.).

WARNING —
Always loosen or tighten axle nuts while the car is on the ground. The leverage required to do this could topple the car from a lift or jack stands.

To remove and install front wheel bearings:

1. Remove the wheel bearing housing from the car as described above.
2. From behind, press the hub out of the wheel bearing housing. Remove the brake disc splash shield.
3. Remove the inner bearing race from the hub using a bearing puller as shown in Fig. 5-30.
4. Remove the circlips from the bearing housing on either side of the bearing outer race, and then press out the rest of the bearing, as shown in Fig. 5-31.

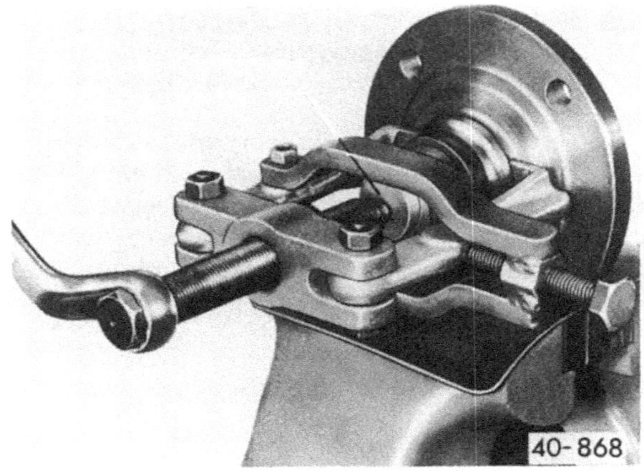

Fig. 5-30. Wheel bearing inner race being removed from hub. Clamping-type puller, as shown, is highly recommended.

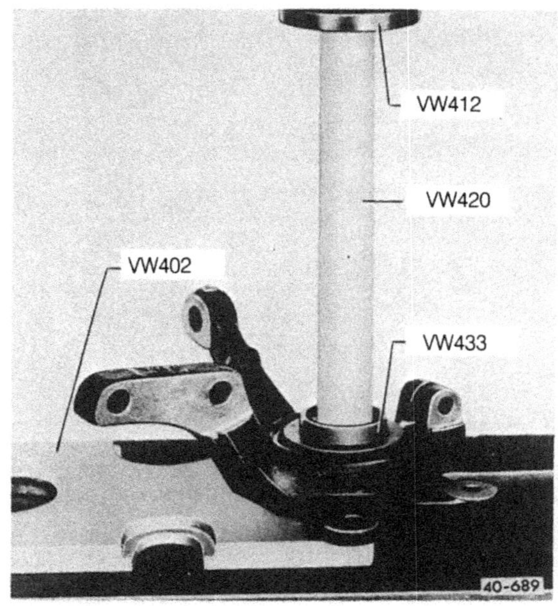

Fig. 5-31. Wheel bearing outer race being pressed out of bearing housing after circlips are removed. Numbers shown identify Volkswagen special tools for press work.

5. Install a new circlip into the innermost groove in the wheel bearing housing. Lubricate the bearing seat with the molybdenum disulfide grease that is supplied with the new bearing, and press in the new bearing, as shown in Fig. 5-32, making sure to press only on the bearing outer race. Then install a new circlip in the outer groove of the housing.

NOTE —
There are two sizes of circlips for the inner groove of the wheel bearing housing, depending on the size of the constant velocity joint. For more information see **5.2 Drive Axles and Constant Velocity Joints**.

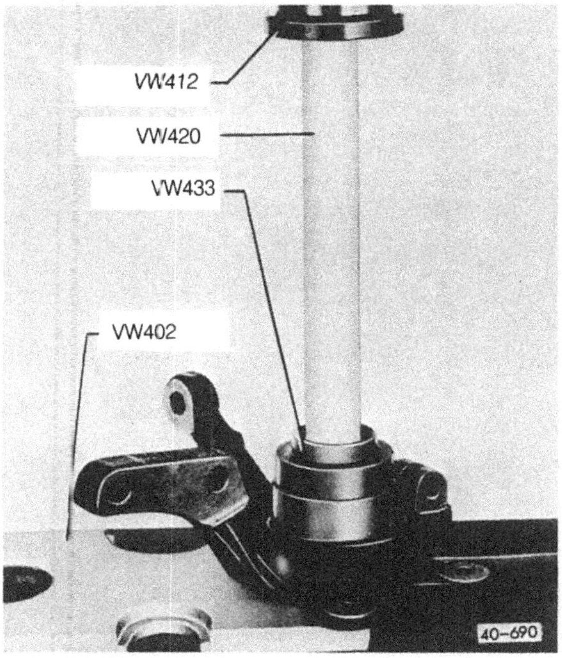

Fig. 5-32. New wheel bearing being pressed into wheel bearing housing. Press on outer bearing race only.

6. Press the hub into the bearing as shown in Fig. 5-33. Be sure to support the bearing on the inner race only.

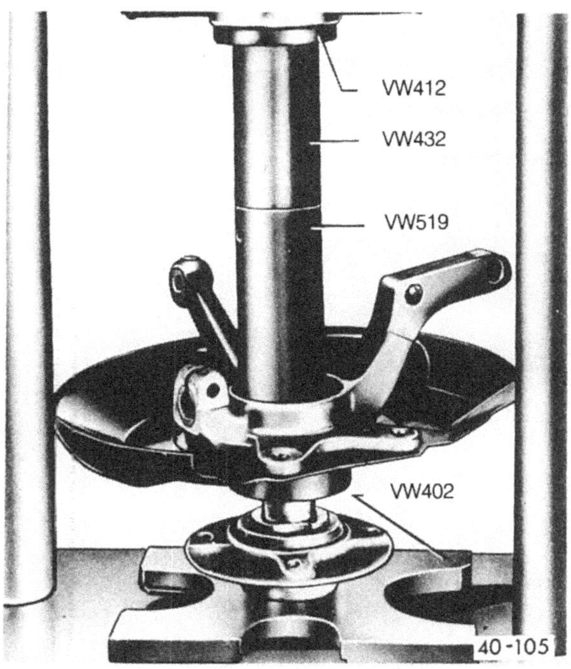

Fig. 5-33. Wheel hub being pressed into bearing. Press tool must support bearing inner race only.

7. Reinstall the wheel bearing housing on the car, as described above.

6. REAR SUSPENSION

The rear suspension, including the axle beam, is shown in Fig. 6-1. The rear shock absorbers and springs are assembled as a unit, much like strut. Unlike the front suspension, the rear springs and shock absorbers can be safely separated without the use of a spring compressor.

The rear wheel bearings do not usually require any routine maintenance, but cleaning, repacking with fresh grease and adjusting the rear wheel bearings are necessary parts of rear brake service. Much of the work described here will also be necessary as a part of work described in **BRAKES**.

The rear suspension components that are most likely to require maintenance or repair are readily accessible and can be removed with standard tools. The wheel bearing outer races are an exception. These fit tightly in the brake drum or rotor and should be removed and installed using a hydraulic press. For any maintenance or repair of the rear suspension or brakes, the following precautions apply:

WARNING —

- *Do not re-use fasteners that are worn or deformed in normal use. Many fasteners are designed to be used only once and become unreliable and may fail when used a second time. This includes, but is not limited to, nuts, bolts, washers, self-locking nuts or bolts, circlips and cotter pins. Always replace these fasteners with new parts.*
- *Do not reinstall bolts and nuts coated with undercoating wax as correct tightening torque cannot be assured on installation. Always clean the threads of removed nuts and bolts with a solvent before installation, or replace them with new parts.*
- *If the axle beam has been removed, or if for any other reason the brake lines have been disconnected, reassembly must include bleeding air from the brake system and readjusting the brakes as described in **BRAKES**.*
- *Do not weld or attempt to straighten a damaged rear axle beam or rear stub axle.*
- *Use only multi-purpose lithium grease to lubricate the rear wheel bearings. Other types of grease will not maintain adequate lubrication and may lead to premature bearing failure.*

6.1 Rear Shock Absorbers and Springs

The rear shock absorbers are removed and installed like any conventional stand-alone shock absorber. In this case, the coil spring is removed or installed at the same time. When the rear suspension is fully extended, as when the car is supported on jack stands, the springs will no longer be compressed and can be removed safely.

For help in determining when the shock absorbers should be replaced, see the information on checking shock absorbers in **5.1 Front Suspension Struts.**

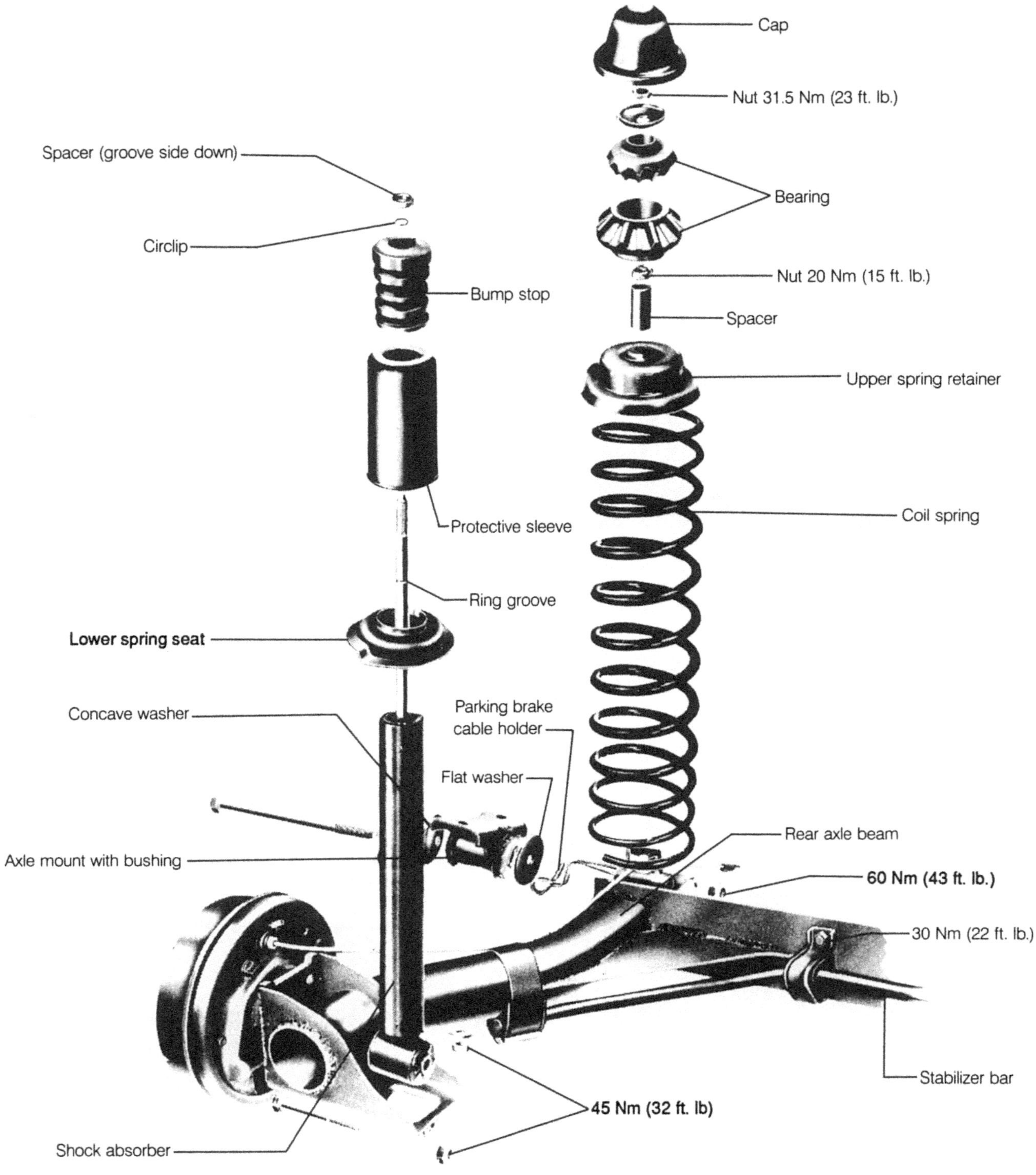

Fig. 6-1. Exploded view of rear suspension, showing shock absorber parts and order of assembly.

Removing and Installing Rear Shock Absorbers and Springs

Remove and install the shock absorber and spring assemblies one side at a time to prevent the rear axle assembly from hanging unsupported and damaging the brake lines.

WARNING —
Do not re-use fasteners that are worn or deformed in normal use. Many fasteners are designed to be used only once and become unreliable and may fail when used a second time. This includes, but is not limited to, nuts, bolts, washers, self-locking nuts or bolts, circlips, cotter pins. Always replace these fasteners with new parts.

To remove and install:

1. Inside the car, snap off the plastic cap that covers the upper shock absorber mount. Loosen and remove the nut and the washer beneath the nut.

2. Raise the car slowly until the coil spring is unloaded and the top of the shock absorber and spring assembly begins to pull away from the body.

NOTE —
When jacking, do not use any part of the suspension for a lifting point. The axle beam and trailing arms must not be supporting the car.

3. Position jack stands to support the car securely. Remove the nut and bolt that mount the bottom of the shock absorber to the trailing arm, and remove the assembly from the car.

4. On installation, torque the bolt and nut at the bottom of the shock absorber to 45 Nm (33 ft. lb.). Torque the top mounting nut to 31.5 Nm (23 ft. lb.).

6.2 Rear Wheel Bearings

The rear wheel bearings are dual tapered roller bearings. The inner bearing races fit on the stub axles. The replaceable outer races are pressed into the brake drum or rotor. Fig. 6-2 is an exploded view of the rear wheel bearings and drum brakes. Cars with rear disc brakes are similar.

Rear wheel bearing service consists of removing the bearings, cleaning and inspecting the bearings and outer races, repacking with fresh grease, and adjusting.

Removing and Installing Rear Wheel Bearings

The rear wheel bearings can be removed, installed, and adjusted using ordinary hand tools. If the bearings are to be replaced, the bearing outer races (which are provided with the new bearings) should be replaced also. The races are most easily and properly installed using a press. The axle nut hardware is locked in place with a cotter pin which should always be replaced rather than reused.

To remove:

1. Raise the rear of the car, support it securely on jack stands which are designed for the purpose, and remove the wheel.

2. On cars with drum brakes, fully back off the brake adjuster (adjust the brake shoes away from the drum) as described in **BRAKES**.

3. On cars with rear disc brakes, remove the rear brake caliper, as described in **BRAKES**, and hang it out of the way without disconnecting the brake line, using a stiff wire or equivalent.

4. Pry off the grease cap in the center of the brake drum or disc, remove the cotter pin and nut lock, and then remove the axle nut, the thrust washer, and the outer bearing.

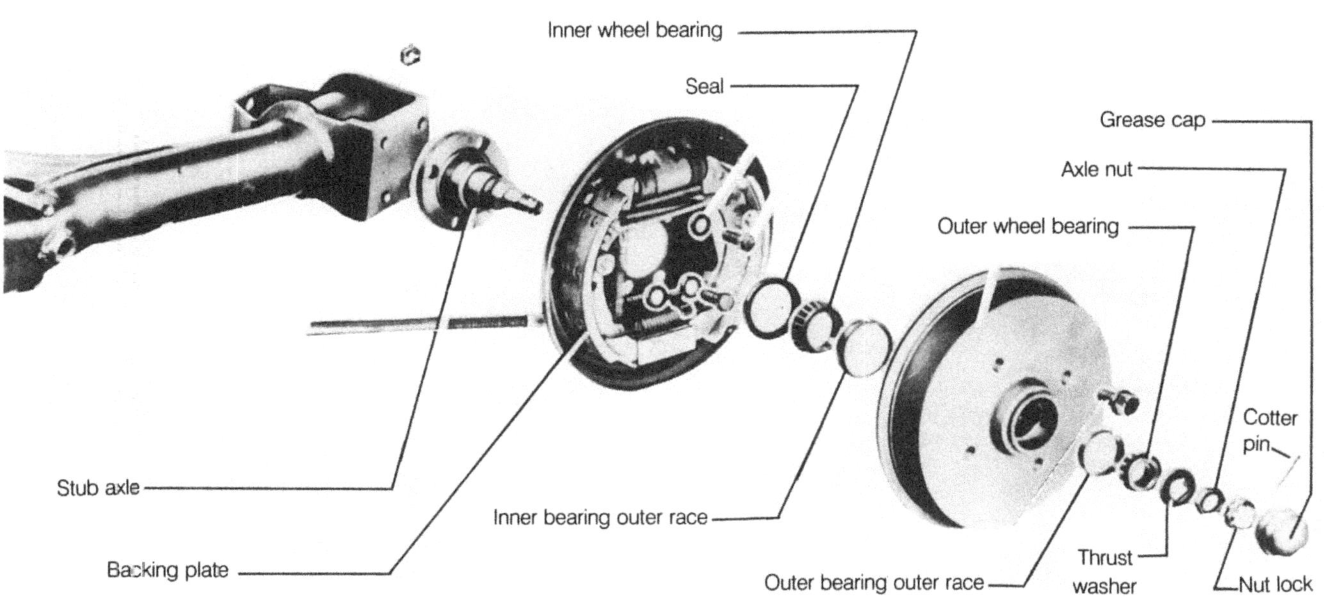

Fig. 6-2. Exploded view of rear axle and wheel bearings. Drum brake model shown. Cars with rear disc brakes are similar.

WARNING —
Do not re-use fasteners that are worn or deformed in normal use. Many are designed to be used only once and may fail when used a second time. This includes self-locking nuts, washers and cotter pins. Always replace these fasteners with new parts.

NOTE —
If the thrust washer and outer bearing are reluctant to move, thread the axle nut back on, leaving a slight gap between it and the thrust washer. A slight jerk on the brake drum, pulling away from the car, will loosen the bearing and thrust washer. The nut will prevent scattering the parts on the ground.

5. Remove the brake drum or rotor. Pull it straight off the stub axle to prevent damage to the bearing races and the threads of the stub axle.
6. Working on the back side of the brake drum or rotor, pry the grease seal out of the recess in the hub. Lift out the inner bearing.
7. If the brake drum or rotor is being replaced, or if the bearings are being replaced, use a brass drift or punch to drive the outer bearing races out of the hub. A brass drift is used to avoid damaging the machined surfaces of the hub.

Fig. 6-3. New outer wheel bearing outer race being pressed into hub. Brake drum shown. Installation in disc brake rotor is similar.

To install:

1. Clean the hub and the bearings in solvent and let dry thoroughly. Dry with compressed air if available.

CAUTION —
Do not use gasoline as a solvent, as it will remove all lubrication. Also, do not let blasts of compressed air spin the unlubricated bearings.

2. Inspect the bearing rollers and races. Replace any that is pitted, burred, rough, or heat-blued.
3. Install the outer bearing races in the hub, using a hydraulic press as shown in Fig. 6-3 and Fig. 6-4.
4. Pack the inner bearing with multipurpose grease. Coat the bearing outer races in the hub with grease, and lightly coat the interior of the hub to prevent corrosion.

CAUTION —
Use only multipurpose (lithium) grease to lubricate the rear wheel bearings. Other greases will not maintain adequate lubrication and may lead to bearing failure.

NOTE —
When packing bearings, cup the grease in the palm of your hand and force it into the bearing all possible ways, especially around the rollers from the ends, until the bearing is completely filled with grease and it oozes out the other side. If in doubt, pack in some more.

Fig. 6-4. New inner wheel bearing outer race being pressed into hub. Brake drum shown. Installation in disc brake rotor is similar.

5. Place the inner bearing in its outer race on the back side of the hub, and press in a new grease seal.
6. Lightly coat the surface of the stub axle with grease, and slide the brake drum or rotor onto the stub axle. Move carefully to avoid damaging the grease seal or the bearing races on the threads of the stub axle.
7. Pack the outer bearing with multipurpose grease, and slide it over the stub axle and into its outer race. Install the thrust washer and the axle nut. Tighten the nut until the bearings just contact their outer races and the assembly just starts to become snug.

8. Adjust wheel bearing end thrust as described below in **Adjusting Rear Wheel Bearings**.

9. Fill the grease cap with approximately 10 grams (about $^3/_8$ oz.) of multipurpose grease. Install the grease cap using a soft-faced mallet.

10. Install the disc brake caliper or adjust the drum brakes as described in **BRAKES**.

11. Install the wheel and tire. Tightening torque for the wheel lug bolts is 110 Nm (81 ft. lb.).

Adjusting Rear Wheel Bearings

The rear wheel tapered roller bearings must be adjusted with the proper pre-load to work smoothly and minimize wear. This adjustment must be made any time the rear brake drum or disc brake rotor has been removed. Other periodic bearing adjustments to compensate for wear may help to increase bearing life or give early warning of excessive wear and the need for replacement. The only materials needed are a little extra multipurpose grease for the grease cap, and new cotter pins.

> ***WARNING —***
> *Do not re-use fasteners that are worn or deformed in normal use. Many fasteners are designed to be used only once and become unreliable and may fail when used a second time. This includes, but is not limited to, nuts, bolts, washers, self-locking nuts or bolts, circlips and cotter pins. Always replace these fasteners with new parts.*

To adjust:

1. Raise the rear of the car and support it securely on jack stands. Remove the wheel.

2. Pry off the grease cap in the center of the brake drum or disc brake rotor, remove the cotter pin and nut lock, then loosen the axle nut.

3. If the wheel bearings, especially new ones, have just been installed, temporarily torque the axle nut to 10 Nm (87 in. lb.), while slowly turning the brake drum or rotor by hand. Then loosen the nut slightly.

> ***CAUTION —***
> *Avoid over-torquing the axle nut, as this may damage the bearings and the bearing races.*

4. Tighten the axle nut again, slowly and in small increments, until the thrust washer can just be moved back and forth with light pressure on from a screwdriver, as shown in Fig. 6-5.

> **NOTE —**
> Do not twist or pry with the screwdriver. Use hand pressure only.

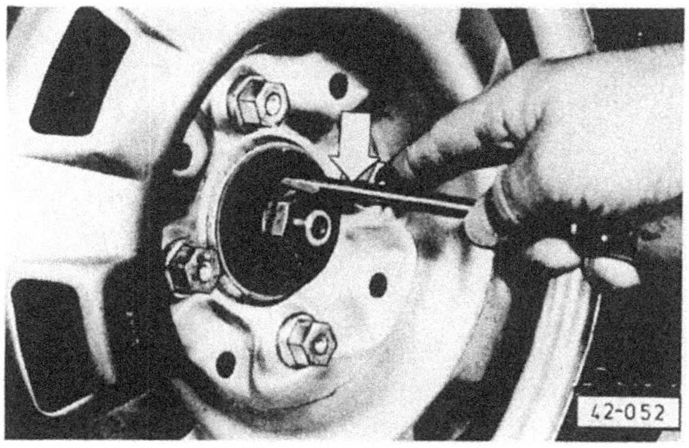

Fig. 6-5. Wheel bearing adjustment being checked. Axle nut is tightened correctly when thrust washer moves with slight pressure.

5. After the wheel bearings are correctly adjusted, install the nut lock so that its projections do not cover the cotter pin hole. If necessary, tighten the axle nut slightly so that the nut lock will align with the hole in the stub axle. Then install a new cotter pin.

6.3 Rear Axle Beam

The rear suspension can be removed and installed as a unit for replacement or for facilitating rear body repairs. The axle beam is mounted to the body by brackets at the front of the axle, cushioned and supported by replaceable rubber bushings. The axle beam bushings can be replaced with the axle installed.

Removing and Installing Rear Suspension as a Unit

The car must be raised high enough to remove the axle assembly, so a lift or a jack and suitable jack stands are required. Removal of the axle assembly also requires disconnecting brake lines to the rear wheels, and bleeding the brakes when reinstallation is complete.

To remove:

1. Disconnect the parking brake cables at the parking brake lever between the front seats. See **BRAKES**.

> **NOTE —**
> If the car can be raised and safely supported high enough, the shock absorber and spring units can stay attached to the axle beam at the bottom and be removed with the axle beam.

2. Remove the rear shock absorbers and springs as described above in **Removing and Installing Rear Shock Absorbers and Springs**.

3. Disconnect the flexible rear brake hoses from the rigid brake lines near the forward edge of the axle beam. Cap both ends of the open connection to prevent contamination.

4. While supporting the front of the axle beam so that it cannot fall, loosen and remove the nuts which hold the axle beam mounting brackets to the car body.

5. Remove the rear suspension assembly by pulling it out to the rear.

To install:

1. If the axle beam mounting brackets have been removed from the axle beam, or if the pivot bolts have been loosened, align the mounting brackets as shown in Fig. 6-6, then torque the pivot bolts and their nuts to 60 Nm (44 ft. lb.).

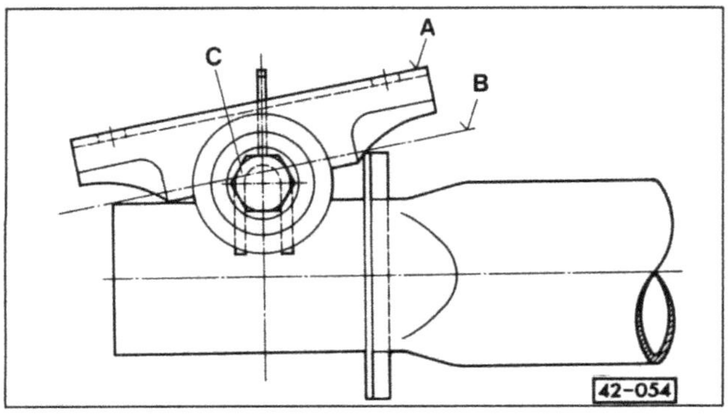

Fig. 6-6. Axle beam mounting brackets being aligned. Align mounting surface **A** with imaginary line **B** and torque nut **C** to 60 Nm (44 ft. lb.).

2. Position the rear suspension assembly under the car and install the mounting nuts. Torque them to 45 Nm (33 ft. lb.).

3. Clean the brake line unions and reconnect the flexible brake hoses to the rigid brake lines. Torque the unions to 15 to 20 Nm (11 to 15 ft. lb.).

4. Remount the rear shock absorbers and springs. Carefully lower the car while guiding the top shock absorber mounts into their mounting holes.

5. Reconnect the top shock absorber mounts as described in **Removing and Installing Rear Shock Absorbers and Springs**. Torque the nuts to 31.5 Nm (23 ft. lb.).

6. Reconnect the parking brake cables.

7. Bleed the brakes and adjust the parking brake as described in **BRAKES**.

Replacing Axle Beam Bushings

Axle beam bushings are replaced by pressing the old bushings out of the axle beam mounting bracket, and pressing the new bushings in. To replace the axle beam bushings without removing the axle beam, support the axle beam and the car equally so that the pivot bolts are not under load. Work on one side at a time. Remove the nuts that hold the axle mounting to the body, and remove the axle pivot bolt. Then remove the mounting and replace the bushing.

Installation is the reverse of removal. Align the axle mounting with the axle as shown above in Fig. 6-6. Torque the pivot bolts and their nuts to 60 Nm (44 ft. lb.), and torque the bracket mounting nuts to 45 Nm (33 ft. lb.).

Removing and Installing Rear Stabilizer Bar

The rear stabilizer bar is held in place by two steel straps and two mounting brackets. See Fig. 6-1 above. To remove the bar, drive off the securing clips at the bottom of the straps using a hammer, and remove the nuts that secure the mounting brackets. Replace any rubber bushings that are cracked or worn. Install the straps by first holding the ends closed with locking pliers, and then slipping on the clips. Torque the nuts on the mounting brackets to 30 Nm (22 ft. lb.)

Checking Rear Stub Axle

To check the rear stub axle for bending or distortion, first remove the wheel bearings as described in **6.2 Rear Wheel Bearings**. Check the stub axle as shown in Fig. 6-7, measuring at least three different places on the axle. Replace the axle if the difference between any two measurements exceeds 0.25 mm (.010 in.). To replace a stub axle, remove the remainder of the brake assembly as described in **BRAKES**, then remove the bolts holding the stub axle to the axle beam. Make sure that the axle mounting surfaces are free of all dirt, corrosion, or grease before installing the new stub axle. Torque the stub axle bolts to 60 Nm (44 ft. lb.).

Fig. 6-7. Rear stub axle being checked for distortion using machinist's square and calipers. Check at least three points between square and outer wheel bearing seat.

7. STEERING

The steering wheel and steering column are connected to the rack and pinion steering gear by a flexible universal joint shaft. When the steering wheel turns, the rack and the tie rods turn the wheels. Power-assisted steering is similar. The power assist comes from hydraulic pressure, used to boost steering response and reduce steering effort. Fig. 7-1 is an exploded view of the basic steering system without power assist. The top end of the universal joint shaft meets the bottom of the steering column. Special procedures and Warnings apply to any repair work on airbag-equipped cars (Cabriolet from 1990).

WARNING —
- *The airbag is operated by an explosive device. Handled improperly or without adequate safeguards, it can be accidently activated and cause serious personal injury.*
- *Thoroughly read all Cautions and Warnings at the front of this manual and in **8. Airbag** before beginning any work on an airbag-equipped car.*

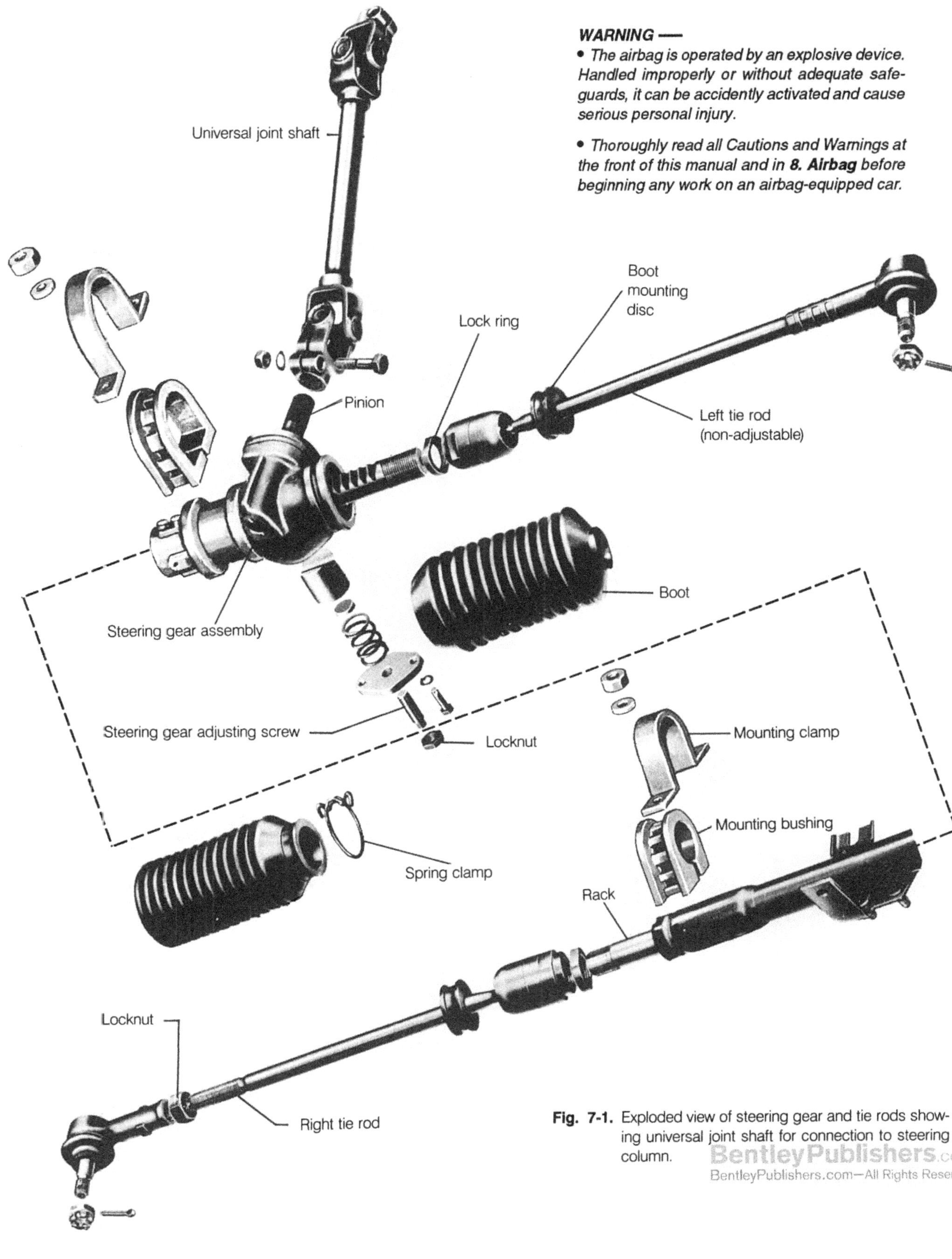

Fig. 7-1. Exploded view of steering gear and tie rods showing universal joint shaft for connection to steering column.

Loose, sloppy or imprecise steering may be caused by wear and excessive play or clearance anywhere in the steering system. The steering gearbox can be adjusted to compensate for internal wear. Otherwise, faults in the steering system must be corrected by replacing worn parts.

Excessive steering effort or imprecise feel may not be caused by faults in the steering system. Faulty front suspension strut bearings or other worn front suspension parts may cause or contribute to the problem. Before doing anything else, check the tires and tire inflation pressures.

7.1 Steering Column

This information applies to 1985-1989 cars with manual or power-assisted steering. The steering column for airbag-equipped cars (Cabriolet from 1990) is covered in **8. Airbag**.

> ***WARNING —***
> - *The airbag is operated by an explosive device. Handled improperly or without adequate safeguards, it can be accidently activated and cause serious personal injury.*
> - *Thoroughly read all Cautions and Warnings at the front of this manual and in **8. Airbag** before beginning any work on an airbag-equipped car.*

Fig. 7-2 is an exploded view of the steering column. The steering column connects the steering wheel to the steering gear. Several accessory switches are also mounted on the tube which houses the steering column. The steering column tube is bolted to the body with shear bolts which are designed to break in the event of a collision.

Removing and Installing Steering Wheel

The steering wheel must be removed for access to the steering column switches and the ignition switch. The splines joining the steering wheel and the steering column also allow the steering wheel position to be adjusted as the final step in the wheel alignment process.

> ***CAUTION —***
> - *The steering wheel should only be repositioned on the splines after the steering rack and tie rods are centered and adjusted as described in **7.2 Steering Gear and Tie Rods**.*
> - *This is especially important for cars with power-assisted steering because the system is sensitive to deviation from the centered, straight-ahead steering wheel position.*

To remove the steering wheel, first disconnect the battery negative (–) terminal. Then pry off the center cover of the steering wheel. On models with four horn pads, use a screwdriver under the center cover to pry it up.

Disconnect the wiring from the horn pad or pads. Hold the steering wheel, remove the nut and washer that hold it to the steering column, and pull it straight off the column.

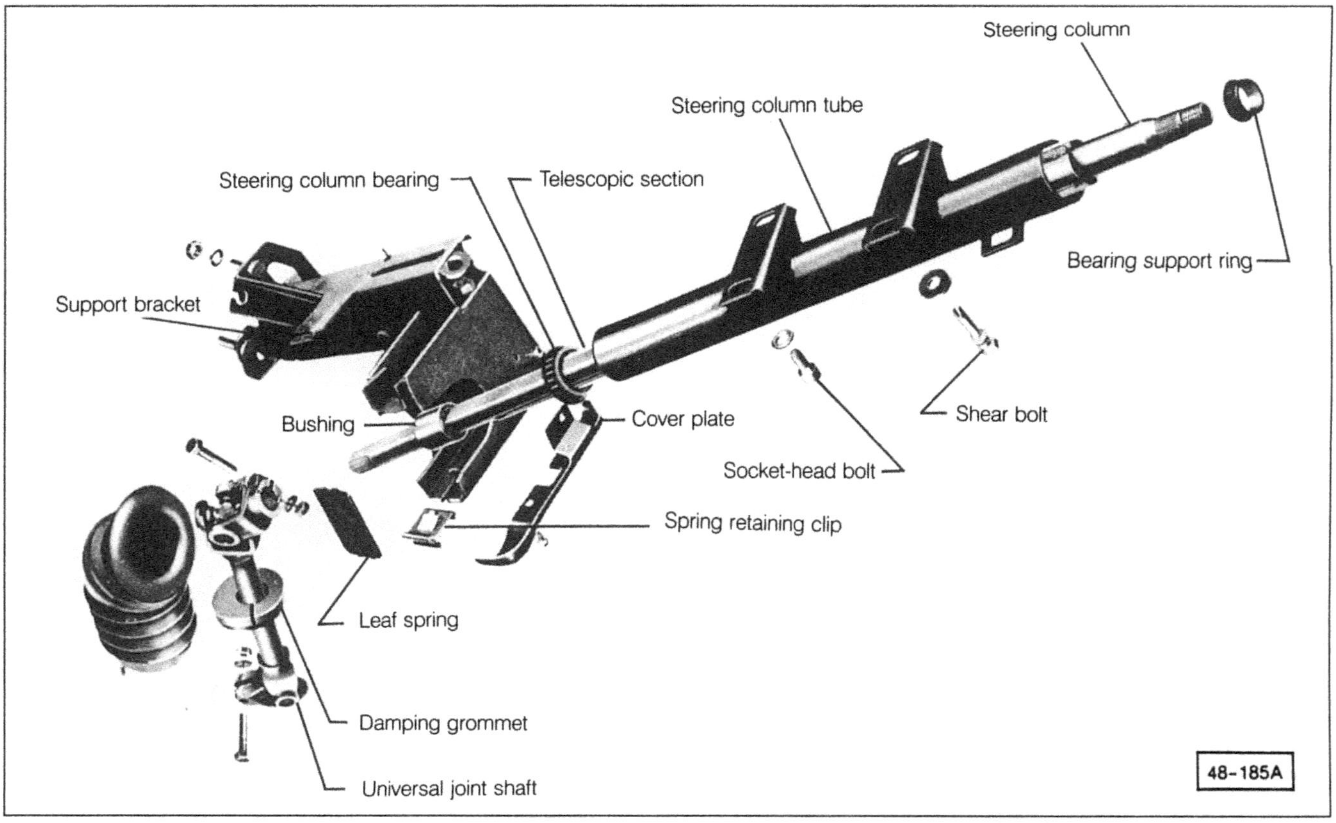

Fig. 7-2. Exploded view of steering column and steering column tube.

NOTE —
Before removing the steering wheel, mark the hub and the shaft so that, unless it is being adjusted, the steering wheel can be reinstalled in exactly the same position.

Installation is the reverse of removal. When aligning the steering wheel, be sure the wheels are pointed straight ahead. Torque the nut on the steering column to 50 Nm (37 ft. lb.).

Removing and Installing Steering Column Switches

Fig. 7-3 is an exploded view of the steering wheel and steering column switches. The turn signal switch and the windshield wiper and washer switch are fastened to the top of the steering column housing by three screws. They can be replaced individually if necessary, but individual parts for repairing the switches are not available.

The ignition switch and lock cylinder also mount to the steering column housing, which in turn clamps to the steering column tube. There is no provision for adjusting the height of the steering wheel or the switches on the steering column.

To remove:

1. Remove the steering wheel as described in **Removing and Installing Steering Wheel**.
2. Remove the Phillips-head screws that hold the two halves of the steering column switch cover together, and remove the lower cover.
3. Remove the three screws that hold the turn signal and windshield wiper switch assemblies to the steering column housing. Disconnect the harness connectors from the switches and remove the switches from the steering column.

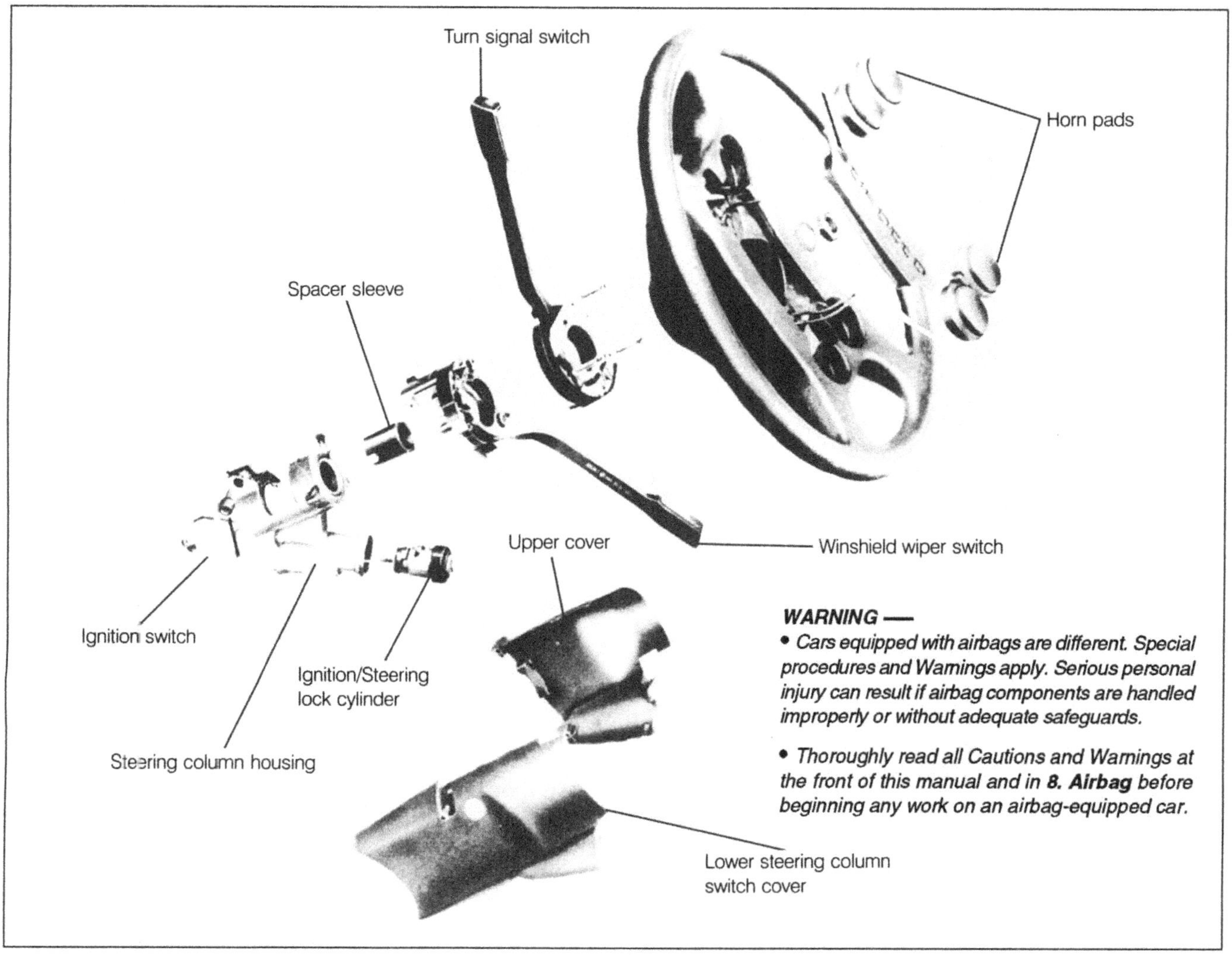

WARNING —
- *Cars equipped with airbags are different. Special procedures and Warnings apply. Serious personal injury can result if airbag components are handled improperly or without adequate safeguards.*
- *Thoroughly read all Cautions and Warnings at the front of this manual and in **8. Airbag** before beginning any work on an airbag-equipped car.*

Fig. 7-3. Exploded view of steering wheel, steering column housing, and switches.

4. Remove the spacer sleeve by twisting it off of the steering column using channel-lock pliers.

5. Remove the lock washer, the spring, and the horn contact ring. Remove the socket head bolt that clamps the steering column housing to the steering column tube.

6. Disconnect the wiring from the ignition switch. Insert the ignition key and unlock the steering. Remove the steering column housing and the ignition/steering lock, together with the upper half of the steering column switch cover.

7. Remove the ignition switch by removing the screw that holds it to the steering column housing.

Installation is the reverse of removal. Drive the spacer sleeve onto the steering column so that the steering wheel clearance is as shown in Fig. 7-4. Use a new lock washer on the steering column. Torque the socket-head steering column housing clamping bolt to 10 Nm (87 in. lb.). Avoid overtightening the three switch mounting screws.

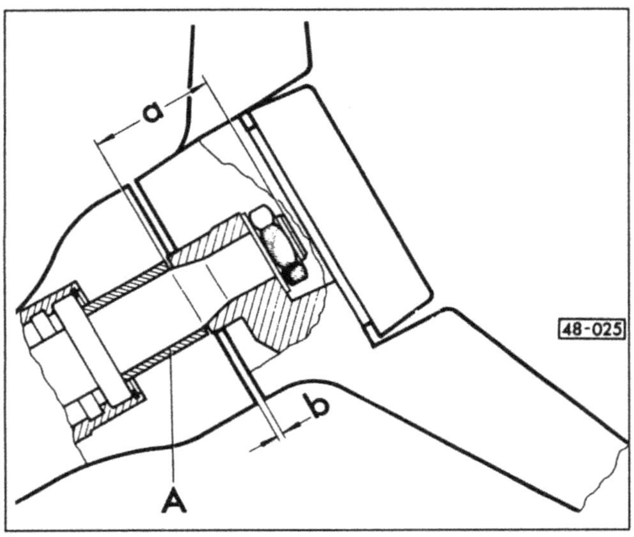

Fig. 7-4. Spacer sleeve (**A**) installed for correct clearance. Drive on sleeve so that dimension **a** is 41.5 mm (1.63 in.). With steering wheel installed, dimension **b** should be 2 to 4 mm (1/16 to 1/8 in.).

Replacing Ignition/Steering Lock Cylinder

The steering column housing does not need to be removed from the steering column in order to replace or repair the lock cylinder, but removal of the entire assembly may make it easier.

To replace the ignition/steering lock cylinder, drill a 3 mm (1/8 in.) hole in the lock cylinder housing at the point indicated in Fig. 7-5. Insert a pin through the hole and press down on the spring that holds the lock cylinder in the housing. Using the ignition key may help by inserting it into the cylinder to pull the cylinder out. To install the cylinder, simply press it in without the key until it snaps into place.

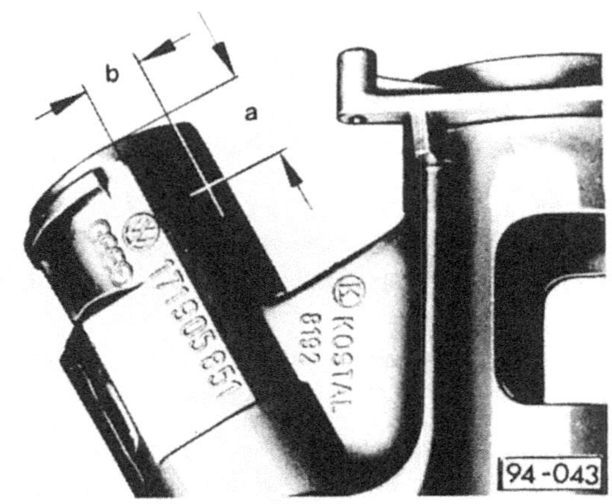

Fig. 7-5. Location of 3 mm (1/8 in.) hole to be drilled in housing for access to ignition/steering lock cylinder retaining spring. Dimension **a** is 12 mm (15/32 in.); dimension **b** is 10 mm (3/8 in.).

7.2 Steering Gear and Tie Rods

In the rack and pinion steering gear, the steering pinion translates rotary motion of the steering wheel into lateral motion of the steering rack. Tie rods connect the ends of the steering rack to the steering arms near the wheels.

The inner and outer tie rod ends are ball-jointed to combine precise steering with flexibility for suspension movement. Rubber boots protect the joints and retain lubricant. Other than periodic inspections for wear or damage, no maintenance or additional lubrication is required for the life of the joints.

Adjustments to the steering rack, and adjustment or replacement of tie rods will virtually always affect wheel alignment. Plan to include a wheel alignment as the final step in any such work.

Inspecting Tie Rod Ends

There are two concerns when inspecting tie rod ends and protective boots. First, wear and excess play in the tie rod ball joints and, second, damage to the protective boot which will accelerate that wear.

Check the condition of the inner and outer tie rod boots. Cracked or ripped inner tie rod boots can be replaced separately, perhaps prolonging the life of the tie rod end. See **Replacing Inner Tie Rod Boots**. Outer tie rod boots are not separately available. Replacement of damaged outer tie rod boots is accomplished only by replacing the tie rod.

To inspect the tie rod ends, look for unwanted play between the steering rack and the tie rods or between the tie rods and the steering arms. At the outer tie rod ends, there should be little or no vertical freeplay. At all four tie rod ends, inner and outer, there should be little or no lateral play. Any turning motion of the wheel should be directly translated to the steering rack with no freeplay in between. Try forcing mating parts in opposite directions and looking for freeplay. Replace any tie rod that fails these tests, as described in **Replacing and Adjusting Tie Rods**.

Replacing Inner Tie Rod Boots

The inner tie rod boots can be replaced without removing the tie rod from the steering rack, and alignment adjustments are not necessary as long as tie rod length is unchanged. Disconnect the outer tie rod end from the steering arm. Use a puller as shown in Fig. 7-6 to avoid damage to the threads. Remove the boot clamps and slide the boot off over the outer tie rod end. Installation is the reverse of removal. Torque the tie rod end stud nut to 30 Nm (22 ft. lb.), and install a new cotter pin.

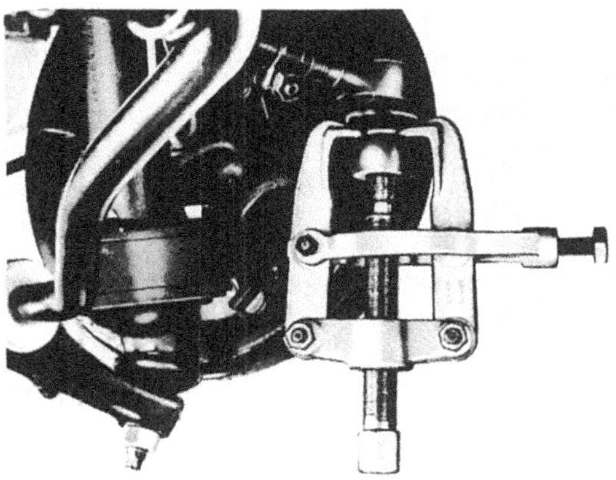

Fig. 7-6. Outer tie rod end being disconnected from steering arm with puller to avoid thread damage (car shown is not one covered by this manual).

Replacing and Adjusting Tie Rods

Only complete tie rods are supplied as replacement parts. All replacement tie rods are adjustable. When replacing a non-adjustable left-side tie rod, the new adjustable tie rod must first be adjusted to a standard length. Use a puller to disconnect outer tie rod ends from the steering arms, as shown in Fig. 7-6 above. Use a thread locking compound such as Volkswagen's D6 when installing tie rods.

On cars with power-assisted steering, tie rod ends are torqued directly to the steering rack. Locknuts are not used. For this reason, the rack itself must be securely held when loosening or tightening the tie rod ends, to avoid damaging the steering gear. The steering gear assembly should be removed from the car and secured in a soft-jawed vise to remove and install the tie rod ends.

To replace and adjust (manual steering only):

1. Remove the nut from the outer tie rod end in the steering arm, and press the stud out using a puller, as shown in Fig. 7-6 above.
2. Remove the clamps that hold the tie rod boots to the steering gear housing and slide the boots back over the tie rod to expose the inner tie rod end.
3. Loosen the locknut on the inner tie rod end and unscrew the tie rod from the steering rack.
4. Center the steering gear housing, as described in **Centering Manual Steering Rack.**
5. Adjust the length of the tie rods by loosening the outer tie rod end locknut and screwing the tie rod in or out. Adjust the length to 379 mm (14.921 in.), as shown in Fig. 7-7.

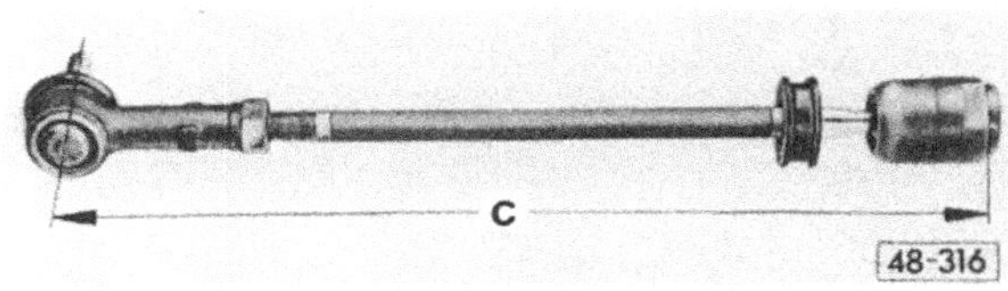

Fig. 7-7. Setting length of adjustable replacement tie rods for manual steering. Dimension **c** is 379 mm (14.921 in.).

6. With a new inner boot in place on the tie rod, thread the new left (driver) side tie rod onto the steering rack, adjusting the position of the tie rod as shown in Fig. 7-8. Tighten the lockring to 50 Nm (37 ft. lb.) to hold this position.

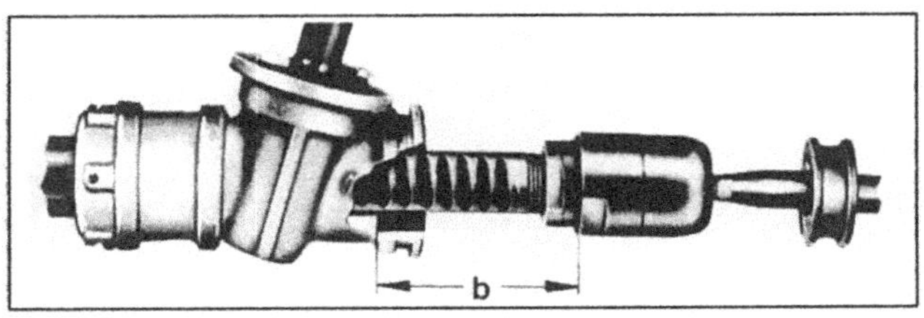

Fig. 7-8. Positioning new left (driver) side ball joint on manual steering rack. With rack centered, dimension **b** is 69.0 mm (2.717 in.).

7. With a new inner boot in place on the tie rod, thread the new right (passenger) side tie rod onto the steering rack, adjusting the position of the tie rod as shown in Fig. 7-9. Tighten the lockring to 50 Nm (37 ft. lb.) to hold this position.

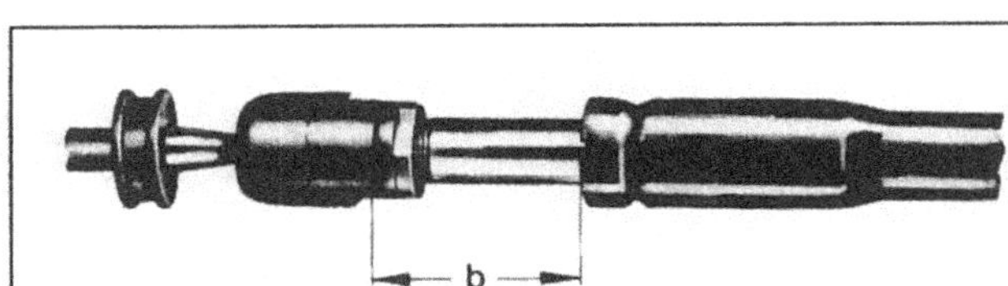

Fig. 7-9. Positioning new right (passenger) side ball joint on manual steering rack. With rack centered, dimension **b** is 67.0 mm (2.638 in.).

8. Reposition the inner boot on the tie rod and the steering gear housing and install the clamps. Make sure that the boot is not twisted.

9. Reconnect the tie rod end to the steering arm. Install the nut, torquing it to 30 Nm (22 ft. lb.), and a new cotter pin.

10. Adjust the toe to the specifications in **4. Wheels, Tires, and Alignment**.

To replace and adjust (power-assisted steering only):

1. Remove the nut from the outer tie rod end in the steering arm, and press the stud out using a puller, as shown in Fig. 7-6 above.

2. Remove the steering gearbox as described below in **Removing and Installing Steering Gearbox**.

3. Remove the clamps that hold the tie rod boots to the steering gear housing and slide the boots back over the tie rod to expose the inner tie rod end.

4. Holding the steering rack securely to prevent steering gear damage, as shown in Fig. 7-10, unscrew the inner tie rod end from the steering rack.

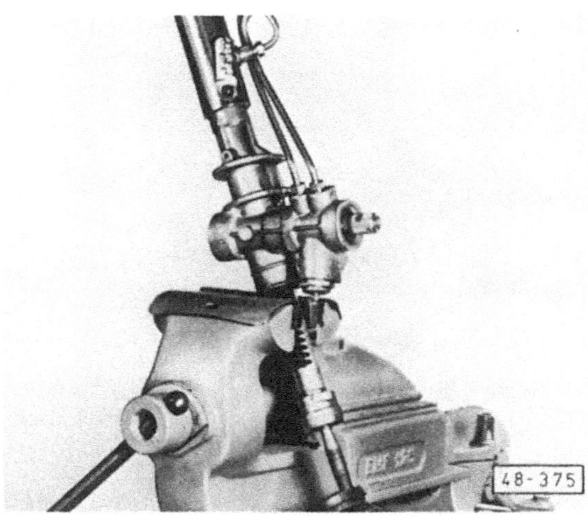

Fig. 7-10. Steering rack for power-assisted steering mounted in soft-jawed vise for removal of tie rod.

5. Clean the threads of the rack, then coat the threads of the new tie rod with locking compound. Thread the tie rod onto the rack and torque to 70 Nm (52 ft. lb.). Make sure that the boot is correctly installed and not twisted.

6. Adjust the length of the adjustable tie rods by loosening the outer tie rod end locknut and screwing the tie rod end in or out. Adjust the length to 362.0 mm (14.252 in.) as shown in Fig. 7-11.

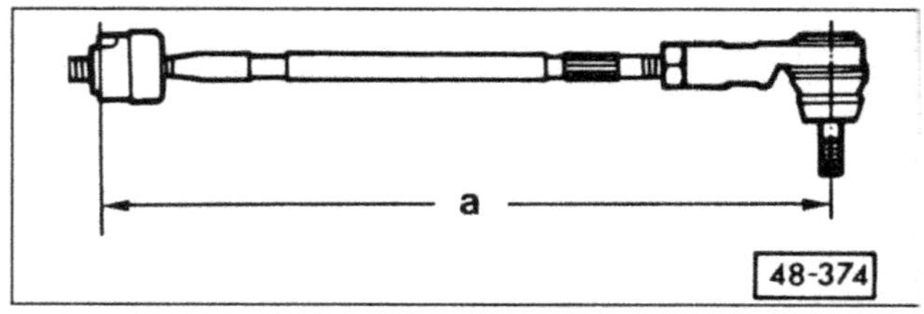

Fig. 7-11. Setting basic length of adjustable tie rods for power-assisted steering. Dimension **a** is 362.0 mm (14.252 in.).

7. Reposition the boot on the tie rod and the steering rack housing and install the clamps. Make sure that the boot is not twisted.

8. Reconnect the tie rod end to the steering arm. Install a new self-locking nut and torque it to 30 Nm (22 ft. lb.).

9. Adjust the toe to the specifications in **4. Wheels, Tires, and Alignment**.

Centering Manual Steering Rack

Centering the steering rack on a manual steering car should be the first step in any mechanical adjustments to the steering system. The rack should be centered whenever the steering gear has been installed and prior to installing tie rods.

Before centering the rack, make sure that the steering gear is firmly mounted and that the universal joint shaft is correctly installed on the pinion shaft of the gearbox.

The steering rack is properly centered when the rack protrudes an equal length from each end of the steering gear housing, as shown in Fig. 7-12. When the steering rack is centered, the steering wheel spokes should be approximately horizontal. If not, see **Removing and Installing Steering Wheel**.

Adjusting Steering Gearbox

The rack and pinion housing of the manual steering gearbox has an adjustment to correct steering play. If the steering rattles or there is play in the steering wheel with the front wheels pointed straight ahead, an adjustment of the steering gearbox may help. If the steering seems too stiff or tends not to center itself after turning, the steering adjustment may be too tight. The power-assisted steering gearbox has a similar adjustment, but it is made using a special tool with the steering gearbox removed from the car.

To adjust the manual steering gearbox, turn the front wheels to the straight ahead position. Loosen the locknut shown above in Fig. 7-1, then turn the adjusting screw clockwise to tighten the steering and counterclockwise to loosen the steering. When finished, hold the screw stationary while tightening the locknut. Make a small change and road test the car. Turn the adjusting screw no more than 20° (approximately 1/16 turn) at any one time. If the steering becomes too stiff, back off the adjusting screw in small steps until the steering effort feels comfortable.

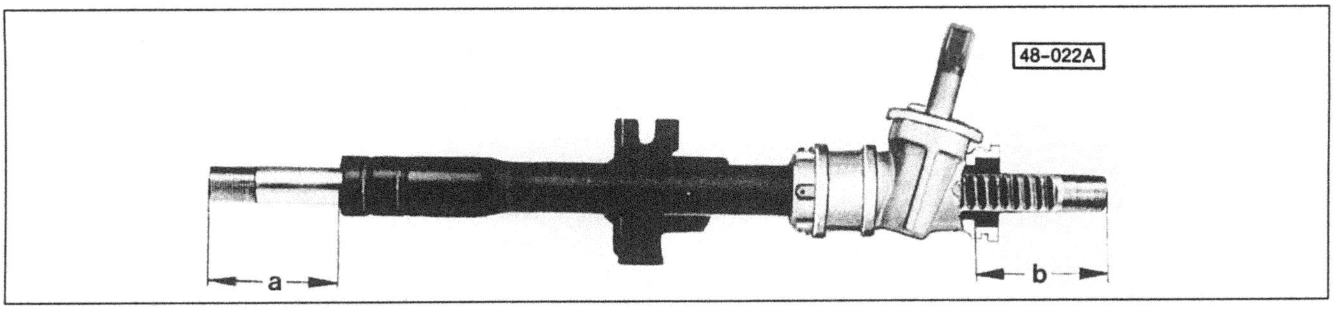

Fig. 7-12. Centering manual steering rack. Dimensions **a** and **b** must be equal on both ends. **b** is measured from first inside shoulder of housing.

To adjust the power-assisted steering gearbox, remove it from the car, as described in **Removing and Installing Steering Gearbox**. Loosen the locknut and, using Volkswagen special tool VW 524 (order no. TV0 524 000 40 ZEL) as shown in Fig. 7-13, turn the hex socket head adjusting bolt until the steering rack can be moved in the housing by hand without binding. Tighten the locknut and reinstall the steering gearbox in the car.

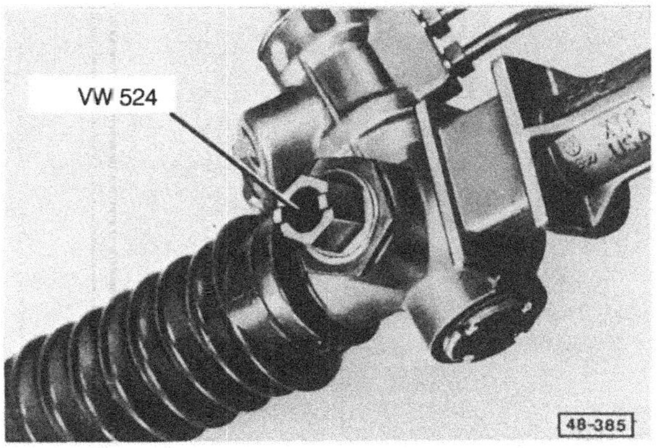

Fig. 7-13. Power-assisted steering gearbox being adjusted using Volkswagen special tool no. VW 524.

Removing and Installing Steering Gearbox

No special tools are required for removing and installing the steering gearbox, although for cars with power-assisted steering the procedure does require removing one of the transaxle mounts.

It is not necessary to remove the steering gearbox to replace the mounting bushings. The bushings are split at the back and the clamps can be removed from the front.

To remove:

1. Disconnect the tie rod ends from the steering arms, as described in **Replacing and Adjusting Tie Rods**.

2. On cars with power-assisted steering, drain the power steering fluid as described in **7.3 Power-assisted Steering**.

3. On cars with manual transmission, remove the gearshift relay shaft bracket from the steering gear housing. See **MANUAL TRANSMISSION**.

4. On cars with power-assisted steering, clean the pressure hose and return hose unions at the steering gearbox and remove them. Do not bend the rigid tubing.

5. On cars with power-assisted steering, remove the rear transaxle mount and its bracket. Also remove the exhaust manifold. If necessary, consult **ENGINE** and **EXHAUST AND EMISSION CONTROLS**.

6. Remove the boot from the steering pinion housing to expose the universal joint shaft. Remove the clamp bolt and nut from the lower end of the shaft and remove the universal joint.

7. Remove the U-shaped steering gear housing mounting clamps. Free the steering gear housing from the mounting studs on the body, push the housing to the right (passenger) side, then pivot the left tie rod down and draw out the housing from the left side.

Installation is the reverse of removal. If the rubber mounting bushings are hard and cracked or oil-soaked, they should be replaced before installation of the steering gear. Make sure that the universal joint shaft lines up with the splined shaft on the steering gearbox as it is being installed. On cars with manual steering, center the steering rack as described in **Centering Manual Steering Rack**. Torque the mounting nuts to 30 Nm (22 ft. lb.).

Torque the universal joint clamping bolt to 30 Nm (22 ft. lb.). Torque the power steering line unions to 20 Nm (15 ft. lb.). Install the relay shaft bracket and check the gearshift linkage as described in **MANUAL TRANSMISSION**. Fill the power steering system with new fluid as described in **7.3 Power-assisted Steering**. Torque the outer tie rod end nuts to 30 Nm (22 ft. lb.) and install new cotter pins. When the installation is complete, a check of front wheel toe as described in **4. Wheels, Tires, and Alignment** is highly recommended, particularly if the steering gear and tie rods have been disassembled.

7.3 Power-Assisted Steering

This section covers the hydraulic portion of the power steering system, including the pump, the fluid reservoir and connecting hoses, and draining and filling the system. The mechanical part of the steering system is covered in **7.2 Steering Gear and Tie Rods**. Information on tightening or replacing the V-belt is found in **LUBRICATION AND MAINTENANCE**.

The power-assisted steering gear is available only as a complete assembly. Service of the steering gear, including repair of housing seal leaks, is by replacement only.

The pump is engine-driven by a V-belt, and provides the hydraulic pressure which is used to assist the movement of the steering rack. Fluid is drawn from the reservoir and pressurized by the pump. Then, depending on the position of the spool valve in the steering gearbox, fluid is directed to either side of the work piston to help move the rack. Fluid which is not doing work flows back to the fluid reservoir through the return line. Fig. 7-14 is a schematic representation of the power-assisted steering system. The system components are shown in exploded view in Fig. 7-15.

Draining and Filling Power Steering Fluid

The fluid should be drained any time the system is to be opened or fluid-bearing components removed. Refilling with new fluid is recommended. The system is essentially self-bleeding, although a simple bleeding step when refilling the system will help remove air from the fluid. Air in the system causes noisy and inconsistent operation.

To drain the fluid from the system, disconnect the suction hose from the pump, as shown in Fig. 7-16. Drain the fluid and dispose of it.

Refill the system with new power steering fluid, Volkswagen part no. ZVW 239 902 or equivalent, until the fluid in the reservoir is at the MAX mark. See **LUBRICATION AND MAINTENANCE** for more information on Volkswagen-approved fluids.

To bleed the system, make sure the fluid level in the reservoir is at the MAX mark and start the engine. With the help of an assistant, turn the steering wheel back and forth from full left to full right, adding fluid to the reservoir as the level drops. When the fluid level remains constant, continue turning the steering wheel until no more air bubbles appear in the reservoir. Add fluid until the fluid level is at the MAX mark. Replace the cap gasket if it is damaged.

Pressure Testing Power Steering System

The pump is tested by measuring its output pressure, as shown in Fig. 7-17. This test requires a pressure gauge with a minimum range of 0-100 bar (0-1500 psi) and a valve between the gauge and the pressure line to the steering gear. Connect the gauge, with the valve open, between the pump and the pressure hose. Center the steering. Start the engine and let it idle.

To test the pump, close the valve on the gauge and quickly read the pressure. Then, open the valve. The pressure should be 70 to 80 bar (1015 to 1160 psi). If not, replace the pump.

CAUTION —

The valve should be closed for no more than 5 seconds. A longer test risks damaging the pump and/or the V-belt.

To test system pressure, leave the gauge valve open and have a helper turn the steering wheel full right and full left. At each position, the pressure with the valve open should be 70 to 80 bar (1015 to 1160 psi). If not, the steering gear should be replaced.

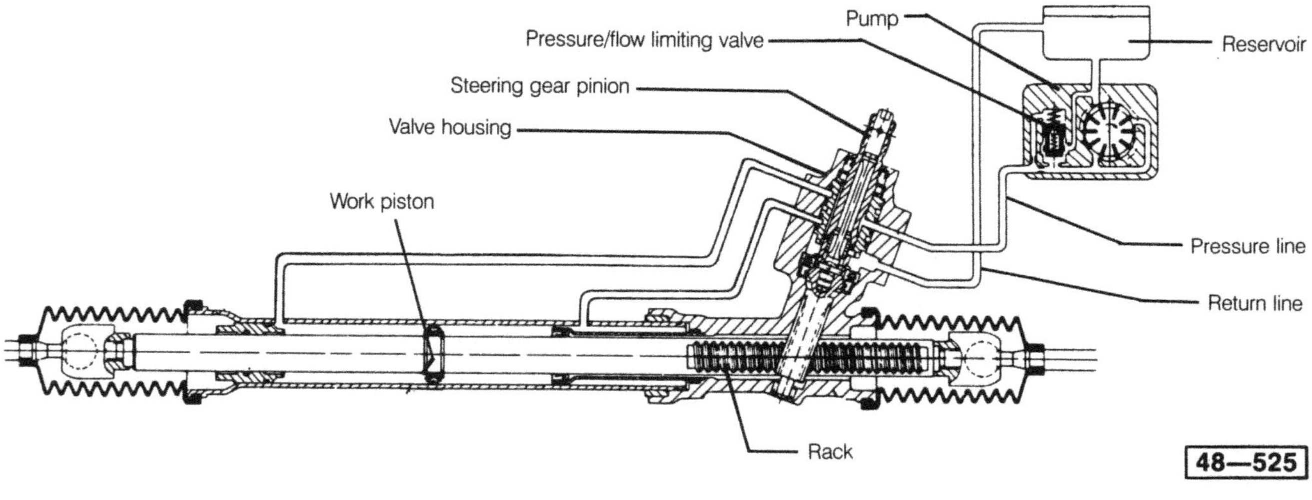

Fig. 7-14. Power steering system operation.

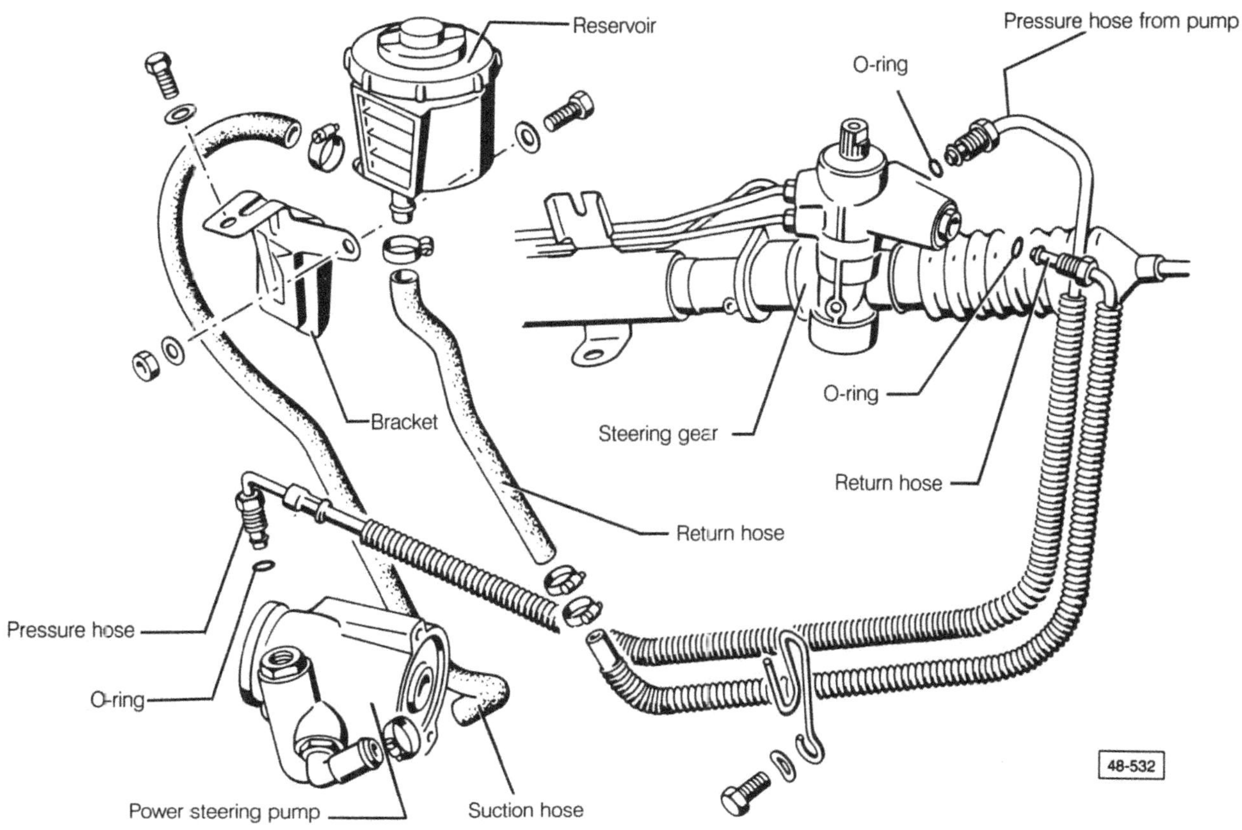

Fig. 7-15. Power-assisted steering system components.

CAUTION —
Hold the steering at full left or right for no more than 5 seconds.

Fig. 7-16. Suction hose (arrow) to be removed from pump for draining fluid.

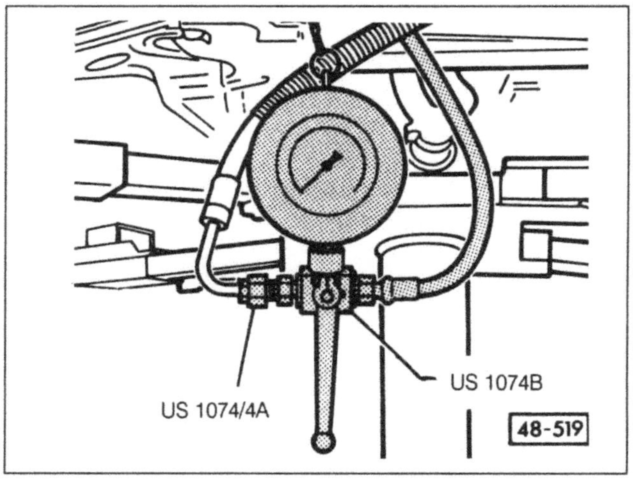

Fig. 7-17. Pressure gauge connected for measuring power steering pump pressure. Gauge shown is Volkswagen special tool no. US 1074B with US 1074/4A adapter.

Power Steering Pump

The power steering pump is mounted on a bracket at the lower front side of the engine, below the coolant pump. The pump and mounting and belt-tensioning brackets are shown in exploded view in Fig. 7-18. For information on V-belt tensioning or replacement, see **LUBRICATION AND MAINTENANCE**.

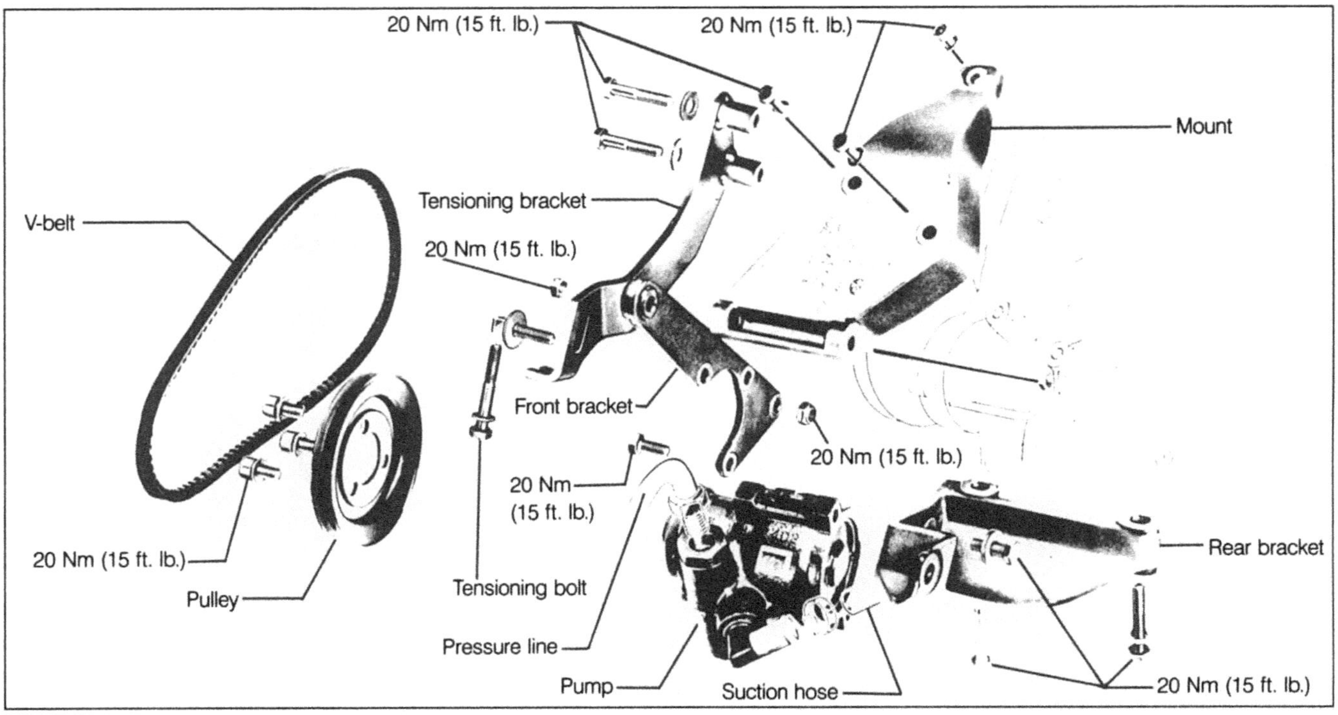

Fig. 7-18. Exploded view of power steering pump and mounting brackets.

8. AIRBAG

Starting with 1990 models, a driver-side airbag is installed as standard equipment on the Cabriolet. The airbag is a supplemental restraint system (SRS), designed to work together with the seat belts and shoulder belts. It will be activated in a frontal collision if the force of impact exceeds a predetermined level. It is not designed to work in all impact situations.

WARNING —
- *The airbag can be inflated only once. It cannot be re-used. After it as been inflated, the airbag unit and all system components must be replaced.*
- *To prevent personal injury or airbag system failure,* ***only trained Volkswagen Service technicians*** *should test, disassemble or service the airbag system.*

Airbag System Warnings

WARNING —
- *Disconnect the power supply before working on the airbag system, or when doing repairs that require removing airbag system components. Disconnect the battery negative (–) terminal and cover the battery. Also disconnect the red single-pin harness connector (marked "airbag") behind the instrument panel.*
- *The airbag unit should not be removed, nor should any other work be performed on the airbag system, until at least 20 minutes after the battery is disconnected.*

WARNING —
- *Do not use a computer memory saver tool to maintain battery power to an anti-theft radio. Use of such a tool will keep the airbag system charged and may cause accidental activation of the airbag.*
- *Always disconnect the battery before doing any electric welding on the car.*
- *Never apply stickers or any other coverings on the steering wheel.*
- *Do not let chemical cleaners, oil or grease come into contact with the vinyl covering of the airbag unit.*
- *Never use a test light on the airbag system. The airbag system must only be tested using the Volkswagen special tool VAG 1551 diagnostic test unit or its Volkswagen-approved equivalent.*
- *The airbag unit must never be electrically tested while it is not installed in the car.*
- *Never open or otherwise attempt to repair airbag system parts. Always use new parts.*
- *Never leave airbag parts or the partially disassembled airbag system unattended.*
- *Do not expose the airbag unit to temperature above 194°F (90°C), even for brief periods. Keep clear of heat sources such as hot plates, soldering irons, heat lamps and welding equipment.*

Airbag Component Warnings

WARNING —
- *The airbag unit must always be stored with the vinyl cover facing up and the metal housing facing down when it is not installed.*
- *Airbag components that have fallen or been dropped from a height greater than 18 inches should not be used.*
- *The airbag control unit and the airbag system wiring harness are a single assembly. If necessary, these items must be replaced together.*
- *The airbag must be deployed (airbag inflated) before disposing of the airbag unit. This can only be done safely and properly while the unit is installed in the car. For more information, see an authorized Volkswagen dealer Service department. Improper activation may cause serious injury.*

CAUTION —
- *Do not turn or release the steering column spiral spring assembly when the steering wheel has been removed.*
- *The turning range for the spiral spring assembly is limited. If it is to be replaced, make sure that the wheels are pointing straight ahead when the steering wheel is removed, and that they remain that way throughout the procedure.*

When to Replace Airbag Components

The airbag system must be replaced if it has been actuated in an accident, or otherwise impaired or damaged. In addition, Volkswagen specifies that the airbag system must be replaced 10 years after the date of manufacture. The date when replacement of the original system is due (year and month) appears on a sticker on the driver-side sun visor. See Fig. 8-1.

WARNING —
- *The airbag can be inflated only once. It cannot be re-used. After it as been inflated, the airbag unit and all system components must be replaced.*
- *Replace parts of the airbag system that are visibly damaged (dented, split, etc.) or have been exposed to oil, grease or chemicals.*
- *Replace all system components (airbag unit, spiral spring assembly, control unit, left and right deceleration sensors) if the airbag system has been triggered in an accident.*
- *Replace system components that have been subjected to previous unauthorized repair.*

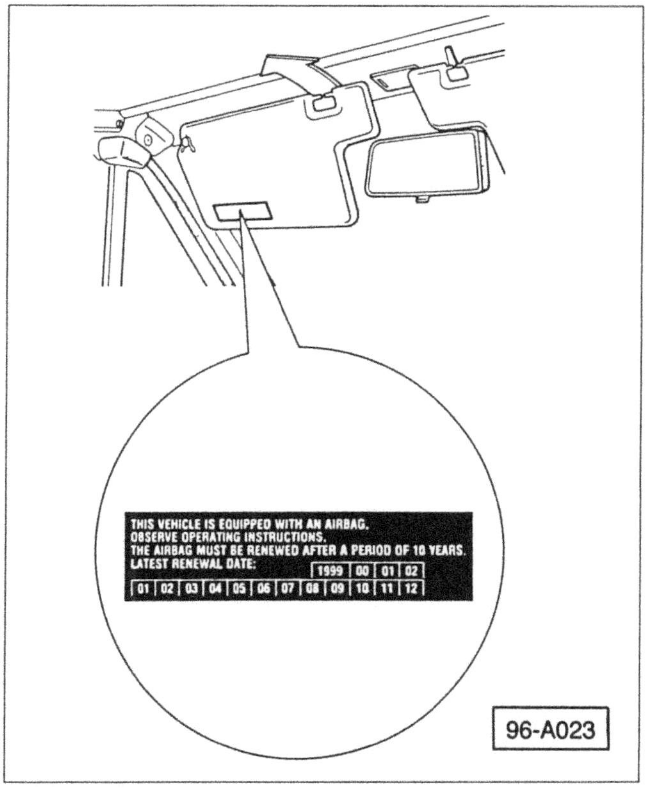

Fig. 8-1. Manufacturer's sticker on driver's side sun visor indicates date (year and month) for replacement of airbag system components.

Steering Column

Fig. 8-2 on the next page is an exploded view of the steering column assembly for airbag-equipped cars, Cabriolet from 1990. If handled improperly or without adequate safeguards, removal of the airbag unit, removal of the steering wheel and any other disassembly of the steering column present a risk of serious personal injury. Review all the Warnings in this section, **8. Airbag**, and at the front of the manual before beginning any work on the steering column.

WARNING —
*To prevent personal injury or airbag system failure, **only trained Volkswagen Service technicians** should test, disassemble or service any part of the airbag system. See the additional airbag system Warnings elsewhere in this section.*

WARNING —
Do not re-use fasteners that are worn or deformed in normal use. Many fasteners are designed to be used only once and become unreliable and may fail when used a second time. This includes, but is not limited to, nuts, bolts, washers, self-locking nuts or bolts, circlips and cotter pins. Always replace these fasteners with new parts.

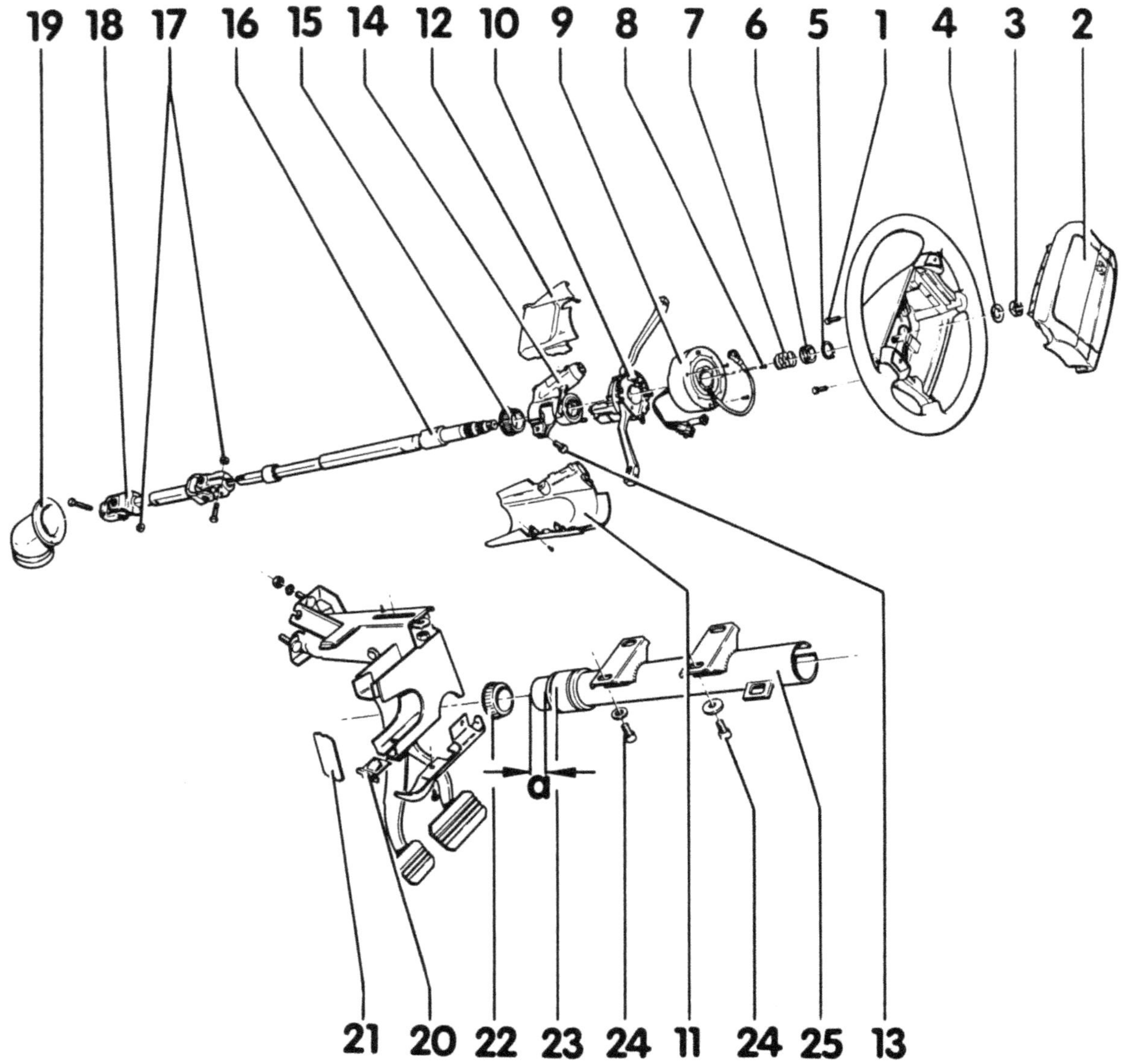

Fig. 8-2. Steering column assembly for airbag-equipped cars, Cabriolet from 1990. Items marked ◆ must be bench-assembled.

1 – Torx®-head screws (T30), captive 10 Nm (89 in. lb.), always replace
2 – Airbag unit see **WARNING**s
3 – 40 Nm (30 ft. lb.)
4 – Spring washer
5 – Locking ring ◆
6 – Spacer ring ◆
7 – Spring ◆
8 – Tapping screw
9 – Spiral spring assembly do not turn or release when steering wheel has been removed
10 – Steering column switch ◆
11 – Trim, lower
12 – Trim, upper
13 – Screw
14 – Steering lock housing ◆
15 – Support ring, ◆ push onto steering column
16 – Steering column, ◆ withdraw from top
17 – 30 Nm (22 ft. lb.)
18 – Universal joint shaft
19 – Boot
20 – Retaining spring
21 – Leaf spring
22 – Steering column bearing, drive out downward to remove
23 – Rubber bushing, warm in hot water, push onto steering column until dimension **a** = 31 mm (1 7/32 in.)
24 – 10 Nm (89 in. lb.)
25 – Steering column tube

WARNING —
To prevent personal injury or airbag system failure, ***only trained Volkswagen Service technicians*** *should test, disassemble or service any part of the airbag system. See the additional airbag system Warnings elsewhere in this section.*

9. TECHNICAL DATA

I. Tightening Torques

Ball joint	
to control arm	25 Nm (18 ft. lb.)
to wheel bearing housing, clamping bolt	50 Nm (37 ft. lb.)
Brake caliper mounting bolts	
front to wheel bearing housing	40 Nm (29 ft. lb.)
rear	65 Nm (48 ft. lb.)
Constant velocity (CV) joint to drive flange	
socket-head bolt	45 Nm (33 ft. lb.)
Control arms to body	
front mounting point	70 Nm (52 ft. lb.)
rear mounting point	45 Nm (33 ft. lb.)
Drive axle to front hub	
self-locking axle nut	240 Nm (177 ft. lb)
Front suspension strut	
top mounting flange nut	20 Nm (15 ft. lb.)
to wheel bearing housing	80 Nm (59 ft. lb.)
Power steering	
hose unions	20 Nm (15 ft. lb.)
pump to mounting bracket	20 Nm (15 ft. lb.)
pulley to shaft/flange (socket-head bolt)	20 Nm (15 ft. lb.)
Rear axle beam	
mounting bracket to body	45 Nm (33 ft. lb.)
pivot bolt	60 Nm (44 ft. lb.)
Rear stub axle (with brake backing plate or splash shield)	
to trailing arm	60 Nm (44 ft. lb.)
Shock absorber, rear	
lower mount	45 Nm (33 ft. lb.)
upper mount	31.5 Nm (23 ft. lb.)
Spring retainer to rear shock absorber	20 Nm (15 ft. lb.)
Stabilizer bar to control arm	25 Nm (18 ft. lb.)
Steering column housing	
clamping socket-head bolts	10 Nm (89 in. lb.)
shear bolts	tighten until heads shear
Steering gear housing to body	30 Nm (22 ft. lb.)
Steering wheel to splined steering column shaft	
cars without airbag	50 Nm (37 ft. lb.)
cars with airbag	40 Nm (30 ft. lb.)
Tie rod to steering rack	
manual steering	50 Nm (37 ft. lb.)
power steering	70 Nm (52 ft. lb.)
Tie rod end, outer	
to steering arm	30 Nm (22 ft. lb.)
to tie rod	30 Nm (22 ft. lb.)
Universal joint shaft	
to steering column	30 Nm (22 ft. lb.)
to steering gear pinion	30 Nm (22 ft. lb.)
Wheel lug bolts	110 Nm (81 ft. lb.)

WARNING —
Do not re-use fasteners that are worn or deformed in normal use. Many fasteners are designed to be used only once and become unreliable and may fail when used a second time. This includes, but is not limited to, nuts, bolts, washers, self-locking nuts or bolts, circlips and cotter pins. Always replace these fasteners with new parts.

Section 11

BRAKES

Contents

TABLES

Brakes

Introduction

All cars covered by this manual are equipped either with front disc brakes and rear drum brakes, or with disc brakes at all four wheels. The brake system is a diagonal, dual-circuit design, hydraulically actuated by the master cylinder, with a vacuum-powered booster to reduce braking effort. Proportioning valves help prevent the rear brakes from locking during heavy braking.

A properly functioning brake system is essential to safe driving. If the red brake/parking warning indicator lights up while driving, indicating low pressure in one of the brake circuits, then it is imperative that the system be given a thorough check, even if braking action still seems satisfactory. The brakes should be regularly inspected, and all brake work must be done with extreme cleanliness, careful attention to specifications, and proper working procedures.

All information needed for performing routine inspection and maintenance is given here, although some of the information may only be of value to the professional mechanic. If you lack the skills, the tools, or a clean workplace for servicing the brake system, we suggest you leave these repairs to an authorized Volkswagen dealer or other qualified shop. We especially urge you to consult your authorized Volkswagen dealer before beginning repairs on a car which may be subject to Volkswagen warranty coverage.

1. GENERAL DESCRIPTION

Fig. 1-1 is a schematic diagram of the dual-diagonal hydraulic brake system. The master cylinder, operated mechanically by the brake pedal, creates pressure in the hydraulic system. At the wheels, the hydraulic pressure acts on the calipers or wheel cylinders to mechanically apply the brakes. The use of hydraulics makes it possible for the driver to generate high braking forces with comparatively little effort.

> ***WARNING —***
> *• Friction materials such as brake linings or brake pads may contain asbestos fibers. Do not create dust by grinding, sanding, or cleaning the pads with compressed air. Avoid breathing any asbestos fibers or dust. Breathing asbestos may result in serious diseases such as asbestosis or cancer, or may result in death.*

> ***WARNING —***
> *• Brake fluid is poisonous. Wear safety glasses when working with brake fluid, and wear rubber gloves to prevent brake fluid from entering the bloodstream through cuts or scratches. Do not siphon brake fluid by mouth.*

> ***CAUTION —***
> *Brake fluid is very damaging to paint. Immediately wipe up any brake fluid that spills on the vehicle.*

Master Cylinder, Vacuum Booster, and Brake Lines

The master cylinder has two separate chambers for two separate hydraulic circuits. The front chamber operates the brakes on the right front and left rear wheels, while the rear chamber operates the brakes on the left front and right rear wheels. In the event of a loss of pressure in one circuit, the other will still supply approximately half of the normal braking force.

The vacuum booster uses engine vacuum to assist braking effort. The rigid brake lines transmit hydraulic pressure to the wheels and resist expansion and pressure loss. Flexible hoses between the rigid lines and the brakes accommodate wheel movement due to steering and suspension action.

Disc Brakes

The disc brakes use a cast-iron caliper with a single hydraulic cylinder to clamp the rotor (disc) between two brake pads. Front rotors are mounted to the wheel hub. Rear rotors are

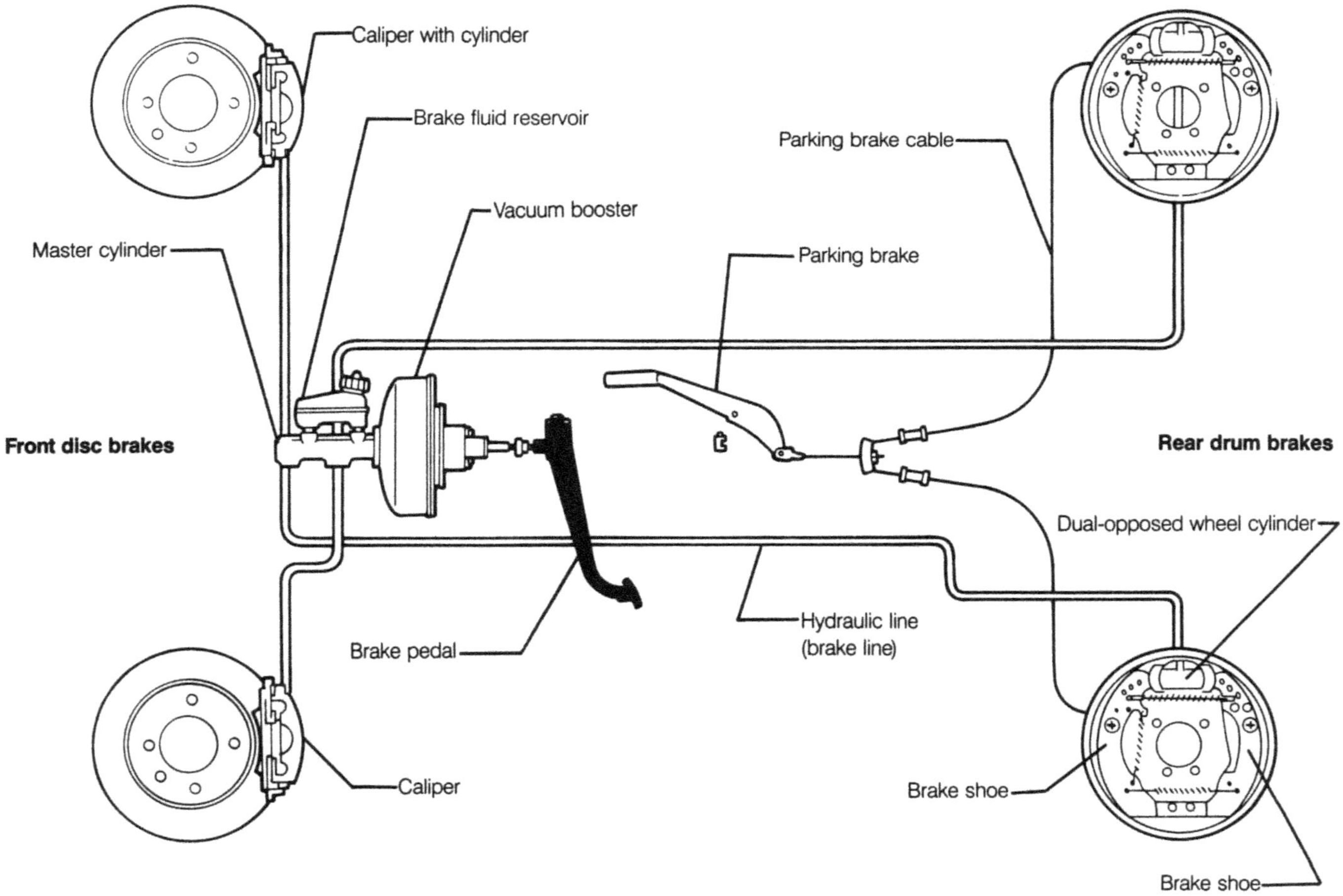

Fig. 1-1. Schematic view of brake system. Front disc and rear drum brakes shown. Four-wheel disc brake system is similar.

integral with the wheel hub; they house the wheel bearings and are held in place by the axle nut. Disc brakes automatically adjust for brake pad and rotor wear.

Rear Drum Brakes

The self-adjusting rear drum brakes use a dual-opposed hydraulic wheel cylinder to apply pressure to the brake shoes, forcing the friction linings against the inside of the brake drum.

Proportioning Valves

To prevent rear-wheel lock-up during hard stops, all cars covered by this manual are equipped with pressure regulating devices, called proportioning valves, that limit the amount of hydraulic pressure applied to the rear brakes. See Fig. 1-2.

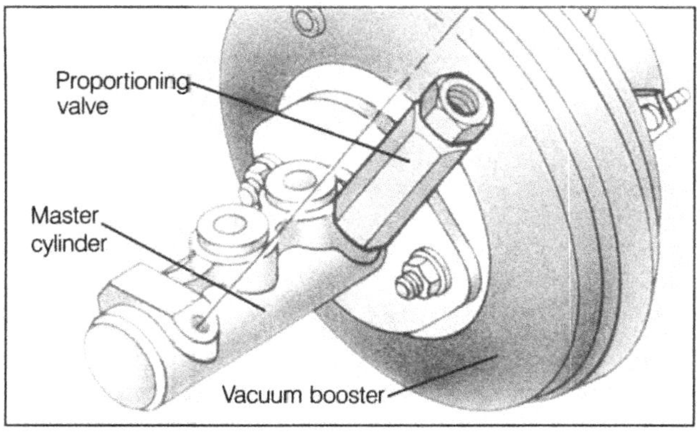

Fig. 1-2. Installed position of one brake proportioning valve. A second valve (not shown) is also mounted on the master cylinder.

Parking Brakes

The cable-operated parking brake mechanically actuates either the rear brake shoes or the rear caliper pistons. The position of the parking brake lever is held by a thumb-release ratchet.

1.1 Brake Applications and Identification

All models except 16-valve are equipped with front disc brakes and rear drum brakes. 16-valve models are equipped with four-wheel disc brakes. The front brake rotors of all models are internally-ventilated for better cooling.

2. MAINTENANCE

The following routine maintenance operations should be performed at the time or mileage intervals listed in **LUBRICATION AND MAINTENANCE**. Instructions can be found in **LUBRICATION AND MAINTENANCE**, or in this section under the listed numbered headings.

1. Checking brake fluid level.
2. Inspecting brake hoses and lines for leaks.
3. Checking brake pad and/or lining wear.
4. Replacing the brake fluid every two years. **4.2**

3. TROUBLESHOOTING

This heading describes symptoms of trouble with the brake system, and suggests possible causes and appropriate corrective actions.

The brake system has the sole function of generating the required friction to stop the car from any speed. Brake problems are usually obvious because they affect the way the car slows and stops. Problems which affect the car's handling or create noise may also be caused by faults in the suspension or steering systems. Information on troubleshooting these systems can be found in **SUSPENSION AND STEERING**. For more help in selecting the appropriate section, see the discussion of troubleshooting in **FUNDAMENTALS** at the front of the manual.

3.1 Basic Troubleshooting Principles

Table a lists symptoms of problems commonly associated with the brakes, their probable causes, and suggested corrective actions. The numbers in bold type in the corrective action column refer to numbered headings in this section where the suggested repairs are described.

As with any troubleshooting, analysis of the observed symptoms is the key to isolating and identifying braking problems. Reliable brake system performance depends on creating and applying hydraulic pressure in the system. Brake performance is affected by three things: the brake fluid, the system's ability to maintain pressure, and the friction components.

Air in the brake fluid, because it is compressible, will make the brake pedal feel spongy during braking, or will increase the force required at the brake pedal to stop the car. Moisture- or dirt-contaminated fluid will dramatically increase internal corrosion and wear.

Seals are used throughout the system to maintain hydraulic pressure. Faulty seals or wear and corrosion on the sealing surfaces will reduce braking efficiency. A symptom of this condition is the need to pump the pedal to get good braking. Simple leaks at the brake line or hose unions may cause the same problems.

Table a. Brake Troubleshooting

Symptom	Probable cause	Corrective action
1. Brake squeak or squeal	**a.** Normal condition **b.** Incorrectly installed brake pads or brake shoes **c.** Brake pad carriers or backing plates dirty or corroded **d.** Brake shoe return springs weak **e.** Pad anti-rattle springs faulty or missing **f.** Brake pads or linings heat-glazed or oil-soaked **g.** Wheel bearings misadjusted	**a.** See **3.3** **b.** Check installation. **5.1, 6.1, 7.1** **c.** Clean calipers and/or backing plates. **3.3, 5.1, 6.1, 7.1** **d.** Replace return springs. **6.1** **e.** Install/replace anti-rattle springs. **5.1, 7.1** **f.** Replace brake pads/linings. Clean rotors/drums. Replace leaking calipers/wheel cylinders as required. **5.2, 6.1, 7.1** Replace wheel bearing grease seals as required. See **SUSPENSION AND STEERING** **g.** See **SUSPENSION AND STEERING**
2. Pedal goes to floor when braking	**a.** Insufficient brake fluid **b.** Master cylinder faulty	**a.** Check fluid level. See **LUBRICATION AND MAINTENANCE** Check for leaks. Fill and bleed system. **4.2** **b.** Replace master cylinder. **4.3**
3. Low pedal even after brake adjustment and bleeding	**a.** Master cylinder faulty	**a.** Replace master cylinder. **4.3**
4. Pedal feels spongy or brakes work only after pedal is pumped	**a.** Insufficient brake fluid **b.** Air in brake fluid **c.** Master cylinder spring weak **d.** Leaking line and hose unions	**a.** Check fluid level. See **LUBRICATION AND MAINTENANCE** Check for leaks. Fill and bleed system. **4.2** **b.** Bleed system. **4.2** **c.** Replace master cylinder. **4.3** **d.** Repair or replace lines and hoses. Bleed system. **4.2**
5. Excessive braking effort with little effect	**a.** Brake pads or linings wet **b.** Brake pads or linings heat-glazed or oil-soaked **c.** Vacuum booster or vacuum line faulty	**a.** Use light pedal pressure to dry brakes while driving **b.** Replace brake pads/linings. Clean rotors/drums. Replace leaking calipers/wheel cylinders as required. **5.2, 6.1, 7.1** Replace wheel bearing grease seals. See **SUSPENSION AND STEERING** **c.** Inspect vacuum lines. Test booster and replace as required. **4.3**
6. Brakes pulsate, chatter, or grab	**a.** Warped disc brake rotors **b.** Brake drums worn or out-of-round **c.** Brake pads or linings worn **d.** Brake pads or linings heat-glazed or oil-soaked	**a.** Recondition or replace rotors. **5.1, 7.1** **b.** Recondition or replace brake drums.**6.1** **c.** Recondition brakes. **5.1, 6.1, 7.1** **d.** Replace brake pads/linings. Clean rotors/drums. Replace leaking calipers/wheel cylinders as required. **5.2, 6.1, 7.1** Replace wheel bearing grease seals as required. See **SUSPENSION AND STEERING**
7. Uneven braking, car pulls to one side, rear brakes lock	**a.** Incorrect tire pressures **b.** Brake pads or linings heat-glazed or oil-soaked **c.** Proportioning valves faulty **d.** Caliper/wheel cylinders or brake pads/shoes binding **e.** Drum brake self-adjusters faulty **f.** Brake fluid contaminated **g.** Worn tires or suspension components	**a.** Check and correct tire pressures. See **LUBRICATION AND MAINTENANCE** **b.** Replace brake pads/linings. Clean rotors/drums. Replace leaking calipers/wheel cylinders as required. **5.2, 6.1, 7.1** Replace wheel bearing grease seals as required. See **SUSPENSION AND STEERING** **c.** Replace proportioning valves **4.4** **d.** Clean and recondition brakes. **5.1, 6.1, 7.1** **e.** Clean, repair, or replace self-adjusting mechanism. **6.1** **f.** Drain, flush, refill, and bleed brake system. **4.2** **g.** See **SUSPENSION AND STEERING**
8. Brakes drag, bind, or overheat	**a.** Brake shoe return springs weak **b.** Brake caliper/wheel cylinder or brake pad/shoe binding **c.** Master cylinder faulty	**a.** Replace springs. **6.1** **b.** Clean and recondition brakes. **5.1, 6.1, 7.1** **c.** Replace master cylinder. **4.3**

Worn or contaminated brake linings are an obvious cause of poor braking performance. The friction material is slowly consumed by use of the brakes and must be periodically replaced. Also, pads and linings which are oil-contaminated or glazed due to overheating cannot produce as much friction, and stopping distances will increase.

When troubleshooting the brake system, consider how quickly the problem occurred. Sudden brake system failure is most likely caused by the failure of one component, such as a burst hose or a disconnected vacuum booster vacuum line. A gradual decline in braking efficiency probably indicates the general state of wear in the system, and more than one component may need to be repaired or replaced.

3.2 Diagnostic Tests

Some general checks of brake system function and component inspection can help isolate problems.

Test the vacuum booster first by pumping the brake pedal a few times with the engine off, and then holding the pedal down. When the engine is started, the pedal should fall slightly. The booster diaphragm rarely fails, but leaks in the vacuum line or a faulty check valve will decrease the vacuum assist. Check the valve by removing it from the booster vacuum line and blowing through it. Air should pass through in the direction of the arrow on the check valve, but not through the opposite way.

Test the rear brakes by applying the parking brake. It should be able to hold the car, perhaps even stalling the engine, when you try to drive away. If not, the brakes probably need adjustment or reconditioning.

Test the master cylinder by holding the pedal down hard with the car stopped and the engine running. The pedal should feel solid and stay solid. If the pedal slowly falls to the floor, then the master cylinder is faulty and should be replaced.

3.3 Brake Noise

Occasional groaning or squealing sounds coming from the brakes, especially disc brakes, are usually caused by vibrations transmitted through the brake pads. The problem is not unique to Volkswagens. New brake friction materials, using little or no asbestos for health reasons, may contribute to an increase in brake noise problems in all makes. These noises, though discomforting, are normal and rarely indicate a problem in brake system function or effectiveness.

Although there is no permanent solution, proper maintenance and repair can help minimize brake noise. Disc brake caliper assemblies include anti-rattle springs, designed to minimize vibration and noise. As much as possible, brake pads and calipers should be kept free of foreign matter and corrosion which may inhibit smooth operation. Always resurface or replace brake rotors when changing brake pads.

A reduction in brake noise may be possible by installing a shim kit with the pads. The kit, Volkswagen part no. 171 698 993, contains self-adhesive shims which attach to the backs of the brake pads to damp vibration. The backs of the pads must be cleaned down to the metal for good adhesion, and the shims can only be installed on thinner, partially worn pads. The combined thickness of the pad and shim must not exceed the thickness of a new pad. See **5.1 Checking Brake Pad Wear** for specifications.

Rear drum brake squeak may be more easily corrected. A lubricant is available to coat the points on the backing plate where the brake shoes make contact with the plate (raised or embossed points). The lubricant is Wolfracote grease, part number G 000 650, and is available from authorized Volkswagen dealers.

CAUTION —
Part numbers are subject to change. Always rely on an authorized Volkswagen dealer parts department for the latest and most accurate information.

To eliminate squeaking drum brakes, remove the brake shoes as described in **6.1 Reconditioning Rear Drum Brakes**, and apply a small amount of Wolfracote grease to the contact points of the brake backing plate, then reinstall the brake shoes.

4. BRAKE SYSTEM SERVICE

This section provides the requirements for the brake fluid, the methods for bleeding the brake system, and the servicing of the master cylinder, vacuum booster, and proportioning valves.

4.1 Brake Fluid

Brake fluid is the heart of the hydraulic system. It transmits the mechanical braking force at the pedal to the brake mechanisms at the wheels. Any contamination of the fluid will have a direct influence on braking efficiency. Dirt in the system will clog small passages and wear rubber parts.

Brake fluid absorbs moisture. Moisture in the fluid lowers its boiling point, allowing it to be boiled more easily by the heat generated during braking. Boiling the fluid will create air in the system and, since air is compressible, will reduce braking efficiency. Moisture in the system also causes corrosion.

When adding brake fluid, use only new fluid from unopened containers. Do not reuse brake fluid that has been bled from the system, even if it is brand new. The bleeding operation aerates the fluid. See **LUBRICATION AND MAINTENANCE** for brake fluid specifications. For best performance and system reliability, use only Volkswagen-approved brake fluid.

Because it readily absorbs moisture, brake fluid should be changed every two years, regardless of the number of miles driven annually. The brake system should also be bled of air after any repairs in which brake lines are disconnected. See **Replacing Brake Fluid** for more information.

> ***WARNING —***
> *Brake fluid is poisonous. Wear safety glasses when working with brake fluid, and wear rubber gloves to prevent brake fluid from entering the bloodstream through cuts or scratches. Do not siphon brake fluid by mouth.*

> ***CAUTION —***
> *Brake fluid is very damaging to paint. Immediately wipe up any brake fluid that spills on painted surfaces.*

4.2 Bleeding Brakes

Bleeding brakes is the process of purging aerated or contaminated brake fluid from the system and replacing it with new, clean fluid. Bleeding brakes periodically is good preventive maintenance, but it is also essential if brake lines or components have been disconnected, or any time air is introduced into the system. Vague or spongy brakes may indicate the need to bleed air from the system.

There are three widely used methods of bleeding brakes. Each employs some means of forcing the fluid through the system to the calipers or wheel cylinders, and releasing contaminated fluid through the bleeder valves. The system can be bled using a pressure bleeder, using a hand-held vacuum bleeder, or manually using the brake pedal and a helper. Each method accomplishes the same result. Pressure bleeding, if available, is fastest. Manual bleeding requires no special tools.

For the best possible results, bleeding should start at the wheel farthest from the master cylinder and end at the wheel closest to the master cylinder. This bleeding sequence is as follows:

1. right rear caliper or wheel cylinder
2. left rear caliper or wheel cylinder
3. right front caliper
4. left front caliper

For best access to the bleeder valves when bleeding the brakes, the car should be raised and firmly supported on jack stands designed for the purpose. The engine should be off, and the vacuum hose and check valve should be disconnected from the vacuum booster. The bleeder valves on both disc brake calipers and drum brake wheel cylinders are opened and closed using an 8 mm (5/16 in.) box wrench. Use a clear container to catch the expelled fluid and a piece of 4 mm (5/32 in.) inside diameter clear, flexible tubing connected to the bleeder valve, as shown in Fig. 4-1, so that outgoing air bubbles are visible.

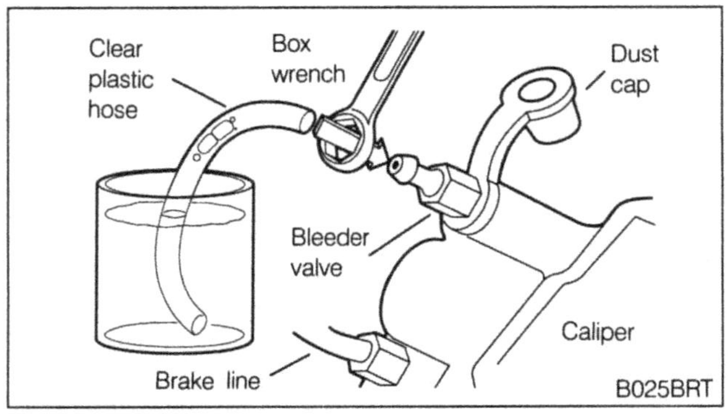

Fig. 4-1. Container (partially filled with brake fluid) connected to caliper bleeder valve with clear tubing for bleeding brakes. Bleeding rear drum brake wheel cylinder is similar.

Pressure Bleeding

Pressure bleeding, using special equipment and compressed air, is the method used by most commercial repair shops. It is the quickest method and the best way to prevent contaminants and air from entering the brake system. The pressure bleeder is connected to the brake fluid reservoir and fills the system with fluid under pressure. Follow the instructions supplied by the equipment manufacturer.

Vacuum Bleeding

Vacuum bleeding uses an inexpensive hand pump connected at the bleeder valve to draw the fluid through the system. It is a practical, do-it-yourself alternative to pressure bleeding, is faster than manual bleeding, and can be accomplished by only one person. Pumps of this general type are available from auto parts and supply outlets.

Follow the pump manufacturer's instructions very carefully. Make sure that the hose connection to the brake bleeder valve is secure and air-tight. Check the brake fluid level in the master cylinder often to guard against emptying the reservoir.

Manual Bleeding

Bleeding the brake system by pumping the brake pedal is the most economical method. The only disadvantage is that it requires a helper.

When manually bleeding the brakes, first fill the brake fluid reservoir on top of the master cylinder until the level is well above the MAX mark, and replace the cap to prevent contamination of the fluid.

Starting with the right rear caliper or cylinder, clean the area around the bleeder valve and remove the dust cap. Fit the box wrench to the bleeder valve, then slip one end of the hose onto the valve and submerge the other end in clean brake fluid in the clear container. Have a helper slowly pump the brake pedal about three times and, on the last time, hold it down. Slowly open the bleeder valve approximately one-half turn. When the brake pedal goes to the floor and fluid stops flowing from the bleeder valve, close the bleeder valve, then release the brake pedal.

CAUTION —

Until some pressure has built up in the system, only depress the pedal approximately halfway. This will prevent overrunning the master cylinder's normal stroke, which may damage the master cylinder piston seals.

CAUTION —

Be sure that the bleeder valve is fully closed before releasing the brake pedal to avoid drawing air back into the bleeder valve.

Repeat the procedure until clear fluid with no air bubbles is coming from the bleeder valve, then fully tighten the valve and replace the dust cap. Using the sequence described above, follow the same procedure at the other three wheels. Check the fluid level in the master cylinder frequently, and add more to keep the level at the MAX mark.

Replacing Brake Fluid

Remove the filter/strainer from the brake fluid reservoir and clean it in new, unused brake fluid. Reinstall it, and refill the reservoir with new, unused brake fluid. Expel the old fluid from the system using the brake bleeding procedure described above. Keep the reservoir refilled, and remove at least 1 pint (about 500 ml) of brake fluid at each caliper or wheel cylinder to expel the old fluid and replace it with new. See the brake fluid specifications in **LUBRICATION AND MAINTENANCE**. Use only brake fluid of the type approved by Volkswagen.

Flushing the Brake System

Do not rely on flushing alone to clean a brake system contaminated with dirt or corrosion. The flushing procedure may actually force dirt in the lines into the wheel cylinders. The brake system must be disassembled and the parts individually cleaned. Use only brake fluid to flush the lines. Alcohol must not be used since it will encourage the accumulation of water in the system.

4.3 Master Cylinder and Vacuum Booster

The master cylinder, its vacuum booster, and the brake pedal assembly are shown in Fig. 4-2. The vacuum booster and master cylinder are mounted to the firewall in the engine compartment. The pedal operates the master cylinder via a pushrod which passes through the firewall. The master cylinder piston acts on the brake fluid to create pressure in the system. The brake fluid reservoir is mounted atop the master cylinder.

The vacuum booster, mounted between the pedal mechanism and the master cylinder, uses the difference between engine vacuum and atmospheric pressure to help actuate the master cylinder and reduce pedal effort.

When the engine is running, vacuum is applied equally to both sides of the booster diaphragm. When the brake pedal is pressed, atmospheric pressure reaches the pedal side of the diaphragm and creates the brake-boosting pressure differential. A check valve in the vacuum hose holds vacuum in the booster when the engine is stopped. The pedal pushrod is directly connected to the master cylinder so that a failure of the vacuum booster, although it will increase braking effort, will not result in total brake failure.

Removing and Installing Master Cylinder

Master cylinder parts or rebuild kits are not available. A faulty master cylinder must be replaced. It can be removed and installed with ordinary hand tools. On installation, use a new gasket between the cylinder and the vacuum booster, and new self-locking mounting nuts. Make sure to have an adequate supply of new, unopened brake fluid on hand, as bleeding the brakes will be necessary after installation. See **4.2 Bleeding Brakes**.

To remove and install:

1. Disconnect the battery negative (–) terminal. Disconnect the electrical connector from the brake fluid reservoir cap.

2. Remove the reservoir cap and, using a clean syringe, empty the brake fluid reservoir.

   WARNING —

   Brake fluid is poisonous. Do not siphon the brake fluid with your mouth.

3. Disconnect the brake lines from the master cylinder. Remove the two nuts holding the master cylinder to the vacuum booster, then remove the master cylinder.

4. If necessary, remove the brake fluid reservoir from the top of the master cylinder by prying it out of the rubber mounting grommets, and install the reservoir on the new master cylinder.

5. Using a new gasket and self-locking nuts, install the new master cylinder. Torque the nuts to 21 Nm (15 ft. lb.). Connect the brake lines and torque the brake line unions to 15 to 20 Nm (11 to 15 ft. lb.).

   CAUTION —

   Do not exceed recommended mounting torque of master cylinder mounting nuts. Overtightening may cause damage to the vacuum booster.

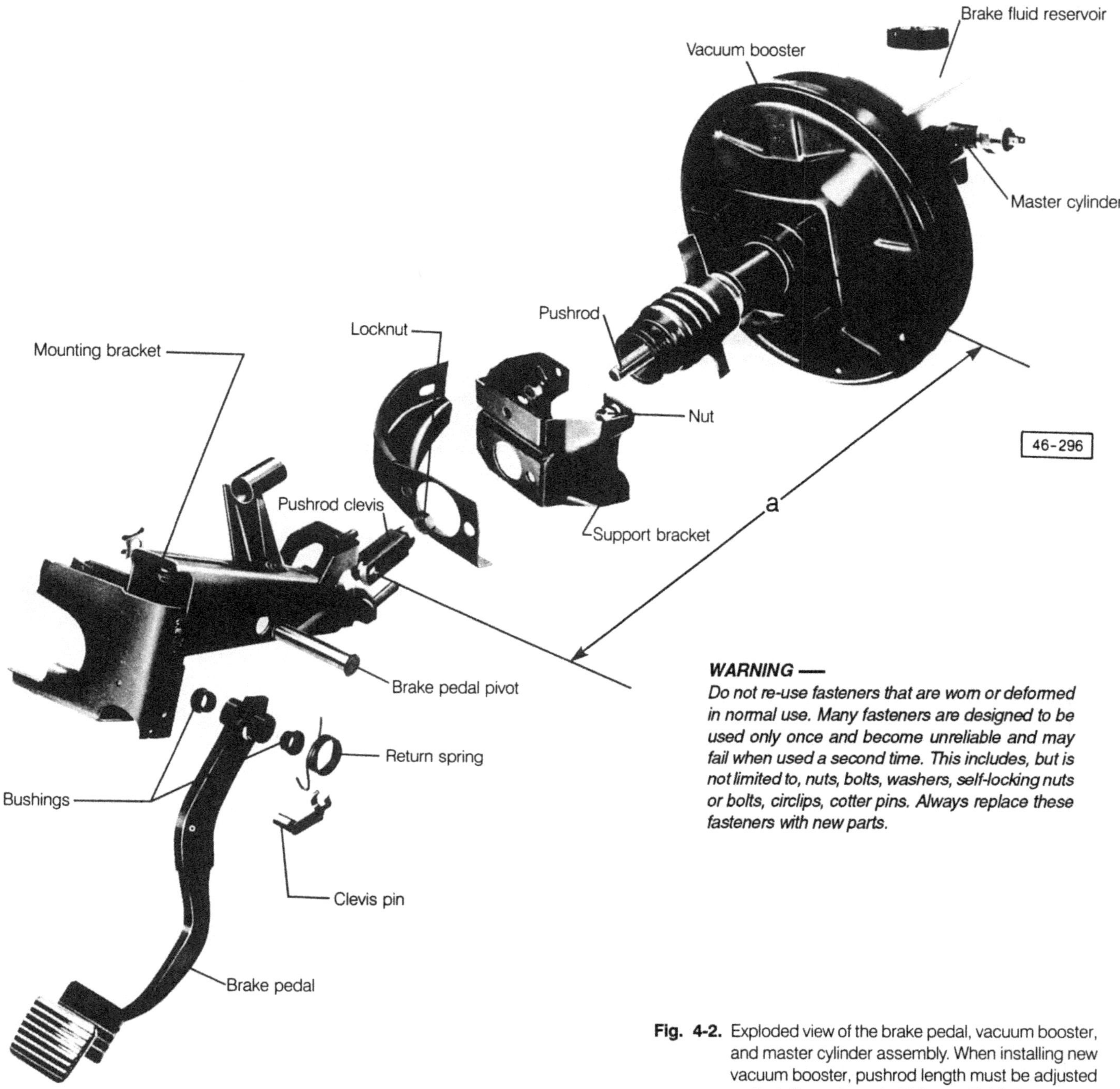

WARNING —
Do not re-use fasteners that are worn or deformed in normal use. Many fasteners are designed to be used only once and become unreliable and may fail when used a second time. This includes, but is not limited to, nuts, bolts, washers, self-locking nuts or bolts, circlips, cotter pins. Always replace these fasteners with new parts.

Fig. 4-2. Exploded view of the brake pedal, vacuum booster, and master cylinder assembly. When installing new vacuum booster, pushrod length must be adjusted so that dimension **a** is 206 mm (8⅛ in.).

NOTE —
The master cylinder pushrod does not require adjustment prior to installation of the master cylinder, even if a new vacuum booster has been installed.

6. Fill the reservoir with new brake fluid and bleed the system as described in **4.2 Bleeding Brakes**. Reconnect the battery negative (−) terminal and the fluid reservoir cap connector. Road test the car and repeat the brake bleeding operation if necessary.

Removing and Installing Vacuum Booster

A faulty vacuum booster is not repairable, and must be replaced. It can be removed and installed with ordinary hand tools. Replacing the vacuum booster requires removing and installing the master cylinder. See **Removing and Installing Master Cylinder** above.

To remove the vacuum booster, disconnect the vacuum hose from the booster, and disconnect the pushrod from the brake pedal. Working in the engine compartment, remove the four nuts that mount the vacuum booster to its support bracket, and remove the booster. Installation is the reverse of removal.

Use new self-locking nuts when installing the booster. Before attaching the pushrod to the brake pedal, adjust the pushrod to the length indicated above in Fig. 4-2 by turning the pushrod clevis.

4.4 Proportioning Valves

Sciroccos and Cabriolets use proportioning valves, mounted in the brake lines at the master cylinder, to control the pressure to the rear wheels. The proportioning valves are not adjustable.

Pressure Testing Proportioning Valves

Testing the pressure-regulating function of the system requires measuring brake system pressure at each wheel, using two pressure gauges with a range of at least 0 to 160 bar (0 to 2320 psi) connected in place of the brake bleeder valves. Because of the need for this specialized equipment, we recommend having this test performed by an authorized Volkswagen dealer or other qualified repair shop.

The gauges are connected to one front and the diagonally opposite rear brake to test one hydraulic circuit, and then to the other front brake and its diagonally opposite rear brake to test the other hydraulic circuit. The system must be bled at the gauges each time they are connected. Have a helper depress the brake pedal while observing the pressure readings on the gauges.

Table b lists the brake pressure specifications for all models. Replace proportioning valves that are not within specifications.

Table b. Brake Pressure Specifications

Front caliper pressure	Rear wheel cylinder or caliper pressure
55 bar(798 psi) 100 bar(1450 psi)	39–43 bar(565–624 psi) 51–57 bar(740–825 psi)

5. FRONT DISC BRAKES

An exploded view of a front disc brake is shown in Fig. 5-1. The single-piston caliper is a floating design. When the brakes are applied, the caliper pushes the inside brake pad against the rotor. This moves the caliper body at the same time, pulling the outside pad against the rotor.

Because the front brakes do more of the work of stopping the car, they are likely to wear faster and require service more frequently than the rear brakes. The brake pads and rotors are subjected to the greatest wear and are the components which most often need attention.

The brake pads are designed to be routinely replaced as they wear out. While it is possible to restore the brakes by replacing only the brake pads, full braking performance and minimum pad wear can only be achieved if the rotors are resurfaced or replaced at the same time.

WARNING —

Brake pad friction materials may contain asbestos fibers. Do not create dust by grinding, sanding, or cleaning the pads with compressed air. Avoid breathing any asbestos fibers or dust. Breathing asbestos may result in serious diseases such as asbestosis or cancer, or may result in death.

Checking Brake Pad Wear

The inspection procedure and specifications for checking the brake pads are found in **LUBRICATION AND MAINTENANCE**. For more complete inspection of pad condition, the pads must be removed as described in **5.1 Reconditioning Front Disc Brakes**.

5.1 Reconditioning Front Disc Brakes

Reconditioning of the front disc brakes typically includes replacing the brake pads and resurfacing or replacing the brake rotors. Unless the calipers and hoses are damaged, extremely worn, or leaking fluid, more extensive front brake repairs are not normally required.

If the brake pads are soaked with oil, grease, or brake fluid, the cause of the contamination must be found and corrected before new pads are installed. Brake rotors must be replaced when resurfacing exceeds specified limits. See **Removing, Reconditioning, and Installing Brake Rotors** for rotor thickness and runout specifications.

Removing and Installing Brake Pads

Although the calipers must be removed to change the brake pads, this can be done without disconnecting the flexible brake hose. Keeping the hydraulic system sealed eliminates the need to bleed the brakes afterward.

Brake pads and the surfaces of the rotors wear slightly differently from one another. Always replace brake pads in complete sets. If old pads are to be reinstalled, such as after inspecting them, always make sure they are refitted in their original locations.

To remove:

1. Raise the front of the car and support it securely on jack stands. Remove the front wheels.

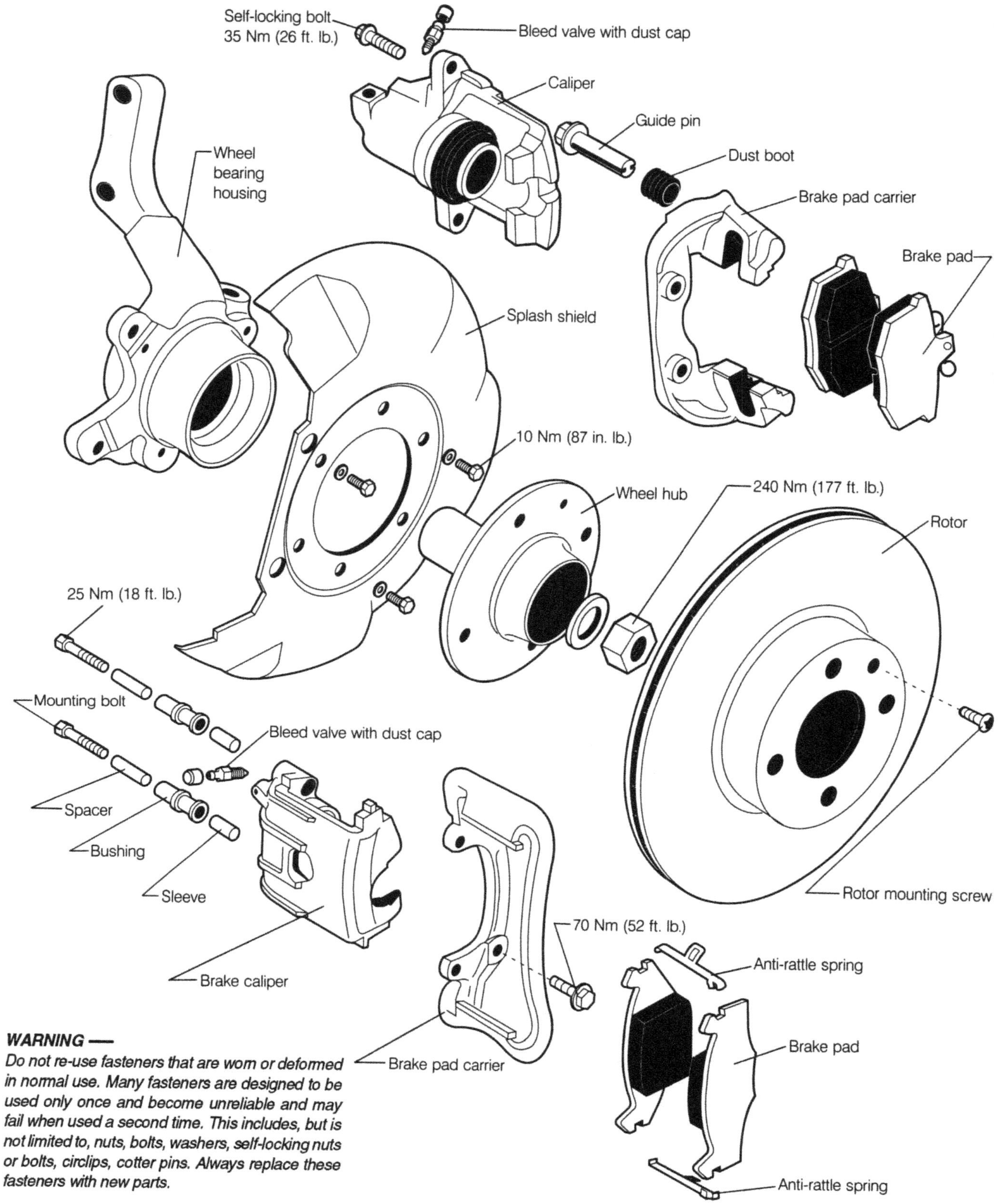

WARNING —
Do not re-use fasteners that are worn or deformed in normal use. Many fasteners are designed to be used only once and become unreliable and may fail when used a second time. This includes, but is not limited to, nuts, bolts, washers, self-locking nuts or bolts, circlips, cotter pins. Always replace these fasteners with new parts.

Fig. 5-1. Exploded view of front disc brake assembly. Calipers for both 16-valve models (top) and all except 16-valve (bottom) are shown.

2. On all except 16-valve models, remove the anti-rattle springs, then use a hex wrench to remove the two brake caliper mounting bolts shown in Fig. 5-2. Also remove the spacers and the sleeves. Pull the caliper away from the brake pad carrier and suspend it from the car body with a stiff wire to avoid stretching the brake hose.

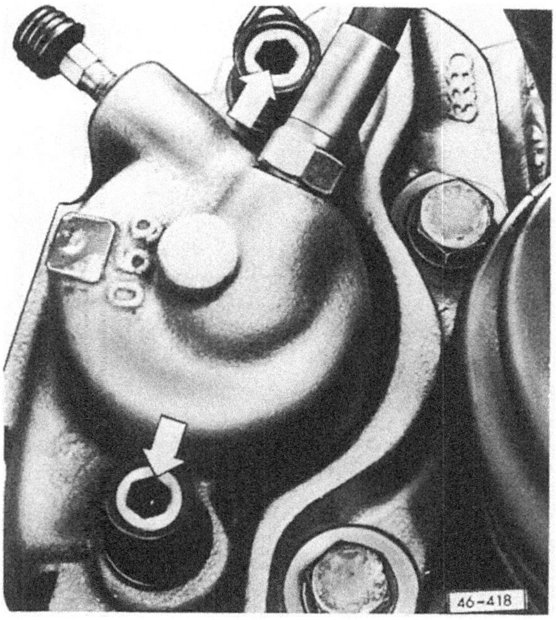

Fig. 5-2. Caliper mounting bolts (arrows) to be removed.

3. On 16-valve models, hold the caliper guide pin with an open end wrench, then remove the lower brake caliper mounting bolt. See Fig. 5-3. Swing the caliper up and away from the brake pads.

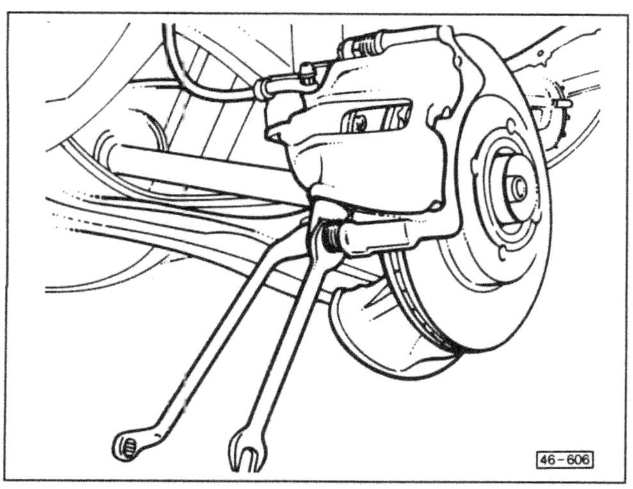

Fig. 5-3. Lower caliper mounting bolt being removed. Hold guide pin with open end wrench when removing mounting bolt.

NOTE —
If the piston creeps out of the caliper due to residual pressure in the brake lines, place a rubber band around the piston and caliper to hold the piston.

4. Lift out the pads from the pad carrier.

CAUTION —
If the pads are to be reused, mark the pads so that they can be reinstalled in their original positions.

To install:

1. Install the brake pads on the pad carrier.

2. When installing new brake pads, slowly push the piston back into the caliper to provide clearance for the thicker new brake pads.

NOTE —
Pushing in the caliper pistons to gain clearance for the thicker new brake pads will cause brake fluid to overflow the master cylinder fluid reservoir. To prevent this, use a clean syringe to first remove some fluid from the reservoir.

3. On all except 16-valve models, engage the caliper with the pad carrier and install the spacers, sleeves, and mounting bolts, torquing the bolts to 25 Nm (18 ft. lb.). Then install the anti-rattle springs.

4. On 16-valve models, swing the caliper down and install the lower mounting bolt. Hold the guide pin stationary as shown above in Fig. 5-3 while torquing the mounting bolt to 35 Nm (26 ft. lb.).

CAUTION —
Make sure that all caliper mounting hardware is clean and free of corrosion.

5. Mount the wheel and loosely install the lug bolts. Lower the car and then torque the lug bolts to 110 Nm (81 ft. lb.).

6. Depress the brake pedal several times to adjust the caliper and brake pads to the rotor.

7. Check the level of brake fluid in the reservoir. If necessary, add new brake fluid to fill the reservoir to the MAX mark.

WARNING —
New brake pads require some break-in. Allow for slightly longer stopping distances for the first 100 to 150 miles of city driving, and avoid hard stops.

Removing, Reconditioning, and Installing Front Brake Rotors

To remove the brake rotors, first remove the caliper and brake pads as described above, then remove the brake pad carrier mounting bolts and the brake pad carrier. On 16-valve models, it will be necessary to suspend the caliper and pad carrier assembly from the body with a stiff wire to avoid stretching the brake hose. Remove the countersunk flathead screw between two of the wheel bolt holes, and pull the rotor off the wheel hub. Use a soft-faced mallet to free a stuck rotor.

> ***CAUTION —***
> *Do not try to remove the rotor without first removing the caliper. Excessive force may damage the rotor or the wheel bearing housing.*

> **NOTE —**
> Be careful when suspending the caliper assembly. The guide pins are lubricated and will allow the caliper to separate from the pad carrier.

Installation is the reverse of removal. Use new self-locking bolts for the brake pad carrier, and torque them to 70 Nm (52 ft. lb.). Make sure that the rotor and wheel hub mounting surfaces mate properly and are free of dirt and corrosion.

> **NOTE —**
> New replacement brake rotors should be cleaned with a grease-free solvent, such as a commercially available brake cleaner, before installing the caliper and brake pads.

Brake rotors should always be resurfaced in pairs, with an equal amount of material removed from both sides of each rotor. **Table c** lists brake rotor specifications, including the minimum thickness to which a rotor can be resurfaced. Rotors which fail to meet the minimum thickness are too thin to be reliably used and should be replaced. Brake rotors can be resurfaced by most local automotive machine shops.

Table c. Front Rotor Reconditioning Specifications

Thickness new	20.0 mm (.787 in.)
Minimum thickness	18.0 mm (.709 in.)
Maximum runout	0.06 mm (.002 in.)

5.2 Calipers

Fig. 5-4 is an exploded view of the front disc brake caliper. The piston seal holds the hydraulic pressure in the system as the piston moves, and also provides the caliper's self-adjusting action.

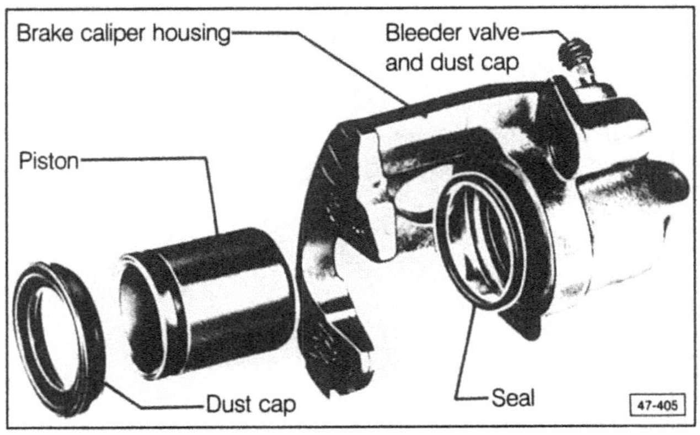

Fig. 5-4. Exploded view of front disc brake caliper. (Calipers used on models covered by this manual differ slightly from one shown.)

The caliper is self-adjusting. When the piston moves and presses on the brake pads, the seal is distorted slightly. When the pressure is removed, the seal pulls the piston back slightly, creating a small gap to minimize brake pad wear. As the pads wear, the seal is repositioned on the piston, so that the gap remains approximately the same.

Brake fluid leaks around the brake caliper piston are the result of a failed or damaged piston seal, perhaps caused by corrosion, scoring, or pitting of the piston or caliper bore. The seal can be replaced, but a damaged piston, which is not available as a replacement part, will promptly destroy the new seal. To remedy a leaking caliper piston seal and avoid future problems, complete replacement of the caliper is recommended. A damaged caliper dust seal can be replaced separately.

To remove front brake calipers, remove the front brake pads as described in **Removing and Installing Brake Pads**. On 16-valve models, also remove the second bolt that holds the caliper to the brake pad carrier. Disconnect the brake hose from the caliper and drain the brake fluid into a container. Cap the end of the brake line to prevent dirt and moisture from entering the brake system.

Installation is the reverse of removal. Torque the brake hose union to 15 Nm (11 ft. lb.). Bleed the brake system as described in **4.2 Bleeding Brakes**. Do not let brake fluid contaminate the brake pads or brake rotor surface.

6. REAR DRUM BRAKES

Fig. 6-1 is an exploded view of the rear drum brake assembly. Retaining springs hold the brake shoes in place. The shoes rest against a fixed bracket at the bottom of the backing plate. When the brakes are applied, the wheel cylinder pushes against the tops of the shoes to force them against the brake drum. Return springs retract the brake shoes when the brakes are released.

The rear brakes are equipped with a self-adjusting mechanism which compensates for wear. At the top of the backing plate, the brake shoes rest against the slotted adjusting bar. As

WARNING —
Do not re-use fasteners that are worn or deformed in normal use. Many fasteners are designed to be used only once and become unreliable and may fail when used a second time. This includes, but is not limited to, nuts, bolts, washers, self-locking nuts or bolts, circlips, cotter pins. Always replace these fasteners with new parts.

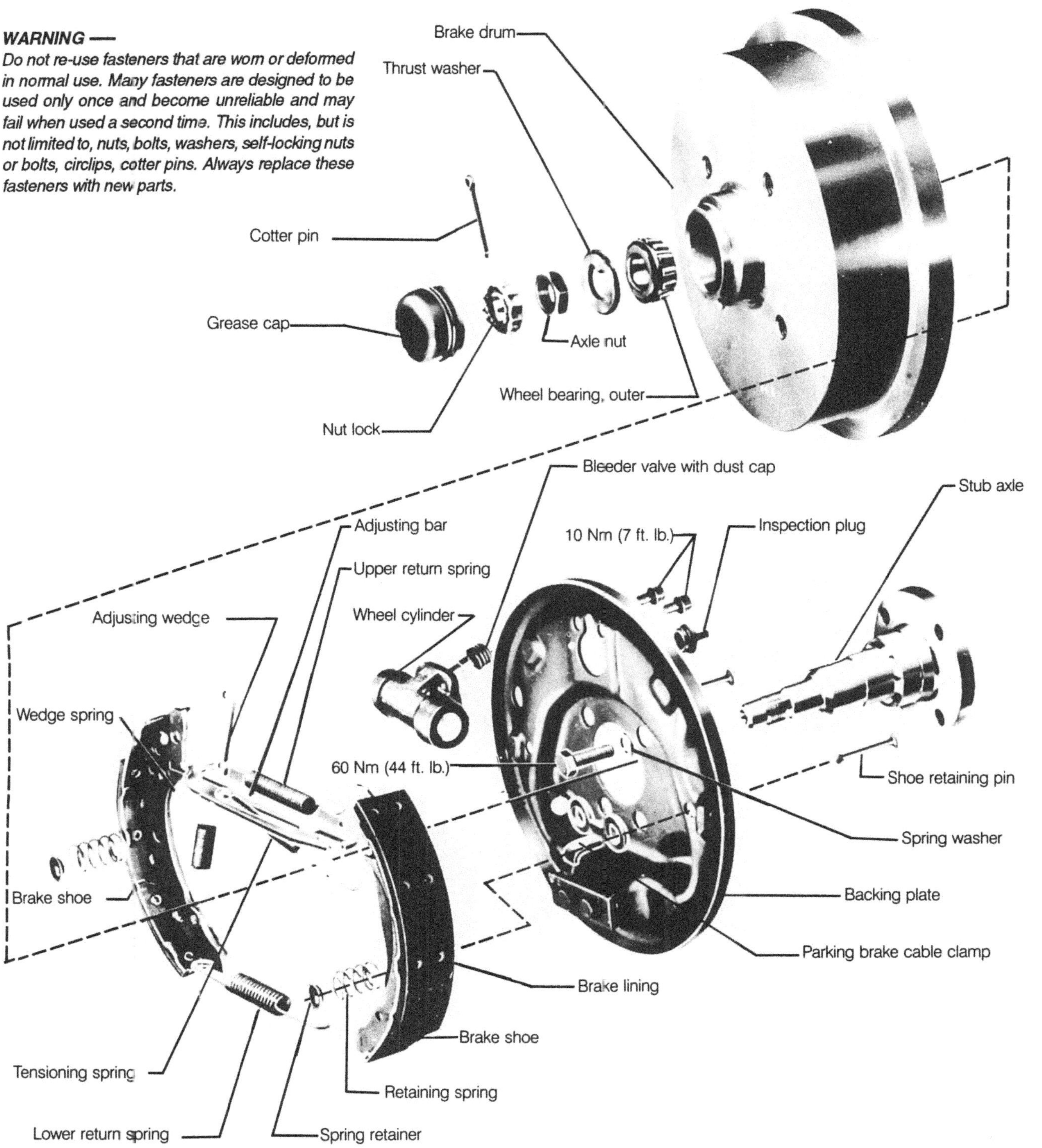

Fig. 6-1. Exploded view of rear drum brake assembly.

the brake lining wears and brake shoe travel increases, the spring-loaded adjusting wedge fills the gap between the brake shoe and the adjusting bar, essentially lengthening the adjusting bar and keeping the brake shoes from retracting too far from the drum.

Because the rear brakes do less of the work of stopping the car, they are not likely to wear as quickly or require service as frequently as the front brakes. Unless problem symptoms are actually observed at the rear brakes, troubleshooting of brake problems should always include the front brakes. If the rear brakes are worn and require service, then at the very least the front brakes should also be thoroughly inspected. The brake shoes, the drums, and the attaching and adjusting parts are subjected to the greatest wear and are the components that most often need attention.

The brake shoes are designed to be routinely replaced as they wear out. While it is possible to restore the brakes by replacing only the brake shoes, full braking performance and minimum lining wear can only be achieved if the drums are resurfaced or replaced at the same time. Removal of the brake drums requires cleaning, repacking, and adjusting the wheel bearings during reassembly, as described in **SUSPENSION AND STEERING**.

> ***WARNING —***
> *Brake pad friction materials may contain asbestos fibers. Do not create dust by grinding, sanding, or cleaning the pads with compressed air. Avoid breathing any asbestos fibers or dust. Breathing asbestos may result in serious diseases such as asbestosis or cancer, or may result in death.*

Checking Brake Lining Wear

The inspection procedure and specifications for checking the brake lining wear are found in **LUBRICATION AND MAINTENANCE**. For more complete inspection of lining condition, the drums must be removed, as described in **6.1 Reconditioning Rear Drum Brakes**.

6.1 Reconditioning Rear Drum Brakes

Reconditioning of the rear drum brakes typically includes replacing the brake shoes, resurfacing or replacing the brake drums, replacing the attaching and adjusting hardware, and cleaning and repacking the wheel bearings. Unless the wheel cylinders or hoses are damaged or leaking fluid, more extensive rear drum brake repairs are not normally required.

If the brake linings are soaked with grease or brake fluid, the cause of the contamination must be found and corrected before new brake shoes are installed. Brake drums must be replaced when resurfacing exceeds specified limits. See **Reconditioning Brake Drums** for specifications.

Removing and Installing Brake Shoes

The brake shoes only need to be removed when they are being replaced. If the backing plate must be removed for other repairs, the brake shoes can remain attached to it. Bleeding the brake system will not be necessary as long as the brake lines or hoses are not disconnected. The brake drums can be removed with or without the wheel in place.

To remove:

1. Raise the rear of the car and support it securely on jack stands. Remove the wheels, or simply remove one of the wheel lug bolts from each side.

2. With the parking brake released, rotate the brake drum until the adjusting wedge can be seen through a wheel bolt hole. Using a screwdriver as shown in Fig. 6-2, push the adjusting wedge up as far as possible.

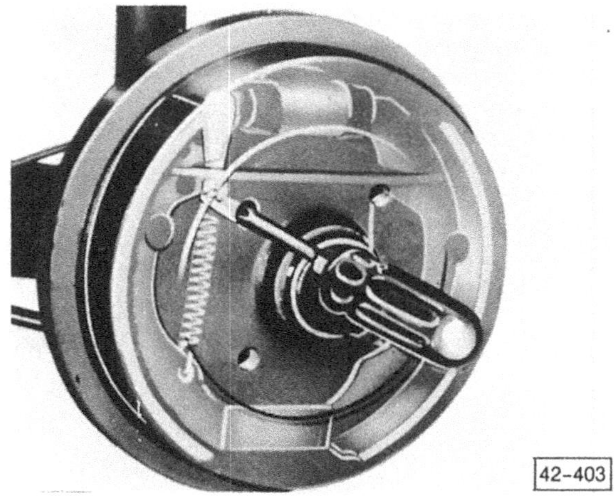

Fig. 6-2. Phantom view of rear drum brake showing adjusting wedge being pushed up with screwdriver to back off adjustment.

3. Pry off the wheel bearing grease cap and remove the cotter pin, the nut lock, and the axle nut. Tugging on the brake drum slightly if necessary, remove the thrust washer and the outer wheel bearing. Remove the wheel and the brake drum.

4. While supporting the head of one shoe retaining pin from behind the backing plate, push the spring retainer in and rotate it 90° to remove it. See Fig. 6-3. Repeat the step for the other retainer.

Fig. 6-3. Rear brake assembly showing brake shoe retaining pins and spring retainers (light arrows), and lower brake shoe support (dark arrows).

5. Disengage the brake shoes from the lower brake shoe support at the bottom of the backing plate, and remove the lower return spring.

6. Unhook the parking brake cable from the lever that is attached to the rear brake shoe. See Fig. 6-4.

Fig. 6-4. Parking brake cable attachment to lever on rear brake shoe (arrow).

7. Unhook the spring from the bottom of the adjusting wedge, and remove the upper return spring. Remove the brake shoes.

CAUTION —
Used brake shoes should not be interchanged. If the shoes are to be reused, mark each shoe's position for reinstallation.

8. Clamp the adjusting bar in a vise, and use pliers to remove the tensioning spring.

To install:

1. Clamp the front brake shoe in a vise. Lubricate the points where the brake shoe, the adjusting bar, the tensioning spring, and the adjusting wedge contact each other. Use Wolfracote grease, Volkswagen part no. G 000 650.

2. Install the adjusting bar and the tensioning spring as shown in Fig. 6-5, then install the adjusting wedge. The lug on the adjusting wedge should face the backing plate.

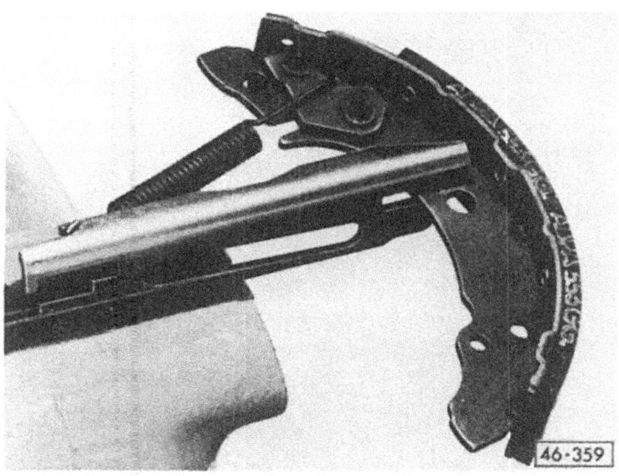

Fig. 6-5. Adjusting bar and tensioning spring being installed on brake shoe. Brake shoe and spring attachment are shown from behind.

3. Attach the rear brake shoe (with parking brake lever) as shown in Fig. 6-6, then install the upper return spring.

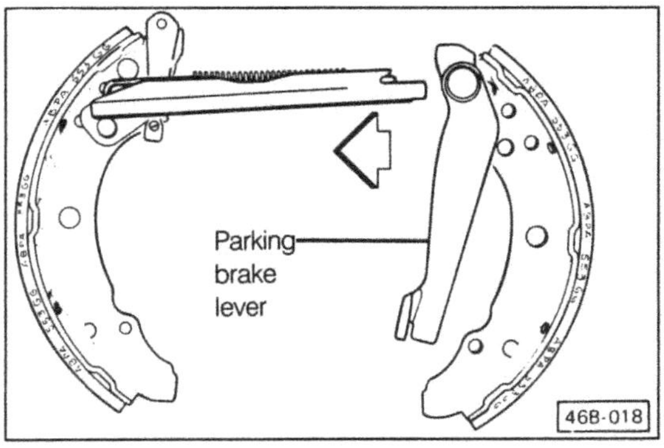

Fig. 6-6. Rear brake shoe with parking brake lever being attached to front shoe and adjusting bar. Shoes are held together by upper return spring.

4. Holding the brake shoes near the backing plate, attach the parking brake cable.

5. Using a small amount of Wolfracote grease, lubricate the points on the backing plate where the edges of the brake shoes make contact.

WARNING —
Do not allow grease to contaminate the brake linings.

6. Place the upper ends of the brake shoes around the wheel cylinder piston, install the lower return spring, and engage the bottom ends of the shoes on the lower support.

7. Install the shoe retaining pins, the springs, and the spring retainers. Turn the retainers 90° to lock them in place.

8. Reinstall the adjusting wedge spring, then push the wedge up to fully retract the brake shoes for installing the drum.

9. Install the brake drum and repack and adjust the wheel bearings as described in **SUSPENSION AND STEERING**. Use a new cotter pin to secure the axle nut lock.

10. If the wheel was removed from the brake drum, mount the wheel and install the wheel lug bolts. Lower the car and then torque the lug bolts to 110 Nm (81 ft. lb.).

11. After completing the installation on both rear brakes, firmly apply the brake pedal once to set the self-adjusting wedge.

WARNING —
New brake linings require some break-in. Allow for slightly longer stopping distances for the first 100 to 150 miles of city driving, and avoid hard stops.

Reconditioning Brake Drums

Both rear brake drums should be resurfaced whenever new brake shoes are installed. **Table d** lists the specifications for resurfacing brake drums, including maximum diameter for the resurfaced drums. Drums which exceed the maximum diameter are too thin to be reliably used and should be replaced.

Brake drums can be resurfaced by most local automotive machine shops. Torquing a wheel to a brake drum changes its shape slightly. If possible, the drums should be resurfaced with the wheel mounted to the drum and the bolts torqued to 110 Nm (81 ft. lb.).

Table d. Brake Drum Reconditioning Specifications

Inside diameter, new	180.0 mm (7.087 in.)
Maximum diameter after resurfacing	180.5 mm (7.106 in.)
Wear limit	181.0 mm (7.126 in.)
Radial runout at friction surface	0.05 mm (.002 in.)
Lateral runout at wheel contact surface	0.2 mm (.008 in.)

Inspecting and Replacing Wheel Cylinder

The wheel cylinder is mounted at the top of the backing plate between the brake shoes. It is reached by removing the brake drum as described in **Removing and Installing Brake Shoes**. Inspect the wheel cylinder by observing the action of the brake shoes while an assistant slowly depresses the brake pedal. Replace the cylinder if the pistons stick, or if the brake shoes do not move out the same distance at a uniform rate. Also replace the cylinder if there are signs of brake fluid leakage under the dust boot. To replace a wheel cylinder the brake line must be disconnected, and it will be necessary to bleed the brakes. See **4.2 Bleeding Brakes**.

> **NOTE —**
> When the brake drum is removed, insert two screwdrivers behind the backing plate flange and press them against the brake shoes to limit their travel.

To replace the wheel cylinder, first disconnect the brake line from the cylinder behind the backing plate. Seal the brake line to prevent contamination. Unhook the upper return spring and the adjusting bar tensioning spring. Push the adjusting wedge down to spread the brake shoes apart. Remove the two bolts that hold the wheel cylinder to the backing plate, and then remove the wheel cylinder.

Installation is the reverse of removal. Torque the wheel cylinder mounting bolts to 10 Nm (87 in. lb.), and torque the brake line union to 15 Nm (11 ft. lb.). Install the brake drum as described in **Removing and Installing Brake Shoes**, and bleed the brake system as described in **4.2 Bleeding Brakes**.

6.2 Parking Brake (Drum Brakes)

The parking brake lever and its cable connections are shown in Fig. 6-7. The mechanical parking brake operates only on the rear wheels, and is independent of the hydraulic brake system. Each of the two parking brake cables is connected to a lever on one of the rear brakes shoes. See Fig. 6-6 above. When the parking brake lever is pulled, the brake shoes are forced out against the brake drum.

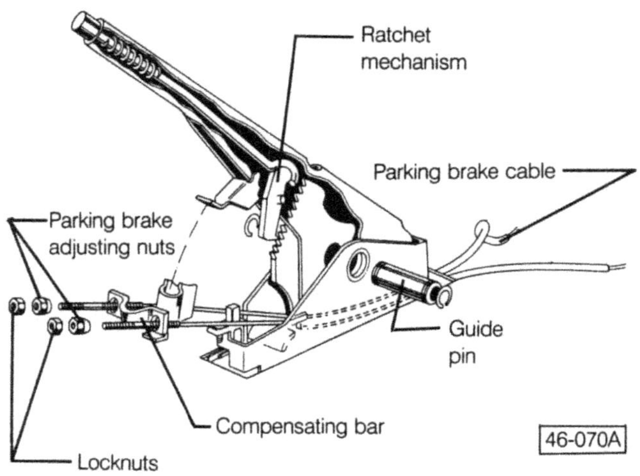

Fig. 6-7. Cut-away view of parking brake lever assembly showing cable connections and internal ratchet.

Adjusting Parking Brake

Adjusting the parking brake should only be necessary if the cable is replaced. Failure of the brake to hold the car most likely indicates worn brake shoes or a faulty self-adjusting mechanism. Check to see that the parking brake functions properly whenever the cables have been detached for rear brake service.

To adjust:

1. Raise the rear of the car and support it securely on jack stands.
2. With the parking brake fully released, firmly apply the brakes one time.
3. Move the parking brake lever to the second detent (second click) from the bottom.
4. Loosen the locknuts and tighten the adjusting nuts by equal amounts just until neither rear wheel can be rotated by hand.
5. Check that when the parking brake is fully released, both rear wheels rotate freely. Readjust if necessary.
6. Tighten the locknuts.

Replacing Parking Brake Cables

The cables can be replaced separately. Remove the locknut and the adjusting nut that hold the cable to the compensating bar. Remove the rear brake drum as described in **Removing and Installing Brake Shoes**, and then unhook the parking brake cable from the lever on the rear brake shoe. It may be necessary to partially disassemble the brakes. Pull the cable and conduit out of the backing plate and remove the cable from the guides and hangers which secure it to the trailing arm.

Pull the old cable out of the body from the rear. Lubricate the new cable with multipurpose grease before inserting it. Feed the cable into the backing plate without kinking it, and hook the end of the cable to the lever on the rear brake shoe. Reinstall the rear brake drum as described in **Removing and Installing Brake Shoes**. At the parking brake lever inside the car, install the compensating bar, the adjusting nut, and the locknut on the cable and adjust the parking brake as described above in **Adjusting Parking Brake**.

7. REAR DISC BRAKES

The components of the rear disc brakes are shown in Fig. 7-1. The rear disc brakes operate in much the same way as the front disc brakes described in **5. Front Disc Brakes**, except that the rear disc rotor doubles as the wheel hub and wheel bearing housing, and the rear calipers can be mechanically activated by the parking brake cables. The wheel lug bolts thread directly into the rotor, and the inner diameter of the rotor hub houses the wheel bearings.

Because the rear brakes do less of the work of stopping the car, they are not likely to wear as quickly or require service as frequently as the front brakes. Unless problem symptoms are actually observed at the rear brakes, troubleshooting of brake problems should always include the front brakes. If the rear brakes are worn and require service, then at the very least the front brakes should also be thoroughly inspected. The brake pads and rotors are subjected to the greatest wear, and are the components that most often need attention.

The brake pads are designed to be routinely replaced as they wear out. While it is possible to restore the brakes by replacing only the brake pads, full braking performance and minimum pad wear can only be achieved if the rotors are resurfaced or replaced at the same time. Removal of the brake rotors requires cleaning, repacking, and adjusting the wheel bearings during reassembly, as described in **SUSPENSION AND STEERING**.

> ***WARNING —***
> *Brake pad friction materials may contain asbestos fibers. Do not create dust by grinding, sanding, or cleaning the pads with compressed air. Avoid breathing any asbestos fibers or dust. Breathing asbestos may result in serious diseases such as asbestosis or cancer, or may result in death.*

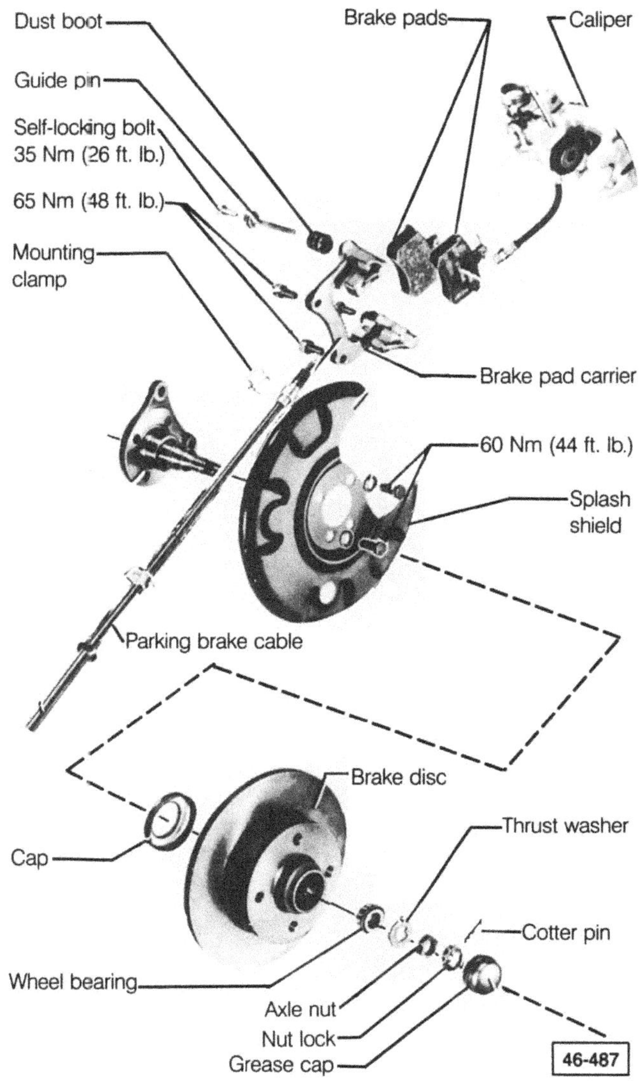

Fig. 7-1. Exploded view of rear disc brake assembly.

Checking Brake Pad Wear

The inspection procedure and specifications for checking the brake pads are found in **LUBRICATION AND MAINTENANCE**. For more complete inspection of pad condition, the pads must be removed, as described in **7.1 Reconditioning Rear Disc Brakes**.

7.1 Reconditioning Rear Disc Brakes

Reconditioning of the rear disc brakes typically includes replacing the brake pads and resurfacing or replacing the brake rotors. Unless the calipers and hoses are damaged, extremely worn, or leaking fluid, more extensive rear brake repairs are not normally required.

If the brake pads are soaked with oil, grease, or brake fluid, the cause of the contamination must be found and corrected before new pads are installed. Brake rotors must be replaced when resurfacing exceeds specified limits. See **Removing, Reconditioning, and Installing Rear Brake Rotors** for rotor thickness specifications.

Removing and Installing Rear Brake Pads

Although the calipers must be removed to change the brake pads, this can be done without disconnecting the flexible brake hose. Keeping the hydraulic system sealed eliminates the need to bleed the brakes afterward.

Brake pads and the surfaces of the rotors wear slightly differently from one another. Always replace brake pads in complete sets. If old pads are to be reinstalled, such as after inspecting them, always make sure they are refitted in their original locations.

When replacing the brake pads, the two upper self-locking caliper mounting bolts should be replaced.

To remove:

1. Raise the rear of the car and support it securely on jack stands. Remove the rear wheels. Release the parking brake.
2. Remove the wire clip that secures the parking brake cable to the brake caliper, as shown in Fig. 7-2, and unhook the cable from the actuating lever. Slide the cable out of the cable guide.
3. Remove the upper self-locking caliper mounting bolt from the brake caliper. See Fig. 7-2. Hold the guide pin with an open-end wrench to remove the bolt. Pivot the caliper downward off the brake pads, as shown in Fig. 7-3.

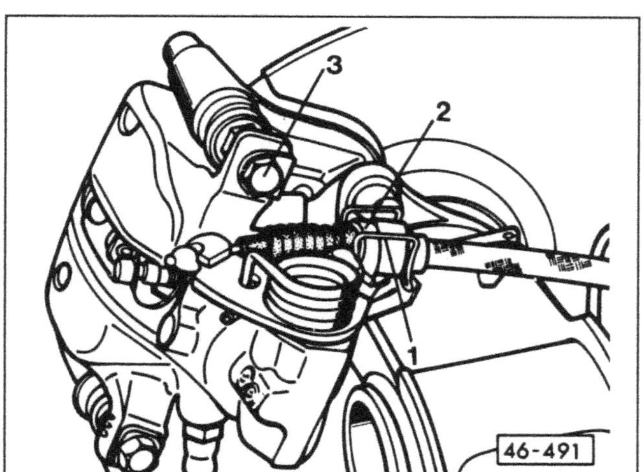

Fig. 7-2. Wire clip **1** to be removed when disconnecting parking brake cable from cable guide **2**. Upper caliper mounting bolt is at **3**.

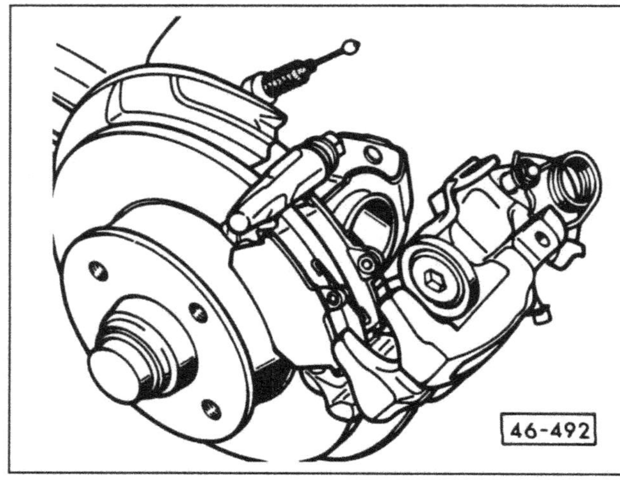

Fig. 7-3. Upper caliper mounting bolt removed and caliper pivoted downward on lower guide pin for removal of brake pads.

4. Lift out the brake pads.

CAUTION —
Do not interchange used pads. If the pads are to be reused, mark them for reinstallation in their same positions.

To install:

1. Place the brake pads in position on the brake pad carrier.
2. When installing new brake pads, adjust the caliper to provide clearance for the thicker new brake pads. Use a hex wrench in the piston socket to turn the caliper piston clockwise all the way into the caliper housing, as shown in Fig. 7-4.

NOTE —
Turning in the caliper pistons to gain clearance for the thicker new brake pads will cause brake fluid to overflow the master cylinder fluid reservoir. To prevent this, use a clean syringe to first remove some fluid from the reservoir.

3. Pivot the caliper back into position and install a new upper self-locking mounting bolt. Hold the guide pin with an open-end wrench, and torque the self-locking bolt to 35 Nm (26 ft. lb.).
4. If new brake pads were installed, adjust the parking brake as described in **7.2 Parking Brake (Disc Brakes)**.
5. Mount the wheel and loosely install the lug bolts, then lower the car and torque the bolts to 110 Nm (81 ft. lb.).

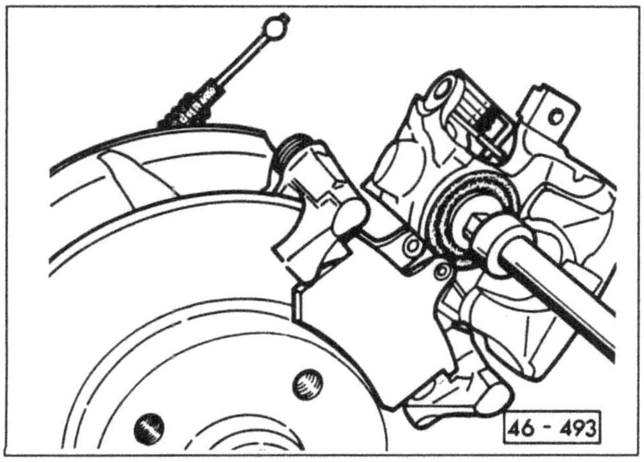

Fig. 7-4. Rear caliper piston being turned into caliper with hex wrench.

6. Before running the engine or moving the car, pump the brake pedal at least 40 times to reset the automatic adjustment of the rear calipers.

> ***CAUTION —***
> *Running the engine and providing vacuum boost to the brakes while resetting the automatic adjustment may jam the caliper.*

7. Check the level of brake fluid in the reservoir. If necessary, add new brake fluid to fill the reservoir to the MAX mark.

> ***WARNING —***
> *New brake pads require some break-in. Allow for slightly longer stopping distances for the first 100 to 150 miles of city driving, and avoid hard stops.*

Removing, Reconditioning, and Installing Rear Brake Rotors

To remove the rear brake rotors, first remove the brake pads as described above. Remove the two bolts holding the brake pad carrier to the stub axle, and suspend the caliper and pad carrier from the suspension or body. Avoid stretching or kinking the hose. If the brake hose remains connected, it will not be necessary to bleed the brakes.

> **NOTE —**
> Be careful when suspending the caliper assembly. The guide pins are lubricated and will allow the caliper to separate from the pad carrier.

Remove the wheel bearing grease cap from the center of the brake rotor, and then remove the cotter pin, the nut lock, and the axle nut from the stub axle. Tugging on the brake rotor slightly if necessary, remove the thrust washer and the outer wheel bearing. Remove the brake rotor. Use a screwdriver to pry out the dust cap and grease seal from the back of the rotor hub, and remove the inner bearing parts.

Installation is the reverse of removal. See **SUSPENSION AND STEERING** for information on cleaning the wheel bearings, packing them with new grease, and adjusting them after installation. Install a new inner grease seal. Torque the brake pad carrier mounting bolts to 65 Nm (48 ft. lb.). Reinstall the brake pads as described in **Removing and Installing Rear Brake Pads**.

> **NOTE —**
> New replacement brake rotors should be cleaned with a grease-free solvent, such as a commercially available brake cleaner, before installing the caliper and brake pads.

Brake rotors should always be resurfaced in pairs, with an equal amount of material removed from both sides of each rotor. **Table e** lists the minimum requirements for resurfacing brake rotors—the minimum thickness of a rotor which can still be resurfaced, and the minimum allowable thickness after resurfacing. Rotors which fail to meet either minimum are too thin to be reliably used and should be replaced. Brake rotors can be resurfaced by most any local automotive machine shop.

Table e. Rear Rotor Reconditioning Specifications

Specification	Value
Thickness new	10.0 mm (.394 in.)
Minimum thickness	8.0 mm (.315 in.)
Maximum runout	0.06 mm (.002 in.)

7.2 Parking Brake (Disc Brakes)

The parking brake operates only on the rear wheels. It is mechanically actuated, and is independent of the hydraulic brake system. Fig. 6-7 above is a cut-away view of the parking brake lever assembly. The lever operates two separate, replaceable cables connected to the rear brake calipers. When the lever is pulled, the actuating lever on the caliper applies pressure to the brake caliper piston.

Adjusting Parking Brake

The parking brake should only have to be adjusted if the cable, the brake rotor, the brake caliper, or the brake pads are replaced. Failure of the brake to hold the car most likely indicates worn brake pads or a faulty caliper. Check to see that the parking brake functions properly whenever the cables have been detached for rear brake service.

> ***CAUTION —***
> *Before adjusting the parking brake cables, adjust the rear disc brakes by retracting the caliper pistons and pumping the brake pedal to reset the automatic adjustment, as described in* **Removing and Installing Rear Brake Pads**.

To adjust:

1. Raise the rear of the car and support it securely on jack stands. Remove the rear wheels. Release the parking brake if it is applied.

2. Loosen the parking brake cable locknuts and adjusting nuts under the parking brake lever just until tension on the cables is relieved.

NOTE —
The actuator levers on the calipers should be against their stops as shown in Fig. 7-5. If not, then the caliper's automatic adjustment mechanism may be faulty.

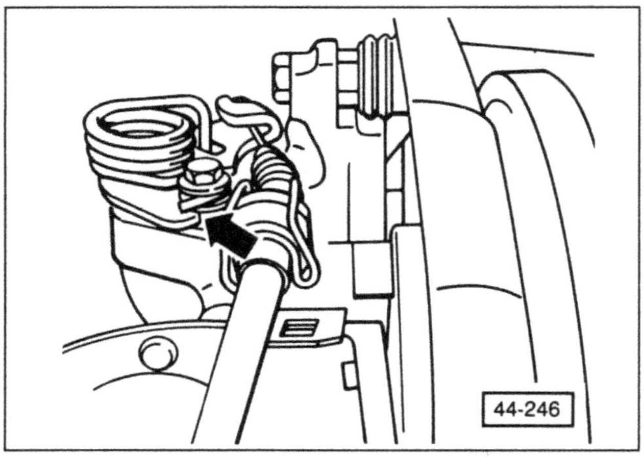

Fig. 7-5. Actuator lever against stop (arrow) on brake caliper.

3. Set and release the parking brake three times to stretch and seat the cables.

4. Tighten the adjusting nuts so that when the parking brake is applied, the following conditions are met: at the first detent (one click) the rear rotors should turn by hand with moderate drag; at the second detent the rotors should turn only with considerable effort; at the third detent it should not be possible to turn the rotors by hand.

NOTE —
When the parking brake is released, the rotors should again turn freely.

5. Check the gap between the actuator lever and the stop on the caliper. When the parking brake is correctly adjusted, it should be no greater than 1.0 mm (.039 in.). If it is, loosen the cable adjusting nuts in small increments until the gap is correct.

6. Tighten the parking brake cable locknuts. Mount the wheels and loosely install the lug bolts. Lower the car and torque the lug bolts to 110 Nm (81 ft. lb.).

Replacing Parking Brake Cable

The cables can be replaced separately. Remove the locknut and the adjusting nut that hold the cable to the compensating bar. Remove the wire clip that holds the parking brake cable to the caliper. Unhook the cable from the actuator lever on the caliper and remove the cable from the guides and hangers which secure it to the trailing arm.

Pull the old cable out of the body from the rear. Lubricate the new cable with multipurpose grease before inserting it. Feed the cable into place without kinking it, through the cable guide on the caliper. Hook the cable to the caliper actuating lever and secure it with the wire clip. At the parking brake lever inside the car, install the compensating bar, the adjusting nut, and the locknut on the cable, and adjust the parking brake as described above in **Adjusting Parking Brake**.

7.3 Calipers

Fig. 7-6 is an exploded view of the rear disc brake caliper. The rear calipers differ from those used at the front, mainly in the addition of the automatically adjusting parking brake actuating mechanism.

Brake fluid leaks around the brake caliper piston are the result of a failed or damaged piston seal, perhaps caused by corrosion, scoring, or pitting of the piston or caliper bore. The seal can be replaced, but a damaged piston, which is not available as a replacement part, will promptly destroy the new seal. To remedy a leaking caliper piston seal and avoid future problems, complete replacement of the caliper is recommended. A damaged caliper dust seal can be replaced separately.

To replace a caliper, first remove the parking brake cable from the caliper, as described in **7.2 Parking Brake (Disc Brakes)**. It may be necessary to loosen the cable adjusting nuts under the parking brake lever. Remove the two self-locking caliper mounting bolts and pull the caliper away from the brake pads and carrier. Hold the guide pins with an open-end wrench to remove the bolts. Disconnect the brake hose from the caliper and drain the brake fluid into a container. Cap the end of the brake line to prevent dirt and moisture from entering the brake system.

Installation is the reverse of removal. Pre-bleed the caliper by laying it on its side, as shown in Fig. 7-7. Fill the piston chamber with brake fluid through the bleeder valve until clear fluid runs out of the brake hose connection, then close the bleeder valve. Torque the brake hose union to 15 Nm (11 ft. lb.). Use new self-locking bolts to mount the caliper to the brake pad carrier, and torque them to 35 Nm (26 ft. lb.). Bleed the brake system as described in **4.2 Bleeding Brakes**. Reconnect the parking brake cable and adjust it as described in **Adjusting Parking Brake**. Do not let brake fluid contaminate the brake pads or brake rotor surface.

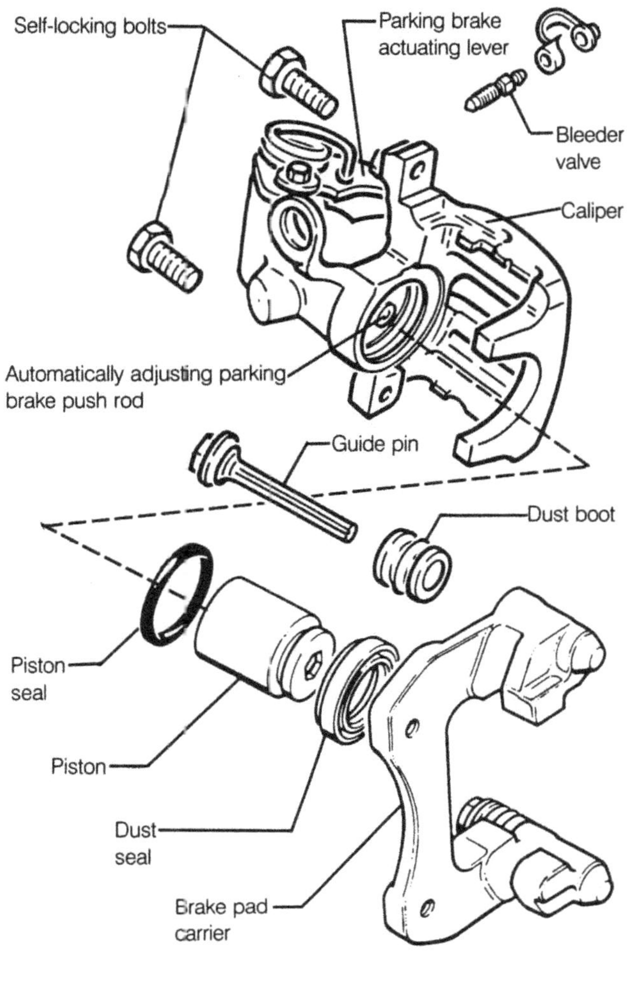

Fig. 7-6. Exploded view of rear brake caliper assembly.

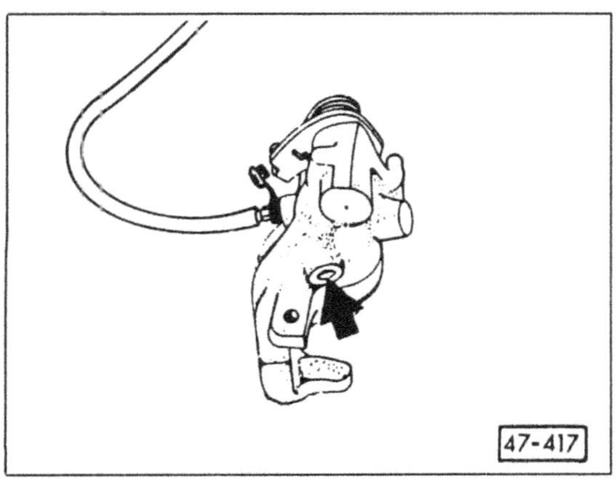

Fig. 7-7. Rear brake caliper being pre-bled prior to installation. Piston faces down. Bleeder valve and brake hose connection (arrow) face up. Fill until fluid escaping is free of air.

8. TECHNICAL DATA

I. Tolerances, Wear Limits, and Settings

Brake disc, front ventilated	
thickness, new part	20.00 mm (.787 in.)
thickness, minimum	18.00 mm (.709 in.)
runout, maximum	0.06 mm (.002 in.)
Brake drum, rear	
inside diameter, new drum	180.00 mm (7.087 in.)
inside diameter, maximum after resurfacing	180.50 mm (7.106 in.)
inside diameter, maximum	181.00 mm (7.126 in.)
runout, lateral at wheel mounting surface	0.2 mm (.008 in.)
runout, radial at friction surface	0.05 mm (.002 in.)
Brake disc, rear	
thickness, new part	10.00 mm (.394 in.)
thickness, minimum	8.00 mm (.315 in.)
runout, maximum permissible	0.06 mm (.002 in.)

II. Tightening Torques

Brake caliper (front) to wheel bearing housing (bolt) except 16-valve	25 Nm (18 ft. lb.)
16-valve	35 Nm (26 ft. lb.)
Brake disc splash shield to wheel bearing housing (bolt)	10 Nm (87 in. lb.)
Brake pad carrier (front) to wheel bearing housing (bolt)	70 Nm (52 ft. lb.)
Rear brake backing plate and stub axle to axle beam (bolt)	60 Nm (44 ft. lb.)
Rear wheel cylinder to backing plate (bolt)	10 Nm (87 in. lb.)
Brake caliper (rear) to brake pad carrier (self-locking bolt)	35 Nm (26 ft. lb.)
Brake pad carrier to stub axle flange (bolt)	65 Nm (48 ft. lb.)
Brake hose to brake caliper	15 Nm (11 ft. lb.)
Brake hose to wheel cylinder	15 Nm (11 ft. lb.)
Brake hoses – except to caliper or wheel cylinder	15–20 Nm (11–15 ft. lb.)
Brake line unions (union nuts)	15–20 Nm (11–15 ft. lb.)
Brake lines to master cylinder	15–20 Nm (11–15 ft. lb.)
Proportioning valves to master cylinder	15–20 Nm (11–15 ft. lb.)
Master cylinder to vacuum booster	21 Nm (15 ft. lb.)
Vacuum booster to support bracket (nut)	21 Nm (15 ft. lb.)
Wheel to brake disc or brake drum (lug bolt)	110 Nm (81 ft. lb.)

WARNING —

Do not re-use fasteners that are worn or deformed in normal use. Many fasteners are designed to be used only once and become unreliable and may fail when used a second time. This includes, but is not limited to, nuts, bolts, washers, self-locking nuts or bolts, circlips, cotter pins. Always replace these fasteners with new parts.

Section 12

BODY AND INTERIOR

Contents

TABLES

Body and Interior

Introduction

Scirocco and Cabriolet unitized-body construction yields body structures that are exceptionally strong and light. This light weight contributes to the outstanding combination of performance and fuel economy. During manufacture, the various body panels and subassemblies, plus a number of smaller pressed-steel panels and plates, are joined by electric welding. Because few screws and other fasteners are used, fewer rattles are likely to develop. The ride is quieted further by the application of sound-dampening material to the floor plates and the body panels.

It is important to ensure that a car suffering collision damage is repaired properly. Special equipment and methods are required to properly straighten the unitized body structure and retain or restore its designed-in structural integrity. Although all body panels are available as replacement parts, replacement panels must be butt-welded to the undamaged parts of the body after the damaged panels have been cut away. Proper assembly procedures and materials must be used to ensure that the repaired sections are still adequately protected against corrosion. Refinishing requires the correct materials and an experienced professional to apply them under the proper conditions. Because of these special considerations, detailed information on body repair and assembly is beyond the scope of this manual.

This section covers a variety of body and interior components that may require mechanical attention during the life of the car, either for repair or for access to other components and systems. Also see **ELECTRICAL SYSTEM**. If you lack the skills and experience, tools and equipment, or a suitable workplace for servicing or repairing any parts of the body and interior, we suggest that you leave these repairs to an authorized Volkswagen dealer or other qualified and properly equipped shop. We especially urge you to check with your authorized Volkswagen dealer before beginning any repairs on a car which may be covered wholly or in part by any of the extensive warranties issued by Volkswagen United States, Inc.

WARNING —

- *Beginning with 1990 models, the Cabriolet is equipped with a driver-side airbag.*
- *Serious personal injury can result if airbag components are handled improperly or without adequate safeguards.*
- *Thoroughly read the Cautions and Warnings at the front of this manual and in **SUSPENSION AND STEERING** before beginning any work on an airbag-equipped car.*

1. GENERAL DESCRIPTION

1.1 Instrument Panel and Instrument Cluster

The padded instrument panel houses the ventilation and heating system. If necessary it can be removed using ordinary hand tools. The instrument cluster can be removed separately. For electrical information see **ELECTRICAL SYSTEM**.

1.2 Seats, Seat Belts and Interior

The front seats are individually mounted to the floor. Folding bench-style or split rear seats are bolted to the floor. Both styles are easily removed for access to the rear seat belt mountings and the fuel gauge sending unit. All interior trim is designed to be removed using ordinary hand tools.

1.3 Ventilation, Heating and Air Conditioning

Fresh air enters the flow-through ventilation system through hood vents and the fresh-air plenum in front of the windshield. It passes through the heater core and, on cars with air conditioning, the evaporator. Interior air exits through vents in the body. A three-speed or four-speed fresh-air blower assists air flow.

The ventilating air can be heated in order to warm the car's interior or, on cars with air conditioning, cooled and de-humidified. In very hot conditions the interior air can be recirculated for maximum cooling by the air conditioning system.

1.4 Body

Unitized-body construction means that there is no separate frame. The sheet metal structure forms a rigid passenger compartment, with large front and rear crumple zones for energy absorption in the event of a collision. See Fig. 1-1.

With front wheel drive, there is no driveshaft to the rear. Instead the longitudinal hump in the front floor serves to strengthen the body. It also houses the shift linkage, exhaust system, parking brake cables and brake lines.

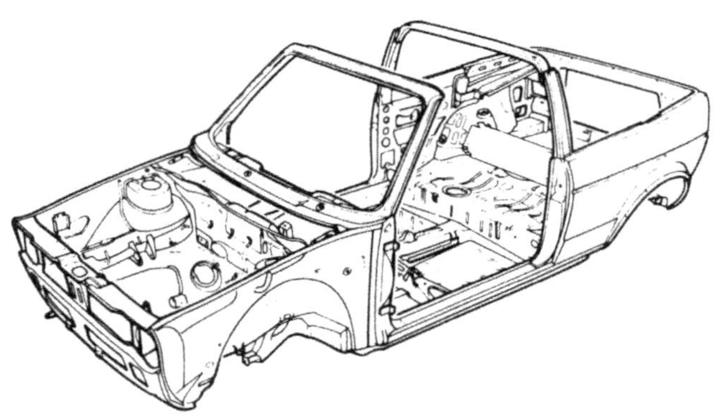

Fig. 1-1. Unitized construction of Cabriolet body. Scirocco is similar.

For corrosion protection, the body steel is specially treated and given a multi-layer finish. Seams are sealed using a PVC compound. The body is undercoated and all interior cavities are flooded with a hot wax.

2. MAINTENANCE

Volkswagen specifies checking the operation of the heating and ventilation system as a part of routine maintenance. See **LUBRICATION AND MAINTENANCE** for the prescribed maintenance intervals.

3. TROUBLESHOOTING

This section covers a variety of components and systems, so specific troubleshooting is included with repair information. For help when troubleshooting electrical problems, including wiring diagrams, see **ELECTRICAL SYSTEM**. Also see the discussion of electrical troubleshooting in **FUNDAMENTALS**.

4. INTERIOR

This section covers removing and installing interior parts. Repairs to interior electrical components are described in **ELECTRICAL SYSTEM**. Information concerning interior door trim, door locks, window controls and other mechanical parts of the door can be found in **5. Doors**.

> ***WARNING —***
> - *Beginning with 1990 models, the Cabriolet is equipped with a driver-side airbag.*
> - *Serious personal injury can result if airbag components are handled improperly or without adequate safeguards.*
> - *Thoroughly read the Cautions and Warnings at the front of this manual and in **SUSPENSION AND STEERING** before beginning any work on an airbag-equipped car.*

4.1 Instrument Cluster and Instrument Panel

Removing and Installing Instrument Cluster

The instrument cluster can be removed for service or repair using ordinary hand tools, without removing the instrument panel.

To remove instrument cluster (Scirocco only):

1. Disconnect the negative (–) battery terminal.
2. Remove the steering wheel as described in **SUSPENSION AND STEERING**.
3. Remove the instrument cluster trim by removing the two screws shown in Fig. 4-1.

Fig. 4-1. Screws (arrows) to be removed for removal of instrument cluster trim.

4. Free the instrument cluster by removing the screw shown in Fig. 4-2. Tip forward, but do not fully remove, the cluster, and remove the speedometer cable and any vacuum lines from the rear of the cluster. To remove the cable, pinch the retaining tabs shown in Fig. 4-3. Disconnect the two instrument cluster wiring harness connectors and remove the instrument cluster.

Fig. 4-2. Screw (arrow) to be removed to free instrument cluster.

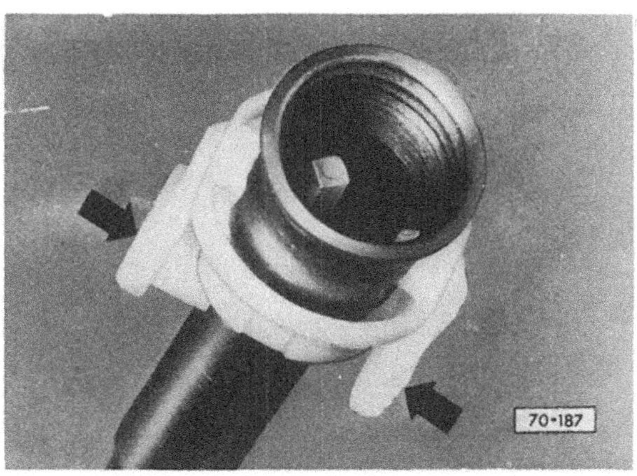

Fig. 4-3. Speedometer cable and retaining tabs (arrows) to be squeezed to remove cable from instrument cluster.

Installation is the reverse of removal.

To remove instrument cluster (Cabriolet only):

1. Disconnect the negative (−) battery terminal.
2. Remove the steering wheel as described in **SUSPENSION AND STEERING**.
3. Remove the heating and ventilation control knobs from the control head by pulling them straight out, and then remove the control head trim plate.
4. Remove the instrument panel storage tray by snapping it out, or, on models so equipped, remove the radio as described below in **4.2 Radio and Antenna**. Pry out the two switch trim rings shown in Fig. 4-4.

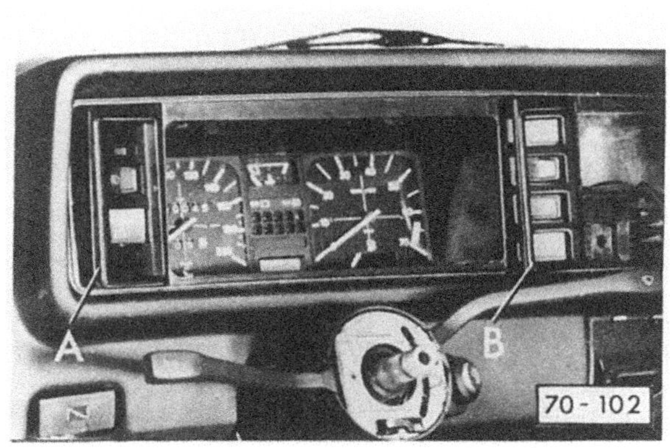

Fig. 4-4. Instrument panel switch trim rings (**A, B**) to be removed.

5. Remove the six screws shown in Fig. 4-5 that secure the instrument panel trim plate. Pull the trim out slightly, press out the panel switches, and then remove the trim.

Fig. 4-5. Screws (arrows) that secure instrument panel trim plate to instrument panel.

6. Free the instrument cluster by removing the two screws on either side of the cluster. Tip forward, but do not fully remove, the cluster, and remove the speedometer cable

from behind by pinching the retaining tabs. See Fig. 4-3. Then disconnect the harness connectors and remove the instrument cluster. Installation is the reverse of removal.

Removing and Installing Instrument Panel

After first removing the instrument cluster as described above, the instrument panel can be removed for access to the heating and ventilation system and other components.

> ***WARNING —***
> *Serious injury can result if airbag components are handled improperly or without adequate safeguards. See the Cautions and Warnings at the front of this manual and in **SUSPENSION AND STEERING** before beginning any work on an airbag-equipped car.*

To remove instrument panel (Scirocco only):

1. Remove the left storage bin. Tilt it down and remove the screws shown in Fig. 4-6. Remove the left lower trim panel by removing the clips and screws. See Fig. 4-7.

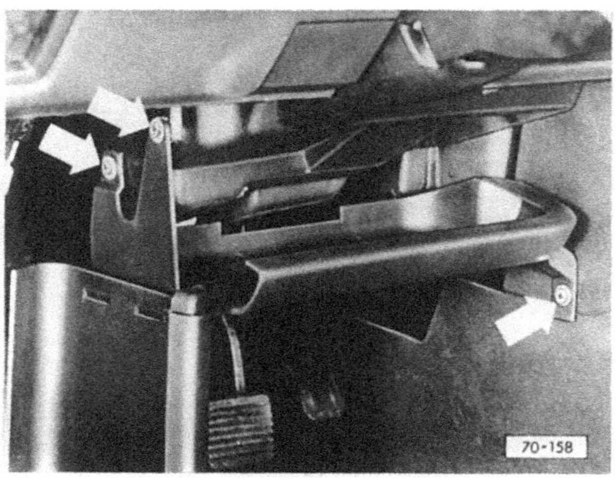

Fig. 4-6. Screws (arrows) to be removed for removal of left storage bin.

Fig. 4-7. Clips (**B**) and screws (arrows) to be removed for removal of left lower trim panel.

2. Remove the right lower shelf by removing the screws shown in Fig. 4-8. Remove the right lower trim panel by removing the clips shown in Fig. 4-9 and then pulling the panel off of its guides. Remove the ventilation air ducts from the instrument panel vents.

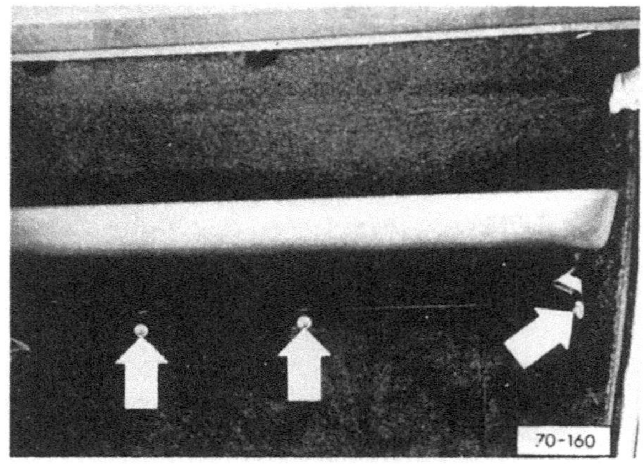

Fig. 4-8. Screws (arrows) to be removed for removal of right lower shelf.

Fig. 4-9. Clips (**B**) to be removed for removal of right lower trim panel. Slide panel off guides (**C**).

3. Remove the control knobs from the heating and ventilation control head by pulling them straight out, and remove the trim plate. Remove the screws shown in Fig. 4-10 and pull the lower console back slightly. Remove the screws for the upper console shown in Fig. 4-11, pull the console out slightly, disconnect any electrical connections to the console, and then remove the upper console.

4. Disconnect the heater cable as shown in Fig. 4-12.

5. Remove the instrument panel mounting screws. One screw is behind the panel as shown in Fig. 4-13, two are inside the glove box as shown in Fig. 4-14, and four are at either end of the panel as shown in Fig. 4-15. To remove the instrument panel, slide it to the right off of its mounting guides.

Installation is the reverse of removal

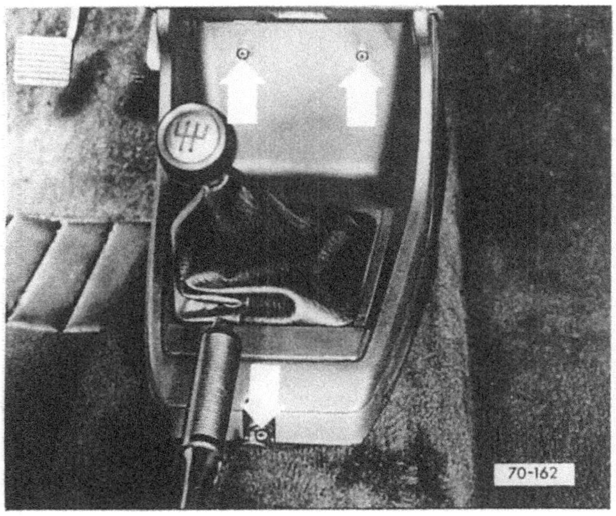

Fig. 4-10. Screws (arrows) to be removed for lower console. Pull console back slightly but do not remove.

Fig. 4-11. Screws (**1, 2**) to be removed for removal of upper console.

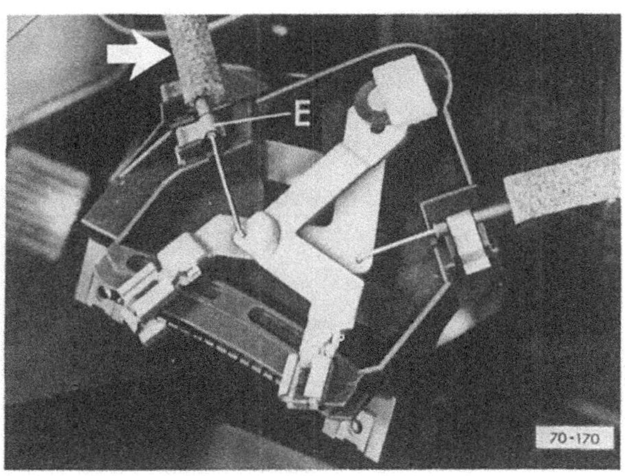

Fig. 4-12. Heater cable to be disconnected (arrow) after removing clip **E**.

Fig. 4-13. Instrument panel mounting screw, located behind the panel.

Fig. 4-14. Instrument panel mounting screws located in glove box.

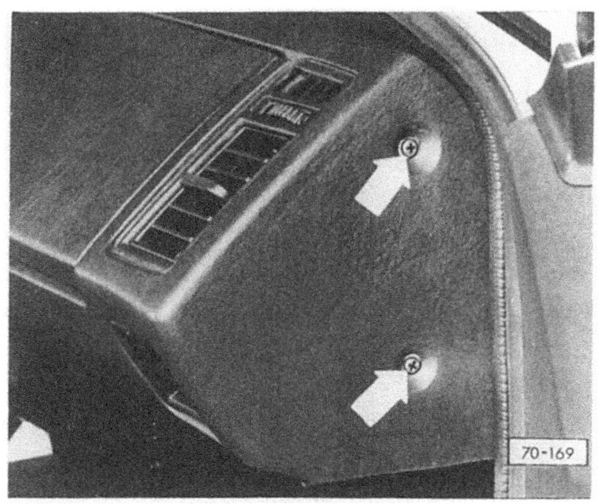

Fig. 4-15. Instrument panel mounting screws located at right end of instrument panel. Two more screws are located at left end.

To remove instrument panel (Cabriolet only):

1. Remove the screws that secure the heating and ventilation control head, shown in Fig. 4-16.

Fig. 4-16. Retaining screws (**1**) for heating and ventilation control head.

2. Remove the ashtray. Remove the four retaining screws for the center console and move the console back slightly.

3. Remove the two instrument panel air vents. Use a screwdriver to press in the retaining springs shown in Fig. 4-17, and pry the vent out.

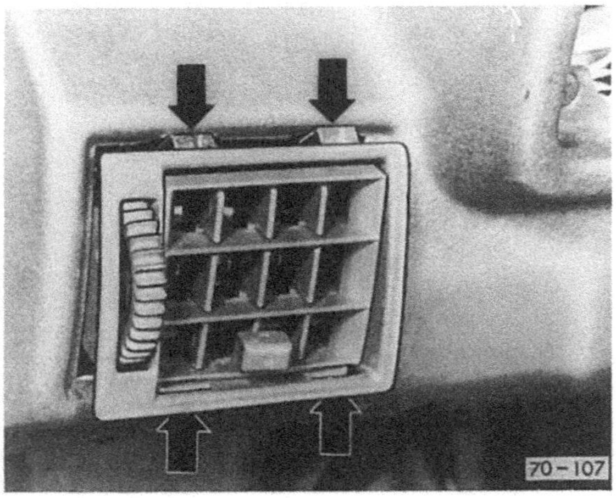

Fig. 4-17. Instrument panel air vent being removed. Press down on retaining springs (arrows) and pry vent out.

4. Remove the kneebar by removing the eight nuts and bolts at either end. See Fig. 4-18. Remove the left and right lower trim panels from underneath the instrument panel by unclipping them.

5. Disconnect the wiring harness connector and the screws shown in Fig. 4-19 and Fig. 4-20. Pull the instrument panel off of its retaining clips and remove it to the right.

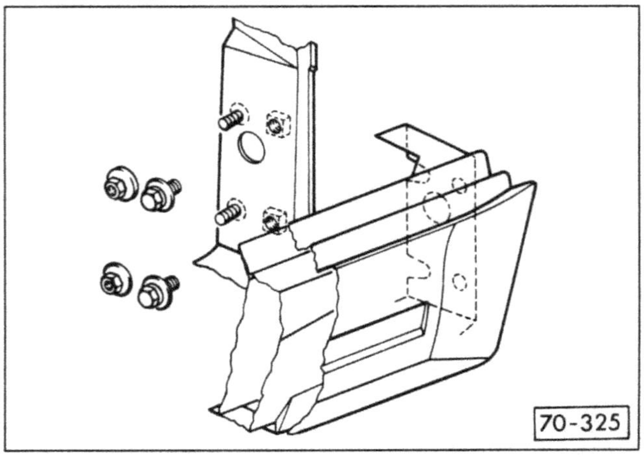

Fig. 4-18. Bolts and nuts that hold kneebar to right "A" pillar of body. Left side of kneebar is similar.

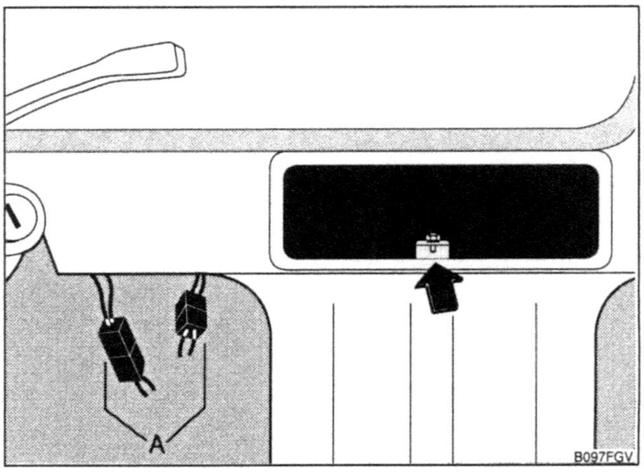

Fig. 4-19. Instrument panel harness connector (**A**) to be disconnected and screw (arrow) to be removed.

Fig. 4-20. Instrument panel mounting screws to be removed at either end of panel (**1**) and from below panel (**2**).

Installation is the reverse of removal. Torque the bolts and nuts for the kneebar to 21 Nm (15 ft. lb.).

Removing and Installing Speedometer Cable and Drive Gear

On automatic transmission cars, the cable mounts directly to the transmission with a union nut. On manual transmission cars, the cable and drive gear mount as shown in Fig. 4-21. Cars with an oxygen sensor (OXS) mileage counter have two cables, one from the transmission to the counter on the firewall and one from the counter to the speedometer.

To remove the cable, first disconnect the battery negative (–) terminal. Remove the left lower instrument panel trim as described in **Removing and Installing Instrument Cluster**. Disconnect the cable from the speedometer as shown earlier in Fig. 4-3.

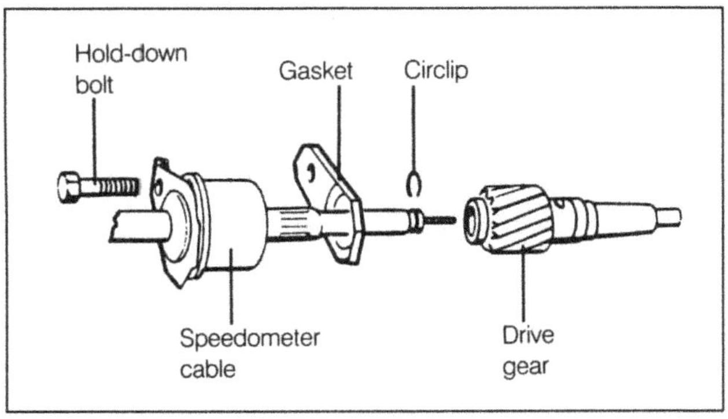

Fig. 4-21. Speedometer cable used on cars with manual transmission.

When replacing the cable on a manual transmission car, pull the drive gear off of the old cable, install a new circlip, and push the drive gear onto the cable until the circlip is visible through the hole in the gear. Position the circlip gap 180° from the hole in the gear. To prevent premature failure of the new cable, position it so that it is free of sharp bends. Do not grease the speedometer end of the cable.

4.2 Radio and Antenna

Factory-installed radios are removed with Volkswagen special tool VW 160 radio removal clips (order no. TV0 160 000 90 ZEL). First, disconnect the battery negative (–) terminal. Push the clips into the holes on either end of the radio face plate until they click into place, then use them to remove the radio. Remove the tools before installing the radio. Make sure that the rubber mount at the rear of the radio is properly engaged.

WARNING —
Do not use a computer memory saver to maintain battery power to an anti-theft radio. Using such a tool will keep the airbag system charged and may cause accidental activation of the airbag.

CAUTION —
With an anti-theft radio, make sure that you know the correct radio activation code before disconnecting the battery or removing the radio. Entering the wrong code when power is restored may lock up the radio and render it inoperable, even if the correct code is later entered.

On cars equipped with the factory-installed radio prep kit, the wiring is accessible after removing the radio trim plate. See the wiring diagrams in **ELECTRICAL SYSTEM** for help in identifying the wires and connectors.

To replace the roof-mounted antenna on Scirocco 16V models, the headliner must be removed as described in **4.9 Headliner**.

4.3 Interior Lights

A lighting failure may be caused by a blown fuse, especially if more than one bulb appears to have failed. Check and, if necessary, replace fuses as described in **ELECTRICAL SYSTEM**. **Table a** lists interior bulb applications and their part numbers. Dashboard lights and their replacement are covered in **ELECTRICAL SYSTEM**.

Table a. Bulb Application and Part Numbers

Bulb	VW Part No.	U.S. Trade No.
Interior dome	N 17 723 2	211
Luggage compartment	N 17 723 2	211
Center stop	N 17 753 2	168

Replacing Interior Light Bulbs

Always turn off the ignition before replacing a bulb. Avoid leaving fingerprints on the bulb and reflector. Fingerprints evaporate when the glass gets hot and can dim the light or reflector.

To replace the dome light bulb, press against the spring clip indicated in Fig. 4-22 and withdraw the assembly. Press the assembly back into position after installing the new bulb.

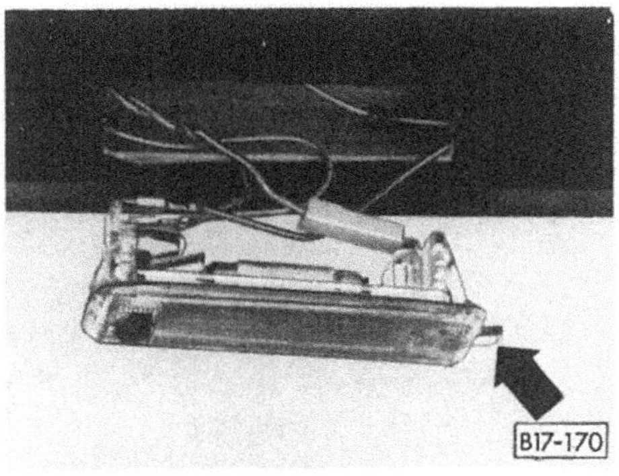

Fig. 4-22. Spring clip (arrow) that holds interior dome light in headliner.

To replace the luggage compartment light bulb, use a small flat-blade screwdriver to pry out the light assembly as shown in Fig. 4-23. Install the end with the wiring connector first, and press the assembly back into position.

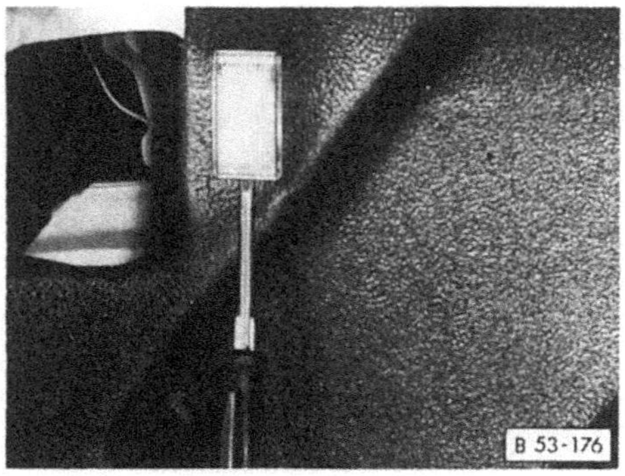

Fig. 4-23. Luggage compartment light assembly being removed with screwdriver.

To replace any one of the three center stop light bulbs, remove the bulb housing by pressing the small clips on either side of the assembly inwards to separate the housing from the lens. After installing the new bulb(s), press the bulb housing into position.

4.4 Heater and Controls

(without air conditioning)

The components and controls of the heating and ventilation system without air conditioning are shown in Fig. 4-24. The heater box contains the heater core, as well as two flaps operated by cables connected to the top two levers of the control head in the instrument panel. One flap regulates air flow

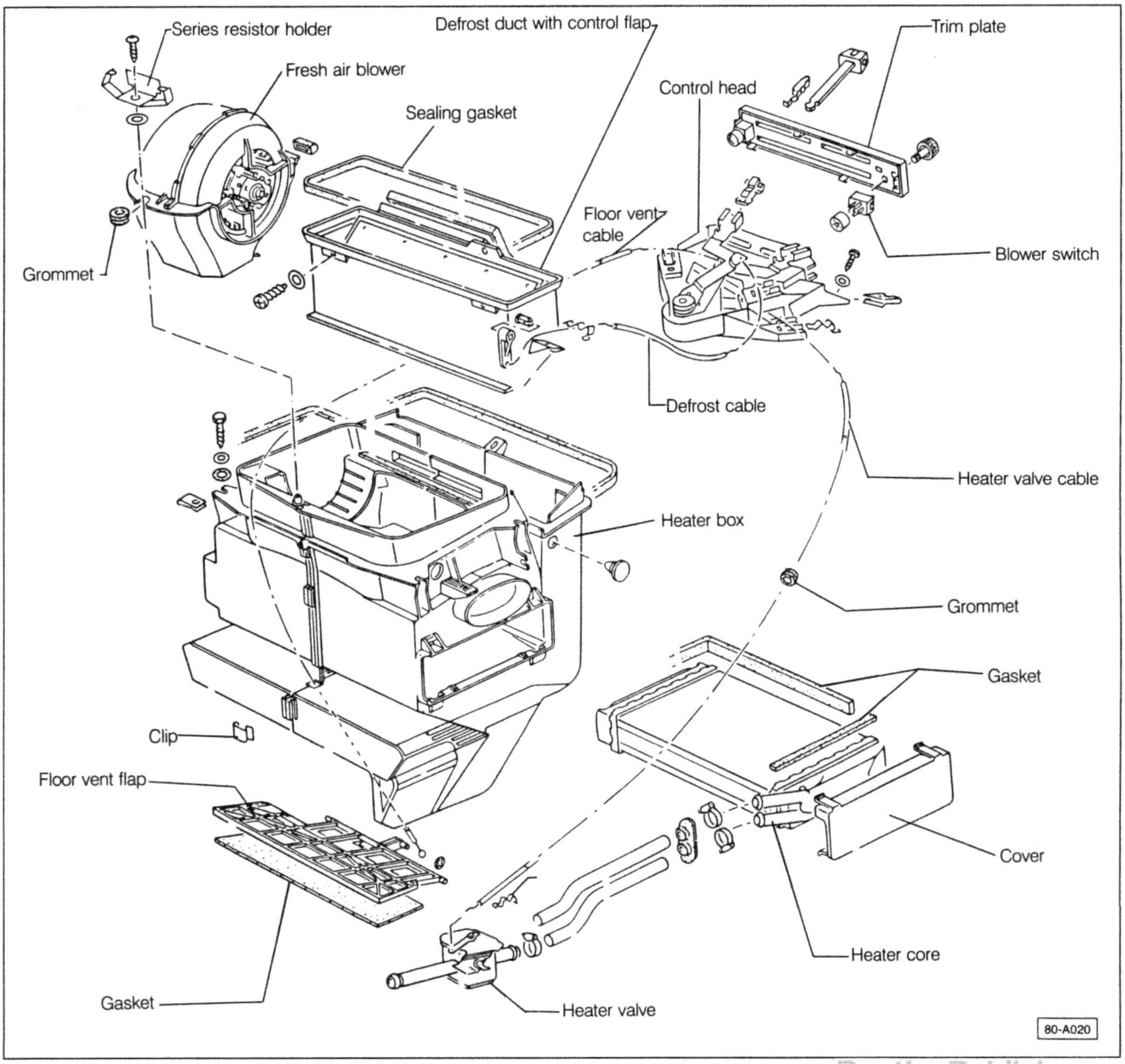

Fig. 4-24. Exploded view of ventilation and heating system used on cars without air conditioning.

to the floor vents, and the other regulates flow to the defrost vents. A third cable attached to the bottom lever of the control head operates the heater valve. When the valve is open, engine coolant flows through the heater core and heats the incoming air. See **ELECTRICAL SYSTEM** for electrical tests of the blower motor.

If moisture is leaking into the front foot well area of the passenger compartment, check the heater core as described below. Also, check the plastic drip tray at the base of the windshield in the engine compartment. It may be cracked, allowing water to leak into the ventilation system fresh air inlet.

Heater Core

Coolant leaking visibly into the passenger compartment is a sign of a faulty core or heater hose. A sweet, anti-freeze odor in the car's interior, or a constantly fogged windshield may also indicate a faulty core. Inspect the carpet and the area near the floor vents for any signs of moisture or coolant. A leaking heater core may be caused by overpressurization of the cooling system. See **COOLING SYSTEM** for more information.

When replacing the heater core or a hose, the coolant must first be drained as described in **COOLING SYSTEM**. The heater core can be pulled out of the side of the heater box after the cover is unclipped and the hose clamps are loosened to remove the heater hoses.

Replacing and Adjusting Heater Controls

The heating and ventilation control head is shown in Fig. 4-25.

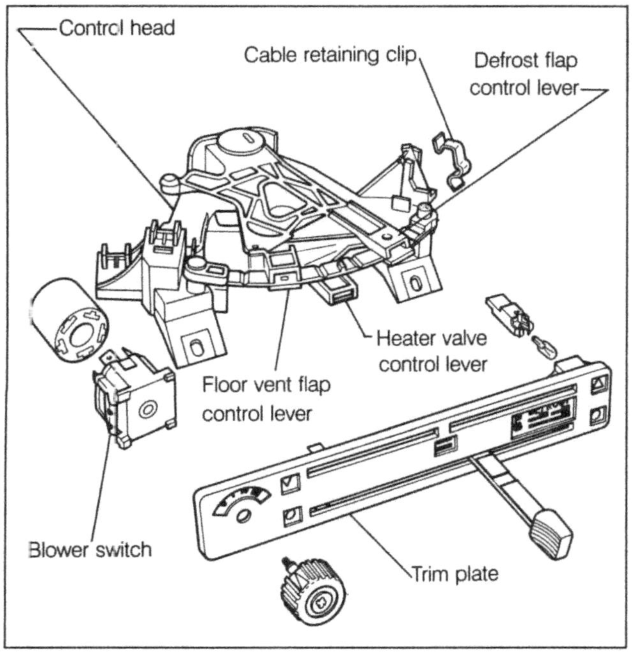

Fig. 4-25. Heating and ventilation control head used on cars without air conditioning.

To remove control cables, first remove the control head trim plate and the lower instrument panel trim panels as described in **4.1 Instrument Cluster and Instrument Panel**. For defrost or floor vent cables, remove the retaining clip that holds the cable conduit to the control head, then remove the cable from the flap and the control lever. For the heater valve cable, remove the retaining clips from the control head and the heater valve, and then remove the cable.

To install and adjust defrost or floor vent cables, place the control levers in the OFF position. Install the cable onto the control lever and the control flap. Making sure that the flap is in the closed position, install the cable retaining clip.

When replacing the heater valve, make sure that the engine is fully cool. Clamp the heater hoses close to the valve to prevent coolant loss, and place a container under the valve to catch any coolant that may leak out when the hoses are disconnected. Disconnect the control cable from the valve. After the new valve is installed, check the coolant level as described in **LUBRICATION AND MAINTENANCE**.

To install and adjust the heater valve control cable, place the control lever in the closed position. Install the cable onto the control head and install the retaining clip. Then install the cable onto the heater valve and, making sure that the valve is closed, install the other retaining clip. See Fig. 4-26.

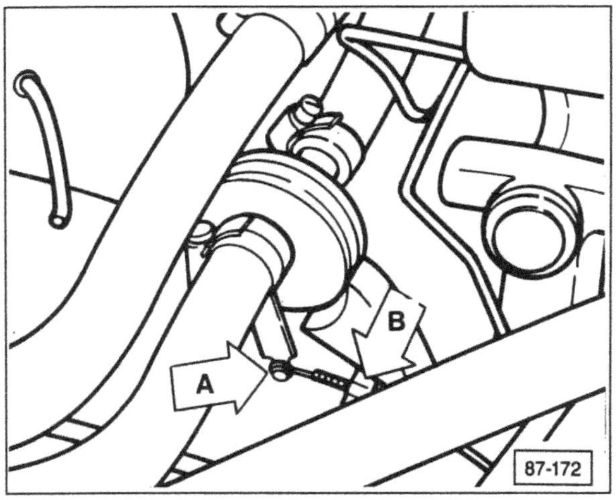

Fig. 4-26. Heater valve control cable being installed and adjusted. With cable secured to control head, move lever to off position (**A**) and install retaining clip (**B**).

Fresh Air Blower

The fresh air blower motor is located atop the heater box. The motor is reached from the engine compartment by removing the plastic drip tray at the base of the windshield, and then removing the plastic screen covering the fresh air inlet. To remove the motor, unclip the series resistor from its bracket and remove the screw under the resistor. See Fig. 4-27. Tilt the motor up and out and unclip the resistor wiring. See **ELECTRICAL SYSTEM** for electrical tests of the series resistor.

Fig. 4-27. Installed position of fresh air blower and series resistor. Motor mounting screw (arrow) is removed after resistor is unclipped.

4.5 Heater and Controls
(with air conditioning)

The components and controls of the heating and ventilation system with air conditioning (A/C) are shown in Fig. 4-29 (facing page). The heater box contains the fresh air blower, the heater core, and the evaporator for the A/C system. The control flaps for air distribution are operated by vacuum servos, which are controlled by the vacuum distributor attached to the top lever in the control head. The top lever also actuates the A/C thermostat switch to turn the air conditioning on and off. The bottom lever of the control head is attached to the heater control valve by a cable. When the valve is open, engine coolant flows through the heater core to warm the incoming air.

If there is moisture leaking into the front foot well area of the passenger compartment, check the heater core as described below. Also, check the plastic drip tray at the base of the windshield for cracks that may allow water to leak into the heater box fresh air inlet. Finally, check the evaporator drain hose for clogging and correct installation. The hose routes A/C system condensation from the bottom of the heater box to underneath the car. It is mounted behind a small cover on the left side of the passenger foot well.

Insufficient heater or air conditioner output, or air flow that does not correspond to control lever position may be caused by problems with the vacuum system. Check the vacuum system as described below before assuming that another component is faulty. For a description of the air conditioning, and for testing procedures, see **10. Air Conditioning**. See **ELECTRICAL SYSTEM** for electrical tests of the blower motor.

Heater core

Coolant leaking visibly into the passenger compartment is a sign of a faulty core or heater hose. A sweet, anti-freeze odor in the car's interior, or a constantly fogged windshield may also indicate a faulty core. Inspect the carpet and the area near the floor vents for any signs of moisture or coolant. A leaking heater core may be caused by overpressurization of the cooling system. See **COOLING SYSTEM** for more information.

When replacing the heater core or a hose, the coolant must first be drained as described in **COOLING SYSTEM**. On some vehicles, the heater core can be pulled out of the side of the heater box after the cover is removed and the hose clamps are loosened to remove the heater hoses.

Replacing and Adjusting Heater Controls

The control head for the heating and ventilation system with air conditioning is shown in Fig. 4-28. The top control lever is connected to the vacuum distributor to control ventilation air flow. The bottom lever is also connected to the vacuum distributor, and to the A/C thermostat switch and the bowden cable for the heater valve. Replacement and adjustment of the heater valve and cable are covered in **4.4 Heater and Controls (without air conditioning)**. Test the vacuum distributor as described below in **Checking Vacuum System**.

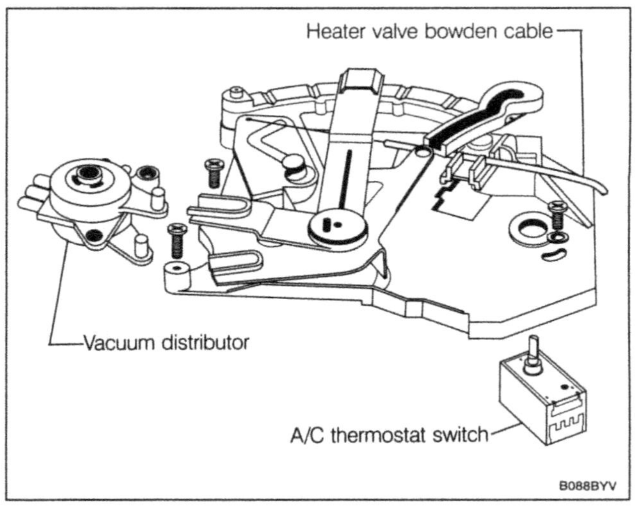

Fig. 4-28. Exploded view of control head for heating and ventilation system with air conditioning.

To free the control head for replacement of the vacuum distributor or A/C thermostat switch, pry off the control head trim plate and then remove the control head mounting screws. Make note of vacuum hose connections when replacing the vacuum distributor.

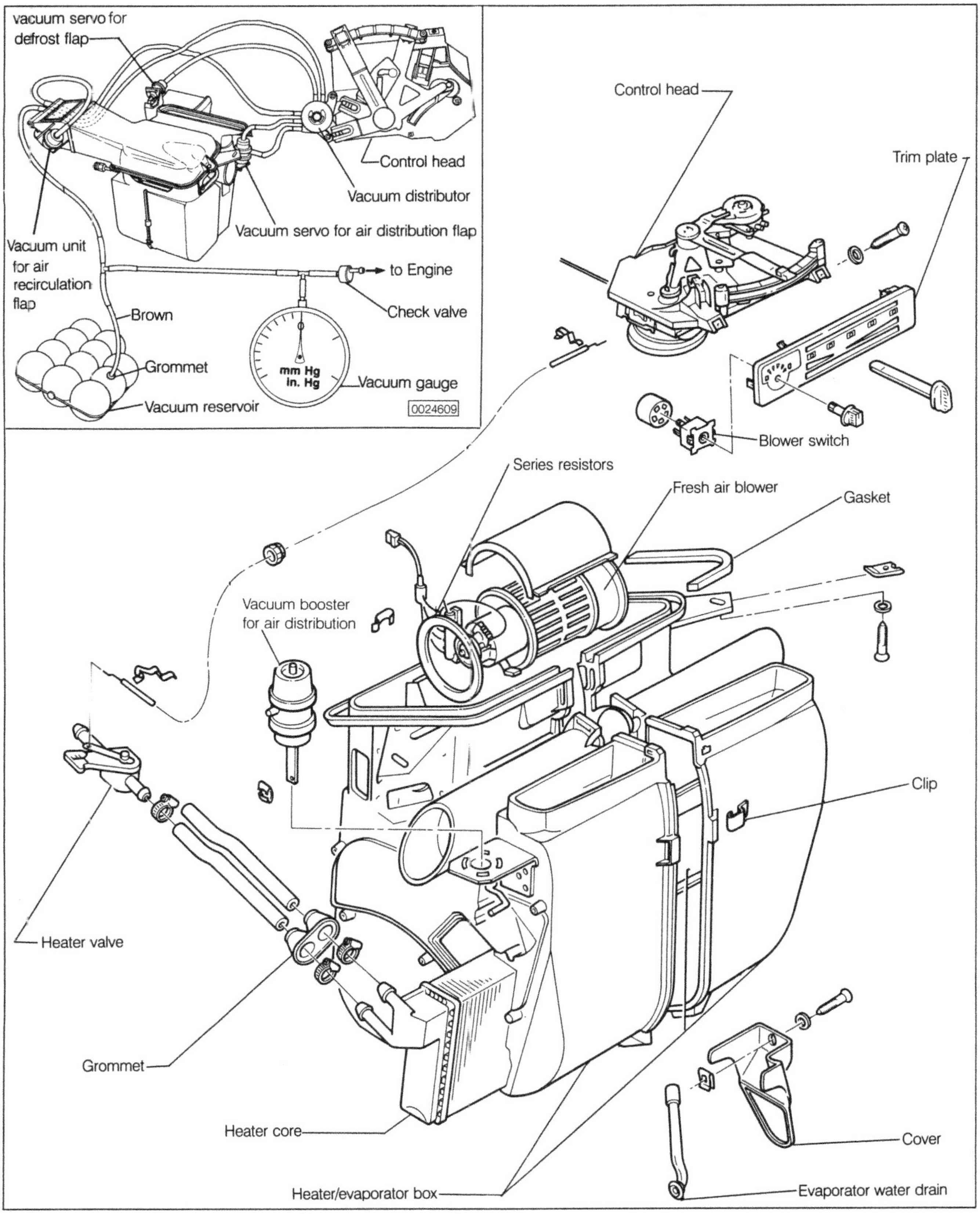

Fig. 4-29. Heater box and vacuum system layout (inset) of heating and ventilation system with air conditioning. Vacuum gauge is used for testing vacuum system.

To replace the A/C thermostat switch, first pull the capillary tube from the evaporator. Then remove the switch from the control head. Install the new switch and adjust it as shown in Fig. 4-30. Insert the capillary tube from the new switch into the evaporator to a depth of approximately 180 mm (about 7 in.).

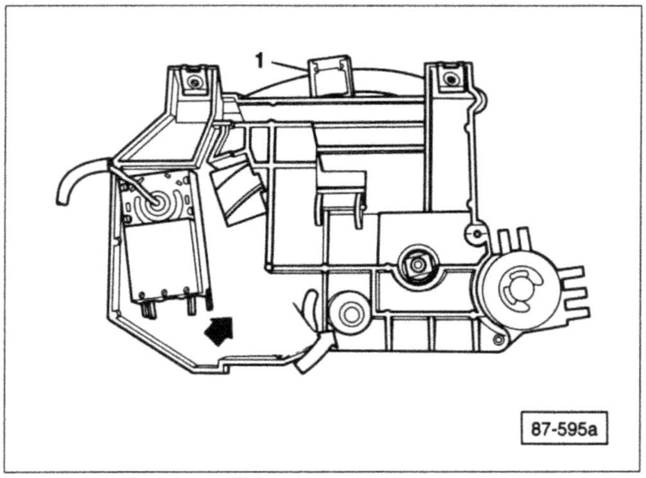

Fig. 4-30. Bottom view of A/C control head showing installation of A/C thermostat switch. Push lever **1** toward switch to stop, turn switch counterclockwise to stop (black arrow), then tighten mounting screw.

Checking Vacuum System

Engine vacuum operates the vacuum servos. If the control head cannot properly direct air flow, check for a faulty servo or leaks in that part of the vacuum system. The servos can be reached by removing the left and right lower instrument panel trim panels. If none of the servos respond properly, there is a problem with the vacuum supply or the vacuum distributor.

To test the vacuum system, install a vacuum gauge as shown earlier in Fig. 4-29. Run the engine until the gauge reads at least 15 inches Hg. Move the air distribution lever and the temperature control lever all the way to the right, and stop the engine. If the gauge reading falls within one minute, check for leaks in the vacuum hoses, the vacuum reservoir, the check valve or the vacuum distributor. If the gauge reading stays the same but one or more of the vacuum servos has retracted, then the vacuum distributor or the control head is faulty. See **Replacing and Adjusting Heater Controls**.

4.6 Seat Belts

Seat belts should be periodically inspected. If they have to be removed for any reason, it is important that they are reinstalled correctly. For problems with the seat belt warning light/buzzer system see **ELECTRICAL SYSTEM**.

Inspecting Seat Belts

Inspect front and rear belt webbing. Replace belts that have broken or pulled threads, cut loops at the belt edge, bowed or twisted webbing, faded areas, or cuts. On airbag-equipped cars (Cabriolet from 1990), the front belts are designed to elongate and absorb energy in a crash. Check these belts to determine whether or not they have been elongated. See Fig. 4-31.

Pull each belt out fully and let it retract. It should move smoothly in both directions. Check for dirt, grease or other contaminants on the webbing. Clean belts using only a mild soap solution recommended for cleaning upholstery or carpet. If this is not adequate to clean the belt, it should be replaced.

> ***WARNING —***
> - *Seat belts should always be replaced as a set if they have been stressed in a collision.*
> - *Do not bleach or dye seat belt webbing. Webbing that is severely faded will not meet strength requirements and must be replaced.*

Replace the belt if the buckle cover is cracked, if the latch release button is loose, or if it does not latch securely. Replace corroded mounting hardware.

Check the belt retractor mechanisms to see that they lock. At the front, tug sharply on the belt while extending it. At the rear, extend the belt, let several inches feed back into the retractor, then tug sharply on the belt. In both cases, the retractor should lock. Replace a belt if the retractor does not lock.

> **NOTE —**
> For front automatic belts, it is difficult with this test to apply enough force to the actuate the inertia reel and lock the retractor.

For front automatic belts, the retractor should also lock with a change in vehicle motion, such as while braking from a speed of 5 to 15 mph. Replace a belt if the retractor does not lock.

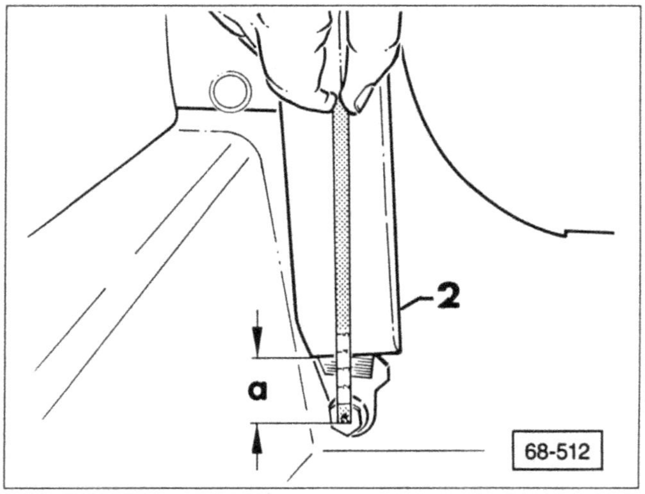

Fig. 4-31. On airbag-equipped cars, measure from mounting to edge of trim **2**. If **a** is greater than 40 mm (1.6 in.) the belt has elongated and must be replaced.

Installing or Replacing Seat Belts

The retractors must be securely anchored. Belt mounting points only must be free to swivel when the bolts are tight. Torque mounting bolts at 35 to 47 Nm (26 to 35 ft. lb.).

WARNING —

For maximum protection from injury, do not interchange buckle and retractor assemblies with those designated for other seating positions or other models.

Child Restraint Tethering

For child restraint systems with a tethering strap, Volkswagen recommends the anchoring points shown in Fig. 4-32 for Sciroccos, and Fig. 4-33 for Cabriolets. Pull back the carpet strip to expose the anchoring point before drilling an 8.5 mm (11⁄32 in.) hole. Cut a 20 mm (13⁄16 in.) hole in the carpet and glue it back. Use the mounting hardware shown in Fig. 4-34 or Fig. 4-35 to mount the strap. Seats can be mounted on either the left or the right side.

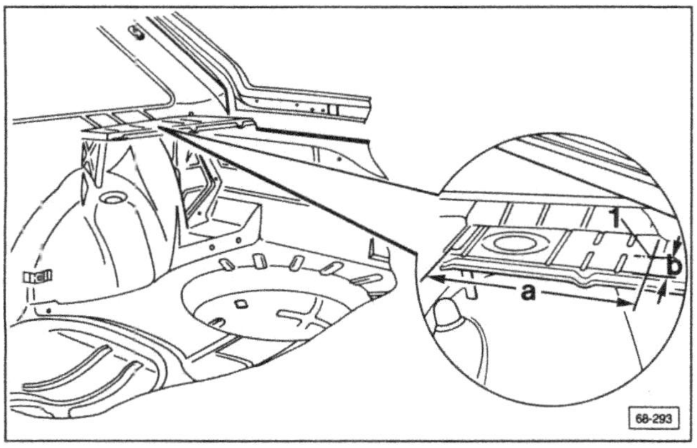

Fig. 4-32. Scirocco anchoring point for child restraint tethering strap. Drill hole where two lines intersect (**1**). Dimension **a** is 451 mm (17 ¾ in.). Dimension **b** is 30 mm (1 3⁄16 in.) from inner edge of shelf.

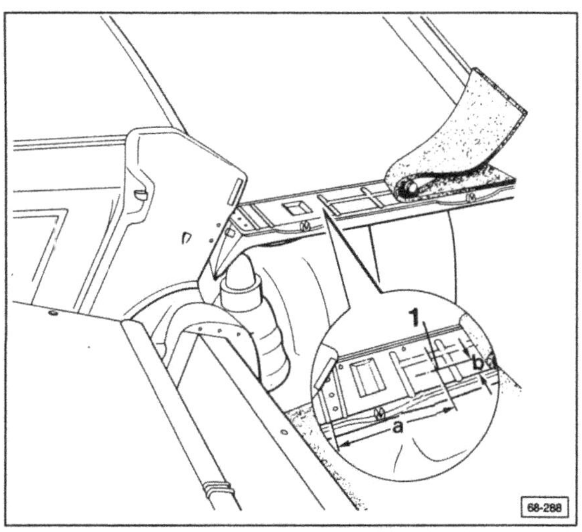

Fig. 4-33. Cabriolet anchoring point for child restraint tethering strap. Drill hole where two lines intersect. (**1**). Dimension **a** is 225 mm (8 ⅞ in.). Dimension **b** is 35 mm (1 ⅜ in.) from inner edge of shelf.

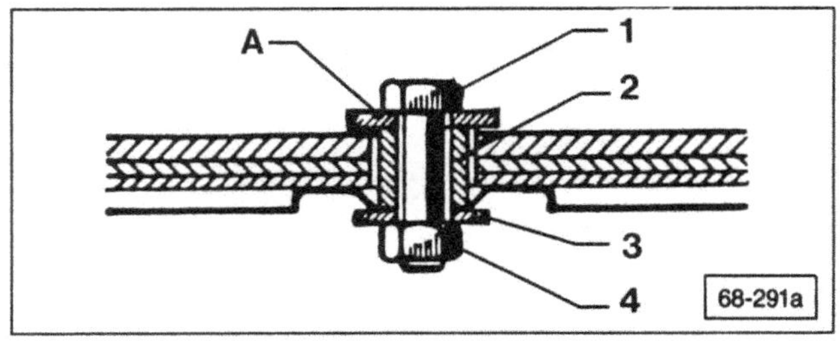

Fig. 4-34. Scirocco mounting hardware for child restraint anchor **A**. Hex bolt **1** is part no. N 010 244 11. Spacer ring **2** is part no. 811 857 813. Washer **3** is part no. N 011 670 11. Hex nut **4** is part no. N 011 008 8.

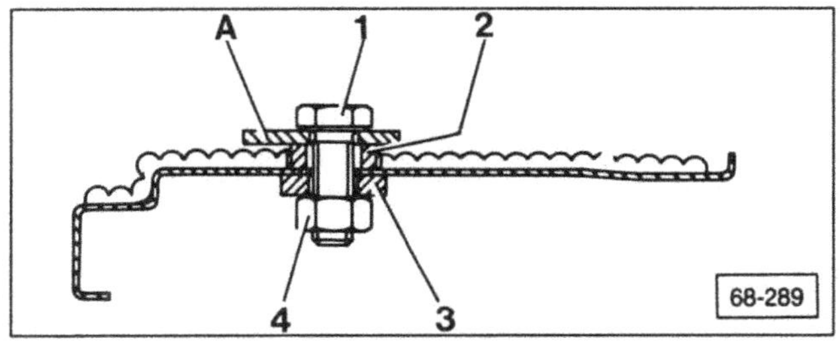

Fig. 4-35. Cabriolet mounting hardware for child restraint anchor **A**. Hex bolt **1** is part no. N 900 245 01. Spacer **2** is part no. 803 019 825 B. Washer **3** is part no. N 015 401 4. Hex nut **4** is part no. N 011 008 18.

WARNING —

To assure maximum protection, follow manufacturers' instructions when installing a child restraint system.

CAUTION —

Part numbers are for reference only. Always consult an authorized Volkswagen dealer for the latest parts information.

NOTE —

If the mounting hardware provided with the child seat does not fit the anchor mounting, contact the child seat manufacturer for suitable parts. Volkswagen specifications are for maximum occupant protection.

4.7 Front Seats

To remove a front seat, first slide the seat all the way forward. Pry up the end of the track cover or remove the screw and slide the cover off of the track as shown in Fig. 4-36. Remove the track stop from the rear of the other rail. On models with heated seats or electric height adjust, disconnect any electrical connectors beneath the seat. Remove the acorn nut, the washer, and the bolt from the front guide shown in Fig. 4-37, and release the retaining rod from the seat while sliding the seat towards the rear of the car and off its tracks. Installation is the reverse of removal.

Fig. 4-36. Front seat track cover removed from retaining clip. Slide cover off track in direction of arrow.

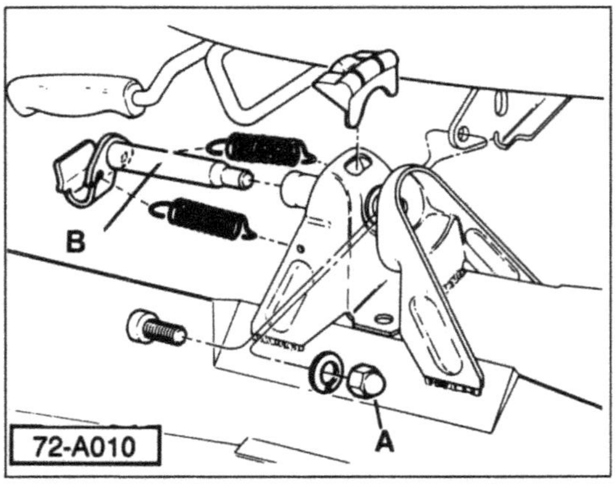

Fig. 4-37. Acorn nut (**A**) and retaining rod (**B**) to be removed when removing seat.

Heated Seats

On models with heated seats, check the electrical circuit to the seats as described in **ELECTRICAL SYSTEM** before assuming that a heating element is faulty.

To replace the heating elements, remove the seat as described above. Separate the backrest wire from the seat wiring connector. See Fig. 4-38. Remove the backrest from the seat cushion by removing the left and right trim pieces and the hinge screws. Remove both seat covers and heating elements. Lay the new heating elements against the seat without wrinkling the elements. Install the seat covers and route the wire harness as shown in Fig. 4-38 and Fig. 4-39.

4.8 Rear Seats

Remove the rear bench seat cushion by removing the two screws shown in Fig. 4-40 and then pulling the cushion out toward the front.

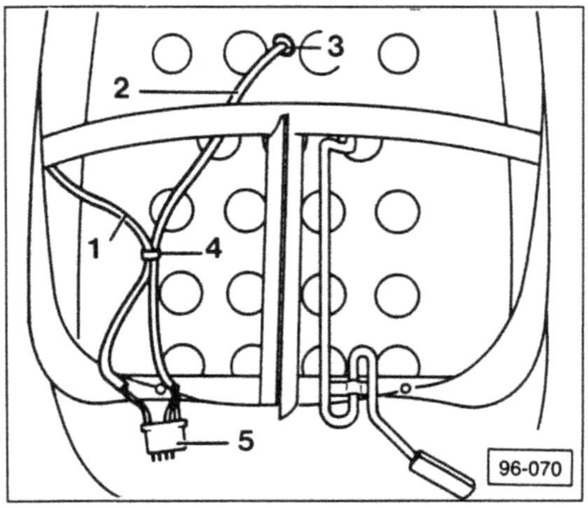

Fig. 4-38. Routing of wires for heated seats beneath driver's seat cushion. Passenger seat is mirror image. Wire **1** is backrest element wire, **2** is seat element wire, **3** is wiring grommet, **4** is wiring clip, **5** is connector.

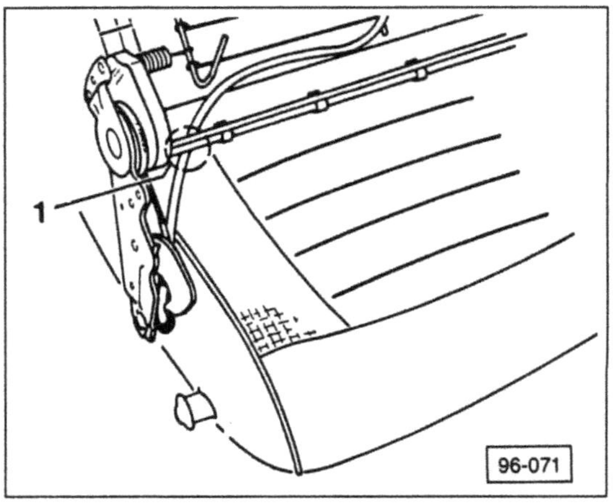

Fig. 4-39. Routing of backrest wire for heated seat. Wire runs between seat cover tensioning wires and seat frame (**1**).

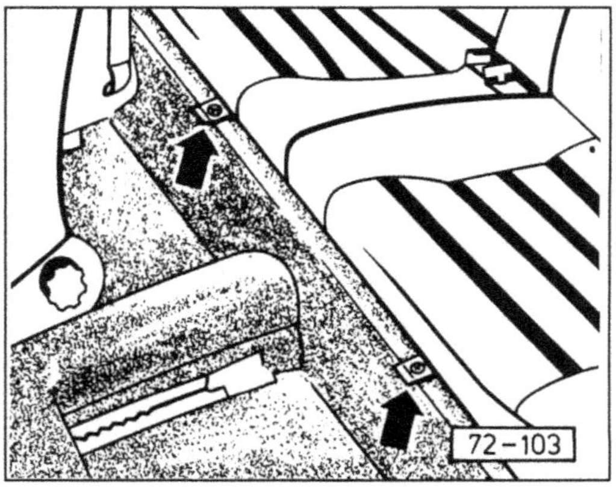

Fig. 4-40. Rear bench seat cushion screws (arrows) to be removed. Slide cushion out toward front.

To remove the Cabriolet rear seat backrest, fold the backrest forward and then remove the two backrest mounting bolts.

To remove the Scirocco rear bench seat backrest, fold the backrest forward, unclip the locking lug shown in Fig. 4-41, and remove the backrest.

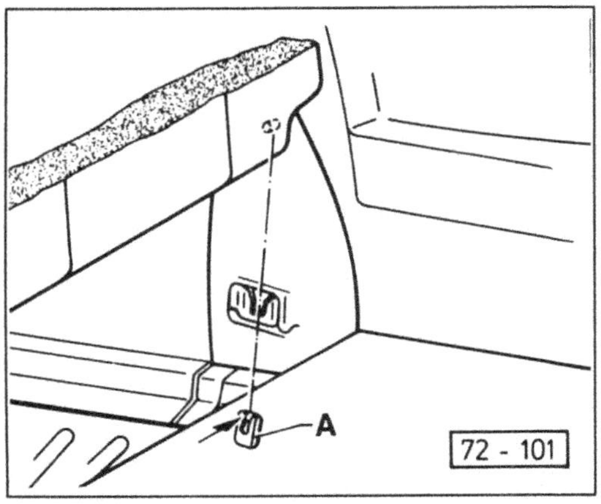

Fig. 4-41. Locking lug **A** to be pushed in direction of arrow for removal of Scirocco rear seat backrest.

To remove the Scirocco split rear seat backrest, fold the backrest forward and unclip the release cable shown in Fig. 4-42. Remove the mounting screws from one side of the backrest, and remove the locking lug from the other as shown above in Fig. 4-41, then remove the backrest.

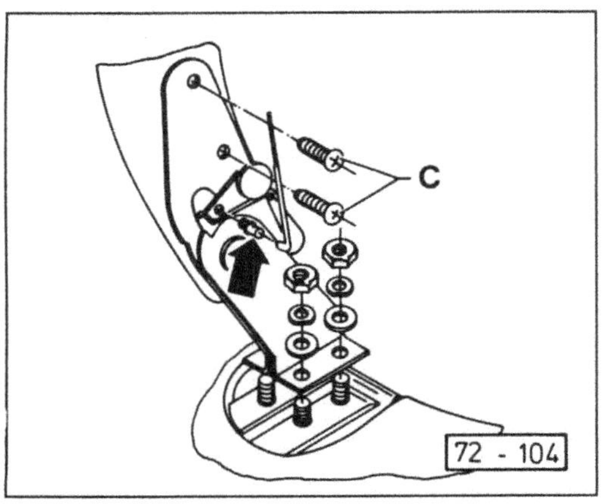

Fig. 4-42. Scirocco rear split seat backrest release cable (arrow) and mounting screws (**C**).

4.9 Headliner
(Scirocco only)

The headliner is a one-piece molded panel that is clipped in place. It is removed after removing the headliner trim panel. Clean the headliner using only mineral spirits, isopropanol, or a mild cleaning solution.

CAUTION —
To avoid damaging the headliner, do not let acetone, paint thinners, or metal primer/polyurethane (a solvent used to install the windshield) contact it.

To remove the headliner trim panel, first remove the sun visors by removing the mounting screws from the swivels mounts. Next, remove the sun visor retaining clips as shown in Fig. 4-43. Unclip the dome light and disconnect the wiring. On cars with a sunroof, remove the sunroof crank handle and trim panel. The headliner trim panel can now be removed by sliding it out toward the front.

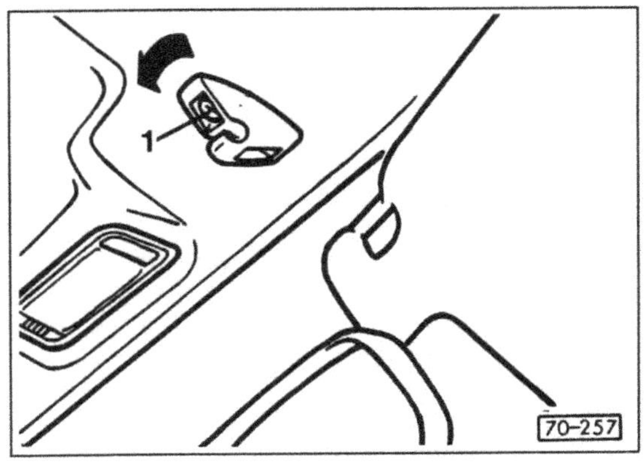

Fig. 4-43. Retaining clip being removed. Remove mounting screw (**1**), then twist clip 90° in direction of arrow.

5. DOORS

The rear half of each door hinge is welded to the door. The front hinge half is bolted to a bracket welded to the car body. Hinge pins are available as replacement parts. The only adjustment point of the door hinge is the center mounting bolt. When removing a door, mark the hinge halves so that the door can be reinstalled in its original position.

To remove a door, remove the mounting screws for the door check rod bracket. Use an impact driver to loosen the hinge bolts. Support the door, then remove the bolts and the door. On models with electric mirrors or central locking system, first remove the mirrors as described in **5.3 Side Mirrors**, and remove the vacuum hose and wiring for the lock actuator as described in **5.2 Central Locking System**.

CAUTION —
If the door is not supported when the first bolt is removed, the second hinge may bend or break.

Install the door using the marks made prior to removal. If a replacement door is being installed, it should contact the weather seal evenly, and the lock should work smoothly. Set the hinge bolts with an impact driver.

5.1 Door Assembly

The interior trim, glass, handles, locks, mirrors and internal components can all be reached without removing the door. Make sure that the drain holes at the bottom of the door are open. On reassembly the plastic moisture barrier should be well sealed, without creases. If necessary, reseal it using a non-hardening adhesive. Also see **11. Cabriolet Power Windows**.

Interior Trim

The interior trim must be removed for access to the internal door components. No special tools are required for the procedure.

To remove trim:

1. Unscrew and remove the lock button.
2. Pry out the trim plate for the inside door lever.
3. Remove the trim for the remote control rear view mirror.
4. Remove the window winder handle by unclipping the trim and removing the mounting screw.
5. Remove the pull handle or armrest. On Sciroccos, remove the upper and lower trim and remove the two screws, then pull the handle down. On Cabriolets, remove the two screws, turn the armrest 90°, and then remove it.
6. On Sciroccos, remove the loudspeaker trim panel.
7. Remove the door trim panel. On Sciroccos, remove any additional screws at the edges and face of the trim panel. On Cabriolets, check for any additional screws that may secure the panel and remove them if necessary. Then gently pry the panel off of its retaining clips.

Installation is the reverse of removal. Apply thread locking compound to the window winder handle screw before installing it.

Window Regulator

To remove the window regulator, first remove the interior trim panel and the plastic moisture barrier from the door as described above.

On Sciroccos, remove the glass mounting bolts and lower the window into the door. The bolts are reached through an access hole near the top of the door. See Fig. 5-1 for the location of the mounting bolts for Scirocco window regulators. When installing the regulator, install but do not fully tighten the window glass mounting bolts, raise the window fully, then lower it and tighten the bolts.

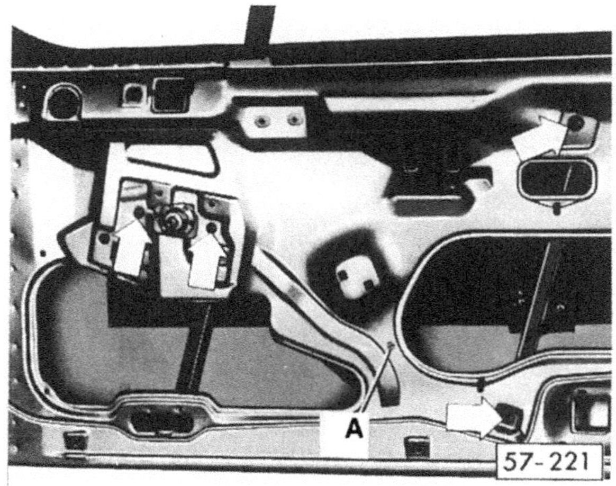

Fig. 5-1. Scirocco window regulator being removed. Remove four bolts (arrows), then unclip regulator from point **A**.

On Cabriolets, lower the window and remove the screw that holds the regulator to the front bottom edge of the window. Remove the seven bolts that hold the regulator to the door, then slide the regulator out while disengaging the rear regulator guide from the window. Adjust the window regulator stop after the regulator is installed. Align the top of the window as described in **Convertible Top**. Then, with the window raised, press the regulator stop against the regulator and tighten the screw.

Window Guides

Both Scirocco and Cabriolet door window guides are replaceable if there are air or water leaks. In either case it is necessary first to remove the door trim as described above for access to the guide mounting screws.

To remove Scirocco rear guides, remove the screw and bolt shown in Fig. 5-2. To remove Scirocco front guides, remove the screws and bolt shown in Fig. 5-3.

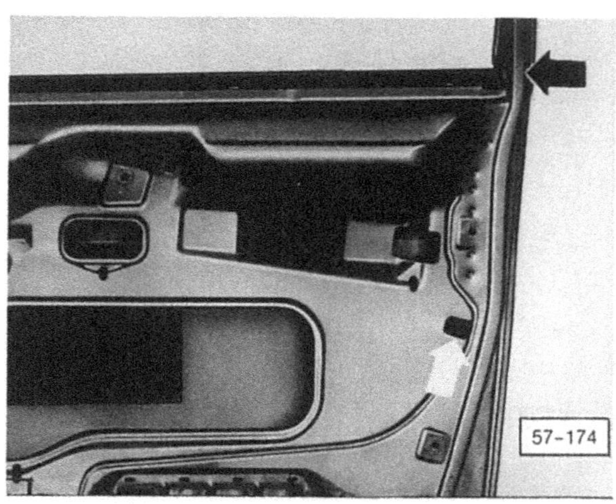

Fig. 5-2. Screw (black arrow) and bolt (white arrow) to be removed for removal of Scirocco window rear guide.

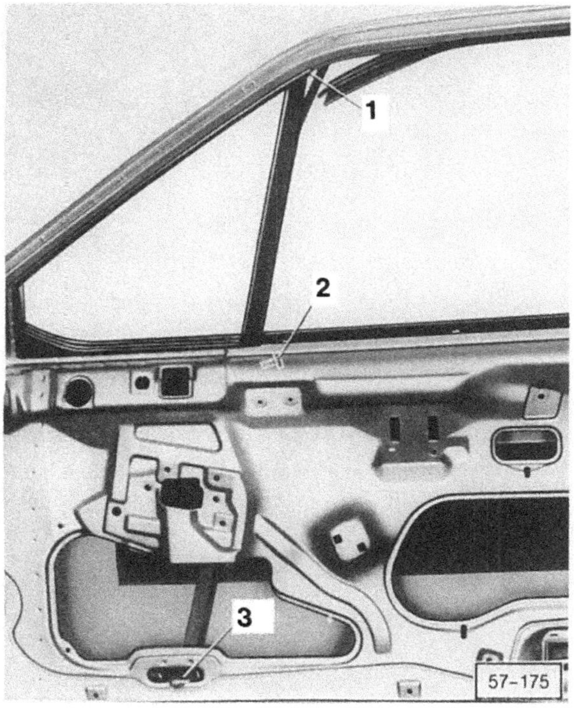

Fig. 5-3. Screws (**1, 2**) and bolt (**3**) to be removed for removal of Scirocco window front guide.

To remove Cabriolet window guides, first remove the top part of the vent window weatherstripping as described below in **Weatherstripping**. Remove the screw shown in Fig. 5-4, then remove the two hex head screws in the door panel.

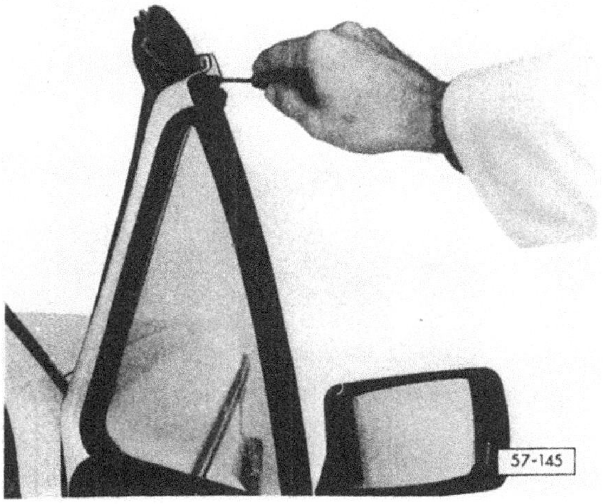

Fig. 5-4. Screw underneath weatherstripping to be removed for removal of Cabriolet window guide.

Weatherstripping

To remove Cabriolet vent window weatherstripping, first remove the screw shown in Fig. 5-5, then pull off the weatherstripping. If the upper part of the door weatherstripping is in poor shape, it is not necessary to replace the entire piece. The upper part of the weatherstripping is available as a separate piece, part no. 155 837 915A/916A, and is installed as shown in Fig. 5-6.

> ***CAUTION —***
> *Part numbers are for reference only. Always consult an authorized Volkswagen dealer for the latest parts information.*

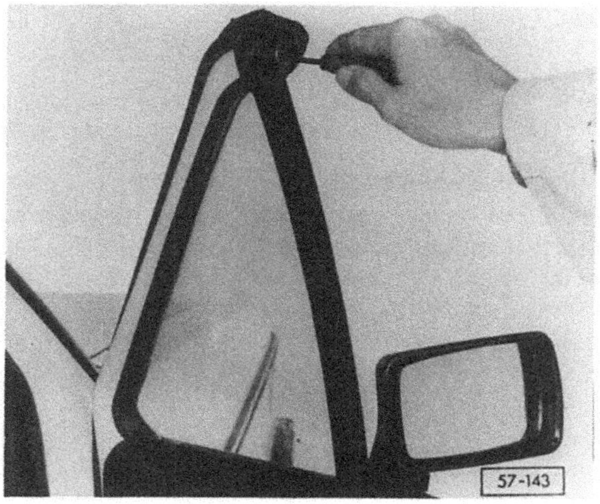

Fig. 5-5. Screw for upper part of Cabriolet vent window weatherstripping being removed.

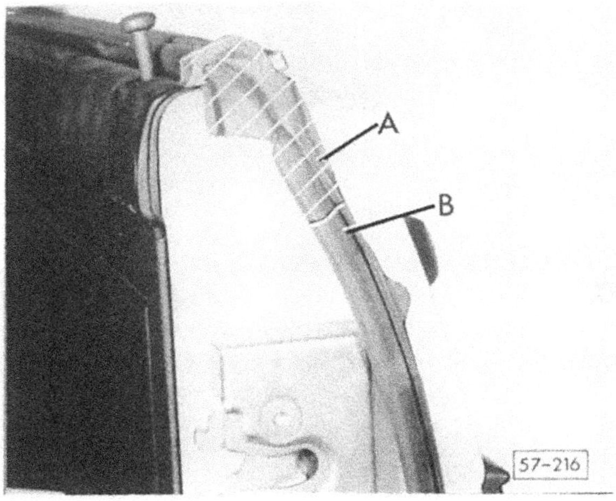

Fig. 5-6. Replacement door weatherstripping (**A**) being installed. Butt together new part with old weatherstripping (**B**).

Handles and Locks

The outside door handle must first be removed for access to the door lock cylinder. To remove the handle, pry out the trim strip, then remove the two retaining screws and slide the handle off in the direction shown in Fig. 5-7. Installation is the reverse of removal.

Fig. 5-7. Exploded view of door handle mechanism. For removal, handle slides off in direction of arrow.

To remove the door lock, lock the door lock with the door open, then remove the two lock mounting screws and pull the lock part-way out. Hold the lock actuating lever in its turned-out position as shown in Fig. 5-8, then detach the lock from the pullrod and sleeve.

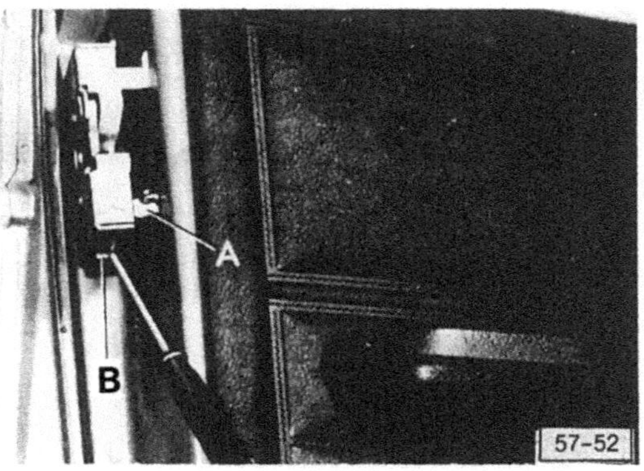

Fig. 5-8. Door lock being removed. Actuating lever (**A**) is held in position by screwdriver inserted through access hole (**B**).

To install the lock, use a screwdriver to hold the actuating lever at a 90° angle as shown in Fig. 5-9, and reattach the lock to the locking knob sleeve and the pullrod as shown in Fig. 5-10. Then install the mounting screws.

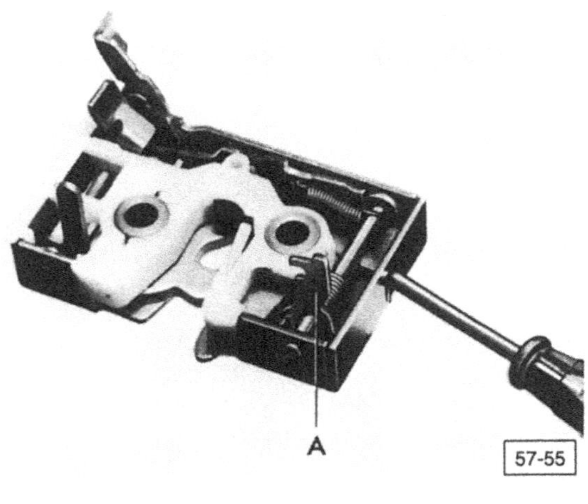

Fig. 5-9. Lock actuating lever **A** being held at 90° angle using screwdriver for installation.

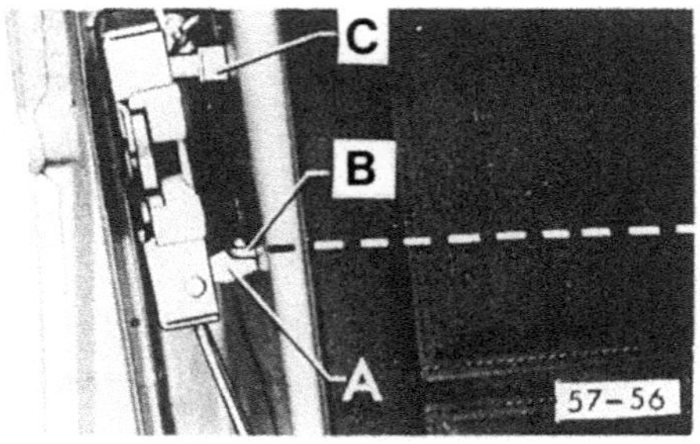

Fig. 5-10. Door lock being installed. Actuating lever (**A**) is attached to pullrod (**B**), and locking knob lever is attached to sleeve (**C**).

To remove the inside door actuating lever, first remove the door trim panel and door lock as described above. Remove the strap from the panel and slide the door lever assembly from the door as shown in Fig. 5-11. Installation is the reverse of removal.

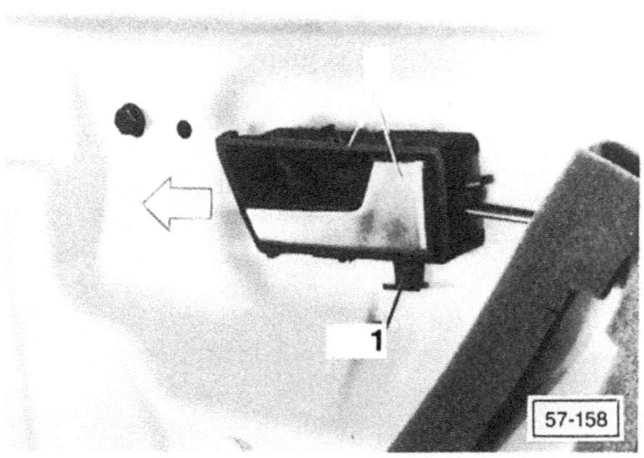

Fig. 5-11. Door actuator lever being removed. Remove strap (**1**) and slide assembly in direction of arrow.

5.2 Central Locking System

The central locking system consists of a special bi-pressure pump in the luggage compartment, connected by vacuum hose to lock activators at the passenger door, at the rear lid, and at the gas tank flap. When the key is turned in the driver's door lock, the master activator turns on the pump to supply pressure to open the locks, or vacuum to close them.

This heading covers only the removal and installation of the mechanical parts, and troubleshooting of the bi-pressure system. If the pump does not run when key is turned, the problem may be electrical. Troubleshoot the pump and the master activator as described in **ELECTRICAL SYSTEM**.

> **NOTE —**
> If the central locking system has not been operated for an extended period, the key may have to be turned several times before the system activates.

Bi-pressure System Troubleshooting

When the key is turned in the master activator, all locks should be activated within 2 seconds. A fault is indicated if the pump runs longer than 5 seconds, or if the pump runs, but none of the locks activate. Troubleshooting should be carried out in the sequence given here, even if only one activator is suspected of being faulty. The replacement of faulty components is covered below under separate headings.

To troubleshoot:

1. Remove the bi-pressure pump from its housing as described below, but do not disconnect it.

2. Clamp the hose as shown in Fig. 5-12. Activate the pump using the door key. If the pump runs longer than 5 seconds, either the hose from the pump is leaking or the pump is faulty. If the pump shuts off within 5 seconds, remove the clamp and continue testing.

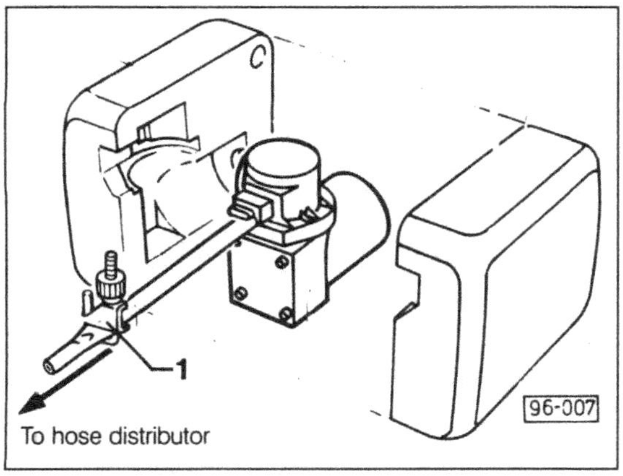

Fig. 5-12. Bi-pressure pump being tested with hose clamp at **1**.

3. Clamp the hose as shown in Fig. 5-13. If the pump runs longer than 5 seconds when activated, the right door activator or the hose to it is faulty. If the pump turns off within 5 seconds, remove the clamp and continue testing.

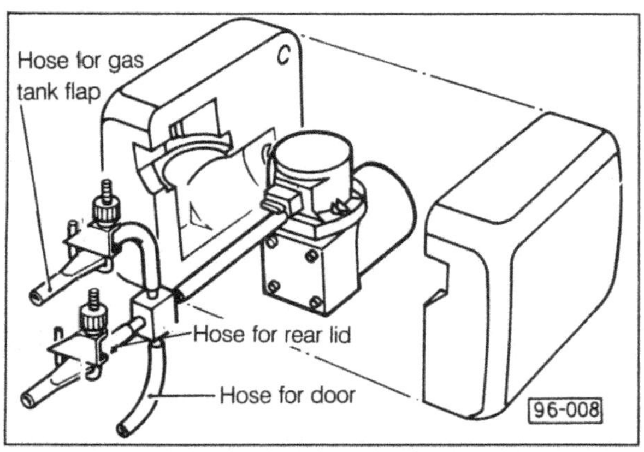

Fig. 5-13. Right door activator being tested with hose clamps installed.

4. Clamp the hose as shown in Fig. 5-14. If the pump runs longer than 5 seconds when activated, then the gas tank flap activator or the hose to it is faulty. If the pump turns off within 5 seconds, then there is a fault either in the activator for the rear lid, or in the off-set grommet that connects the rear lid and the body, or in the vacuum hose.

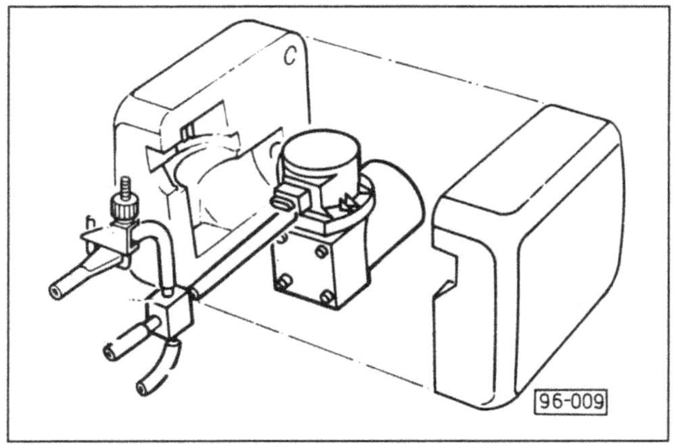

Fig. 5-14. Gas tank flap being tested with hose clamp installed.

Removing and Installing Door Activators

To remove a faulty activator, first remove the door trim panel and moisture barrier as described in **5.1 Door Assembly**. Remove the activator retaining screws and unhook the connecting rod shown in Fig. 5-15. When removing the master activator from the driver's door, first disconnect the battery ground cable, and then disconnect the electrical connector. When removing the passenger door activator, disconnect the vacuum hose. Installation is the reverse of removal.

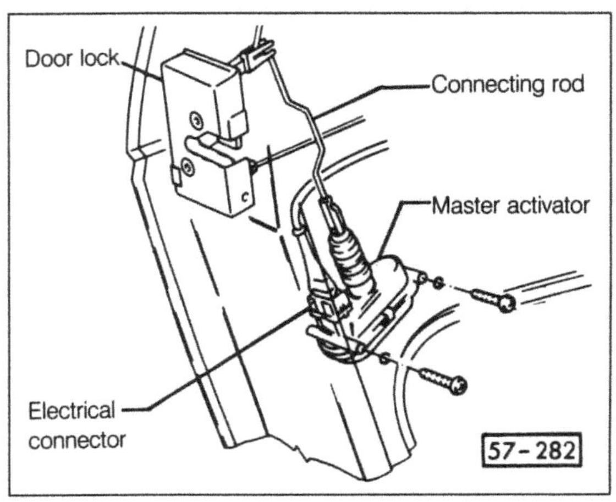

Fig. 5-15. Phantom view of master activator installation. Passenger door activator is similar.

Bi-pressure Pump

The bi-pressure pump is located in the luggage compartment on the right side, behind the trim panel and the container for the washer fluid. When removing the pump, first disconnect the negative (−) battery terminal. Remove the retaining strap, and then remove the pump from its housing. Disconnect the electrical connector and the hose, and remove the pump. See Fig. 5-16. Installation is the reverse of removal.

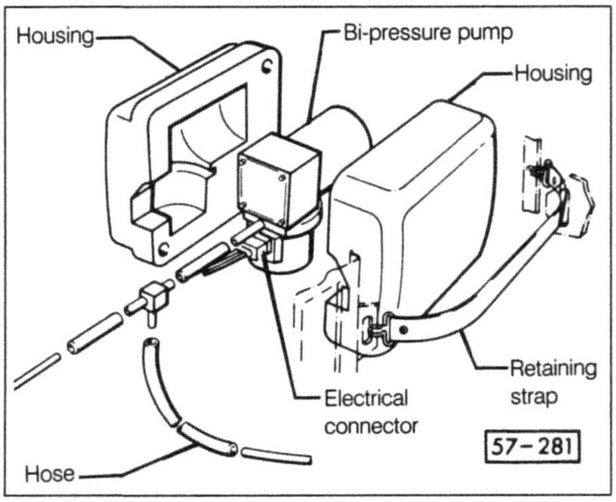

Fig. 5-16. Bi-pressure pump and related components.

Rear Lid Activator

The rear lid activator is shown in Fig. 5-17. To remove the activator, open the lid and remove the inside trim panel by turning the fasteners 90°. Loosen, but do not remove, the activator mounting bolts. It is necessary to unscrew the rear lid lock and the mounting bracket for the rear wiper for clearance to remove the activator. Disconnect the vacuum hose from the activator and disconnect the connecting rod at the lock. Installation is the reverse of removal.

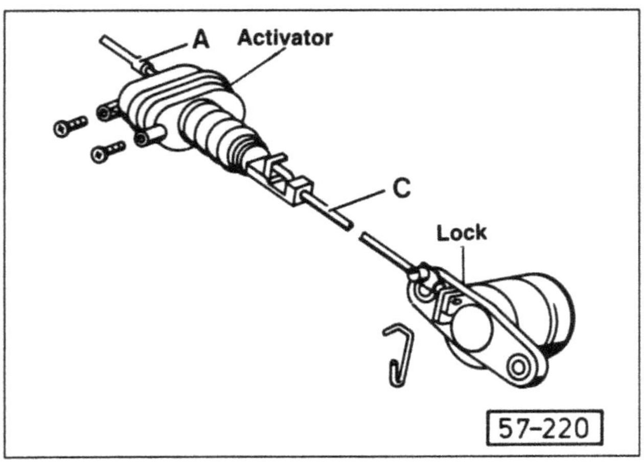

Fig. 5-17. Rear lid activator being removed. Remove vacuum hose (**A**) and connecting rod at **C**.

Gas Tank Flap Activator

The gas tank flap activator is shown in Fig. 5-18. It is located in the right-hand side of the luggage compartment, behind the trim panel and the container for the washer fluid. To remove the activator, drill out the three rivets and disconnect the vacuum hose. Installation is the reverse of removal. New rivets are available from an authorized Volkswagen dealer.

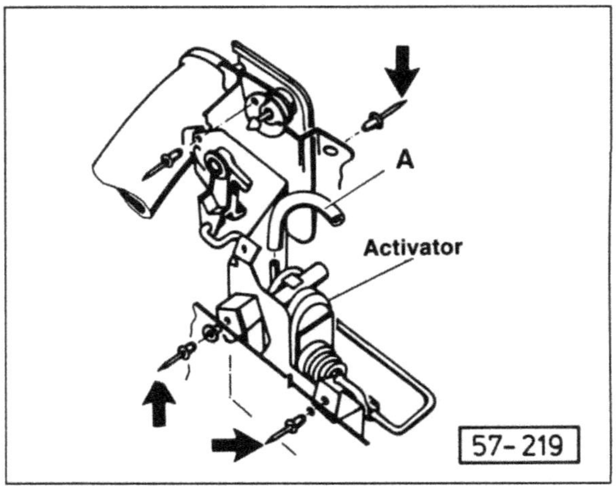

Fig. 5-18. Gas tank flap activator being removed. Drill out rivets (arrows) and disconnect vacuum hose.

Off-set Grommet

The off-set grommet is the vacuum connection between the car body and the rear lid for the rear lid activator. If the grommet must be replaced, install the new part as shown in Fig. 5-19.

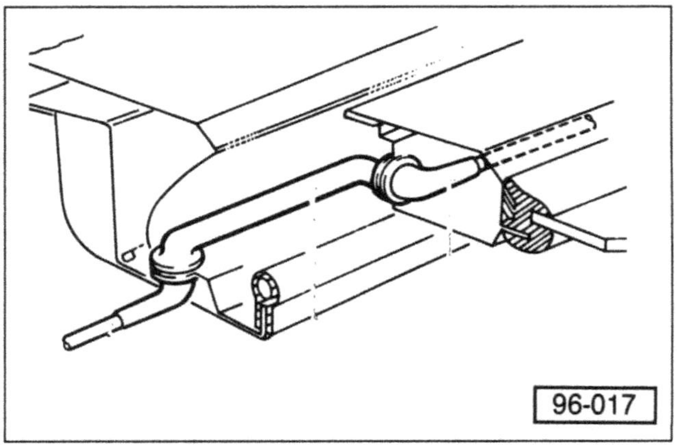

Fig. 5-19. Off-set grommet, showing correct installation.

5.3 Side Mirrors

Both manual remote-control side mirrors and electric side mirrors are easily removed using ordinary hand tools. In either case, the interior door trim panel must first be removed and the moisture barrier peeled back as described in **5.1 Door Assembly**.

Manual Remote-Control Mirrors

The manual remote-control mirrors for Scirocco and Cabriolet are shown in Fig. 5-20 and Fig. 5-21. With the door trim *panel removed*, remove the lock nut for the remote control knob and remove the mounting screws for the mirror. Lift the mirror away from the door, guiding the knob and its cables through the door. On Sciroccos, the mounting screws for the mirror are located behind the small trim panel at the front corner of the door window. Installation is the reverse of removal.

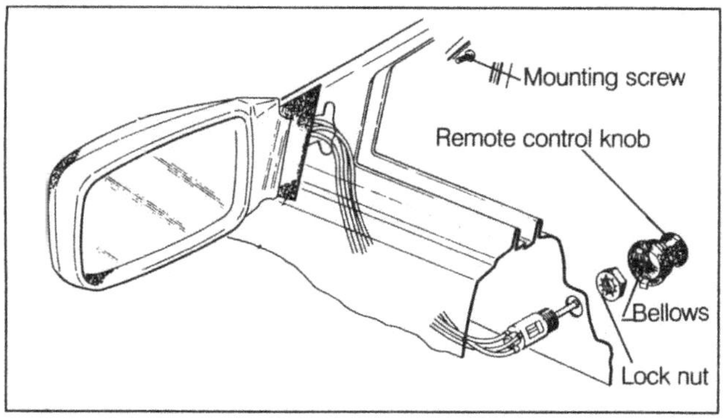

Fig. 5-20. Exploded view of Scirocco manual remote-control mirror.

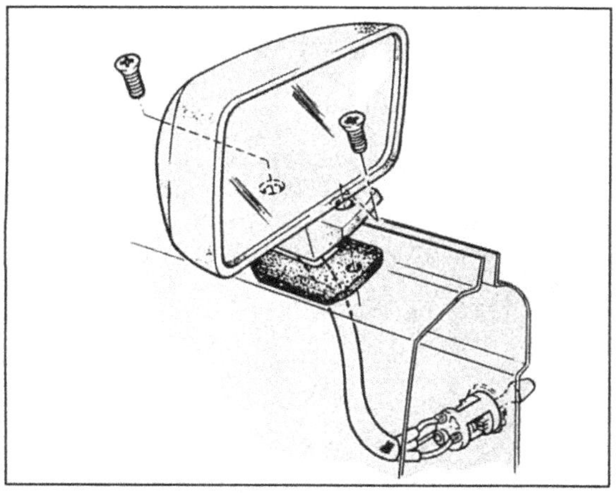

Fig. 5-21. Phantom view of Cabriolet manual remote-control mirror.

Electric/Heated Remote-Control Mirrors

The procedure for removal of electric remote-control mirrors is similar to that of manual mirrors described above. The only difference is that instead of a remote control knob, it is the mirror wiring inside the door that must be disconnected.

If the mirror will not operate, check the electrical circuit as described in **ELECTRICAL SYSTEM** before assuming that the motor is faulty. It is not necessary to remove the mirror to replace the motor.

To replace the mirror motor, first remove the mirror glass as shown in Fig. 5-22. Take note of the wiring connections to the glass. With the glass removed, remove the motor mounting screws shown in Fig. 5-23. Take note of the wiring connections. Installation is the reverse of removal. With the mirror glass in place, insert the screwdriver and turn the lock ring the opposite way to engage the detent.

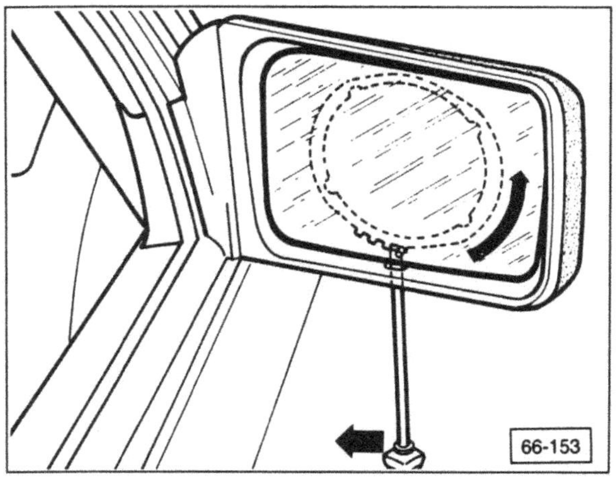

Fig. 5-22. Electric mirror glass being removed. Insert screwdriver to engage notch in lock ring, then move screwdriver in direction of arrow to unlock glass.

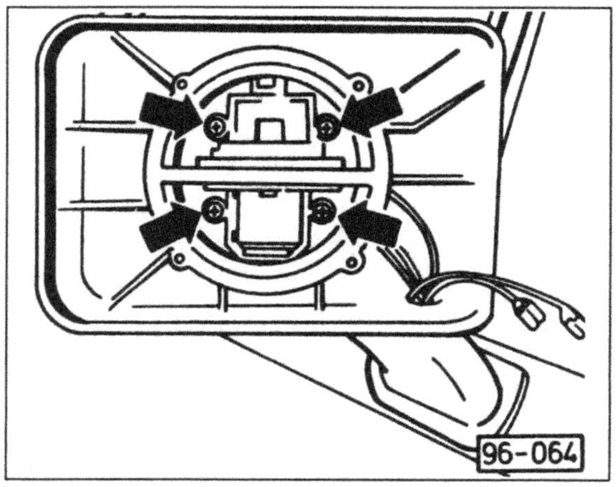

Fig. 5-23. Motor mounting screws (arrows) to be removed.

6. EXTERIOR

The exterior body parts and trim discussed under this heading are fastened with conventional fasteners, and are easily removed using ordinary hand tools. A thorough knowledge of automobile body repair is not needed to perform the procedures.

6.1 Headlights

The Scirocco and Cabriolet use sealed beam headlights. See Fig. 6-1 and Fig. 6-2. A retainer holds the headlight to the adjusting frame, and two screws are used to aim the headlights. If a headlight fails, the entire sealed beam unit must be replaced.

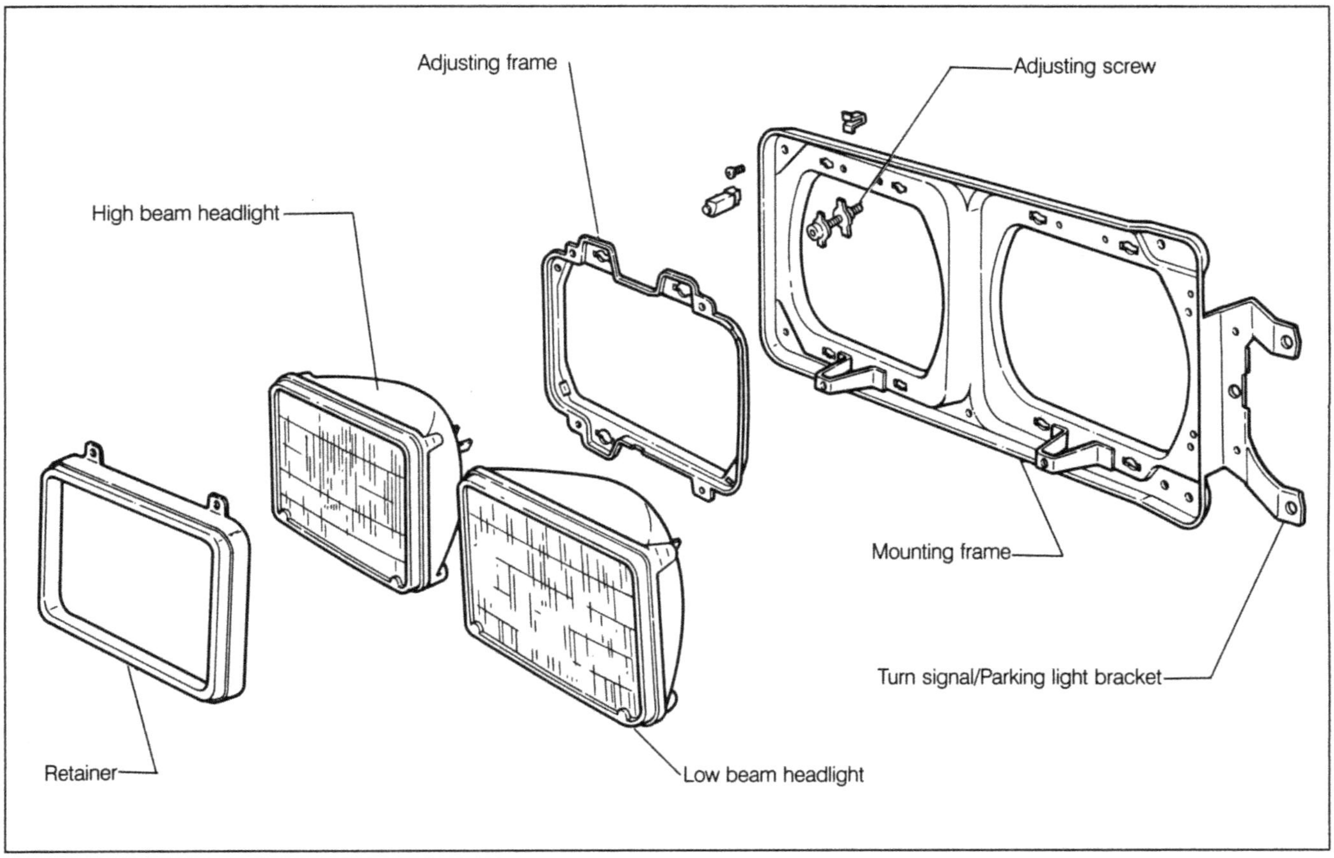

Fig. 6-1. Scirocco headlight assembly.

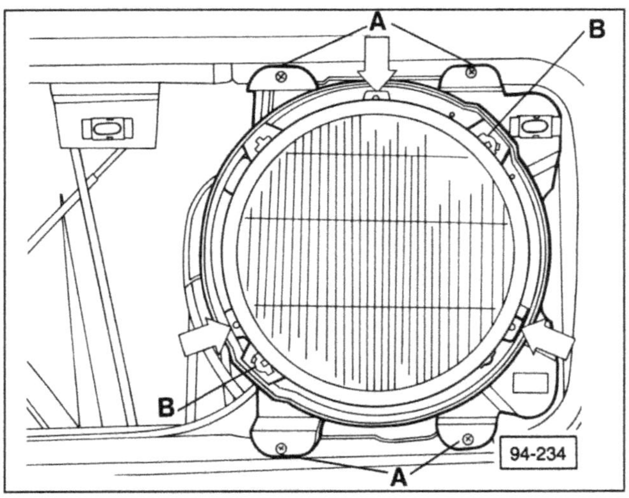

Fig. 6-2. Cabriolet headlight assembly. Screws for retainer are shown by arrows. Screws for mounting frame are at **A**, and screws for adjustment are at **B**

Replace a sealed beam unit with the headlights off. Remove the radiator grille as described in **6.4 Trim, Bumpers, and Body**. On Sciroccos there is an additional molding below the headlights, attached with two screws, that must be removed. Remove the screws from the retainer and remove it. The screws for the Scirocco are shown in Fig. 6-3. The screws for the Cabriolet are shown above in Fig. 6-2. Pull the headlight forward and disconnect the wiring. Orient the new lens such that the part of the lens marked UP or TOP, or else the single adjusting lug, is uppermost. It should not be necessary to aim the headlight if the adjustment screws have not been touched. However, it is a good idea to check headlight aim as described below.

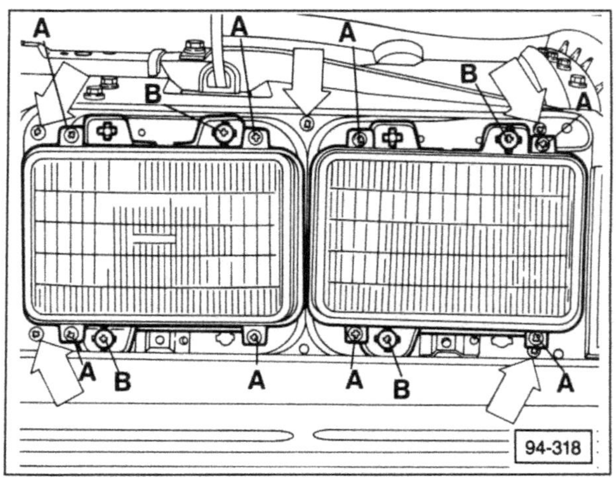

Fig. 6-3. Scirocco headlight retainer screws (**A**). Screws also shown for mounting frame (arrows) and adjustment (**B**).

Aiming Headlights

Special adaptors are supplied with the car's tool kit to aim the headlights. Headlight aiming using the adaptors should be carried out by an authorized Volkswagen dealer or other qualified shop. Headlight aiming can also be done as described below without any special tools, but the results will not be as exact.

NOTE —
Check your state laws to see if headlight adjustments must be made by a licensed shop. Some states may also have different aiming specifications from those described here.

To aim the headlights, position the car on a level surface, 25 feet (approx. 7.6 meters) from a vertical wall. The gas tank should be half full and the tire pressures should be correct. Turn the headlights to low beam. With a person in the driver's seat, the low beams must be in the areas indicated in Fig. 6-4. If re-aiming is necessary, remove the radiator trim grille and use a Phillips head screwdriver on the height and side-to-side adjusting screws to move the headlight.

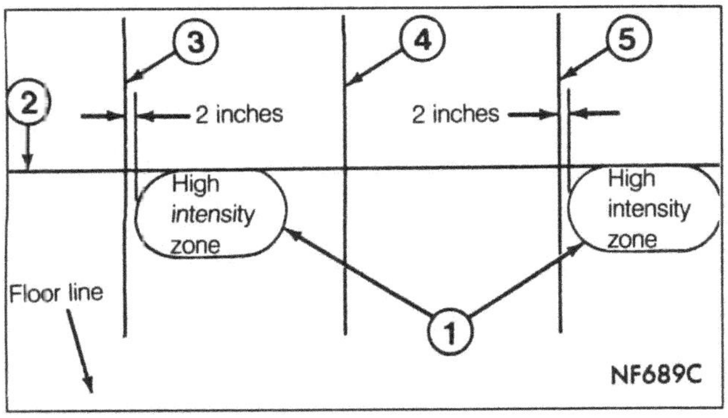

Fig. 6-4. Headlight aiming target on vertical wall. Low beam light intensity areas are at **1**. **Line 2** is at height of headlight centers. Line **4** is car centerline. Lines **3** and **5** are distance from centerline to centers of headlights.

6.2 Taillight Assembly

All rear bulbs and the taillight lens are replaced from the luggage compartment. Replacement of the center brake light is covered in **4.3 Interior Lights**.

On Sciroccos, replace a faulty taillight bulb by first removing the bulb holder. Pinch the two clips on the top and bottom of the holder shown in Fig. 6-5 and remove the holder. Press the bulb in and twist to remove and install it. Make sure that both clips are engaged and that no wires are pinched when the holder is installed.

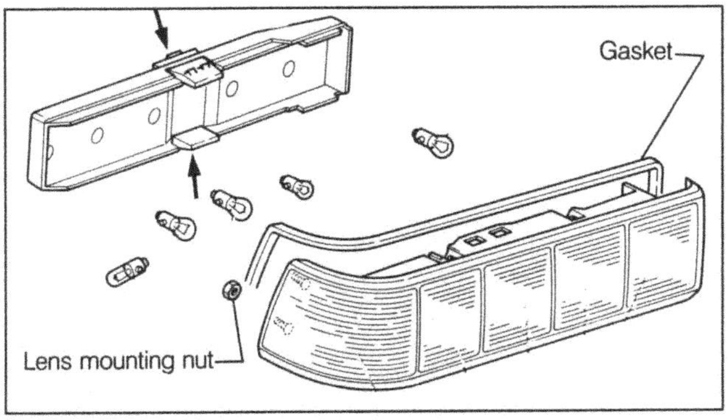

Fig. 6-5. Scirocco taillight bulb holder. Pinch clips (arrows) to remove. Replacement turn-signal, stop-light and back-up light bulbs are part no. N 017 732 2 (U.S. trade no. 1073). Taillight bulbs are part no. N 017 719 2 (U.S. trade no. 67).

On Cabriolets, replace a faulty bulb by first removing the bulb holder. Depress the tab to remove the bulb cover, and then depress the tab to remove the holder. See Fig. 6-6. Press the bulb in and twist to remove and install it.

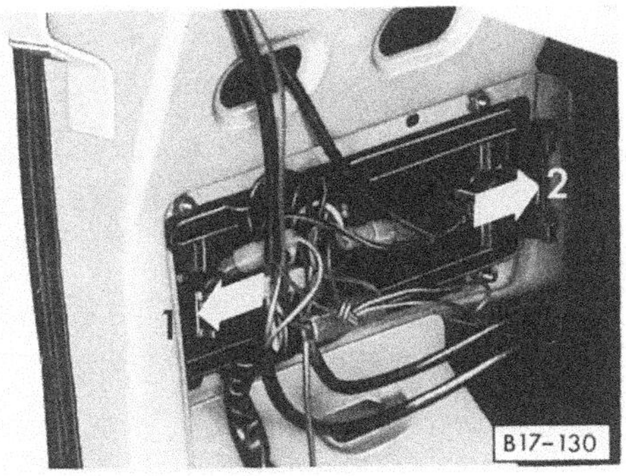

Fig. 6-6. Cabriolet bulb holder. Press tab **1** to remove bulb cover. Press tab **2** to remove holder. Replacement bulbs are same as for Scirocco. See Fig. 6-5.

The taillight lens is replaced by removing the four nuts in the luggage compartment that hold the lens to the body. Pull the lens off the car with its gasket.

6.3 Side Marker and Front Turn Signal Lights

To replace a Scirocco front side marker or turn signal bulb or lens, open the hood. Pull back the rubber cap from the light and squeeze the lug to remove the bulb holder. See Fig. 6-7. Press the bulb in and twist to remove and install it. To remove the lens, first remove the bulb holder, then remove the lens mounting screws.

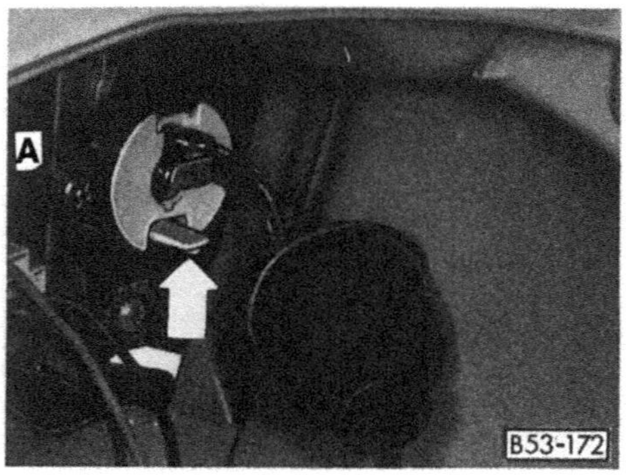

Fig. 6-7. Scirocco front side marker and turn signal light. Arrow points to bulb holder lug. Lens mounting screws are at **A** (one screw not visible in illustration). Replacement side marker/turn signal bulb is Volkswagen part no. N 017 738 2 (U.S. trade no. 1034).

To replace a Cabriolet side marker bulb, remove the two mounting screws and pull the light assembly away from the fender. Pull back the rubber cap, then press the lugs on the bulb holder outward to release the holder. To replace a turn signal bulb or lens, remove the lens mounting screws and the lens. Do not overtighten the lens mounting screws, or the lens may crack.

6.4 Trim, Bumpers, and Body

For corrosion protection, the fenders, hood, and trunk lid are assembled with zinc plates which act as anodes, and with a sealing tape between mating surfaces to prevent moisture from becoming trapped. See Fig. 6-8. The plates are part no. AKL 381 035 50, and the tape is part no. D 001 900.

NOTE —

The plates and tape are installed as part of the corrosion warranty. If in doubt about warranty protection, consult an authorized Volkswagen dealer.

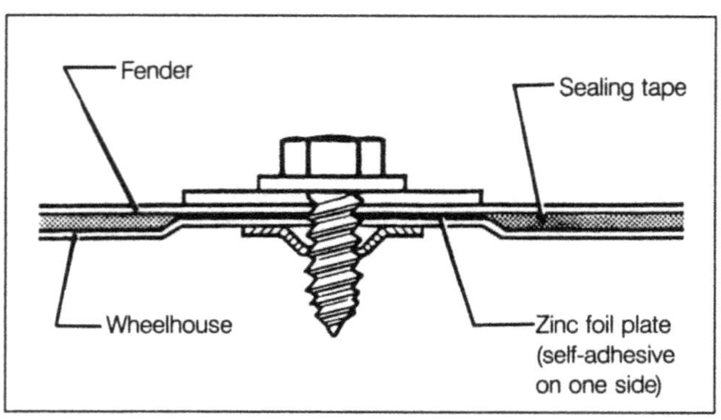

Fig. 6-8. Correct installation of bolt-on body parts for maximum corrosion protection.

Removing Front Body Parts and Grille

The front radiator grille is held in place by a combination of screws and spring clips. Scirocco grilles and 1985-1987 Cabriolet grilles are in 3 pieces. 1988 and later Cabriolet grilles are one-piece.

To replace a fender, first remove the wheelhouse liner, bumper, and bumper trim as described below. On Cabriolets, remove the side marker light. Carefully use a heat gun and razor knife to cut through the PVC coating along the A-pillar where the fender meets the body, then remove the mounting bolts. Scirocco fenders are held by 13 bolts, while Cabriolet fenders are held by 9 bolts. If the fender does not pull away with a moderate amount of effort, check again for any fasteners still installed. Excessive force will damage the fender. During installation, install the zinc plates and sealing tape as shown above in Fig. 6-8.

CAUTION —

Use extreme care when applying heat to the body to avoid blistering the paint or discoloring the PVC body coating.

To replace a wheelhouse liner, support that corner of the car securely on a jack stand and remove the wheel. Remove the mounting screws and pull the liner down and out. See Fig. 6-9. Installation is the reverse of removal. Torque the wheel bolts to 110 Nm (80 ft. lb.) after lowering the car to the ground.

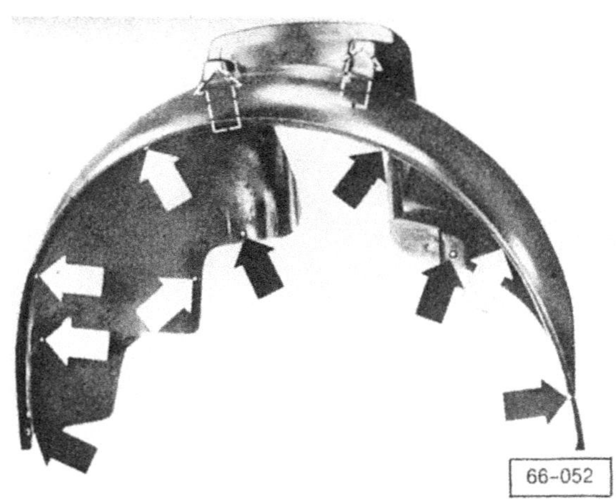

Fig. 6-9. Location of fasteners (arrows) that secure wheelhouse liner on Sciroccos. Cabriolet wheelhouse liners are similar, but are secured with only seven screws.

To remove the hood, first mark the hinge plate locations on the hood for reinstallation. Loosen the mounting bolts, then with a helper supporting the opposite side, remove the bolts. Install the original hood by aligning the matching marks made earlier. If installing a new hood, loosely install the mounting bolts and adjust the hood so that it contacts the body evenly. During installation, install the zinc plates and sealing tape as shown above in Fig. 6-8.

Removing and Installing Body Trim

Body trim includes the wheelhouse trim, body side moldings, stone guards, and front and rear spoilers. A rivet gun is necessary when installing some wheelhouse trim. Replacement rivets are available from an authorized Volkswagen dealer.

There are two types of wheelhouse trim on Sciroccos. The basic trim is secured with screws and rivets. Full-length wheelhouse trim on 16-valve models is secured with screws and adhesive.

To replace Scirocco 16-valve wheelhouse trim:

1. Remove the bumper. Remove the wheel.

NOTE —
The illustrations used here show the front wheelhouse trim only. The procedure is similar for the rear trim.

2. Remove the wheelhouse liner as described above, then remove the trim mounting screws.

3. Partially remove the spoiler trim where it meets the wheelhouse trim by cutting through the adhesive. See Fig. 6-10. Then partially remove the rocker panel trim in the same manner.

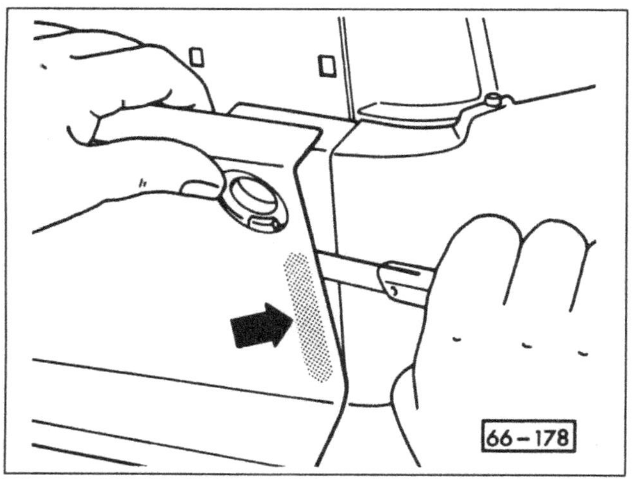

Fig. 6-10. Spoiler trim being partially removed from wheelhouse trim by cutting adhesive with knife (arrow).

4. Remove the wheelhouse molding by cutting it away with a knife as shown in Fig. 6-11. With the molding removed, trim the adhesive on the fender to a height of 1.0 mm (3/64 in.). Remove all other traces of adhesive using 3M solvent.

5. Install a bead of adhesive on the wheelhouse trim, using adhesive part no. AKD 476 KD5 05 or similar. Then install the molding onto the fender as shown in Fig. 6-12.

CAUTION —
Part numbers are for reference only. Always rely on an authorized Volkswagen dealer for the latest and most accurate parts information.

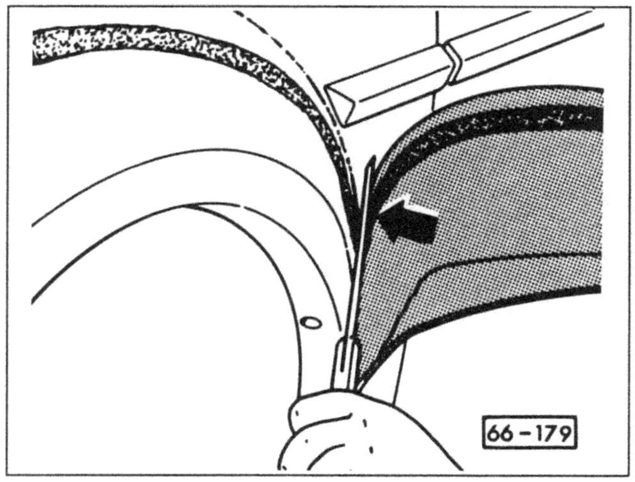

Fig. 6-11. Wheelhouse trim being removed with knife (arrow)

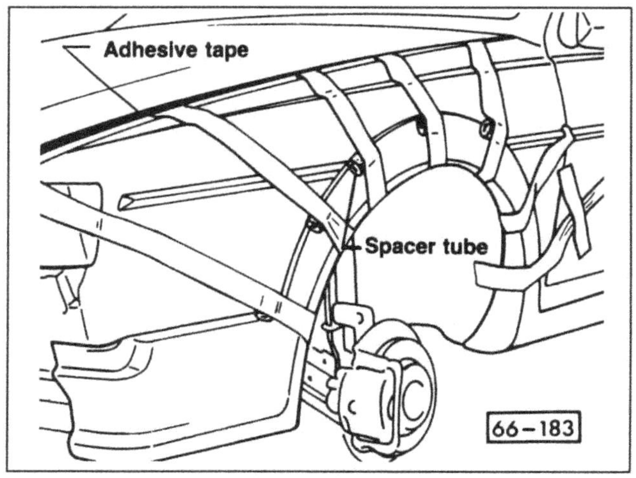

Fig. 6-12. Wheelhouse trim being installed. Tubing presses against upper edge of trim.

6. The remainder of installation is the reverse of removal. Use the same adhesive to join the wheelhouse trim to the spoiler and the rocker panel trim. Allow the adhesive to set for at least ten hours before removing the tape.

There are two types of wheelhouse trim on Cabriolets. Trim on 1985 through 1987 models is secured by screws. Trim on 1988 and later models is secured by screws and rivets. To remove 1988 and later Cabriolet wheelhouse trim, first remove the body side molding as shown in Fig. 6-13, then remove the trim as shown in Fig. 6-14.

To remove the stone guard, pry the fastening studs on the guard away from the retention clips as shown in Fig. 6-15.

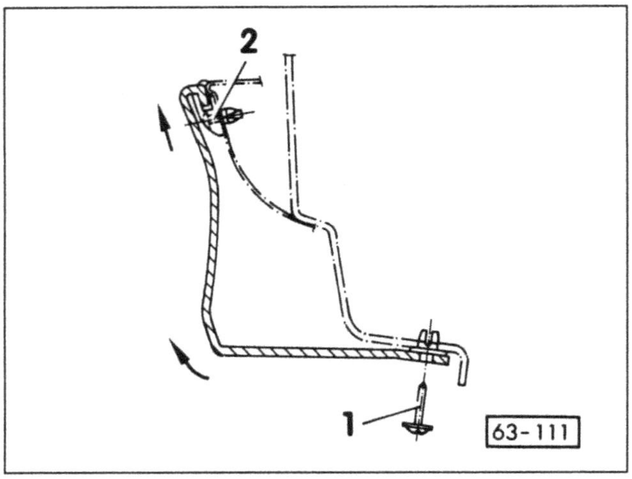

Fig. 6-13. Cabriolet body side molding being removed. Remove screws (**1**), then tilt molding up and out of clips (**2**).

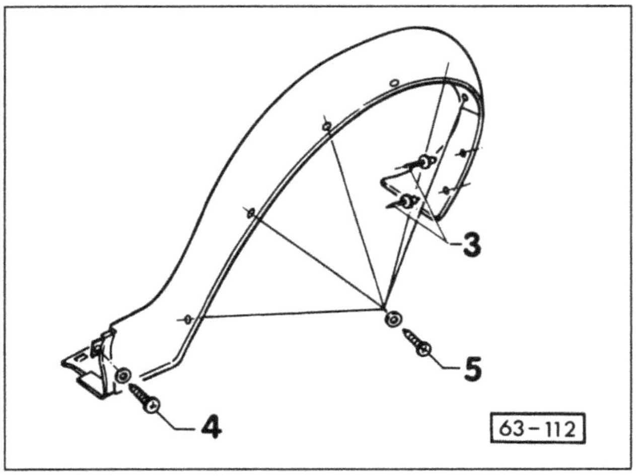

Fig. 6-14. Wheelhouse trim being removed. Drill out rivets (**3**) and remove screws (**4, 5**).

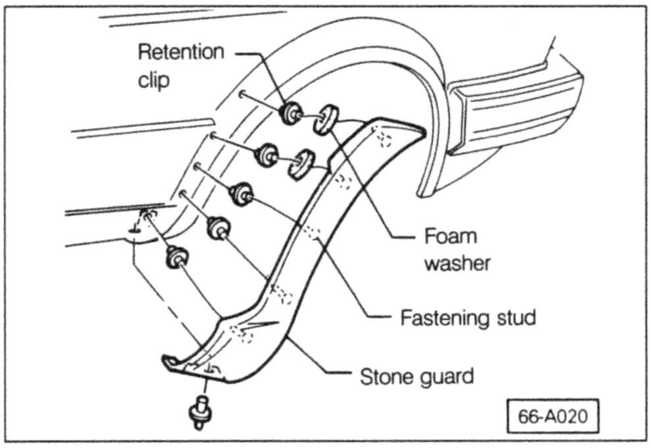

Fig. 6-15. Stone guard attachment to fender.

The spoiler for the Scirocco rear lid is shown in Fig. 6-16. To remove the spoiler, open the lid and remove the bolts from the holding angles. Use a putty knife to cut the bonding adhesive and sealant. Remove the remaining sealant on the lid to a level of 1.0 mm (3/64 in.). Apply sealant (part no. D 009 100) around the holding angles for water-proofing, and apply adhesive (part no. D 001 200) to the bottom of the spoiler. If the spoiler is being installed on a new piece of glass, first use the spoiler as a template to locate the areas on the glass on which to apply sealant primer (part no. D 009 200).

CAUTION —

Part numbers are for reference only. Always rely on an authorized Volkswagen dealer for the latest and most accurate parts information.

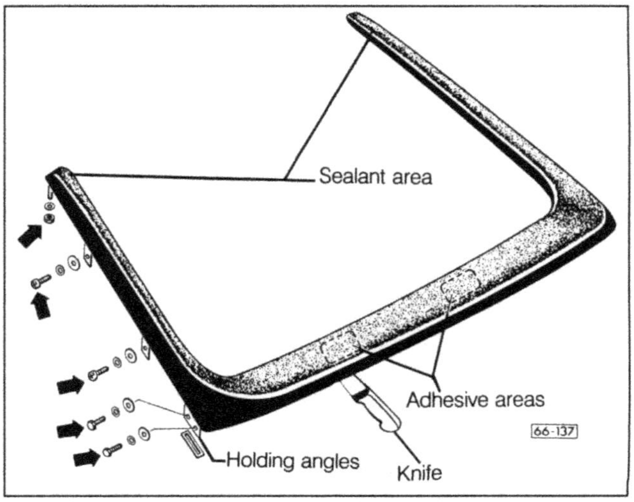

Fig. 6-16. Scirocco rear lid spoiler.

Removing and Installing Bumpers

The front and rear bumpers are mounted on solid brackets bolted to the frame, with guide brackets attached to the body. To remove a bumper, simply unbolt the brackets from both sides of the car frame and pull the bumper straight off. See Fig. 6-17, Fig. 6-18, Fig. 6-19, or Fig. 6-20. On 1988 and later Cabriolet models, it is necessary first to remove the fender mountings for the bumper cover as shown in Fig. 6-21. Installation is the reverse of removal. Torque the mounting bolts to 90 Nm (66 ft. lb.).

Fig. 6-17. Location of mounting bolts (arrows) for Scirocco front bumper (shown with radiator grille and headlights removed for clarity). Opposite side is similar.

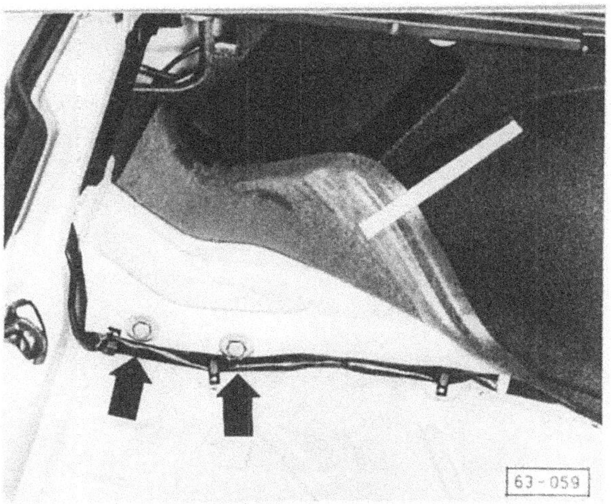

Fig. 6-18. Location of mounting bolts (arrows) for Scirocco rear bumper, inside luggage compartment. Opposite side is similar.

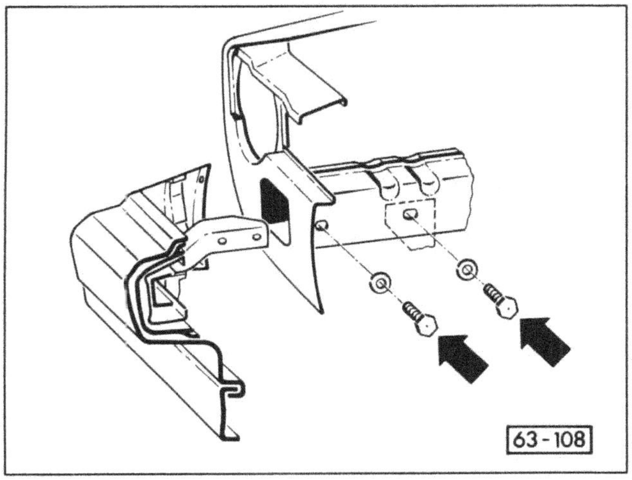

Fig. 6-19. Location of right-side mounting bolts for Cabriolet front bumper (arrows). Left side is similar.

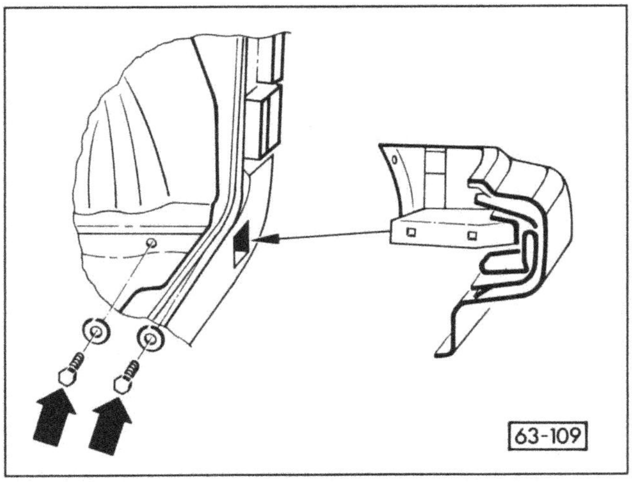

Fig. 6-20. Location of right-side mounting bolts for Cabriolet rear bumper (arrows). Left side is similar.

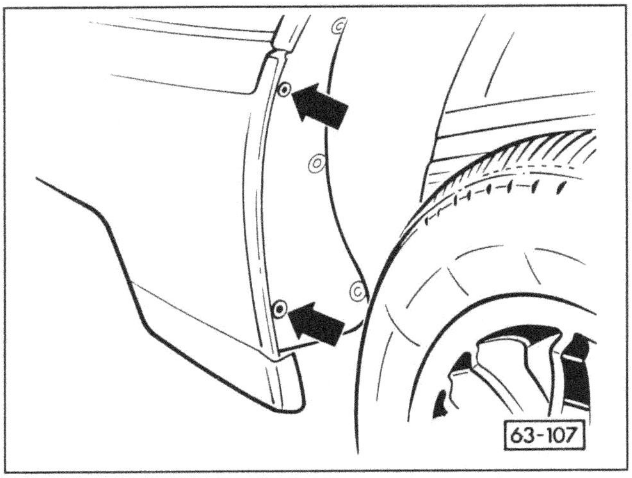

Fig. 6-21. Location of bumper cover mounting rivets in wheelhouse.

Hood Lock and Cable

The hood lock is held in place by rivets, and is easily replaced once the cable is removed from the lock. If the lock and lockpin are not aligned, adjust the lockpin plate on the hood by loosening the plate mounting bolts. The hood lock cable can be adjusted by loosening the cable lock screws as shown below.

Fig. 6-22 shows the routing of the Scirocco hood lock cable. To replace the cable, disconnect it from the lock by loosening the cable lock screw shown in Fig. 6-23, and remove the cable from its retaining clips. Remove the mounting screws for the cable release lever in the passenger compartment, then draw the cable through the firewall. Installation is the reverse of removal. Connect the cable to the lock without any tension, tighten the lock screw, and wind any excess cable around the screw.

Fig. 6-22. Scirocco hood lock cable routing with guide sleeves (**1, 2**).

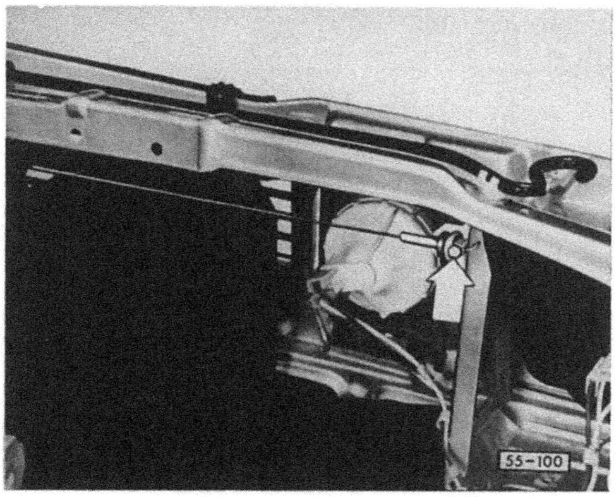

Fig. 6-23. Lock screw (arrow) for Scirocco hood lock cable.

Fig. 6-24 shows the routing of the Cabriolet hood lock cable. To replace the cable, disconnect it from the lock by loosening the cable lock screw and removing the clip shown in Fig. 6-25. Remove the cable from its mounting clips. Remove the mounting screws for the cable release lever in the passenger compartment, then draw the cable through the firewall. Installation is the reverse of removal. Connect the cable to the lock without any tension.

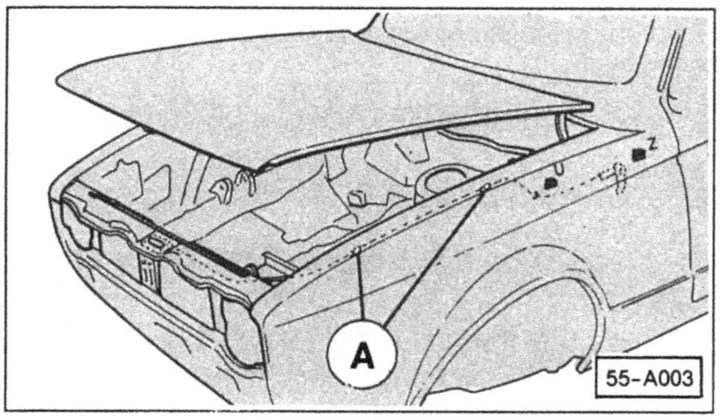

Fig. 6-24. Cabriolet hood lock cable routing and mounting clips (**A**).

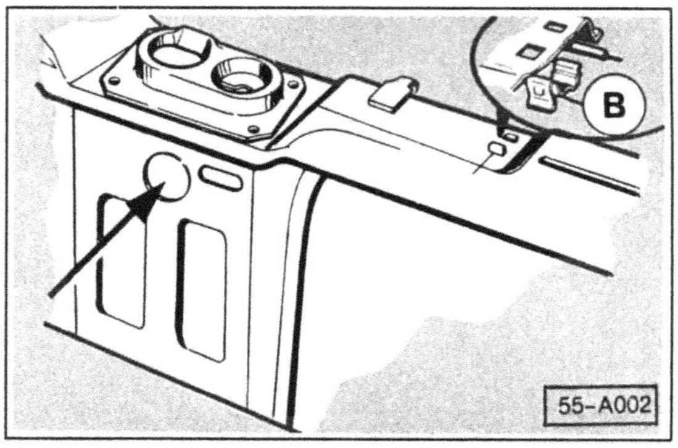

Fig. 6-25. Location of locking screw for Cabriolet hood lock cable (arrow), and clip to be removed (**B**).

Rear Lid

The rear lid locks for Scirocco and Cabriolet are shown in Fig. 6-26 and Fig. 6-27. Both locks are secured to the lid with screws. The lock and the position of the bottom edge of the rear lid is adjusted by moving the striker plate. To adjust, loosen the striker plate mountings screws just enough so that plate can be moved by hand, but with some effort. Close the lid and adjust it until the bottom edge is even, then open the lid and torque the striker plate mounting screws to 10 Nm (87 in. lb.).

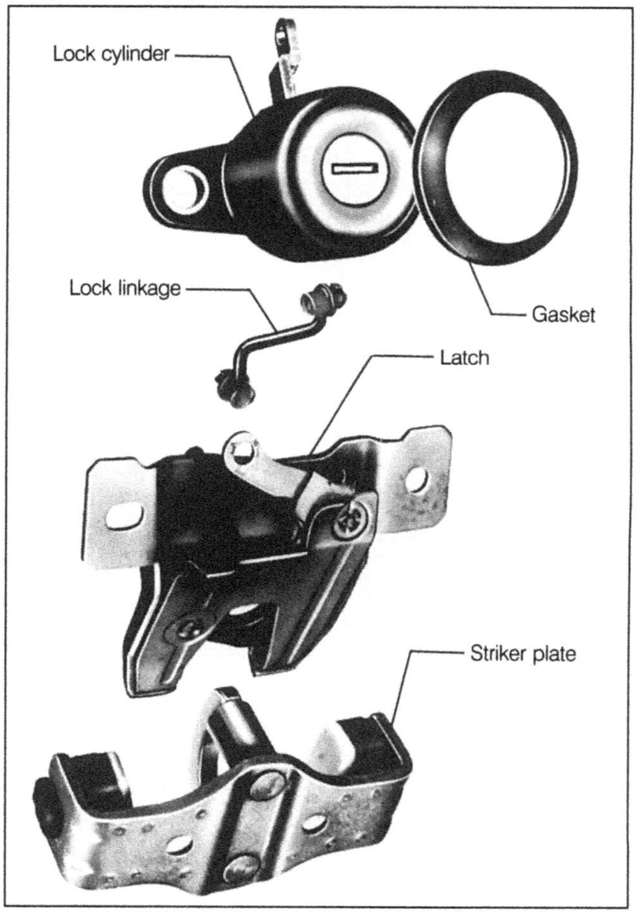

Fig. 6-26. Exploded view of Scirocco rear lid lock.

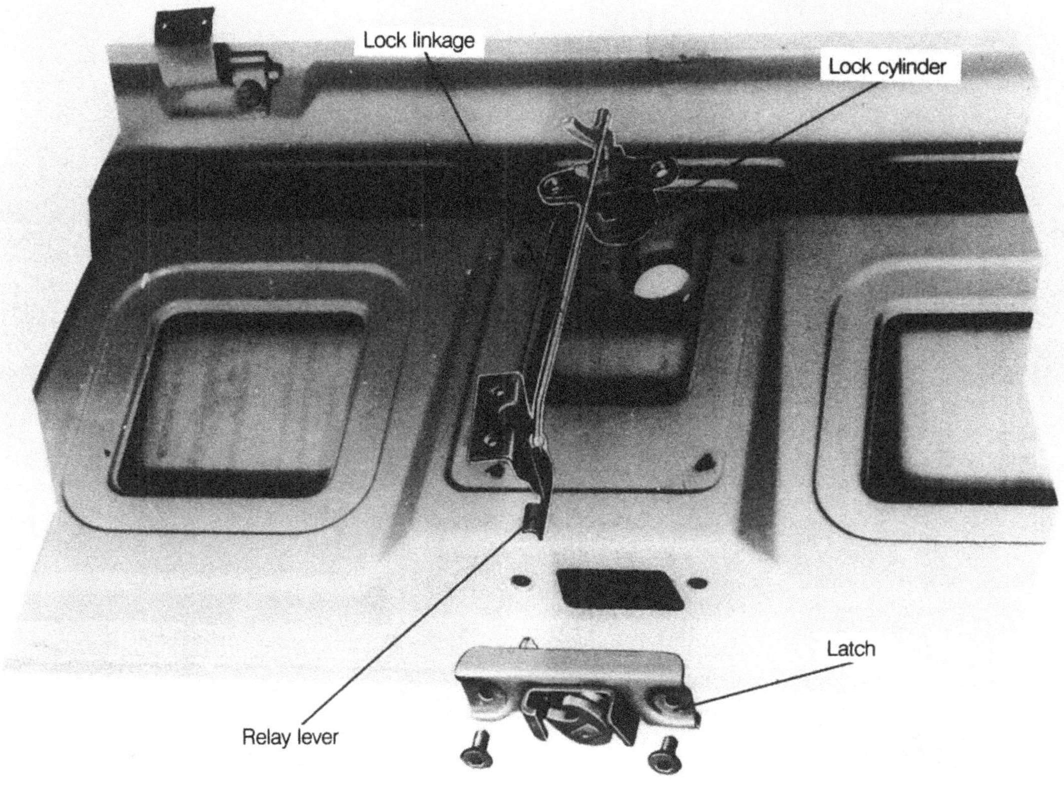

Fig. 6-27. Exploded view of Cabriolet rear lid lock.

To replace a license plate bulb, remove the screws that hold the lens to the lid. Press in and twist the bulb to remove and install it. Tighten the lens screws just until they are snug to avoid cracking the lens. Replacement bulbs are part no. N 017 717 2 (U.S. trade no. 1816).

7. SUNROOF

The sunroof is operated by a set of cables which pull the sunroof panel along guide rails when the crank is turned. The sunroof can be adjusted without removing it from the car, but replacement of the sunroof liner or components such as the cable assembly require sunroof removal. If the sunroof leaks or sticks, adjust it as described in **7.2 Adjusting Sunroof Fit** before deciding that it needs to be removed from the car. Leaks may also be caused by clogged drain hoses. The front hoses lead from the A-pillar into the space between the front fender and the door. The rear hoses lead to the rear section of the side panels. Clean the hoses with a length of flexible cable, such as an old speedometer cable. Work down from the top of the front hoses, and up from the bottom of the rear hoses.

7.1 Removing and Installing Sunroof

The only special tool needed to remove the sunroof is a length of wire, bent into the position shown in Fig. 7-1. It is used to unhook the sunroof trim spring.

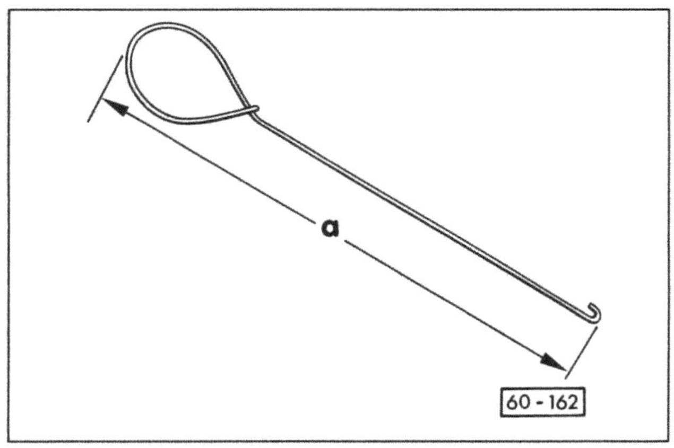

Fig. 7-1. Special tool for detaching sunroof trim spring. Dimension **a** is 300 mm (12 in.).

To remove:

1. Open the sunroof. Using the tool shown above in Fig. 7-1, detach the spring from the sunroof trim as shown in Fig. 7-2, then close the sunroof.

2. Push the sunroof trim back from the front guide, then remove the trim rod from the left and right guide rails. See Fig. 7-3.

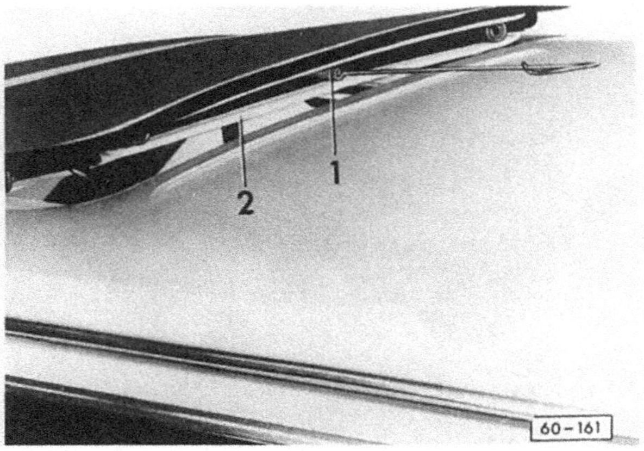

Fig. 7-2. Spring (**1**) being detached from sunroof trim (**2**) with special tool.

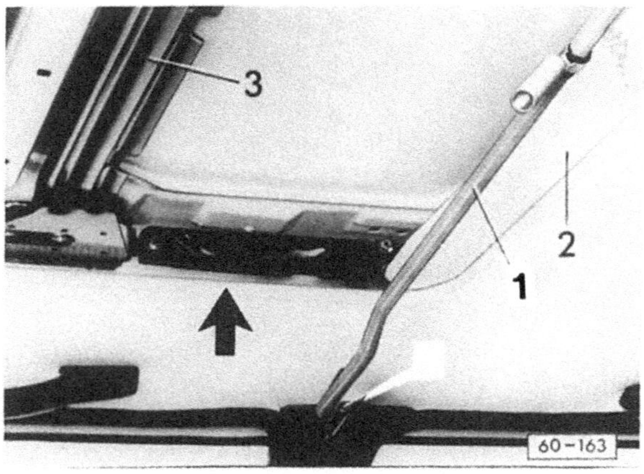

Fig. 7-3. Trim rod (**1**) being removed from guide rail (arrow). Trim (**2**) is shown pulled back from guide (**3**).

3. Disconnect the water deflector plate lever from the left and right guides as shown in Fig. 7-4.

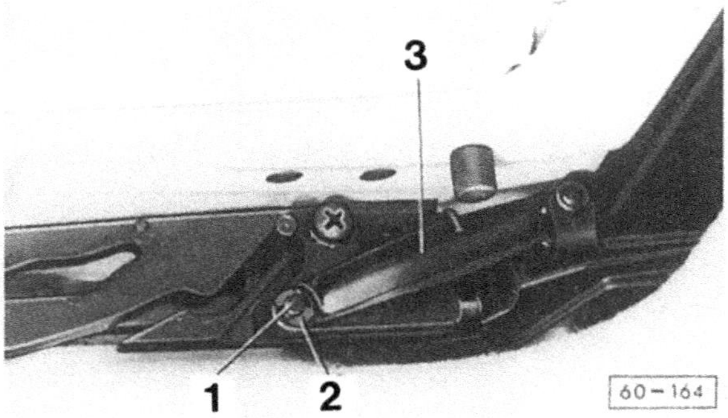

Fig. 7-4. Water deflector plate being disconnected. Remove circlip (**1**) from pin (**2**) and pry deflector lever (**3**) out of guide. Procedure is same for both sides.

4. Remove the right and left sunroof mounting screws from the front guide as shown in Fig. 7-5.

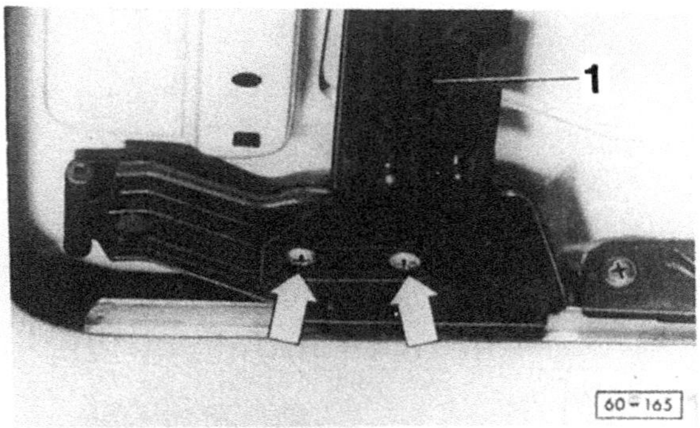

Fig. 7-5. Location of sunroof mounting screws (arrows) on front guide (**1**). Remove screws from both sides.

5. Crank the sunroof forward until the guide pins for the rear guide slide out of the guide plate, and then remove the sunroof. See Fig. 7-6.

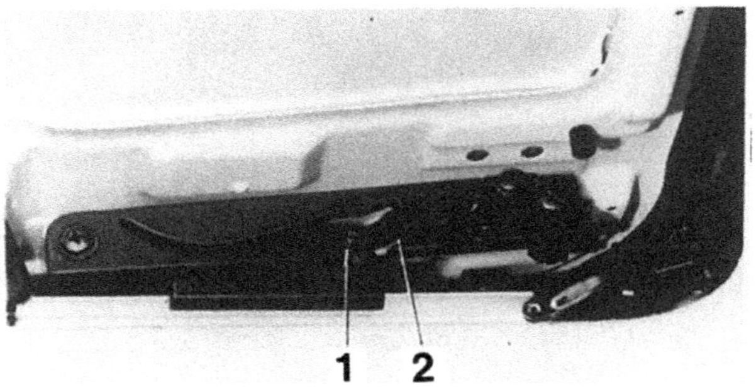

Fig. 7-6. Rear guide pin (**1**) being cranked out of guide plate (**2**).

Installation is the reverse of removal. It may be necessary to loosen the front guide plate screws as shown in Fig. 7-7 for installation of the water deflector plate lever. Adjust the sunroof as described in **7.2 Adjusting Sunroof Fit**, then attach the trim panel as shown in Fig. 7-8.

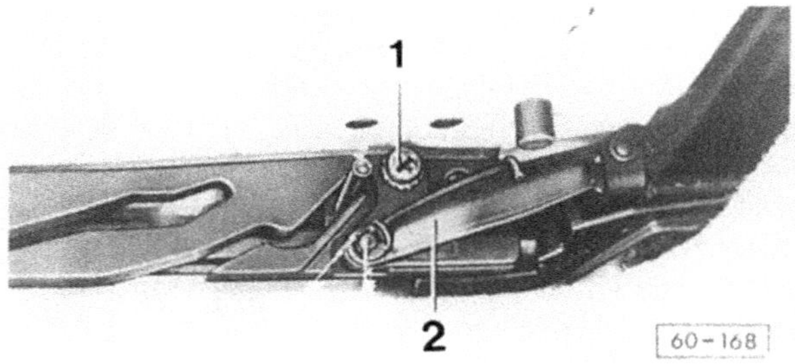

Fig. 7-7. Mounting screw for front guide plate (**1**) to be loosened when installing water deflector plate lever (**2**). Tighten screw when finished.

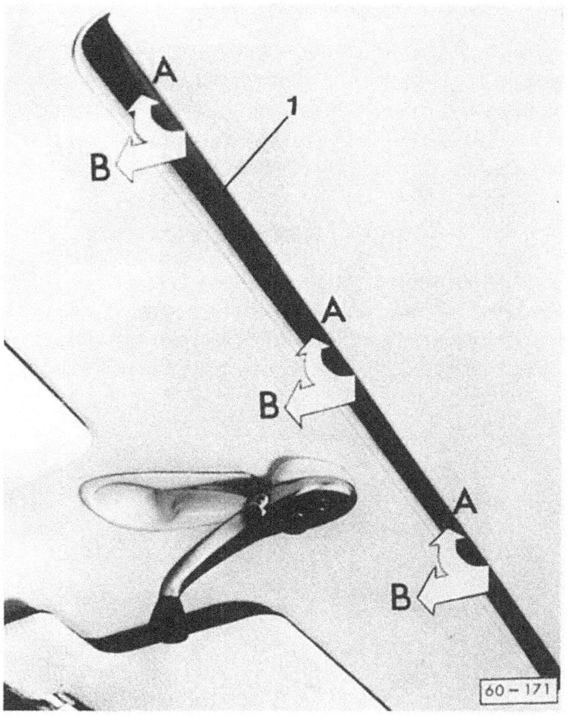

Fig. 7-8. Sunroof trim being installed. With sunroof open, close trim panel until there is a gap of approximately 10-15 mm (3/8-9/16 in.) at **1**, then push trim up (**A**) and forward (**B**). Engage trim spring using hook when finished.

Removing and Installing Sunroof Guides, Cables, and Trim

This heading covers the removal of the sunroof parts including the wind deflector, the front guides, the front cable guides, the rear guides with cables, the water deflector plate, and the guide rails.

To remove:

1. Remove the sunroof as described above.
2. Move the water deflector plate to the rear after pushing down on the detent spring shown in Fig. 7-9.
3. Crank the rear guides to the rear and push the front guides back for access to the wind deflector actuating rod mounting screws. See Fig. 7-10. Remove the actuating rod screws and the deflector mounting screws, and remove the deflector.
4. Remove the sunroof crank and cable drive as shown below in Fig. 7-18.

Fig. 7-9. Release detent spring (arrow) for water deflector.

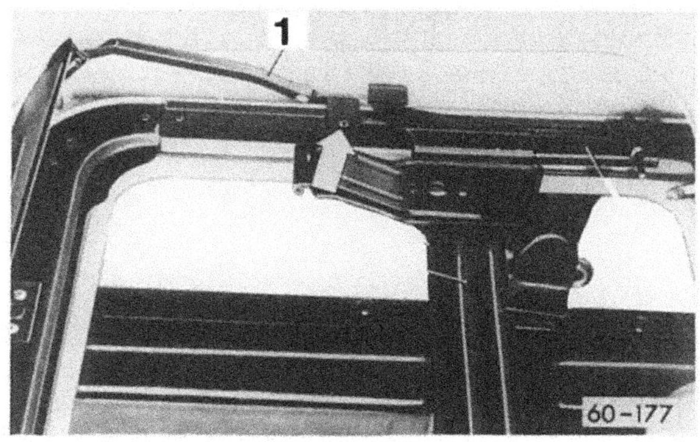

Fig. 7-10. Mounting screw (arrow) for wind deflector actuating rod (**1**). Remove screw for other side also.

5. Remove the front cable guides by removing the screws shown in Fig. 7-11, then remove the two cover plates from either side as shown in Fig. 7-12.

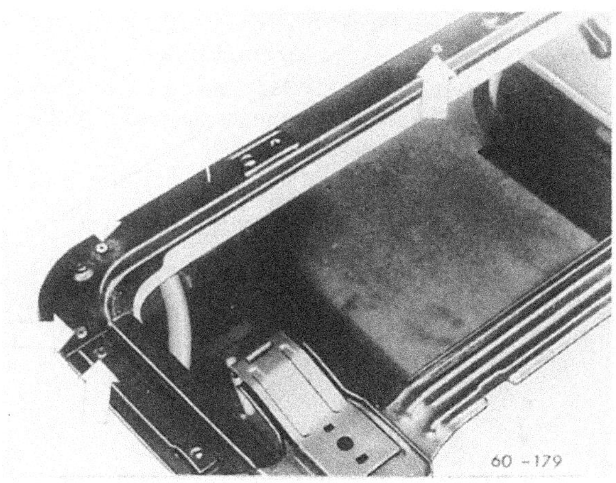

Fig. 7-11. Front cable guide mounting screws (arrows) to be removed.

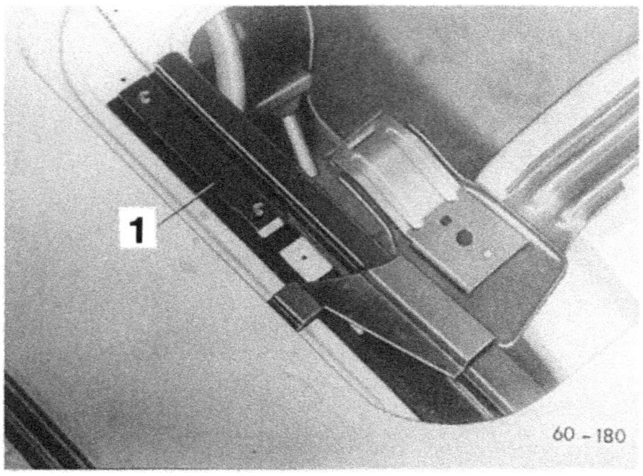

Fig. 7-12. Cover plate (**1**) to be removed from both sides.

6. Remove the guide rail mounting screws shown in Fig. 7-13, then remove the remaining sunroof parts in the following order: the front guide, the rear guides with cables, the water deflector plate, the trim panel, and the guide rails.

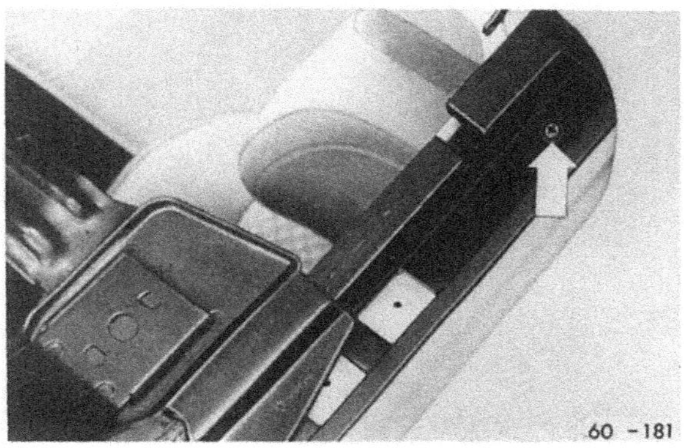

Fig. 7-13. Guide rail mounting screw (arrow) to be removed from either side.

Installation is basically the reverse of removal. When the cover plates and front cable guides are installed, loosely install the screws, then push the cover plates against the guides as shown in Fig. 7-14 and tighten the screws. Adjust the sunroof as described below when finished.

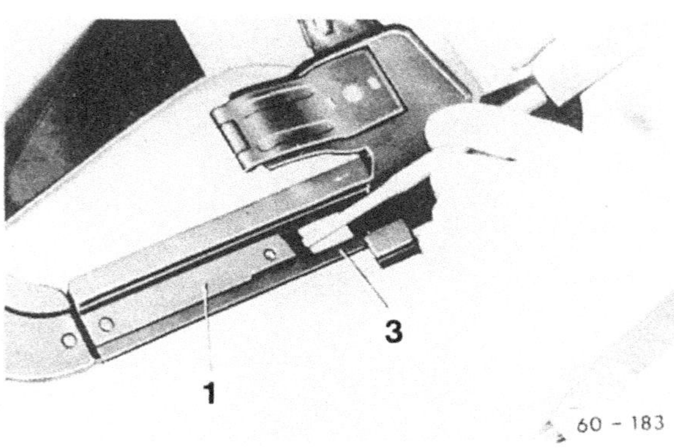

Fig. 7-14. Cover plates and front cable guides being installed. Lightly push guides (**3**) against cover plates (**1**) and tighten cover plate screws.

7.2 Adjusting Sunroof Fit

There are two adjustments for the sunroof: height adjustment and parallel adjustment. Incorrect height adjustment may cause wind noises at high speed, water leaks, or scratching of the sunroof panel. Incorrect parallel adjustment may make the sunroof difficult to operate, or prevent it from closing squarely.

Sunroof Height Adjustment

Fig. 7-15 shows the correct height adjustment of the sunroof. Height is adjusted at the front and rear guides after the sunroof trim has been detached from the sunroof and pushed back as described in **7.1 Removing and Installing Sunroof**.

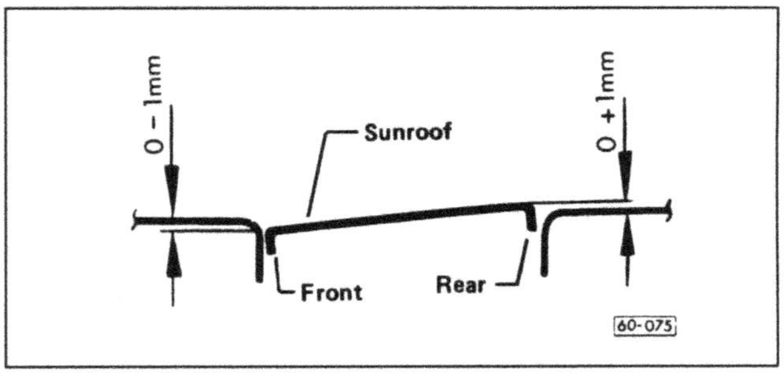

Fig. 7-15. Correct height adjustment of sunroof.

Adjust the front height at both sides of the front guide member as shown in Fig. 7-16. To adjust the rear height, first remove the water deflector plate levers from both sides as shown above in Fig. 7-4, then adjust the height as shown in Fig. 7-17. Reinstall the deflector lever plate as shown in Fig. 7-7, and reinstall the trim panel.

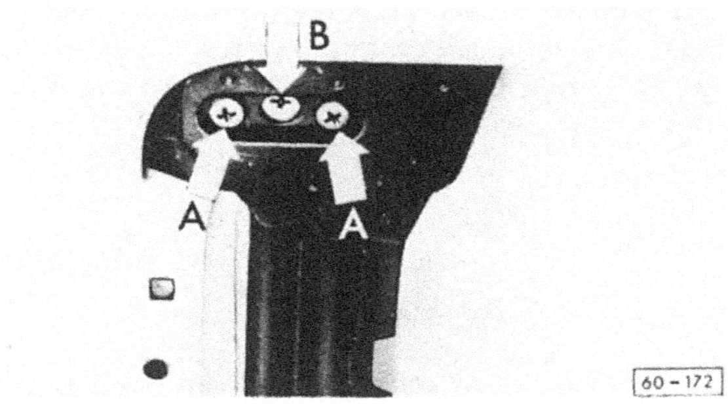

Fig. 7-16. Sunroof front height adjustment for both sides. Loosen screws **A** and turn screw **B** to adjust height. Tighten screws **A** when finished.

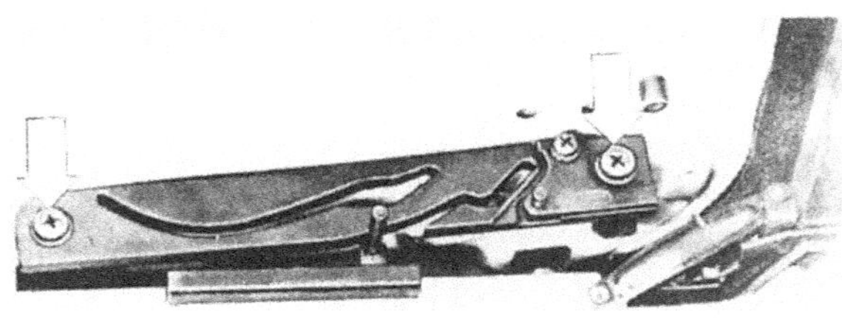

Fig. 7-17. Sunroof rear height being adjusted by loosening screws (arrows). Tighten screws when finished.

Sunroof Parallel Adjustment

The sunroof parallel adjustment aligns the rear guides so that they operate parallel to each other. The adjustment is made using a pair of inexpensive tools, Volkswagen special tool no. 3060 (order no. T03 060 000 60 ZEL). The sunroof panel must be removed as described in **7.1 Removing and Installing Sunroof** to make the adjustment.

To adjust sunroof parallel operation, turn the sunroof crank handle until it locks, then remove the handle and the cable drive. See Fig. 7-18. Use the special tools to adjust the position of the rear guides as shown in Fig. 7-19, then reinstall the cable drive and recheck the measurement. Install the sunroof panel when finished.

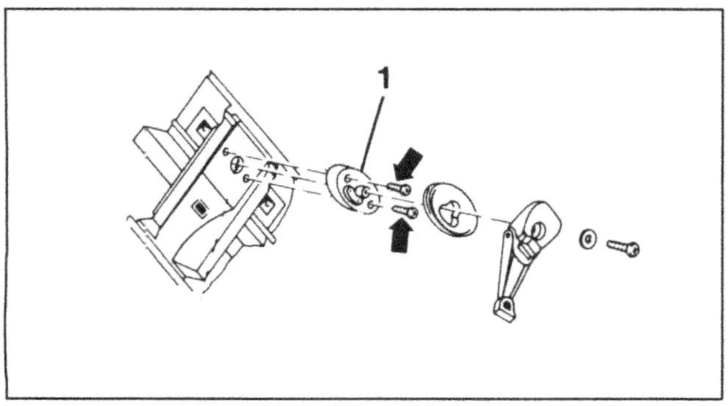

Fig. 7-18. Sunroof crank handle and cable drive (**1**) with mounting screws (arrows).

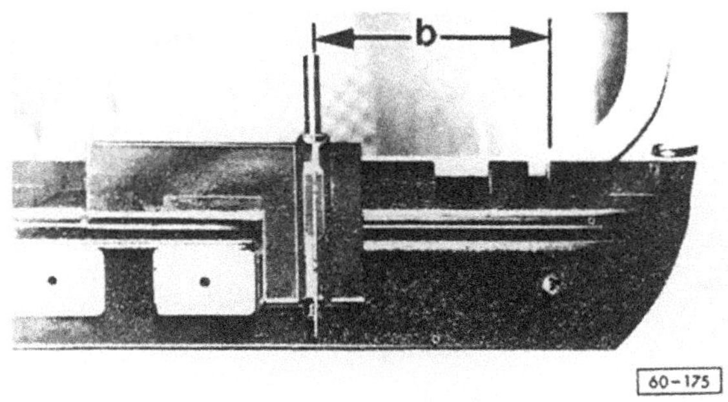

Fig. 7-19. Rear sunroof guides being adjusted using Volkswagen special tool 3060. Dimension **b** should be 62 ± .05 mm (2 7/16 in.).

CAUTION —
Always move the rear guides together to the front or rear when adjusting them. Do not move one forward and one back, or vice versa.

8. CONVERTIBLE TOP

The Cabriolet is a modification of a basic two-door unitized body. When the convertible top and its mounting frame are added, extra body-strengthening members and a steel rollbar are also added to the body structure. The configuration of the convertible top is shown in Fig. 8-1. For information about the power-operated top available since 1991, see **12. Power Convertible Top**.

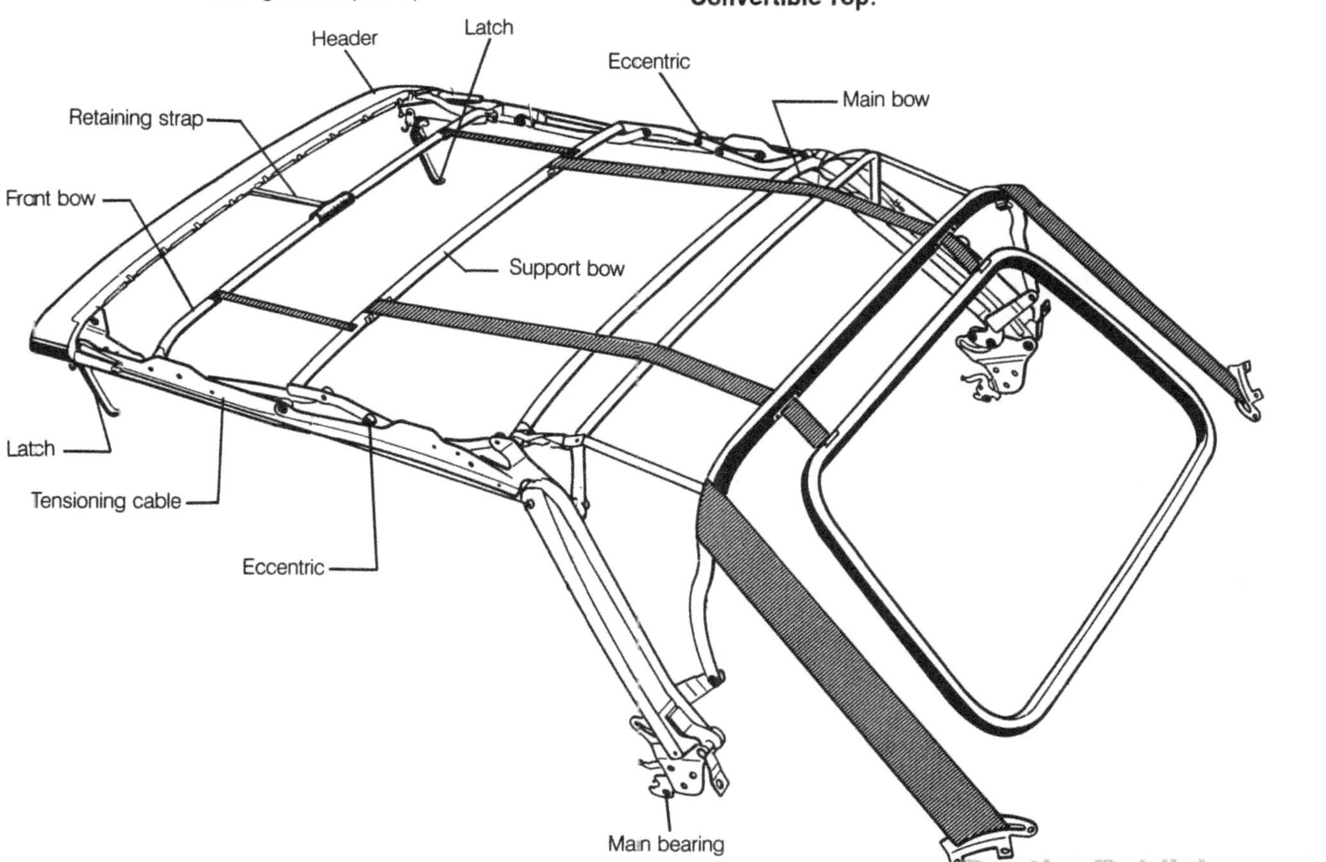

Fig. 8-1. Convertible top mounting frame. Top adjustment points are at front latches and left and right eccentrics.

A number of adjustments can be made to the convertible top and to the windows in order to ensure proper weather sealing. In general, if the top is difficult to open or close, or if drafts, water leaks, or fluttering noises come from around the top's front header, the front part of the top needs to be adjusted. If there are rattles or air leaks around the roll bar, then either the frame eccentric or the windows must be adjusted. Make all adjustments with the car on level ground.

8.1 Front Header

If there are air or water leaks at the front edge of the top, adjusting the front latches may cure them. If not, then the top seal may need replacing.

The latches are adjusted after loosening the adjustment locknuts shown in Fig. 8-2. With the latches undone, turn the latch hooks clockwise or counterclockwise so that the top contacts the header evenly and with reasonable pressure when the latches are closed. Tighten the locknuts when finished.

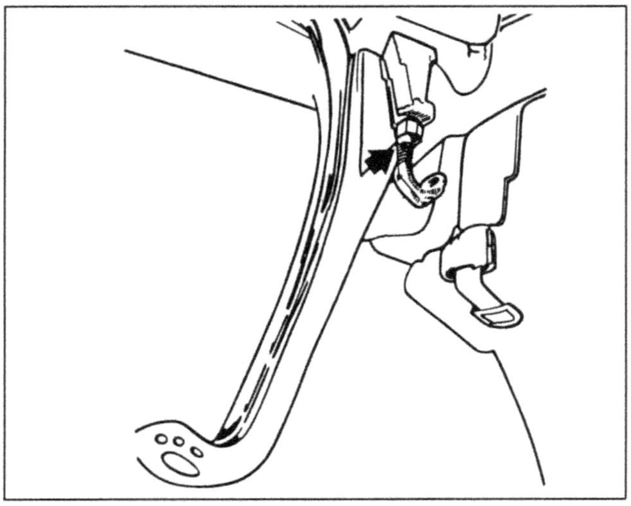

Fig. 8-2. Locknut (arrow) to be loosened for convertible top header latch adjustment.

To replace the top seal, drill out the two rivets at either end of the seal and pry the seal off. Clean off any old adhesive. Apply adhesive as shown in Fig. 8-3, then push the new seal into place and install new hollow rivets. Finally, seal the area shown in Fig. 8-4 with a silicone sealer.

8.2 Eccentric Adjustment

The two eccentrics on either side of the convertible top frame are used to adjust the fit of the top against the roll bar. To adjust the fit, open the top halfway and use a wrench as shown in Fig. 8-5 to turn the eccentric one way or the other. Close the top to check the fit of the weatherseal against the roll bar. Readjust as necessary.

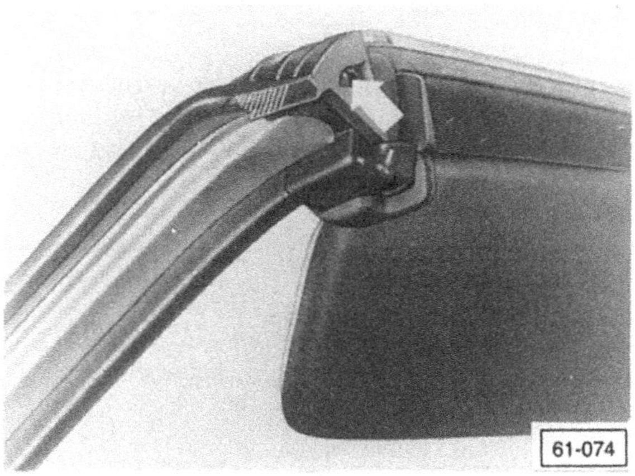

Fig. 8-3. Convertible header seal being replaced. Apply adhesive in shaded area on both sides. Replace rivet (arrow) on both sides.

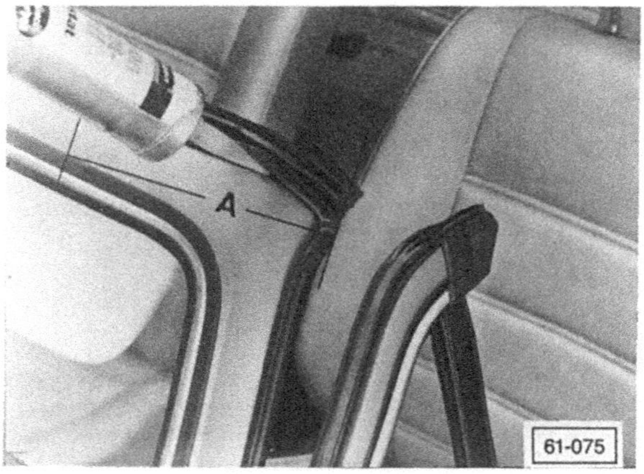

Fig. 8-4. Silicone sealer being applied to header seal. Dimension **A** on both sides of frame is approximately 200 mm (7 7/8 in.).

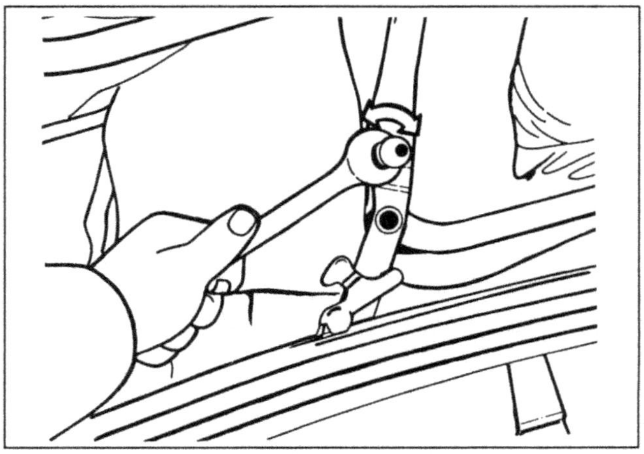

Fig. 8-5. Eccentric being turned with wrench for top adjustment.

8.3 Window Adjustments

Three adjustments are made to the front and rear side windows to ensure a weather tight fit and smooth operation. The adjustments include the window fit against the convertible top seal, the parallel fit of the window in its frame, and the fore-and-aft fit of the windows against the side seals. The correct fit of the windows against the convertible top seal is shown in Fig. 8-6.

Front window adjustments are made with the door trim panel removed as described in **5.1 Door Assembly**. Rear window adjustments are made with the rear seat removed as described in **4.8 Rear Seats**, and with the side trim panel unclipped.

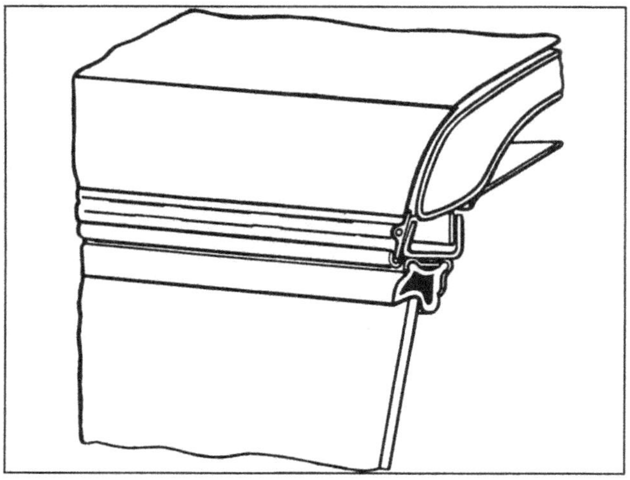

Fig. 8-6. Cutaway view of proper fit of window glass against convertible top seal.

Adjusting Front Window Fit

Front window fit against the seal is adjusted by loosening the bolts shown in Fig. 8-7 and moving the channel in or out as necessary. Tighten the bolts when finished.

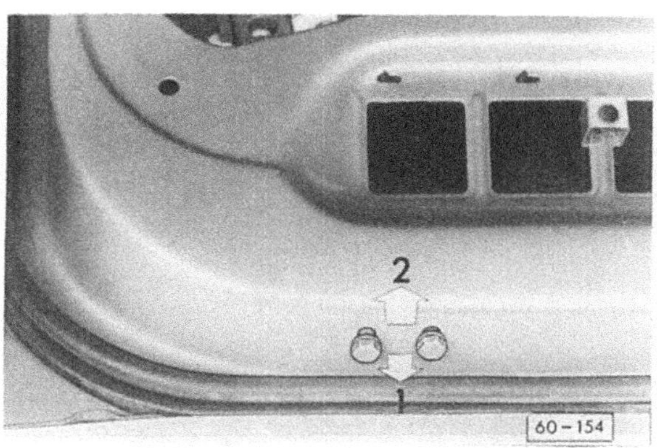

Fig. 8-7. Bolts (arrow) to be loosened for adjustment of front window fit against top seal.

Front window parallel fit is adjusted by loosening the bolts shown in Fig. 8-8 to move either the front or the rear of the glass up or down. Tighten the bolts when finished.

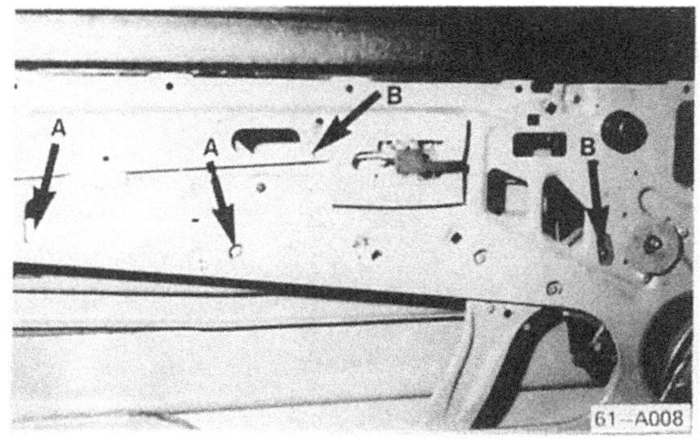

Fig. 8-8. Bolts to be loosened for adjusting front window parallel fit. Loosen bolts **A** to adjust rear of glass, and bolts **B** to adjust front of glass.

The fore-and-aft fit of the front windows against the front guide channel is adjusted after the bolt shown in Fig. 8-9 is loosened. Check the movement of the window glass after adjustment, and readjust if necessary.

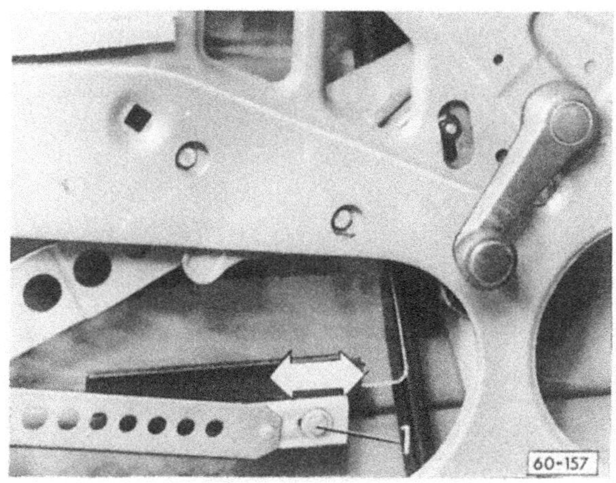

Fig. 8-9. Bolt to be loosened (**1**) for fore-and-aft adjustment of front window.

Adjusting Rear Window Fit

Rear window fit against the top seal is adjusted by loosening the bolt shown in Fig. 8-10 and moving the glass up or down as necessary. Tighten the bolt when finished.

To adjust rear window parallel fit, first remove the convertible top hinge cover as shown in Fig. 8-11. Adjust the fit by loosening the bolts shown in Fig. 8-12 and moving the rear of the glass up or down as necessary. Tighten the bolts and replace the cover when finished.

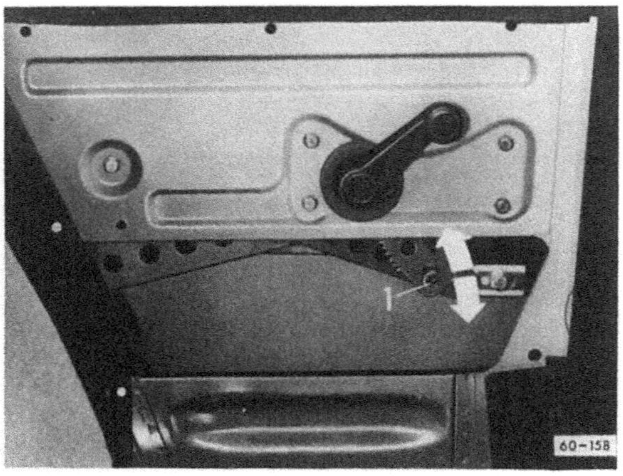

Fig. 8-10. Bolt to be loosened (**1**) for adjustment of rear window fit against top seal. Move glass up or down (arrows) as necessary.

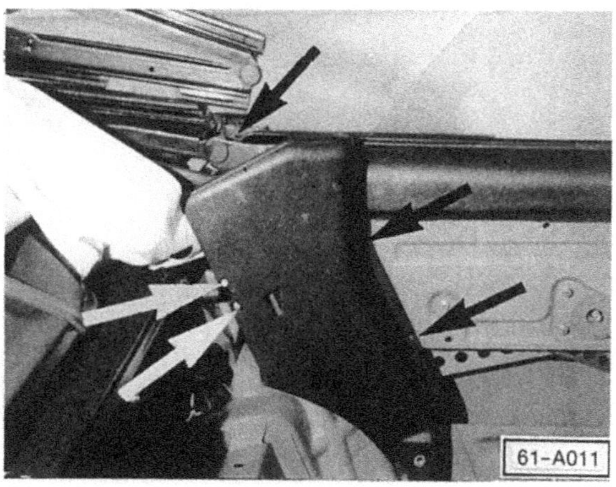

Fig. 8-11. Screws to be removed (arrows) for removal of convertible top hinge cover.

Fig. 8-12. Bolts to be loosened (**1, 2**) for parallel adjustment of rear window.

The fore-and-aft fit of the rear window is adjusted after loosening the bolt shown in Fig. 8-13. Roll the window up and down a few times to center it in the channel, then tighten the bolt.

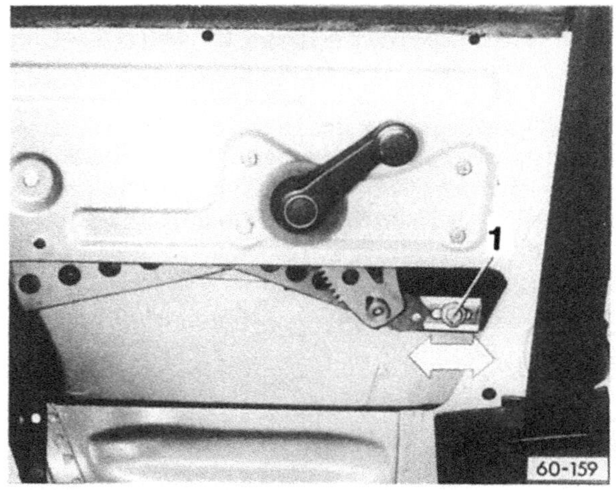

Fig. 8-13. Bolt to be loosened (**1**) for adjustment of rear window fore-and-aft fit. Move window as necessary (arrows) for smooth operation.

9. BODY ASSEMBLY MATERIALS

The assembly materials listed in **table b** are specified by Volkswagen for use in body and interior repairs.

CAUTION —
Volkswagen part numbers listed in this manual are provided for reference only. Always check with an authorized Volkswagen dealer Parts department for the most accurate and up-to-date information.

Table b. Body Assembly Materials

Application	Type, Volkswagen part number
Fastening rear view mirror to windshield	Glass to metal adhesive (part no. D 000 702) Adhesive activator (part no. D 000 701)
Gluing interior leatherette	Synthetic adhesive (part no. D 001 101)
Gluing carpet, rubber, felt and fabric to painted metal	Universal adhesive (part no. D 001 200)
Gluing door seals	Profile rubber adhesive (part no. D 002 100)
Fastening protective liner to door	Two-sided tape (part no. AKL 440 025)
Adhesive cleaning solution	Cleaning solution (part no. D 009 400 1)
Primer for painting	Paint primer (part no. D 009 300)
Sealing gaps in bolts, cables and wires to panels	Original plastic sealant (part no. D 001 400)

10. AIR CONDITIONING (A/C)

The subassemblies and lines of the air conditioning system are filled with difluorodichloromethane (CF_2Cl_2) refrigerant, known by the commercial names Frigen R-12 or Freon R-12.

> **WARNING —**
> - *The A/C system is filled with R-12 refrigerant gas under pressure. When other work makes it necessary to remove A/C components, try to do so without disconnecting the refrigerant lines. Otherwise the system will have to be recharged.*
> - *R-12 automotive refrigerant is hazardous to the earth's atmosphere. The A/C system should be serviced only by trained technicians using approved refrigerant recovery/recycling equipment, trained in related safety precautions, and familiar with regulations governing the discharging and disposal of automotive chemical refrigerants.*
> - *Do not expose any part of the A/C system to high temperatures such as open flame. Excessive heat will increase system pressure and may cause the system to burst.*
> - *Always wear eye protection when working around the A/C system.*

> **WARNING —**
> - *Always be careful to ensure that refrigerant does not come in contact with skin or eyes.*
> - *If refrigerant has come in contact with skin or eyes, do not rub skin or eyes. Immediately flush with cool water for at least 15 minutes. Seek medical attention from a doctor or hospital. Do not attempt to treat yourself.*
> - *R-12 refrigerant becomes a poisonous gas if it comes into contact with any open flame. Do not smoke if refrigerant gasses are present or if their presence is suspected.*

This section of the manual contains information about the operation of the air conditioning system, as well as some basic system checks. Beyond these basic checks, all A/C testing and service should be left to an authorized Volkswagen dealer Service department or other qualified repair shop. For information on A/C instrument panel controls, see **4.5 Heater and Controls**.

10.1 System Description

Fig. 10-1 is a schematic view of the typical Volkswagen air conditioning (A/C) system. The refrigerant in the evaporator absorbs heat from the passenger compartment and boils,

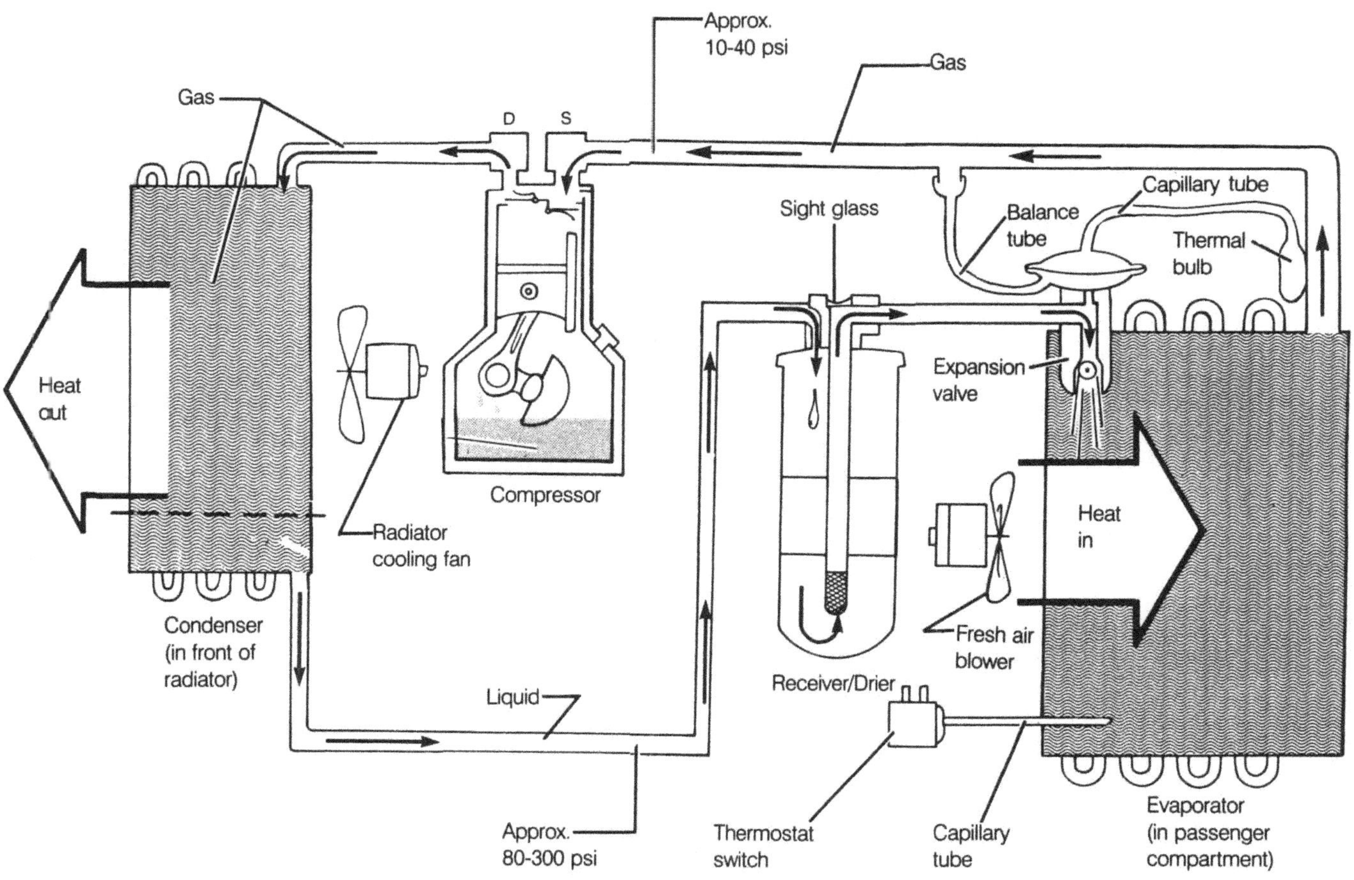

Fig. 10-1. Schematic view of typical air conditioning system.

turning from liquid to gas. The system also dehumidifies as the moisture in the air condenses on the cold evaporator and is routed outside. This is why a small puddle often accumulates under the car when the A/C is operating. The expansion valve meters the flow of refrigerant based on its temperature as it leaves the evaporator.

The A/C compressor forces the refrigerant through the system and at the same time pressurizes it, raising its boiling point to make it condense more readily. When the circulating refrigerant gets to the condenser, near the radiator, heat is transferred to the atmosphere. The refrigerant is cooled and condenses back into a liquid. The receiver/drier removes moisture and dirt from the system.

The compressor is engaged by an electro-magnetic clutch when the A/C is switched on. A thermostat switch disengages the clutch when the temperature in the passenger compartment reaches the level set on the operating controls.

Safety Features

To prevent system freeze-up, the fresh air blower and the radiator cooling fan come on at low speed whenever the A/C is turned on. A low-pressure switch prevents engagement of the compressor clutch if the R-12 charge is too low. A high-pressure switch runs the radiator cooling fan at high speed A/C system pressure is too high. A blow-out plug in the receiver/drier releases if system pressure exceeds approximately 500 psi.

10.2 Inspections and Tests

The condenser fins may become bent or blocked by debris, reducing air flow and raising system pressure. With the engine off, clean the debris from the front of the condenser, and straighten any bent fins using a fin comb. Check the fresh air intake for obstructions. It is on the passenger side of the car, behind the firewall and underneath the plastic drip tray. Inspect the compressor, the refrigerant hoses, and all visible system components for oil or refrigerant leaks. Evidence of leaks is most commonly visible at the bottom side of the fittings and components. Inspect the wiring to the pressure switches and compressor clutch. Check the compressor mountings to be sure they are tight, and check the condition and tension of the V-belt as described in **LUBRICATION AND MAINTENANCE**.

Checking Refrigerant Charge

There are almost always small leaks in the system that will eventually require that it be recharged. Inspect the refrigerant charge with the engine and the system set to "MAX A/C" and while the compressor is running (clutch cycled on).

The sight glass, located in-line near the condenser in the engine compartment, should be nearly free of bubbles. Constant foaming indicates that the system charge is low. Streaks on the interior of the glass may mean that the system is totally discharged. To guard against component damage, a completely discharged system must be evacuated (also referred to as pulling a vacuum), using special equipment to remove any moisture before recharging.

WARNING —
- *Be careful that refrigerant does not come in contact with skin or eyes. Always wear eye protection when working around the A/C system.*
- *The A/C system should be serviced only by technicians using approved refrigerant recovery/recycling equipment and trained in related safety precautions.*

10.3 Air Conditioning Specifications

Fig. 10-2 and the specifications in **table c** are intended only for use by trained and properly equipped A/C service professionals.

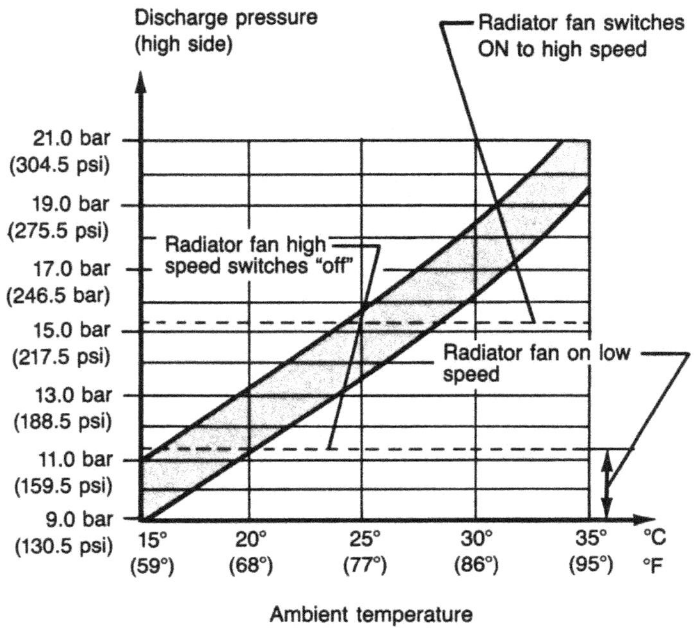

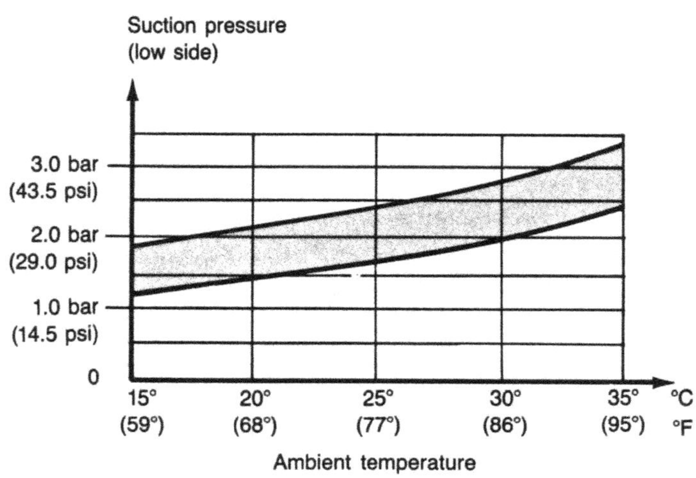

Fig. 10-2. Low and high side system pressures used when checking refrigerant charge and servicing the system. The system pressures should be in the shaded areas according to the ambient air temperatures.

Table c. Air Conditioning Specifications

Refrigerant capacity (R-12)	1150 ± 50 grams (40 ± 2 oz.)
Refrigerant oil capacity (total)	175 cc (6 oz.)
Oil distribution through system	
Condenser	20 cc (⅔ oz.)
Evaporator	30 cc (1 oz.)
Receiver/Drier	20 cc (⅔ oz.)
Discharge hose	15 cc (½ oz.)
Suction hose	15 cc (½ oz.)
Compressor	75 cc (2½ oz.)
High pressure switch	
Closes	14.5 ± 0.7 bar (210.0 ± 10.2 psi)
Opens	12.0 ± 0.7 bar (174.0 ± 10.2 psi)
Low pressure switch	
Closes	3.0 ± 0.3 bar (43.5 ± 4.3 psi)
Opens	2.0 ± 0.3 bar (29.0 ± 4.3 psi)

WARNING —
Adding refrigerant or servicing the system without the proper tools and equipment risks serious personal injury and may damage the system.

11. CABRIOLET POWER WINDOWS

Fig. 11-1 is an exploded view of the window regulator and related Cabriolet power window parts. The scissors-type regulator and front guide rail help ensure the proper position of the glass in the window opening.

A power window control unit is mounted behind the glovebox. A 20-amp power window circuit breaker is located on the main fuse/relay panel. For system wiring diagrams and more information on the locations of fuses and circuit breakers see **ELECTRICAL SYSTEM**.

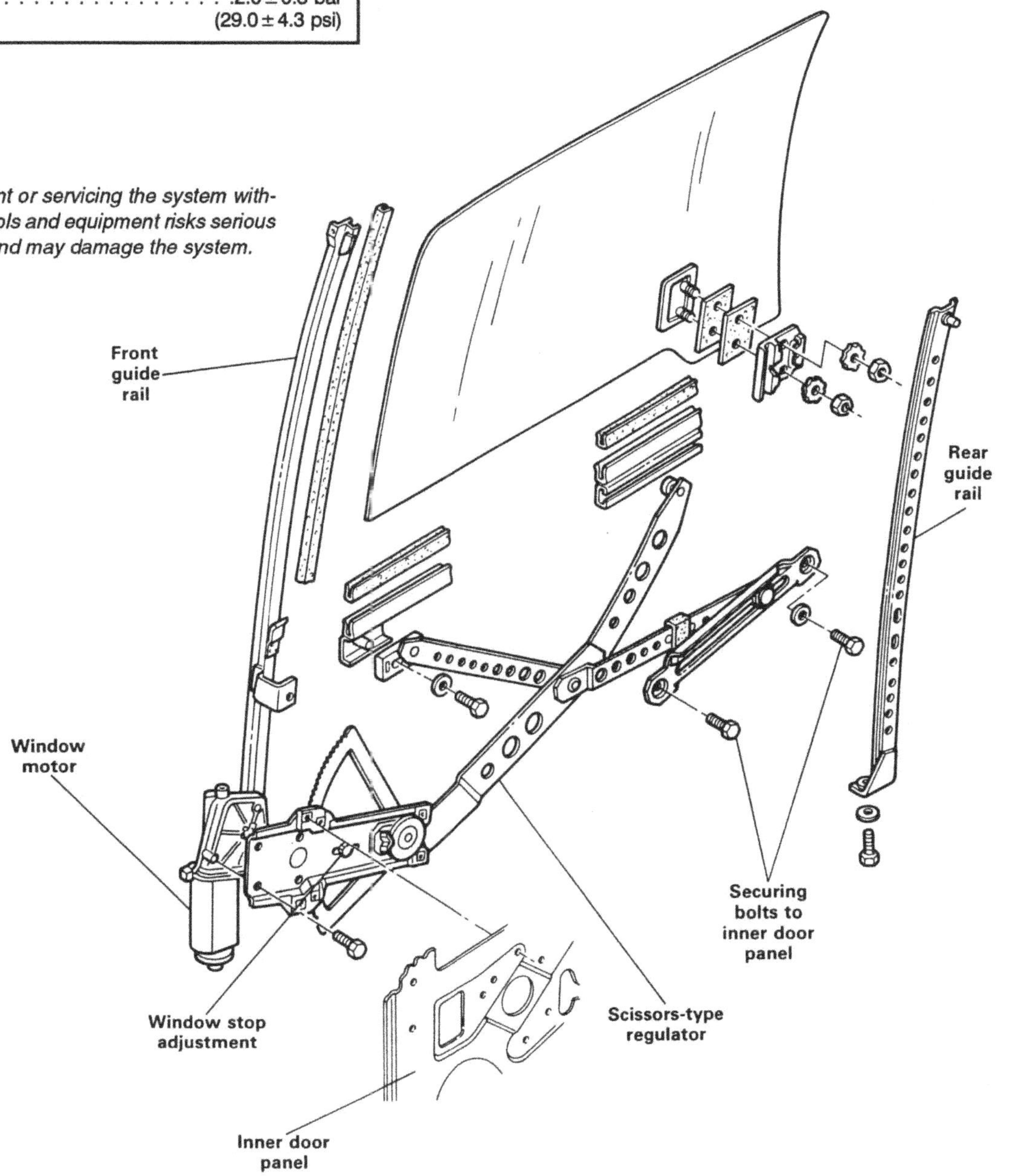

Fig. 11-1. Cabriolet power window regulator assembly.

12. Power Convertible Top

The electro-hydraulic power top system is shown in Fig. 12-1. A bi-directional electric motor and pump unit supplies pressure to two hydraulic cylinders that operate the top mechanism. A system control unit provides power and ground for the motor. The motor-pump unit, hydraulic cylinders and control unit are mounted in the luggage compartment.

Power is provided by the ignition switch's "SU" circuit. This means that the system is powered when the key is inserted in the ignition switch, but without switching on the ignition. Wires in the convertible top boot complete a ground circuit to inform the control unit when the boot is installed. Then the control unit interrupts power to the hydraulic pump motor. When the top is down, the boot should always be installed. For more electrical information see the wiring diagrams in **ELECTRICAL SYSTEM**.

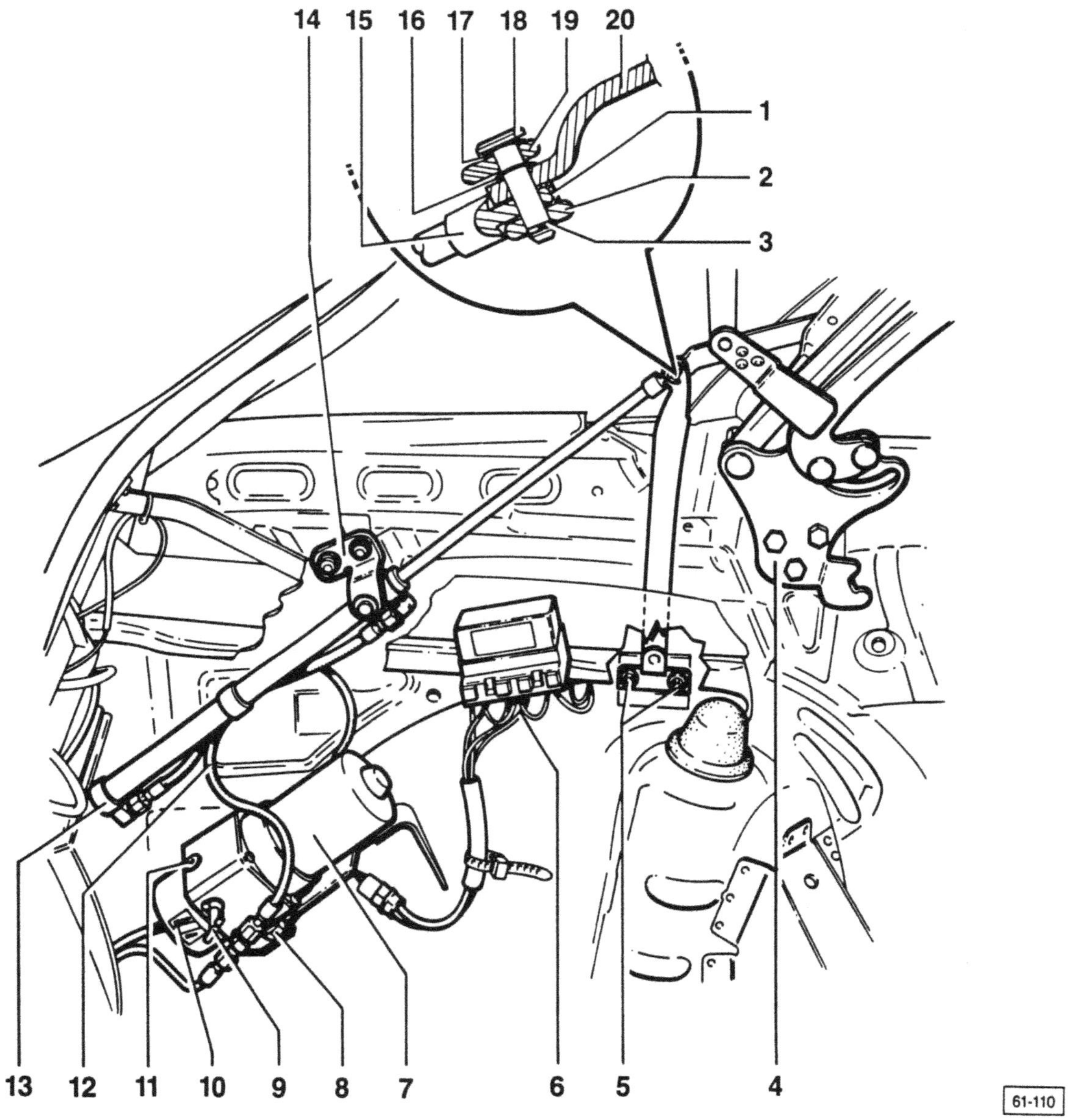

Fig. 12-1. Electro-hydraulic system to operate power top.

1 – Slide
2 – Guide rod
3 – Lock washer
4 – Spring washer
5 – Nuts
6 – Control unit
7 – Hydraulic pump
8 – Bonded rubber bushing
9 – Thumb-screw
10 – Hydraulic oil level indicator
11 – Hydraulic oil filler plug open/close with $^{3}/_{16}$ in. hex wrench
12 – High-pressure hydraulic hoses
13 – Hydraulic cylinder (one per side)
14 – Bracket for hydraulic cylinder
15 – Hydraulic cylinder piston rod
16 – Corrugated washer
17 – Corrugated washer
18 – Retaining bolt
19 – Guide rod
20 – Guide rod

A circuit breaker above the main fuse/relay panel protects the circuit against over-load if, for example, someone trying to lower the top has neglected to unlock the front latches. It will interrupt power to the system, then cool and reset after another 30 to 35 seconds.

Emergency Operation

The boot must not be installed. To operate the top manually open the luggage compartment, find the access hole in the trim panel, and locate the T-shaped thumb-screw just inside the hole. See Fig. 12-2. The trim panel and the access hole are *also visible* in Fig. 12-3.

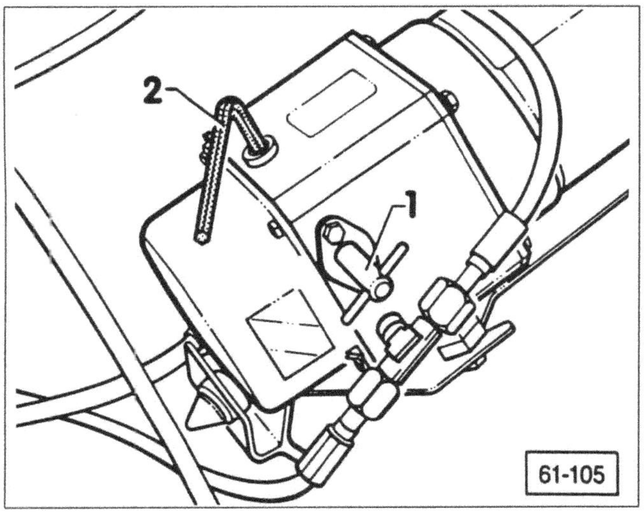

Fig. 12-2. Detail view of hydraulic pump unit, with trim panel removed, showing thumb-screw (**1**) and $^3/_{16}$ in. hex wrench (**2**) used to remove filler plug and check hydraulic oil level.

Turn the thumb-screw all the way to the left, counterclockwise until it stops. This opens a bypass port in the hydraulic system. The top can now be raised or lowered manually.

> ***CAUTION —***
> - *Take care to raise or lower both sides evenly.*
> - *When raising the top, be sure to lock it at the front.*

Bleeding the Hydraulic System

Bleeding may be required to purge air from the hydraulic system, if the top does not go up and down smoothly, or if lines have been disconnected. For access, unclip and remove the left side luggage compartment trim as shown in Fig. 12-3. The thumb-screw used to bleed the system is shown above in Fig. 12-2.

To bleed the system:

1. Turn the thumb-screw on the hydraulic pump housing all the way to the left, counterclockwise until it stops.
2. By hand, lower the convertible to its open position, raise it to its closed position, then lower it again.

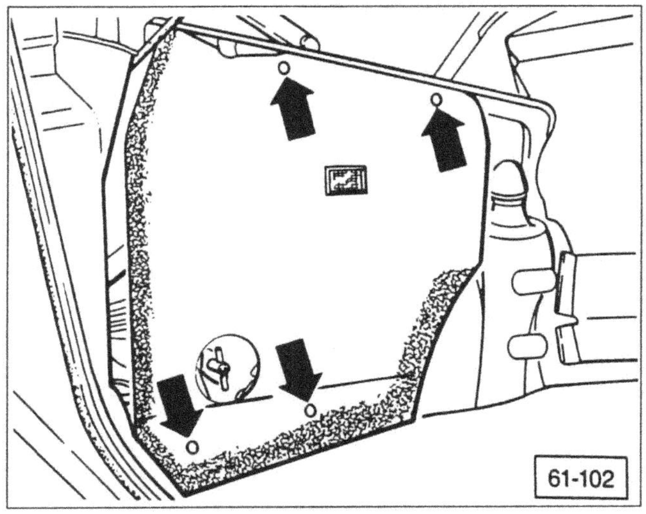

Fig. 12-3. Luggage compartment trim clips (arrows) to be removed for access to power top components. Notice location of thumb-screw access hole.

3. Turn the thumb-screw on the hydraulic pump housing all the way to the right, clockwise until it stops.
4. Check the hydraulic oil level and add oil as necessary, as described below.

> ***CAUTION —***
> *Hydraulic oil level must be checked with the top lowered, in the open position. Oil level is slightly lower with the top closed.*

Checking and Correcting Hydraulic Oil Level

Check hydraulic oil level at the indicator on the side of the hydraulic pump housing, as shown in Fig. 12-4. The level should be between **MIN** and **MAX** marks. If the oil level is below the **MIN** mark, add oil according to the procedure that follows.

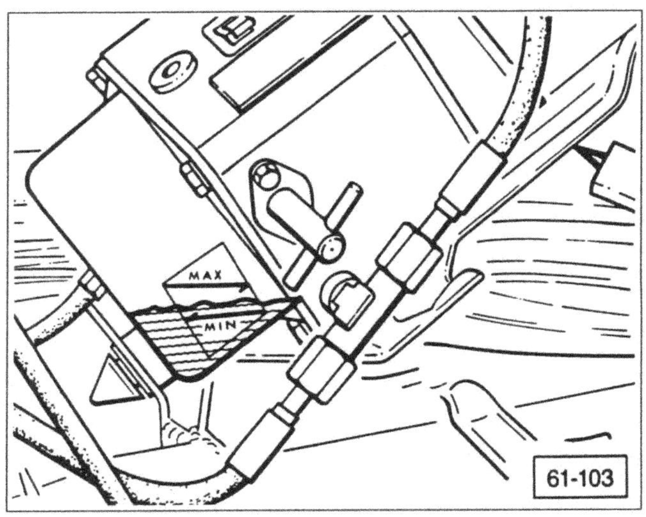

Fig. 12-4. Hydraulic oil level being checked at level indicator on pump housing. Check with top down (open).

To add hydraulic oil:

1. Use a hex wrench to open the oil filler plug, as shown earlier in Fig. 12-2.
2. Add oil through the filler hole as shown in Fig. 12-5 until the level indicated is between the **MIN** and **MAX** marks.

   Hydraulic oil specification:
   - Volkswagen part no. G 002 500 or equivalent

CAUTION —
- *Hydraulic oil level must be checked with the top lowered, in the open position. Oil level is slightly lower with the top closed.*
- *Part numbers listed in this manual are provided for reference only. Always check with an authorized Volkswagen dealer Parts department for the most accurate and up-to-date information.*

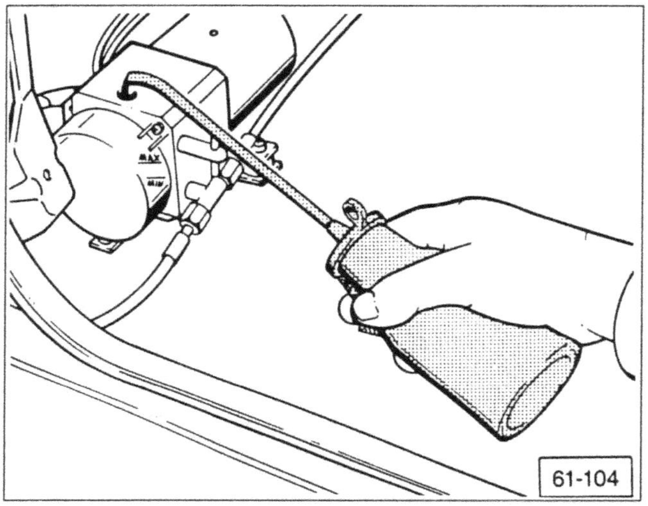

Fig. 12-5. Add hydraulic oil, Volkswagen part no. G 002 500 or equivalent, until level is between marks.

3. Reinstall the filler plug. Leave it slightly loose, just enough to allow air to slowly escape.
4. Insert the key into the ignition switch, but do not switch the ignition on.
5. Actuate the power top switch to raise and lower the top four or five times. End this step with the top lowered, in the open position.
6. Recheck the fluid level and correct it as necessary.
7. Tighten the oil filler plug.

Component Removal and Installation

There are three things to bear in mind any time the hydraulic system has to be opened. First, the system operates at very high pressure. It must be assumed that there is some residual pressure in the system at all times. Next, cleanliness is very important. Care must be taken to keep dirt or other contaminant from entering the system. Finally, it is necessary to check the hydraulic oil level and bleed the system of air any time it has been opened.

Before loosening any hydraulic fitting:

1. Disconnect the battery negative (–) terminal.
2. Slowly loosen the oil filler plug, just enough to allow air to slowly escape. There may be a hissing sound as air escapes.
3. Have a suitable container handy and be prepared to catch any oil that may escape as fittings are loosened.
4. Carefully note the location and routing of hoses before disconnecting any fittings.

CAUTION —
- *Clean all connection points before loosening or removing components of the hydraulic system.*
- *Dirt and dust should not be allowed to enter the system.*
- *Cover parts that have been removed. Use plastic or paper. Do not use rags.*

Section 13

ELECTRICAL SYSTEM

Contents

TABLES

Electrical System

Introduction

The electrical system is basically an efficient means for transmitting power from the engine to the electrical components of the car. It does this with the help of the alternator, which converts some of the engine's mechanical energy into electrical energy. The electrical energy is then carried by wires to the various electrical components such as motors, light bulbs, or electronic control units. The battery in the system supplies electrical power when the engine is not running, and also supplies power to start the engine.

The electrical system is based on negative (–) ground. In other words, the negative terminal of the battery is connected to the car body, and any electrical connection to the body is a connection to ground. This greatly reduces the amount of wire required in the system. Voltage from the battery to the various electrical components is carried by the wiring harness. Most components are then grounded through either direct mounting to the car body or a ground wire leading to the car body. It is very important to the trouble-free operation of the electrical system that the ground connections, including the negative battery cable and the body ground straps, remain sound and free of corrosion.

Most terminals are identified with a number stamped next to the pins on the components. On relays, the terminals are also identified by numbers on the relay. Throughout this section, sockets on the relay panel are identified by referring to the terminal which fits into them. The terminal numbers and the location of all the major electrical connections are given in the current flow diagrams near the end of this section.

All electrical circuits except those required for starting and operating the engine are protected by fuses. To prevent accidental shorts that might blow a fuse or damage wires and electrical components, the ground strap from the negative terminal of the battery should always be disconnected before working on the electrical system. If you lack the skills or the equipment needed for testing and repairing the electrical system, we suggest you leave this work to an authorized Volkswagen dealer or other qualified repair shop. We especially urge you to consult an authorized Volkswagen dealer before beginning repairs on a car that may be subject to Volkswagen's warranty coverage.

1. GENERAL DESCRIPTION

A brief description of the principal parts of the electrical system is presented here for familiarization with the system. The components are each discussed in greater detail in later parts of this section.

Voltage and Polarity

The cars covered by this manual have a 12-volt, direct current (DC), negative-ground electrical system. The voltage regulator maintains the voltage in the system at approximately the 12-volt rating of the battery, and all circuits are grounded by direct or indirect connection to the negative (–) terminal of the battery.

Battery

The six-cell, 12-volt lead-acid battery is located on the driver's side of the engine compartment. The battery capacity is determined by the amount of current needed to start the car, and by the amount of current consumed by the electrical system. Volkswagen batteries are rated by ampere/hours (Ah) and cold cranking amps (CCA) rating. The Ah rating is determined by the average amount of current the battery can deliver over time without dropping below a specified voltage. The CCA is determined by the battery's ability to deliver starting current at 0°F (–18°C).

Starter

The starter and its attached solenoid are located at the front of the engine and transaxle assembly on cars with manual transmissions, and at the rear of the engine and transaxle assembly on cars with automatic transmissions. To maximize the amount of current available to the starter, a load-reduction relay interrupts the voltage to many of the electrical accessories when the ignition key is in the START position.

Charging System

The charging system consists of a belt-driven alternator and a voltage regulator. The voltage regulator is an integral part of the alternator brush holder. An alternator manufactured by either Bosch or Motorola is installed on cars covered by this manual. The alternator may be one of three capacities—55 amperes, 65 amperes, or 90 amperes—depending on the type and number of electrical accessories fitted to the car.

Ignition System

The transistorized ignition system is covered in **IGNITION**. To avoid personal injury, as well as damage to the sensitive electronic components of the ignition system, please review the cautions and warnings in that section before troubleshooting or repairing any ignition-related part of the electrical system.

Wiring, Fuses, and Relays

Nearly all parts of the wiring harness connect to components of the electrical system with keyed, push-on connectors which lock into place. Notable exceptions are the heavy battery cables. The wiring is color-coded for circuit identification. Wire colors are indicated in the current flow diagrams near the end of the section.

With the exception of the battery charging system, all electrical power is routed from the ignition switch or the battery through the fuse/relay panel, located behind the instrument panel to the left of the steering column. Fuses prevent excessive current from damaging components and wiring. Fuses are color coded to indicate their different current capacities.

The relays are electromagnetic switches that operate on low current to switch a high-current circuit on and off. The relays are mounted on the fuse/relay panel. For information concerning relay and fuse locations, see **13. Fuse/Relay Panel** and **14. Current Flow Diagrams**.

Lights

The lighting system includes the parking lights, the side marker lights, the turn signals, the back-up lights, the interior lighting, and the headlights. The headlight dimmer switch is on the steering column. There is no relay in the headlight circuit. High/low beam switching is handled directly from the contacts in the dimmer control. Information on changing bulbs and lenses can be found in **BODY AND INTERIOR**.

Heating and Ventilation

The heating and ventilation system includes a three-speed fresh air blower controlled by a dashboard-mounted switch. Cars with air conditioning have a four-speed blower. The control levers for the air conditioning system switch power to the air conditioning compressor, the fresh air blower, and the radiator cooling fan. Removal and installation of the heating and ventilation components is covered in **BODY AND INTERIOR**.

Windshield Wipers and Washers

The blades of the two-speed windshield wiper system with intermittent operation come to rest in the "park" position automatically when the wiper switch is turned off. The wiper switch includes a windshield washer control. A motor-driven pump supplies the washers with fluid. Some Scirocco models have an additional rear wiper and washer.

Instruments

With the exception of the speedometer, which is gear-driven by a flexible cable from the transaxle, the dashboard instruments are electrical. All differences in instrumentation among models are covered in the current flow diagrams. Removal and installation of the instrument cluster are described in **BODY AND INTERIOR**.

2. MAINTENANCE

No routine lubrication of the alternator, the starter, or other motors is required. The following electrical system maintenance is described in **LUBRICATION AND MAINTENANCE**.

1. Checking the battery.

3. TROUBLESHOOTING

This heading describes general procedures for electrical system circuit and component troubleshooting, and provides specific troubleshooting information for the battery, the starter, and the charging system. The later heading **14. Current Flow Diagrams** contains information such as wire colors, terminal identification, and ground and connector locations which is invaluable to electrical troubleshooting.

3.1 How To Read Volkswagen Current Flow Diagrams

The circuit diagrams presented in this manual are organized to indicate current flow, from positive to negative, in the electrical system. This is different from other types of wiring diagrams that actually show the routing of the wiring harness.

As a general rule, current flows from positive (+) at the top of the diagram to negative (−) at the bottom (which represents ground). One exception to this is the fuse/relay panel's internal ground circuit, labeled with the terminal number 31. See **Terminal and Circuit Identification** for further information.

Fig. 3-1 is an example of a Volkswagen current flow diagram, showing the meanings of the various markings. This is a general example. It does not show all of the schematic symbols used. For a complete listing of the symbols, see **14. Current Flow Diagrams**.

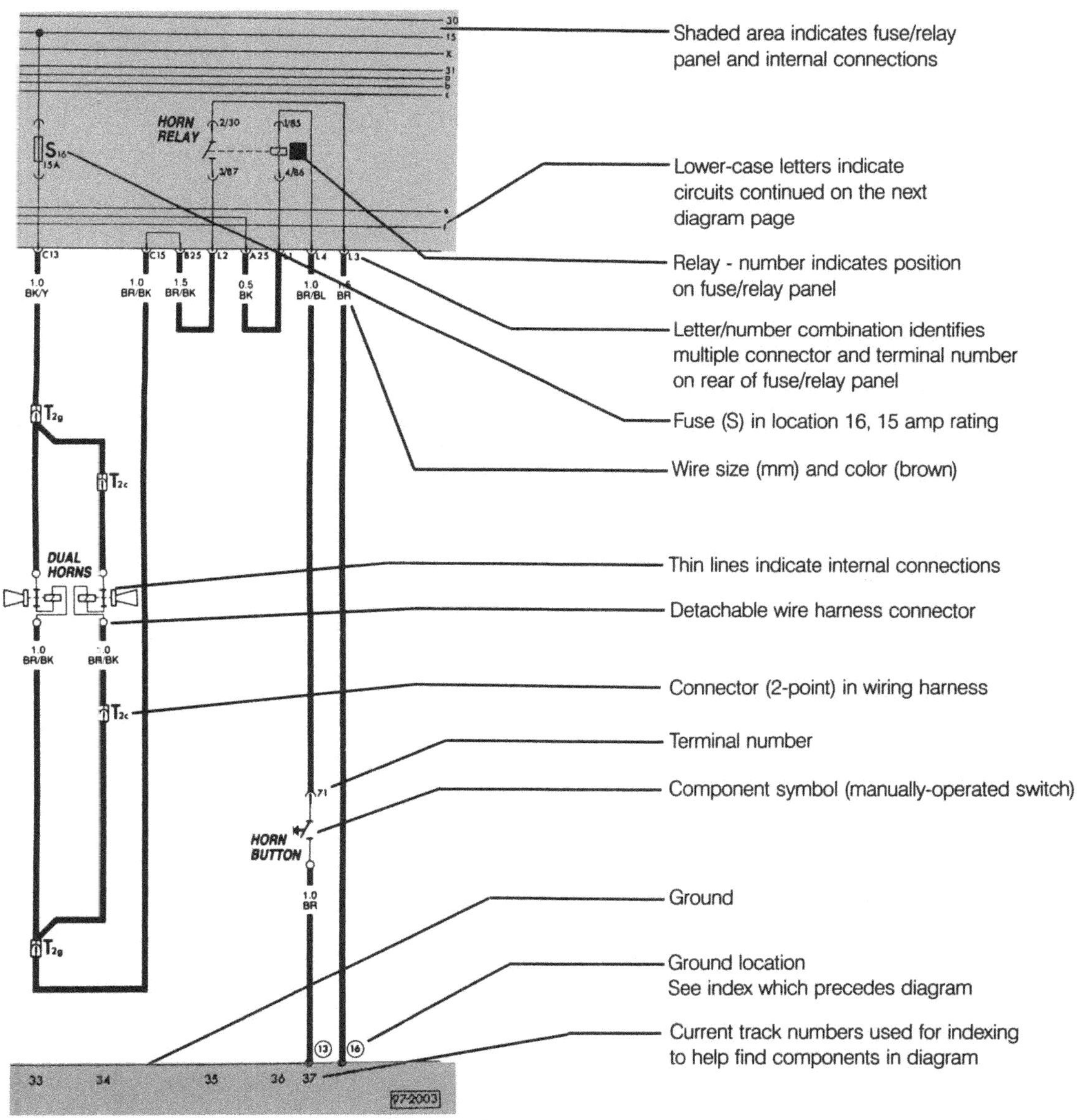

Fig. 3-1. Sample current flow diagram for Scirocco and Cabriolet models.

NOTE —
Current tracks are numbered for use with the diagram index. If a wire ends at a number inside a square box, that number indicates the current track where the wire continues.

NOTE —
A key to the wire color codes is found on each diagram. The colors shown in the diagrams are as accurate as possible. Some wire colors on a specific car may differ from the colors shown, however, due to running production changes.

NOTE —
Thin black lines are not always actual wires. When used to show current flow to ground, they may indicate a direct connection such as a switch threaded into the engine block.

Terminal and Circuit Identification

On the wiring diagrams at the end of this section, terminals are usually identified by numbers or letters. These terminal designations are often found on the components and harness connectors too. If in doubt refer to the wire color, which also can be found on the wiring diagrams.

Terminal numbers are often unique to a particular terminal, but some appear in many places throughout the electrical system to identify certain common circuits. A letter suffix may be added to the number to distinguish between two different circuits, or between two parts of the same circuit. **Table a** lists several of the most common circuit designations, the circuit type and the wire color normally used for that circuit.

3.2 Battery, Starter and Charging System Troubleshooting

The causes of a dead battery or an inoperative starter are often interrelated and hard to pinpoint. **Table b** lists symptoms of trouble with the battery, starter, and charging system, their probable causes, and suggested corrective actions. The bold numbers in the corrective action column refer to the numbered headings in this section.

3.3 Basic Electrical Troubleshooting Principles

Four things are required for current to flow in any electrical circuit: a voltage source, wires or connections to transport the voltage, a consumer or device that uses the electricity, and a connection to ground. Most problems can be found using only a relatively inexpensive multimeter (volt/ohm/amp meter) to check for voltage supply, for breaks in the wiring (infinite resistance or no continuity), or for a path to ground that completes the circuit. In addition, test leads with flat male connectors are needed to prevent damaging terminal connections. For information on using a multimeter or constructing test leads, see **FUNDAMENTALS** at the front of the manual.

Table a. Terminal and Circuit Numbers

Number	Circuit description	Most common wire color
1	Low voltage (−) terminal of coil	Green
4	High voltage center terminal of coil	Black
15	Originates at ignition switch. Supplies power when ignition switch is in ON or START position	Black
X	Load-reduction relay circuit. Originates at ignition switch. Supplies power only when ignition switch is in ON position. Circuit is interrupted when starter is actuated	Black/yellow
30	Battery positive (+) voltage. Supplies power whenever battery is connected. (Not dependent on ignition switch position)	Red
31	Ground; battery negative (−)	Brown
50	Supplies power from battery to starter solenoid when ignition switch is in START position	Red/black
B+	Originates at alternator and supplies voltage to battery	Red
D+	Alternator warning light and field energizing circuit	–
85	Ground side (−) of switching relay	Brown
86	Power-in side (+) of switching relay	–
87	Relay change-over contact	–

Table b. Battery, Starter and Charging System Troubleshooting

Symptom	Probable cause	Corrective action
1. Engine cranks slowly or not at all, solenoid clicks when starter is operated	**a.** Battery cables loose, dirty, or corroded	**a.** Clean or replace cables. See **LUBRICATION AND MAINTENANCE**
	b. Battery discharged	**b.** Charge battery; test and replace if necessary. **4.**
	c. Body ground strap loose, dirty, or corroded	**c.** Inspect ground strap; clean, tighten, and replace as necessary
	d. No voltage reaching starter motor	**d.** Test for voltage at starter. **6.** Test neutral/park/reverse light switch. **10.1**
	e. Starter motor or solenoid faulty	**e.** Test starter. **6.1**
2. Battery will not stay charged (more than a few days)	**a.** Shorted circuit draining battery	**a.** Test for current drain. **5.2**
	b. Alternator V-belt loose or damaged	**b.** Inspect alternator V-belt. **5.3**
	c. Battery faulty	**c.** Test battery and replace if necessary. **4.1**
	d. Battery cables loose, dirty, or corroded	**d.** Clean or replace cables. See **LUBRICATION AND MAINTENANCE**
	e. Alternator or voltage regulator faulty	**e.** Test alternator and regulator. **5.2**
3. Battery losing water	**a.** Battery being overcharged	**a.** Test voltage regulator. **5.2**
4. Lights dim, light intensity varies with engine speed	**a.** Alternator V-belt loose or damaged	**a.** Inspect alternator V-belt. **5.3**
	b. Alternator or voltage regulator faulty	**b.** Test alternator and voltage regulator. **5.2**
	c. Body ground straps loose, dirty, or corroded	**c.** Inspect ground straps; clean, tighten, and replace as required

Electric current is logical in its flow, always moving from the voltage source toward ground. Keeping this in mind, electrical faults can be located through a process of elimination. Fig. 3-2 is used to illustrate the logic used in locating a sample fault: back-up lights which do not light.

To troubleshoot a circuit:

1. Check to see that all the connections in the circuit are tight and free of corrosion. Especially make sure that all ground connections are clean, dry, tight, and corrosion-free.

2. Check the fuse. In the circuit example shown in Fig. 3-2, the circuit receives power through fuse 15. Fuse and relay locations are given in **13. Fuse/Relay Panel.** A fuse that repeatedly fails indicates an unusually high current flow, probably caused by a damaged wire, a faulty component, or a short directly to ground.

3. Check for voltage reaching the circuit. Connect a voltmeter or test light between a point in the circuit and ground (a clean, unpainted metal part of the car). In the example, the ignition must be switched on since fuse 15 only receives power when the ignition is on. See **3.1 How To Read Volkswagen Current Flow Diagrams.**

4. Pick a point where the diagram indicates that there should be voltage. In the example, the terminals "RF" at the connection to the back-up lights should be getting voltage when the ignition is on and the transmission is in reverse. Voltage indicated by the voltmeter (or by the test light lighting up) means that the circuit is doing its job at least as far as that point.

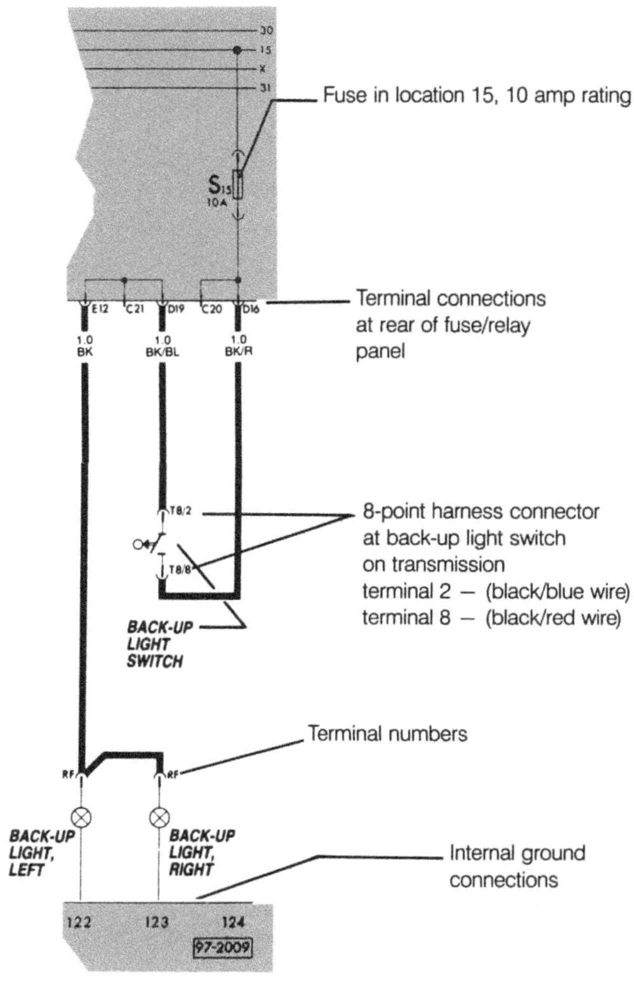

Fig. 3-2. Sample current flow diagram. See procedure **To troubleshoot a circuit** for example of troubleshooting a circuit.

5. If there is no indication of voltage, something has interrupted the circuit between the power source and the test point. Pick a new test point farther "upstream" in the current path. In the example, terminal T8/8 at the back-up light switch would be a good point. If there is voltage at this point, then the fault lies somewhere between the two test points.

6. Check circuit integrity between two points with a continuity test using an ohmmeter to measure resistance between the two points. Continuity (little or no resistance) between two points indicates that that part of the circuit is complete and would allow current flow. No continuity (infinite resistance) indicates that part of the circuit is interrupted and would not allow current flow. In the example, operation of the back-up light switch could be checked by disconnecting the harness connector from the switch and testing for continuity across terminals 2 and 8. There should be continuity only when in reverse gear—otherwise the switch is faulty and should be replaced.

4. BATTERY

The battery does not actually store electricity—it stores energy chemically. The battery contains a number of negative and positive lead plates, along with a liquid solution of sulfuric acid and water called the electrolyte. When an electrical load is placed across the battery, the sulfuric acid combines with the lead in the plates to produce lead sulfate and water. A great number of electrons are released as electrical current flow by this chemical process. Charging the battery reverses the reaction —lead sulfate is released from the plates and sulfuric acid is formed.

Charging the battery does not remove all of the sulfate from the plates. The battery will gradually become sulfated and will no longer hold a sufficient enough charge to start the car. Replacement of the battery is the only remedy. Temperature also affects the efficiency of the battery. The current-producing capacity of a battery at 5°F (−15°C) is only half its capacity at 68°F (20°C), and partly-discharged batteries can freeze due to the higher proportion of water in the electrolyte. A frozen battery produces no current, but can usually be restored when thawed.

Both conventional and maintenance-free batteries are used in cars today. Conventional batteries have filler caps and require that distilled water be added periodically to maintain the electrolyte level. Maintenance-free batteries have no filler caps, and require no maintenance of the electrolyte.

The low-maintenance battery installed by Volkswagen in new cars covered by this manual is a combination of these two types. Although similar in appearance to a conventional battery, the caps on a low-maintenance battery are generally used only for access during hydrometer testing. Under normal operation the battery is maintenance-free, and normal charging can be done without removing the caps.

For more information on checking the battery and on battery maintenance, see **LUBRICATION AND MAINTENANCE**. For information on starting with jumper cables, see **FUNDAMENTALS**.

WARNING —
Wear goggles, rubber gloves, and a rubber apron when working with battery electrolyte. Battery electrolyte contains sulfuric acid, and can cause severe burns and damage to clothing. If electrolyte is spilled on your skin or clothing, flush the area at once with large quantities of water. If electrolyte gets into your eyes, bathe them with large quantities of clean water for several minutes and call a physician.

WARNING —
Batteries generate explosive gasses. Keep sparks and open flame away. Do not smoke.

CAUTION —
Replace batteries with cracked or leaking cases. Leaking electrolyte can damage the car. If acid is spilled on the car, clean the area with a solution of baking soda and water.

4.1 Testing Battery

Battery testing determines the state of battery charge. On conventional or low-maintenance batteries the most common method of testing the battery is that of checking the specific gravity of the electrolyte using a hydrometer. On maintenance-free batteries, because of the sealed case, an open-circuit voltage test is used.

Hydrometer Testing

The hydrometer consists of a glass cylinder with a freely moving float inside. When electrolyte is drawn into the cylinder, the level to which the float sinks indicates the specific gravity of the electrolyte. The more dense the concentration of sulfuric acid in the electrolyte, the less the float will sink, resulting in a higher reading (and state of charge).

Electrolyte temperature affects the reading, so a thermometer should be used to determine electrolyte temperature before making a hydrometer test. Add .004 to the hydrometer reading for every 10°F (6°C) that the electrolyte is above 80°F (27°C); subtract .004 from the reading for every 10°F (6°C) that the electrolyte is below 80°F (27°C).

Before checking the specific gravity of a battery, load the battery with 15 amps for one minute. This can be done by turning on the headlights without the engine running if the battery is installed in the vehicle. **Table c** lists the percent of charge based on specific gravity values.

Table c. Specific Gravity of Electrolyte at 80°F (27°C)

Specific gravity	State of charge
1.265	Fully charged
1.225	75% charged
1.190	50% charged
1.155	25% charged
1.120	Fully discharged

The battery is in satisfactory condition if the average specific gravity of the six cells is at least 1.225. If the specific gravity is above this level, but the battery lacks power for starting, determine the battery's service condition with a load voltage test, as described below. If the average specific gravity of the six cells is below 1.225, recharge the battery. If, after recharging, the specific gravity varies by more than .050 between any two cells, replace the battery.

Open-circuit Voltage Test

This test is mainly for maintenance-free batteries, but it can also be performed on conventional and low-maintenance batteries if a hydrometer is not available. An open-circuit voltage test checks battery voltage by connecting an accurate digital voltmeter to the battery posts after disconnecting the battery ground cable. Before making an open-circuit voltage test on a battery, first load the battery with 15 amps for one minute. If the battery is installed in a car, this can be done by turning on the headlights without the engine running. Open-circuit voltage levels and their corresponding percentages of charge are given in **Table d**.

Table d. Open-circuit Voltage and Battery Charge

Open-circuit Voltage	State of Charge
12.6 v or more	Fully charged
12.4 v	75% charged
12.2 v	50% charged
12.0 v	25% charged
11.7 v or less	Fully discharged

The battery is in satisfactory condition if the open-circuit voltage is at least 12.4 volts. If the open-circuit voltage is at this level or above, but the battery lacks power for starting, have a load voltage test performed to determine the battery's service condition. If the open-circuit voltage is below 12.4 volts, recharge the battery. If the battery cannot be recharged to at least 75%, it should be replaced. New maintenance-free batteries should not be installed unless they are at least 75% charged.

Load Voltage Testing

A load voltage battery test should be performed by an authorized Volkswagen dealer or qualified repair facility if, after performing the tests described earlier, the battery appears to be in good condition but still can't provide enough current for starting.

CAUTION —
Do not give a maintenance-free battery a load voltage test unless it is at least 75% charged.

4.2 Battery Charging

Discharged batteries can be recharged using a battery charger, but a battery can never be charged to a voltage in excess of that which it is capable of producing electrochemically. During charging, the battery's voltage builds to a peak called the terminal voltage. If charging is continued beyond the terminal voltage, the water in the electrolyte begins to decompose into hydrogen and oxygen. This condition, called gassing, will evaporate the electrolyte to a level that can damage the battery.

Thoroughly read the instructions provided by the battery charger's manufacturer and always follow the instructions carefully. Do not use a charger if the instructions are not available. **Table e** lists charging rates and times that should be followed when charging batteries.

WARNING —
The gasses given off by the battery during charging are explosive. Do not smoke. Keep open flames away from the top of the battery, and prevent electrical sparks by turning off the battery charger before connecting or disconnecting it.

CAUTION —
Always allow a frozen battery to thaw before attempting to recharge it.

Table e. Battery Charging Specifications

Charging rate (conventional or low-maintenance batteries only)	Specific gravity	Approximate charging time
Fast charge (at 80% to 90% of battery's capacity, example: 43 to 48 amperes for a 54-ampere hour battery)	1.150 or less 1.150 to 1.175 1.175 to 1.200 1.200 to 1.225	1 hour 3/4 hour 1/2 hour 1/4 hour
Slow charge (at 10% of battery's capacity, example: 5.4 amperes for a 54-ampere hour battery)	Above 1.225	Slow charge only, to a specific gravity of 1.250 to 1.265

WARNING —
Maintenance-free or heavily sulfated batteries should only be charged at the slow rate. There is risk of explosion if charged at a high rate.

5. CHARGING SYSTEM

The charging system provides the current necessary to keep the battery charged. The system includes an alternator driven from the engine crankshaft by a V-belt to generate the charging current, and a voltage regulator to control the rate at which the battery is charged.

The charging capacity of the alternator installed on each car depends on the type and number of electrical accessories. The alternator rating can be found on its housing as shown in Fig. 5-1. The voltage regulator is mounted on the end of the alternator housing. Voltage regulators and alternators are available as replacement parts from an authorized Volkswagen dealer. Volkswagen remanufactured alternators are also available as an alternative.

Fig. 5-1. Location of rating stamp on alternator housing. Example shown is 65-ampere alternator.

To prevent damage to the alternator or regulator when making tests or repairs, make all connections with negative (−) to negative, and positive (+) to positive unless directed otherwise. Even momentary contact with a conductor of the wrong polarity can cause damage to the alternator's diodes. Make certain that the battery ground strap is connected to the battery's negative (−) terminal and that the cable to terminal 30 on the starter is connected to the battery's positive (+) terminal. Never operate the engine with the battery disconnected. Never operate the alternator with its output terminal (B+) disconnected and the other terminals connected. Never short, bridge, or ground any terminals of the charging system except as specifically described in **5.1 Charging System Troubleshooting**.

5.1 Charging System Troubleshooting

Charging system trouble is indicated by an illuminated alternator warning light on the instrument panel, or by an undercharged or overcharged battery.

The alternator generates electrical current by electrical induction. That is, a magnetic field is placed in motion to induce a current in a stationary coil. When the engine is running and the alternator is spinning, part of the current it produces is used to energize its electromagnetic field. When starting, some other current must be provided to initially energize the field and begin the current generating process. This current is provided by the battery.

The battery current to the alternator rotor is routed through the alternator warning light in the instrument cluster. As soon as the alternator's output equals the battery's voltage, the light goes out. Normally, the warning light should be off when the ignition is off and the engine is stopped. The light should only come on when the ignition is turned on (current to the rotor) and go out again when the engine is started and the battery is being charged. **Table f** describes symptoms of trouble indicated by the warning light, lists tests and probable causes for the problem, and suggests corrective actions. The bold numbers in the corrective action column refer to the numbered headings in this section where the repair procedures are described.

An undercharged battery is usually associated with starting trouble. Again, make sure that the battery is in good condition and capable of accepting a full charge before blaming the charging system. Causes of an under-charged battery are: the simultaneous use of many electrical accessories for long periods of time, leaving accessories or lights in operation with the engine stopped, frequent long periods of starter usage, and frequent short-trip driving. Broken or frayed charging system wiring or corroded connections at the D+ and B+ terminals of the alternator, as well as worn, corroded, or loose battery cable connections will also prevent adequate charging or increase charging time.

5.2 In-car Testing of Charging System

The tests described here will help determine the cause of charging system trouble. For the tests, the battery should be fully charged and the alternator V-belt should be tensioned correctly. If in doubt about either, test the battery as described in **4.1 Testing Battery**, and adjust the V-belt as described in **5.3 Removing and Installing Alternator and Voltage Regulator**. The charging system warning light should be operating correctly. All electrical connections should be clean and tight. Replace wires that are hard or cracked.

A general test of charging system output can be made with an accurate digital voltmeter. The most accurate testing is done using an alternator and regulator tester, such as the Sun VAT-40, that applies a high-current load to the alternator. Conclusive tests using this equipment can be performed inexpensively by an authorized Volkswagen dealer or other qualified shop.

CAUTION —

An alternator must never be run with the battery disconnected. This will severely damage the alternator, the voltage regulator, or both. The alternator wiring should be secure, with the plug held by its clip.

Table f. Warning Light Troubleshooting

Symptom	Test and Probable cause	Corrective action
1. Ignition off, engine not running, warning light glowing or on	**(TEST)** Disconnect plug from alternator. **a.** Light goes out: Faulty diodes **b.** Light does not go out: Short in wiring harness or wiring connector	**a.** Repair or replace alternator. **5.3** **b.** Repair or replace faulty wiring. **14**
2. Ignition on, engine not running, warning light off	**a.** Battery discharged **(TEST)** Remove and test warning light bulb. **b.** Bulb burned out **(TEST)**—bulb not burned out. Disconnect plug from alternator. Using a voltmeter, test between the plug terminal for the red wires and ground. **c.** No voltage: open circuit between plug and battery positive (+) pole **(TEST)**—bulb not burned out, voltage reaching connector plug. Disconnect the plug from the alternator. With the ignition on, ground the plug terminal for the blue wire to the engine **d.** Light does not come on: Faulty bulb socket, open circuit in blue wire, or open circuit between socket and terminal 15 on the ignition switch **e.** Light comes on: Loose connection between regulator and alternator or loose connection between brushes and regulator **f.** Light comes on, regulator properly connected: Internal alternator faults or faulty regulator	**a.** Charge battery. **4.2** **b.** Replace bulb. **7.1** **c.** Repair wire or connections. **14** **d.** Replace faulty socket. Repair wires or connections. **14** **e.** Correct loose connections. **5.3** **f.** Repair or replace alternator or voltage regulator. **5.3**
3. Engine running at any speed, warning light stays on	**a.** Loose or broken alternator V-belt **b.** Exciter diodes burned out **c.** Faulty regulator or faulty alternator windings **d.** High voltage drop between positive (+) battery terminal and alternator due to broken, loose, or corroded wires	**a.** Replace or adjust V-belt. See **LUBRICATION AND MAINTENANCE** **b.** Repair or replace alternator. **5.3** **c.** Test charging system and replace faulty components as needed. **5.2** **d.** Repair wires or connections. **14**

Testing Alternator and Regulator

Start the engine and run it at about 3000 rpm with all electrical accessories turned off. Set the voltmeter to the DCV scale and measure the voltage between the positive and negative terminals of the battery. Make sure that the tester is connected to clean areas of the terminals. A reading between 13.5 to 14.5 volts indicates a correctly operating system. A reading higher than 14.5 volts most likely indicates a faulty voltage regulator.

A reading below 13.5 volts means that the battery is not being adequately charged because either the regulator or the alternator is faulty. To determine which, keep the car running at about 3000 rpm and turn on all electrical accessories, then check voltage across the battery. A reading that is the same as the first most likely indicates a faulty regulator. A reading that is lower most likely indicates a faulty alternator.

Testing Current Drain

If tests show that the alternator and regulator are operating correctly, but the battery still continually runs down, there may be a short in the electrical system causing a continuous current drain.

To test for current drain, turn off the ignition switch, the radio, and all lights. Disconnect the clock, if so equipped, by removing its fuse. Turn on the switches for the heated rear window and, if applicable, the air conditioning; this will determine if the load reduction relay is faulty. Disconnect the negative (−) cable from the battery and connect a test light between the cable and the negative post. If the test light comes on, some electrical accessory or a short in the electrical system is draining current from the battery. Isolate the faulty circuit by removing and replacing the fuses one at a time. When the test light goes out, the circuit with the short is located. If no fault is found in this way, a fault may exist in the components without fuses, such as the alternator, the starter motor, the ignition system, and the instrument cluster. Disconnect the items one at a time until the test light goes out.

Low level current drain may not be detectable with a test light. If the light does not come on, but a current drain is still suspected, repeat the test using an ammeter set to the 0 to 200 mA range.

Noisy Alternator

Alternator noises are usually mechanical in origin, but a high, soft whistling sound may be produced by an alternator that is overcharging because of a faulty regulator diode. The same sound may be heard if there is a shorted diode that is placing abnormal electrical strain on the alternator. Some alternators make this sound when operating normally at maximum output.

Alternator mechanical noises are usually the result of misalignment between the V-belt and the pulley, a loose or broken pulley, worn bearings, or a bent rotor shaft. Check for bad bearings by removing the V-belt as described below and rotating the alternator pulley by hand. If the pulley or shaft seem to grind or grate, replace the alternator.

5.3 Removing and Installing Alternator and Voltage Regulator

The alternator is connected to battery current even when the ignition is turned off. To prevent shorts and electrical damage, always disconnect the battery ground strap before removing the alternator.

To remove:

1. Disconnect all wiring at the alternator. On some models there is a main connector plug held in place by a wire. On others, there is an insulator and two nuts that hold the wires to the alternator. If applicable, remove the screw that holds the suppression condenser to the regulator so that it is free from the alternator.

2. Loosen the upper and lower alternator mounting bolts. Loosen the bolt that holds the alternator belt adjusting bracket to the cylinder head or the air-conditioning compressor's mounting bracket.

3. Push the alternator as far as possible toward the engine, then take the V-belt off the alternator pulley.

4. Remove the lower alternator mounting bolt, then remove the upper alternator mounting bolt and remove the alternator from the engine.

Installation is the reverse of removal. Torque the nut on the D+ terminal of the alternator to 3 Nm (27 in. lb.), and torque the nut on the B+ terminal to 16 Nm (12 ft. lb.). On models with a toothed-rack adjusting mechanism, use a torque wrench and a 22 mm crowfoot wrench to tension the V-belt. Turn the larger part of the tensioning gear center bolt counterclockwise until the torque wrench reads 8 Nm (70 in. lb.) on used V-belts or 4 Nm (35 in. lb.) on new V-belts. Hold the wrench steady and tighten the center bolt to 35 Nm (26 ft. lb.). Torque the cylinder head bracket bolt to 20 Nm (15 ft. lb.) and the lower mounting bolt to 35 Nm (26 ft. lb.). See Fig. 5-2.

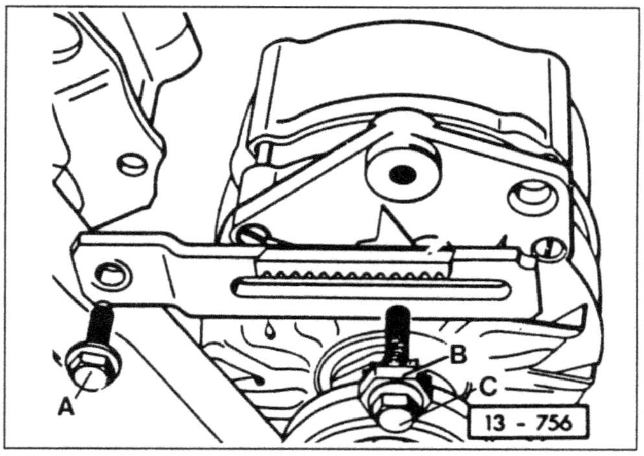

Fig. 5-2. Tensioning alternator V-belt on models with toothed rack adjusting mechanism. Make sure bolt **A** and bolt **C** are loose before adjusting tension. Place crowfoot wrench on outer part of bolt at **B**.

On models without a toothed rack adjusting mechanism, adjust V-belt tension by pushing the alternator away from the engine until the belt can be depressed as shown in Fig. 5-3. When the tension is correct, torque both adjusting bracket bolts to 25 Nm (18 ft. lb.). Torque the lower mounting bolt to 30 Nm (22 ft. lb.).

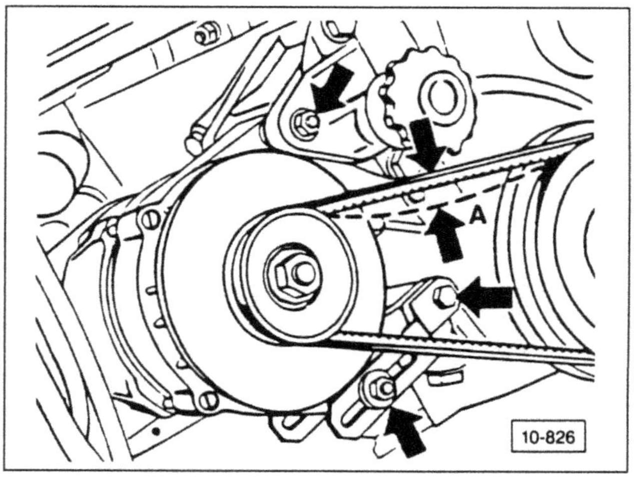

Fig. 5-3. Tensioning alternator V-belt on models without toothed rack adjusting mechanism. Mounting bolts (arrows) should be loose before adjusting tension. Dimension **A** should be between 2 mm and 5 mm (5/64 in. and 3/16 in.)

Brushes and Voltage Regulator

The brushes contact the alternator's slip rings to supply the current that magnetizes the alternator rotor. The voltage regulator maintains a nominal voltage in the electrical system by feeding excess output from the alternator back to ground. In addition, it regulates the amount of current supplied to the alternator rotor, turning it on and off as needed. If either component is faulty, the result will be an over or undercharged battery. The brushes wear under normal use, and will eventually need to be replaced.

The brushes and regulator are mounted on the rear of the alternator and can be replaced with the alternator installed. The regulator and brush holder is removed by removing two screws. Check the brush length as shown in Fig. 5-4. If the brushes are worn to less than 5 mm (7/32 in.), they should be replaced. On some alternator models the brushes are available as separate parts; on others they can only be replaced with the regulator. See an authorized Volkswagen dealer parts department for more information.

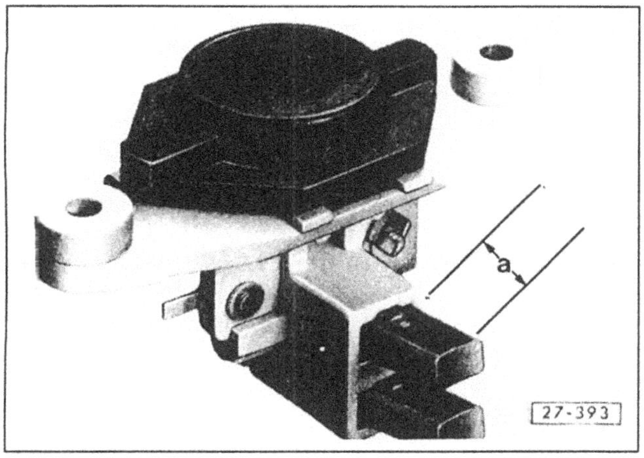

Fig. 5-4. Alternator brush length being checked. Length **a** of new brushes is 13 mm (17/32 in.). Replace brushes worn to 5 mm (7/32 in.) or less. Regulator and brush holder shown is used on Bosch alternators, Motorola is similar.

6. STARTING SYSTEM

When the ignition key is in the START position, a solenoid switch engages the starter's drive pinion with the ring gear on the engine flywheel or drive plate. To minimize wear and stress on the drive pinion and ring gear, the solenoid does not switch current to the starter until the drive pinion is fully engaged. Also, the drive pinion has an overrunning clutch to prevent the starter from being driven by the engine.

A number of different starters are used on the cars covered by this manual, based on the engine and transmission combination. Starters are not interchangeable among these different applications. When replacing the starter, make sure that the replacement part is correct for the application.

6.1 Starting System Troubleshooting

The battery and its cables should be in good condition when troubleshooting the starter. If in doubt about the battery, see **4. Battery**. If the outside temperature is at or below 32°F (0°C), check the viscosity of the engine oil, as described in **LUBRICATION AND MAINTENANCE**. The use of an oil of improper viscosity for cold conditions will increase engine friction and can lead to starting difficulties. Troubleshooting information for the starting system appears in **Table g**. The bold numbers in the corrective action column refer to headings in this section where the repair procedures are described.

6.2 Removing and Installing Starter

To remove the starter, first disconnect the ground strap from the negative (−) battery terminal. On cars with automatic transmission, work beneath the car to remove the two bolts that hold the cover plate to the starter, and remove the plate.

Disconnect the wire to terminal 50 and the positive (+) battery cable from the starter solenoid. Remove the bolts that hold the starter to the transaxle bellhousing and remove the starter. On cars with automatic transmission, it is also necessary to remove the bolt holding the starter mounting bracket to the engine.

Inspect the starter. Replace or repair the starter if the drive pinion teeth are worn or broken. On cars with manual transmission, inspect the starter bushing that is pressed into the transaxle's bellhousing. If the bushing is worn or damaged, it should be replaced as described below.

Installation is the reverse of removal. Lubricate the starter bushing with multipurpose grease. On cars with manual transmission, torque the starter mounting bolts to 46 Nm (34 ft. lb.). On cars with automatic transmission, torque the mounting flange bolts and the starter mounting bracket bolt to 23 Nm (17 ft. lb.). Install the starter wires as shown in Fig. 6-1.

CAUTION —

Connect wires to the proper terminals. Incorrect installation may damage the electrical system.

Removing and Installing Solenoid Switch

A faulty solenoid switch can be replaced separately. To remove the solenoid switch, first remove the starter as described above. Remove the nut from the field-winding connecting strap shown in Fig. 6-1, and remove the strap from the solenoid. Remove the two screws holding the solenoid to the starter, lift the solenoid plunger's pull rod up and off the engaging fork, and remove the solenoid. Installation is the reverse of removal. Apply a sealer to the screw heads after installing the solenoid.

Table g. Starting System Troubleshooting

Symptom	Probable cause	Corrective action
1. Starter does not operate when ignition switch is turned to "Start"	**a.** Ignition switch or wire leading from ignition switch to solenoid faulty (less than 8 volts to solenoid switch)	**a.** Test for voltage at terminal 50 of solenoid switch with ignition switch at "Start." If not at least 8 volts, test for voltage at terminal 50 of ignition switch with switch at "Start". Replace ignition switch (see **SUSPENSION AND STEERING**) or eliminate open circuit between ignition switch and solenoid switch. **14.**
	b. Solenoid switch faulty (less than 8 volts to starter motor)	**b.** Test for voltage at field-winding connecting strap with ignition at "Start." (See Fig. 6-1 below). If not at least 8 volts, replace solenoid. **6.2**
	c. Starter motor faulty.	**c.** Test for voltage at field-winding connecting strap with ignition at "Start." If at least 8 volts, repair or replace starter motor. **6.2**
	d. Neutral/park/reverse switch faulty (auto. trans. only)	**d.** Test switch. **10.1**
2. Starter turns slowly or fails to turn engine	**a.** Dirty, loose, or corroded starter connections	**a.** Remove, clean, and tighten connections. **6.2**
	b. Dirty, loose, or corroded ground strap between body and transmission	**b.** Remove and clean or replace strap. See **MANUAL TRANSMISSION AND CLUTCH** or **AUTOMATIC TRANSMISSION**
	c. Starter faulty	**c.** Repair or replace starter. **6.2**
3. Starter makes unusual noise, turns erratically, or fails to turn	**a.** Drive pinion defective	**a.** Repair or replace starter. **6.2**
	b. Flywheel or driveplate ring gear damaged	**b.** Replace flywheel or driveplate. See **MANUAL TRANSMISSION AND CLUTCH** or **AUTOMATIC TRANSMISSION**
4. Starter operates, but does not turn engine	**a.** Starter drive pinion or armature shaft faulty	**a.** Repair or replace starter. **6.2**
	b. Solenoid switch mechanism faulty	**b.** Replace starter solenoid switch. **6.2**

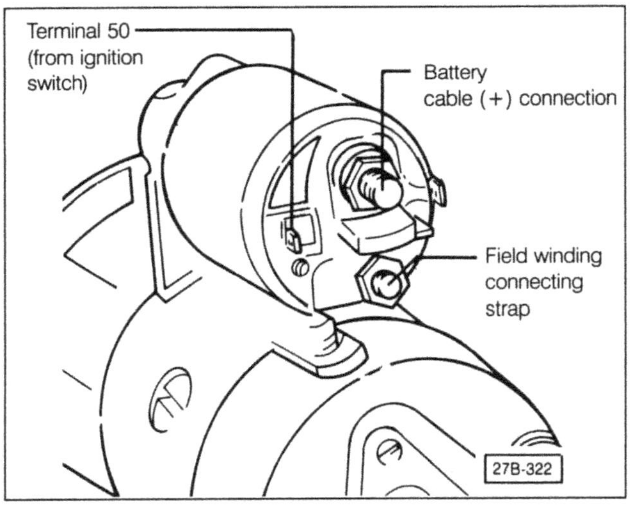

Fig. 6-1. Wiring connections to starter solenoid terminals.

Replacing Starter Bushing

(manual transmission only)

Jamming of the starter is usually a sign of a bad starter bushing. A faulty bushing that is not replaced with the starter will cause the starter bearings to quickly fail. The starter bushing is replaceable with the transmission installed and the starter removed. Fig. 6-2 shows the bushing being removed from the bellhousing. To install the bushing, drive the new bushing into the bellhousing until it is firmly seated.

Fig. 6-2. Starter bushing being removed from bellhousing using puller Volkswagen special tool no. VW 228b (order no. TV0 228 B00 027 ZEL).

7. INSTRUMENTS

Fig. 7-1 is an exploded view of the instrument cluster. A flexible printed circuit is used in place of the numerous wires and connectors that would otherwise be required to connect the instruments and switches. Power and electrical signals to the instruments are supplied to the wire terminal housings from two main wire harnesses.

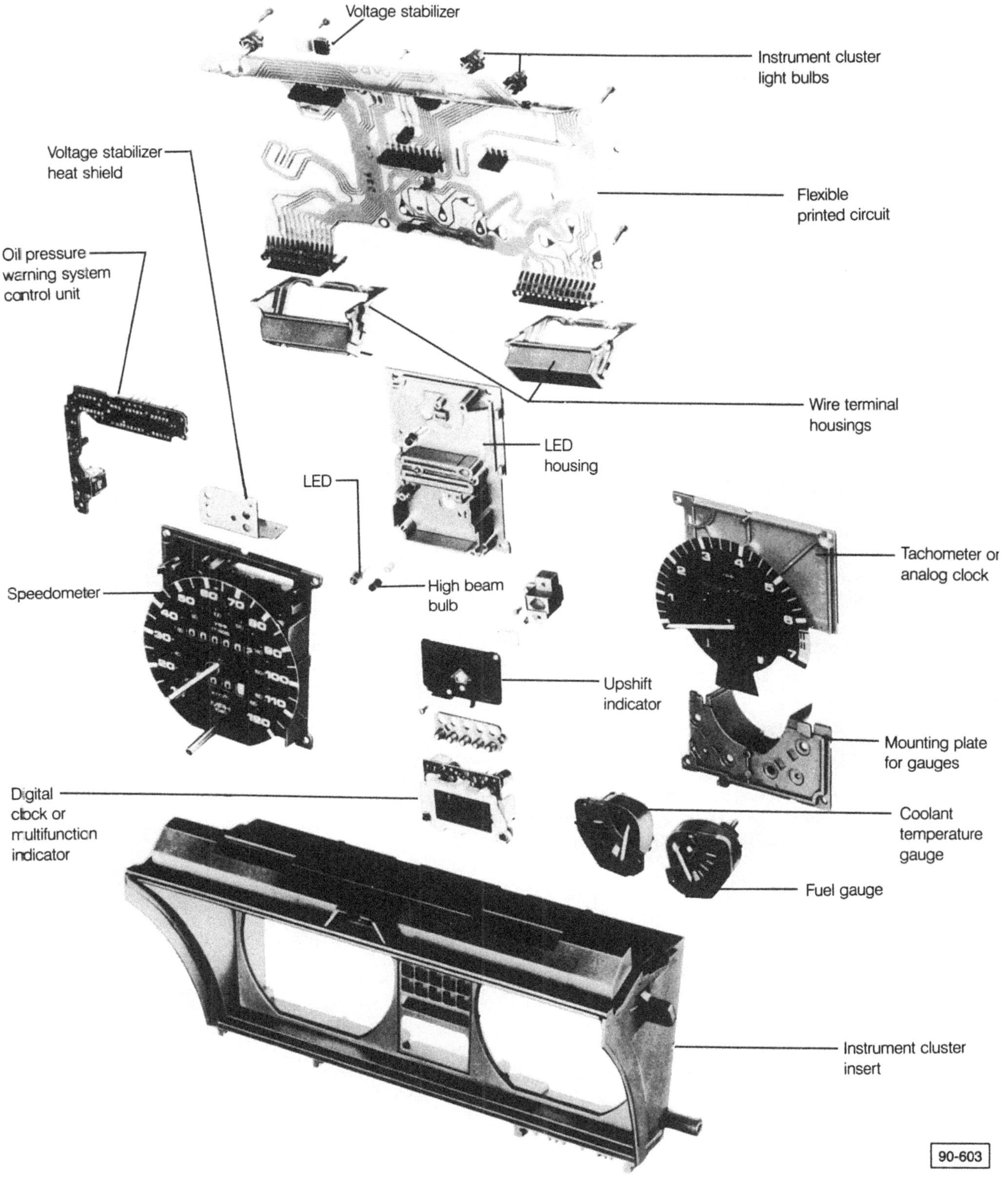

Fig. 7-1. Exploded front view of instrument cluster.

In addition to the instruments, the instrument cluster also houses an electronic control unit for the oil pressure warning system and, on certain models, an electronic control unit for the multifunction indicator. These control units are not repairable and should be replaced if faulty.

Troubleshooting or repair of the instrument cluster is easiest with the cluster removed. Removal of the instrument cluster, as well as replacement of the speedometer drive gear and cable, are described in **BODY AND INTERIOR**.

7.1 Troubleshooting Instrument Cluster and Gauges

Begin diagnosing instrument cluster electrical problems by checking the fuses. Make a thorough check in the engine compartment and under the instrument panel for disconnected or damaged wires or connectors. If the fuses, wires, and connectors are intact, then problems are caused by the correct signal not reaching the instrument cluster, or by faulty instrument cluster components.

For an additional aid to troubleshooting the instrument cluster, refer to the current flow diagrams at the end of this section for terminal and connector locations. For information on how to use the diagrams, see **3. Troubleshooting**.

If only one function is affected, the problem is most likely one which affects only that individual circuit. If more than one function of the instrument cluster is affected, it is logical to begin with the instrument cluster's common wiring. Check the main power supply from the ignition switch, the main ground wire, and the voltage stabilizer as described below.

Testing Instrument Cluster Voltage and Ground

If the warning lamps or gauges do not respond with the ignition on, then either battery voltage is not reaching the instrument cluster or the cluster is not properly grounded. With the battery negative (–) terminal disconnected, remove the instrument cluster as described in **BODY AND INTERIOR**. Make sure that the short black and long white 7-point connectors at the rear of the cluster that connect to the wire terminal housings are also disconnected. Reconnect the battery negative cable.

CAUTION —
Do not allow any unprotected part of the instrument cluster electrical circuitry to touch any metal part of the car while the cluster is removed and the battery is connected.

Use the current flow diagrams at the end of the section to identify the wire and terminal of the short black connector which supplies power from the ignition switch. With the ignition switched on, use an ohmmeter or test light to check for voltage between that connector terminal and ground. If there is no voltage, check for voltage at terminal 15 of the ignition switch. If voltage is reaching the switch but the switch will not complete the circuit, then the ignition switch is faulty and should be replaced as described in **SUSPENSION AND STEERING**.

To test for ground, use the ohmmeter or test light to check for continuity between the car body and the ground terminal of the long white connector that goes to ground. (See the appropriate current flow diagram). If there is no continuity, check the wiring harness for faults between the connector and ground.

Testing and Replacing Voltage Stabilizer

Test the voltage stabilizer with the instrument cluster removed, but with the harness connectors connected. Connect a voltmeter to the ground terminal and the stabilized voltage output terminal on the voltage stabilizer. See Fig. 7-2. Connect the ground cable to the battery and turn the ignition on. The meter should indicate approximately 10 volts. If it is above 10.5 volts or below 9.5 volts, the voltage stabilizer is faulty and should be replaced. Remove the voltage stabilizer by removing its retaining screw and sliding it straight off of its terminals.

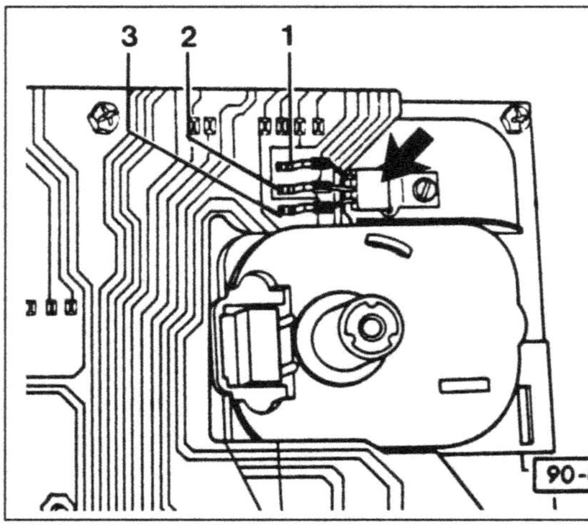

Fig. 7-2. Voltage stabilizer (arrow). Terminal **1** is stabilized voltage output, terminal **2** is ground, terminal **3** is voltage input. Test by connecting voltmeter between terminals **1** and **2**.

Testing and Replacing Light Emitting Diodes (LEDs) and Indicator Bulbs

Test LEDs and indicator bulbs with the instrument cluster removed and the harness connectors disconnected. Use an external 12-volt power source and jumper wires to apply current to the appropriate terminal in the wire terminal housing. See Fig. 7-3 and **14. Current Flow Diagrams** for terminal identification. If the LEDs or bulbs do not light, check the printed circuit for continuity from the wire terminal housing to the LED or bulb. No continuity indicates that the printed circuit is faulty and should be replaced. If there is continuity, replace the LED or bulb.

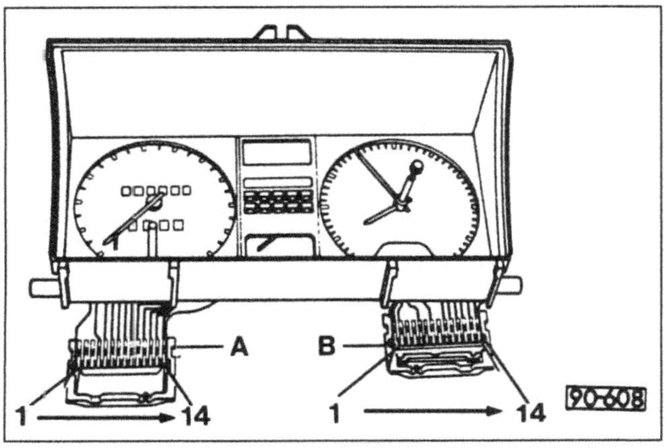

Fig. 7-3. Wire terminal housings **A** and **B** and pins used when testing LEDs or indicator bulbs.

To replace LEDs or indicator bulbs, the instrument cluster must be partially disassembled. Remove the voltage stabilizer as described above, and remove the flexible printed circuit and the LED/bulb holder. Pull the LED or bulb from its holder as shown in Fig. 7-4. When installing new LEDs, take care not to reverse their polarity. The negative (−) terminal of the LED is slightly wider than the positive (+) terminal.

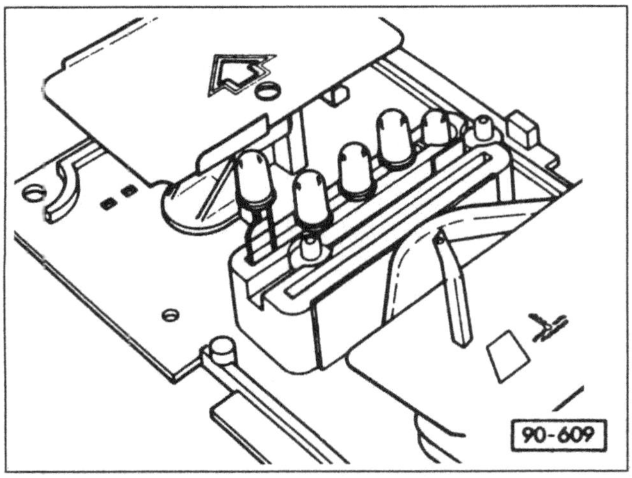

Fig. 7-4. LED being removed from LED holder.

Testing Upshift Indicator

The upshift indicator system on 1985 through 1987 cars with manual transmission indicates the optimum upshift points for maximum fuel efficiency. Fig. 7-5 is a schematic diagram of the system. For more information on the electrical details of the system, see **14. Current Flow Diagrams**.

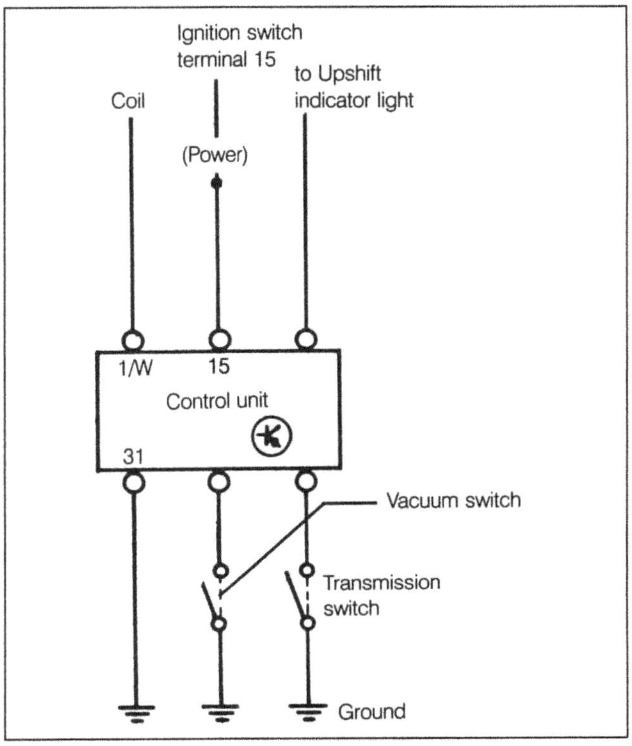

Fig. 7-5. Schematic diagram for upshift indicator light system.

The control unit receives an rpm signal from terminal 1 of the ignition coil. Engine load is relayed by the vacuum switch (located in the vacuum line to the ignition distributor). The upshift/reverse light switch (located on the transmission) disables the system when either high gear or reverse is selected. The control unit is located in the fuse/relay panel. The indicator light is an LED. **Table h** gives troubleshooting information.

Table h. Troubleshooting Upshift Indicator Light System

Symptom	Probable cause	Corrective action
1. Upshift light on continuously	**a.** Upshift/reverse light switch faulty	**a.** Perform continuity test of switch. There should be continuity only in high gear or reverse. Replace faulty switch
	b. Vacuum switch faulty	**b.** Perform continuity test of switch. There should be continuity only when vacuum is applied to switch
	c. No rpm signal from coil reaching control unit	**c.** Check wiring harness and repair as necessary

continued on next page

Table h. Troubleshooting Upshift Indicator Light System (continued)

Symptom	Probable cause	Corrective action
2. Upshift light does not come on	**a.** LED or instrument cluster connection faulty	**a.** Detach coil or alternator wire from control unit and ground it. If LED does not come on (ignition on), LED or LED connection is faulty
	b. Upshift/reverse light switch faulty	**b.** Detach wiring from switch and test switch's continuity. There should be continuity only in high gear or reverse. Replace faulty switch
	c. Vacuum switch faulty	**c.** Detach wiring from switch and test switch's continuity. There should only be continuity when vacuum is applied to switch
	d. Broken or shorted wire in upshift light harness	**d.** Test wires for shorts and open circuits
	e. Control unit faulty	**e.** Replace control unit if none of the above applies
3. Upshift light flickers	**a.** Vacuum switch faulty	**a.** Detach wiring from switch and test switch's continuity. There should be continuity only when vacuum is applied to switch.
	b. Loose connections or broken wires	**b.** Check wiring for secure connections and continuity while wires are being flexed
	c. Control unit faulty	**c.** Replace control unit if none of the above applies
4. Upshift light glows before coming on at shift point	**a.** Control unit faulty	**a.** Replace control unit (atop fuse/relay panel)

To remove and replace the upshift indicator, remove the printed circuit from the back of the instrument cluster and disassemble the cluster. Remove the screws which mount the upshift indicator front cover and the control unit as shown in Fig. 7-6.

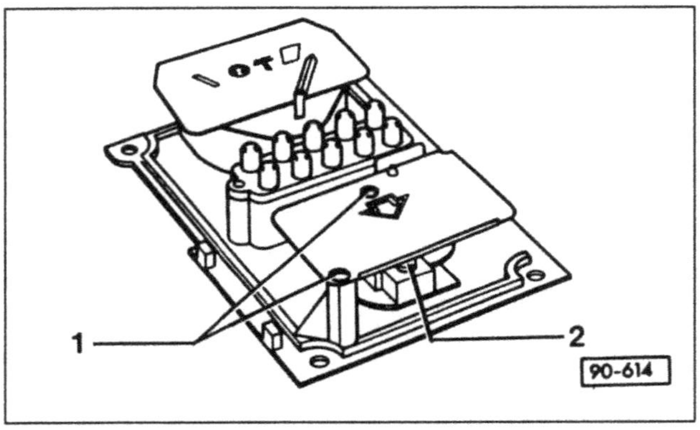

Fig. 7-6. Upshift indicator front cover screws (**1**) and upshift indicator control unit mounting screw (**2**).

Dynamic Oil Pressure Warning System Control Unit

Symptoms of a faulty dynamic oil pressure warning system are most likely caused by a problem with the warning switches, or by an actual oil pressure problem. Before assuming that the warning system control unit is faulty, check the oil pressure and the oil pressure switches as described in **ENGINE**. In addition, check the circuit wiring using the current flow diagrams at the end of this section.

NOTE —

When first turning on the ignition, it is possible for an impulse from the ignition system to briefly set off the low pressure warning buzzer. This condition is normal and does not indicate a fault.

To replace a faulty control unit, remove the instrument cluster as described in **BODY AND INTERIOR**. Remove the instrument illumination bulbs and the voltage stabilizer, and then remove the flexible printed circuit from the instrument cluster. See Fig. 7-1 above. Remove the speedometer and release the locking tabs to lift the control unit printed circuit away as shown in Fig. 7-7. On some models, an additional screw may hold the control unit to the speedometer. Install the new printed circuit, press in the locking tabs, and reassemble and install the instrument cluster.

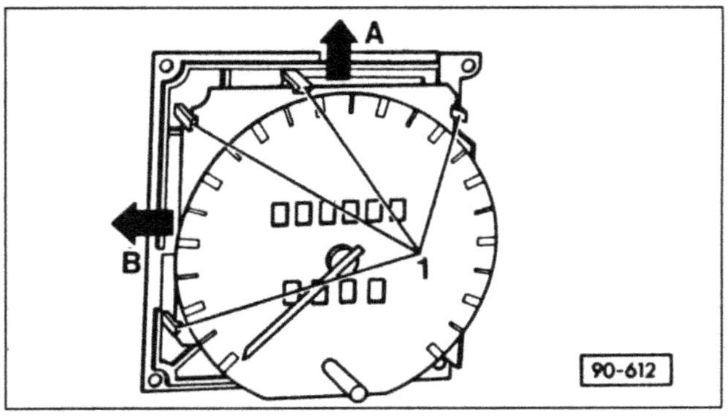

Fig. 7-7. Oil pressure warning system printed circuit being removed. Disengage tabs **1** and pull printed circuit in direction **A** until contacts project, then slide off in direction **B**.

8. WINDSHIELD WIPERS AND WASHERS

The operation of the two-speed windshield wipers and washers is controlled by a lever on the steering column. The wiper blades park automatically when the lever is turned off. The wiper motor is located beneath the drip tray in the engine cowl. A relay on the fuse/relay panel controls the intermittent operation of the wipers, and runs the wipers for several sweeps if the washer is actuated while the wiper switch is in the off position.

On Sciroccos equipped with a rear wiper, the motor is behind the hatchback trim panel. The rear window washer is operated by a separate motor and reservoir in the right side of the luggage compartment.

If the wiper or washer motor fails to operate when the steering column lever is moved, see **8.1 Windshield Wiper and Washer Troubleshooting**. If the wiper motor runs but the wiper arms do not move, see **8.2 Windshield Wiper Motor and Linkage**.

8.1 Windshield Wiper and Washer System Troubleshooting

For best results, these troubleshooting tests should be done in the sequence outlined below. Refer to the current flow diagrams for the most complete model-specific information on terminals and connectors.

Testing Windshield Wiper Motor

The windshield wiper motor has three voltage inputs. Terminal 53a is the main power input coming from the fuse panel. The other two inputs are from the wiper/washer switch and provide signals to operate the motor on either low or high speed.

Check the individual terminals of the harness connector for voltage at the various switch positions according to **Table i**. Also check the appropriate connector terminals for continuity to ground. If voltage is reaching the connector as described in the table and the motor still fails to run on one or more speeds, then the motor is faulty and should be replaced. See **Removing and Installing Windshield Wiper Motor**.

Table i. Windshield Wiper Motor Voltage Tests

Wiper/washer switch position	Test terminal	Correct value
Off	53 53a 53b	No voltage No voltage Battery voltage (approximately 12 volts)
Low speed	53 53a 53b	Battery voltage Battery voltage Battery voltage
High speed	53 53a 53b	No voltage Battery voltage Battery voltage

If voltage is not reaching the terminals as described above, check the continuity of the wires between the motor's harness connector and the connector for the wiper/washer switch on the steering column. If the wiring harness between the motor and the switch is sound, continue testing the wiper/washer switch as described in **Testing Wiper/Washer Switch**.

Testing Rear Wiper Motor

The rear wiper motor has two voltage inputs. One comes directly from the load reduction relay, and one is from the rear wiper relay. For electrical wiring information pertaining to a specific model, refer to the current flow diagrams at the end of this section.

Disconnect the harness connector from the motor and check for voltage. With the ignition on, there should be voltage coming from the load reduction relay. With the rear wiper switch activated, there should be voltage from the relay. Also check for continuity to ground at the third (ground) terminal. If voltage is reaching the connector as described and the motor still fails to run, then the motor is faulty and should be replaced. See **8.2 Windshield Wiper Motor and Linkage.**

If voltage is not reaching the connector as described, check the continuity of the wiring harness. Check for voltage input to the rear wiper relay by removing the relay from the panel and measuring voltage at the appropriate socket terminals with the ignition on. If there is voltage input to the relay but no voltage reaching the motor, then the relay is faulty and should be replaced.

Testing Wiper/Washer Switch

The switch receives power from the ignition switch and the intermittent relay, and routes it according to the position of the lever. To reach the switch, remove the steering-column cover by removing the Phillips-head screws. The switch terminals are identified in Fig. 8-1.

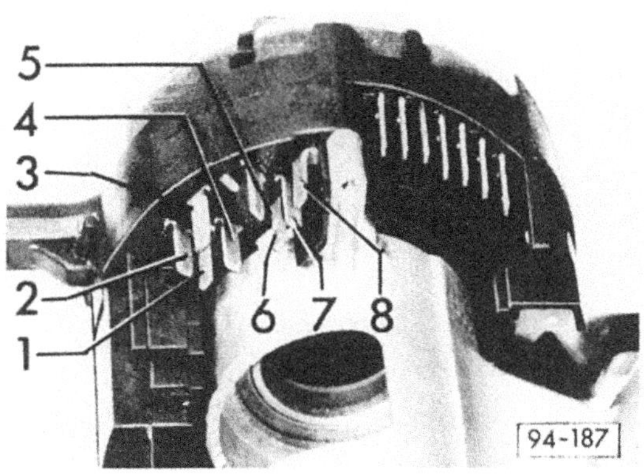

1. Terminal 53e
2. Terminal 53
3. Terminal 53a
4. Terminal J
5. Terminal 53b
6. Terminal L
7. Terminal T
8. Terminal 31

Fig. 8-1. Wiper and washer switch terminals used to test switch. Make tests of wiring harness using corresponding sockets in harness connector.

Use an ohmmeter to test the switch by making continuity checks as indicated in **Table j**. If the switch fails any of these tests, it is faulty and should be replaced as described in **SUSPENSION AND STEERING**.

Table j. Wiper/Washer Switch Continuity Tests

Wiper/washer switch position	Terminals	Test result
Off	53 and 53e	Continuity
Low Speed	53 and 53a	Continuity
High Speed	53a and 53b 53a and 53	Continuity Continuity
Intermittent	53a and J 53e and 53	Continuity Continuity
Washer	53a and T	Continuity
Rear Wiper	53a and L	Continuity

Testing Windshield Wiper Intermittent Relay

Only the low speed function of the wiper motor is wired through the intermittent relay. Before testing the intermittent relay, make sure that the wiper motor and the wiper/washer switch are functioning correctly.

Remove the intermittent relay from the fuse/relay panel. Check for voltage reaching the relay from the ignition switch with the ignition on, and check the ground socket for continuity to ground. By process of elimination, if the relay is getting voltage and has a good ground and no other wiring harness or switch faults can be found, then the relay is faulty and should be replaced.

8.2 Windshield Wiper Motor and Linkage

The front and rear wiper motors and their linkages are readily accessible for replacement. Check with an authorized Volkswagen dealer parts department for information on Volkswagen remanufactured wiper motors. For information on windshield wiper blades, see **LUBRICATION AND MAINTENANCE**.

Removing and Installing Windshield Wiper Motor

Fig. 8-2 and Fig. 8-3 are exploded views of the windshield wiper motors and linkages for Sciroccos and Cabriolets. The wiper motor frame does not need to be removed from the body when removing the motor or linkage. The drive crank should not be removed unless the motor is being replaced.

Disconnect the battery negative (−) terminal and carefully remove the plastic drip tray by unclipping it and pulling it straight up. Disconnect the harness connector from the wiper motor. Disengage the relay rods from the ball pin on the motor's drive crank, then remove the motor mounting bolts and remove the motor from beneath the frame.

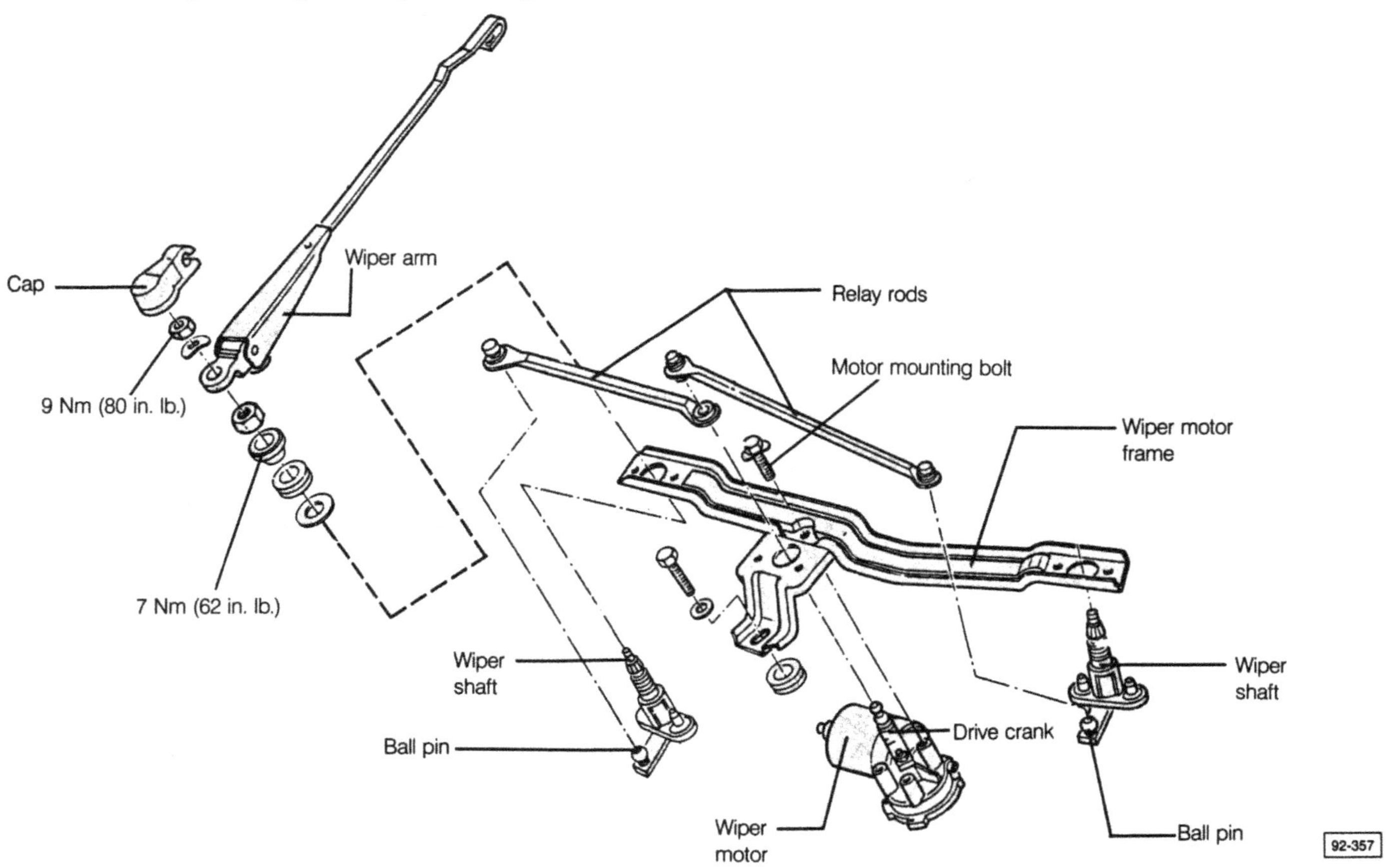

Fig. 8-2. Exploded view of Scirocco windshield wiper components.

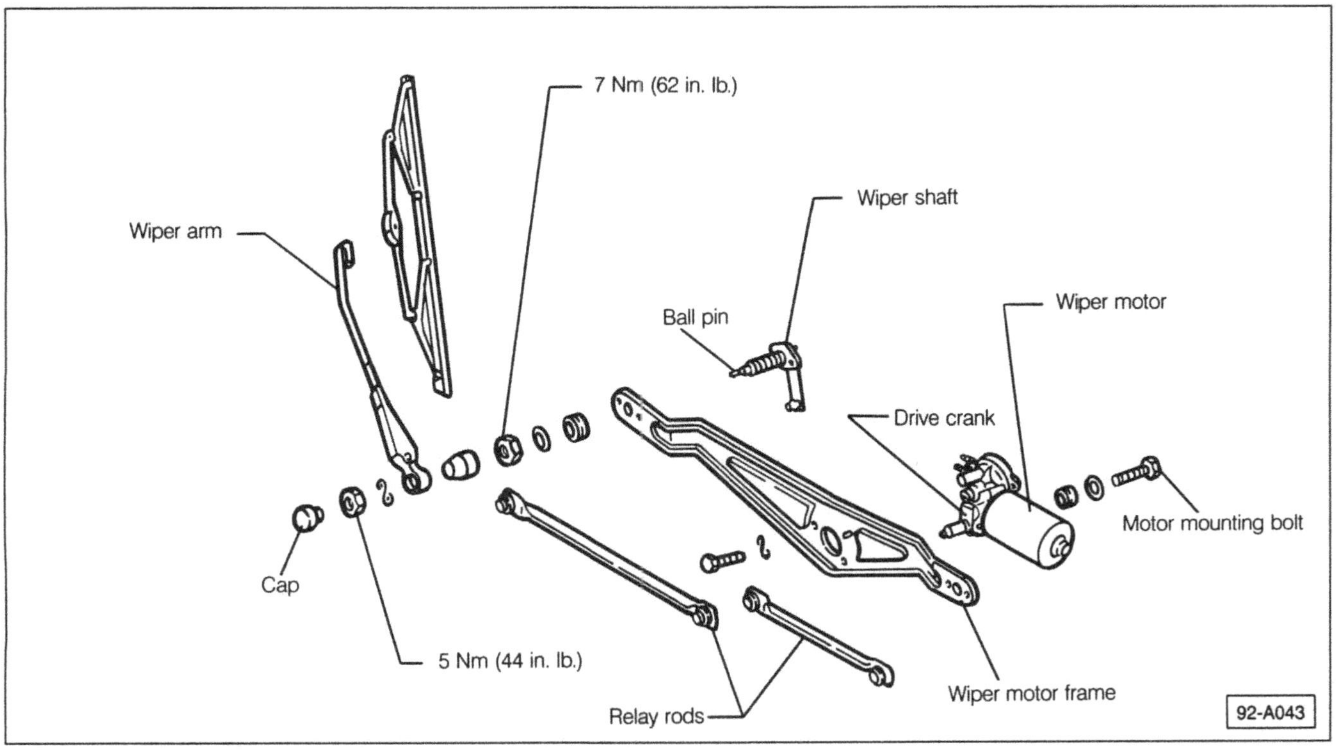

Fig. 8-3. Exploded view of Cabriolet windshield wiper components.

Installation is the reverse of removal. Completely install the motor and connect the battery, but do not connect the relay rods. Turn on the ignition and the wiper switch to run the motor for approximately one minute, then turn off the wiper switch. The motor will stop in its park position. If installing a new motor, transfer the drive crank from the old motor to the new one in the position shown in Fig. 8-4 or Fig. 8-5. Lubricate the ball pins with multipurpose grease and install the relay rods. Check wiper arm position as shown in Fig. 8-6.

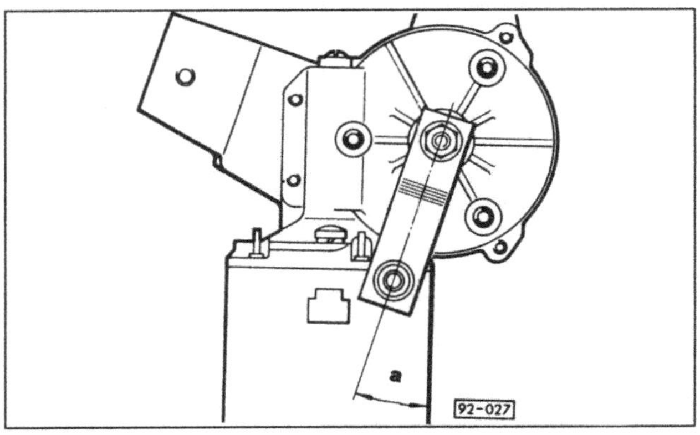

Fig. 8-5. Cabriolet windshield wiper motor drive crank correctly installed in motor (motor in park position). Angle **a** is 20°.

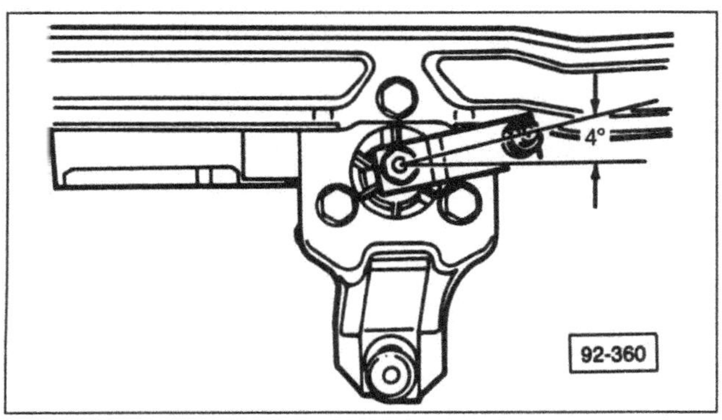

Fig. 8-4. Scirocco windshield wiper drive crank correctly installed in motor (motor in park position).

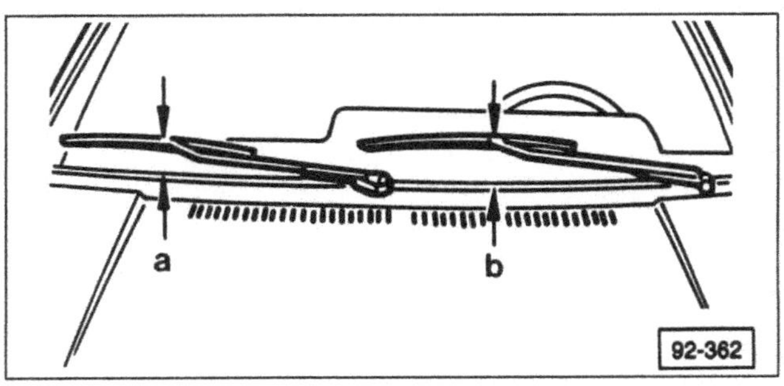

Fig. 8-6. Correct position of windshield wiper arms with relay rods installed and motor in park position. Dimension **a** is 65 mm (2 ½ in.), dimension **b** is 35 mm (1 ⅜ in.).

Removing and Installing Rear Wiper Motor

Fig. 8-7 is a view of the Scirocco rear hatch showing the rear wiper motor and linkage assembly. The rear wiper arm attaches to the wiper shaft in the same manner that the front wiper arms are attached.

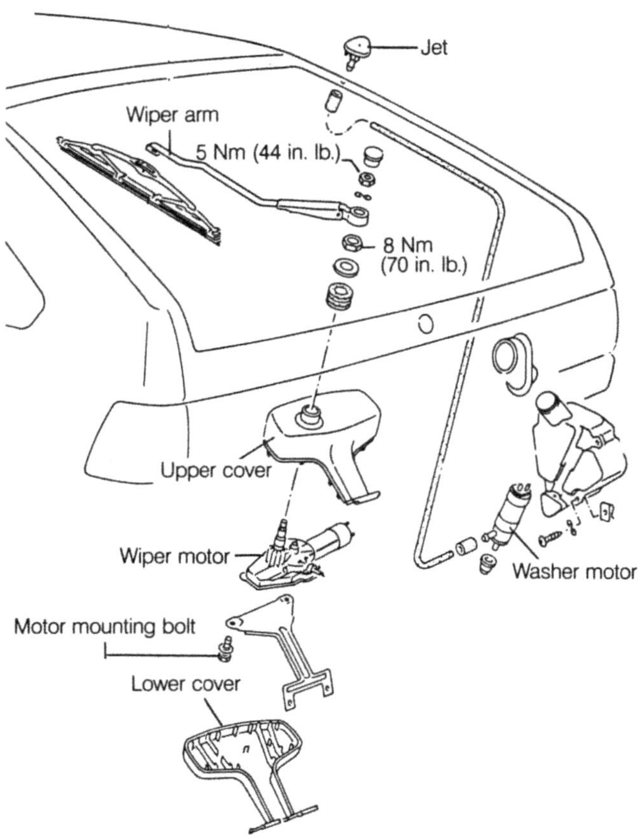

Fig. 8-7. Scirocco rear hatch showing rear wiper motor and linkage components.

To remove the motor, disconnect the battery negative (–) terminal. Remove the interior lower motor cover and disconnect the wiring connector from the motor. Remove the wiper arm mounting nut and the arm. Remove the motor mounting bolts and remove the motor.

Installation is the reverse of removal. Completely install the motor and connect the battery, but do not connect the wiper arm. Turn on the ignition and the wiper switch to run the motor for approximately one minute, then turn off the wiper switch. The motor will stop in its park position. Mount the wiper arm on the shaft as shown in Fig. 8-8.

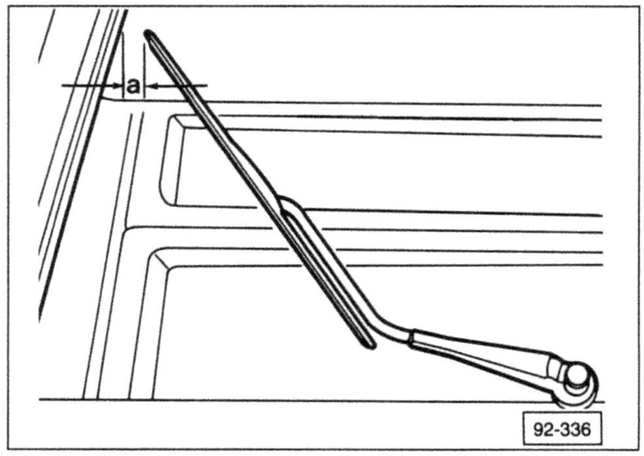

Fig. 8-8. Rear wiper arm installation and washer adjustment. Dimension **a** is 30 mm (1 3⁄16 in.).

9. HORNS

The components of the horn system are the dual horns, the horn relay, and the horn button on the steering wheel. The horns and the relay are connected to positive (+) voltage through a fuse whenever the ignition is on. Depressing the horn button activates the horn relay, which in turn completes the circuit to ground to sound the horns.

The horn button receives voltage through terminal 71 of the turn signal switch. See Fig. 9-1. The button is grounded through the same connector. The connector is accessible by removing the steering wheel as described in **SUSPENSION AND STEERING**. See **14. Current Flow Diagrams** for detailed information on terminal and wire identification.

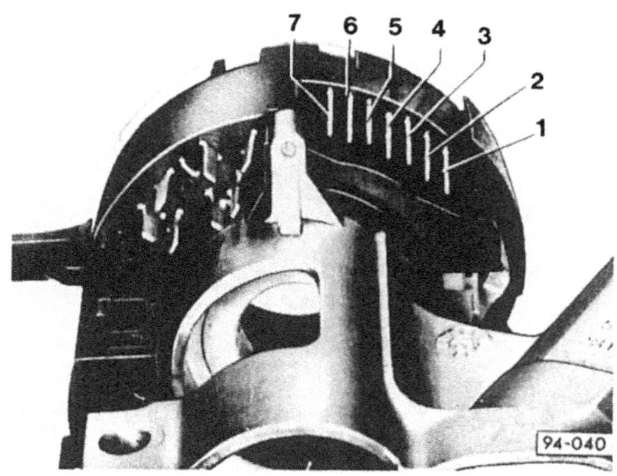

Fig. 9-1. Location of horn and turn signal switch terminals on steering column. Terminals are labeled. (Although terminal numbers shown do not apply to cars covered by this manual, actual terminals on switch are labeled for identification).

9.1 Horn Troubleshooting

Horns that continually blow indicate either a short in the horn circuit—most likely in the switch or steering column—or a faulty relay. If the horns do not sound, check the fuse, the relay, and all the wiring connections. Troubleshoot the circuit using the current flow diagram as a guide.

10. EXTERIOR LIGHTS

This heading covers exterior light electrical circuits. Replacement of bulbs and lenses for the taillights, side marker lights, turn signals, brake lights, back-up lights, and headlights is covered in **BODY AND INTERIOR**.

10.1 Troubleshooting Exterior Lights

Most of the exterior lighting involves pairs of lights, and this fact is an aid to troubleshooting. If only one of a pair of lights is out—one taillight for example—then the problem is most likely due to a failed bulb or some other problem with that particular light socket or its wiring. A simple test is to exchange the bulb with its counterpart which is known to be good. If the same bulb fails to light in a new location, then the bulb is faulty and should be replaced. If the same light fails to light with the other bulb, then the problem is in the socket or wiring. Many lighting problems are due to dirty or corroded sockets, or loosely-fitting bulb contacts or connectors. Check that voltage is reaching the bulb and that the socket has a good connection to ground.

If a pair of bulbs are both out—both taillights for example—then the problem is most likely in some part of the system which is common to both lights. Begin by checking for a failed fuse. Test switches using simple continuity checks made with an ohmmeter. Check the switch connectors for voltage and continuity to ground.

Headlights

Power to the headlight low beams comes only with the ignition on (terminal 15). Power to the high beams comes directly from the battery, and is routed through the high-beam portion of the turn signal switch.

Turn Signals and Emergency Flashers

Power for the turn signal circuit is routed through the emergency flashers switch only when the ignition is on (terminal 15). The emergency flashers circuit receives power from the battery at all times. When the emergency flashers switch is on, the turn signal switch is bypassed.

The turn signals and the emergency flashers share the same flasher relay, but each circuit has its own fuse. One system working indicates that the flasher relay is good. Look for a failed fuse.

If both the flashers and turn signals are inoperative, then it is probably either the flasher relay or the emergency flasher switch that is faulty. Check to see if power is reaching the relay. Check the flasher switch by checking continuity in the ON and OFF positions, using the current flow diagram as a guide. For failed turn signals, test the turn signal switch as described in **10.2 Steering Column Switches**.

Parking Lights, Taillights, Side Marker Lights, and License Plate Lights

The parking, side marker, tail, and license plate lights all receive power directly from the battery through the light switch in the first or second position, independent of the ignition switch. If these lights fail together, first look for a failed fuse, then for problems in the switch.

Brake Lights

The brake lights receive their power directly from the battery through a fuse and through the brake light switch, independent of the ignition switch. The brake light switch closes when the brake pedal is depressed.

Back-up Lights

On cars with manual transmission, the reverse lights receive power through the back-up/upshift light switch when the transmission is in reverse gear. The switch is mounted on the transmission as shown in Fig. 10-1 and is accessible from the engine compartment.

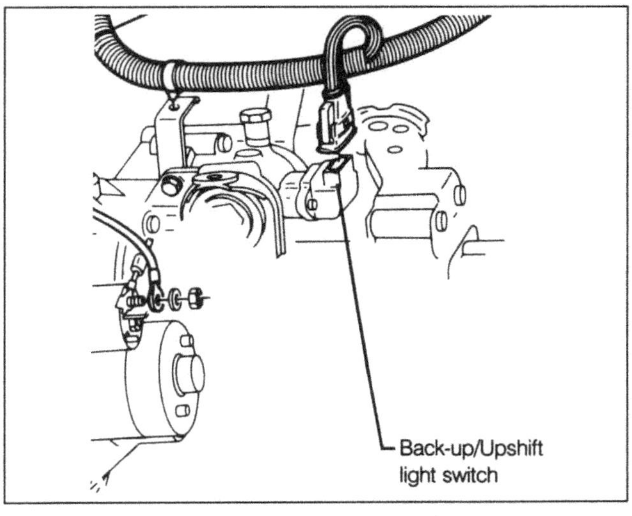

Fig. 10-1. Location of back-up/upshift light switch on manual transaxle housing. Torque mounting bolts to 10 Nm (87 in. lb.).

On cars with automatic transmission, the back-up lights receive their power through the neutral/park/back-up light switch mounted at the base of the shift lever in the passenger compartment. The switch prevents the car from starting if the shift lever is not in the neutral or park position.

10.2 Steering Column Switches

The steering column switches can be tested without being removed simply by removing the upper and lower steering column covers and disconnecting the harness connectors. For testing the windshield wiper switch, see **8. Windshield Wipers and Washers**. Replace faulty switches as described in **SUSPENSION AND STEERING**.

Ignition Switch

Fig. 10-2 identifies the terminals of the ignition switch. Terminal 30 brings power in to the ignition switch from the battery. Terminal 15 provides power to the ignition system and other parts of the electrical system when the ignition key is in the ON or START position. Terminal X supplies power through the load reduction relay. It has power when the ignition switch is in the ON position, but has no power when the switch is moved to the START position. Terminal Su is for the seat belt warning/interlock system. **Table k** lists ignition switch continuity checks. A switch which fails any of these tests is faulty and should be replaced.

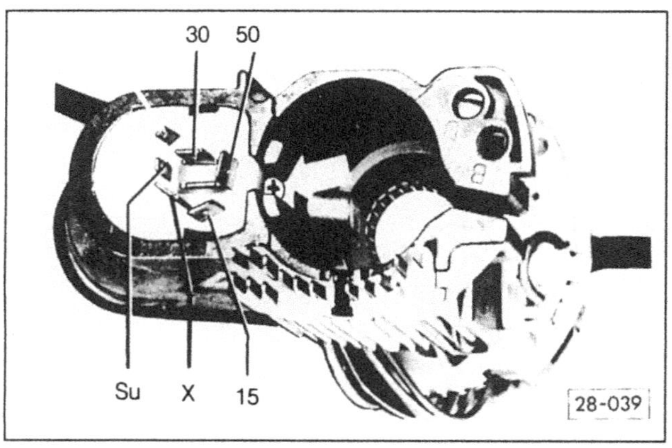

Fig. 10-2. Ignition switch and terminal identification. Switch is held in place by retaining screw (arrow).

Table k. Ignition Switch Continuity Tests

Switch position	Continuity between terminals
Ignition Off	30 and Su
Ignition On	30 and X 30 and 15
Start	30 and 50 30 and 15

11. HEATING, VENTILATION AND AIR CONDITIONING

Fresh Air Blower

The speeds of the fresh air blower motor are controlled by varying resistance. Maximum resistance produces the lowest speed. At high speed, power is supplied directly to the blower motor. On cars with air conditioning, power for high speed is supplied through the air conditioning relay.

Before replacing a suspected faulty blower motor, check for voltage at the blower motor harness connector with the ignition on. If there is no voltage reaching the motor, then the motor may not be at fault. Refer to the current flow diagrams for help in troubleshooting the circuit.

A blower motor that runs on high speed but fails to run on one or more of the lower speeds indicates a faulty series resistor. If voltage is reaching the blower motor connector, remove the blower motor as described in **BODY AND INTERIOR** and check the series resistor. The series resistor is located under the drip tray near the blower motor, as shown in Fig. 11-1.

Fig. 11-1. Series resistor in blower motor circuit regulates blower speeds.

To test the series resistor, connect the test leads of an ohmmeter to measure the resistance values corresponding to the different fan speeds. For cars without air conditioning, the resistance between the yellow and white wires should be slightly higher than the resistance between the yellow and yellow/black wires. For cars with air conditioning, resistance should increase in steps when measuring red/black to white, red/black to red/white, and red/black to black/yellow. If the results are not as described, or if any one test shows an open circuit (infinite resistance), then the series resistor is faulty and should be replaced.

Air Conditioning

The radiator cooling fan and the air conditioning compressor clutch are operated by a relay. The relay is actuated by the air conditioning temperature switch when the lever on the heating control head is moved to one of the A/C positions. The system will not operate unless the blower motor is switched on. Test the blower motor as described above. The temperature switch can be tested and replaced after removing the control head as described in **BODY AND INTERIOR**. Any other testing of the air conditioning system, including the operation of the pressure switches, requires specialized knowledge and equipment, and is beyond the scope of this manual. Servicing by an authorized Volkswagen dealer or other qualified air conditioning shop is highly recommended.

Rear Window Defogger

The rear window defogger is a series of resistance elements that heat up as electrical current passes through them. One or two wires that do not heat up suggests a break in the heating element. Repair material for the element is available from an authorized Volkswagen dealer to repair the breaks as described below. An element that does not operate at all suggests that power is not reaching the heating element. Check the fuse, and check for voltage supply to the window element and for a good connection to ground. Refer to the current flow diagrams at the end of the section for details of the rear window defogger wiring.

To find and repair heating element breaks:

1. Wrap a small piece of aluminum foil around the negative (−) probe of a voltmeter.

2. With the heating element's wiring connected, touch the positive (+) voltmeter probe to the end of the heater element closest to the incoming voltage source.

3. With the ignition and the defogger switch on, place the foil-covered negative probe on the faulty wire near the positive side of the element and slowly slide it toward the negative side. The point at which the voltmeter deflects from zero volts to several volts is where the wire is broken.

4. Apply a strip of masking tape to either side of the break, leaving the break exposed. Apply the repair material over the break and allow it to dry for one hour at room temperature. Remove the tape and retest the wire.

12. POWER OPTIONS AND ACCESSORIES

The power windows, the heated seats, and the heated/power mirrors are all powered through the load reduction relay. If these systems fail together, one likely cause is a faulty relay. The power window circuit has an additional relay.

The central locking system receives its power directly from the battery through a fuse. The cruise control is powered only when the ignition is on.

When troubleshooting these systems, check first for blown fuses. Fuses are simple to check and are a likely cause of trouble. Current flow diagrams for the major Volkswagen power options and accessories, including the radio, are found in **14. Current Flow Diagrams**.

12.1 Heated Seats

The heat control switches are variable resistors (sometimes called rheostats). Current runs through the heating elements, and then to ground. The control units underneath the seats receive temperature information from a temperature sender in each seat. The heating elements, the switches, and the control units are not repairable. For information on replacing the heating elements, see **BODY AND INTERIOR**. Remove the switch by prying it out using a screwdriver.

12.2 Electric/Heated Outside Mirrors

The mirror motors are operated by a door-mounted switch. The motor controls the mirror's movement via a magnetic clutch. The mirror heating elements are wired together with the rear window defogger. For removal and replacement of the mirror assemblies or motors, see **BODY AND INTERIOR**. The door-mounted switch is pressed into place and can be carefully pried out for testing.

12.3 Central Locking System

There are two electrical components of the central locking system—the bi-pressure pump in the luggage compartment and the master activator switch in the driver's door. The master activator switch controls the electrical signal to the bi-pressure pump. If the pump does not run when the door key is turned, test the electrical portion of the system as described below. If the pump runs, but the system does not operate, test the bi-pressure system as described in **BODY AND INTERIOR**.

> **NOTE —**
> If the system has not been operated for an extended period of time, several turns of the key may be required to reactivate the system.

To test the system, remove the bi-pressure pump from its housing as described in **BODY AND INTERIOR**, and disconnect the pump's harness connector. Place the driver's door lock button in the up (open) position. There should be voltage at the pump's harness connector as shown in Fig. 12-1. Place the driver's door lock in the down (locked) position. There should be voltage at the pump's wiring connector as shown in Fig. 12-2. If these conditions are not met, test the switch and the circuit wiring. See **14. Current Flow Diagrams**. The inside door trim panel must be removed as described in **BODY AND INTERIOR** to reach the master activator connector.

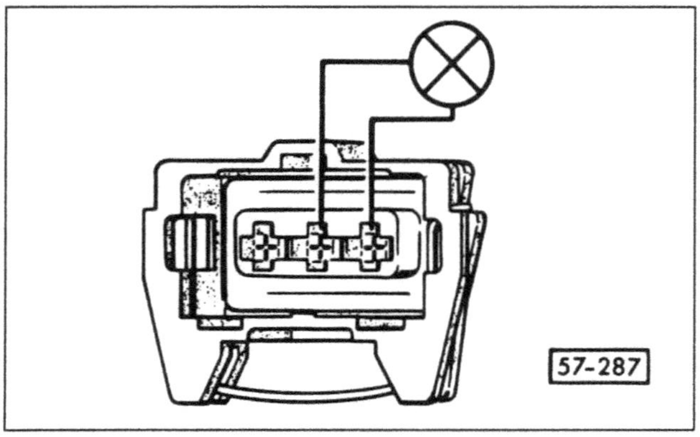

Fig. 12-1. Test light (shown schematically) connected between terminals of central locking system's bi-pressure pump harness connector for checking for voltage from master activator in driver's door (open position).

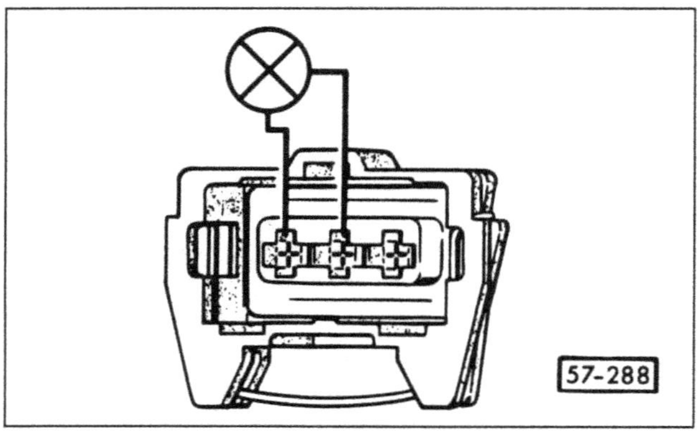

Fig. 12-2. Test light (shown schematically) connected between terminals of central locking system's bi-pressure pump harness connector for checking for voltage from master activator in driver's door (locked position).

12.4 Power Windows

The power windows are operated by polarity-reversing motors. Depending on the position of the rocker switches, current is routed to change the rotation direction of the motor.

A single power window that does not operate can be tested without disassembling the door. Remove the rocker switch and disconnect its harness connector. Turn the ignition on and check for voltage at the connector. Also check ground. If voltage is reaching the switch and there is a good ground, perform continuity checks on the switch.

> **NOTE —**
> A window that is stuck in the raised position may be due to a locked winder mechanism. For more information see **BODY AND INTERIOR**.

12.5 Cruise Control

The components of the cruise control system are shown in Fig. 12-3. The control unit, located under the instrument panel on the passenger side, compares the car's true speed to the speed selected by the driver. True speed is indicated by the induction pickup mounted on the back of the speedometer. The control unit then either turns on the vacuum pump to contract the vacuum servo to open the throttle, or opens a vent at the pump to bleed air into the vacuum lines to release the throttle and slow the car. A combination vacuum valve and electric switch vents air into the system and turns off the cruise control whenever either the brake or clutch pedal is depressed. The cruise control operates at speeds above 22 mph (35 km/h).

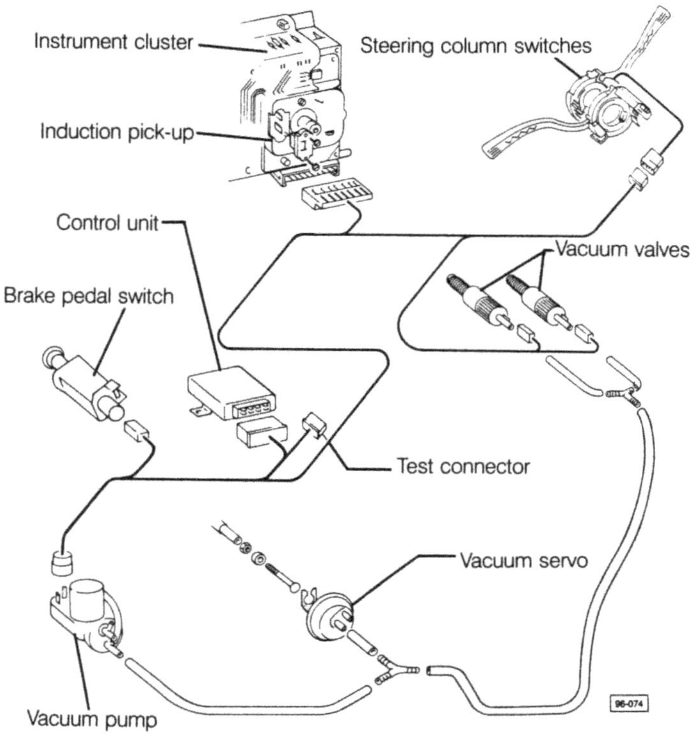

Fig. 12-3. Schematic view of cruise control system and controls.

Troubleshoot a faulty cruise control system using the tests outlined in **Table I**. Disconnect the control unit harness connector and, using a test light or volt/ohmmeter and a jumper wire, test at the appropriate terminals of the connector shown in Fig. 12-4. If the test results are all correct, and the vacuum servo is functioning correctly as described below, then the control unit is most likely faulty and should be replaced.

To leak-test the vacuum servo, remove the vacuum line, squeeze the servo until it is fully depressed, and then plug the vacuum line connection. If the servo does not hold its depressed position, it is faulty and should be replaced.

Replacement of components is fairly straightforward. In particular, remove vent valves using long nose pliers, after disconnecting the vacuum hose and wiring connector. Press new valves all the way in using a 10 mm socket. Used vent valves cannot be reinstalled, as their threads are damaged upon removal. When installing the vacuum servo, make sure that the linkage is adjusted as shown in Fig. 12-5, with the engine fully warmed up.

Table I. Cruise Control Troubleshooting

Component or Function	Test conditions	Correct results	Probable cause if incorrect
Control unit ground	Test light connected between terminal 8 and battery positive (+) terminal	Test light lights	Break in wiring to ground
On/Off switch	Ignition and cruise control on Test light connected between terminal 1 and ground	Test light lights	Break in wiring or faulty switch
Vacuum valves	Ignition and cruise control on Test light connected between terminal 3 and terminal 1 Press brake and clutch pedals in turn	Test light goes out when each pedal pressed	Break in wiring or faulty switch
RESUME function	Ignition and cruise control on Test light connected between terminal 6 and ground Press RESUME	Test light lights	Break in wiring or faulty switch
SET function	Ignition and cruise control on Test light connected between terminal 2 and ground Press SET	Test light lights	Break in wiring or faulty switch
Vacuum pump	Ignition and cruise control on Jumper wire connected to terminal 4 Briefly touch wire to ground	Pump runs	Break in wiring or faulty pump
Pump vent	Ignition and cruise control on Jumper wire connected to terminal 7 Briefly touch wire to ground	Vent clicks	Break in wiring or pump faulty
Induction pick-up	Ignition off Ohmmeter connected between terminal 5 and ground	90-110 ohms	Break in wiring or pick-up faulty

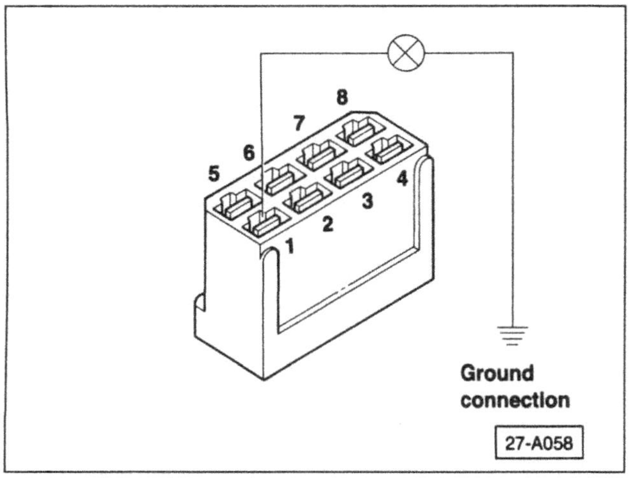

Fig. 12-4. Control unit harness connector used when testing cruise control. Test light (shown schematically) connected to terminals for one of the tests.

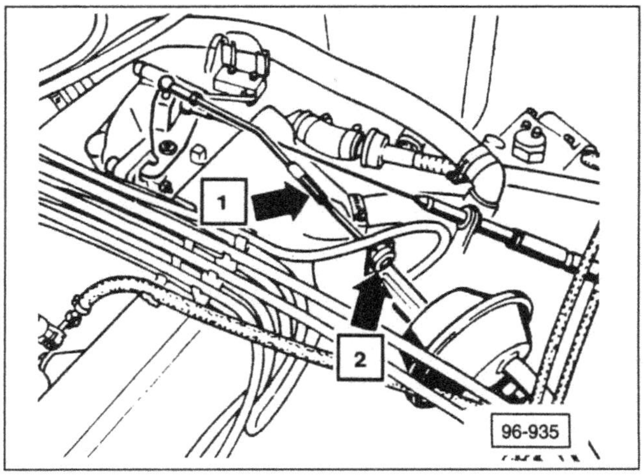

Fig. 12-5. Cruise control vacuum servo linkage being adjusted. Turn nut (**1**) until there is 0.1 to 0.3 mm (.004 to .012 in.) of play between bushing and stop plate (**2**).

13. FUSE/RELAY PANEL

Most fuses and relays are mounted together on the main fuse/relay panel, located below the instrument panel on the driver's side. To reach the Scirocco fuse/relay panel, fold the storage bin down after releasing the locking tab, as shown in Fig. 13-1. On Cabriolets, the panel is up behind the kneebar.

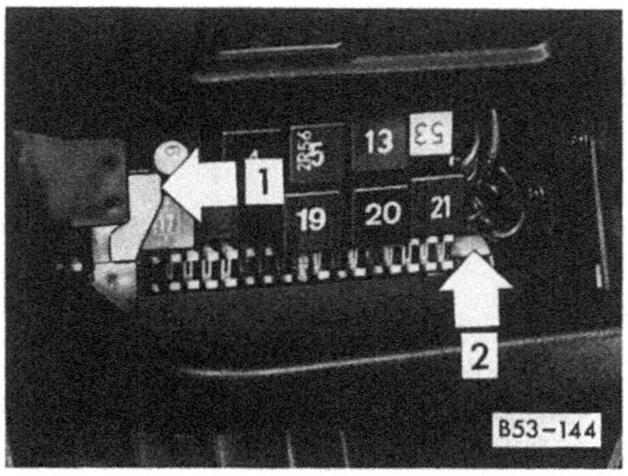

Fig. 13-1. Location of fuse/relay panel on Scirocco models. Press locking tab (**1**) to lower bin. Use special tool (**2**) to remove fuses.

Fuses come in standard colors, and each color corresponds to a different current rating. The fuse for each circuit, with its particular current rating, has been chosen to protect the circuit against current flow that exceeds the rating. When replacing fuses, always use a fuse of the specified current rating. It is never appropriate to substitute a fuse of a higher rating.

CAUTION —
- *Replacement fuses must always have the same current rating as the fuse being replaced.*

- *If the fuse in a particular circuit fails repeatedly there is a problem in that circuit, in the wiring harness or in a component, that is causing the circuit to be overloaded and the fuse to fail. The fault must be repaired. Never substitute a fuse with a higher current rating.*

Fuses and relays are identified in Fig. 13-2 according to their positions on the panel. On the wiring diagrams, relays are identified by a number in a black box that indicates its location on the relay panel. Also, each fuse number is preceded by the letter **S**. For example, **S12** in the circuit diagrams indicates fuse #12 on the fuse panel.

NOTE —
Relay locations may vary slightly from car to car depending on equipment and options, especially for those relays mounted above the main panel. If in doubt, the connectors and wire colors shown on the wiring diagrams may help in identifying the relays. Also remember that an authorized Volkswagen dealer is your best source for the most accurate and up-to-date information.

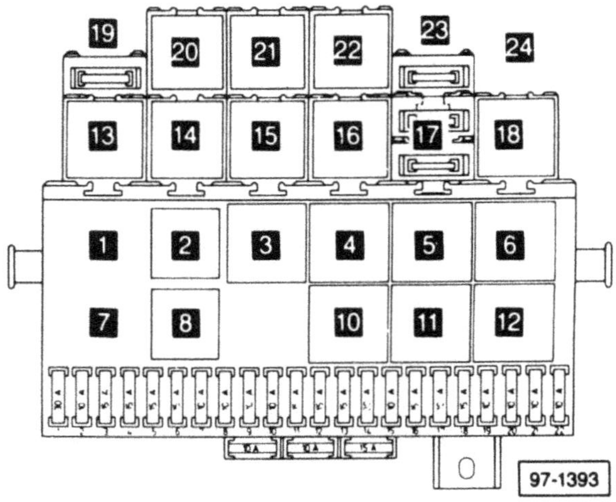

Fig. 13-2. 1985 through 1989 Cabriolet and Scirocco fuse/relay panel.

Fuses

No.	Description	Ampere rating
1	Radiator cooling fan	30
2	Brake lights, cruise control	10
3	Cigar lighter, radio, clock, interior lights, central locking	15
4	Emergency flashers	15
5	Fuel pump Oxygen sensor heater (Cabriolet from 1990)	15
6	Open	
7	Taillights, parking lights, side marker lights (left)	10
8	Taillights, parking lights, side marker lights (right)	10
9	Headlights (high beam, right), high beam indicator	10
10	Headlights (high beam, left)	10
11	Windshield wipers/washers	15
12	Heated seats, power mirrors, rear wiper/washer Multi-function indicator, power windows (Scirocco)	15
13	Rear window defogger, heated mirrors	15
14	Fresh air (heater) blower, glovebox light Air conditioning (1985 Scirocco only)	20
15	Back-up lights, shift console light (auto. trans.) Cruise control (1985-1990 w/automatic trans. only)	10
16	Horn	15
17	Knock sensor control unit (Scirocco 16V only)	10
18	Seat belt warning light, horn relay, coolant level/ temperature warning light, oil pressure gauge, oil temperature gauge, voltmeter Cruise control (except 1985-1990 auto. trans.) Idle speed boost (Scirocco from 7/87 except 16V)	15
19	Turn signals, brake warning light	10
20	License plate lights Luggage compartment light (Cabriolet from 1987) Instrument panel lights (Cabriolet from 1991)	10
21	Headlights (low beam, left)	10
22	Headlights (low beam, right)	10

Separate fuses above main panel

Relay panel position	Description	Ampere rating
17	Air conditioning thermofuse (S23)	30
	Cabriolet convertible top circuit breaker	20
19	Power windows (Scirocco, S37)	30
	Power windows (Cabriolet fr. 1990, S37)	20

Relays

No.	Description
1	Digifant control unit (Cabriolet from 1990)
2	Fuel pump
3	Seat belt warning
4	Upshift indicator control unit
5	Air conditioning
6	Horns
7	open
8	Load reduction
9	open
10	Intermittent wipers
11	Rear wiper/washer (Scirocco)
12	Turn signals, emergency flashers
13	Power windows (Scirocco only)
14	Hot-start pulse relay (Cabriolet to July 1987) Cold-start enrichment (July 1987 to 1989 except 16V) Rad. fan after-run control unit (Cabriolet from 1990)
15	Fuel injection power supply (1985-1989) Starter interlock (Cabriolet w/auto. trans. from 6/90)
16	Idle speed boost/Idle stabilizer control unit (to 1989) Hot-start pulse relay (July 1987 to 1989 except 16V) open (from 1990)
17	A/C thermofuse – 30 amp Cabriolet convertible top circuit breaker – 20 amp
18	Coolant low-level warning control unit Idle boost control unit (July 1987 to 1989 except 16V)
19	Power window fuse (S37) Cabriolet –20 amp; Scirocco – 30 amp
20	open
21	open
22	open
23	open
24	open

14. CURRENT FLOW DIAGRAMS

Volkswagen wiring diagrams are current flow diagrams that indicate current flow, from positive to negative, in the electrical system. Information such as wire size and color is included, but the diagrams indicate how a circuit functions.

As a general rule, the diagrams indicate current flow from positive (+) at the top of the diagram to negative (–) or ground at the bottom. Each diagram features connector descriptions, a list of ground points in the circuit, and a component index referring to the current track numbers at the bottom of each page. For more information on the features of these diagrams, see **3.1 How To Read Volkswagen Current Flow Diagrams** and **Terminal and Circuit Identification** earlier in this section.

The current flow diagrams in this manual are organized in the order listed below. Main Wiring Diagrams are first. They normally cover all the features of the basic electrical system. Following are the sub-system or accessory diagrams in alphabetical order. Parts of the electrical system vary from year to year and from model to model. The titles identify the models and model years covered by that diagram, and any exceptions. References to "Scirocco" and "Cabriolet" apply to all Scirocco and Cabriolet models, unless otherwise specified.

Main wiring diagram – Scirocco 1985, 1986
Main wiring diagram – Scirocco 1987-1989

Main wiring diagram – Cabriolet 1985, 1986
Main wiring diagram – Cabriolet 1987-1989
Main wiring diagram – Cabriolet 1990
Main wiring diagram – Cabriolet from 1991

Airbag – Cabriolet from 1990

Air conditioning – Scirocco 1985-1988
Air conditioning – Cabriolet 1985-1988
Air conditioning – Scirocco, Cabriolet 1989
Air conditioning – Cabriolet from 1990

Automatic shiftlock – Cabriolet from June 1990

Central locking – Scirocco 1985-1988

Cruise control, man. trans. – Scirocco, Cabriolet 1985-1990
Cruise control, auto. trans. – Scirocco, Cabriolet 1985-1990
Cruise control, man. trans. – Cabriolet from 1991
Cruise control, auto. trans. – Cabriolet from 1991

Daytime running lights (Canada) – Cabriolet 1990
Daytime running lights (Canada) – Cabriolet from 1991

Electronic engine controls – Scirocco 16V
Electronic engine controls – Scirocco (except 16V) from 7/87
Electronic engine controls – Cabriolet, July 1987-1989
Electronic engine controls – Cabriolet from 1991 (Calif. only)

Heated seats – Scirocco, Cabriolet to 1990
Heated seats – Cabriolet from 1991

Lights – Scirocco, Cabriolet 1989

Multi-function indicator – Scirocco 1985-1989

Power convertible top – Cabriolet from 1991

Power/heated mirrors – Scirocco 1985-1988

Power windows – Scirocco 1985-1989
Power windows – Cabriolet from 1990

Stereo radio – Scirocco, Cabriolet 1985-1989

Stereo radio w/roof-mounted antenna – 16 valve Scirocco
Stereo radio w/power antenna – Scirocco, Cabriolet 1985-1989
Stereo radio – Cabriolet 1990
Stereo radio w/six amplified speakers – Cabriolet from 1991
Stereo radio w/six passive speakers – Cabriolet from 1991

Main wiring diagram—Scirocco 1985, 1986

Description	Current track
Alternator	1, 2
Alternator warning lite	38
Ashtray lite	67
Auxiliary air regulator	21
Back-up lite, left	120
Back-up lite, right	119
Back-up lite switch	121
Battery	3
Brake and parking brake indicator lite	60, 61
Brake fluid level warning switch	62
Brake lite, left	102
Brake lite, right	101
Brake lite switch	121
CAT elapsed mileage indicator	44
Cigar lighter	73
Cigar lighter lite	71
Clock	46a
Clock lite	65
Cold start valve	25
Coolant low level control unit	46b
Coolant low level switch	46b
Coolant over-temp warning lite	43
Coolant temperature gauge	41, 42
Coolant temperature sending unit	47
Control pressure regulator	20
Door contact/buzzer switch, left	55, 56
Door contact switch, right front	54
Engine oil pressure control unit	33, 34
Engine oil pressure switch (1.8 bar)	48
Engine oil pressure switch (0.3 bar)	49
Engine oil temperature lite	69
Emergency flasher relay	91-93
Emergency flasher switch	90-95
Emergency flasher warning lite	95
Fresh air fan	106
Fresh air fan switch	104-106
Fuel enrichment switch	13
Fuel gauge	40
Fuel gauge sending unit	24
Fuel pump	22
Fuel pump relay	20
Glove compartment lite	104
Hall generator	7, 8
Headlite dimmer/flasher switch	76, 77
Headlite, high beam, left	80
Headlite, high beam, right	83
Headlite, high beam warning lite	37
Headlite, low beam, left	78, 81
Headlite, low beam, right	79, 82
Headlite switch lite	103
Horns	122, 123
Horn button	128
Horn relay	125, 127
Idle speed stabilizer valve	117
Ignition coil	9, 10
Ignition control unit	6-10
Ignition distributor	9, 11
Interior lite, front	53, 54
Instrument panel lite	64
Instrument panel lite switch/dimmer	71

Description	Current track
Junction box (behind fuse/relay panel)	
License plate lites	74, 75
Load reduction relay	65-66
Luggage compartment lite	52
Luggage compartment lite switch	52
Lite switch	66-71
Oil pressure warning lite	35
Oil temperature gauge	125
Oil temperature sensor	126
OXS control unit	12-18
OXS elapsed mileage counter	45, 46
OXS frequency valve	17
OXS thermoswitch	14
Oxygen sensor	15
Parking brake warning lite switch	60
Parking lite, left	88
Parking lite, right	85
Power supply relay, FI	16-19
Radiator cooling fan thermoswitch	129
Radiator fan	129
Rear window defogger indicator lite	130
Rear window defogger switch	131
Rear window defogger switch lite	68
Rear window heater element	132
Rear window washer pump	112
Rear window wiper motor	114-115
Rear window wiper washer relay	112-114
Resistance wire	17
Safety belt switch, left	57
Seat belt warning switch relay	56-58
Series resistance (fresh air blower)	105
Side marker lites, rear	84-89
Spark plugs	9-11
Spark plug connectors	9-11
Starter	4-5
Tachometer	32
Tail lite, left	85
Tail lite, right	86
Transfer fuel pump	22
Transmission switch, upshift indicator	47
Turn signal indicator lite	39
Turn signal lite, left front	96
Turn signal lite, left rear	97
Turn signal lite, right front	98
Turn signal lite, right rear	99
Turn signal switch	92, 93
Upshift control lite	31
Upshift indicator control unit	30-33
Vacuum switch, upshift indicator	51
Voltage regulator	1, 2
Voltage stabilizer	40
Washer/wiper intermittent relay	110, 111
Windshield washer pump	113
Windshield wiper/washer intermittent switch	109-111
Windshield wiper motor	107, 108

Wire connectors

T — junction box, by fuse/relay panel
T1a — single, left of battery
T1b — single, behind instrument panel
T1d — single, by brake master cylinder
T1f — single, engine compartment, left
T2 — double, behind instrument panel by OXS control
T2a — double, next to fuse/relay panel
T2b — double, by brake master cylinder
T2e — double, behind instrument panel
T4 — radio connections
T7 — 7 point, short white housing, instr. cluster
T7a — 7 point, long black housing, instr. cluster
T7b — 7 point, long white housing, instr. cluster
T7c — 7 point, short black housing, instr. cluster
T1c — single, behind instrument panel
T1h — single, behind instrument panel
T2c — engine compart. by trans. switch
T2h — double, luggage compartment, right
T2h — double, engine comp. by trans. switch
T3 — 3 point, behind instrument panel
T4 — 4 point, behind instrument panel
T8 — 8 point, on transmission
T1e — single, next to fuse relay panel
T1g — single, behind instrument panel
T2d — double, behind instrument panel
T2t — double, luggage comp. rear left
T2g — double, behind instrument panel
T2i — double, engine comp., left
T2k — double, below battery

Ground connections

(1) — from battery to trans. via body
(4) — rear panel
(6) — on rear deck lid
(7) — on right door post
(8) — on steering gear
(10) — on fuse/relay panel
(13) — on luggage comp., rear wheel housing
(15) — front wiring harness
(16) — in instrument panel wrapped in tape

Welded wiring connections

(20) — in wiring harness for lite switch

Main wiring diagram—Scirocco 1985, 1986
page 2 of 10

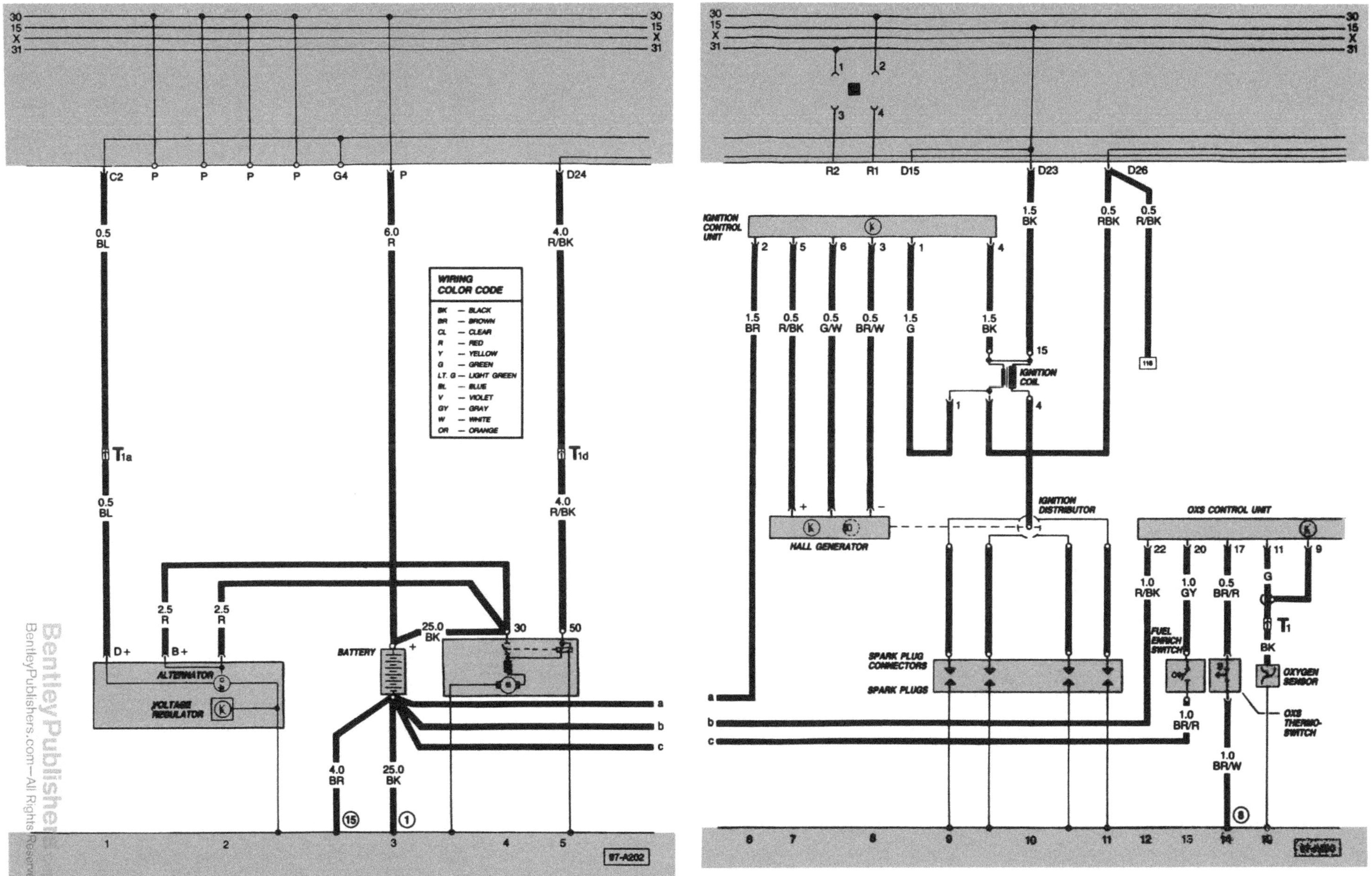

Main wiring diagram—Scirocco 1985, 1986
page 3 of 10

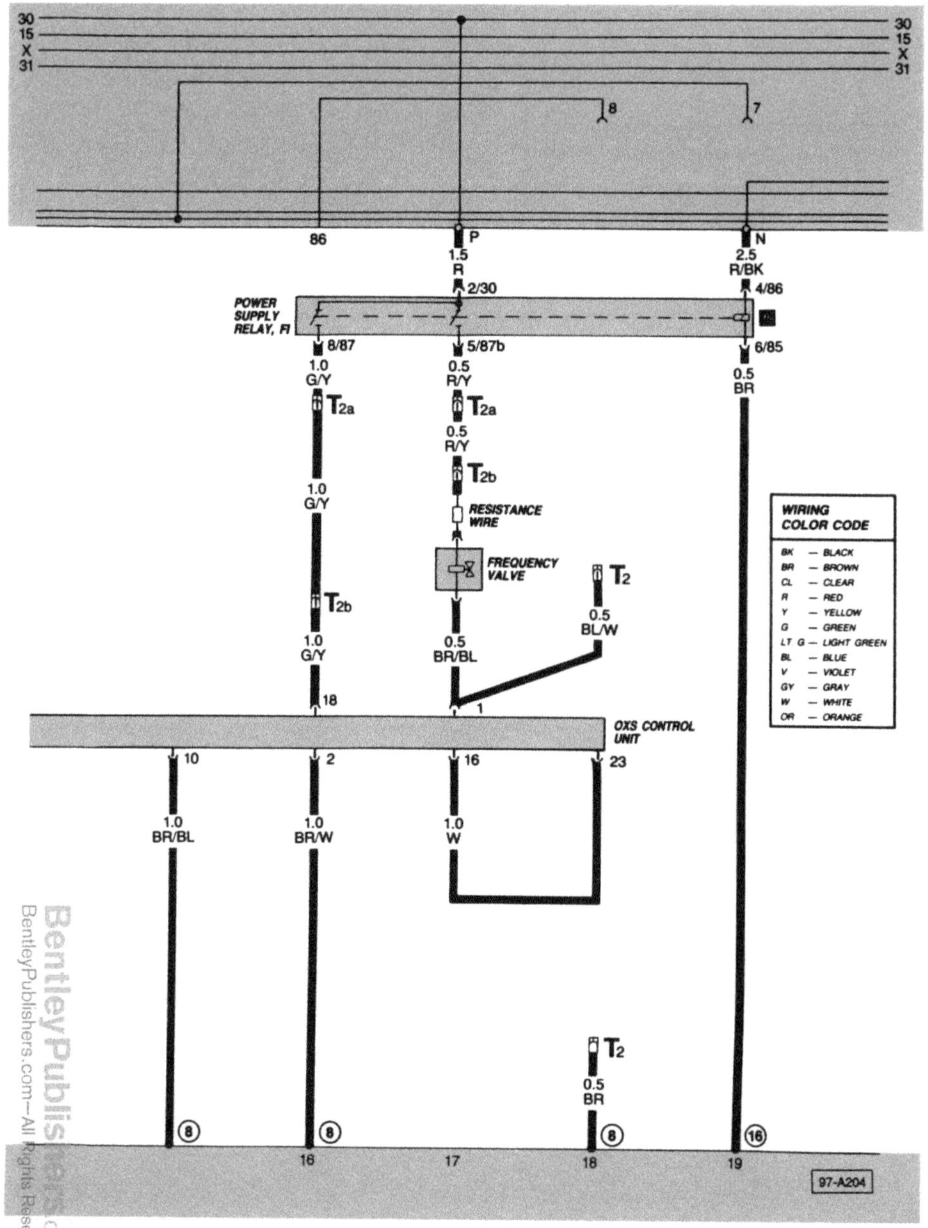

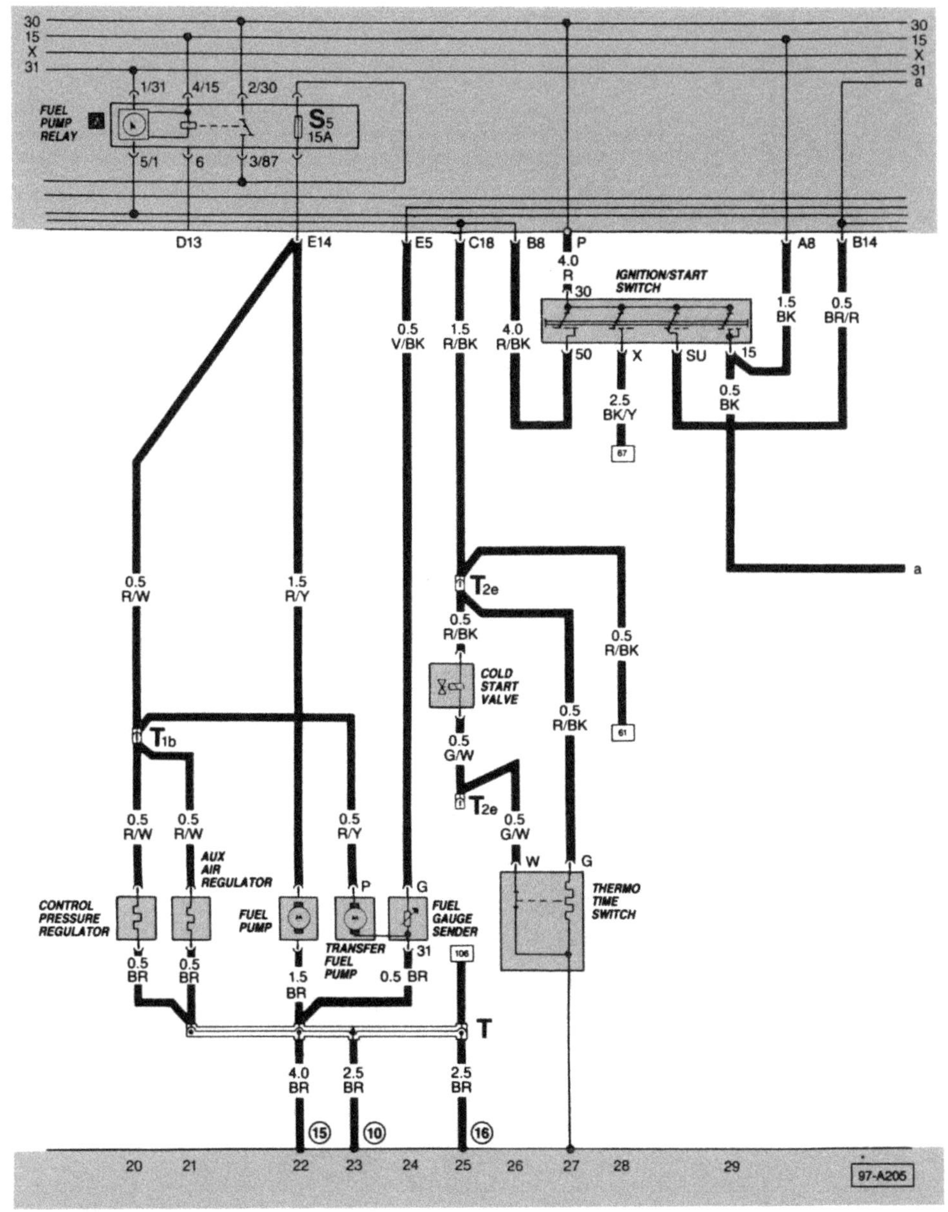

Main wiring diagram—Scirocco 1985, 1986
page 4 of 10

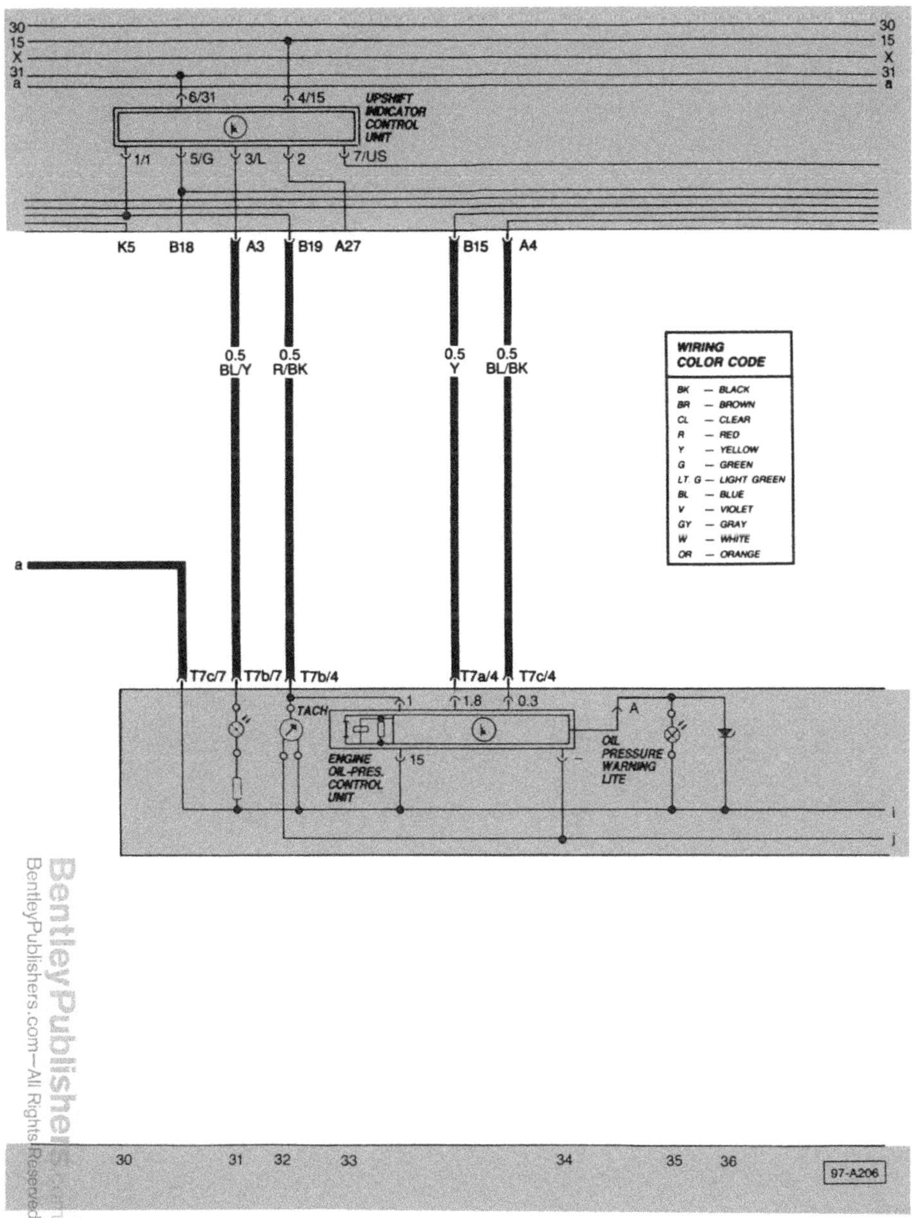

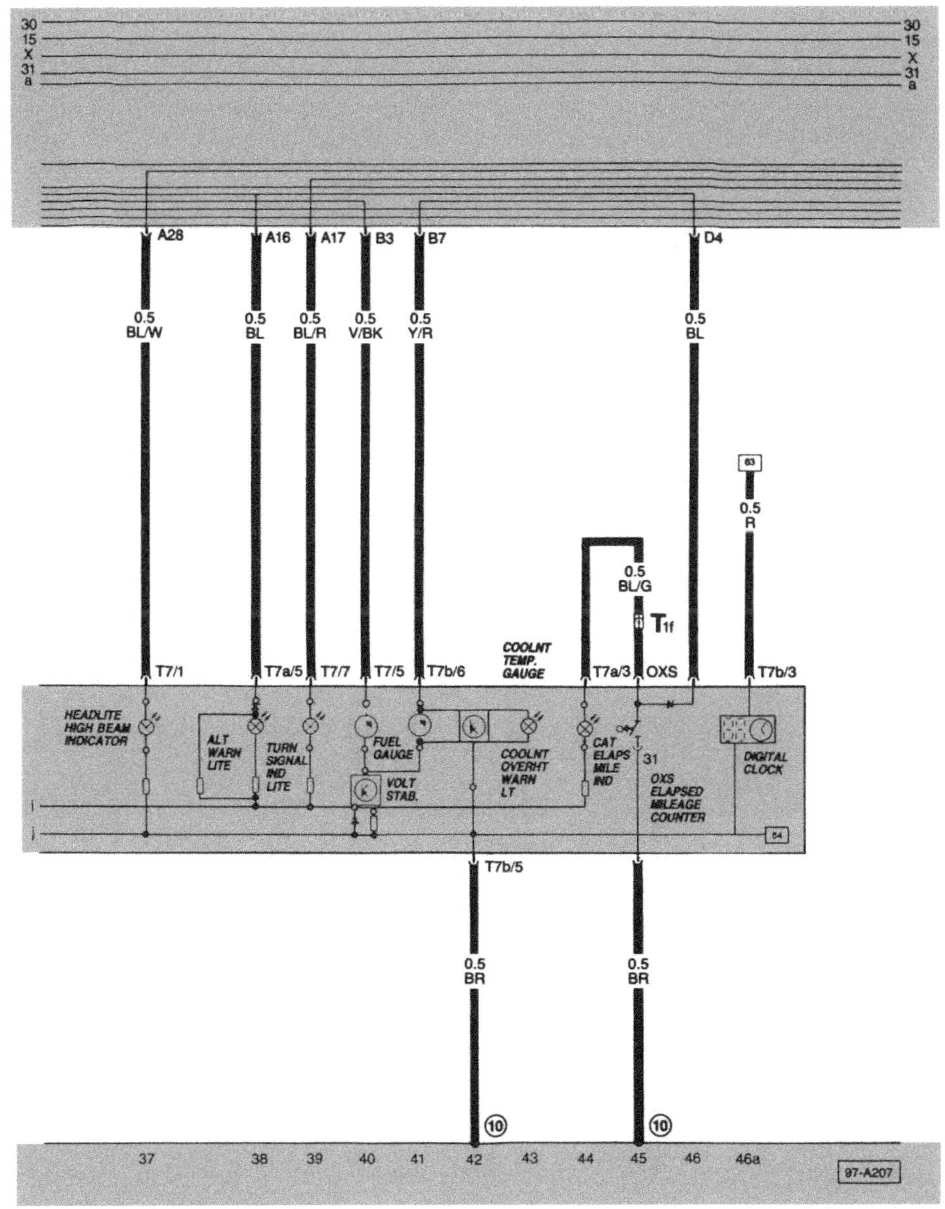

Main wiring diagram—Scirocco 1985, 1986

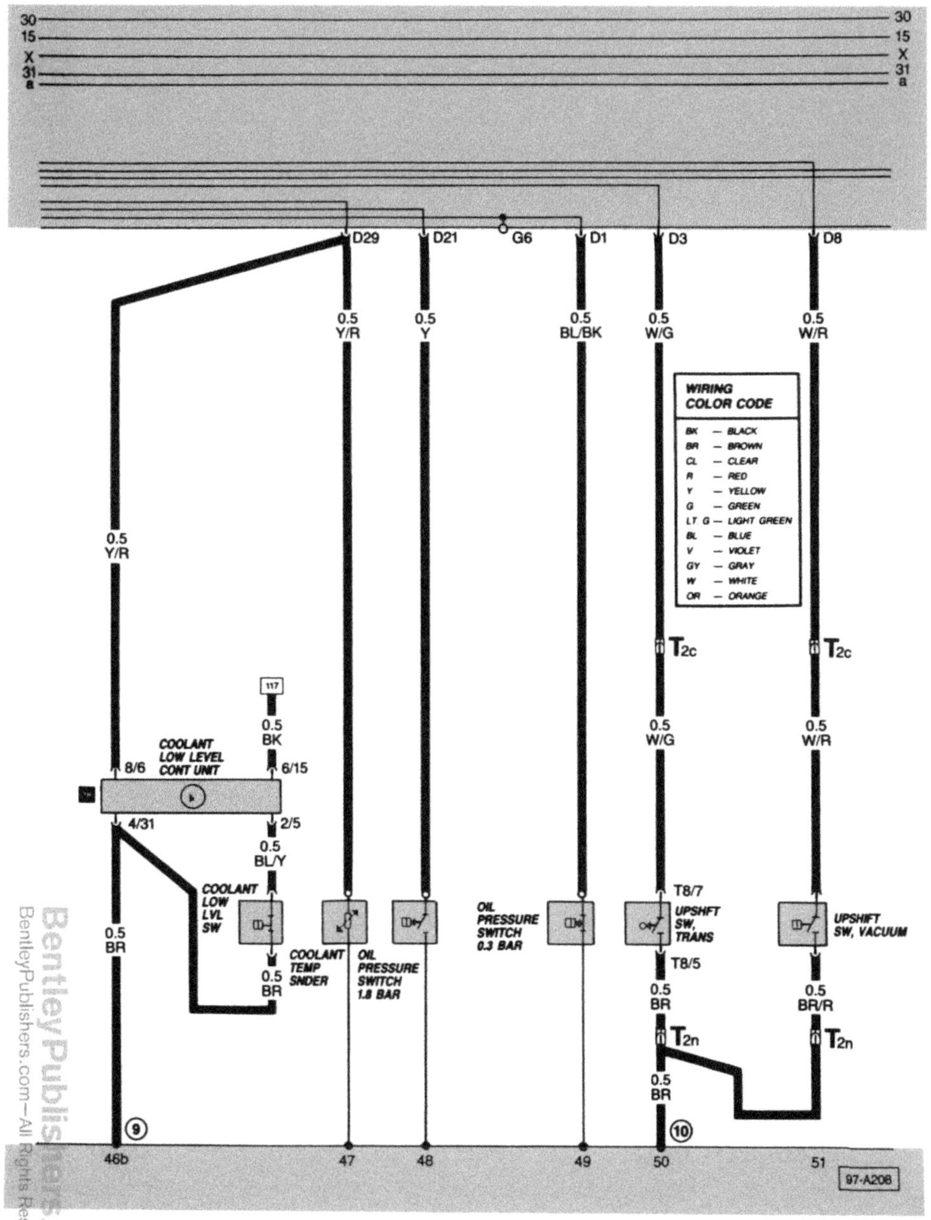

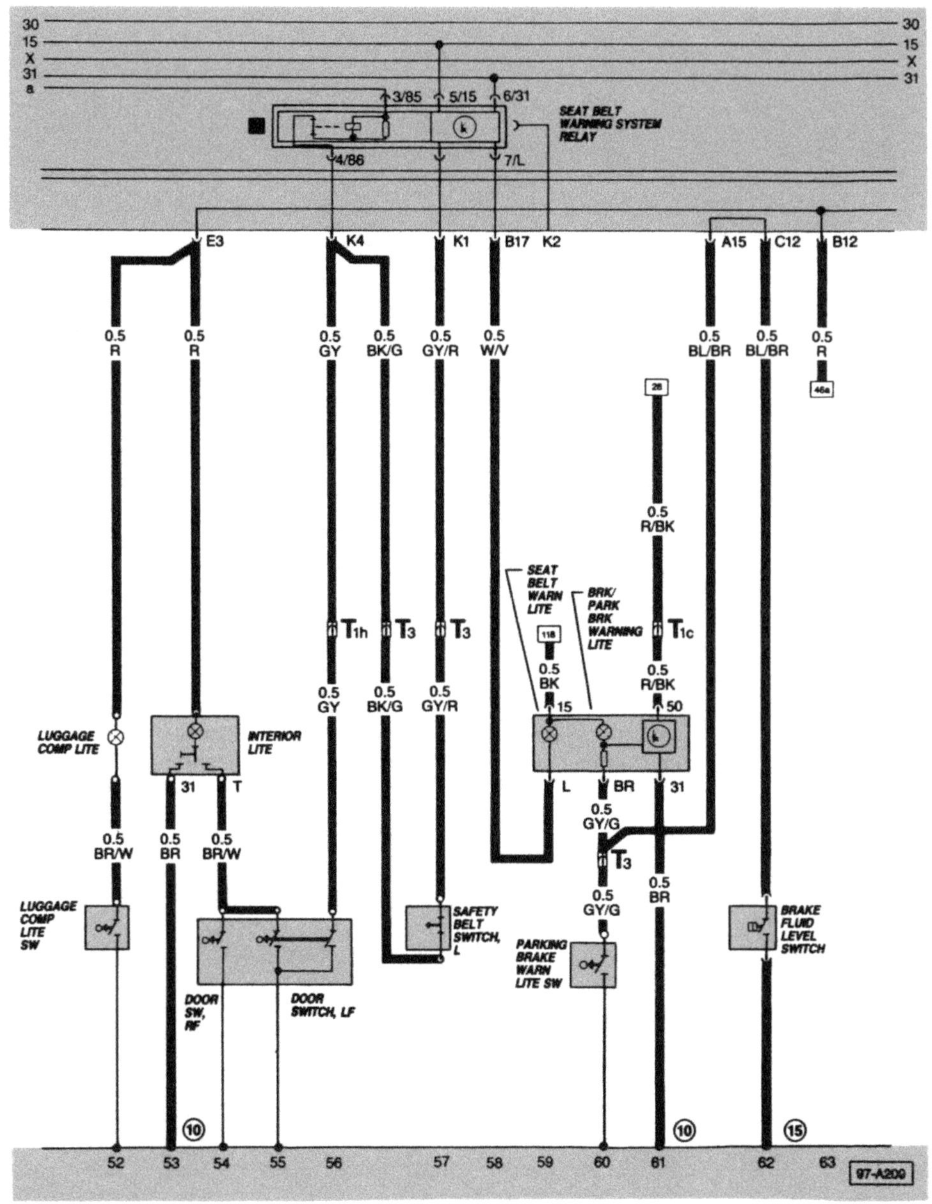

Main wiring diagram—Scirocco 1985, 1986
page 6 of 10

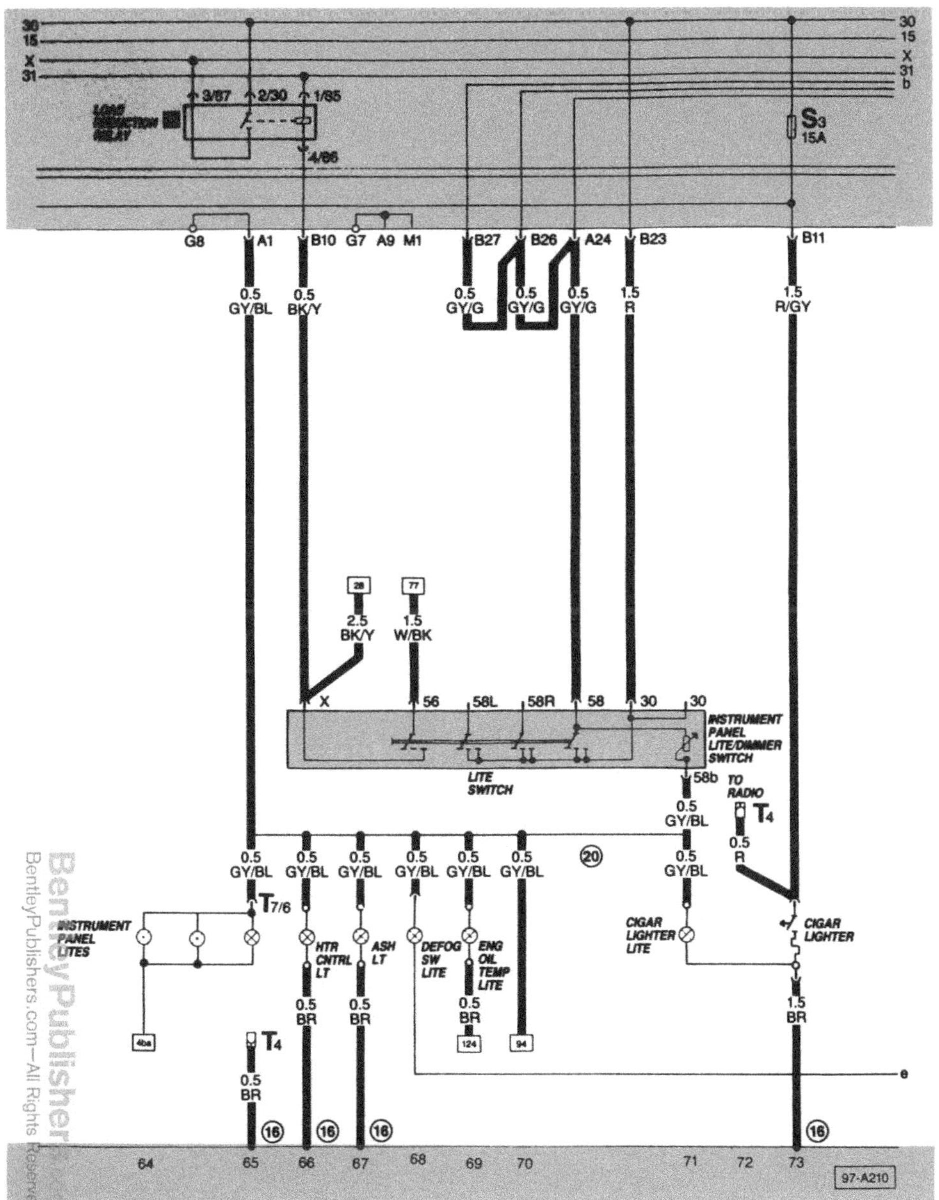

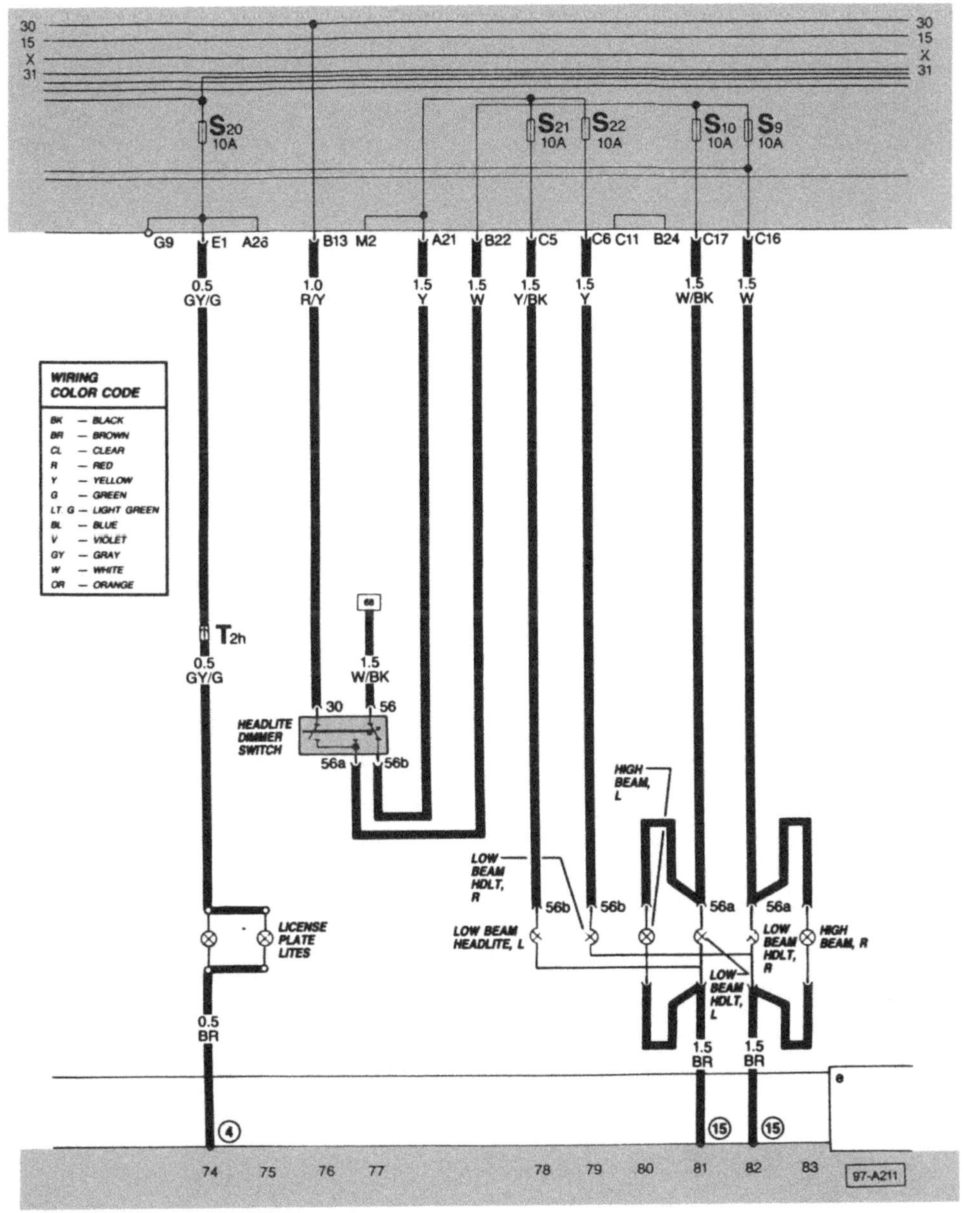

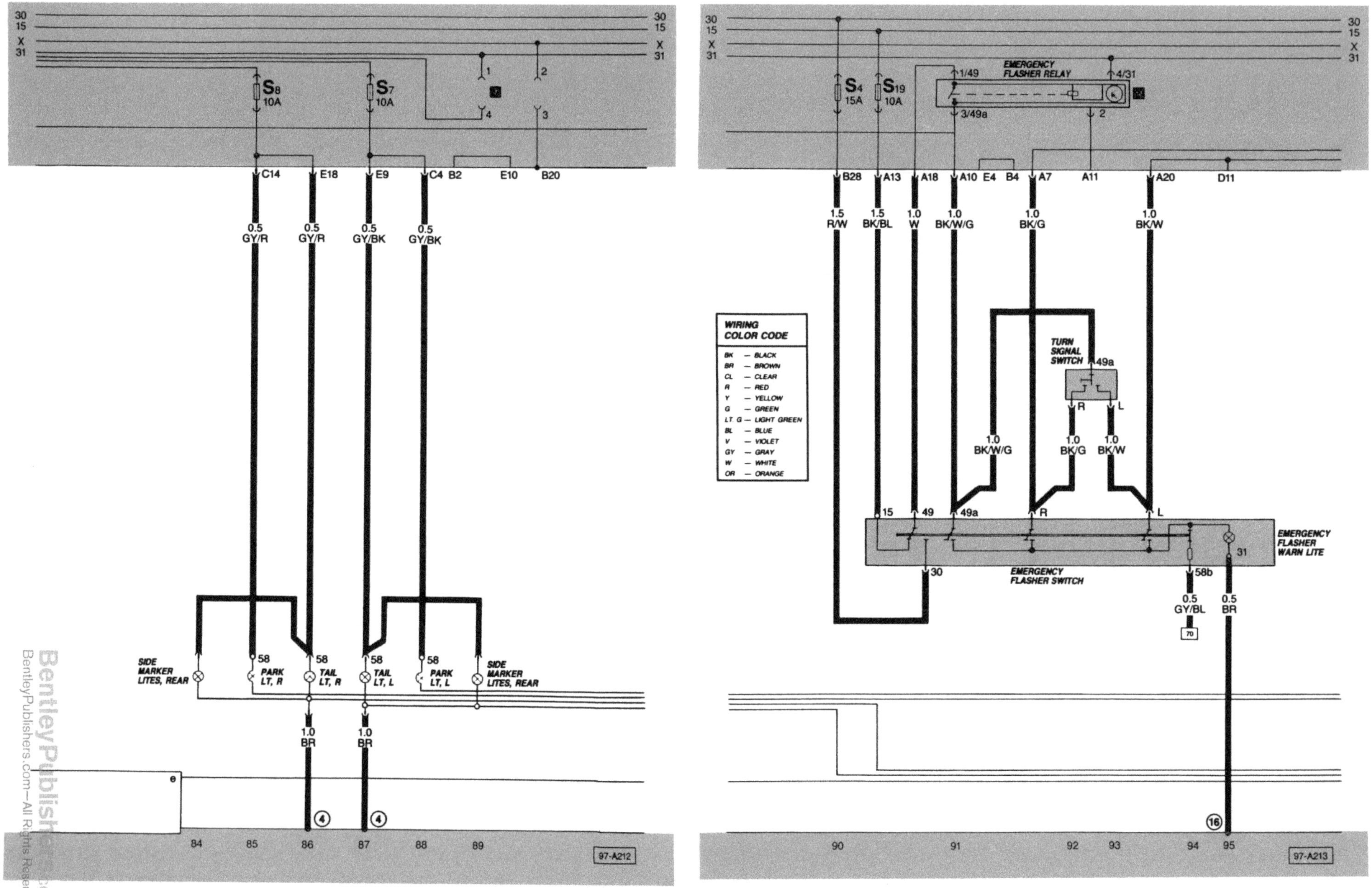

WIRING COLOR CODE

BK	BLACK
BR	BROWN
CL	CLEAR
R	RED
Y	YELLOW
G	GREEN
LT G	LIGHT GREEN
BL	BLUE
V	VIOLET
GY	GRAY
W	WHITE
OR	ORANGE

Main wiring diagram—Scirocco 1985, 1986
page 8 of 10

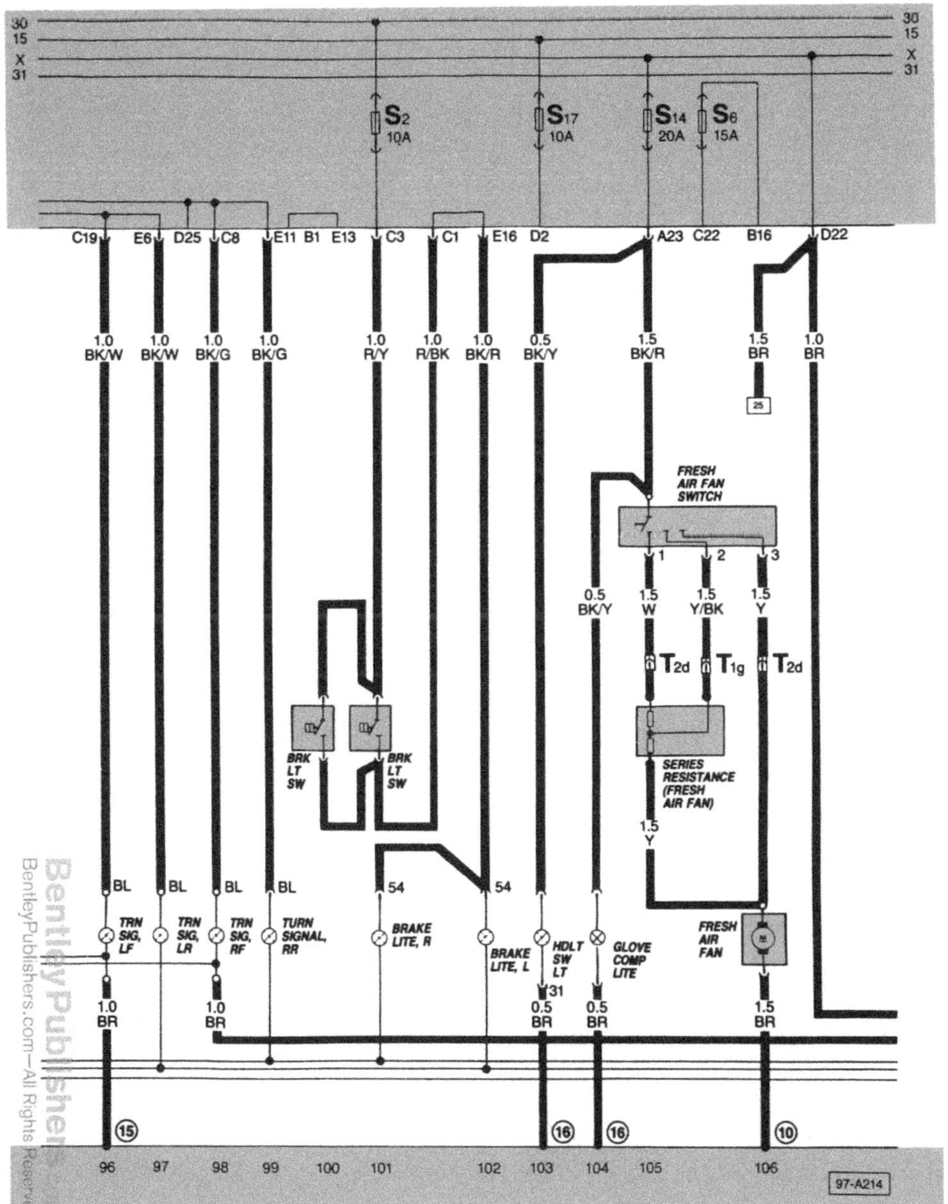

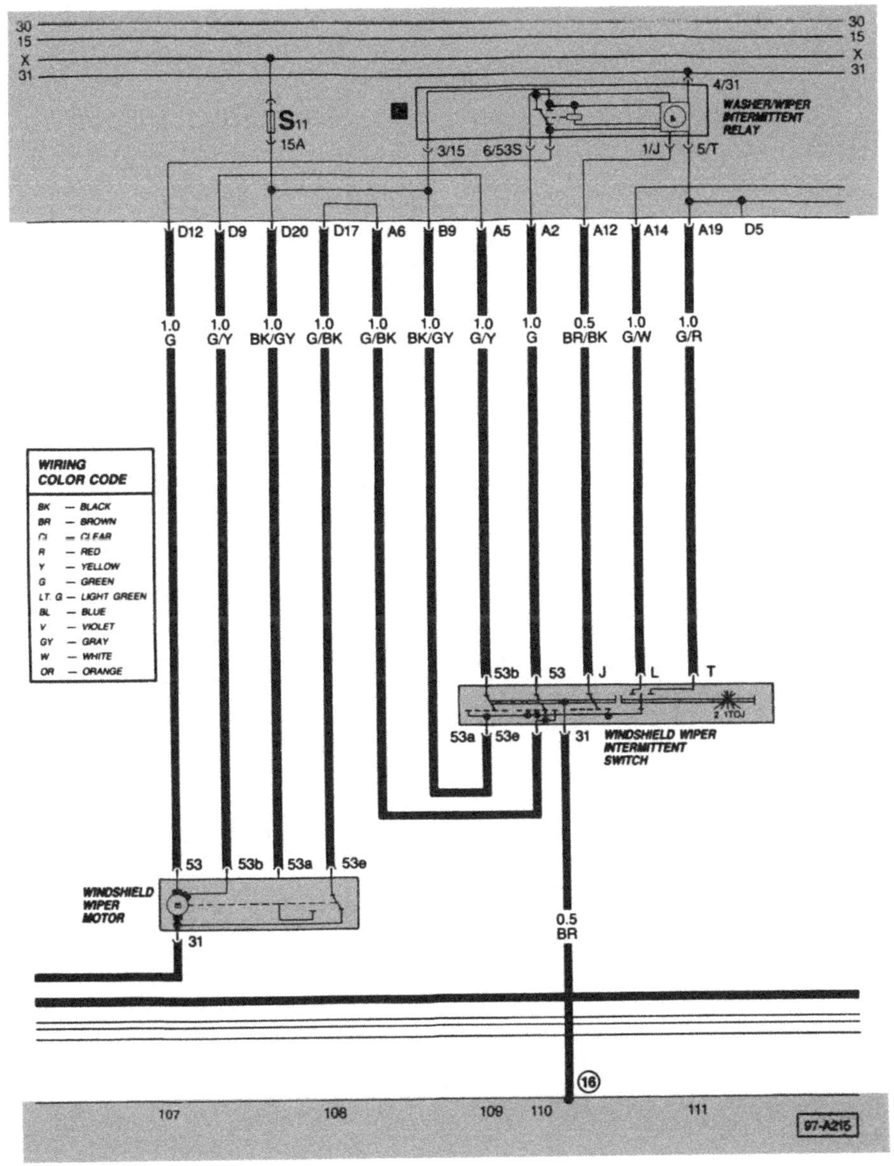

Main wiring diagram—Scirocco 1985, 1986
page 9 of 10

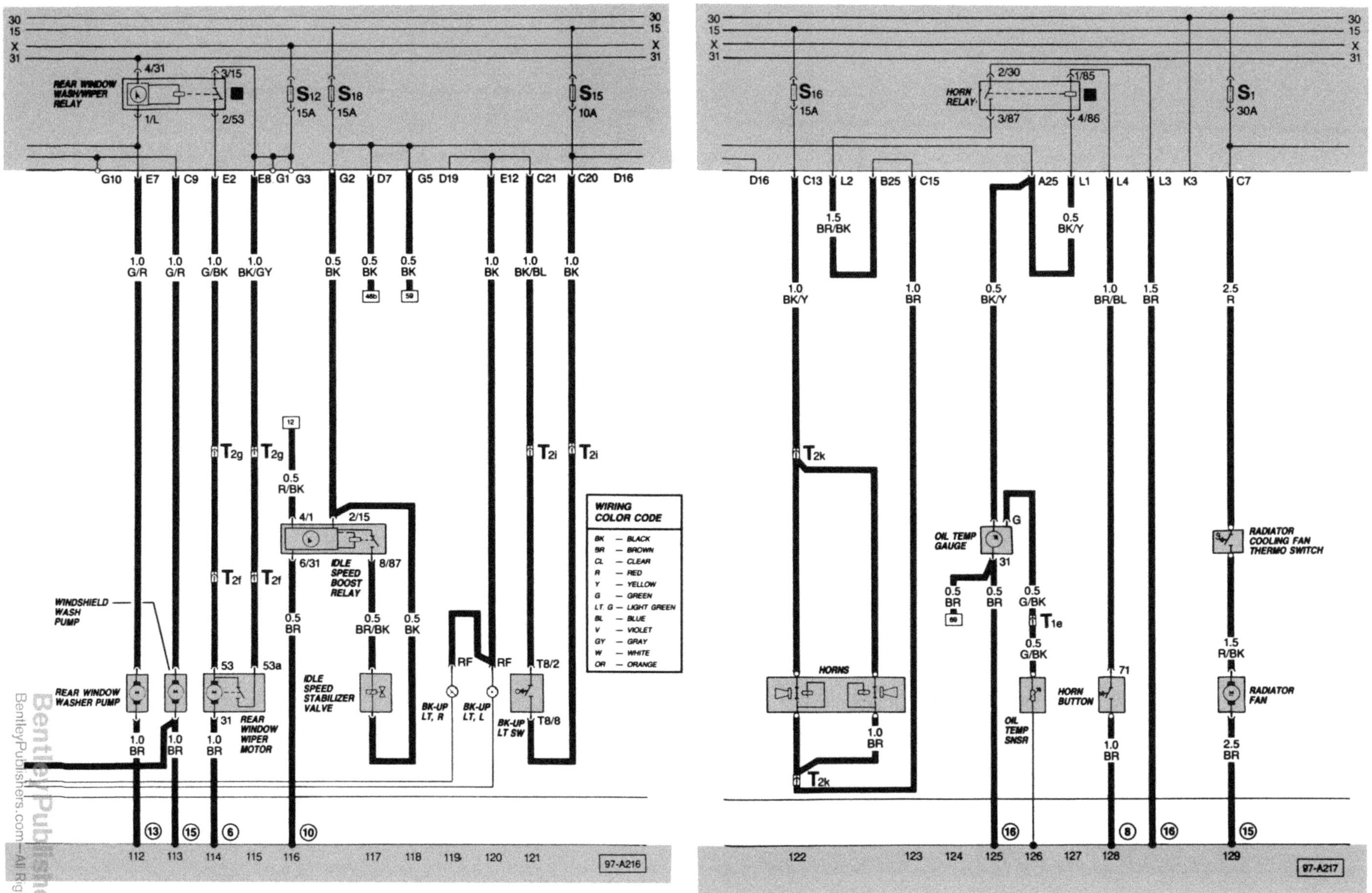

Main wiring diagram—Scirocco 1985, 1986
page 10 of 10

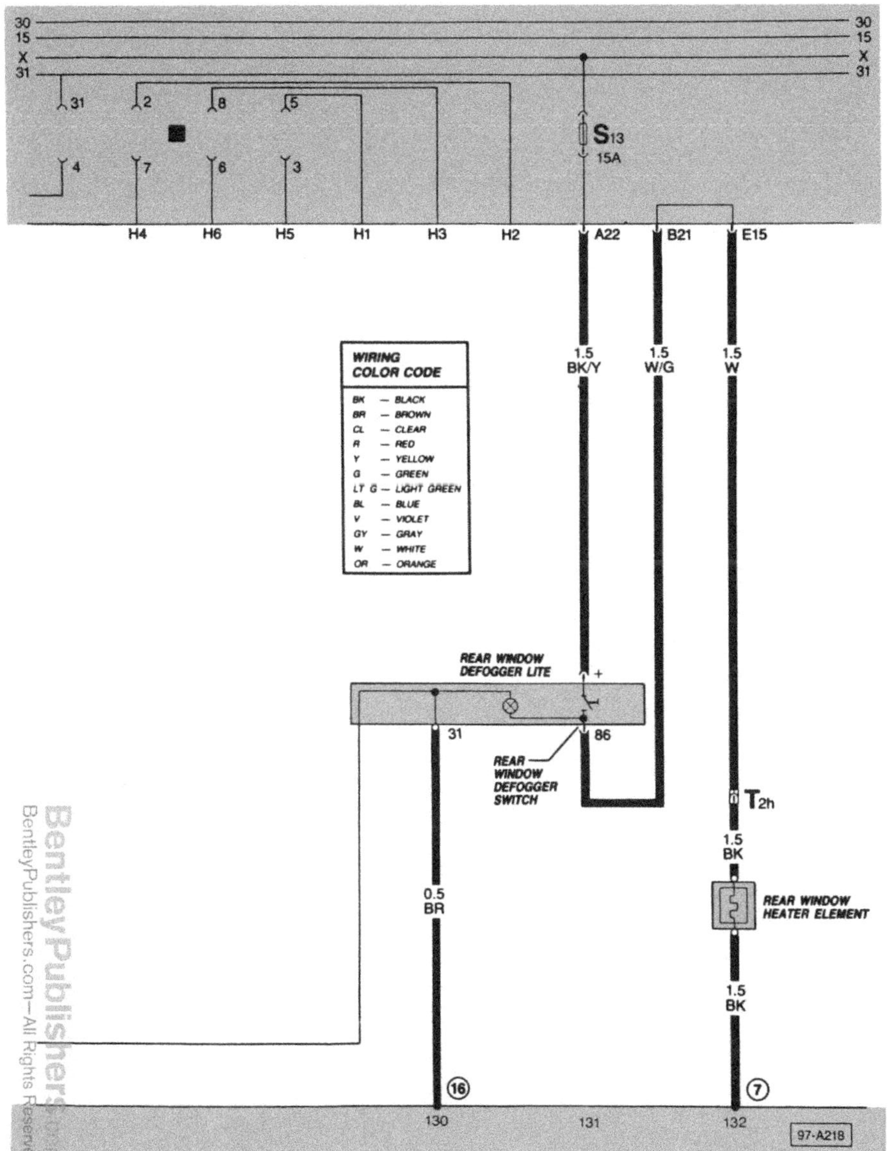

Main wiring diagram – Scirocco 1987-1989

Description	Current track
2-way idle speed boost	128
Alternator	1,2
Alternator warning lite	42
Ashtray lite	69
Auxiliary air regulator	22
Back-up lite switch	131
Back-up lite, left	130
Back-up lite, right	129
Battery	3
Brake and parking brake indicator lite	61-63
Brake fluid level warning contact	64
Brake lite switch	109
Brake lite, left	110
Brake lite, right	109
CAT elapsed mileage counter	48,49
Center stop lite, rear window	108
Cigarette lighter lite	74
Clock lite	67
Cold start valve	28
Control pressure regulator	21
Coolant overheat warning lite	46
Coolant temperature gauge	45
Coolant temperature sending unit	50
Digital clock	65
Door contact switch, left front	58
Door contact switch, right front	57
Emergency flasher relay	95-97
Emergency flasher switch	92-99
Emergency flasher warning lite	99
Engine oil pressure control unit	37-39
Engine oil pressure switch (0.3 bar)	52
Engine oil pressure switch (1.8 bar)	51
Engine oil pressure warning lite	40
Fresh air blower series resistance	113,114
Fresh air control lever lite	68
Fresh air fan	115
Fresh air fan switch	113-115
Fuel enrichment switch	15
Fuel gauge	44
Fuel gauge sending unit	26
Fuel injection power supply relay	18-20
Fuel pump	23
Fuel pump relay	24-26
Glove compartment lite	112
Hall generator	7,8
Headlite dimmer/flasher switch	78,79
Headlite high beam warning lite	41
Headlite, left	80,83
Headlite, right	81,84
Headlite switch lite	111
High-beam headlite, left	82
High-beam headlite, right	85
Horn button	137
Horn relay	134-136
Horns	132,133
Idle boost control unit	127,128
Ignition coil	10-12
Ignition control unit	6-9
Ignition distributor	10-12
Ignition/starter switch	30-33
Instrument panel lite	66
Instrument panel lite dimmer switch	74
Interior lite, front	56,57
License plate lite	76,77
Lite switch	70-74
Load reduction relay	70,71
Luggage compartment lite	55
Luggage compartment lite switch	55
OXS control unit	14-18
OXS elapsed mileage indicator	47
OXS frequency valve	19
OXS thermoswitch	16
Oxygen sensor	17
Parking brake warning lite switch	62
Parking lite, left	91
Parking lite, right	86
Radiator cooling fan	139,140
Radiator cooling fan after run relay	138-140
Radiator cooling fan thermoswitch	138,139
Rear window defogger indicator lite	105
Rear window defogger switch	104-106
Rear window defogger switch lite	104
Rear window heater element	107
Rear window washer pump	123
Rear window wiper motor	125,126
Rear window wiper/washer relay	123-125
Refrigerator 12V connector	75
Resistor	19
Safety belt switch, left	60
Seat belt warning lite	61
Seat belt warning relay	58-60
Side marker lites, rear	87,90
Spark plug connectors	10-12
Spark plugs	10-12
Starter	4,5
Tachometer	36
Tail lite, left	89
Tail lite, right	88
Thermo time switch	29,30
Transfer fuel pump	25
Turn signal indicator lite	43
Turn signal lite, left front	100
Turn signal lite, left rear	101
Turn signal lite, right front	102
Turn signal lite, right rear	103
Turn signal switch	96,97

Description	Current track
Up-shift control lite	35
Up-shift indicator control unit	35-38
Upshift switch, transmission	53
Upshift switch, vacuum	54
Voltage regulator	1,2
Voltage stabilizer	44
Voltmeter	134
Voltmeter lite	70
Washer/wiper intermittent relay	120,121
Windshield washer pump	124
Windshield wiper motor	117,118
Windshield-wiper intermittent switch	119-122

Wire connectors

T — Junction box behind relay panel plate
T1a — Wire connector, single near battery
T1b — Wire connector, single behind relay panel plate
T1c — Wire connector, single
T1d — Wire connector, single
T1e — Wire connector, single near ignition coil
T1f — Wire connector, single
T1g — Wire connector, single behind dash
T1h — Wire connector, single
T1k — Wire connector, single behind dash
T1n — Wire connector, single
T1o — Wire connector, single behind dash
T2a — Wire connector, double behind relay panel plate
T2b — Wire connector, double near ignition coil
T2c — Wire connector, double
T2d — Wire connector, double behind dash
T2e — Wire connector, double behind relay panel plate
T2f — Wire connector, double near ignition coil
T2g — Wire connector, double in luggage compartment, left
T2h — Wire connector, double in luggage compartment, rear part left
T2i — Wire connector, double in engine compartment, left
T2k — Wire connector, double near battery
T2r — Wire connector, double
T2s — Wire connector, double
T2u — Wire connector, double

Wire Connectors

T3a — Wire connector, 3 point in rear lid
T3b — Wire connector, 3 point
T7/ — Wire connector, 7 point on instrument cluster
T7a/ — Wire connector, 7 point on instrument cluster
T7b/ — Wire connector, 7 point on instrument cluster
T7c/ — Wire connector, 7 point on instrument cluster
T8/ — on transmission
T8/5 —
T8a/ — behind dash

Ground connections

(1) — ground strap battery to body
(24) — Ground connection on cold starting valve
(30) — Ground-connection, beside relay plate
(42) — Ground-connection, beside steering column
(51) — Ground-connection, luggage compartment, right
(52) — Ground-connection, in rear lid, left
(53) — Ground-connection, in rear lid, right
(54) — Ground-connection, on rear apron
(81) — Ground-connection, in dashboard wiring harness
(82) — Ground connection -1- in wiring harness front/left
(108) — Ground connection -2- in wiring harness, front/left

NOTE —
Electronic engine controls shown in this diagram are accurate through June 1987 production only. Electronic engine controls for later 1987 cars (from July 1987 production), and all 1988-1989 cars are covered by separate "Electronic Engine Controls" diagrams elsewhere in this section.

Main wiring diagram – Scirocco 1987-1989
page 3 of 9

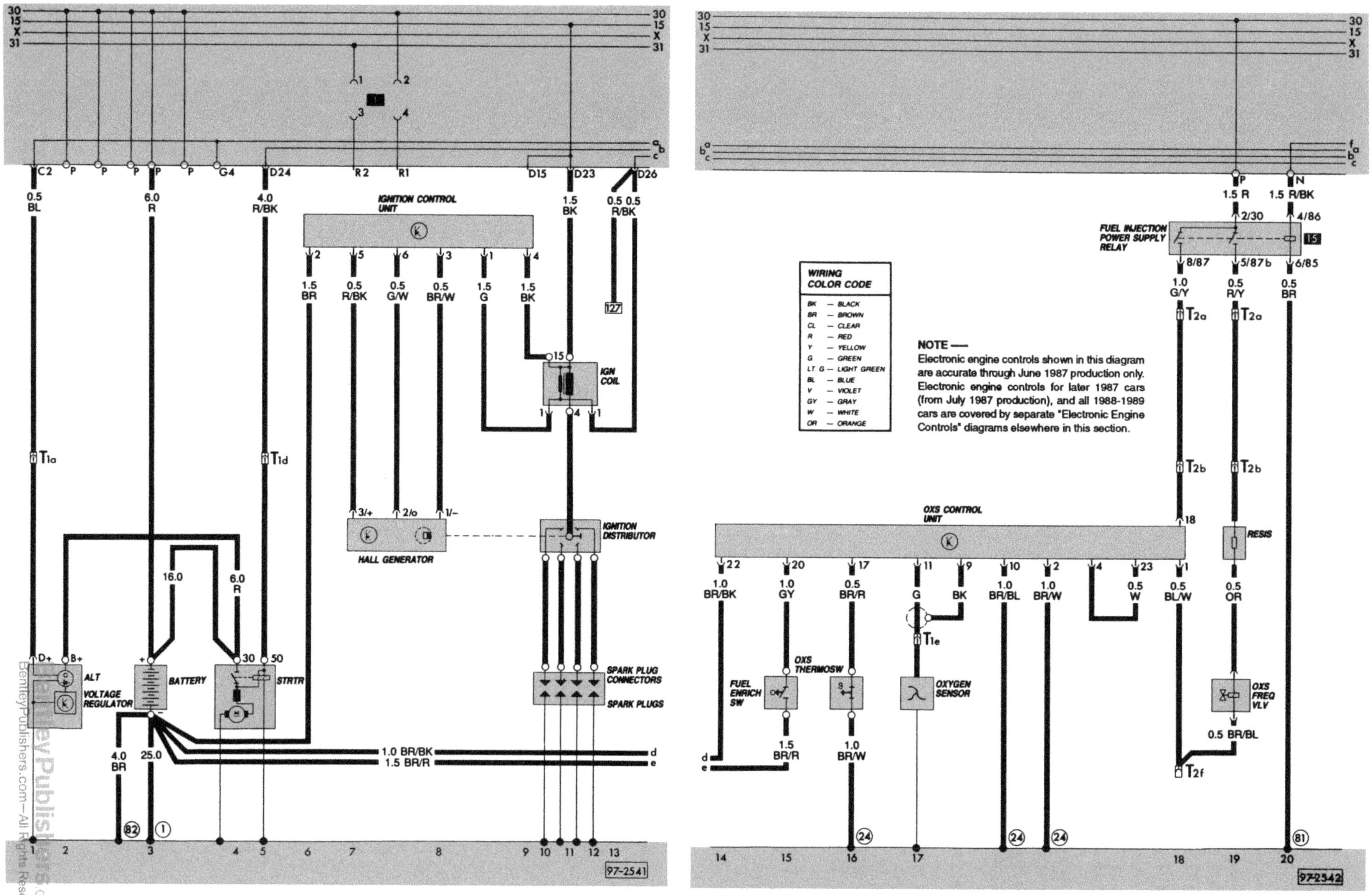

Main wiring diagram – Scirocco 1987-1989
page 4 of 9

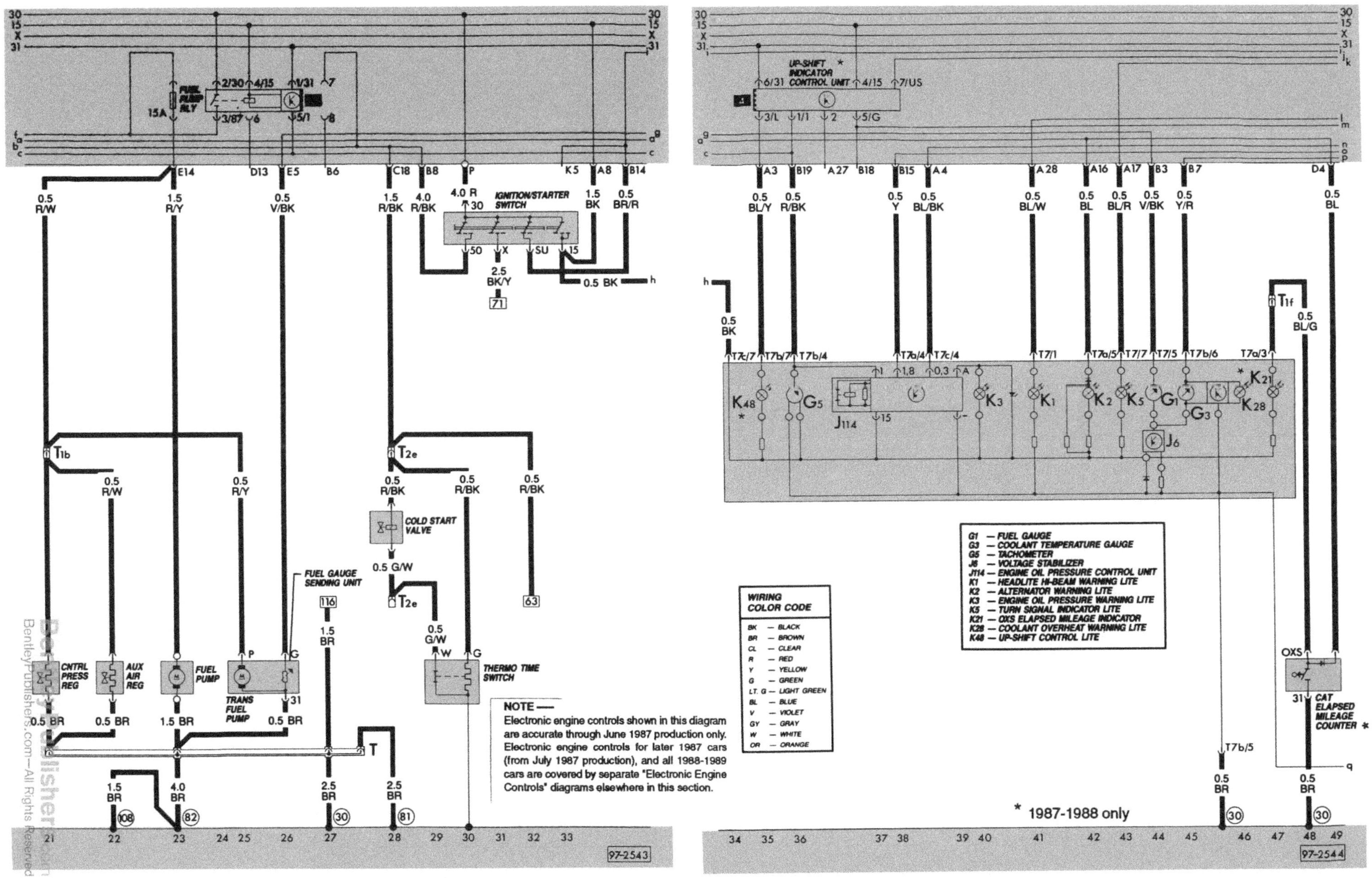

Main wiring diagram – Scirocco 1987-1989
page 5 of 9

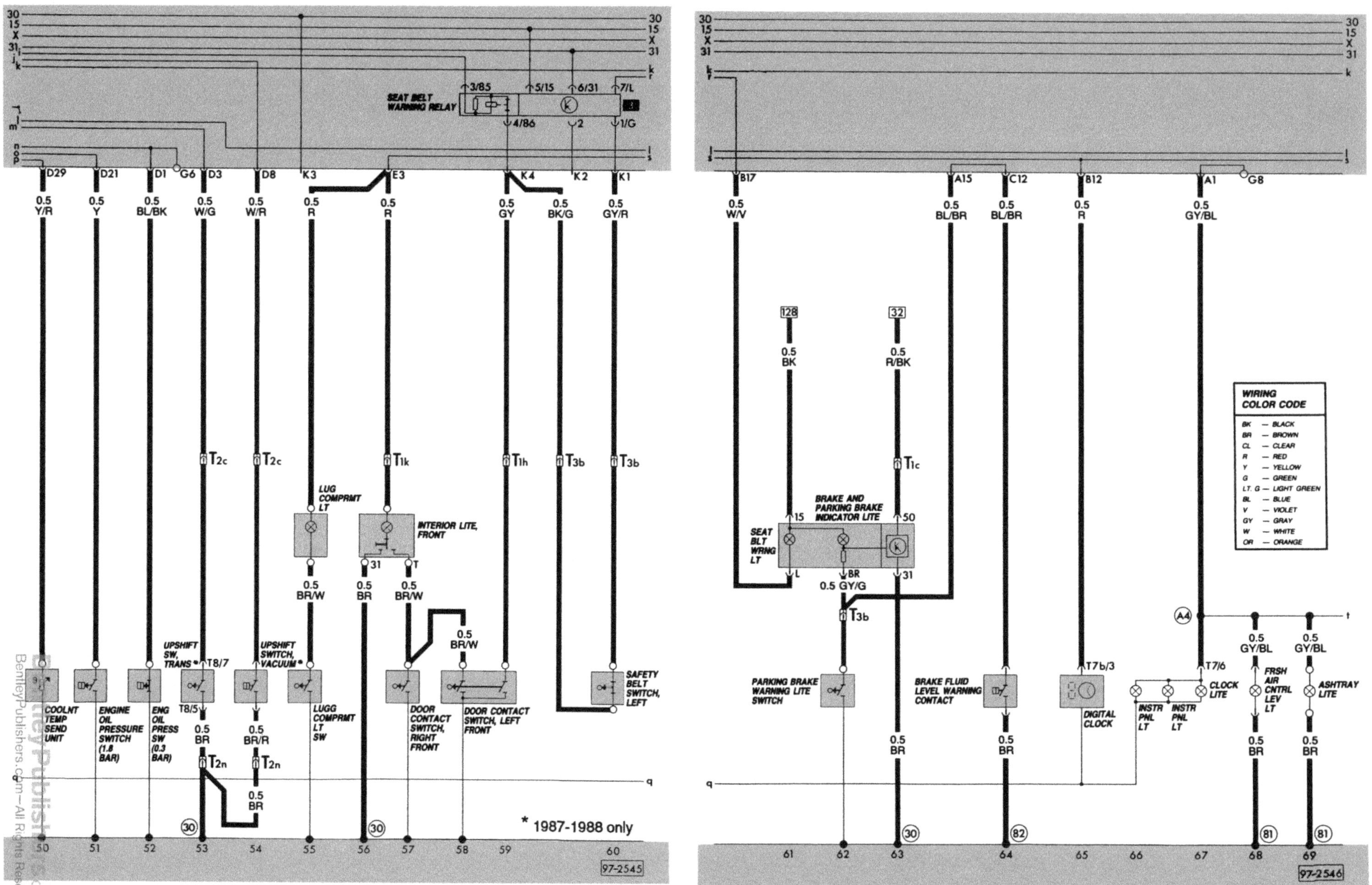

Main wiring diagram – Scirocco 1987-1989
page 6 of 9

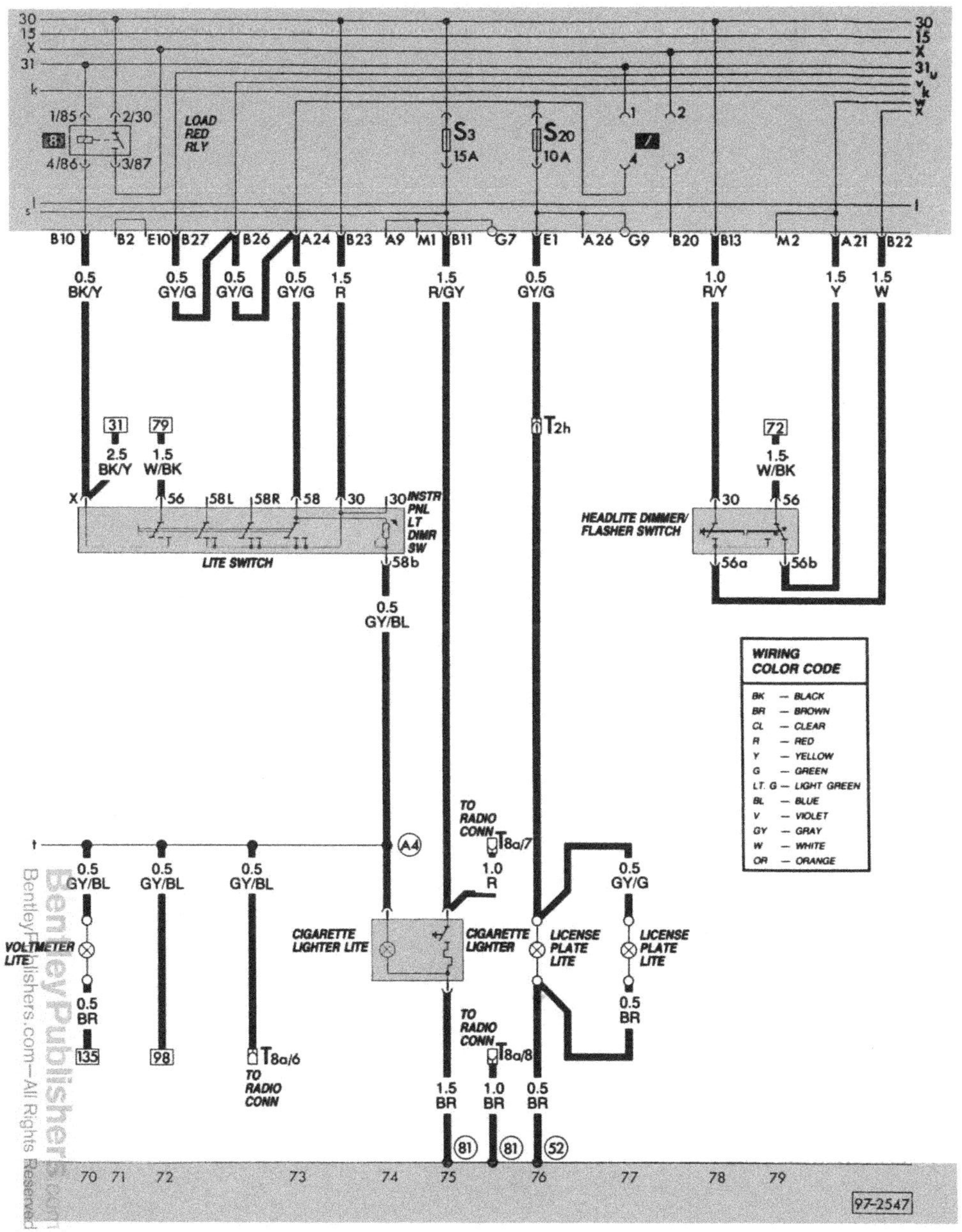

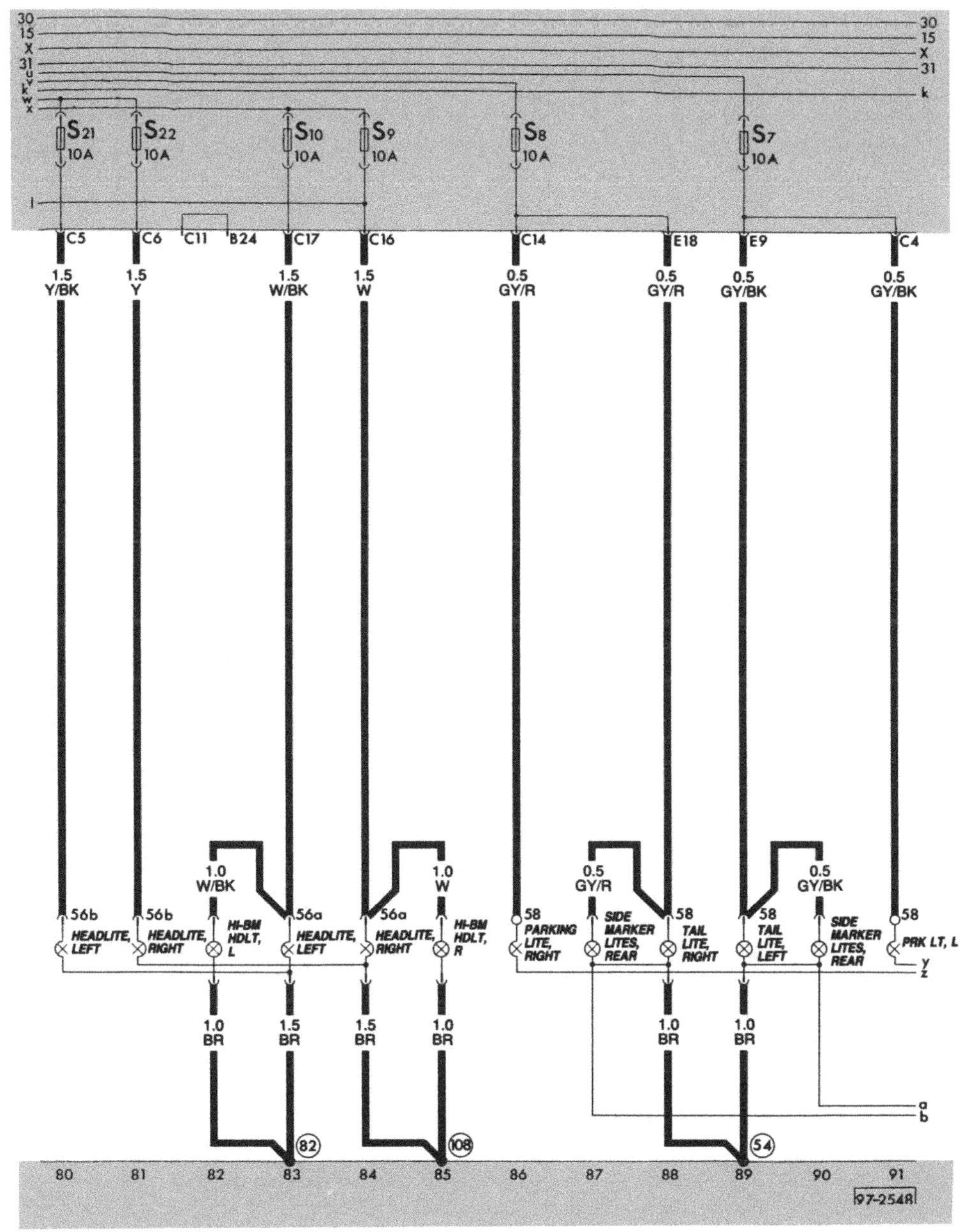

Main wiring diagram – Scirocco 1987-1989
page 7 of 9

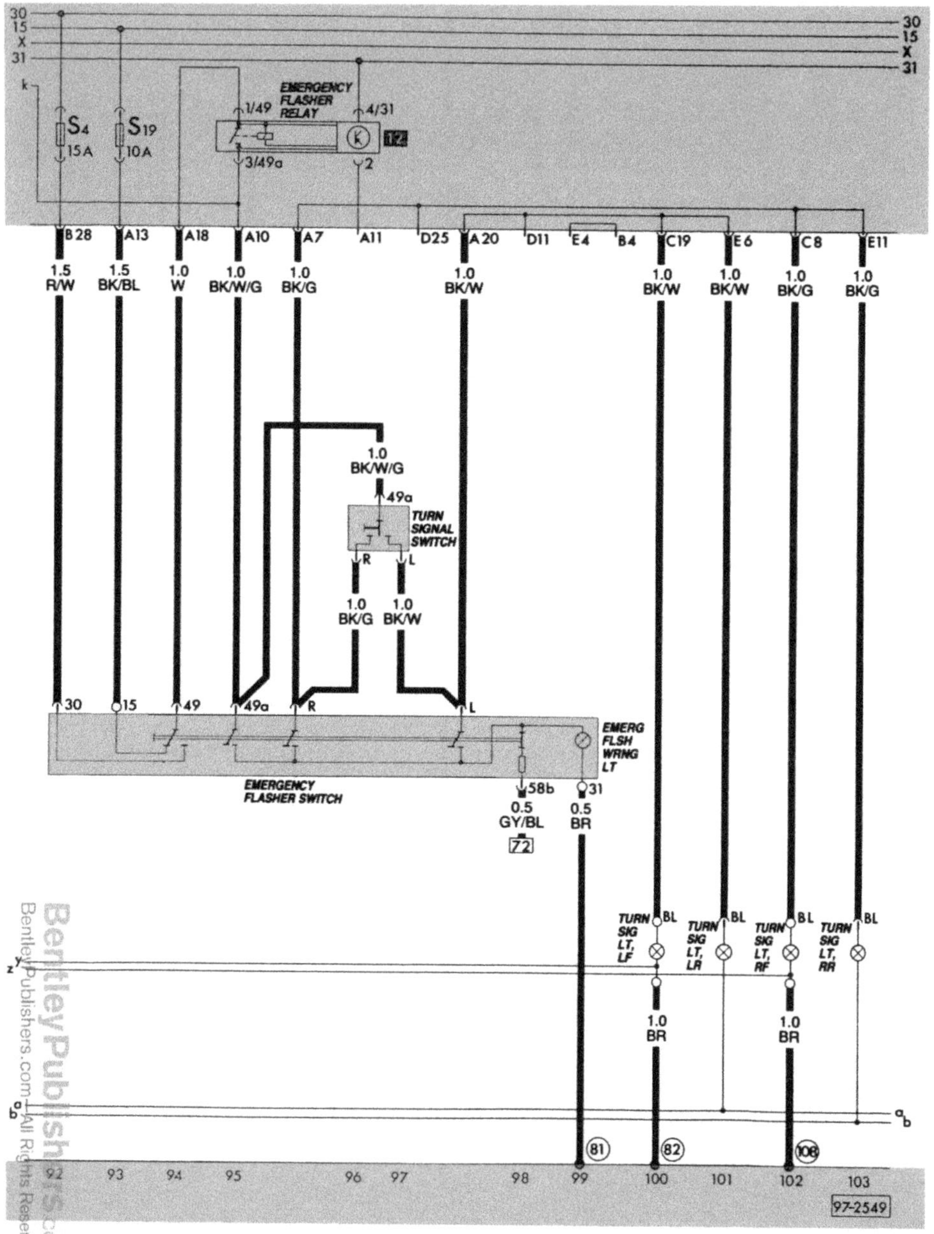

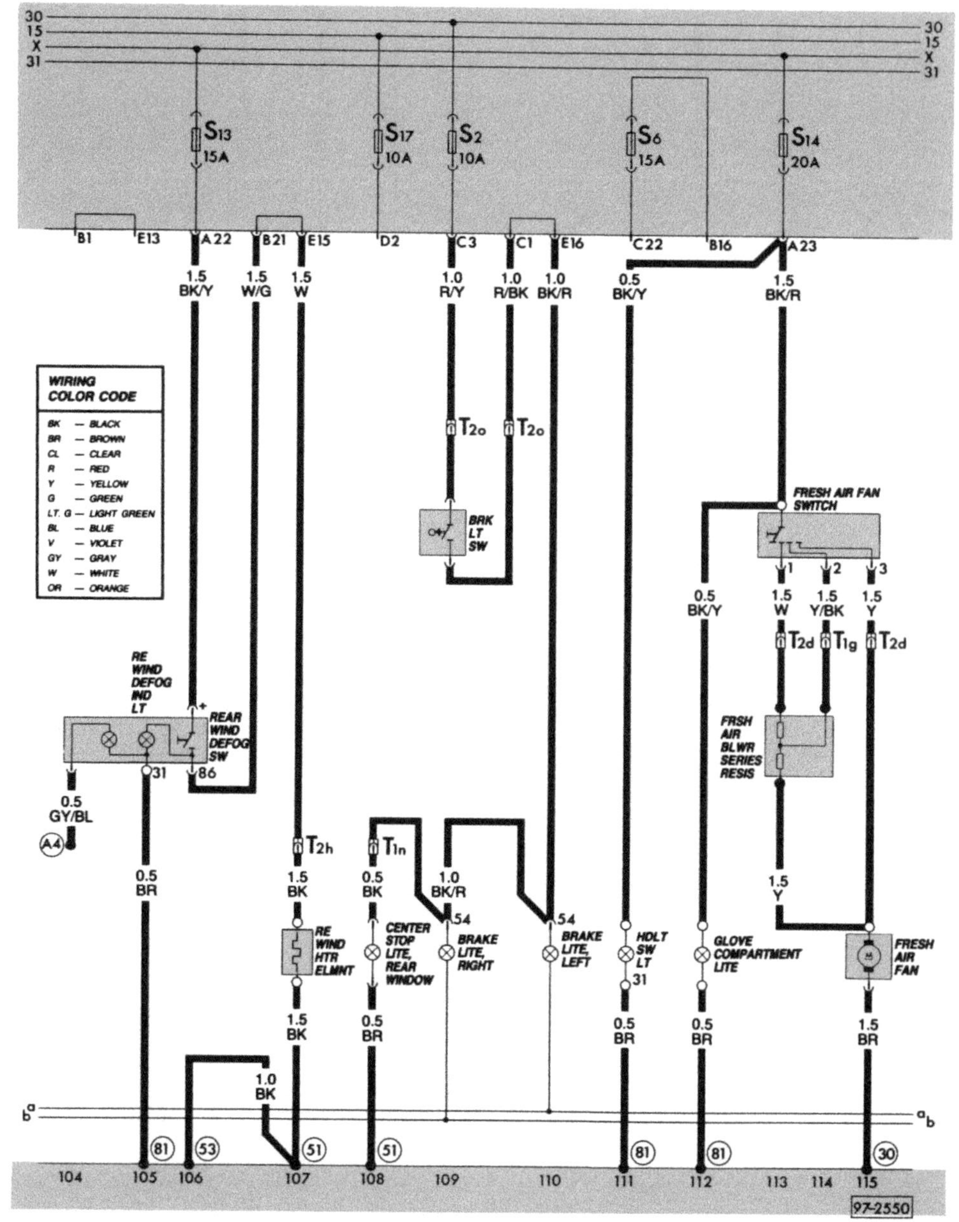

Main wiring diagram – Scirocco 1987-1989
page 8 of 9

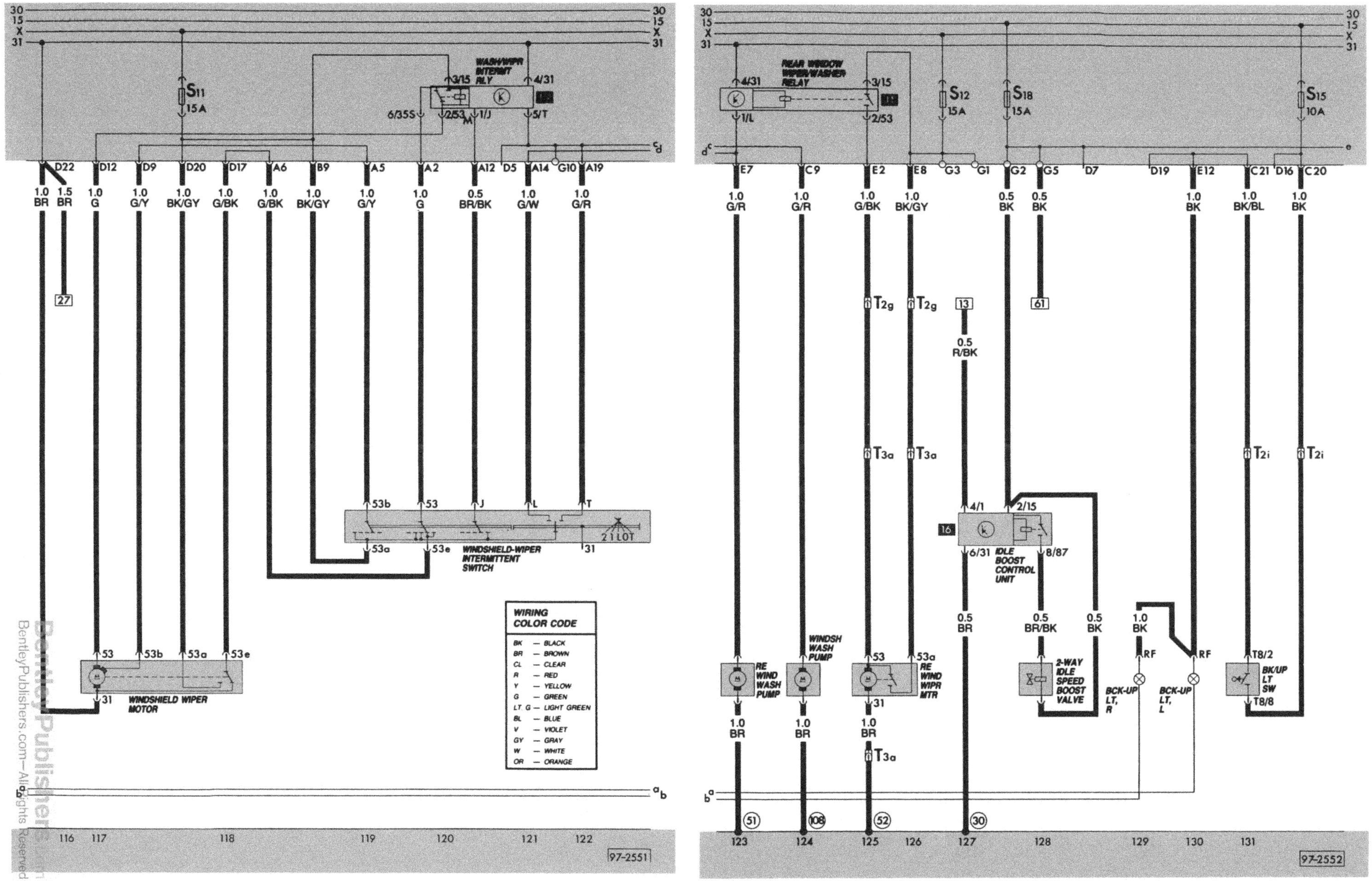

Main wiring diagram – Scirocco 1987-1989
page 9 of 9

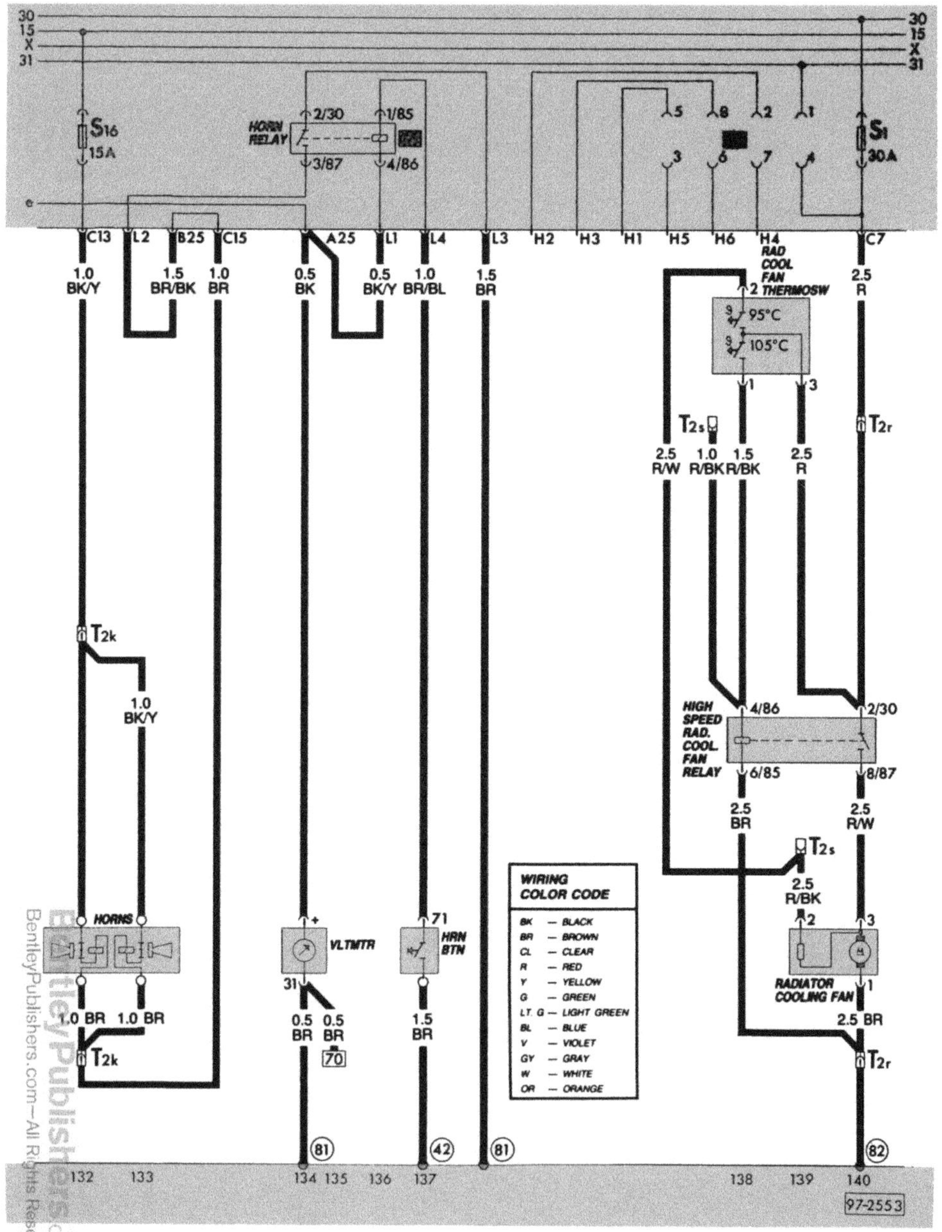

Main wiring diagram—Cabriolet 1985, 1986

Description	Current Track
Alternator	1.2
Alternator warning light	39
Auxiliary air regulator	21
Back-up light, left	115
Back-up light, right	114
Back-up light switch	115
Battery	4
Brake light, left	100
Brake light, right	99
Brake light switch	98, 99
Brake system warning switch	57
C.A.T. warning light	45
Cigarette lighter	73
Cigarette lighter light	70
Clock light	69
Cold starting valve	25
Common ground connector	21-23
Control pressure regulator	20
Coolant indicator control unit	111, 112
Coolant low level warning switch	112
Coolant temperature gauge	42
Coolant temperature sending unit	59
Coolant temperature warning light (overheating red)	44
Defogger cut-out relay (X-contact)	63, 64
Door contact/buzzer switch, left	51, 52
Door contact switch, front right	50
Dual brake & parking brake warning light	55
Electric fuel pump	23
Emergency flasher relay	89, 90
Emergency flasher switch	88-91
Emergency flasher warning light	93
Engine oil pressure control unit	33, 34
Engine oil pressure switch (1.8 bar)	60
Fresh air blower series resistance	102
Fresh air fan	103
Fresh air fan switch	102
Fuel gauge	41
Fuel gauge sending unit	24
Fuel-injection system power supply relay	17-19
Fuel pump relay	19-23
Fuses (S1-S20, S22) on fuse/relay panel	
Hall generator	9, 10
Headlight dimmer/flasher switch	76
Headlight high beam warning light	38
Headlight, left	77, 80
Headlight, right	78, 81
Headlight switch light	101
High beam, left	79
High beam, right	82
Horn button	120
Horn, dual	116
Horn relay, dual horn	118-120
Ignition coil	12
Ignition distributor	11-13
Ignition/starter switch	27-29
Interior light, front	49
Instrument panel light	67, 68
Instrument panel light switch/dimmer	70
License plate light	74-75
Light switch	64-70
Luggage compartment light	48
Luggage compartment light switch	48
Oil pressure switch	34
Oil pressure warning light	36
OXS/AFC-system control unit	14-18
OXS elapsed mileage odometer	46
OXS frequency valve*	18
OXS thermoswitch*	15
Oxygen sensor*	16
Parking light, left	86
Parking light, right	83
Parking brake warning light switch	55
Radiator fan	121
Radio connection	72
Rear window defogger switch	124
Rear window defogger switch light	122
Rear window defogger warning light	123
Rear window heater element	125
Seat belt buzzer contact, left	53
Seat belt system warning light	54
Seat belt warning system relay	52-54
Shift indicator	30
Shift indicator switch	61
Spark plugs	11-13
Spark plug connectors	11-13
Starter	5-7
Tachometer	32
Tail light, left	85
Tail light, right	84
Thermo time switch	27
Transistorized ignition control unit	8-11
Turn signal light, front left	94
Turn signal light, front right	96
Turn signal light, rear left	95
Turn signal light, rear right	97
Turn signal switch	90
Turn signal warning light	40
Up-shift control light	31
Up-shift indicator control unit	30-33
Up-shift/indicator solenoid shut-off	30
Vacuum switch/up-shift indicator	62
Voltage regulator	1, 2
Voltage stabilizer	41
Washer/wiper intermittent relay	107-109
Windshield washer pump	110
Windshield-wiper intermittent switch	107-109
Windshield wiper motor	105-106

Wire Connectors

T1 — single, engine compartment
T1a — single, on battery (+)
T1b — single, behind fuse/relay plate
T1c — single, behind dashboard
T1d — single, behind dashboard
T1e — single, behind fuse/relay panel
T1g — single, behind fuse/relay panel
T1h — single, engine compartment, left
T1i — single, behind dashboard
T1k — single, next to battery
T2 — double, behind dashboard
T2a — double, behind dashboard
T2b — double, behind fuse/relay panel
T2c — double, engine compartment, left
T2d — double, behind dashboard
T2e — double, engine compartment, left
T2f — double, luggage compartment, left
T2g — double, next to dual horn
T2h — double, next to dual horn
T3 — three-point, behind fuse/relay panel
T4 — four-point, behind dashboard
T4a — four-point, engine compartment
T6 — six-point, on dashboard cluster
T8 — eight-point, on transmission
T14 — fourteen-point, on dashboard cluster

Ground Connectors

(1) — battery/body/transmission
(8) — on cold start valve
(10) — on-fuse/relay panel
(13) — steering gear
(14) — roof member center
(15) — on wiring harness front
(16) — in dashboard wiring harness (wrapped with tape)
(18) — rear panel

Main wiring diagram—Cabriolet 1985, 1986
page 2 of 5

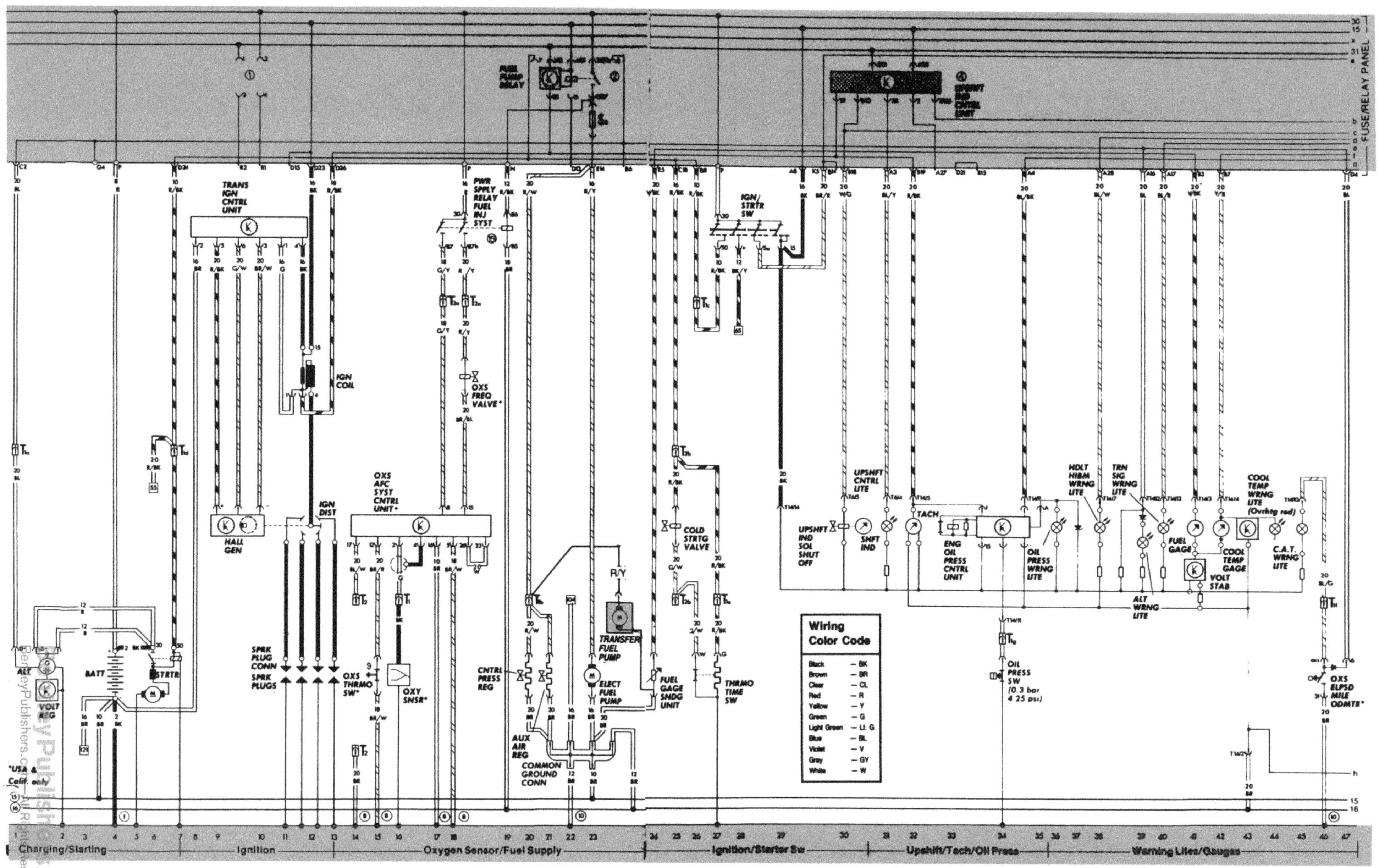

Main wiring diagram—Cabriolet 1985, 1986
page 3 of 5

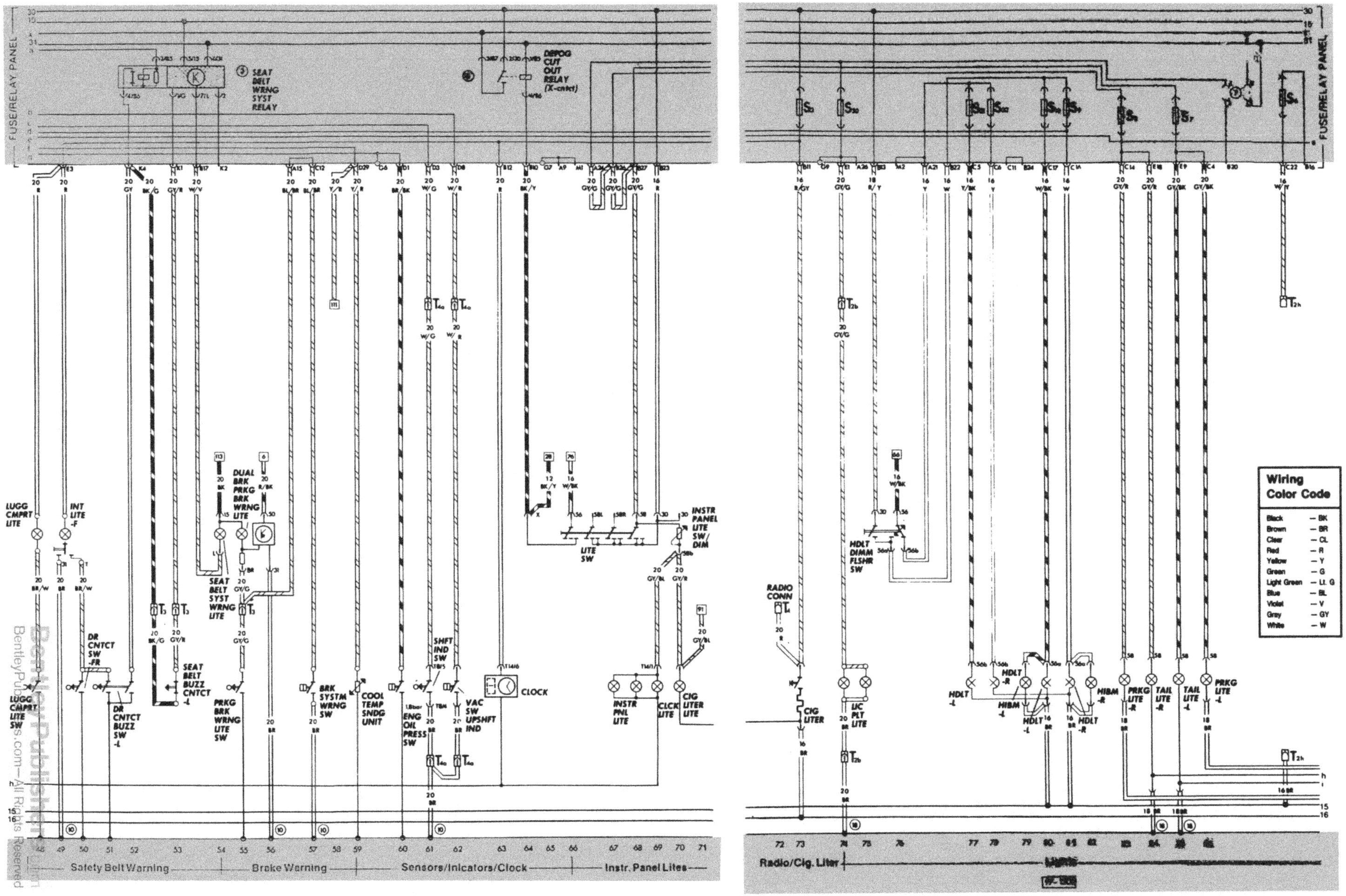

Main wiring diagram—Cabriolet 1985, 1986
page 4 of 5

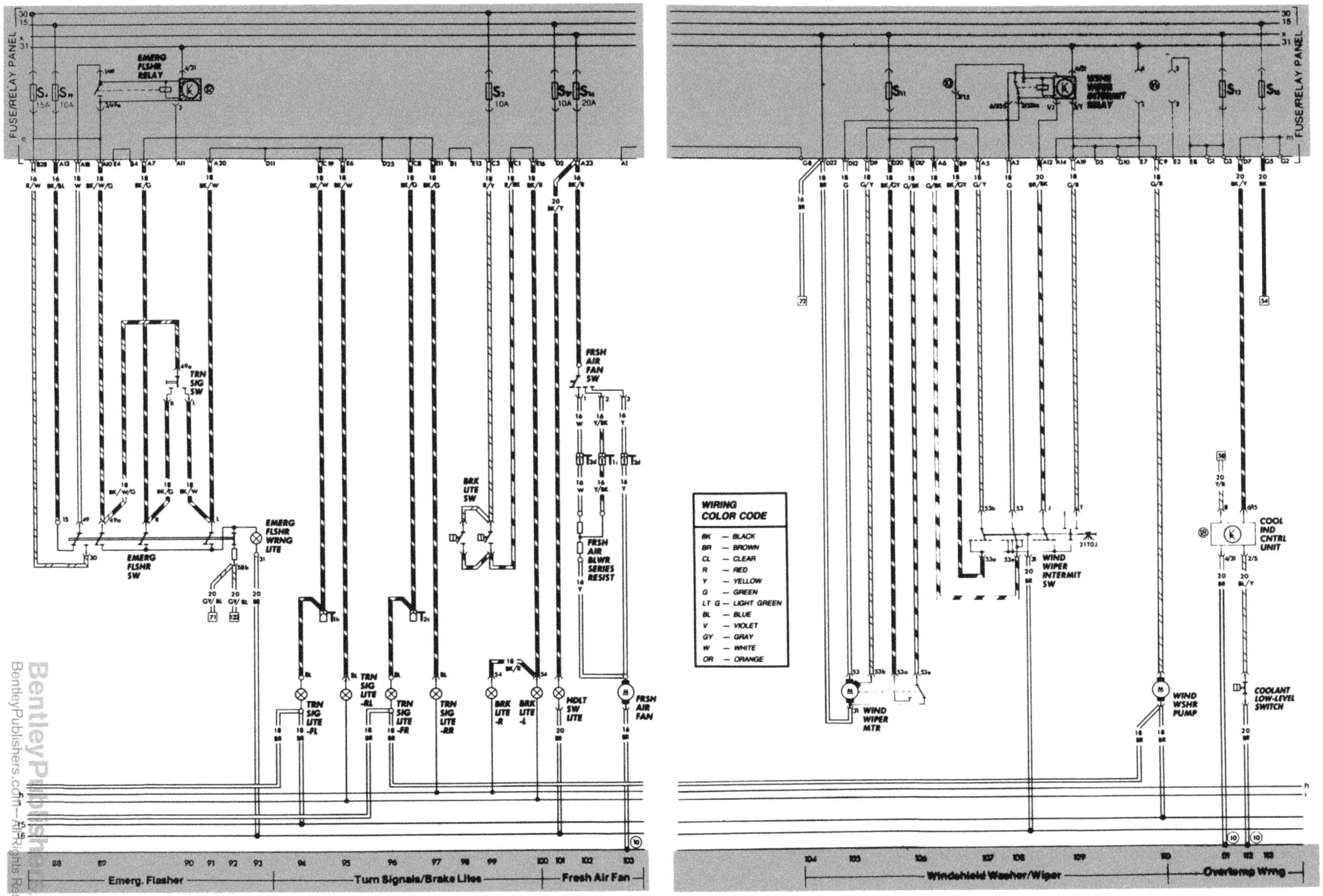

Main wiring diagram—Cabriolet 1985, 1986
page 5 of 5

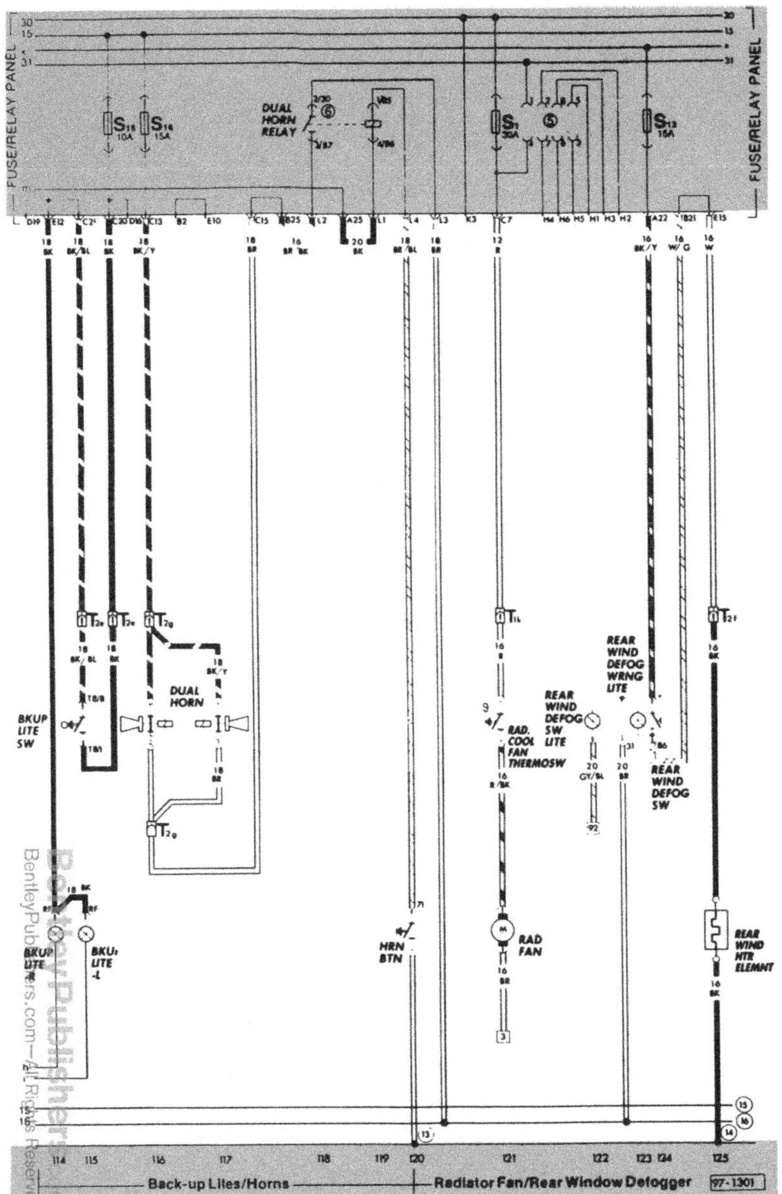

Wiring Color Code	
Black	– BK
Brown	– BR
Clear	– CL
Red	– R
Yellow	– Y
Green	– G
Light Green	– Lt G
Blue	– BL
Violet	– V
Gray	– GY
White	– W

Main wiring diagram—Cabriolet 1987-1989
page 1 of 8

Description	Current track
2-way idle speed boost valve	14
Alternator	1,2
Alternator warning lite	42
Auxiliary air regulator	25
Back-up lite switch	102
Back-up lite, left	101
Back-up lite, right	100
Battery	3
Brake and parking brake indicator lite	65
Brake fluid level warning contact	66
Brake lite switch	112
Brake lite, left	113
Brake lite, right	114
Cigarette lighter lite	72
Clock lite	48
Cold start valve	29
Control pressure regulator	26
Coolant overheat warning lite	46
Coolant temperature gauge	45
Coolant temperature sending unit	55
Digital clock	47
Door contact switch, left front	60,61
Door contact switch, right front	59
Emergency flasher relay	104,105
Emergency flasher switch	103-107
Emergency flasher warning lite	107
Engine oil pressure control unit	38,39
Engine oil pressure gauge	78,79
Engine oil pressure gauge lite	80
Engine oil pressure sensor	78
Engine oil pressure switch (0.3 bar)	53
Engine oil pressure switch (1.8 bar)	54
Engine oil pressure warning lite	40
Engine oil temperature gauge	75,76
Engine oil temperature gauge lite	77
Engine oil temperature sensor	75
Fresh air blower series resistance	116,117
Fresh air control lever lite	71
Fresh air fan	118
Fresh air fan switch	116-118
Fuel enrichment switch	15
Fuel gauge	44
Fuel gauge sending unit	23
Fuel injection power supply relay	17-19
Fuel pump	22
Fuel pump relay	19-21
Hall generator	7,8
Headlite dimmer/flasher switch	86,87
Headlite high beam warning lite	41
Headlite switch lite	115
Headlite, left	88,91
Headlite, right	89,92
High-beam headlite, left	90
High-beam headlite, right	93
Hi-mounted brake lite	112
Horn button	131
Horn relay	128-130
Horns	125,126
Hot pulse relay	30,31

Description	Current track
Idle boost control unit	13,14
Ignition coil	9-12
Ignition control unit	6-10
Ignition distributor	10-12
Ignition/starter switch	32-34
Instrument panel lite	49
Instrument panel lite dimmer switch	70
Interior lite, front	58,59
License plate lite	83,84
Lite switch	67-70
Load reduction relay	67-69
Luggage compartment lite	85
Luggage compartment lite switch	85
OXS control unit	15-18
OXS Elapsed mileage counter	51,52
OXS elapsed mileage indicator	50
OXS frequency valve	18
OXS thermoswitch	16
Oxygen sensor	17
Parking brake warning lite switch	64
Parking lite, left	98
Parking lite, right	95
Radiator cooling fan	132
Radiator cooling fan thermoswitch	132
Rear window defogger indicator lite	134
Rear window defogger switch	133-135
Rear window defogger switch lite	133
Rear window heater element	136
Refrigerator 12V connector	74
Safety belt switch, left	62
Seat belt warning lite.	63
Seat belt warning relay	61-63
Sidemarker lites, front	94,99
Spark plug connectors	10-12
Spark plugs	10-12
Starter	4,5
Tachometer	37
Tail lite, left	97
Tail lite, right	96
Thermo time switch	27,28
Transfer fuel pump	24
Turn signal indicator lite	43
Turn signal lite, left front	108
Turn signal lite, left rear	109
Turn signal lite, right front	110
Turn signal lite, right rear	111
Turn signal switch	105
Up-shift Indicator lite	36
Up-shift indicator control unit	36-38
Upshift switch, transmission	56
Upshift switch, vacuum	57
Voltage regulator	1,2
Voltage stabilizer	44
Voltmeter	81
Voltmeter lite	82
Washer/wiper intermittent relay	122,123
Windshield washer pump	124

Decription	Current track
Windshield wiper motor	120,121
Windshield-wiper intermittent switch	122,123

Wire connectors

T1a — single, near battery
T1b — single, behind relay panel plate
T1c — single, near ignition coil
T1d — single, near brake booster
T1f — single, behind relay panel plate
T1g — single, under fuse/relay panel
T1h — single, under fuse/relay panel
T1i — single, behind dash, center
T1k — single, under fuse/relay panel
T1l — single, under fuse/relay panel
T1m — single, under fuse/relay panel
T1n — single, in luggage compartment, left
T1p — single, near right headlight
T1q — single, near right headlight
T1r — single, near left headlight
T1s — single, in luggage compartment, left

T2a — double, behind dash, right
T2b — double, under fuse/relay panel
T2c — double, near ignition coil
T2d — double, behind dash, center
T2e — double, in engine compartment, left
T2f — double, in eng. compart., left
T2g — double, in eng. compart., left
T2h — double, in eng. compart., left
T2k — double, behind relay panel plate
T2l — double, in rear decklid, left
T2m — double, in engine compartment, left

T3a — three-point, under fuse/relay panel
T3b — three-point, behind dash, center
T3c — three-point, under inst. panel middle
T3d — three-point, near right headlight
T3e — three-point, near left headlight

T6 — 6 point, on instrument cluster

Ground connections

(24) — on cold starting valve
(30) — beside relay plate
(42) — beside steering column
(50) — luggage compartment, left
(51) — luggage compartment, right
(81) — in dashboard wiring harness
(82) — -1- in wiring harness, front/left
(108) — -2- in wiring harness, front/left

Plus connections

(A4) — (58b), in dashboard wiring harness

NOTE —
Electronic engine controls shown in this diagram are accurate through June 1987 production only. Electronic engine controls for later 1987 cars (from July 1987 production), and all 1988-1989 cars are covered by separate "Electronic Engine Controls" diagrams elsewhere in this section.

Main wiring diagram—Cabriolet 1987-1989
page 2 of 8

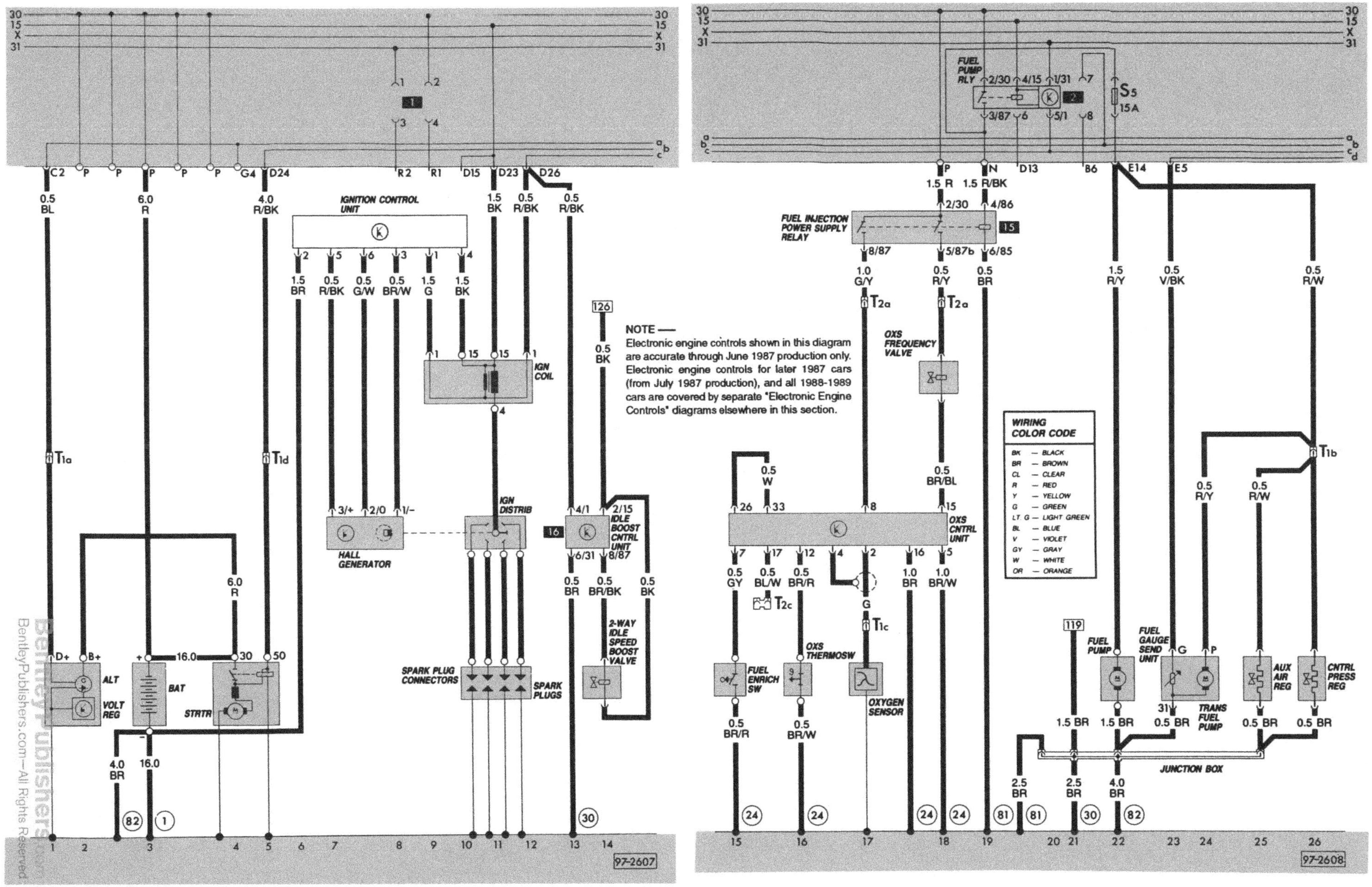

NOTE —
Electronic engine controls shown in this diagram are accurate through June 1987 production only. Electronic engine controls for later 1987 cars (from July 1987 production), and all 1988-1989 cars are covered by separate "Electronic Engine Controls" diagrams elsewhere in this section.

Main wiring diagram—Cabriolet 1987-1989
page 3 of 8

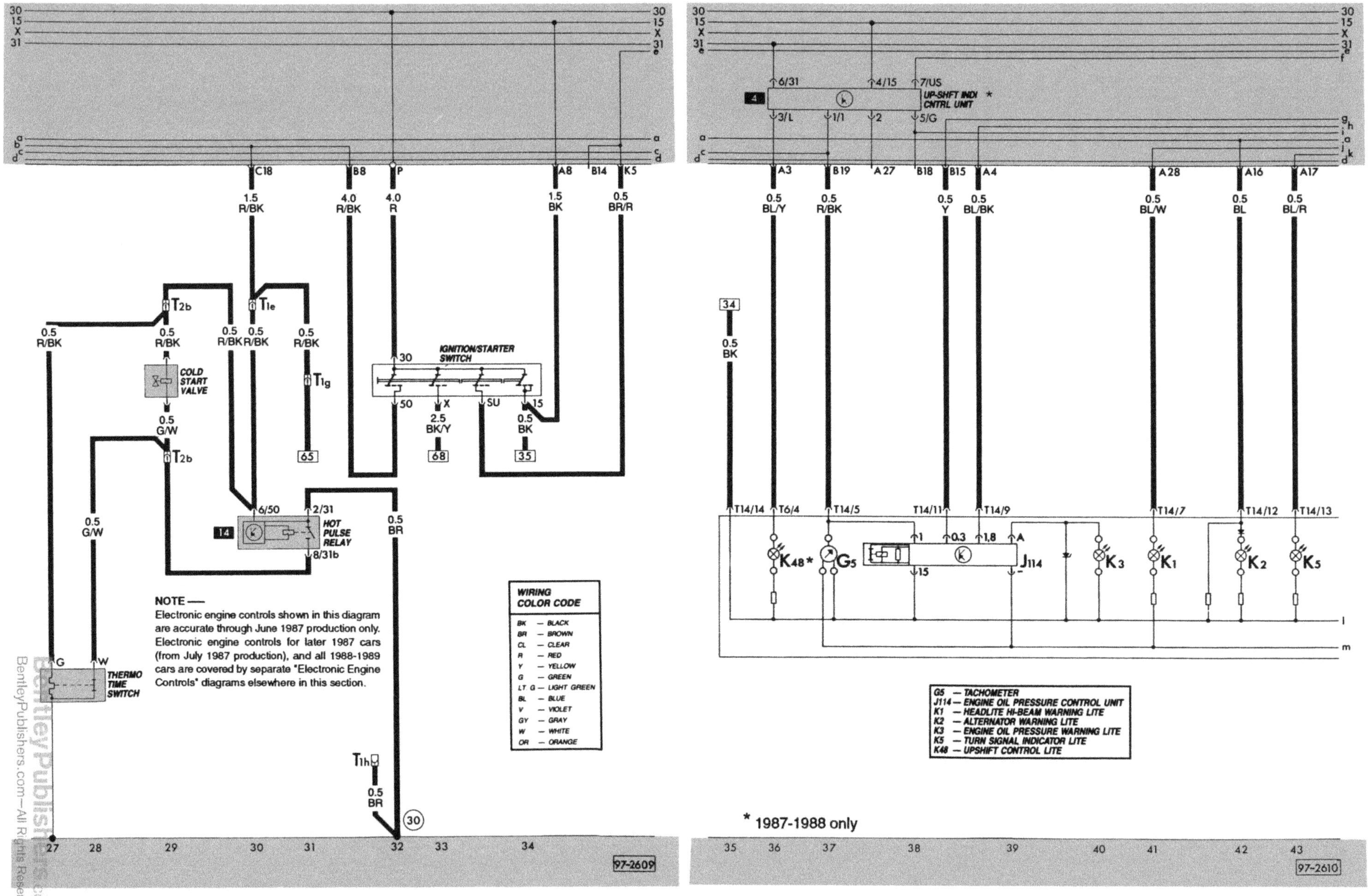

Main wiring diagram—Cabriolet 1987-1989
page 4 of 8

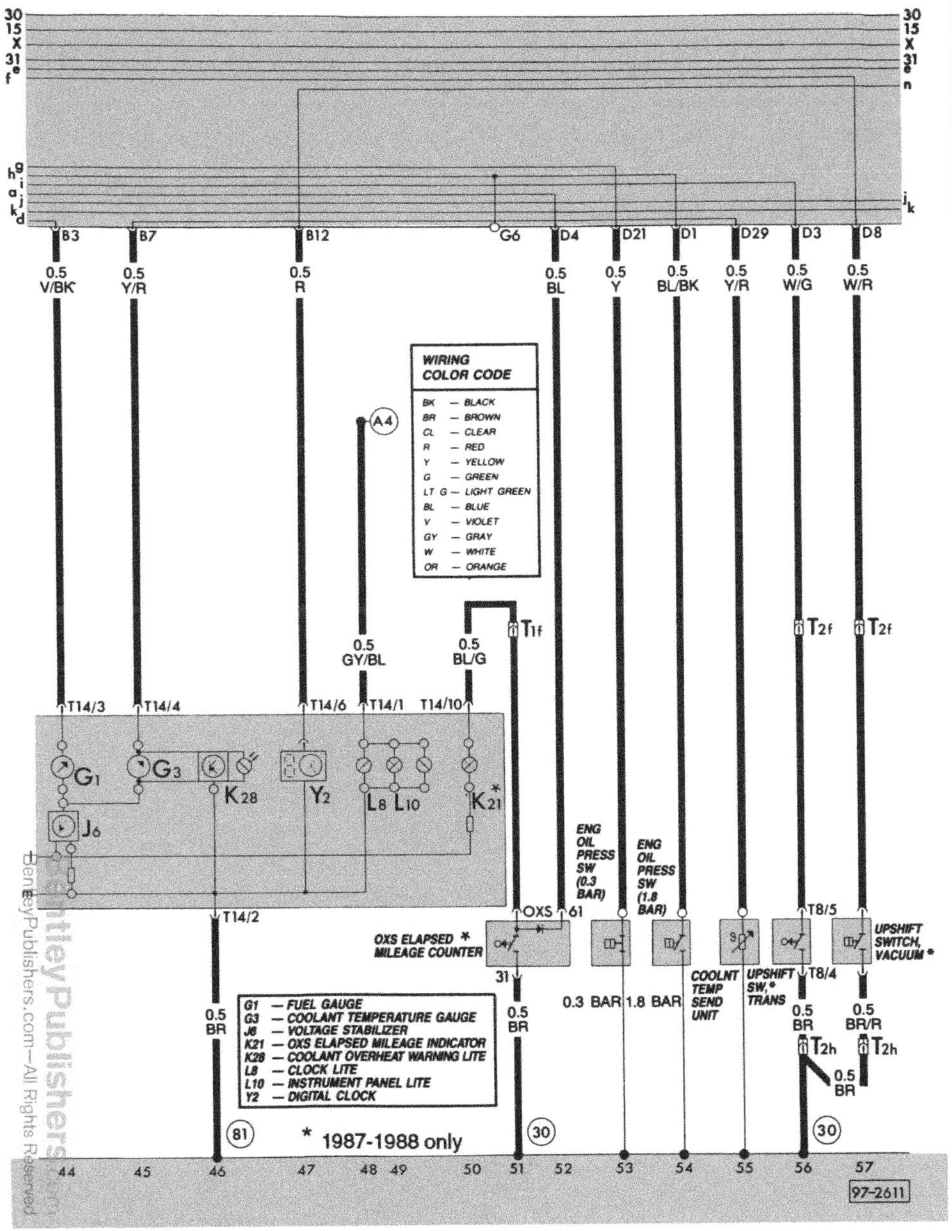

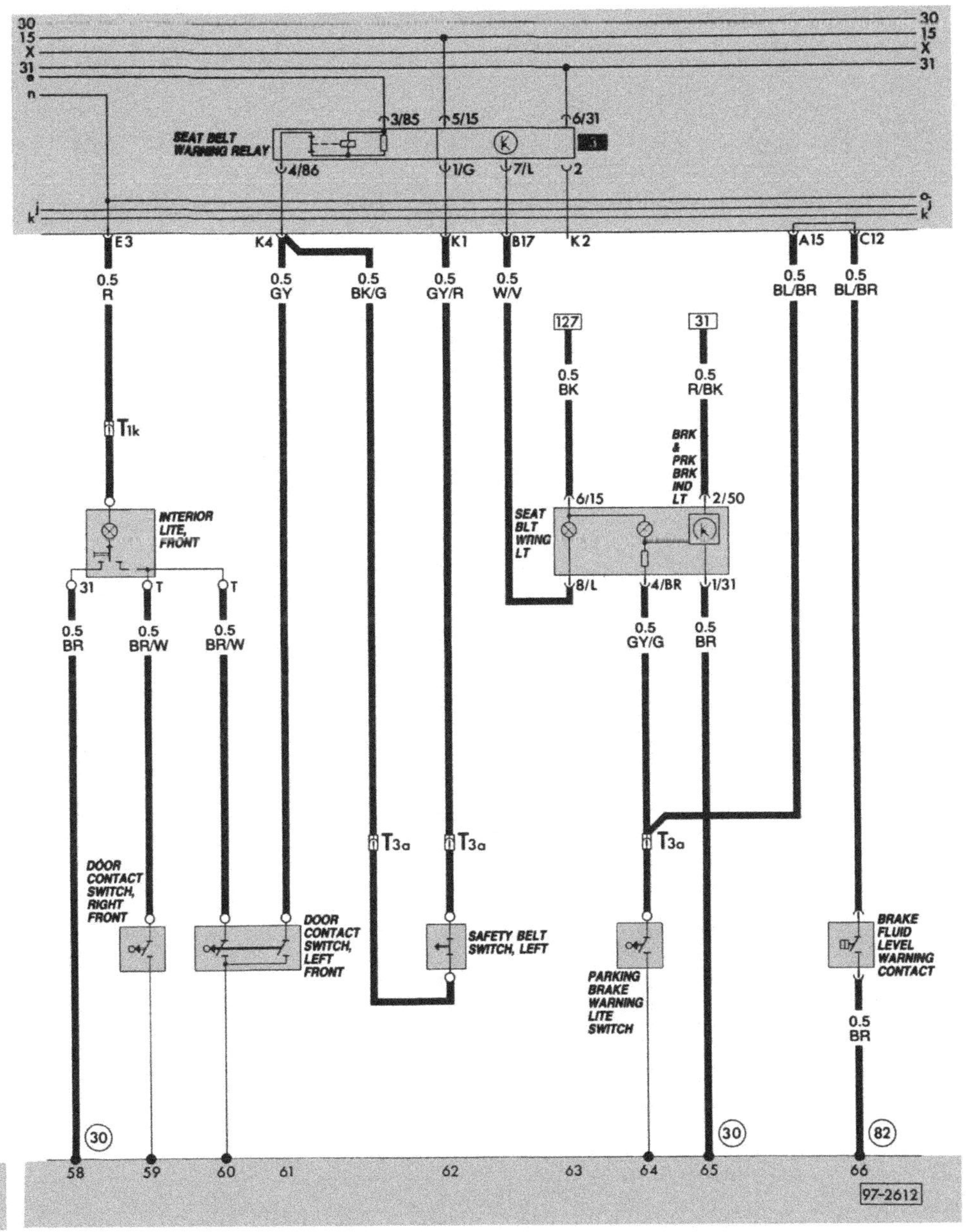

Main wiring diagram—Cabriolet 1987-1989
page 5 of 8

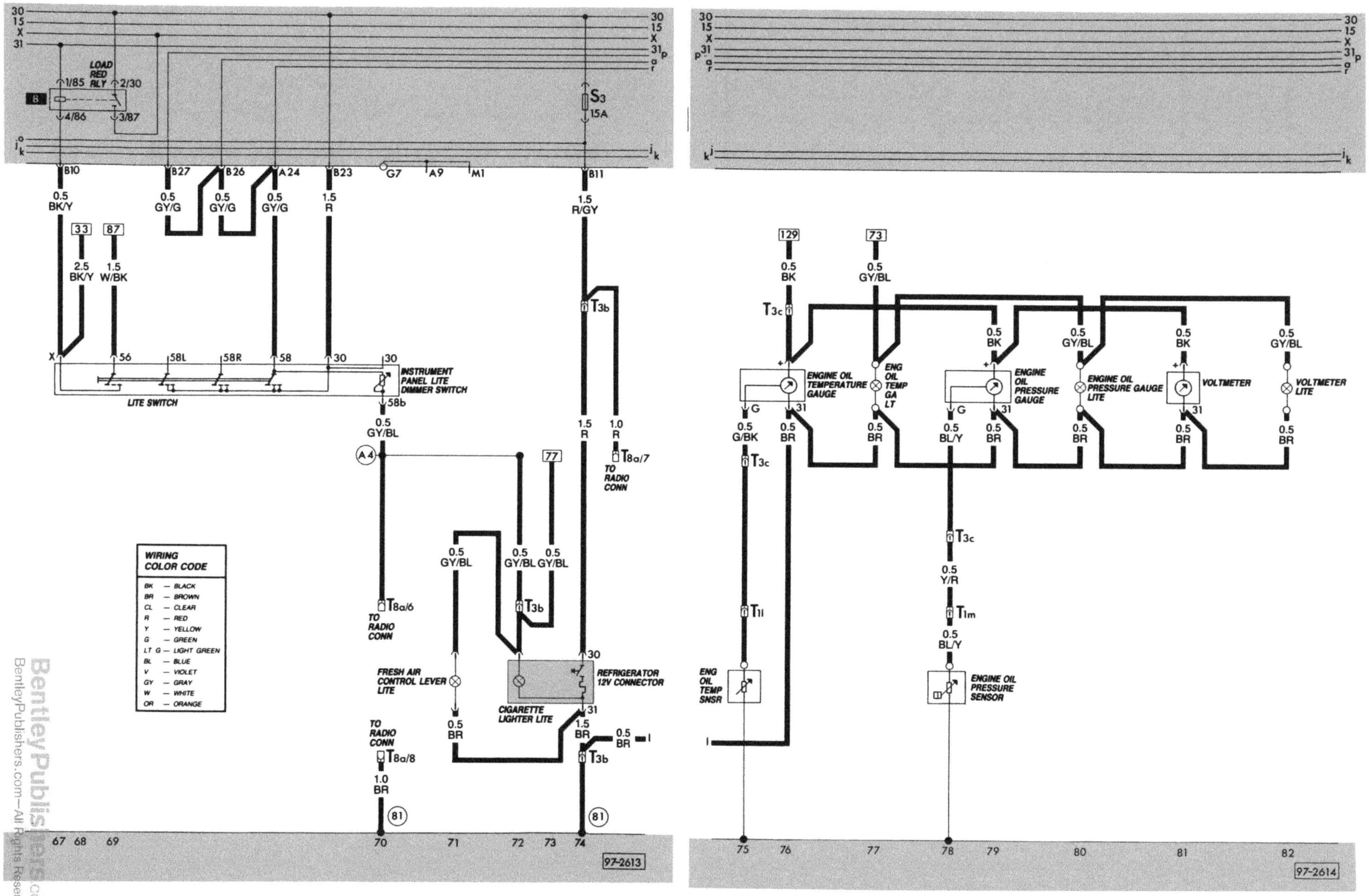

Main wiring diagram—Cabriolet 1987-1989
page 6 of 8

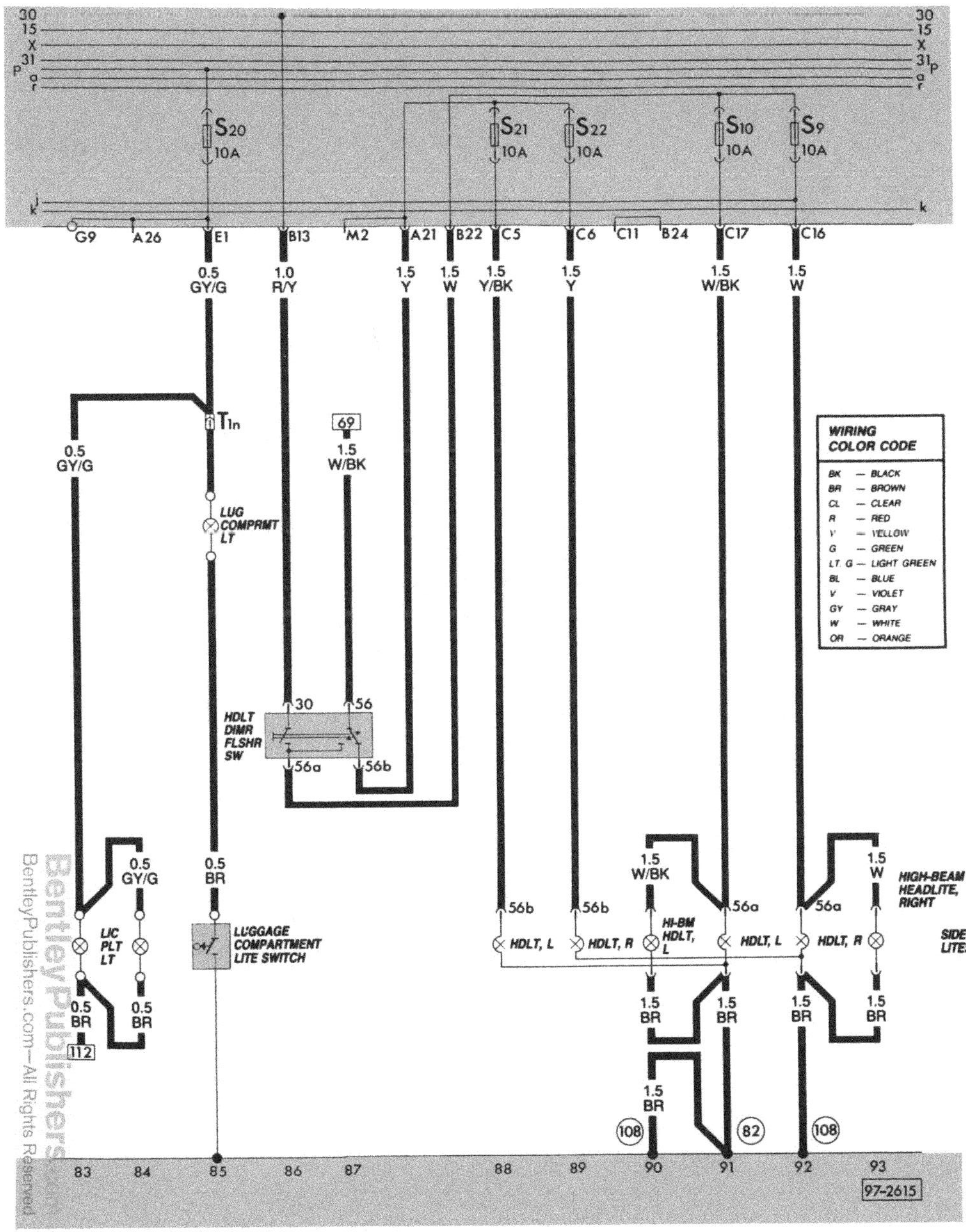

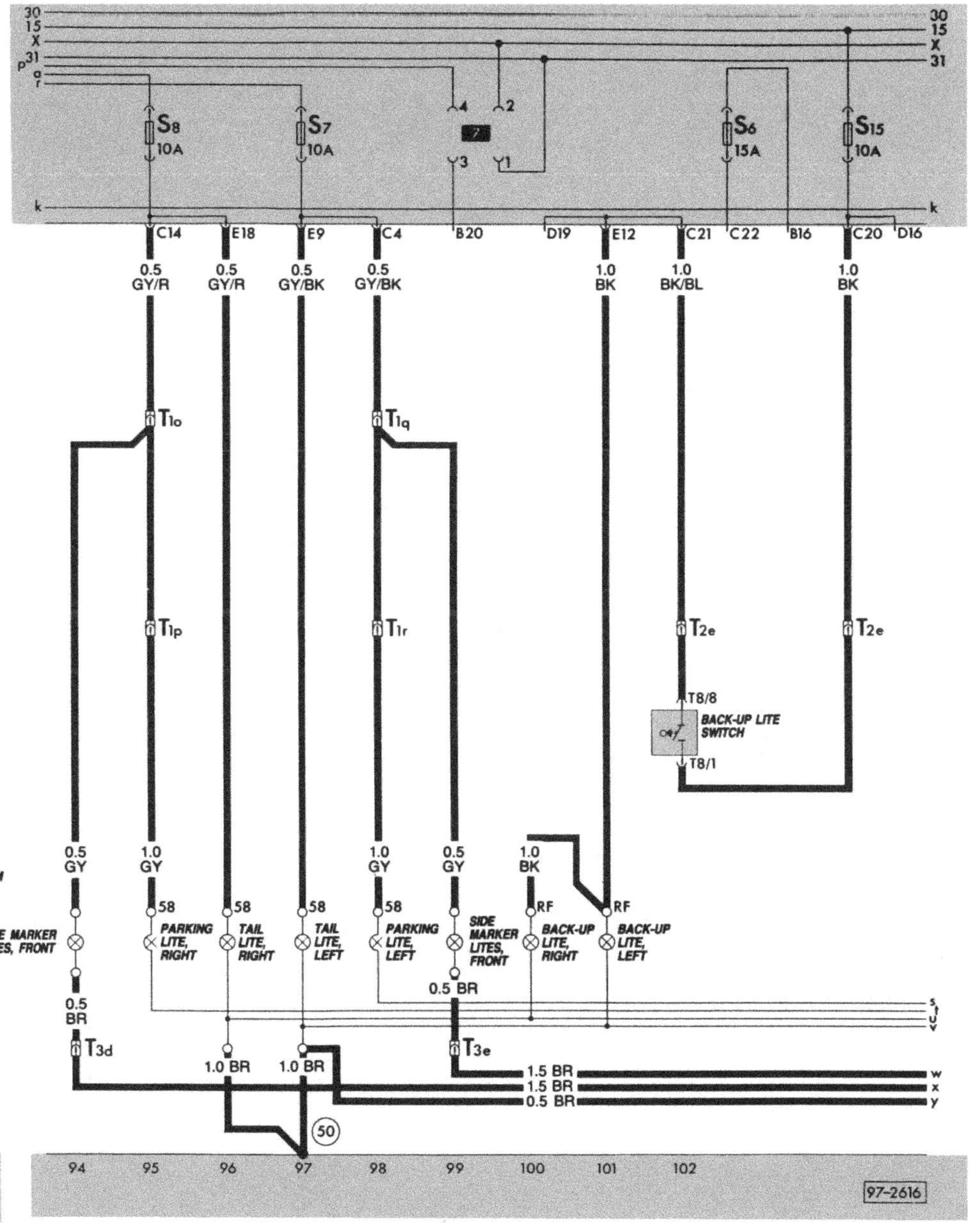

Main wiring diagram—Cabriolet 1987-1989
page 7 of 8

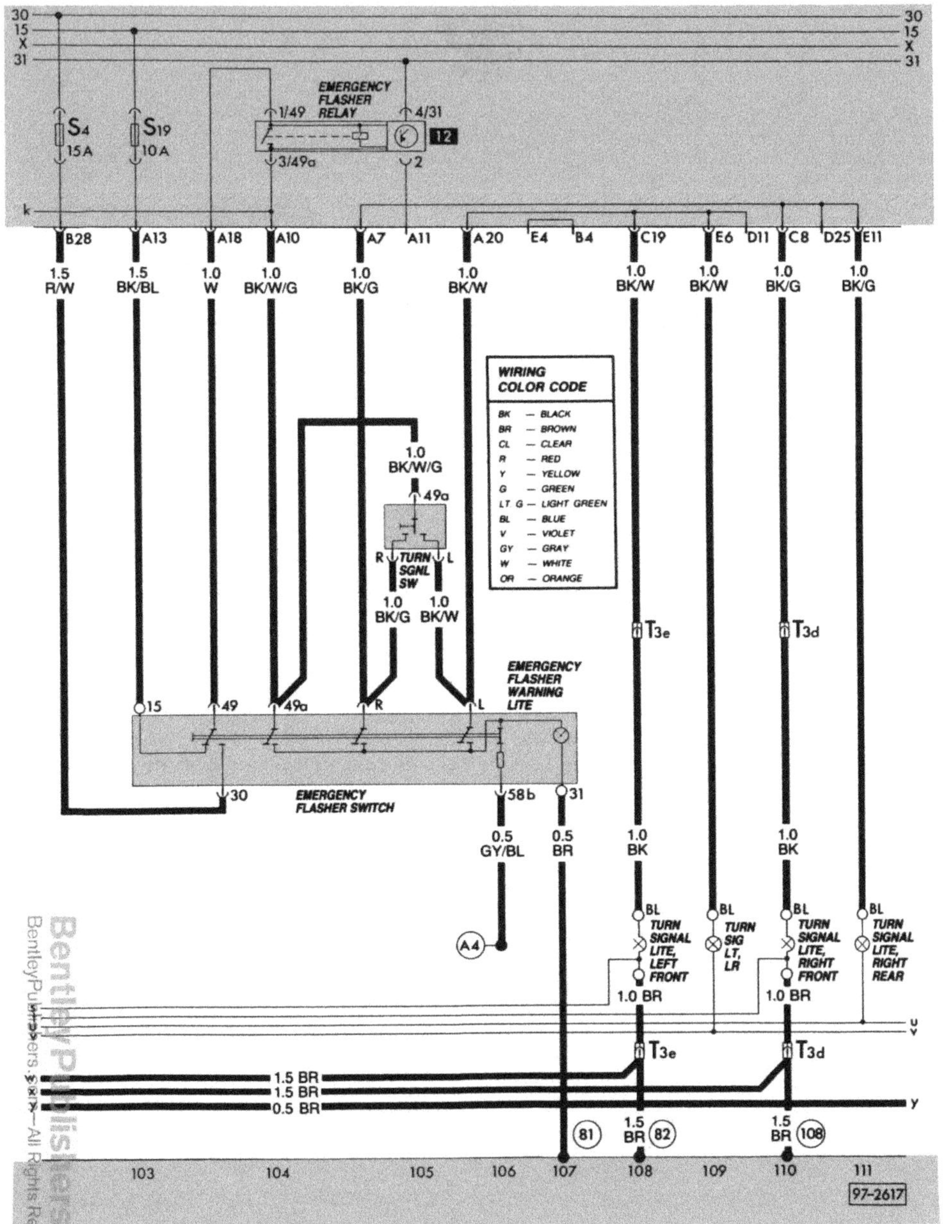

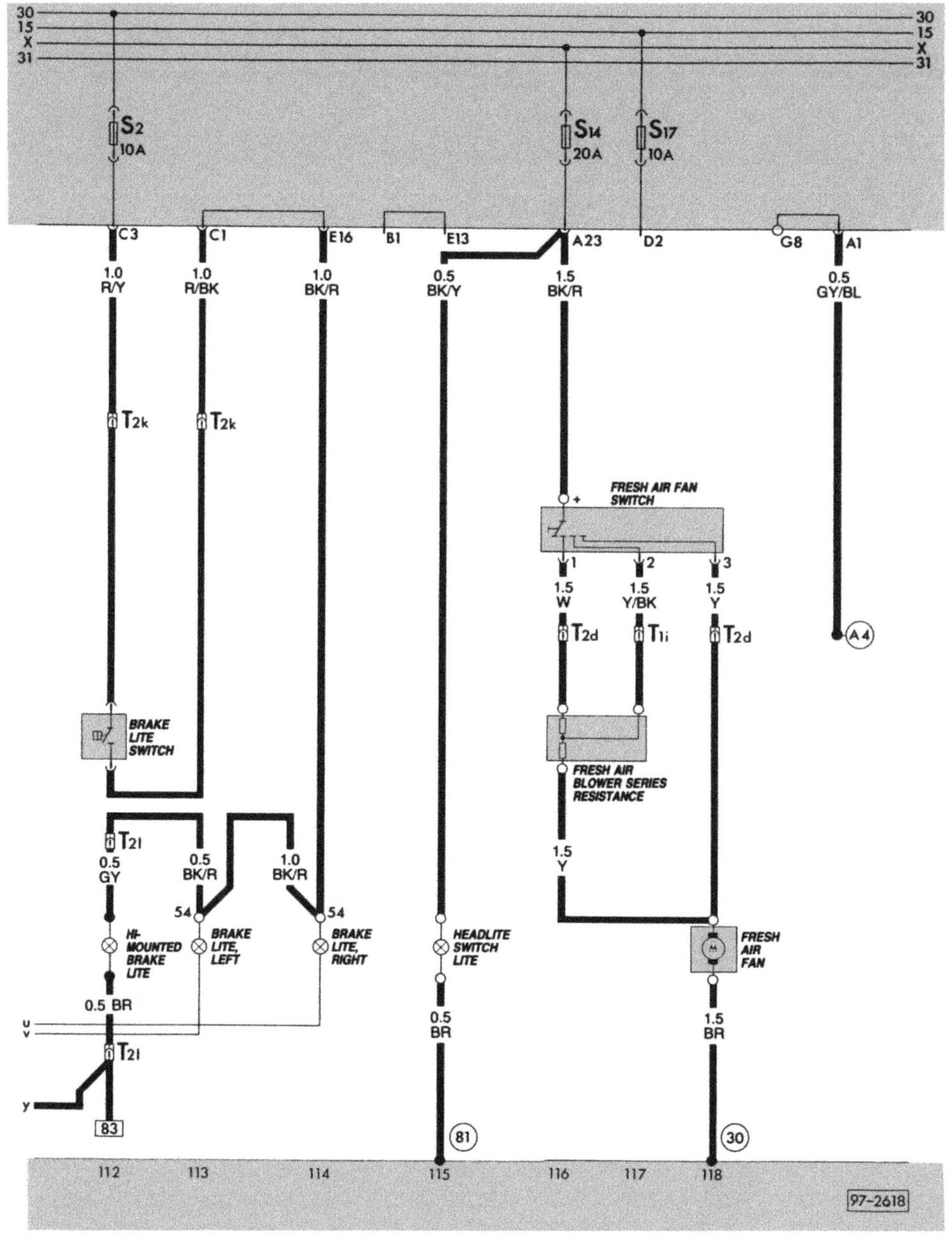

Main wiring diagram—Cabriolet 1987-1989
page 8 of 8

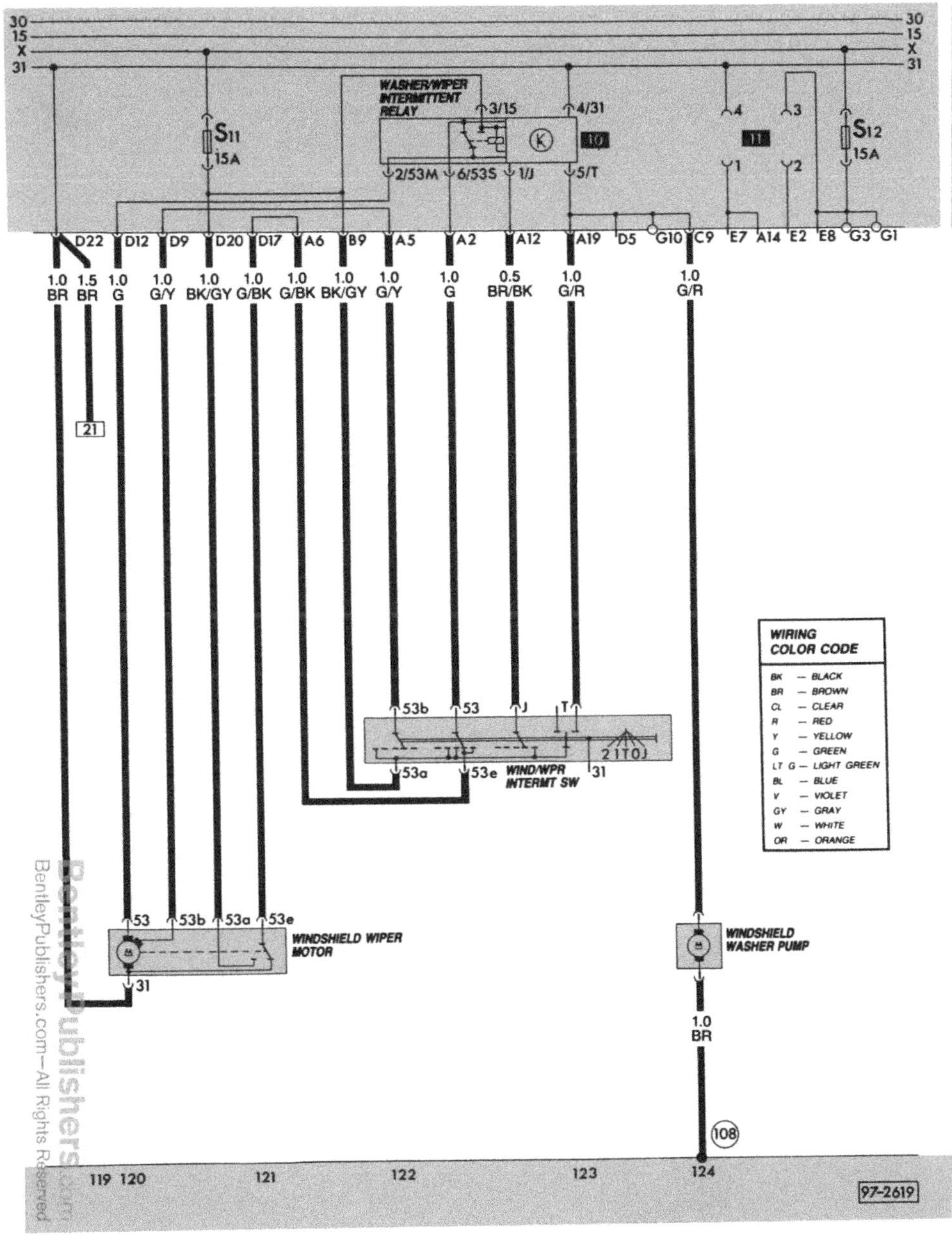

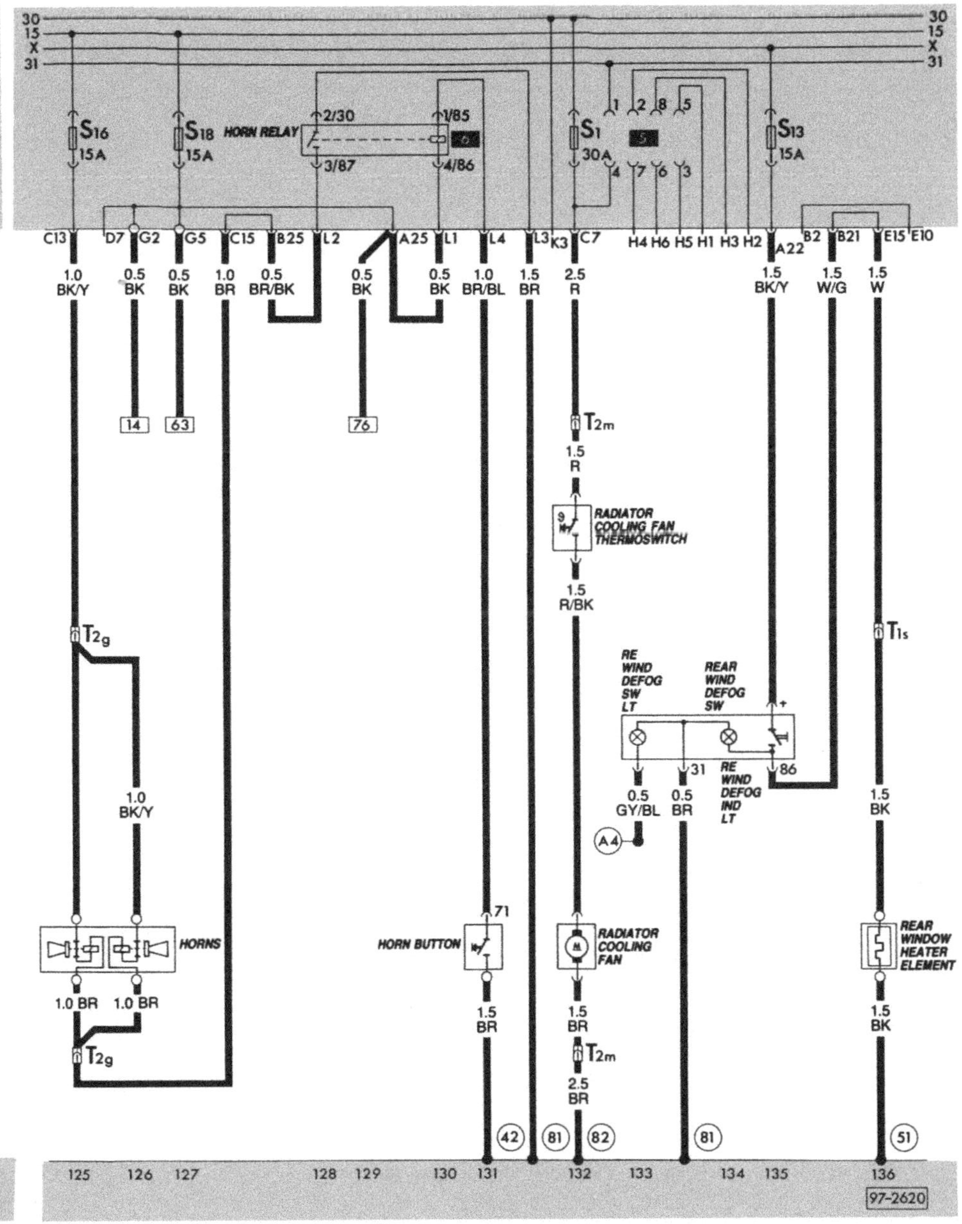

Main wiring diagram – Cabriolet 1990

Description	Current track
Air flow sensor potentiometer, G19	27
Alternator, C	1-2
Alternator warning light, K2	70
Back-up light switch, F4	179
Back-up light, left, M16	171
Back-up light, right, M17	180
Battery, A	4
Brake and parking brake indicator light, K7	93
Brake fluid level warning contact, F34	98
Brake light switch, F	173
Brake light, left, M9	173
Brake light, right, M10	176
Cigarette lighter, U1	104
Cigarette lighter light, L28	102
Clock light, L8	82
Coolant low level control unit, J120	58-59
Coolant low level sender, G32	58
Coolant overheat warning light, K28	79
Coolant temperature gauge, G3	77
Coolant temperature sender, G62	24
Coolant temperature sending unit, G2	57
Digifant control unit, J169	20-38
Digifant control unit relay, J176	37-38
Digital clock, Y2	81
Door contact switch, left front, F2	88-89
Door contact switch, right front, F3	86
Emergency flasher relay, J2	187-189
Emergency flasher switch, E3	184-193
Emergency flasher warning light, K6	193
Engine oil pressure control unit, J114	65-67
Engine oil pressure gauge, G11	116-117
Engine oil pressure gauge light, L27	124
Engine oil pressure sensor, G10	116
Engine oil pressure switch (0.3 bar), F22	61
Engine oil pressure switch (1.8 bar), F1	62
Engine oil pressure warning light, K3	63
Engine oil temperature gauge, G9	113-114
Engine oil temperature gauge light, L24	126
Engine oil temperature sensor, G8	113

Description	Current track
Fresh air blower series resistance, N23	157-158
Fresh air control lever light, L16	99
Fresh air fan, V2	159
Fresh air fan switch, E9	157-159
Fuel gauge, G1	76
Fuel gauge sending unit, G	51
Fuel pump, G6	50
Fuel pump relay, J17	52-54
Full throttle switch, F81	20
Fuse, S1	236
Fuse, S2	173
Fuse, S3	104
Fuse, S4	186
Fuse, S5	50
Fuse, S6	155
Fuse, S7	162
Fuse, S8	165
Fuse, S9	133
Fuse, S10	129
Fuse, S11	215
Fuse, S12	229
Fuse, S13	208
Fuse, S14	157
Fuse, S15	178
Fuse, S16	199
Fuse, S17	158
Fuse, S17	34
Fuse, S18	197
Fuse, S19	184
Fuse, S20	154
Fuse, S21	130
Fuse, S22	134
Hall generator, G40	20-22
Headlight dimmer/flasher switch, E4	137-138
Headlight high beam warning light, K1	75
Headlight switch light, L9	156
Headlight, left, L1	129, 130
Headlight, right, L2	133, 134
Hi-mounted brake light, M25	174
High-beam headlight, left, L13	127
High-beam headlight, right, L14	131
Horn button, H	204
Horn relay, J4	203-204
Horns, H1	199, 201
Idle speed stabilizer valve, N71	36-37
Idle switch, F60	23
Ignition coil, N	16-18
Ignition control unit, N41	17-20
Ignition distributor, O	17-19
Ignition/starter switch, D	10-13
Injector, cyl. 1, N30	42
Injector, cyl. 2, N31	41
Injector, cyl. 3, N32	40
Injector, cyl. 4, N33	39
Instrument panel light, L10	83, 84
Instrument panel light dimmer switch, E20	149
Intake air temperature sensor, G42	26
Interior light, front, W	85-86

Description	Current track
Knock sensor, G61	25-27
License plate light, X	151, 153
Light switch, E1	142-149
Load reduction relay, J59	142-143
Luggage compartment light, W3	154
Luggage compartment light switch, F5	154
Oxygen sensor, G39	34-35
Parking brake warning light switch, F9	94
Parking light, left, M1	163
Parking light, right, M3	166
Radiator cooling fan, V7	236-237
Radiator cooling fan after run control unit, J138	232-235
Radiator cooling fan thermoswitch, F87	232
Radiator cooling fan thermoswitch, F18	236-238
Rear window defogger indicator light, K10	206-208
Rear window defogger switch, E15	208
Rear window defogger switch light, L39	206
Rear window heater element, Z1	210
Seat belt switch, left, E24	91
Seat belt warning light, K19	93
Seat belt warning relay, J34	89-93
Side marker lights, front, M11	162, 165
Spark plug connectors, P	17-19
Spark plugs, Q	17-19
Speed sensor, G54	77
Starter, B	7-9
Tachometer, G5	71
Tail light, left, M4	170
Tail light, right, M2	181
Transfer fuel pump, G23	52
Turn signal indicator light, K5	74
Turn signal light, left front, M5	164
Turn signal light, left rear, M6	169
Turn signal light, right front, M7	167
Turn signal light, right rear, M8	182
Turn signal switch, E2	190-191
Voltage regulator, C1	1-2
Voltage stabilizer, J6	76
Voltmeter, G14	119
Voltmeter light, L25	122
Washer/wiper intermittent relay, J31	219-223
Windshield washer pump, V5	224
Windshield wiper motor, V	213-216
Windshield-wiper intermittent switch, E22	219-223

Wire connectors

T — junction box, behind fuse/relay panel
T1a — single, near ignition coil
T1b — single, behind fuse/relay panel
T1c — single, behind fuse/relay panel
T1d — single, on firewall, right
T1e — single, behind fuse/relay panel
T1f — single, behind fuse/relay panel
T1g — single, behind fuse/relay panel
T1h — single, behind instrument panel
T1i — single, behind instrument panel
T1j — single, behind instrument panel, center
T1k — single, behind fuse/relay panel
T1m — single, behind fuse/relay panel
T1n — single, in luggage compartment, left
T1p — single, near left headlight
T1r — single, near right headlight
T1u — single, behind instrument panel
T1v — single, behind fuse/relay panel
T1w — single, in engine compartment, left front
T1x — single, in luggage compartment, left
T2a — double, behind fuse/relay panel
T2b — double, behind fuse/relay panel
T2c — double, behind instrument panel, center
T2d — double, near intake manifold
T2e — double, in engine compartment, left
T2f — double, behind fuse/relay panel
T2g — double, in luggage compartment, left
T2h — double, in luggage compartment
T2l — double, near left headlight
T3a — three-point, behind fuse/relay panel
T3b — three-point, behind instrument panel, center
T3c — three-point, behind center console
T3d — three-point, near right headlight
T3e — three-point, near left headlight
T3g — three-point, near starter
T3h — three-point, near valve cover
T4a — four-point, in engine compartment, rear
T5a — five-point, on firewall, left
T6 — six-point, on instrument cluster
T6a — six-point, behind fuse/relay panel
T8 — eight-point, on radio
T14 — fourteen-point, on instrument cluster

Ground connections

(1) — ground strap battery to body
(18) — on engine block
(30) — near fuse/relay panel
(42) — beside steering column
(50) — in luggage compartment, left
(51) — in luggage compartment, right
(81) — in instrument cluster wiring harness
(82) — in wiring harness, left front
(94) — in Digifant wiring harness
(108) — in wiring harness, left front

Welded wiring harness points

(A4) — plus connection (58b), in instrument cluster wiring harness

(G3) — plus connection, in fuel injector wiring harness

(G4) — wire connection, in fuel injector wiring harness

Main wiring diagram – Cabriolet 1990
page 3 of 11

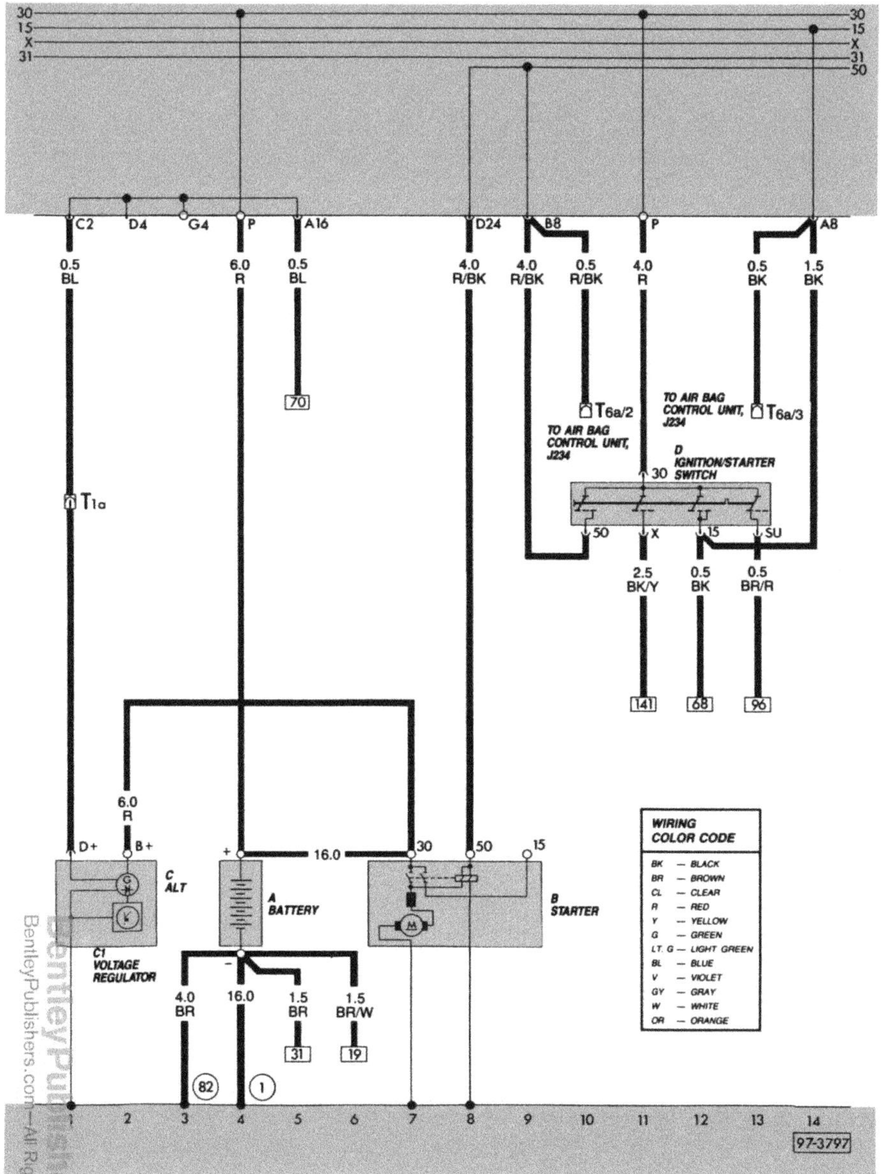

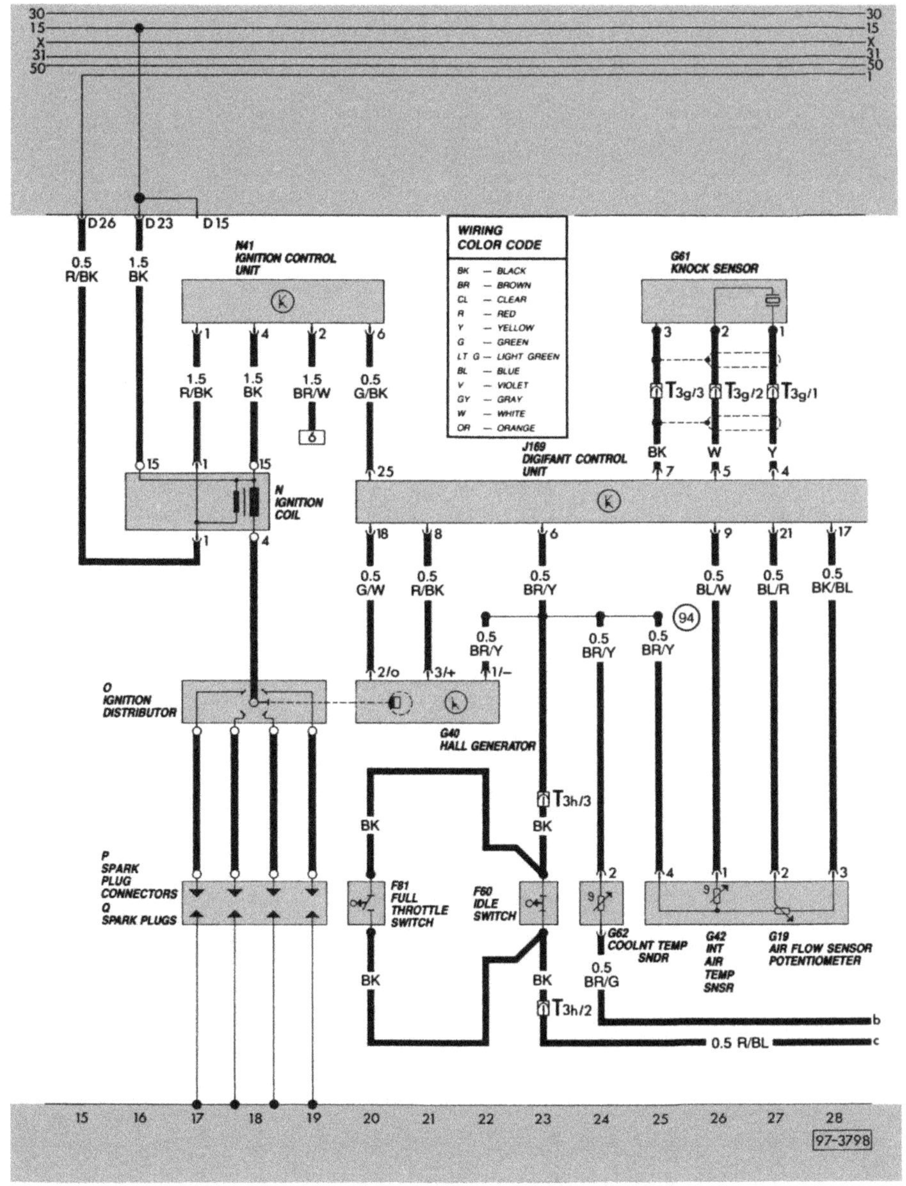

Main wiring diagram – Cabriolet 1990

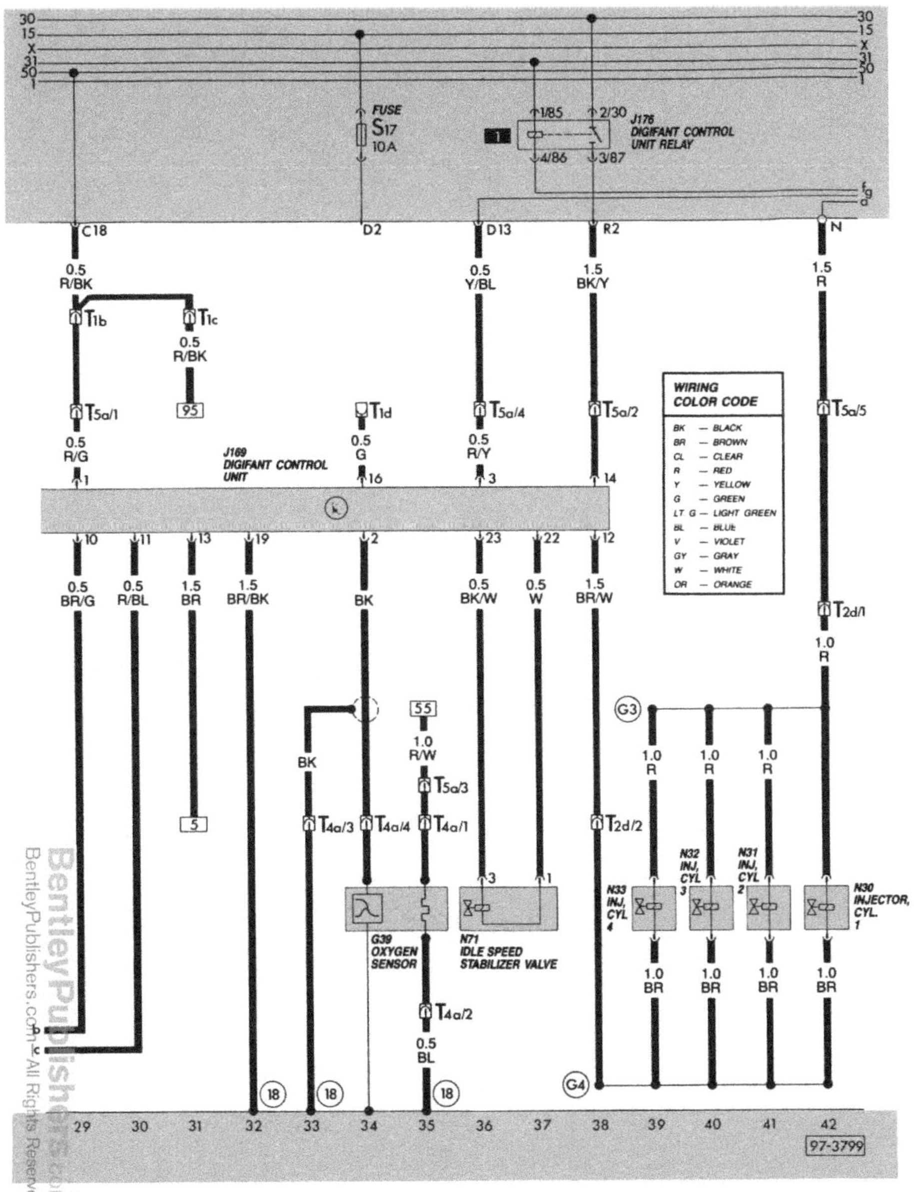

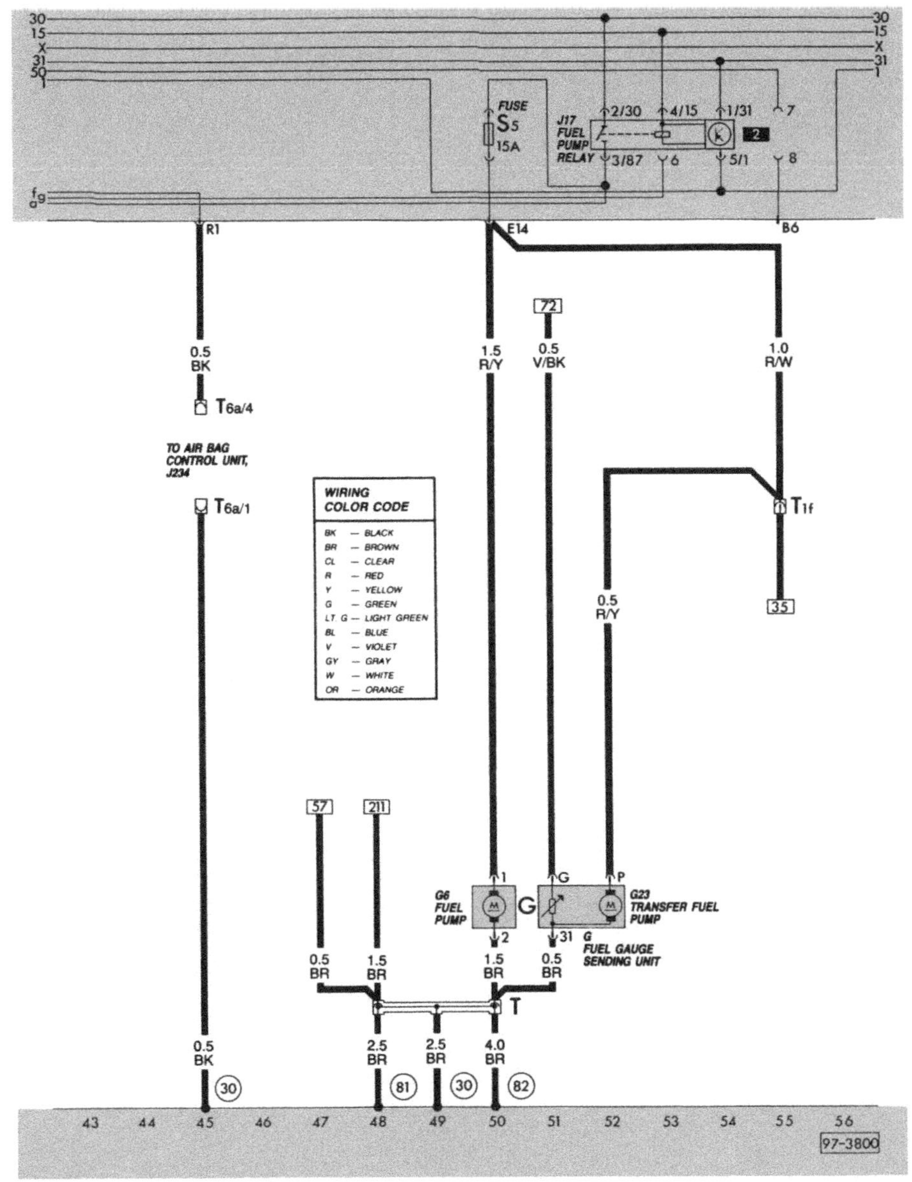

Main wiring diagram – Cabriolet 1990

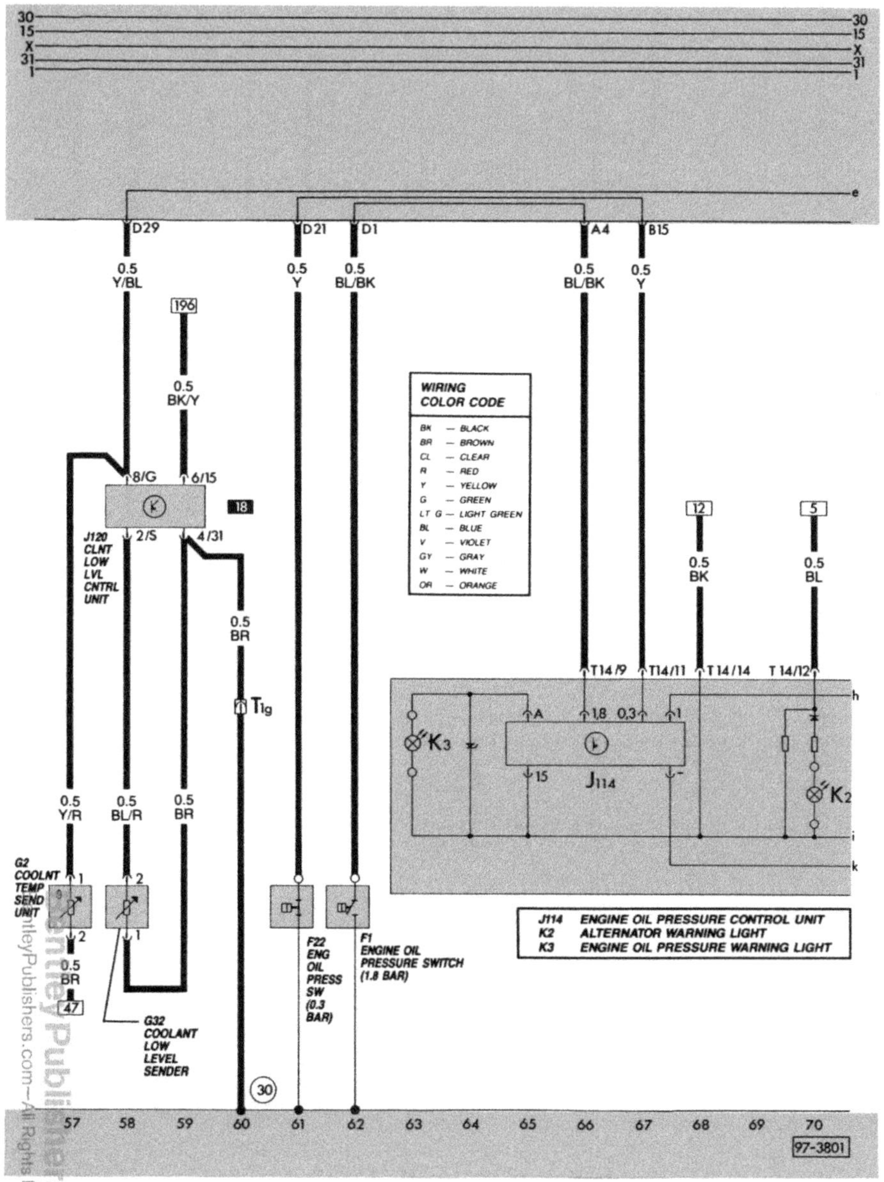

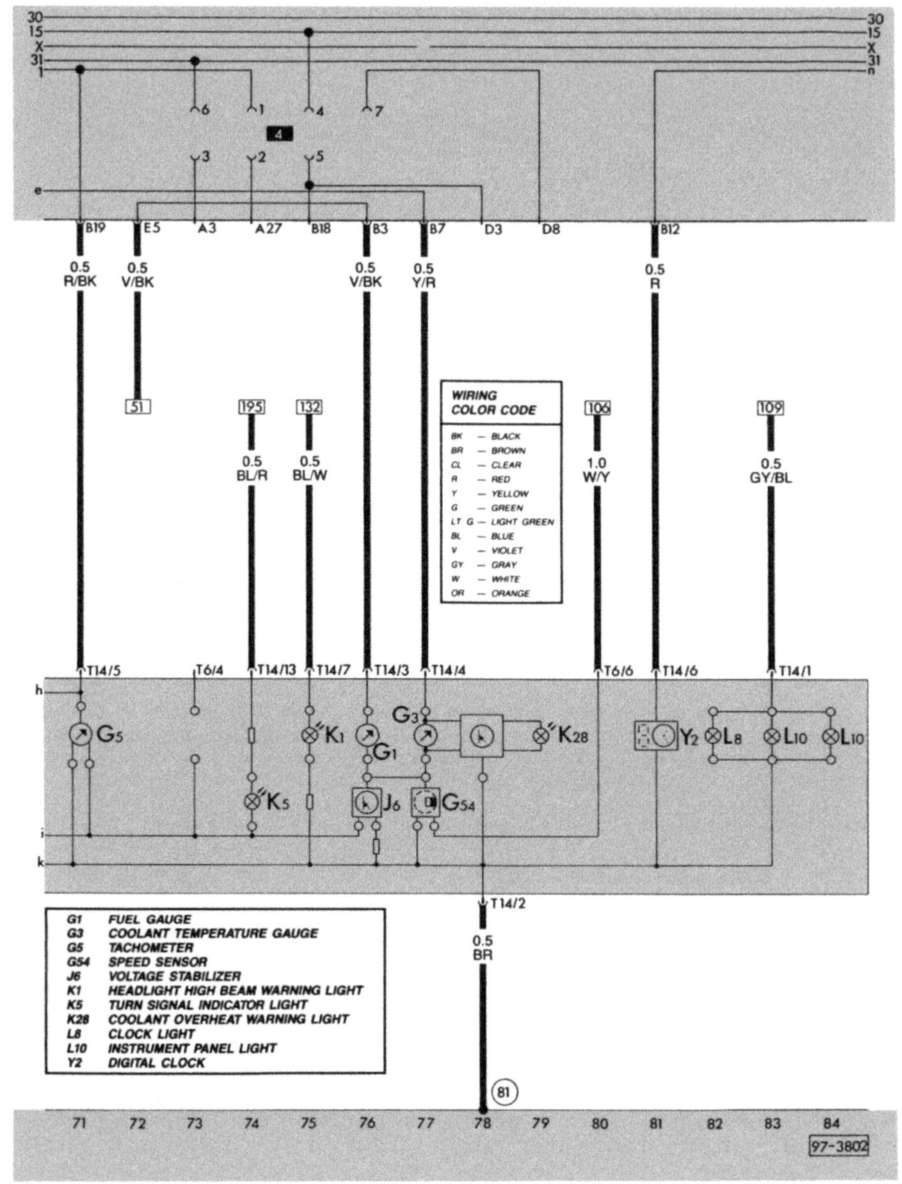

Main wiring diagram – Cabriolet 1990
page 6 of 11

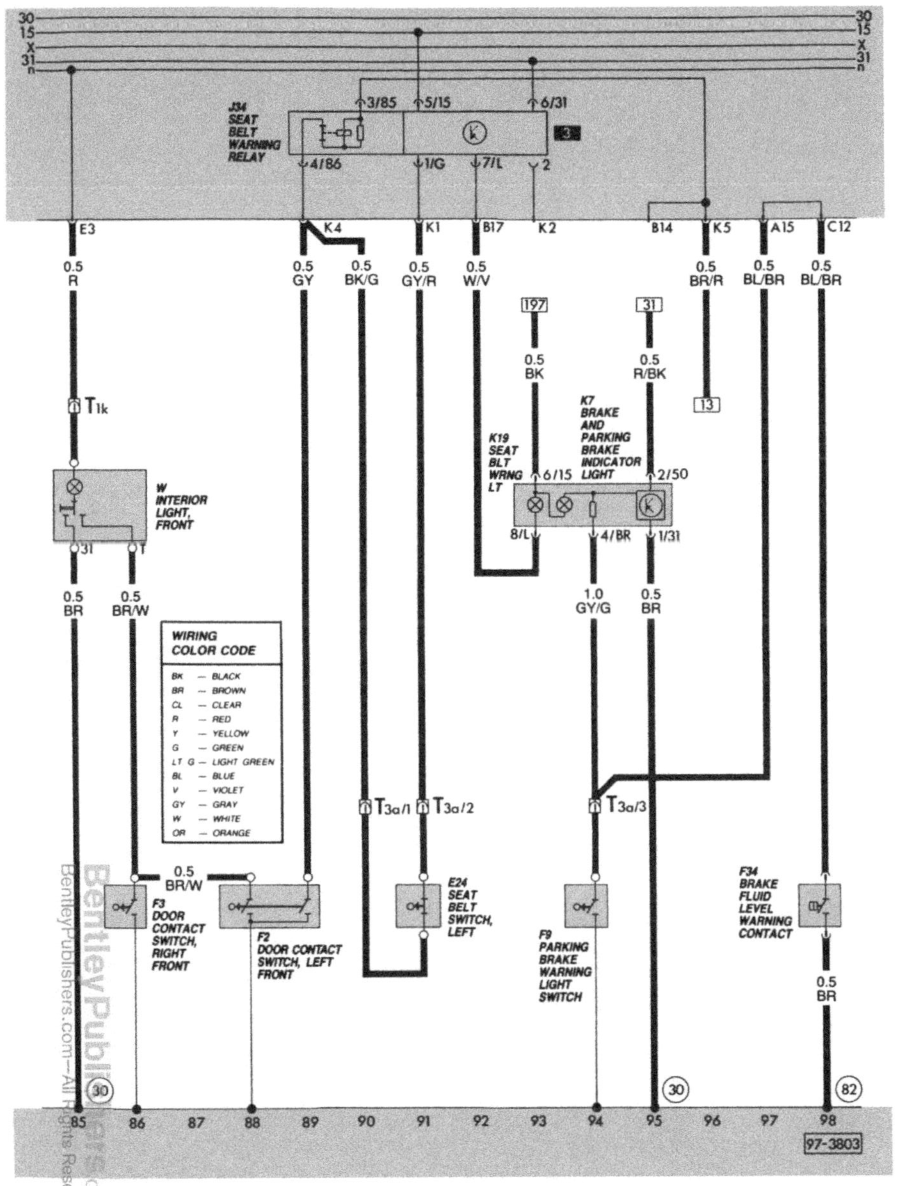

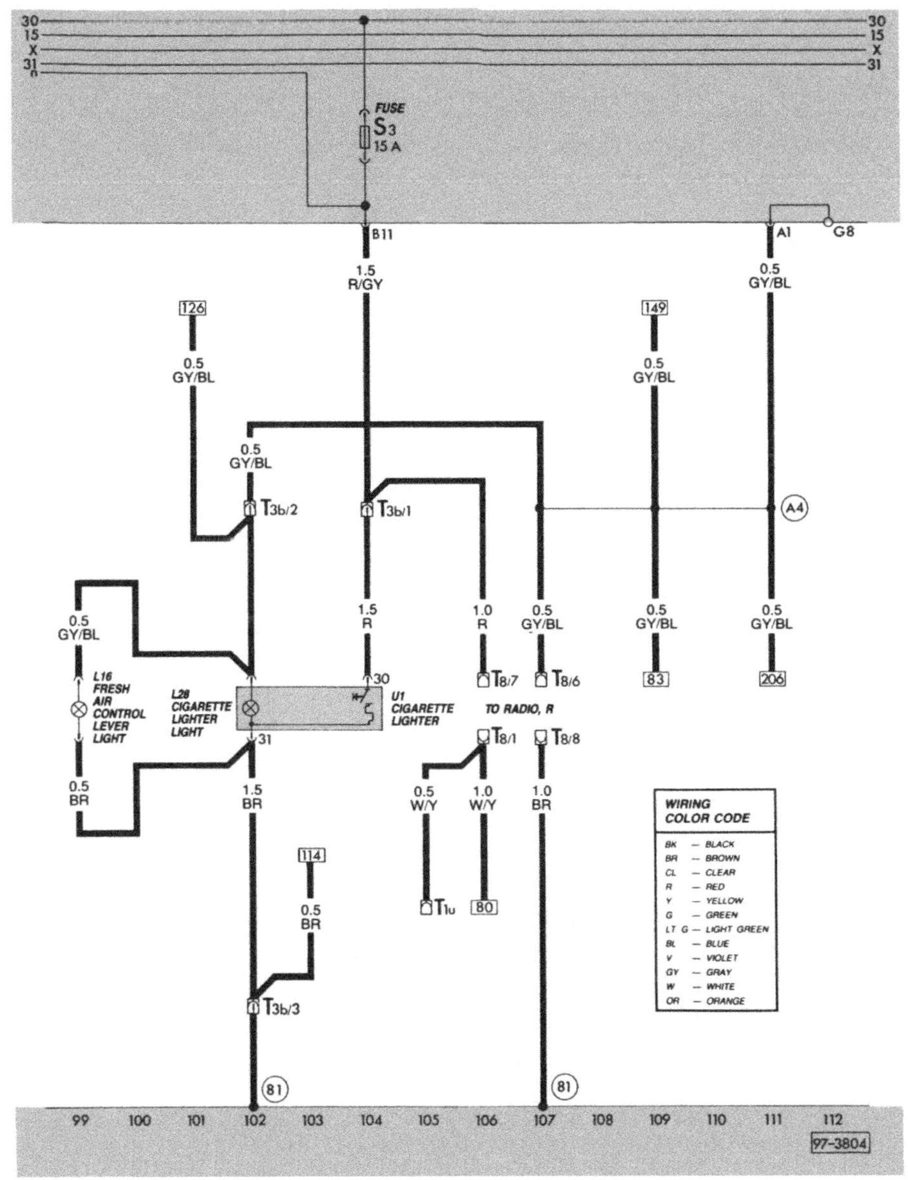

Main wiring diagram – Cabriolet 1990
page 7 of 11

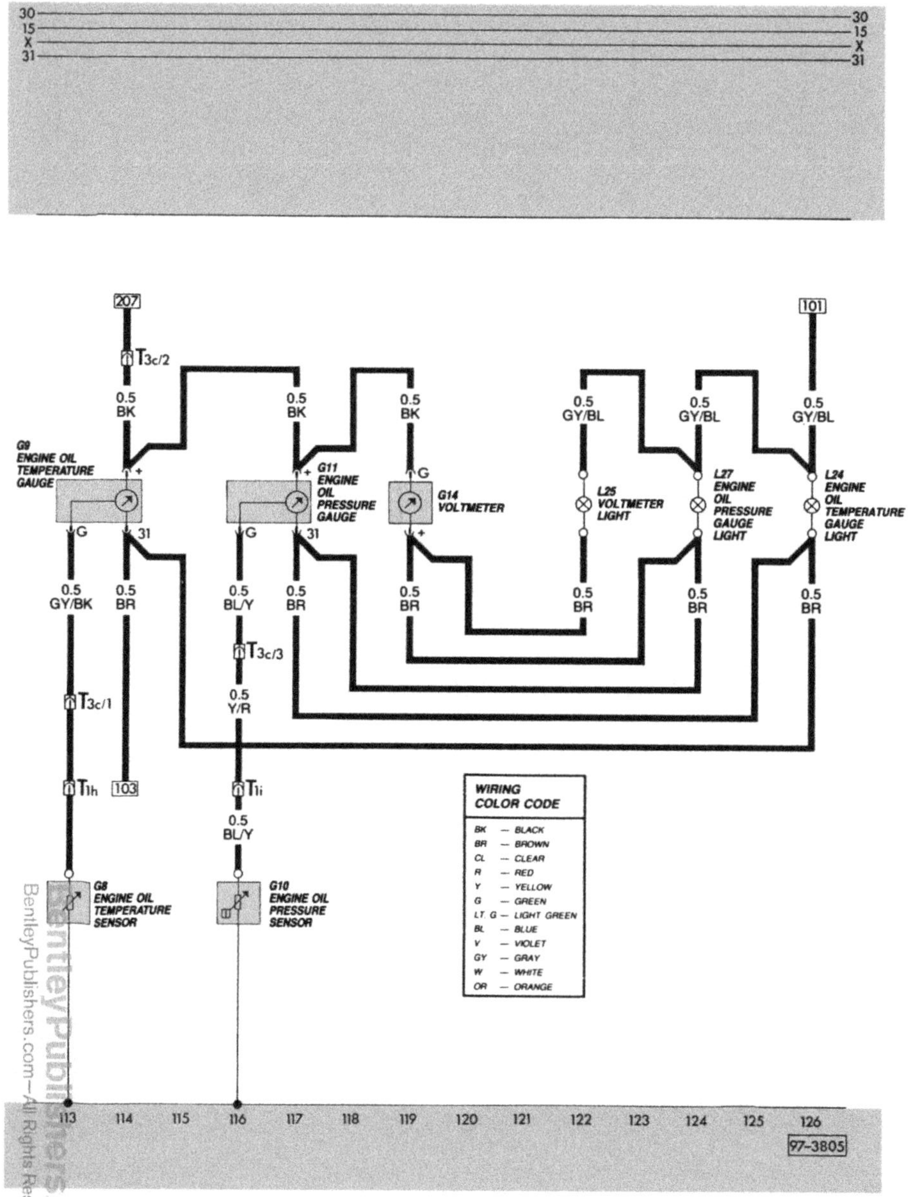

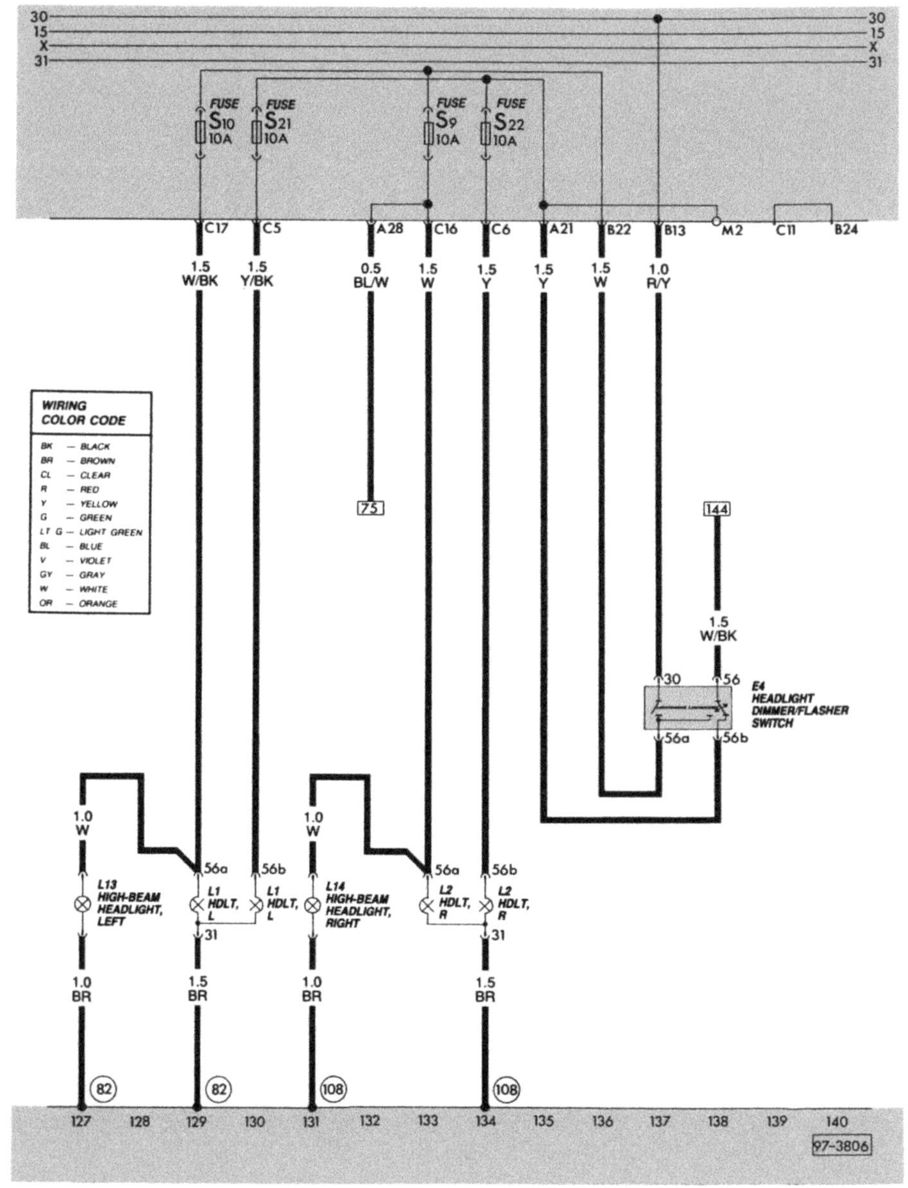

Main wiring diagram – Cabriolet 1990
page 8 of 11

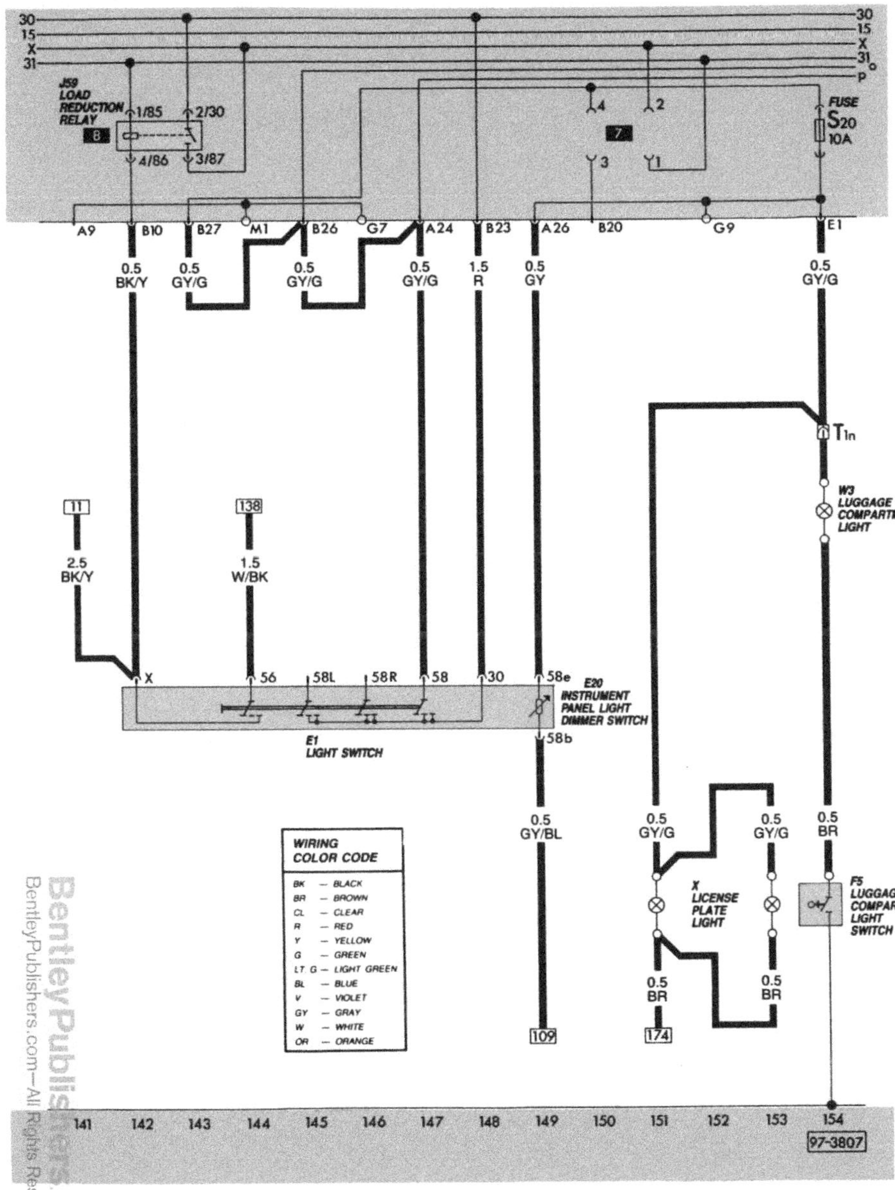

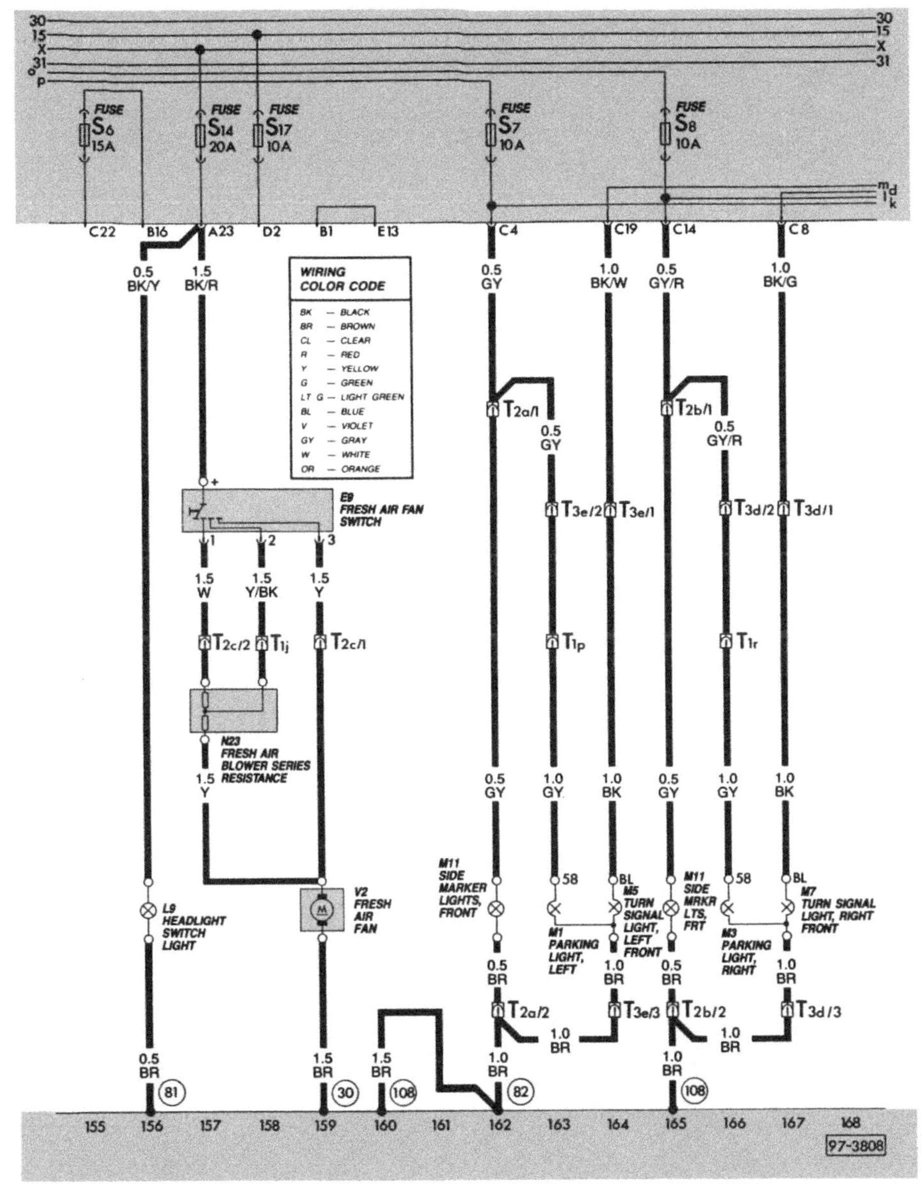

Main wiring diagram – Cabriolet 1990
page 9 of 11

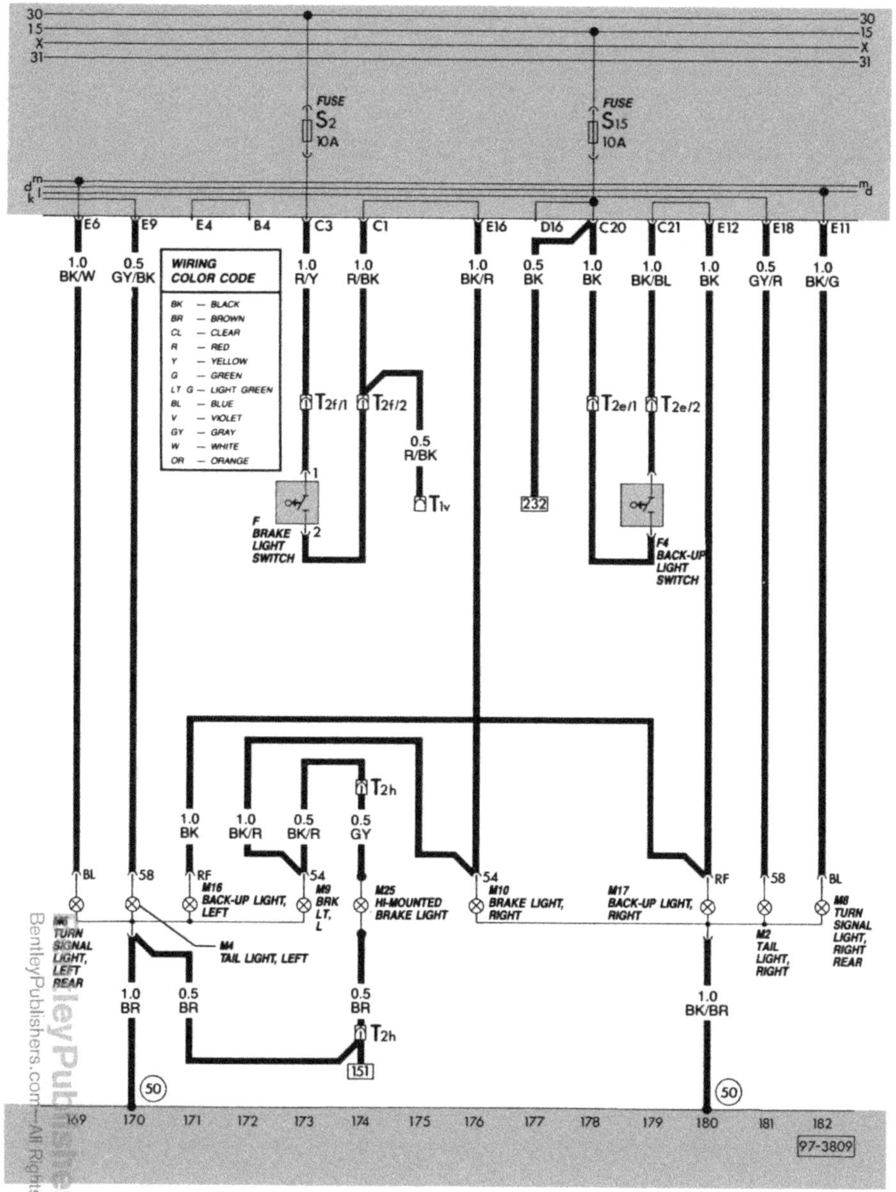

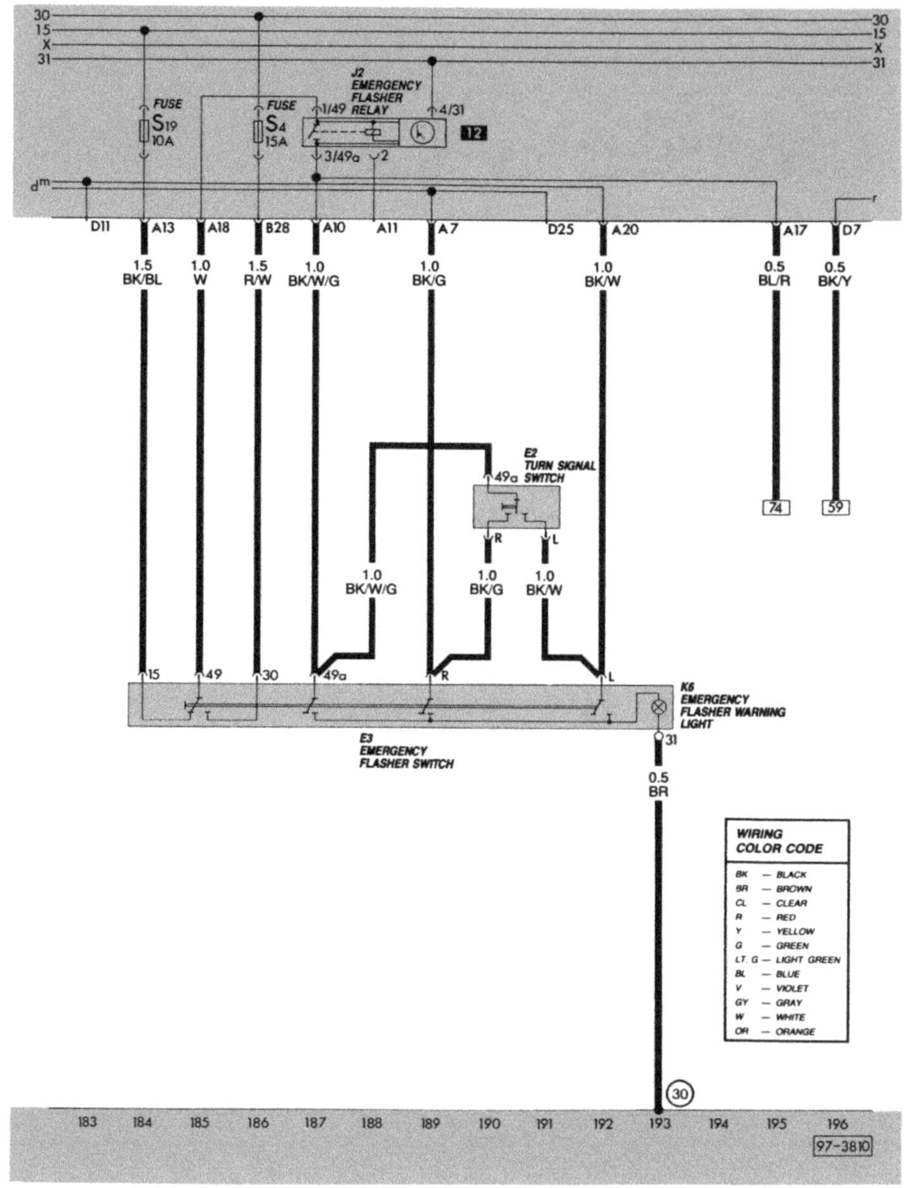

Main wiring diagram – Cabriolet 1990
page 10 of 11

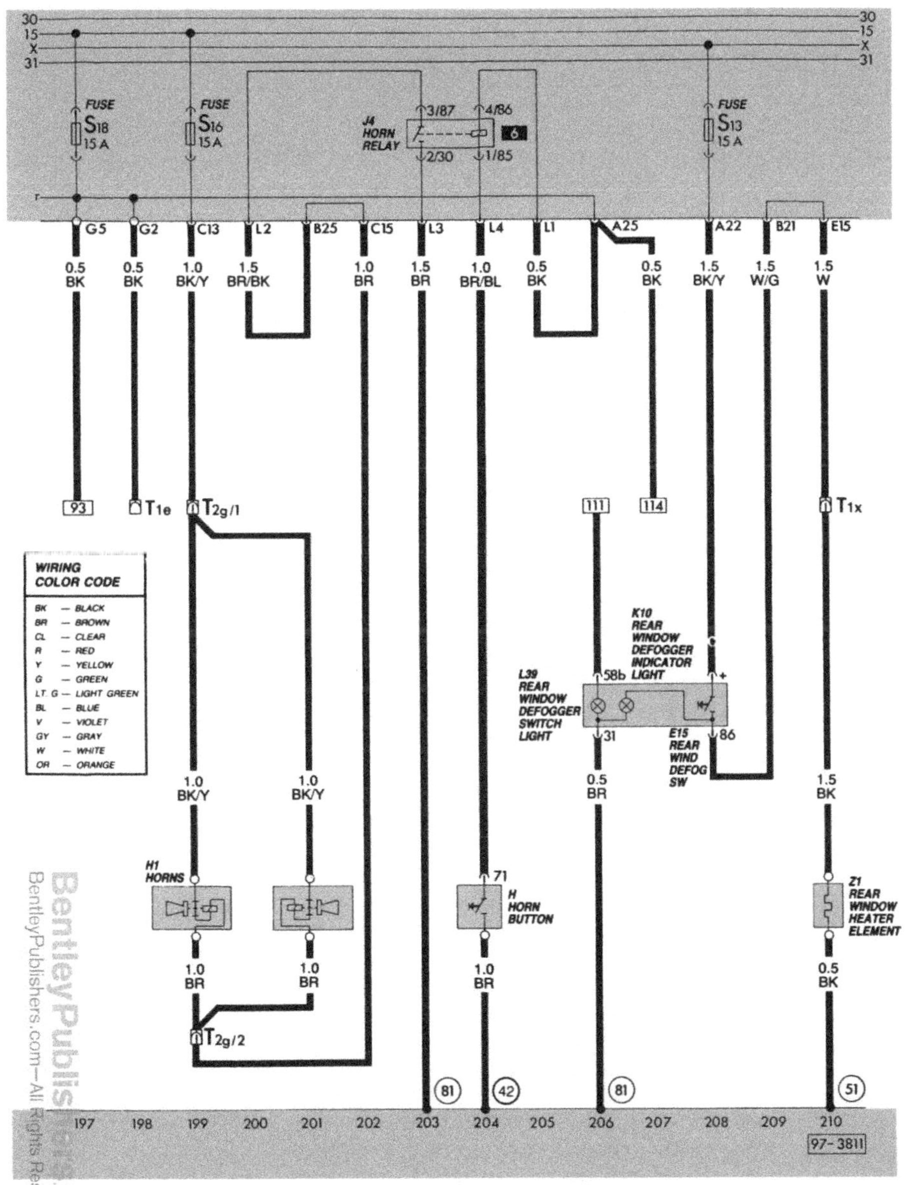

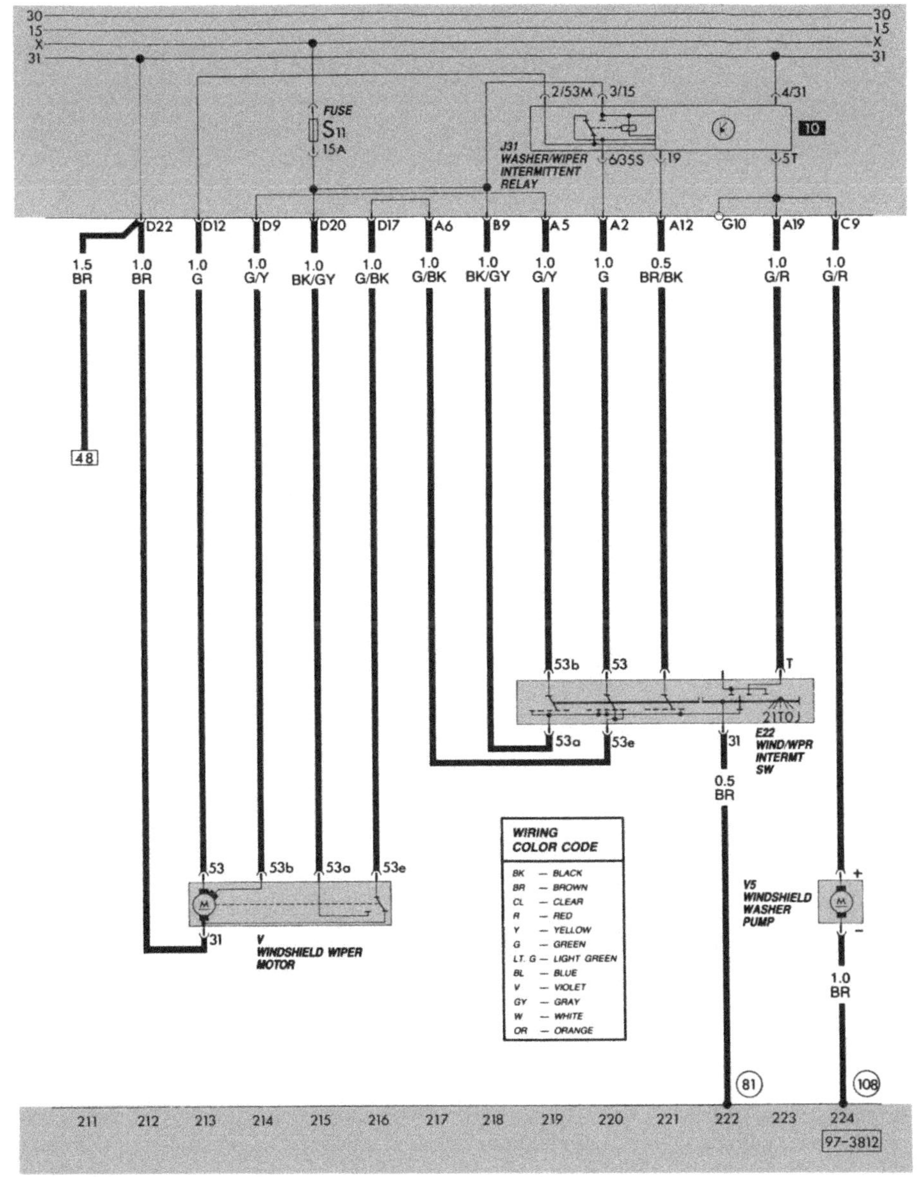

Main wiring diagram – Cabriolet 1990
page 11 of 11

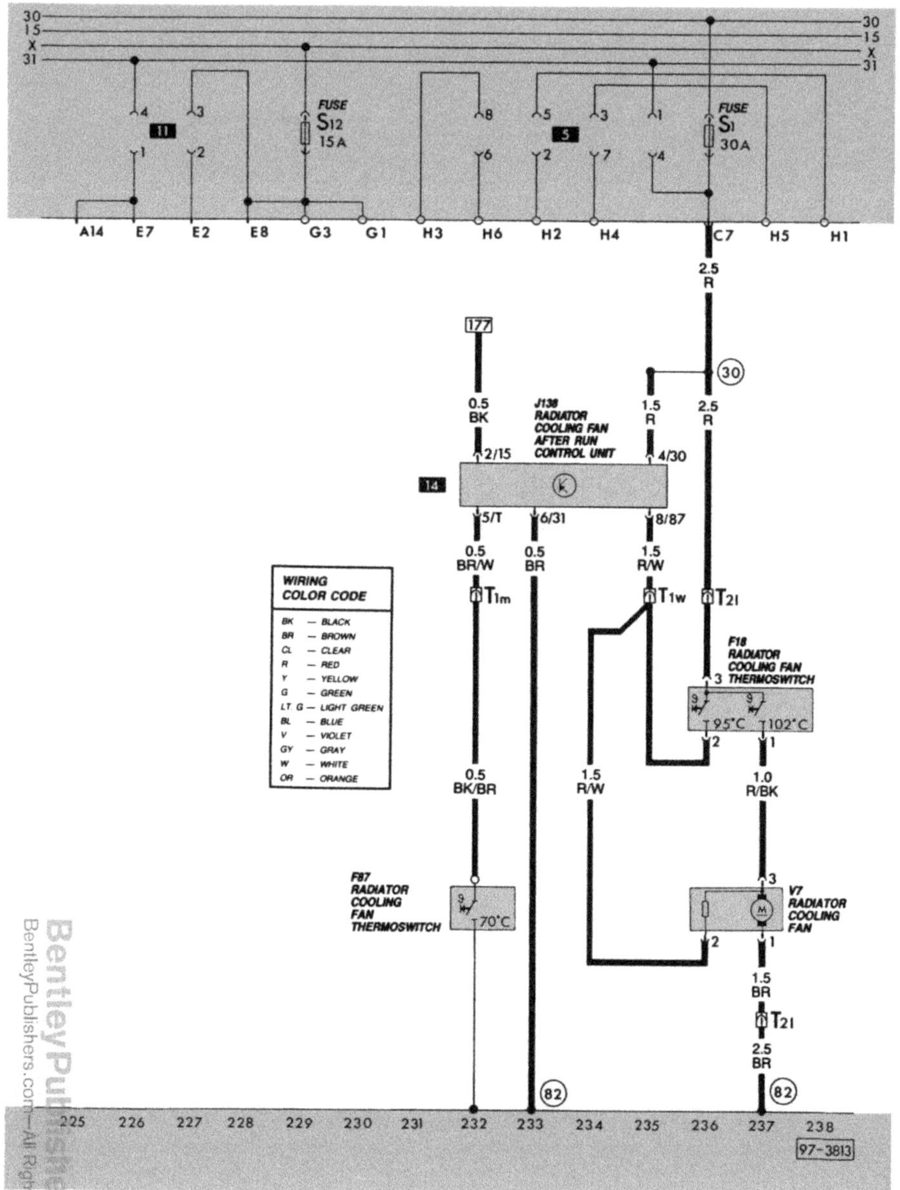

Main wiring diagram – Cabriolet from 1991
page 1 of 10

Description	Current track
Air flow sensor potentiometer, G19	27
Airbag control light, K75	69
Airbag warning light, K90	73
Alternator, C	1-2
Alternator warning light, K2	70
Back-up light switch, F4	173
Back-up light, left, M16	171
Back-up light, right, M17	180
Battery, A	4
Brake and parking brake indicator light, K7	93-95
Brake fluid level warning contact, F34	98
Brake light switch, F	179
Brake light, left, M9	174
Brake light, right, M10	177
Cigarette lighter, U1	104
Cigarette lighter light, L28	102
Clock light, L8	82
Coolant low level control unit, J120	58-59
Coolant low level sensor, G32	58
Coolant overheat warning light, K28	79
Coolant temperature gauge, G3	77
Coolant temperature sending unit, G2	57
Coolant temperature sensor, G62	24
Digifant control unit, J169	20-38
Digifant control unit relay, J176	35-37
Digital clock, Y2	81
Door contact switch, left front, F2	88-89
Door contact switch, right front, F3	86
Emergency flasher relay, J2	187-189
Emergency flasher switch, E3	184-193
Emergency flasher warning light, K6	194
Engine oil pressure control unit, J114	65-67
Engine oil pressure gauge, G11	116-117
Engine oil pressure gauge light, L27	124
Engine oil pressure sensor, G10	116
Engine oil pressure switch (0.3 bar), F22	61
Engine oil pressure switch (1.8 bar), F1	62
Engine oil pressure warning light, K3	63
Engine oil temperature gauge, G9	113-114
Engine oil temperature gauge light, L24	126
Engine oil temperature sensor, G8	113

Description	Current track
Fresh air control lever light, L16	99
Fresh air blower, V2	159
Fresh air blower series resistance, N23	157-158
Fresh air blower switch, E9	157-159
Fuel gauge, G1	76
Fuel gauge sending unit, G	51
Fuel pump, G6	49
Fuel pump relay, J17	51-53
Full throttle switch, F81	20
Fuse, S17	34
Fuse, S12	43
Fuse, S1	47
Fuse, S5	49
Fuse, S3	104
Fuse, S10	129
Fuse, S21	130
Fuse, S9	133
Fuse, S22	134
Fuse, S20	154
Fuse, S6	155
Fuse, S14	157
Fuse, S7	162
Fuse, S8	165
Fuse, S15	174
Fuse, S2	180
Fuse, S19	184
Fuse, S4	186
Fuse, S18	197
Fuse, S16	199
Fuse, S13	208
Fuse, S11	215
Hall generator, G40	20-22
Hall ignition control unit, N41	17-20
Headlight dimmer/flasher switch, E4	137-138
Headlight high beam indicator light, K1	75
Headlight switch light, L9	156
Headlight, left, L1	129, 130
Headlight, right, L2	133, 134
High-beam headlight, left, L13	127
High-beam headlight, right, L14	131
High-mount brake light, M25	175
Horn button, H	204
Horn relay, J4	203-204
Horns, H1	199, 201
Idle stabilizer valve, N71	36-37
Idle switch, F60	23
Ignition coil, N	16-18
Ignition distributor, O	17-19
Ignition/starter switch, D	9-12
Injector, cyl. 1, N30	42
Injector, cyl. 2, N31	41
Injector, cyl. 3, N32	40
Injector, cyl. 4, N33	39
Instrument panel light dimmer switch, E20	149
Instrument panel light, L10	83, 84
Intake air temperature sensor, G42	26
Interior light, front, W	85-86

Description	Current track
Junction box, T	51-53
Knock sensor, G61	25-27
License plate light, X	151-153
Light switch, E1	142-148
Load reduction relay, J59	142-148
Luggage compartment light, W3	154
Luggage compartment light switch, F5	154
Oxygen sensor, G39	34-35
Parking brake warning light switch, F9	94
Parking light, left, M1	163
Parking light, right, M3	166
Radiator cooling fan, V7	47-48
Radiator cooling fan after run control unit, J138	43-46
Radiator cooling fan thermoswitch, F18	47-49
Radiator cooling fan thermoswitch, F87	43
Radio, R	107
Rear window defogger indicator light, K10	206-208
Rear window defogger switch, E15	208
Rear window defogger switch light, L39	206
Rear window heater element, Z1	210
Seat belt switch, left, E24	91
Seat belt warning control unit, J34	89-93
Seat belt warning light, K19	93
Side marker lights, front, M11	162, 165
Spark plug connectors, P	17-19
Spark plugs, Q	17-19
Speed sensor, G54	77
Starter, B	6-7
Tachometer, G5	71
Tail light, left, M4	170
Tail light, right, M2	181
Transfer fuel pump, G23	52
Turn signal indicator light, K5	74
Turn signal light, left front, M5	164
Turn signal light, left rear, M6	169
Turn signal light, right front, M7	167
Turn signal light, right rear, M8	182
Turn signal switch, E2	190-191
Voltage regulator, C1	1-2
Voltage stabilizer, J6	76
Voltmeter, G14	119
Voltmeter light, L25	122
Washer/wiper intermittent relay, J31	219-223
Windshield washer pump, V5	224
Windshield wiper motor, V	213-216
Windshield wiper intermittent switch, E22	219-223

Wire connectors

T — junction box, behind fuse/relay panel
T1 — single, in luggage compartment, left
T1a — single, near ignition coil
T1b — single, behind fuse/relay panel
T1c — single, behind fuse/relay panel
T1d — single, on firewall, right
T1e — single, behind fuse/relay panel
T1f — single, behind fuse/relay panel
T1g — single, in luggage compartment, left
T1h — single, behind instrument panel
T1i — single, behind instrument panel
T1j — single, behind instrument panel, center
T1k — single, behind fuse/relay panel
T1l — single, behind fuse/relay panel
T1m — single, behind fuse/relay panel
T1n — single, behind fuse/relay panel
T1p — single, near left headlight
T1r — single, near right headlight
T1s — single, behind fuse/relay panel
T1u — single, behind instrument panel
T1v — single, behind fuse/relay panel
T2a — double, behind fuse/relay panel
T2b — double, behind fuse/relay panel
T2c — double, behind instrument panel, center
T2d — double, near intake manifold
T2f — double, behind fuse/relay panel
T2h — double, in luggage compartment
T2l — double, near left headlight
T3a — three-point, behind fuse/relay panel
T3b — three-point, behind instrument panel, center
T3c — three-point, behind console
T3d — three-point, near right headlight
T3e — three-point, near left headlight
T3g — three-point, near starter
T3h — three-point, near valve cover
T4a — four-point, in engine compartment, rear
T5a — five-point, on firewall, left
T6 — six-point, behind fuse/relay panel
T6a — six-point, on instrument cluster
T8 — eight-point, on radio
T14 — fourteen-point, on instrument cluster

Ground connections

(1) — battery to body
(12) — in engine compartment, left
(18) — on engine block
(30) — beside fuse/relay panel
(42) — near steering column
(50) — in luggage compartment, left
(51) — in luggage compartment, right
(81) — in instrument panel wiring harness
(82) — in left front wiring harness
(94) — in Digifant wiring harness
(108) — in left front wiring harness
(129) — in horn wiring harness

Welded wiring harness points

(A2) — plus connection (15), in instrument panel wiring harness
(C3) — plus connection (30), in left front wiring harness
(C13) — in horn wiring harness
(G3) — plus connection, in injector valves wiring harness

Wire Size

Wiring Diagrams identify wires by the metric wire size. Metric wires sizes indicate cross-sectional area in square millimeters (mm²). The chart below lists metric wire sizes and their approximate equivalents in American Wire Gauge (AWG) sizes.

Metric size — cross section (mm²)	American Wire Gauge Size (AWG)
0.35	22
0.5	20
0.75	18
1.0	16
1.5	14
2.5	12
4.0	10
6.0	8
16.0	4
25.0	2
35.0	2

Main wiring diagram – Cabriolet from 1991
page 3 of 10

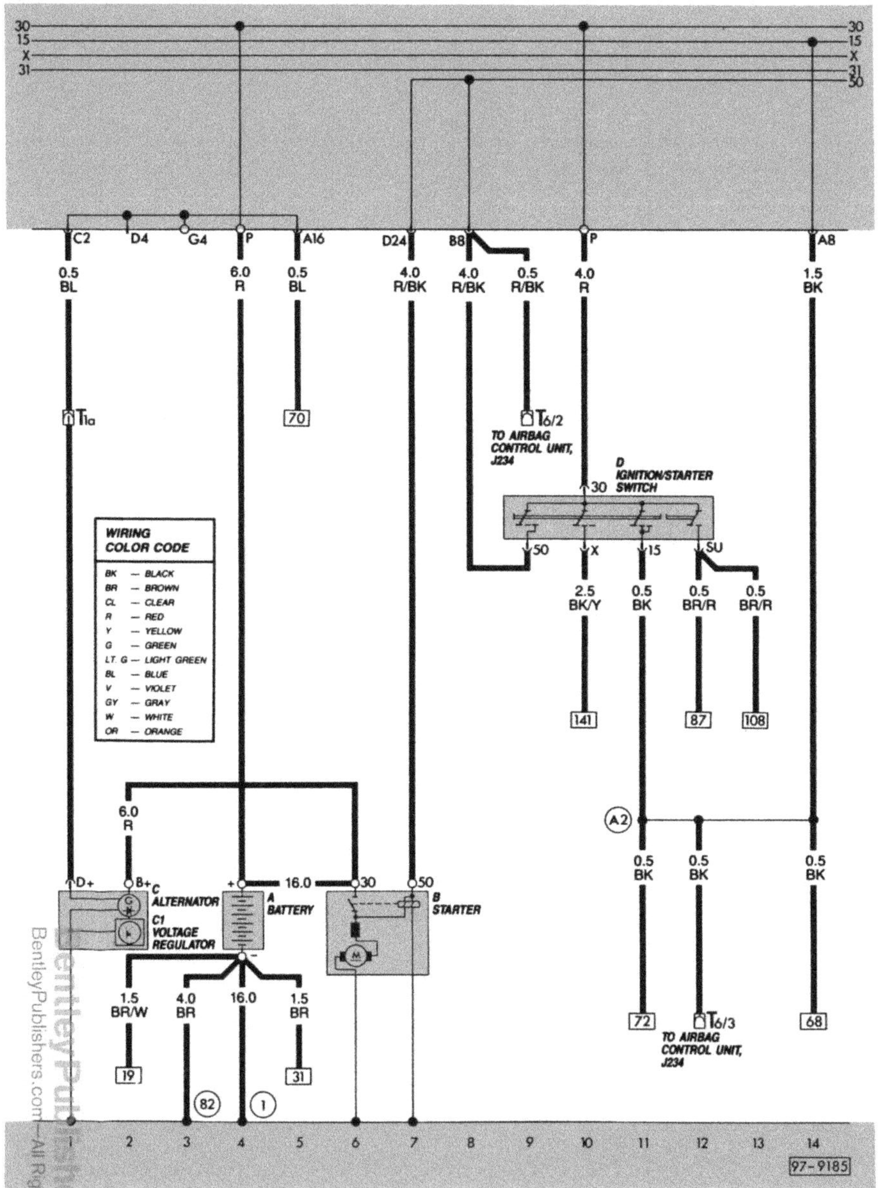

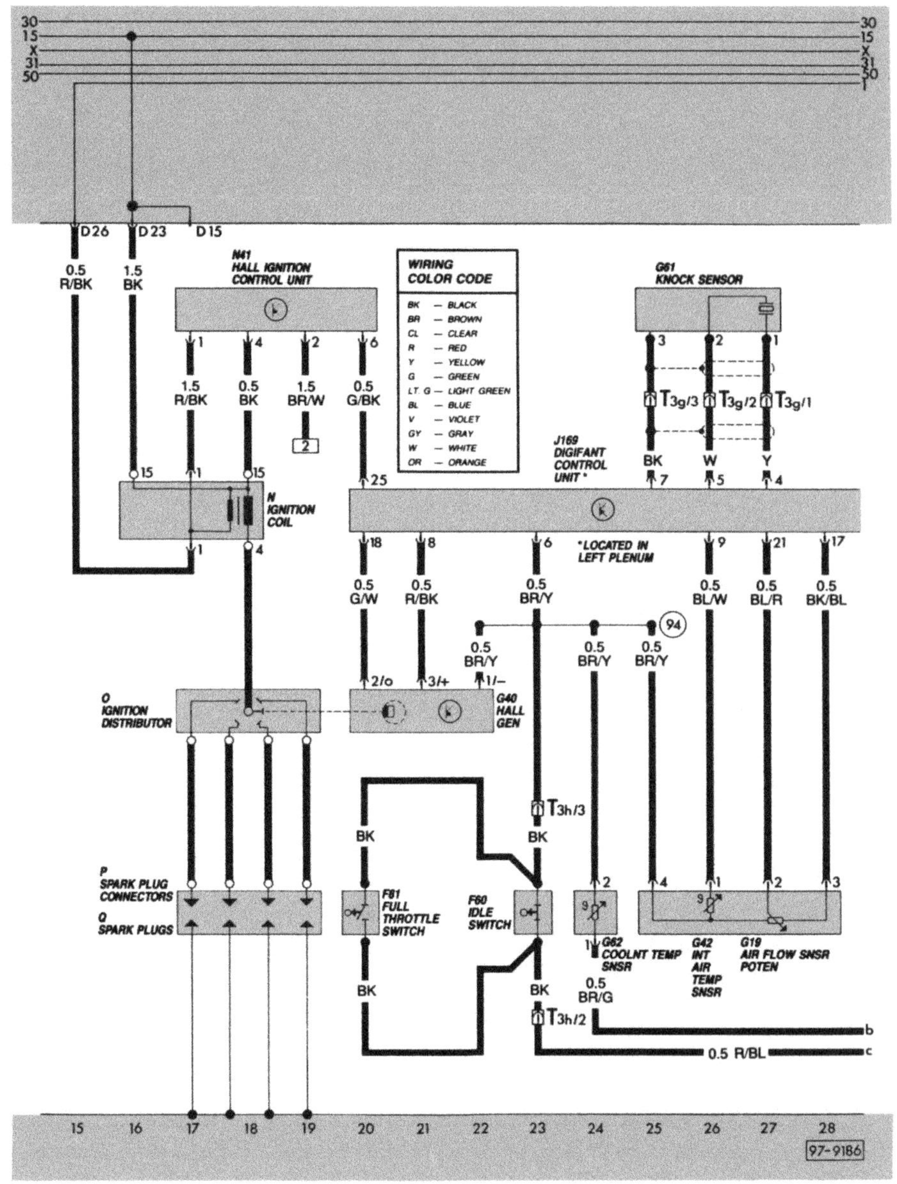

Main wiring diagram – Cabriolet from 1991
page 4 of 10

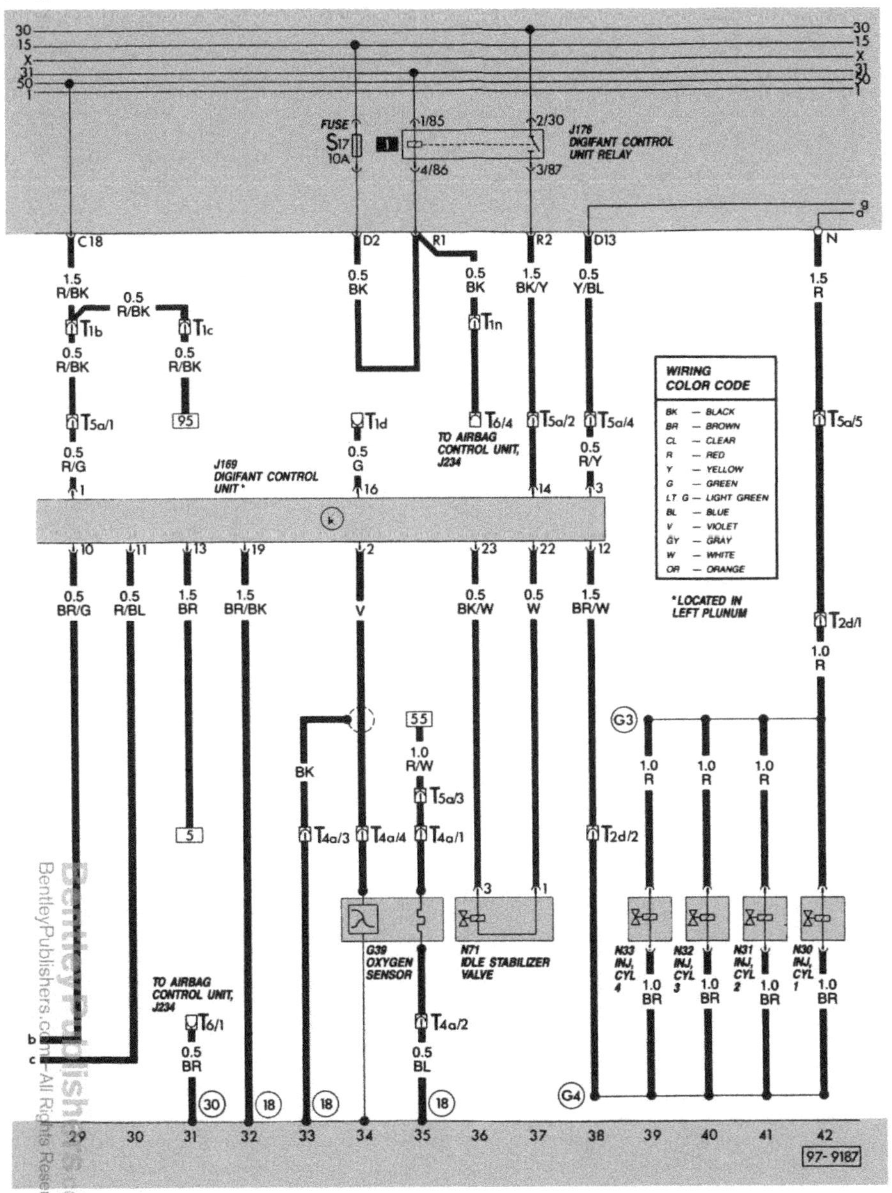

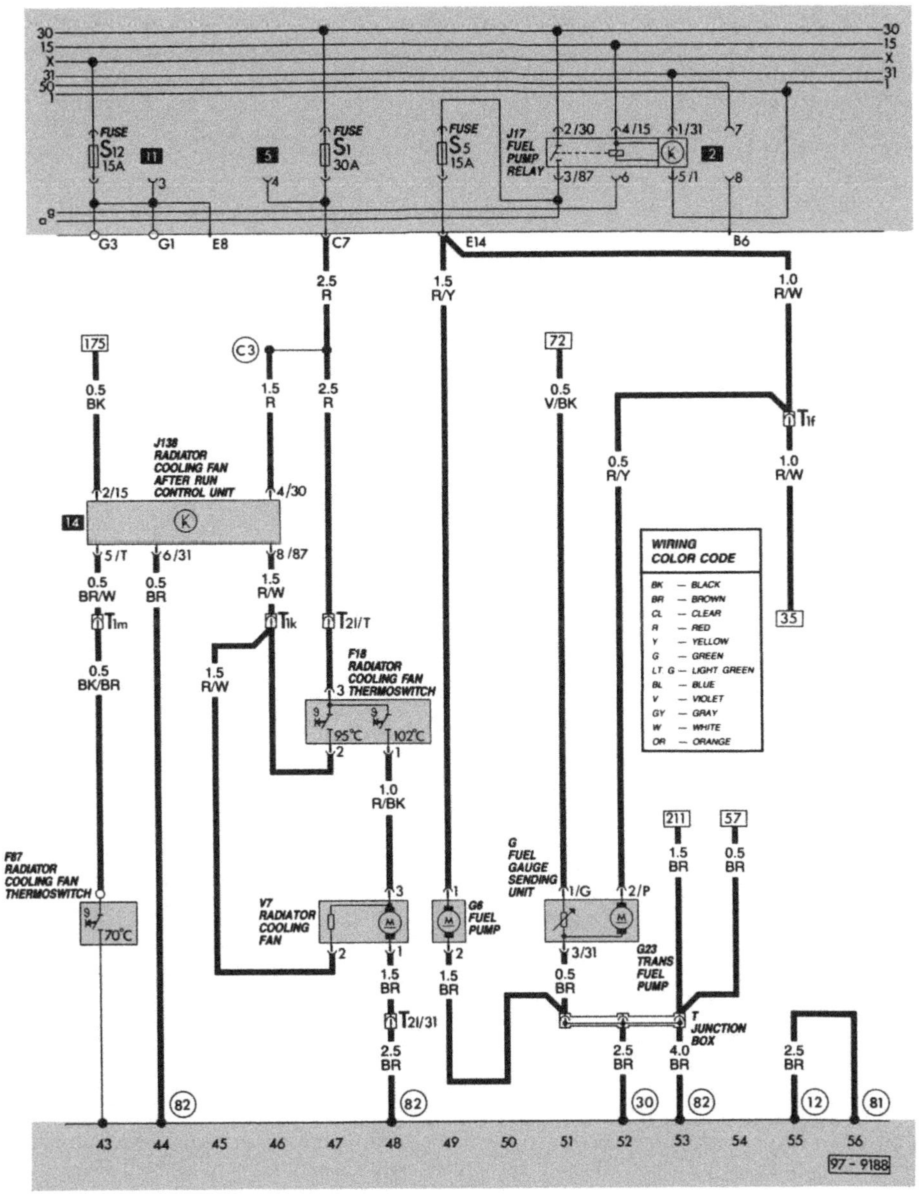

Main wiring diagram – Cabriolet from 1991
page 5 of 10

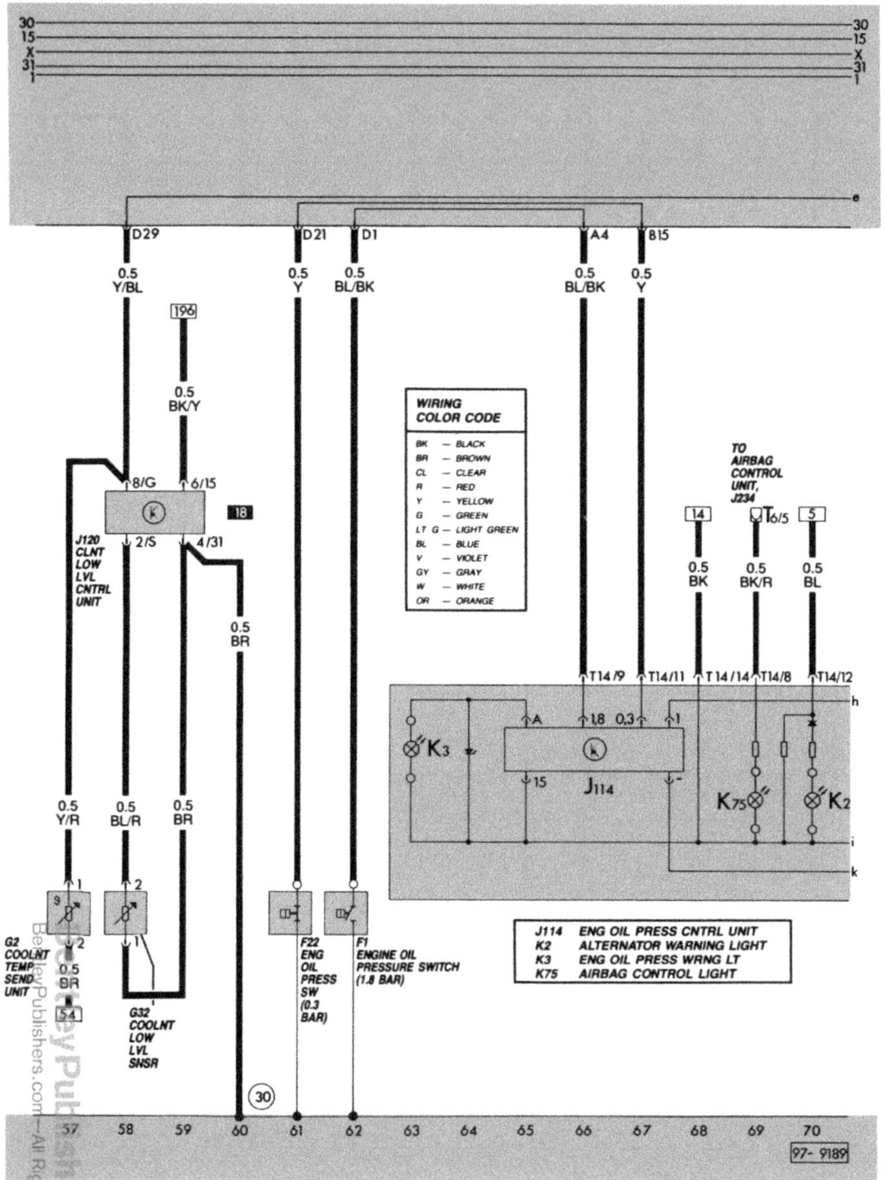

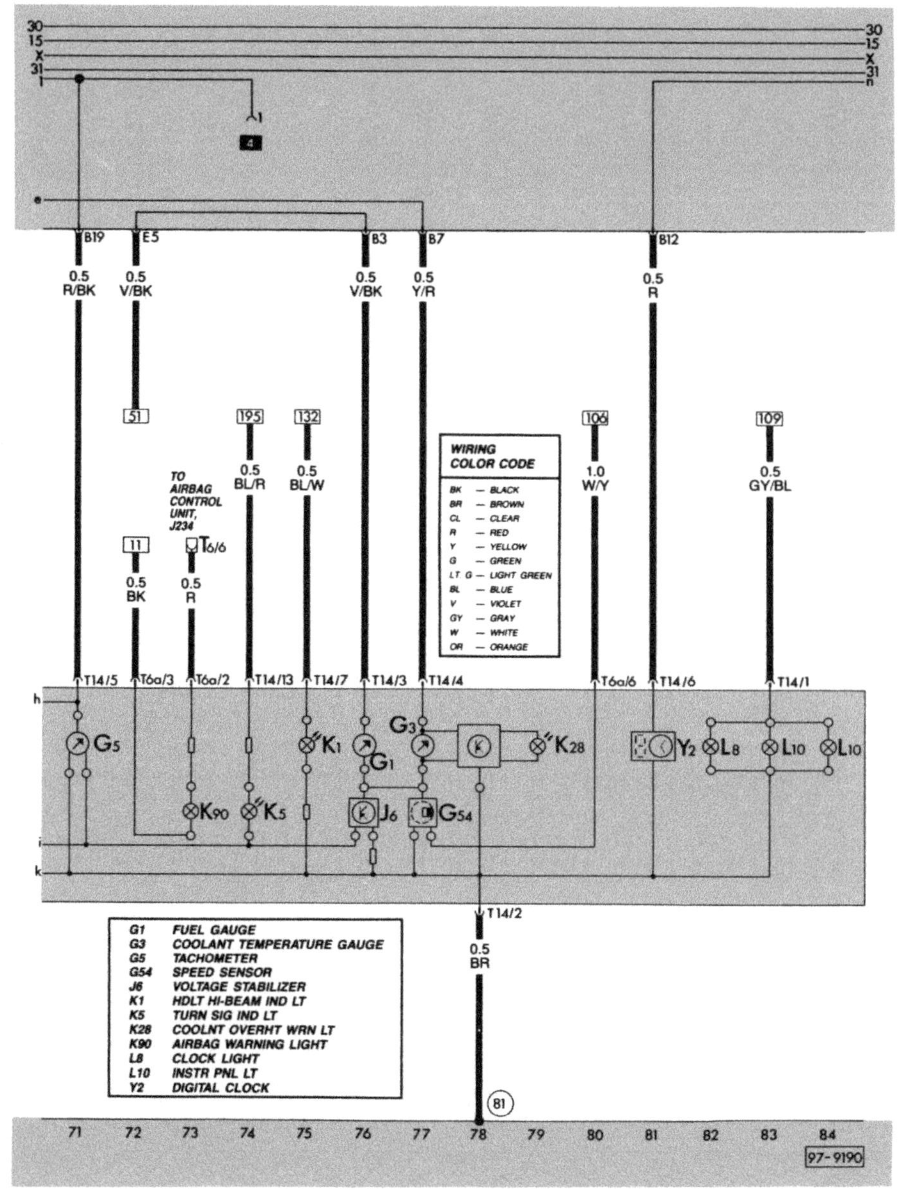

Main wiring diagram – Cabriolet from 1991
page 6 of 10

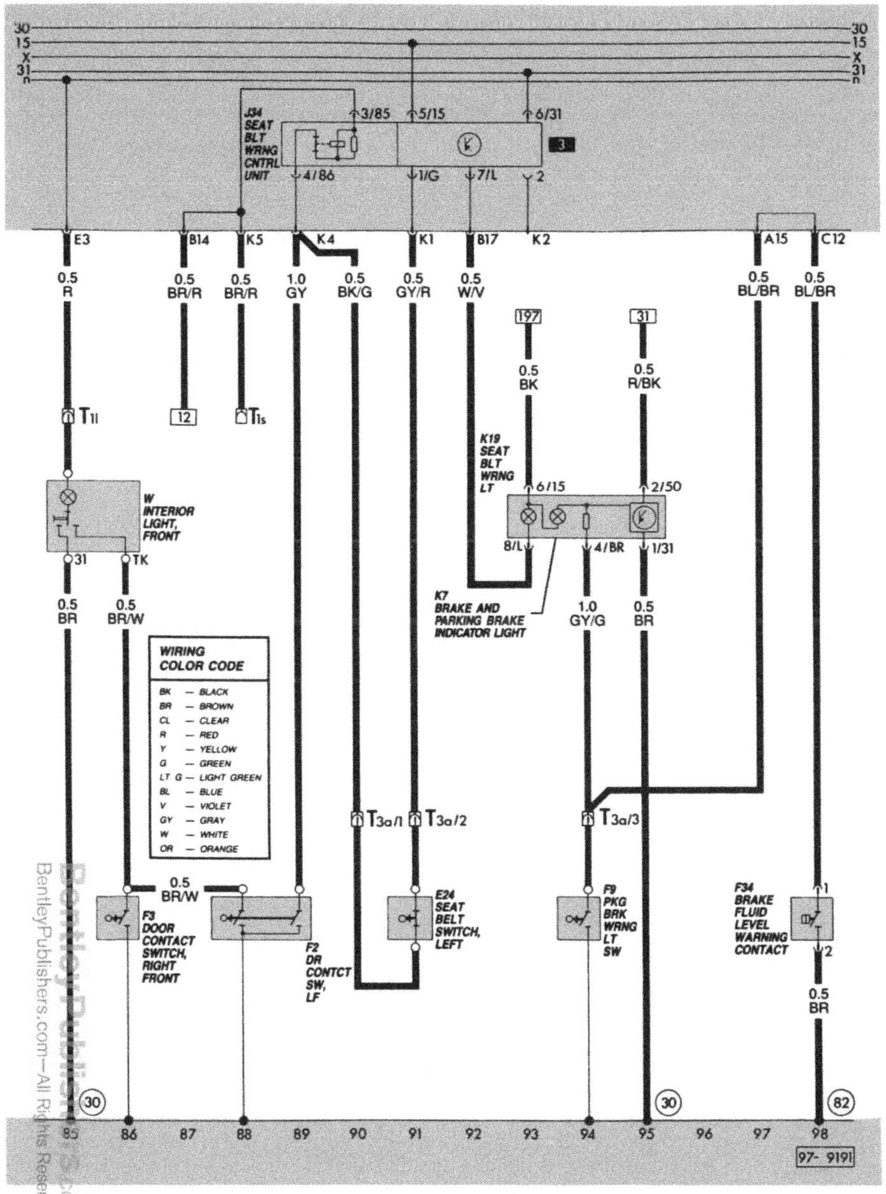

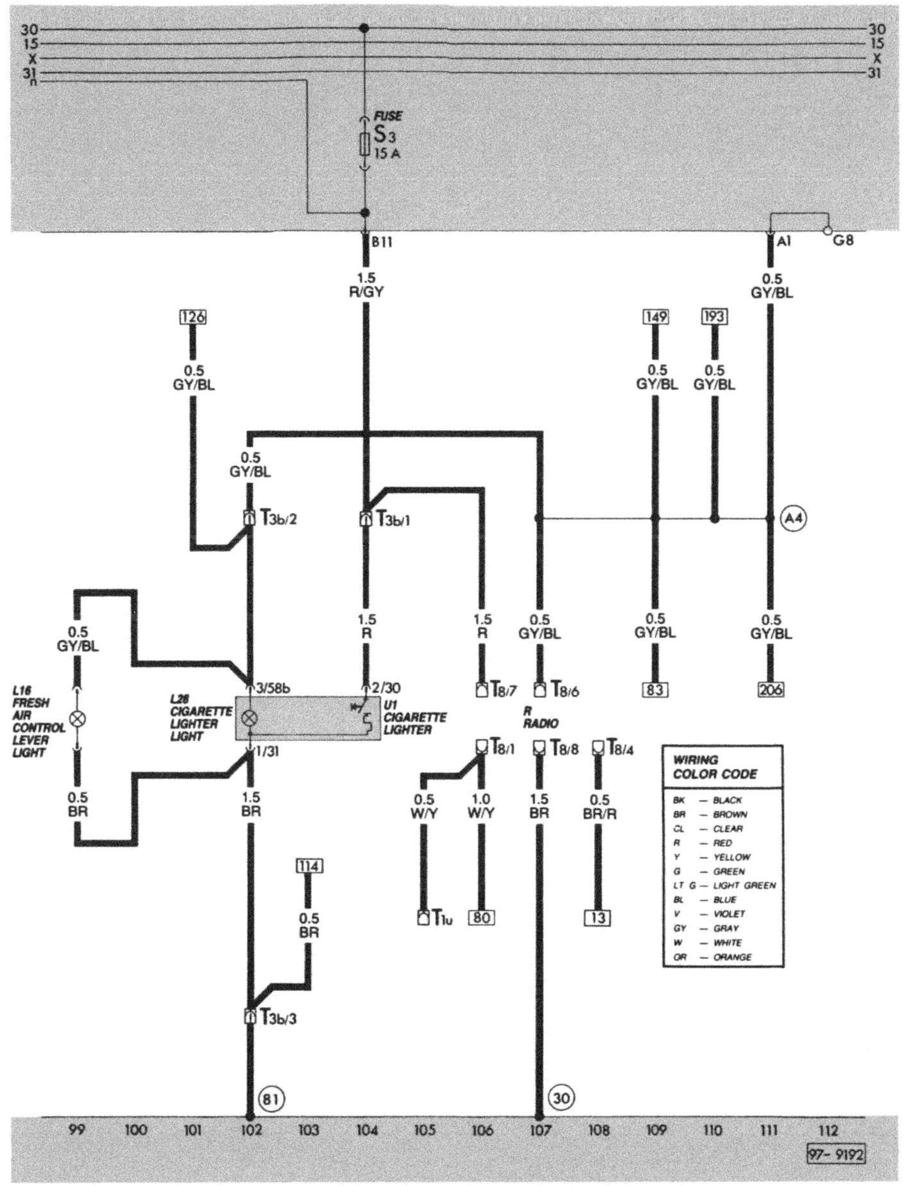

Main wiring diagram – Cabriolet from 1991
page 7 of 10

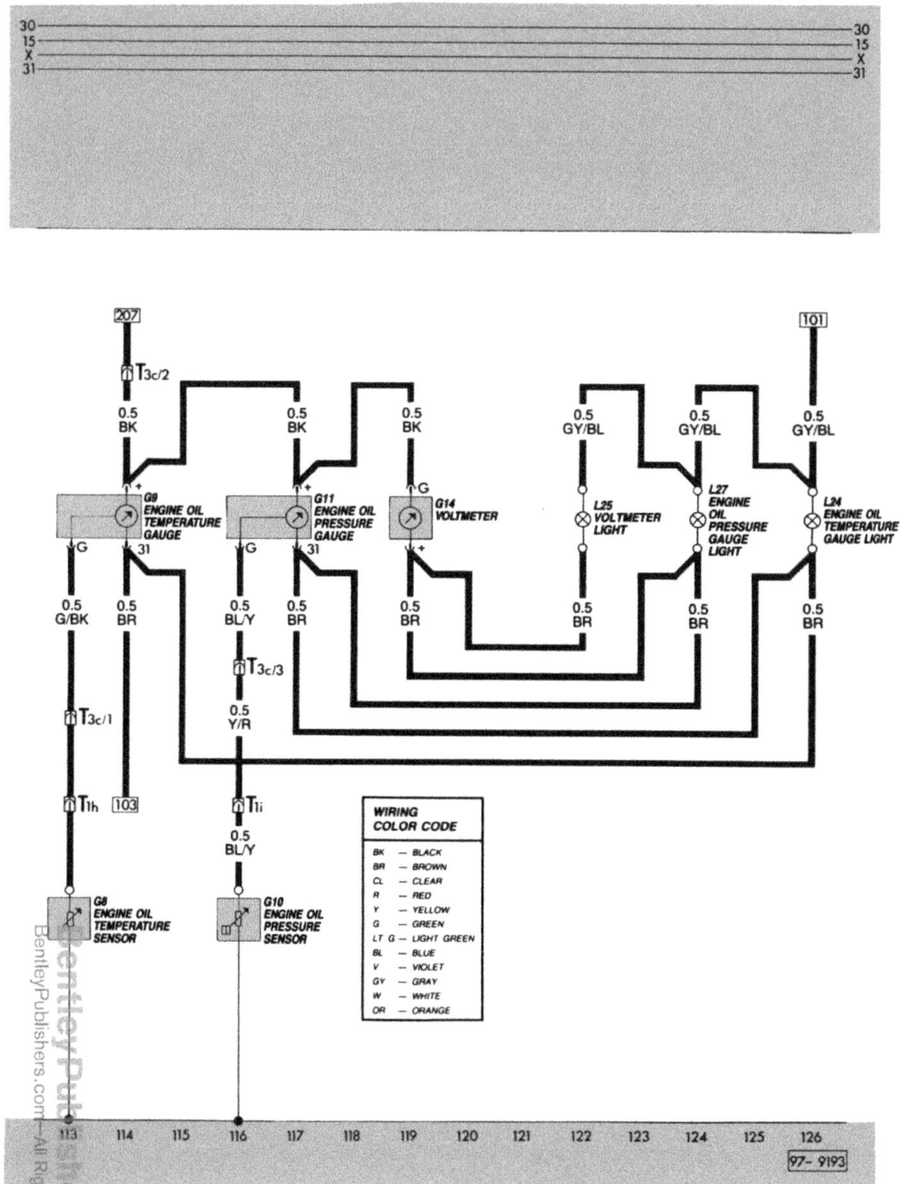

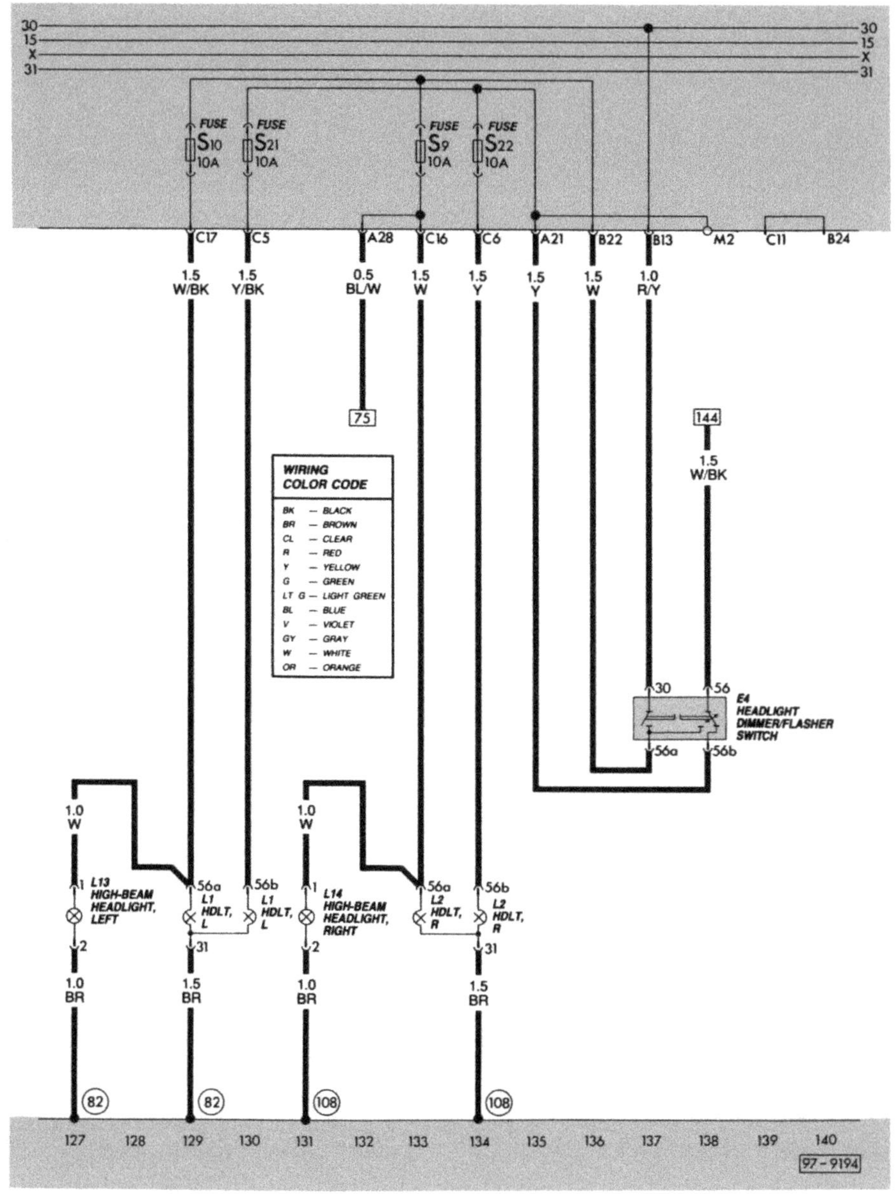

Main wiring diagram – Cabriolet from 1991
page 8 of 10

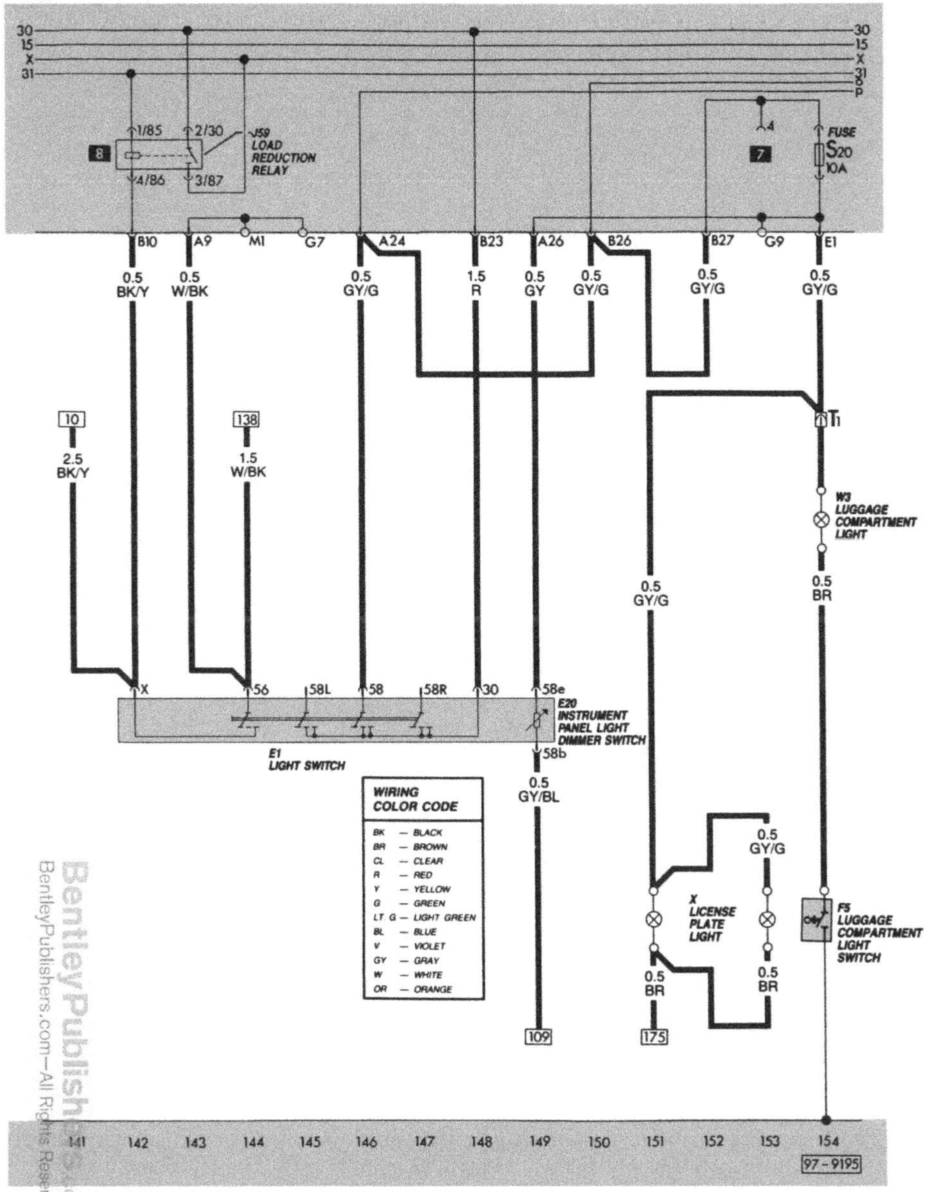

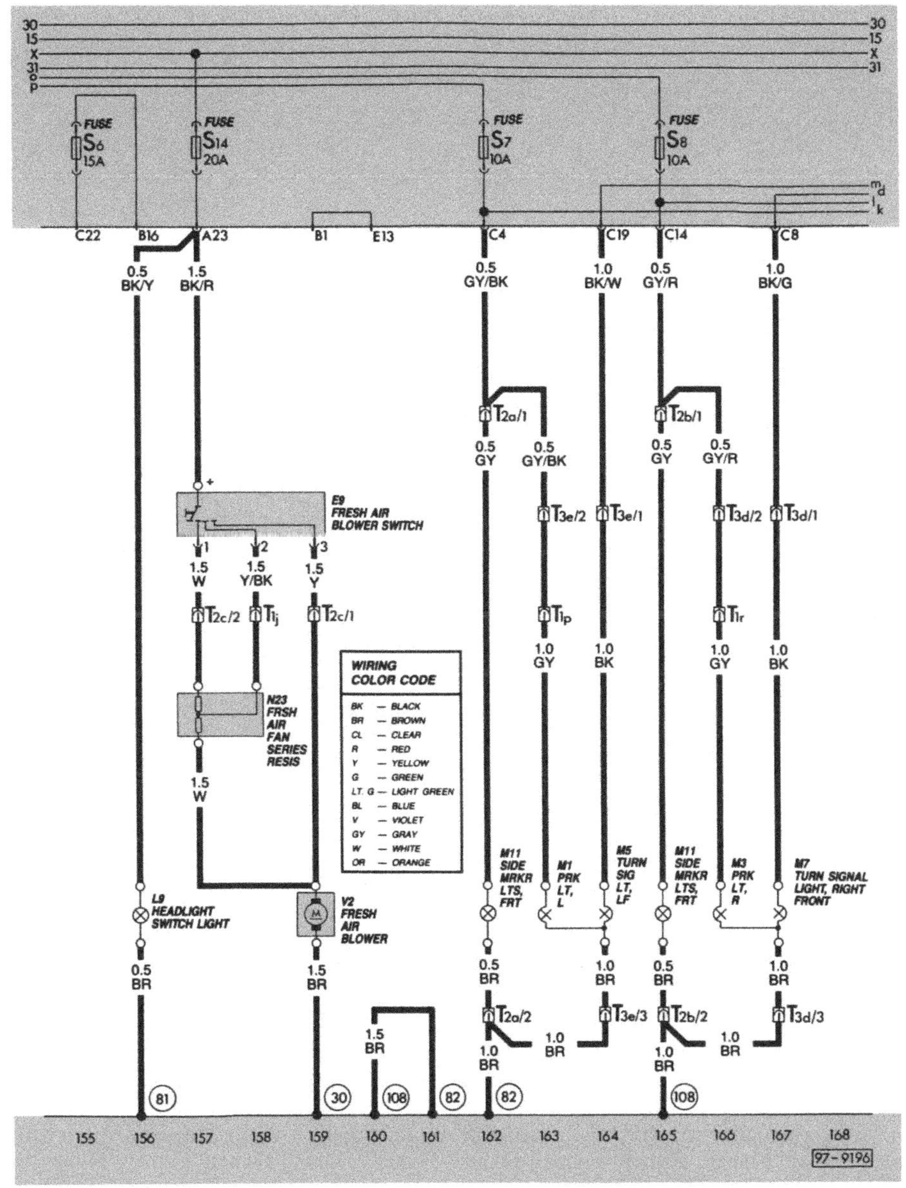

Main wiring diagram – Cabriolet from 1991
page 9 of 10

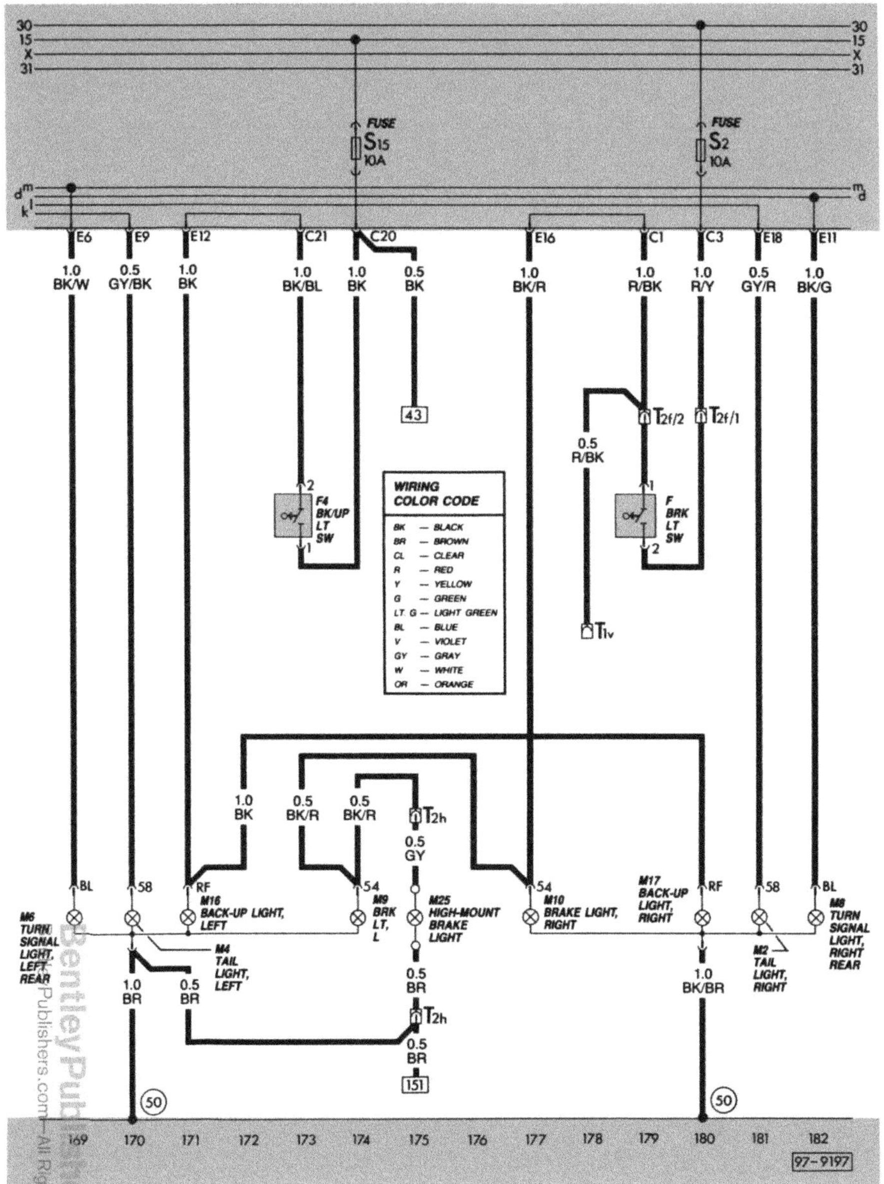

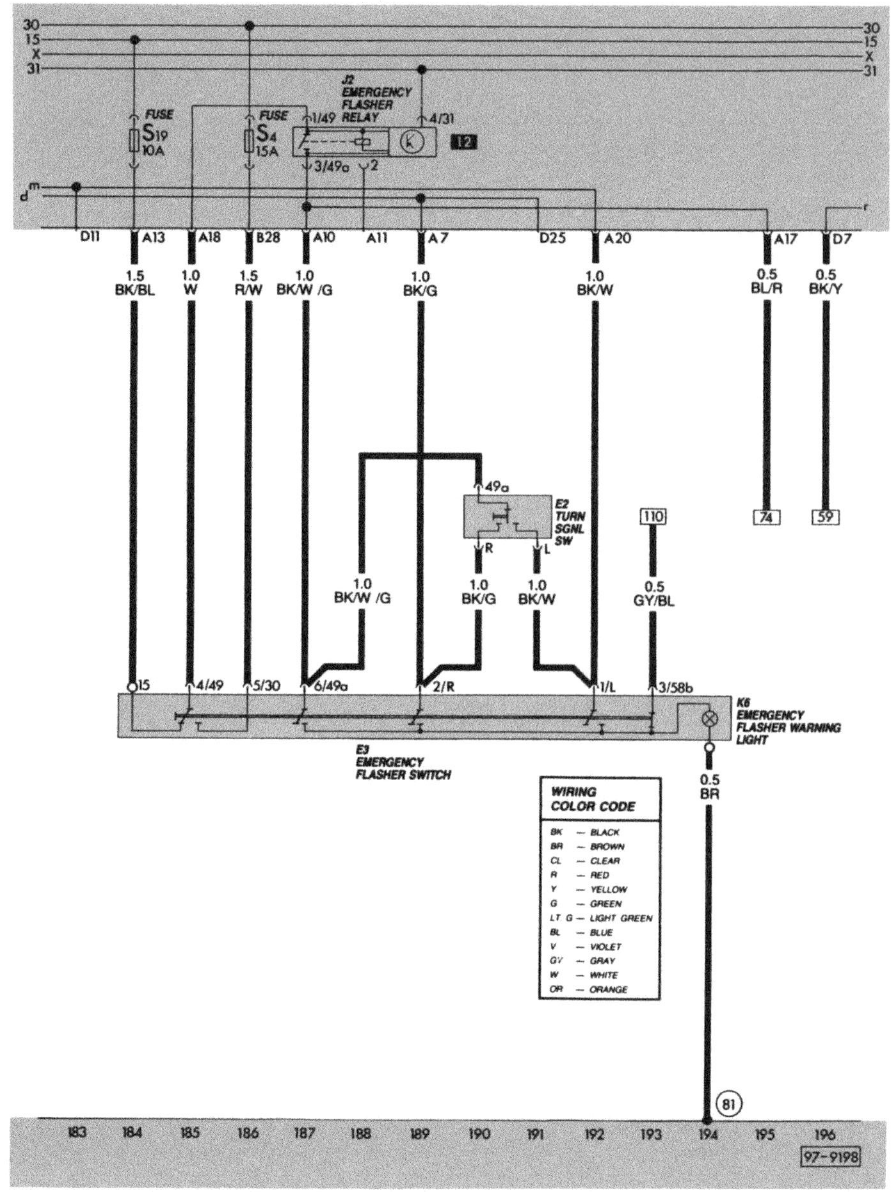

Main wiring diagram – Cabriolet from 1991
page 10 of 10

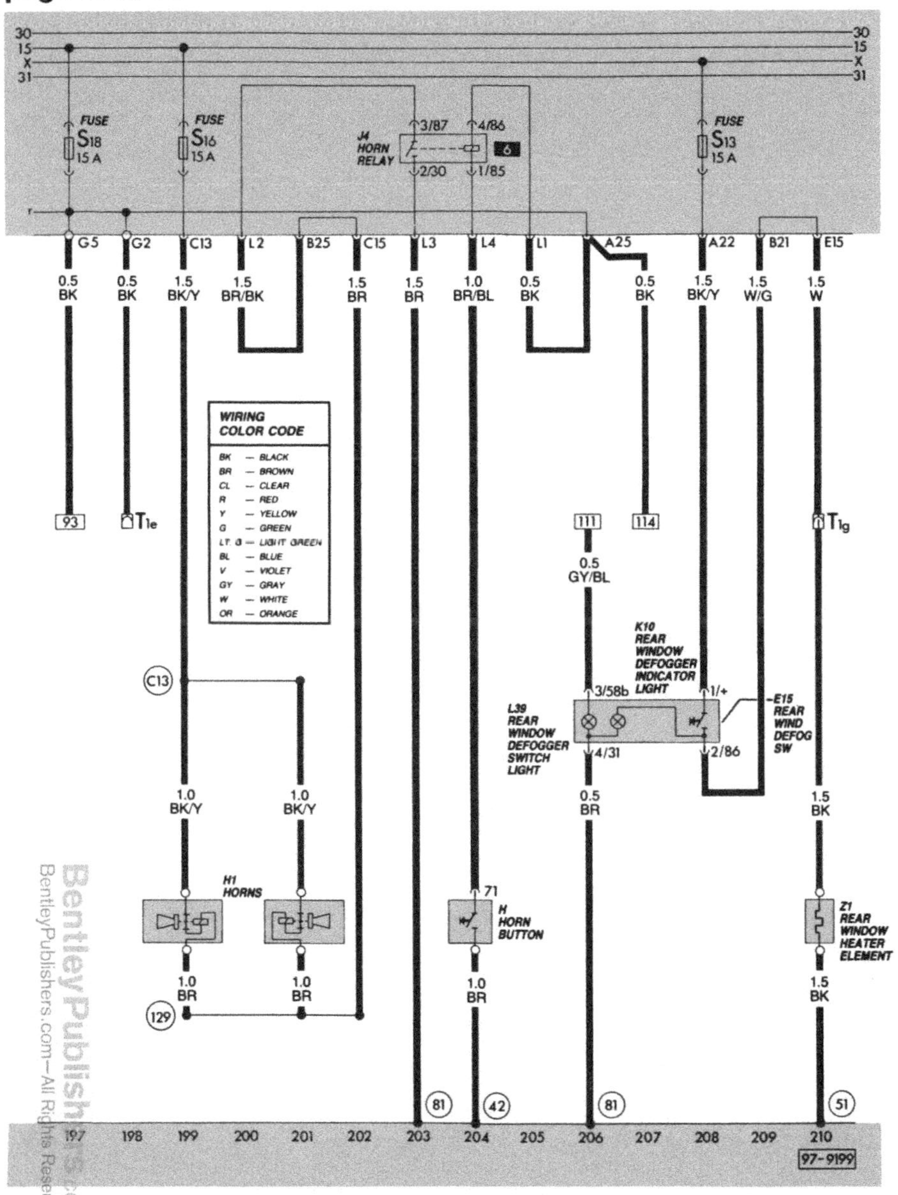

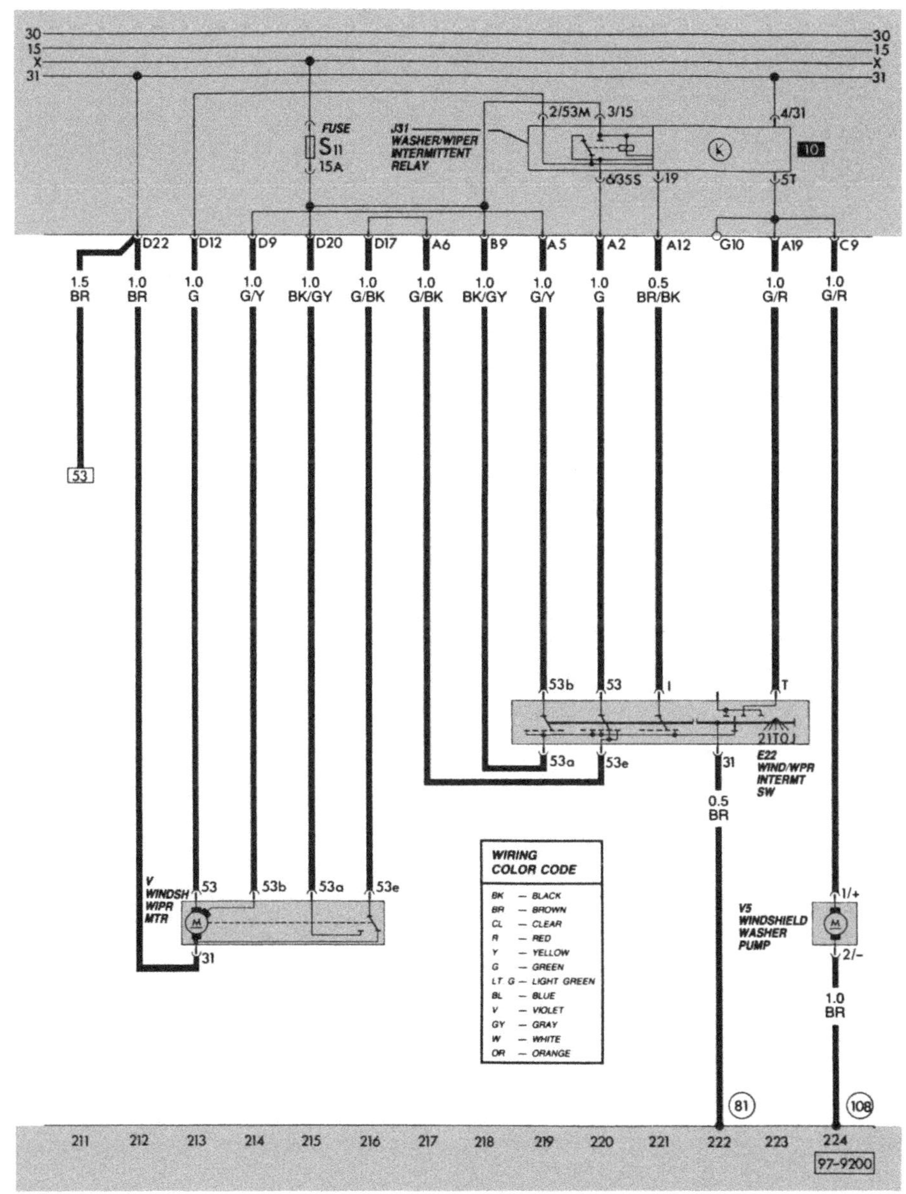

Airbag – Cabriolet from 1990
page 1 of 1

Description	Current track
Airbag control light, K75	10
Airbag control unit, J234	2-15
Airbag deceleration sensor, left, G104	3-5
Airbag deceleration sensor, right, G105	7-9
Airbag ignitor, N95	15
Airbag spiral spring, F138	11-13
Airbag warning light, K90	12
Digital clock, Y2	9
Fuse, S3	6

Wire connectors

T1 — single, behind fuse/relay panel
T1a — single, behind fuse/relay panel
T1k — single, behind fuse/relay panel
T2a — double, behind steering column switch cover
T2x — double, diagnostic connector, black, below transmission shift lever cover
T2z — double, diagnostic connector, red, below transmission shift lever cover
T3 — three-point, behind instrument panel, left
T3a — three-point, behind instrument panel, right
T6 — six-point, behind fuse/relay panel
T6a — six-point, on instrument cluster
T14 — fourteen-point, on instrument cluster

Ground connections

(30) — beside fuse/relay panel

Welded wiring harness points

(T1) — plus connection, in airbag wiring harness

WARNING —
*To prevent personal injury or airbag system failure, **only trained Volkswagen Service technicians** should test, disassemble or service the airbag system.*

WARNING —
- *Disconnect the power supply before working on the airbag system, or when doing repairs that require removing airbag system components. Disconnect the battery negative (–) terminal and cover the battery. Also disconnect the red single-pin harness connector (marked "airbag") behind the instrument panel.*
- *The airbag unit should not be removed, nor should any other work be performed on the airbag system, until at least 20 minutes after the battery is disconnected.*

WARNING —
- *Do not use a computer memory saver tool to maintain battery power to an anti-theft radio. Use of such a tool will keep the airbag system charged and may cause accidental activation of the airbag.*
- *Never use a test light on the airbag system. The airbag system must only be tested using the Volkswagen special tool VAG 1551 diagnostic test unit or its Volkswagen-approved equivalent.*
- *The airbag unit must never be electrically tested while it is not installed in the car.*

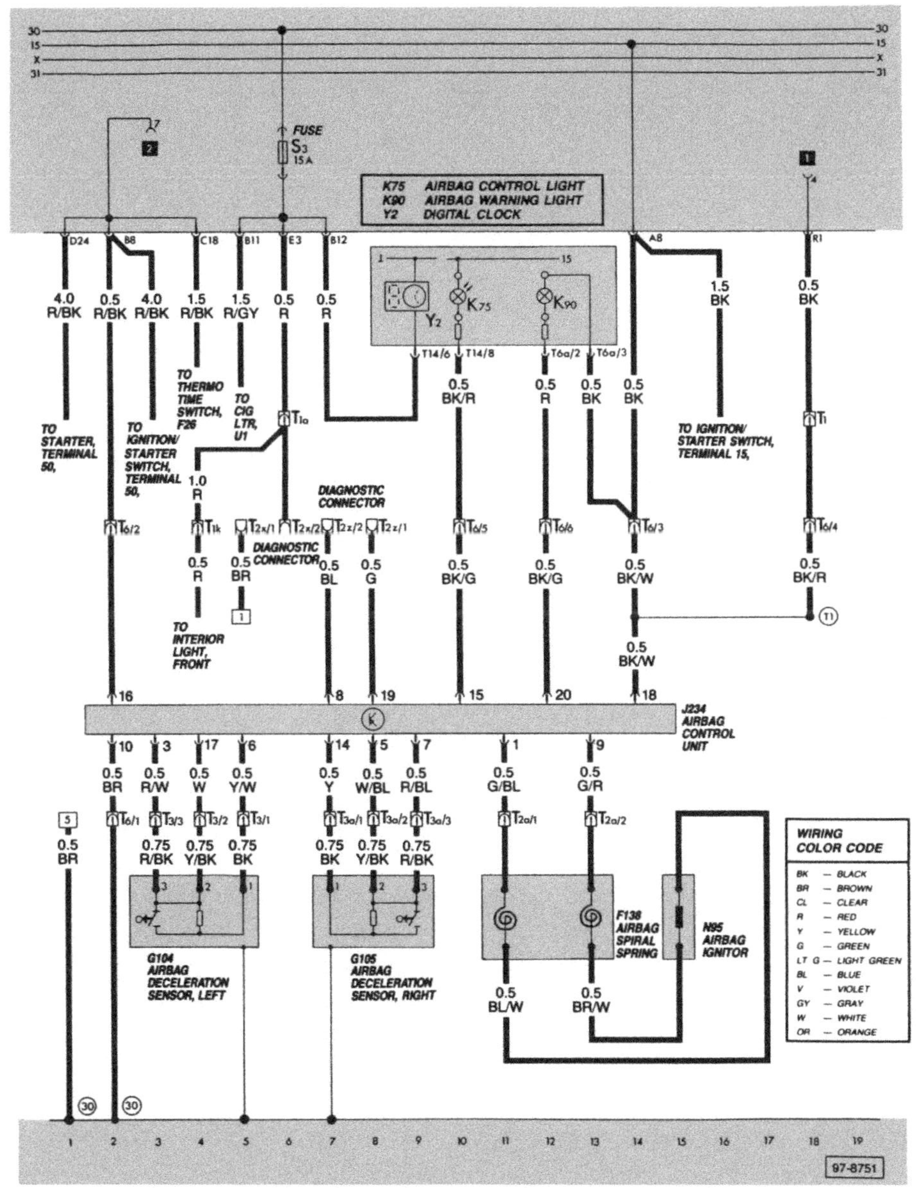

Air conditioning—Scirocco 1985-1988

Description	Current Track
A/C compressor clutch	6
A/C switch	4
A/C thermostat	6
A/C refrig. low press. switch	6
A/C refrig. high press. switch	10
A/C relay	4-9
Battery	7, 11
Fresh air fan	5
Fresh air fan series resistance	3
Fresh air fan switch	1-3
Idle speed stabilizer valve	7
Rad. cooling fan	9
Rad. cooling fan relay	11, 12
Rad. cooling fan resistance	9
Rad. cooling fan thermoswitch	9, 10

Wire connectors

T1a — single, under instrument panel
T1b — single, in engine compartment
T1c — single, under instrument panel
T1d — single, in engine compartment
T5a — five point, under instrument panel
T5b — five point, under instrument panel

Ground connections

(9) — under fuse/relay panel

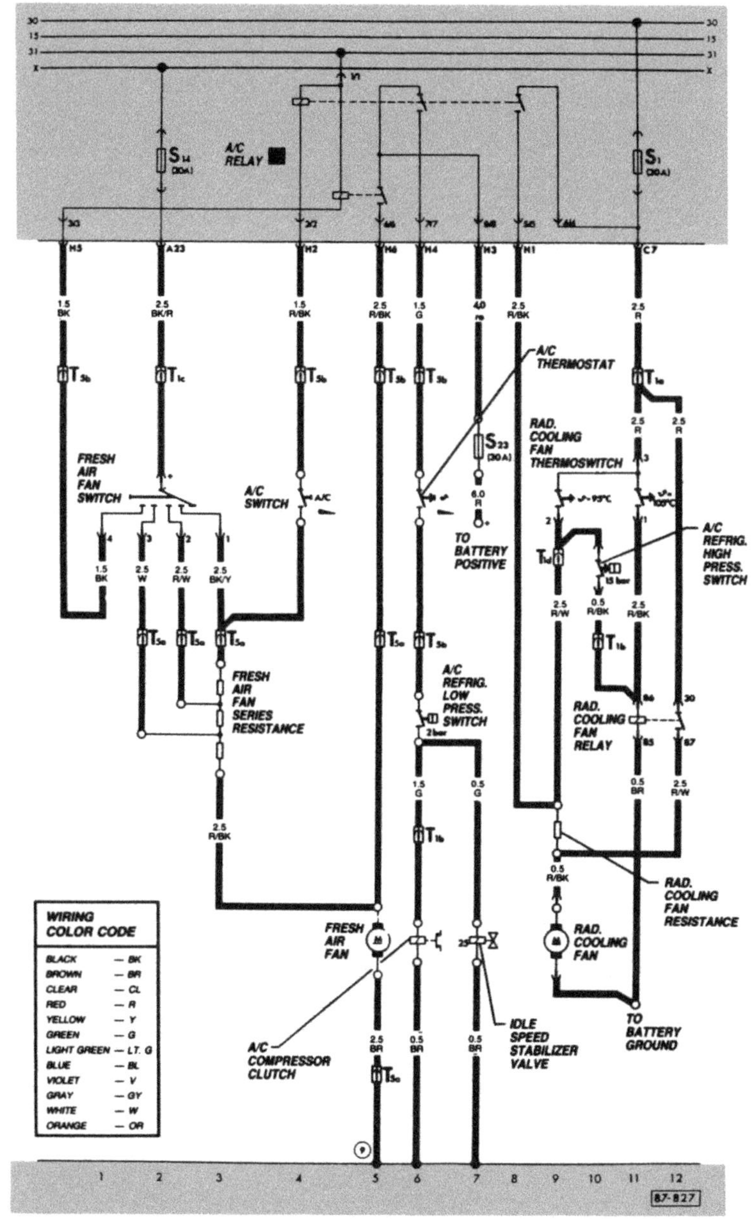

Air conditioning – Cabriolet 1985-1988
page 1 of 1

Description	Current Track
A/C compressor clutch	6
A/C relay	4-8
A/C switch	4
A/C thermostat	6
Battery	7
Fresh air fan	5
Fresh air fan series resistance	3
Fresh air fan switch	1-3
Idle speed stabilizer valve	7
Rad. cooling fan	10
Rad. cooling fan thermoswitch	9

Connectors

T1a — single, under instrument panel
T1b — single, in engine compartment
T1c — single, under instrument panel
T1d — single, in engine compartment
T5a — five point, under instrument panel
T5b — five, point, under instrument panel

Ground Connectors

(1) — at battery

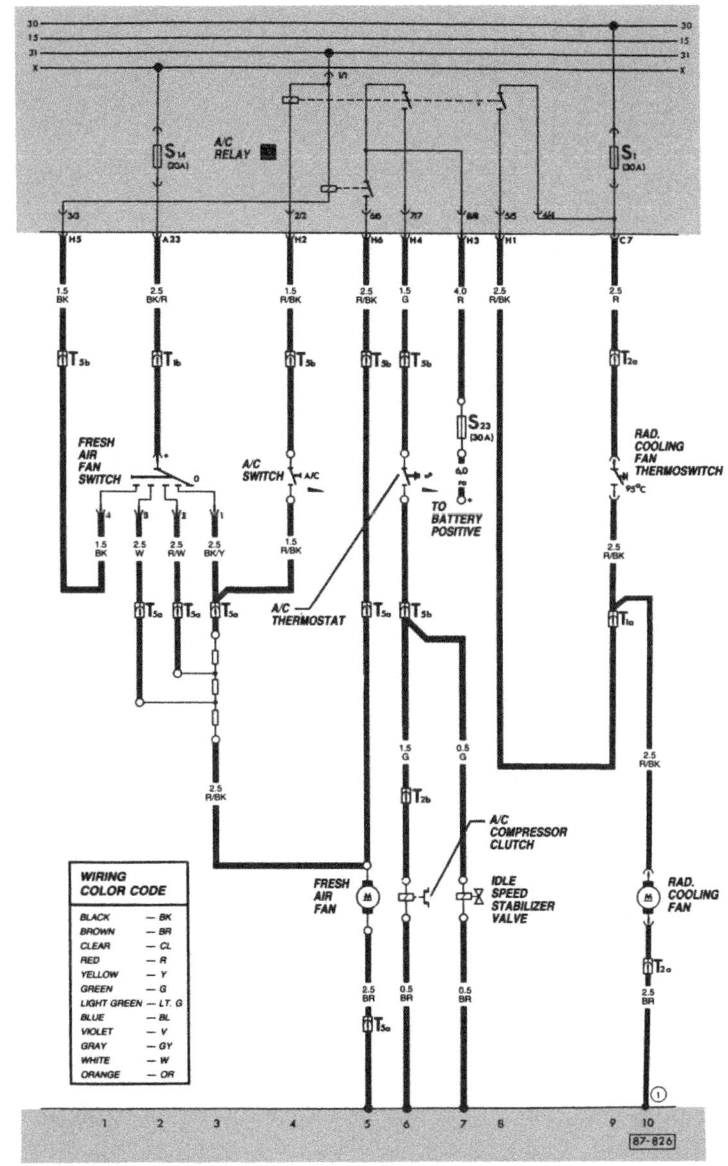

Air conditioning – Scirocco, Cabriolet 1989
page 1 of 1

Description	Current track
A/C compressor clutch	11
A/C refrigerant high pressure switch	13
A/C refrigerant low pressure switch	10
A/C relay	8-10
A/C switch	7
A/C thermostat	10
Fresh air blower series resistance	4-6
Fresh air fan	2
Fresh air fan switch	3-6
Idle enrichment valve	10
Main heater fuse	1
Radiator cooling fan	13-15
Radiator cooling fan after run relay	15-17
Radiator cooling fan thermoswitch	15

Wire connectors

T2 — double, near A/C compressor
T2a — double, on fan shroud
T2b — double, on fan shroud
T2c — double, on firewall, center
T5a — five-point, behind dashboard
T5b — five-point, behind dashboard

Ground connections

(18) — on engine block
(28) — on bulkhead
(30) — beside fuse/relay panel
(82) — in wiring harness, left front

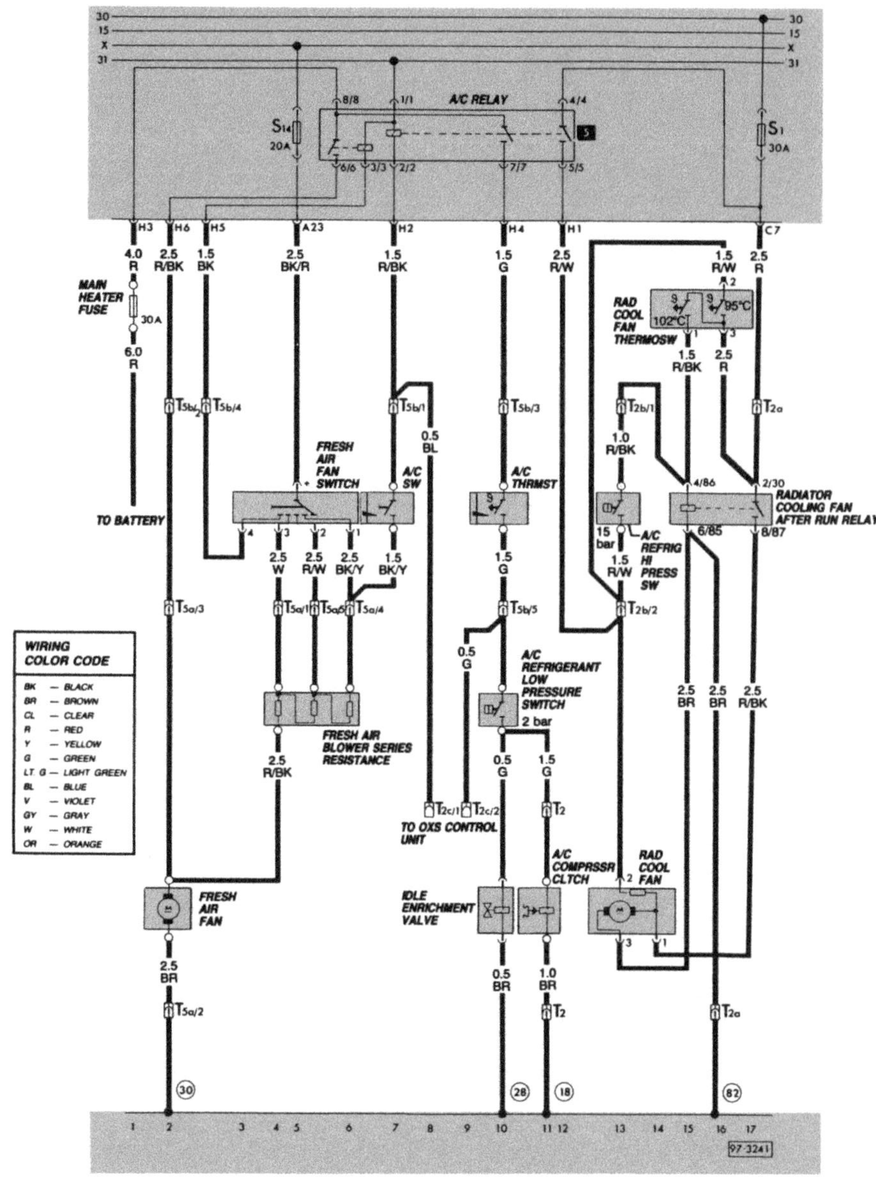

Air conditioning – Cabriolet from 1990
page 1 of 1

Description	Current track
A/C compressor clutch, N25	6
A/C refrigerant high pressure switch, F23	10
A/C refrigerant low pressure switch, F73	6
A/C relay, J32	1-9
A/C switch, E35	5
A/C thermostat, E33	7
Fresh air blower series resistance, N23	2-4
Fresh air fan, V2	1
Fresh air fan switch, E9	2-4
Fuse, S14	12
Fuse, S1	14
Fuse, S23	4
Radiator cooling fan, V7	9-14
Radiator cooling fan relay, J26	12-14
Radiator cooling fan thermoswitch, F18	11-12

Wire connectors

T1d — single, on firewall, center
T1k — single, in engine compartment, left front
T2 — double, on A/C compressor
T2a — double, in engine compartment, left
T2b — double, in engine compartment, left
T5a — five-point, behind instrument panel
T5b — five-point, behind instrument panel

Ground connections

(18) — on engine block
(30) — near fuse/relay panel
(82) — in left front wiring harness

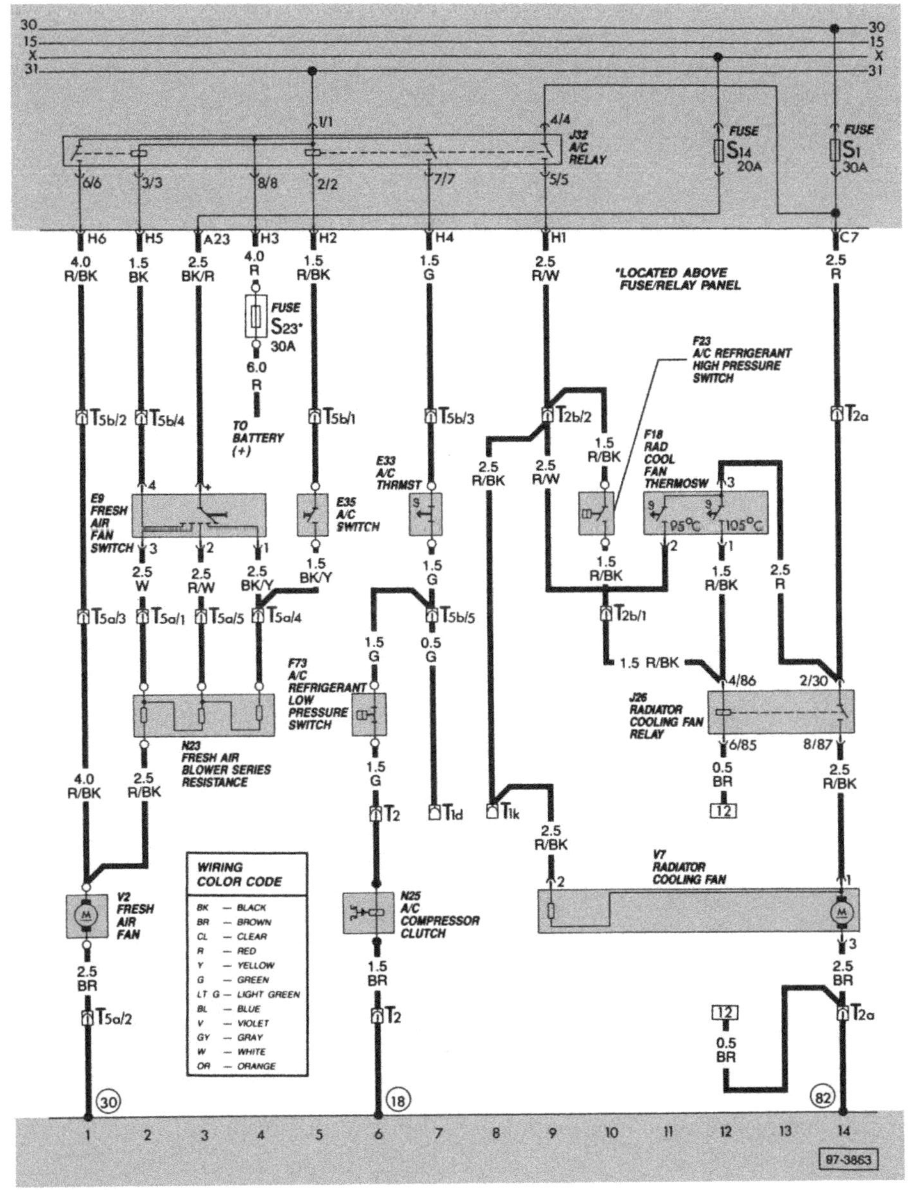

Automatic shiftlock – Cabriolet from June 1990
page 1 of 2

Description	Current track
Automatic transmission console light, L19	7
Back-up light switch, F4	2
Back-up light, left, M16	4
Back-up light, right, M17	6
Fuse, S18	10
Fuse, S15	18
Radiator cooling fan after run control unit, J138	1
Shift lock control unit, J221	19-25
Shift lock solenoid, N110	19
Starter, B	34-35
Starter cut-out/back-up light switch, E17	8-14
Starting interlock relay, J207	31-38

Wire connectors

T1a — single, behind fuse/relay panel
T1b — single, behind fuse/relay panel
T1c — single, behind fuse/relay panel
T1d — single, behind fuse/relay panel
T1e — single, behind fuse/relay panel
T1f — single, behind fuse/relay panel
T1g — single, behind fuse/relay panel
T2 — double, behind fuse/relay panel
T2a — double, below transmission shift lever cover
T2e — double, in engine compartment, left
T2f — double, behind fuse/relay panel
T3 — three-point, below transmission shift lever cover
T4 — four-point, behind fuse/relay panel
T4a — four-point, near transmission shift lever
T6 — six-point, below transmission shift lever cover
T8 — eight-point, on radio
T9 — nine-point, on starting interlock relay

Ground connections

(30) — near fuse/relay panel
(54) — on rear apron
(63) — on tail light bulb holder, left
(64) — on tail light bulb holder, right
(102) — in automatic transmission wiring harness

Welded wiring harness points

(A4) — plus connection (58b), in instrument panel wiring harness
(U2) — plus connection (15), in automatic transmission wiring harness

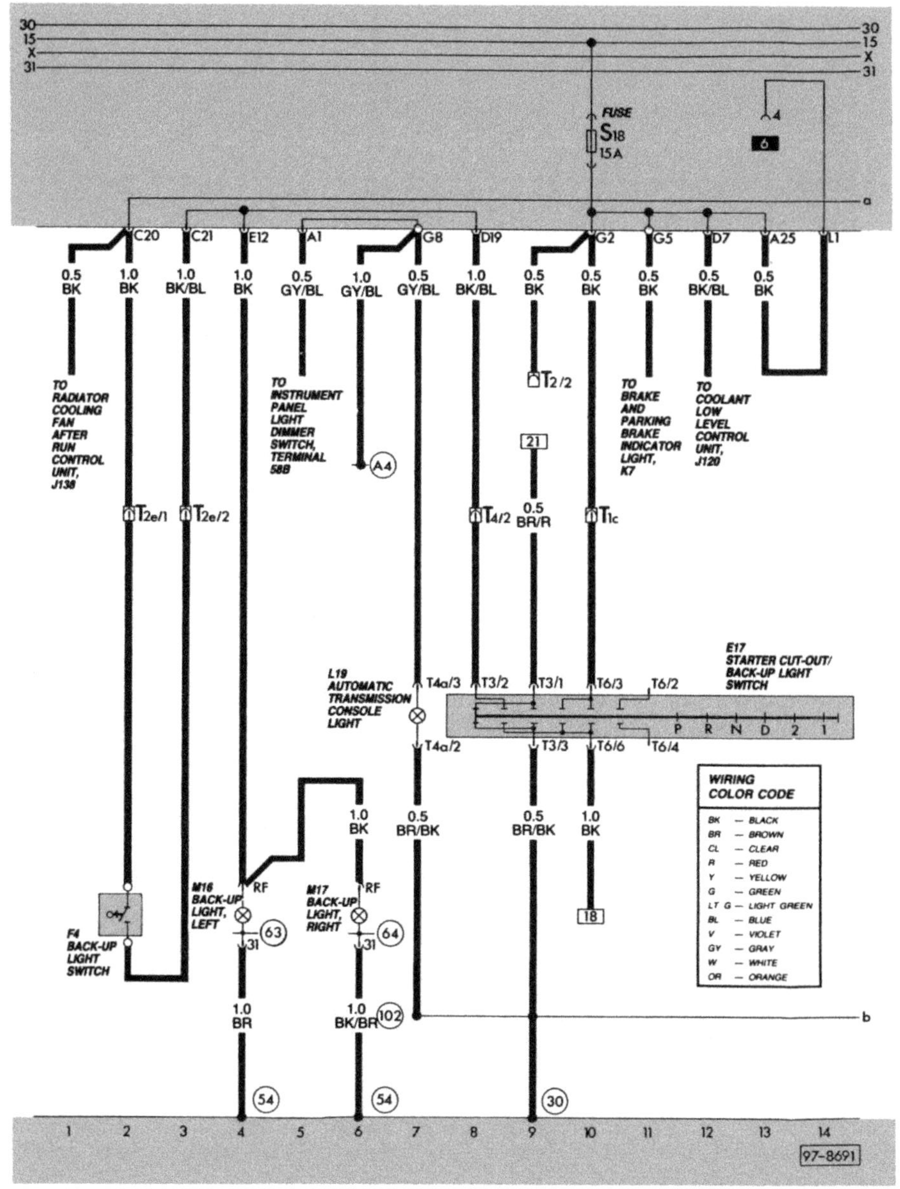

Automatic shiftlock – Cabriolet from June 1990
page 2 of 2

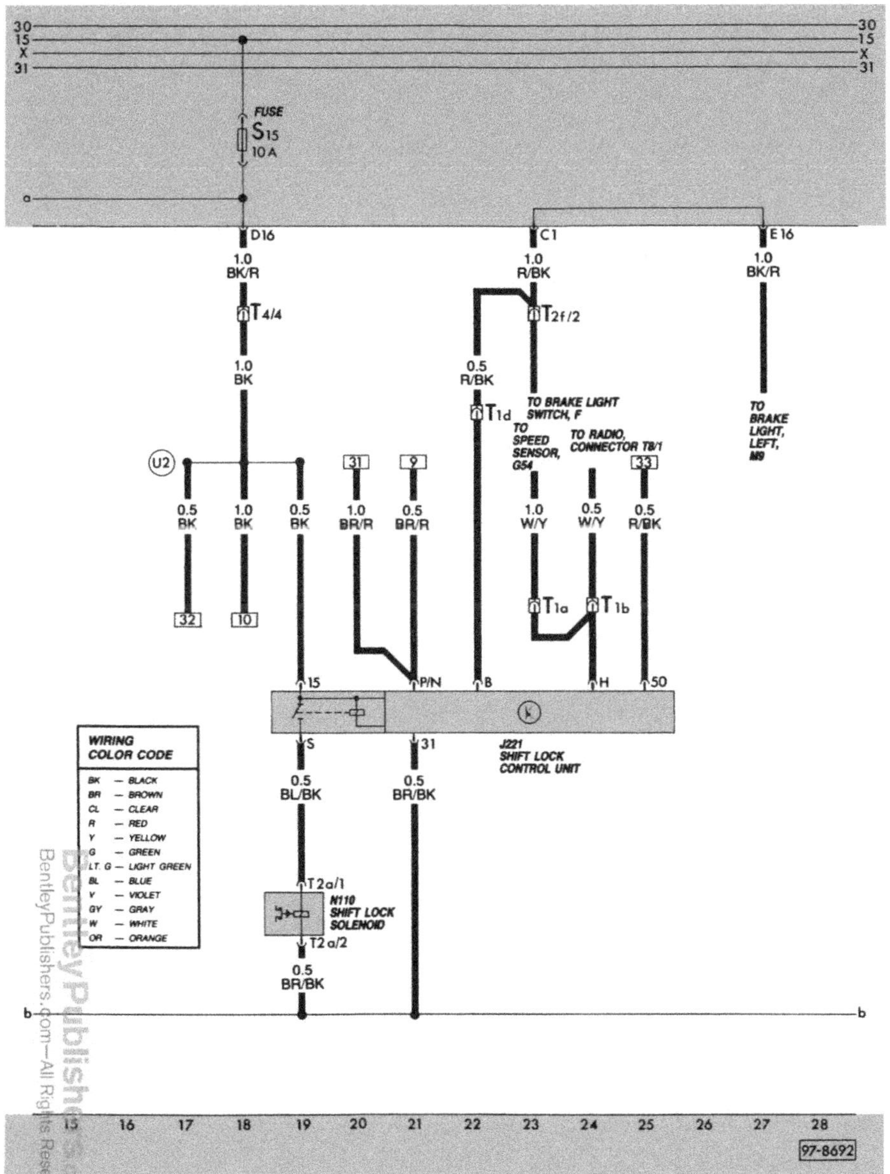

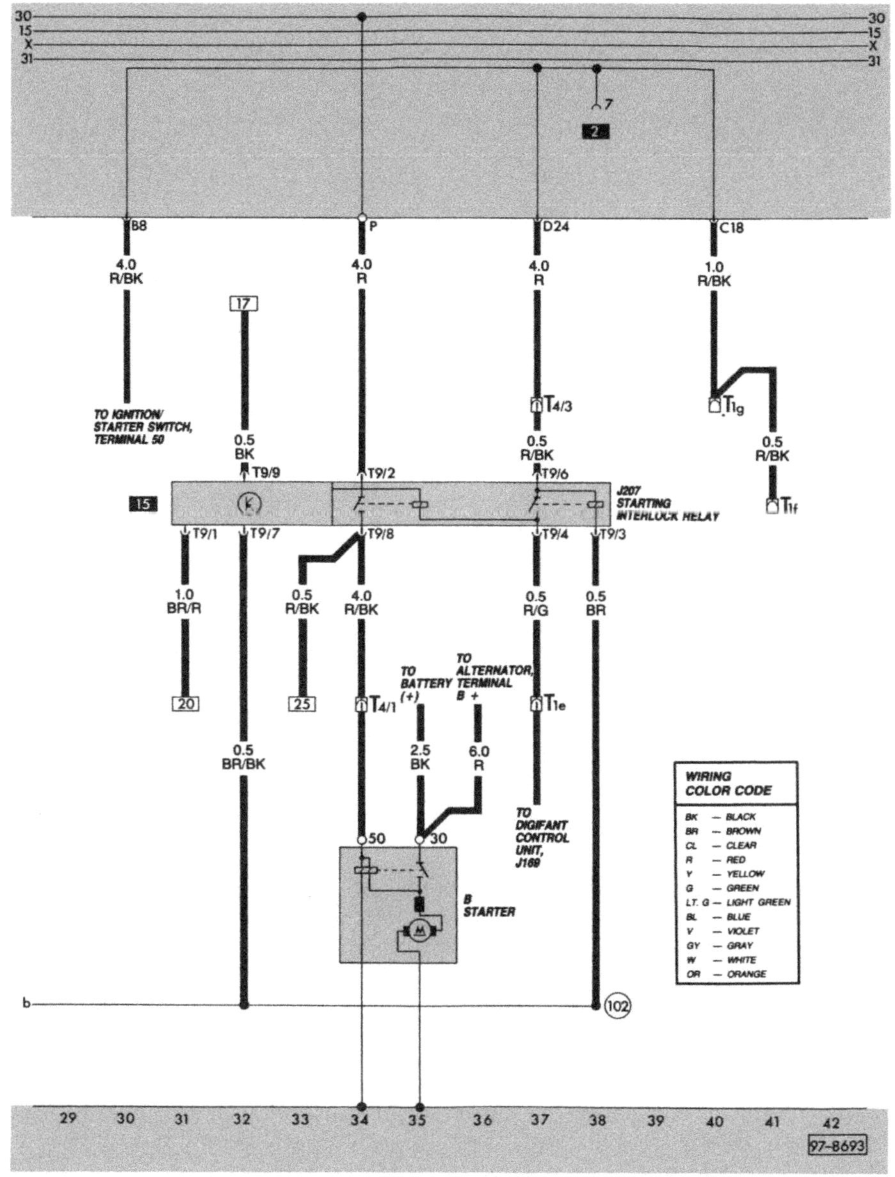

Central locking—Scirocco 1985-1988
page 1 of 1

Description	Current Track
Central locking motor	2, 3
Central locking switch	2, 3
Cigarette lighter	1
Digital clock	2
Interior light, front	3
Fuse S3	1

Wire connectors

T1 — single, next to fuse/relay panel
T3 — 3-point, below dash, left

Ground Connectors

(14) — next to Bi-pressure pump

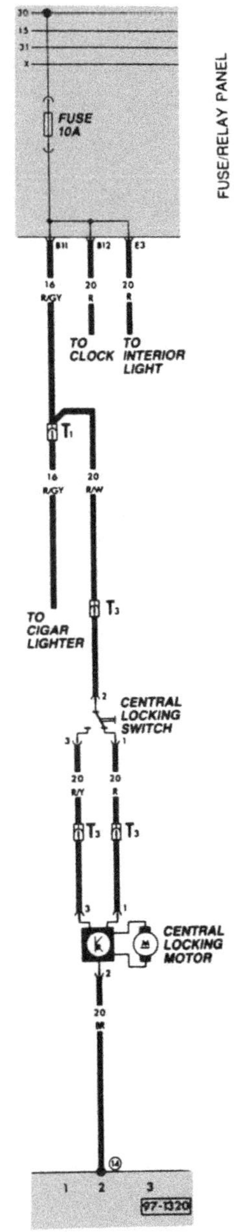

WIRING COLOR CODE	
BLACK	— BK
BROWN	— BR
CLEAR	— CL
RED	— R
YELLOW	— Y
GREEN	— G
LIGHT GREEN	— LT G
BLUE	— BL
VIOLET	— V
GRAY	— GY
WHITE	— W
ORANGE	— OR

Cruise control, automatic transmission - Scirocco, Cabriolet 1985-1990
page 1 of 1

Description	Current track
Automatic transmission console lite	13
Back-up lite, left	15
Back-up lite, right	14
Brake lite switch	2,3
Brake lite, left	1
Brake lite, right	2
Cruise control inductive sender	6
Cruise control switch	5-7
Cruise control, main control unit	4-8
Starter	9-11
Starter cut-out/back-up lite switch	9-11
Vacuum control motor	7-9
Vacuum vent valve, brake	4

Wire Connectors

T1a — single, under relay panel
T1c — single, in eng. compart. near ign. coil
T2a — double, under relay panel
T2c — double, under relay panel
T3 — three point, under relay panel
T4 — four point, under center console
T4a — four point, under relay panel
T4b — four point, under steering column cover

Ground Connections

(10) — near relay panel
(21) — by neutral safety switch

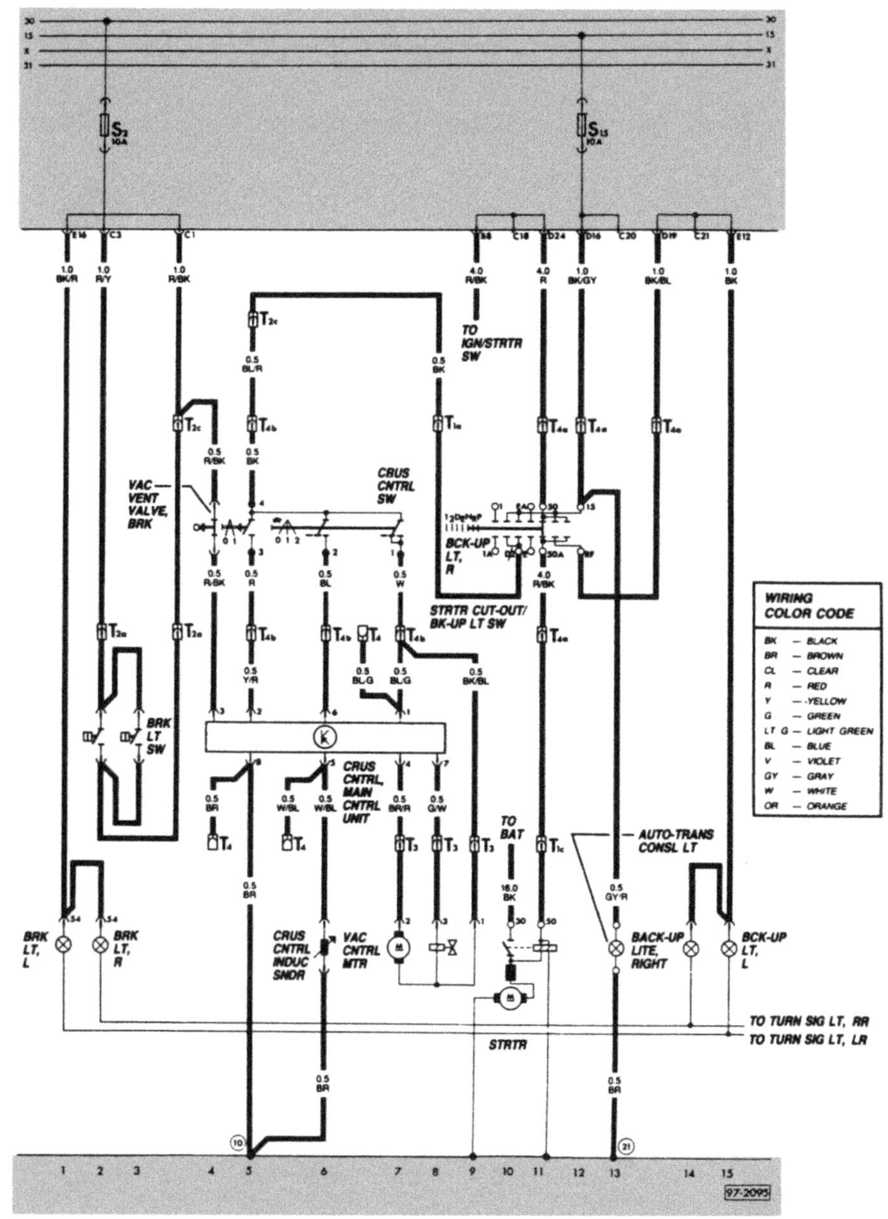

Cruise control, manual transmission - Scirocco, Cabriolet 1985-1990
page 1 of 1

Description	Current track
Brake lite switch	2,3
Brake lite, left	1
Brake lite, right	2
Cruise control inductive sender	6
Cruise control switch	5-7
Cruise control, main control unit	4-8
Horn relay	8
Vacuum control motor	7-9
Vacuum vent valve, brake	4
Vacuum vent valve, clutch	4

Wire connectors

T1a — single, under relay panel
T2a — double, under relay panel
T2c — double, under relay panel
T3 — three point, under relay panel
T4 — four point, under relay panel
T4b — four point, under steering column cover

Ground connections

(10) — near relay panel

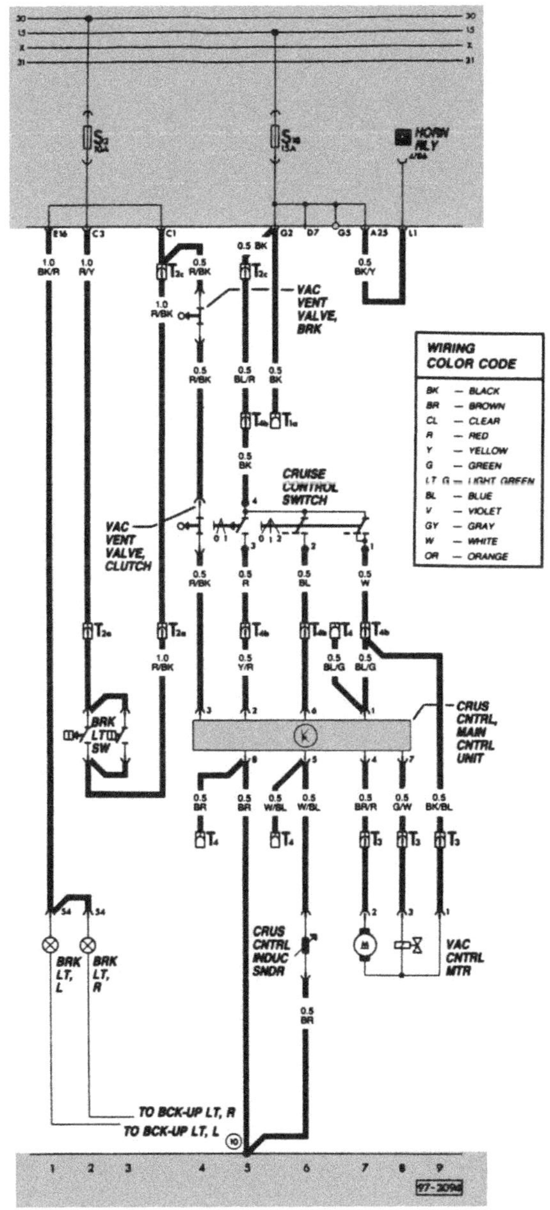

Cruise control, manual transmission – Cabriolet from 1991
page 1 of 1

Description	Current track
Brake light switch, F	14
Cruise control switch, E45	2-4
Cruise control vacuum pump, V18	7-8
Cruise control, control unit, J213	2-11
Fuse, S18	1
Fuse, S2	14
Vacuum vent valve, brake, F47	11
Vacuum vent valve, clutch, F36	11

Wire connectors

T1b — single, behind fuse/relay panel
T1c — single, behind fuse/relay panel
T1v — single, behind fuse/relay panel
T2 — double, behind fuse/relay panel
T2f — double, behind instrument panel, left
T3e — three-point, above fuse/relay panel
T4g — four-point, behind steering column switch cover
T4k — four-point, near steering column
T8 — eight-point, on radio
T8a — eight-point, on control unit for cruise control

Ground connections

(30) — beside fuse/relay panel

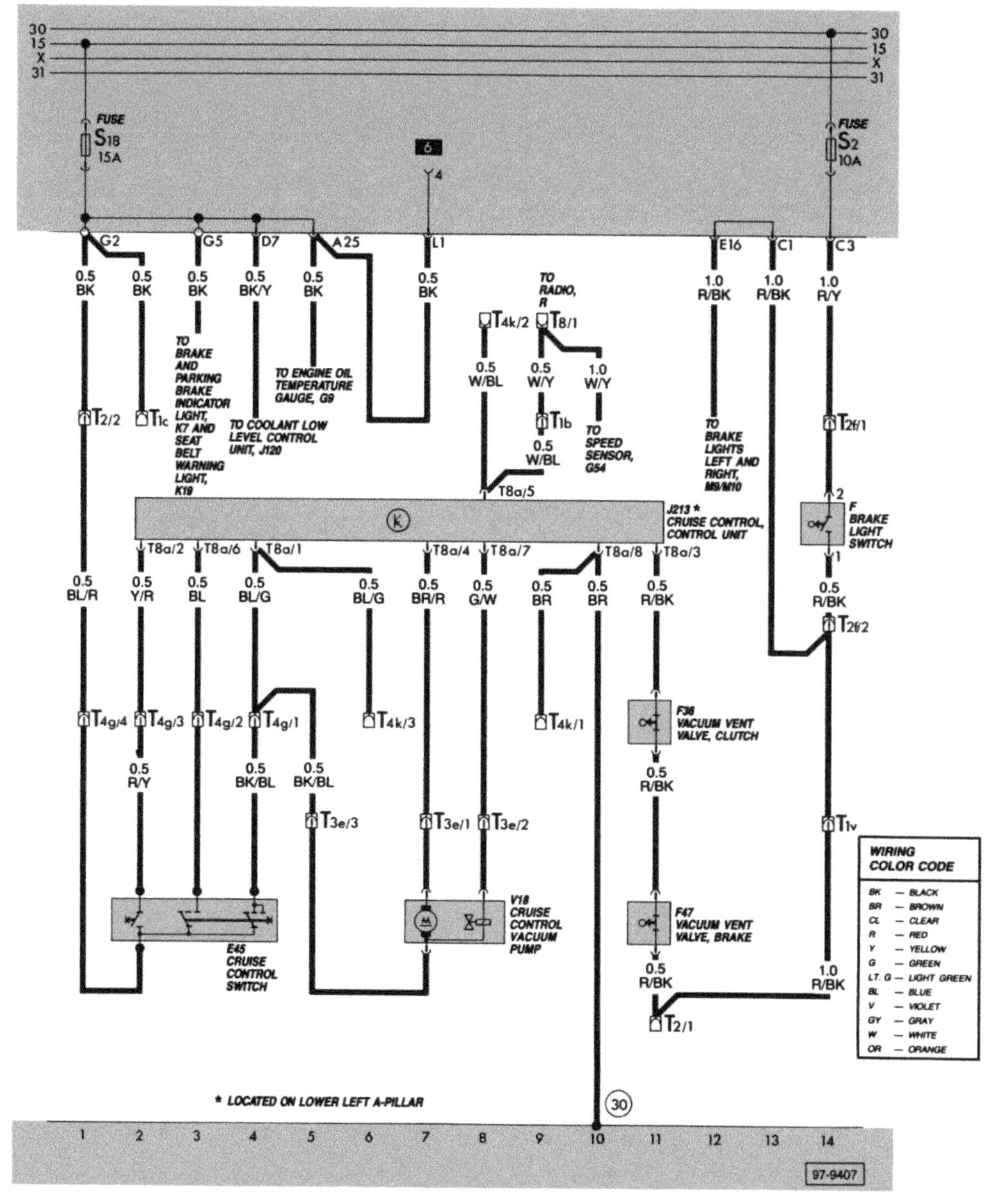

Cruise control, automatic transmission – Cabriolet from 1991
page 1 of 1

Description	Current track
Brake light switch, F	14
Cruise control switch, E45	2-4
Cruise control vacuum pump, V18	7-8
Cruise control, control unit, J213	2-11
Fuse, S18	1
Fuse, S2	14
Vacuum vent valve, brake, F47	11

Wire connectors

T1b — single, behind fuse/relay panel
T1c — single, behind fuse/relay panel
T1v — single, behind fuse/relay panel
T2 — double, behind fuse/relay panel
T2f — double, behind instrument panel, left
T3e — three-point, above fuse/relay panel
T4g — four-point, behind steering column switch cover
T4k — four-point, near steering column
T6 — six-point, on starter cut-out/back-up light switch
T8 — eight-point, on radio
T8a — eight-point, on control unit for cruise control

Ground connections

(30) — beside fuse/relay panel

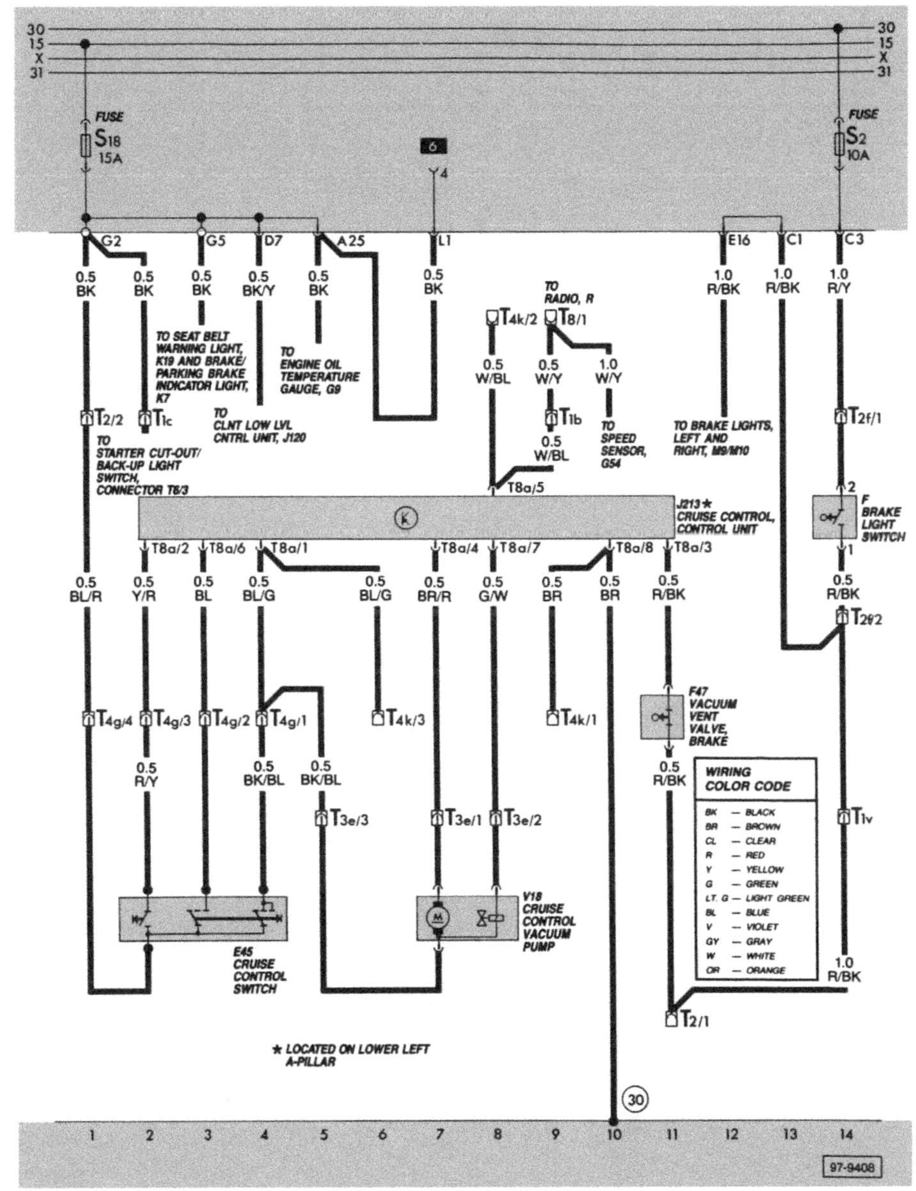

Daytime running lights (Canada) – Cabriolet 1990
page1 of 2

Description	Current track
Daytime driving lights resistor, N53	13
Fuse, S10	3
Fuse, S21	4
Fuse, S9	7
Fuse, S22	8
Fuse, S20	28
Fuse, S7	32
Fuse, S8	38
Headlight dimmer/flasher switch, E4	10-11
Headlight, left, L1	3, 4
Headlight, right, L2	7, 8
High-beam headlight, left, L13	1
High-beam headlight, right, L14	5
Instrument panel light dimmer switch, E20	23
License plate light, X	25, 27
Light switch, E1	16-23
Load reduction relay, J59	16-17
Luggage compartment light, W3	28
Luggage compartment light switch, F5	28
Parking light, left, M1	33
Parking light, right, M3	39
Side marker lights, front, M11	32, 38
Tail light, left, M4	29
Tail light, right, M2	37
Turn signal light, left front, M5	34
Turn signal light, right front, M7	40

Wire connectors

T1n — single, in luggage compartment, left
T1p — single, near left headlight
T1r — single, near right headlight
T2 — double, behind fuse/relay panel
T2a — double, behind fuse/relay panel
T2b — double, behind fuse/relay panel
T3d — three-point, near right headlight
T3e — three-point, near left headlight
T8 — eight-point, on radio
T14 — fourteen-point, on instrument cluster

Ground connections

(50) — in luggage compartment, left
(82) — in left front wiring harness
(108) — in left front wiring harness

Welded wiring harness points

(A4) — plus connection (58b), in instrument cluster wiring harness

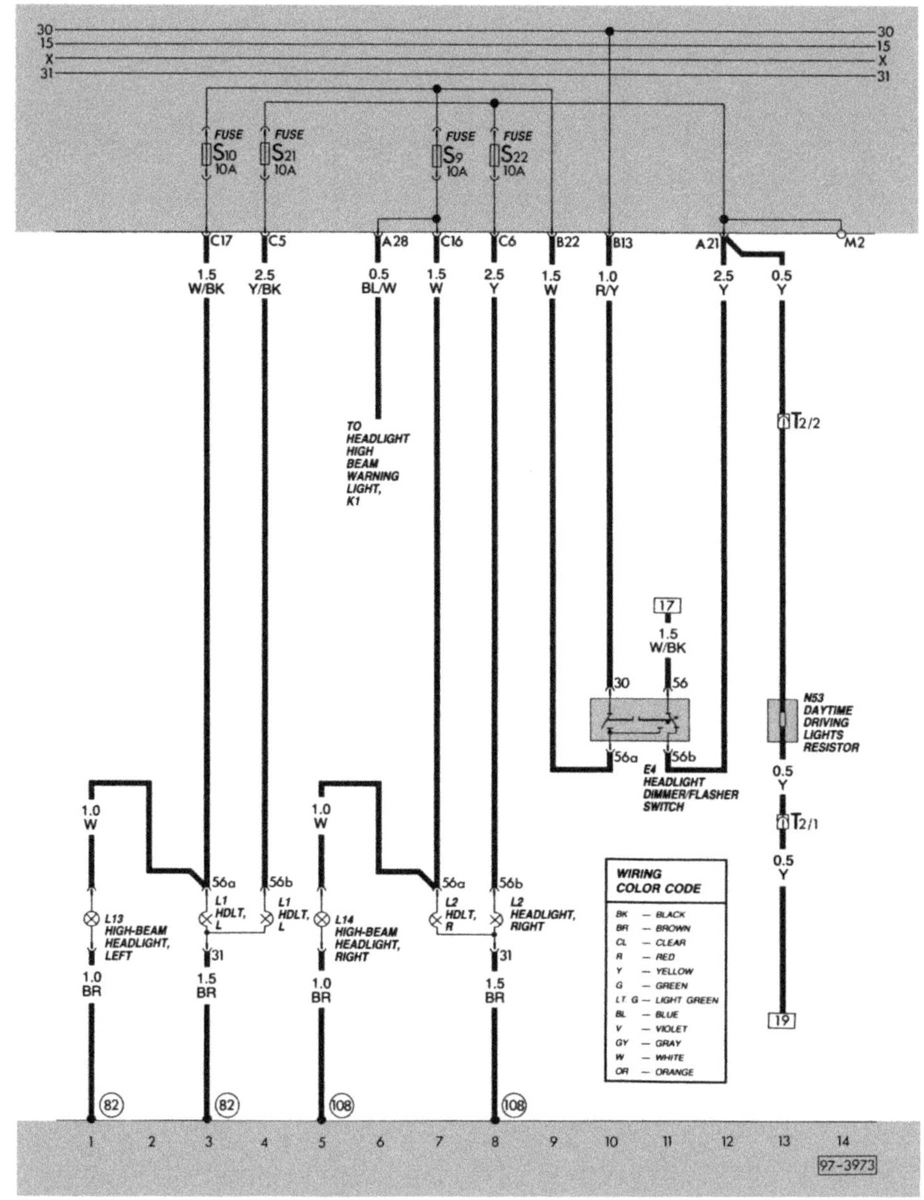

Daytime running lights (Canada) – Cabriolet 1990
page 2 of 2

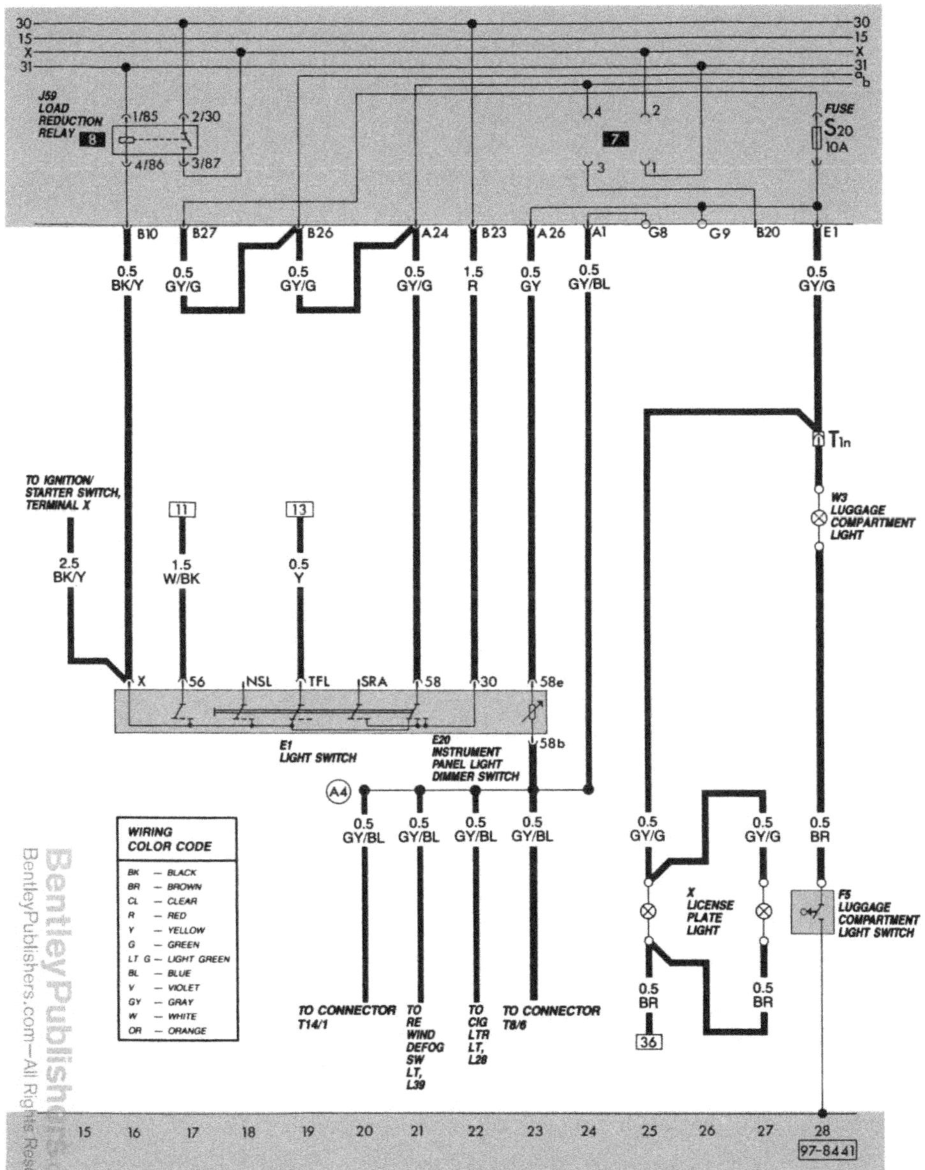

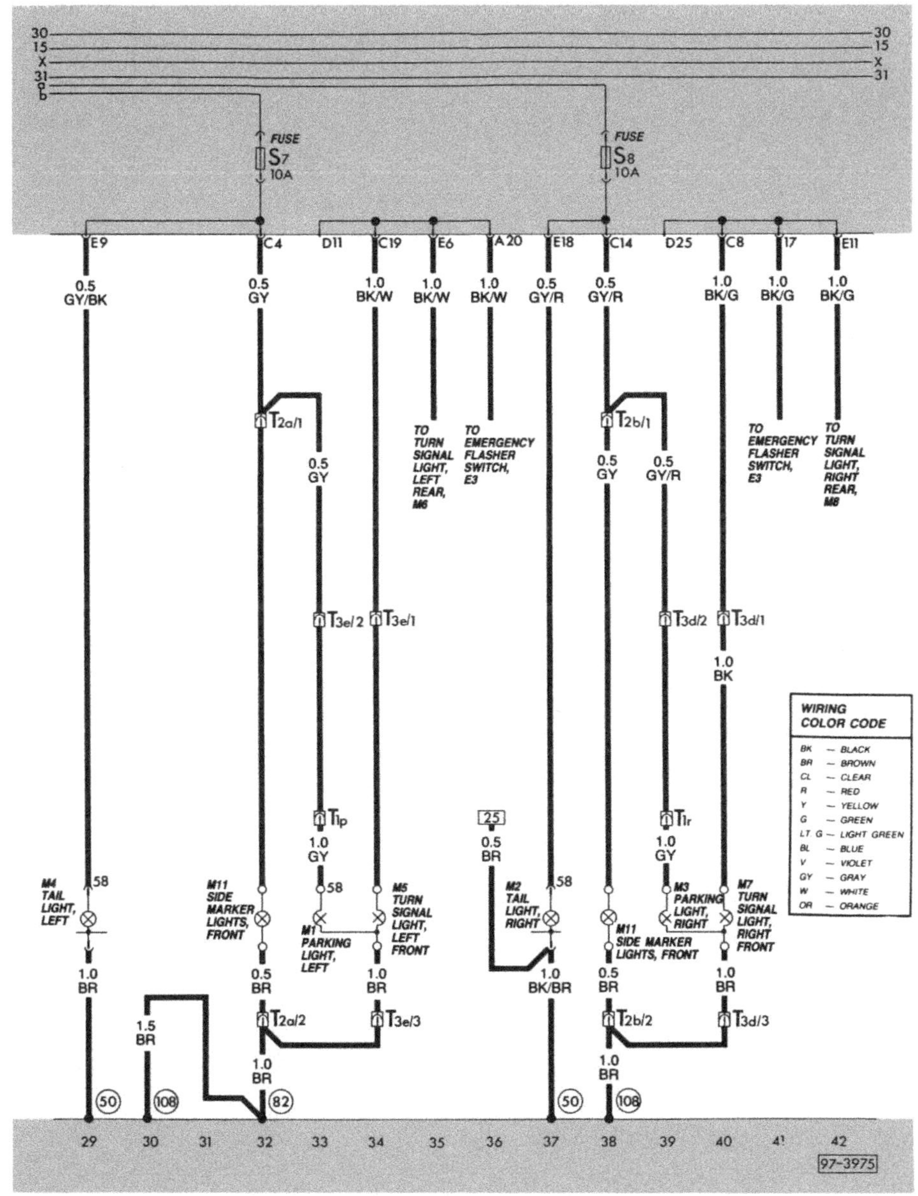

Daytime running lights (Canada) – Cabriolet from 1991 page 1 of 2

Description	Current track
Daytime running light relay, J90	15-17
Daytime running lights resistor, N53	13
Fuse, S10	3
Fuse, S21	4
Fuse, S9	7
Fuse, S22	8
Fuse, S20	28
Fuse, S7	32
Fuse, S8	38
Headlight dimmer/flasher switch, E4	10-11
Headlight, left, L1	3, 4
Headlight, right, L2	7, 8
High-beam headlight, left, L13	1
High-beam headlight, right, L14	5
Instrument panel light dimmer switch, E20	23
License plate light, X	25, 27
Light switch, E1	16-23
Load reduction relay, J59	16-17
Luggage compartment light, W3	28
Luggage compartment light switch, F5	28
Parking light, left, M1	33
Parking light, right, M3	39
Side marker lights, front, M11	32, 38
Tail light, left, M4	29
Tail light, right, M2	37
Turn signal light, left front, M5	34
Turn signal light, right front, M7	40

Wire connectors

T1 — single, behind fuse/relay panel
T1a — single, behind instrument panel, left
T1b — single, behind instrument panel, left
T1c — single, behind instrument pnael, left
T1n — single, in luggage compartment, left
T1p — single, near headlight, right
T1r — single, near headlight, left
T2 — double, behind fuse/relay panel
T2a — double, behind fuse/relay panel
T2b — double, behind fuse/relay panel
T3d — three-point, near headlight, right
T3e — three-point, near headlight, left
T8 — eight-point, on radio
T14 — fourteen-point, on instrument cluster

Ground connections

(50) — in luggage compartment, left
(82) — in left front wiring harness
(108) — in left front wiring harness

Welded wiring harness points

(A4) — plus connection (58b), in instrument panel wiring harness

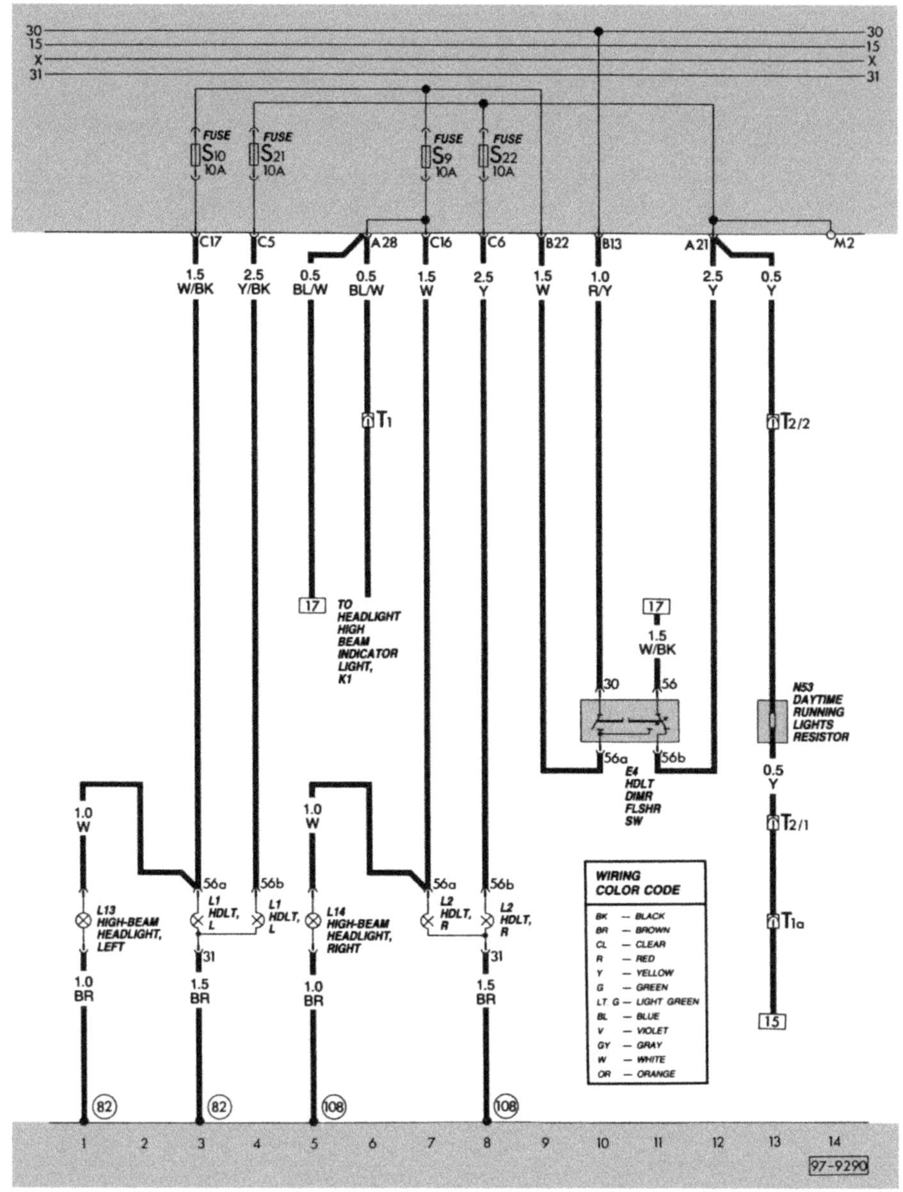

Daytime running lights (Canada) – Cabriolet from 1991 page 2 of 2

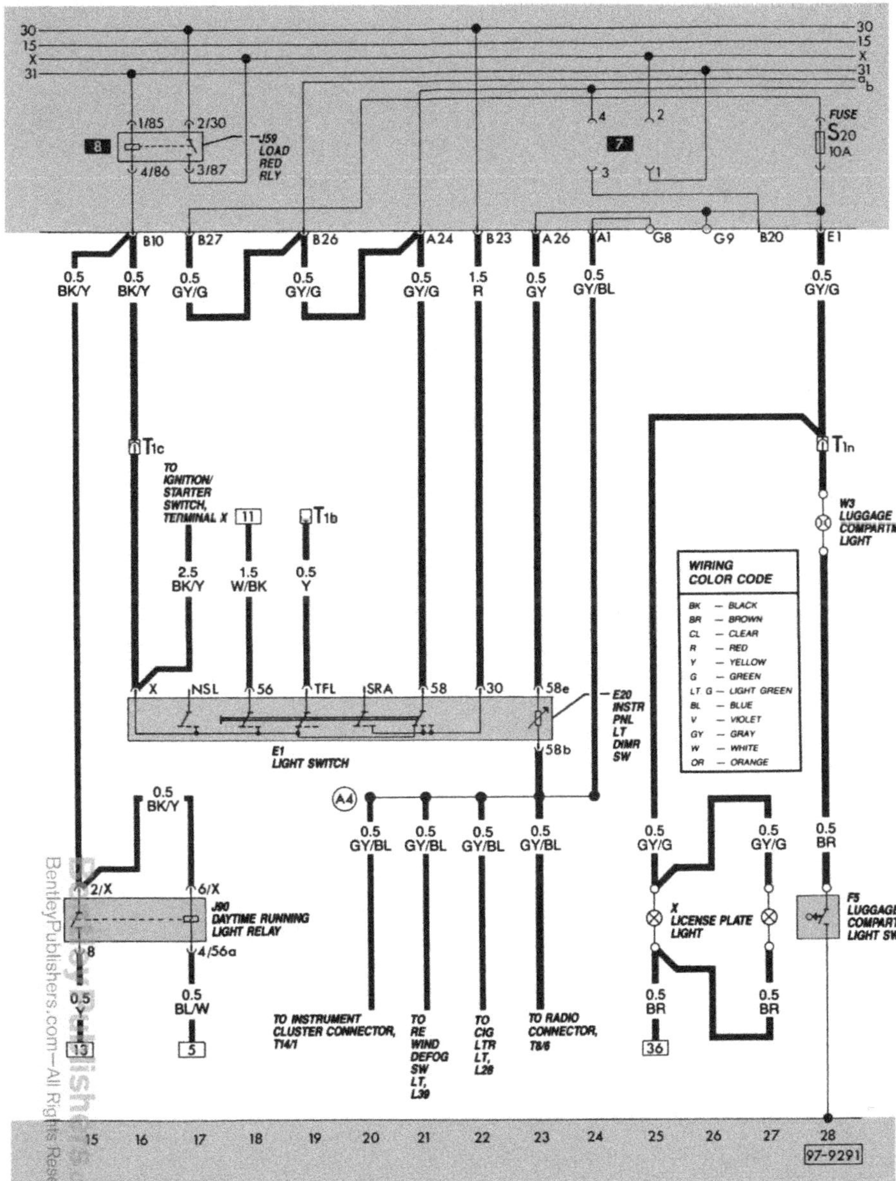

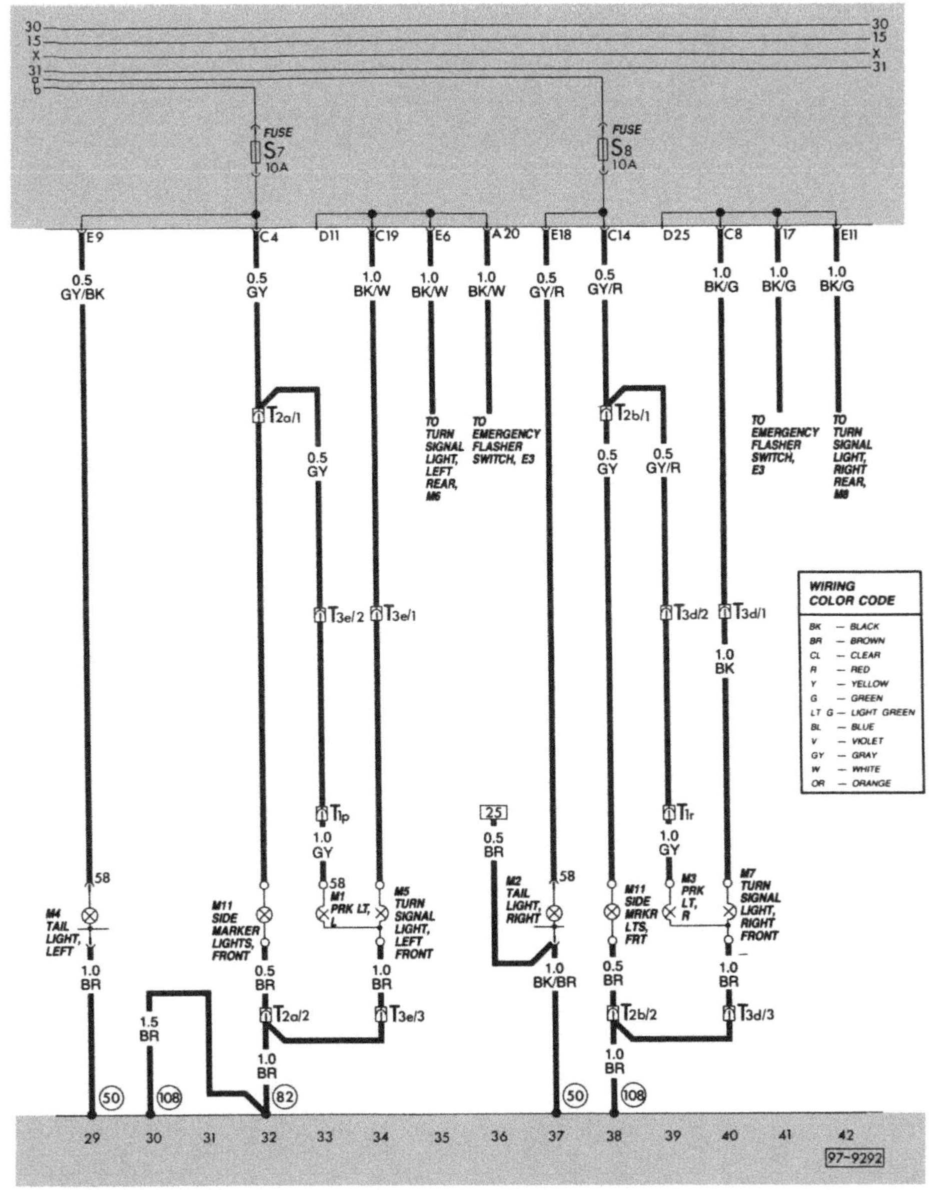

Electronic engine controls – 16-valve Scirocco page 1 of 4

Description	Current track
Air flow sensor potentiometer	28, 29
Alternator	1, 2
Battery	4
Cold start valve	8
Differential pressure regulator	26
Fuel-gauge sending unit	38
Fuel pump	43
Fuel-pump relay	41, 42
Hall generator	17, 18
Idle speed stabilizer valve	24, 25
Ignition coil	12-14
Ignition control unit	10-14
Ignition distributor	13-15
Knock sensor	19, 20
Knock sensor control unit	17-21
OXS control unit	19-32
Oxygen sensor	27
Oxygen sensor heater	44
Spark plug	13-15
Spark plugs	13-15
Starter	5, 6
Temperature sensor (NTC)	32
Thermo time switch	8, 9
Throttle-valve switch (enrichment)	22, 23
Transfer fuel pump	40
Up-shift-indicator control unit	34-37
Upshift switch, transmission	37
Upshift switch, vacuum	35
Voltage regulator	1, 2

Wire connectors

T — wire distributor, under fuse/relay panel
T1a — single, eng. compart., near battery
T1b — single, under fuse/relay panel
T1c — single, eng. compart., near ign. coil
T1d — single, under fuse/relay panel
T1e — single, eng. compart., near ign. coil
T1f — single, eng. compart., near ign. coil
T1g — single, under fuse/relay panel
T2c — double, eng. compart., near ign. coil
T2d — double, eng. compart., near ign. coil
T2f — double, eng. compart., near master brake cylinder
T3a — three-point, eng. compart., near oil cooler
T3b — three-point, eng. compart., near throttle valve
T3c — three-point, in water tray
T3d — three-point, near brake master cylinder
T8/ — eight-point, on transmission

Ground connections

(1) — battery ground cable
(8) — on cylinder head
(10) — near fuse/relay panel
(15) — in front wiring harness
(16) — in instrument panel wiring harness
(25) — in CIS-E wiring harness

Electronic engine controls – 16-valve Scirocco
page 2 of 4

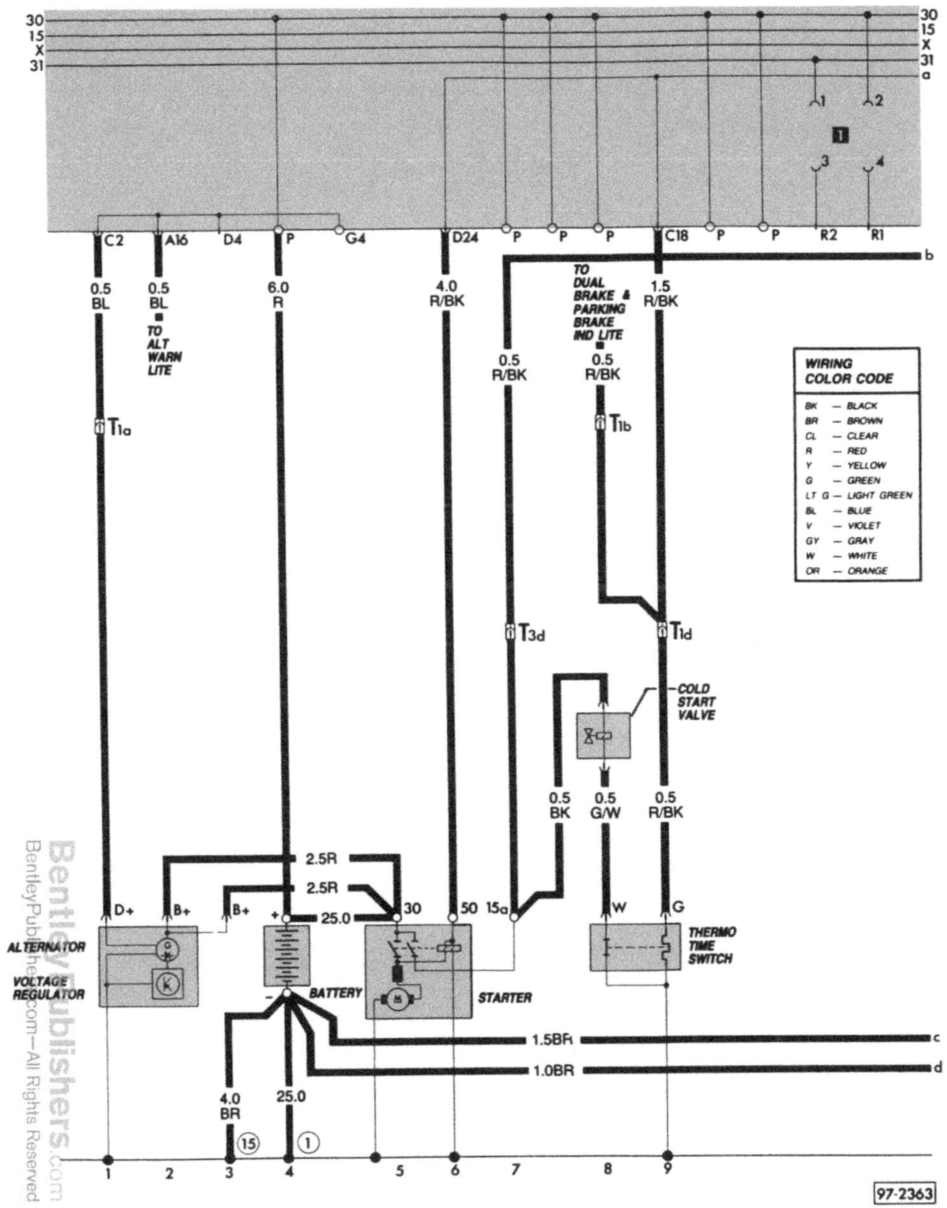

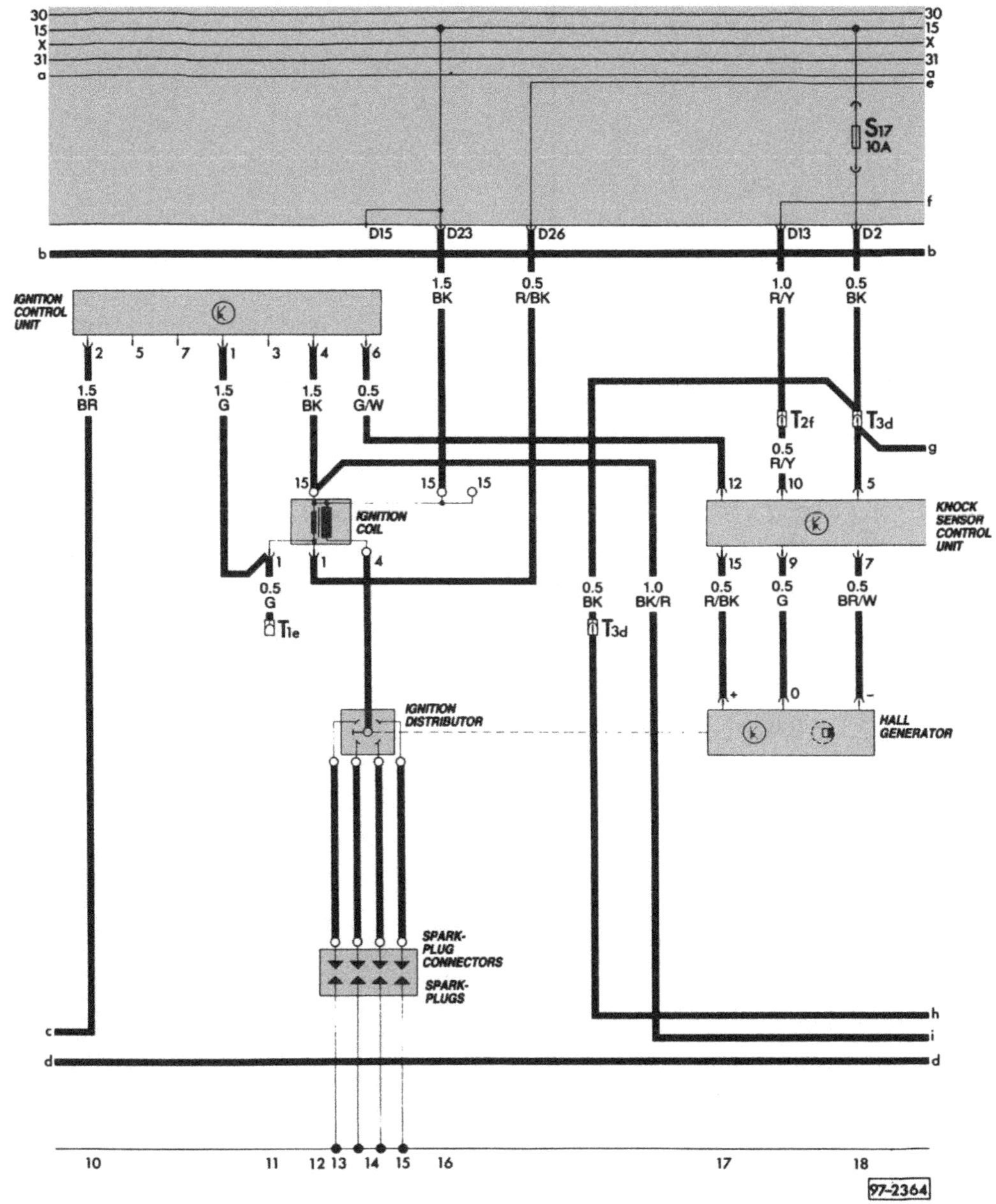

Electronic engine controls – 16-valve Scirocco
page 3 of 4

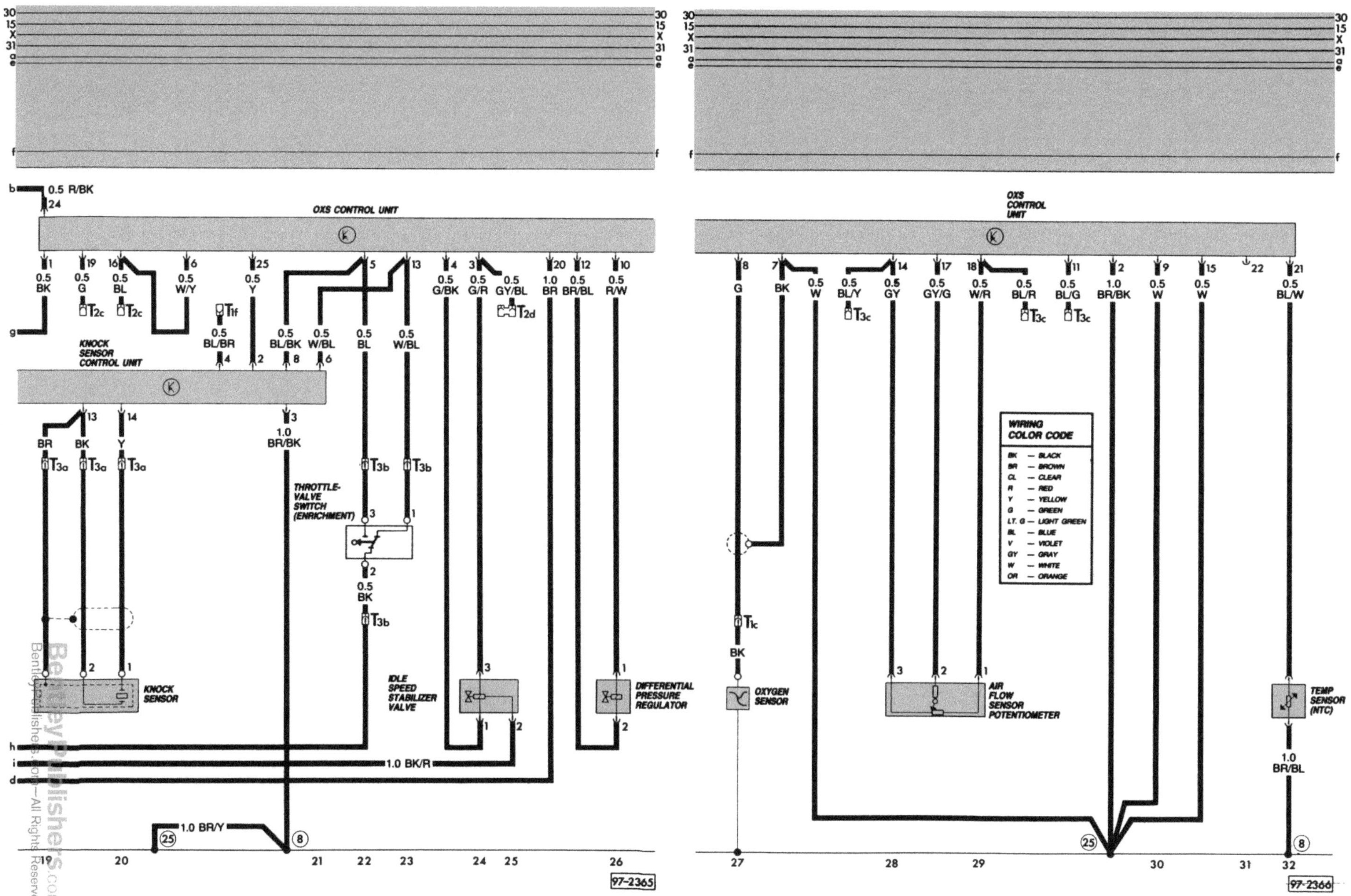

Electronic engine controls – 16-valve Scirocco
page 4 of 4

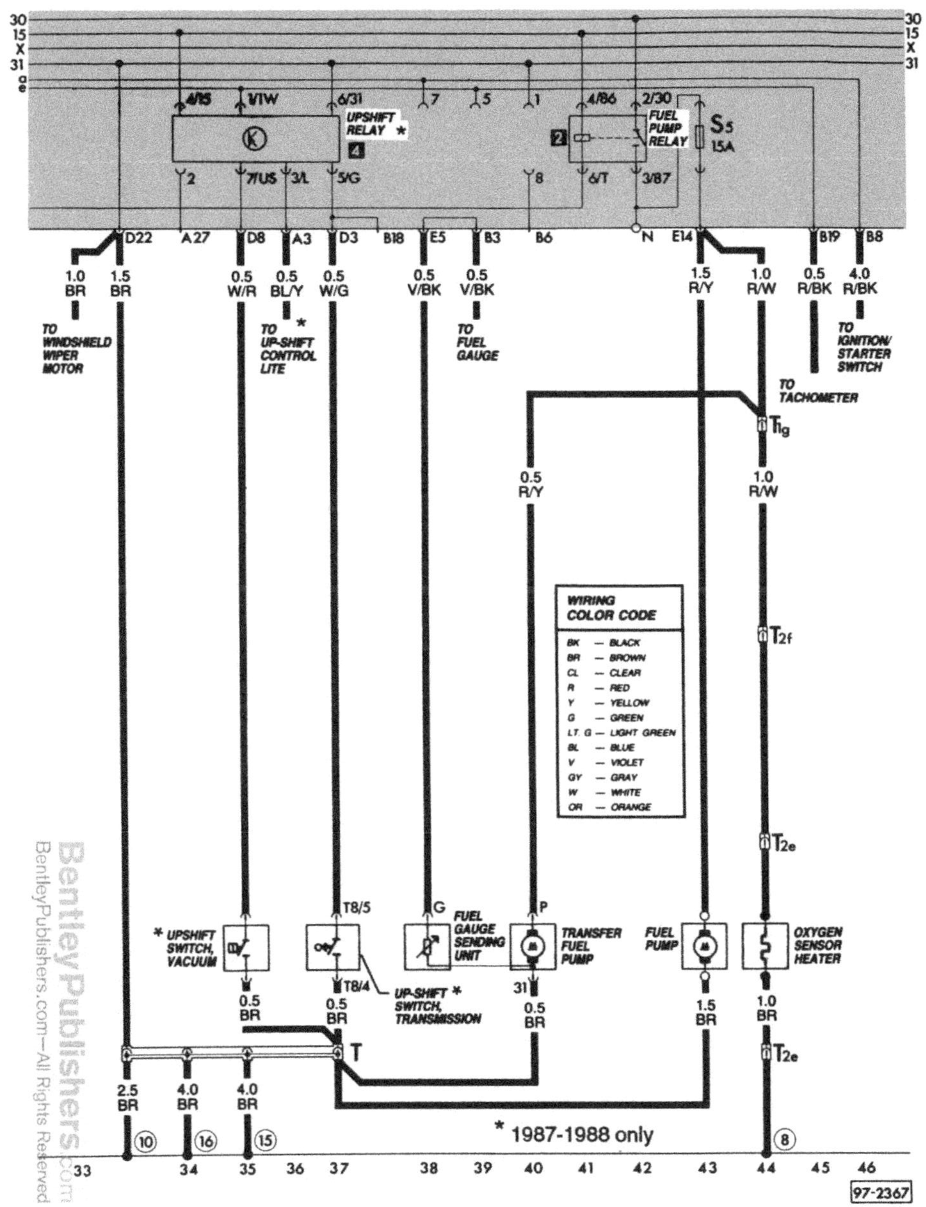

Electronic engine controls – Scirocco (except 16-valve) from July 1987

Description	Current track
2-way idle speed boost valve	17
Alternator	1-3
Auxiliary air regulator	35
Battery	4
Cold start enrichment relay	18,19
Cold start enrichment vacuum switch	25
Cold start valve	19
Control pressure regulator	34
Fuel enrichment switch	27
Fuel gauge sending unit	40
Fuel injection power supply relay	30-32
Fuel pump	36
Fuel pump relay	37,38
Hall generator	9-11
Hot pulse relay	22,23
Idle boost control unit	16,17
Ignition coil	12-14
Ignition control unit	8-13
Ignition distributor	12-14
Junction box	35-43
OXS control unit	27-30
OXS frequency valve	31
OXS thermoswitch	26
Oxygen sensor	28,29
Resistor	31
Spark plug connectors	12-14
Spark plugs	12-14
Starter	6,7
Thermo time switch	20,21
Transfer fuel pump	39
Voltage regulator	1-3

Wire connectors

T1 — single, behind fuse/relay panel
T1a — single, near ignition coil
T1b — single, behind fuse/relay panel
T1d — single, near ignition coil
T2h — double, behind fuse/relay panel
T2m — double, in engine compartment, right
T2w — double, near valve cover
T2x — double, behind fuse/relay panel
T2z — double, behind fuse/relay panel
T14 — fourteen-point, on instrument cluster

Ground connections

(1) — ground strap, battery to body
(10) — in plenum
(15) — on cylinder head
(16) — on valve cover
(24) — on cold start valve
(30) — beside fuse/relay panel
(90) — in fuel injection wiring harness

Electronic engine controls – Scirocco (except 16-valve) from July 1987 page 2 of 3

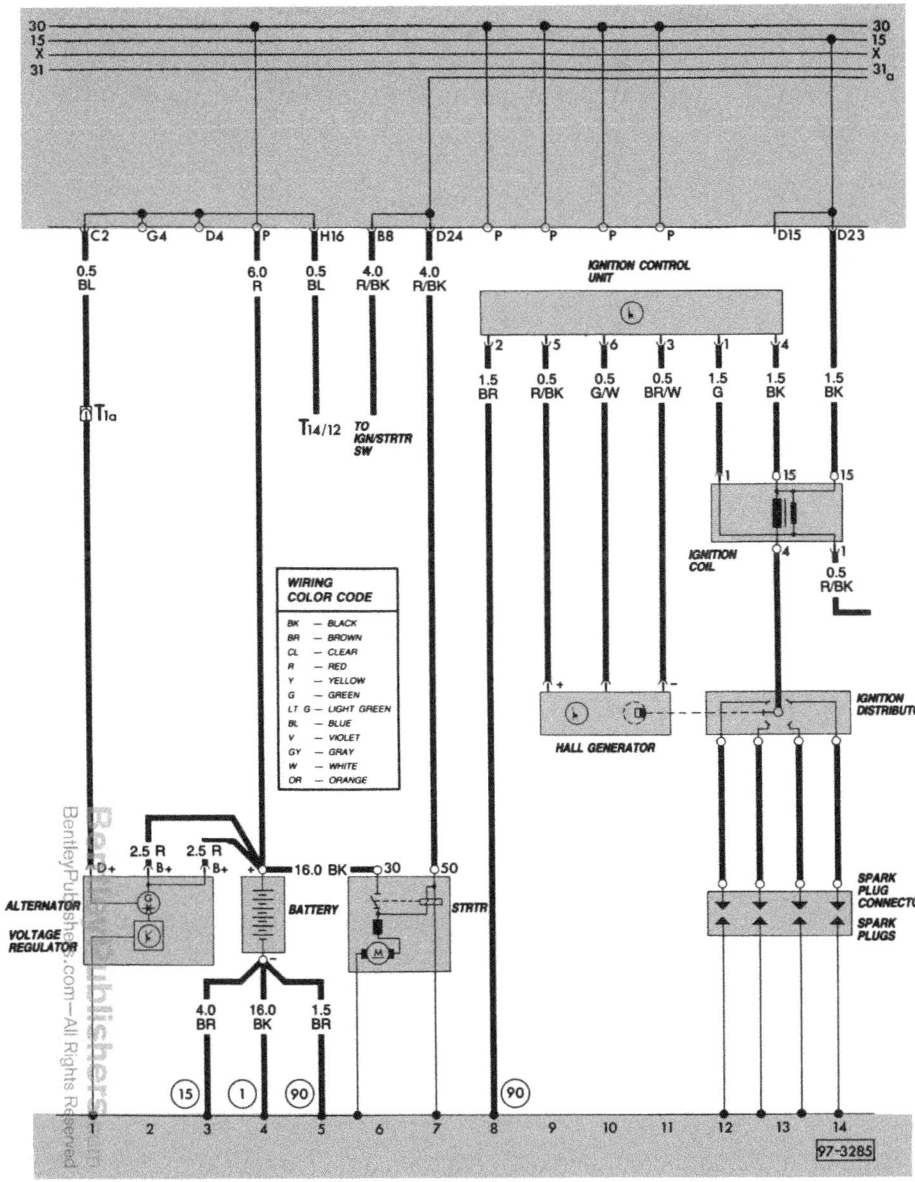

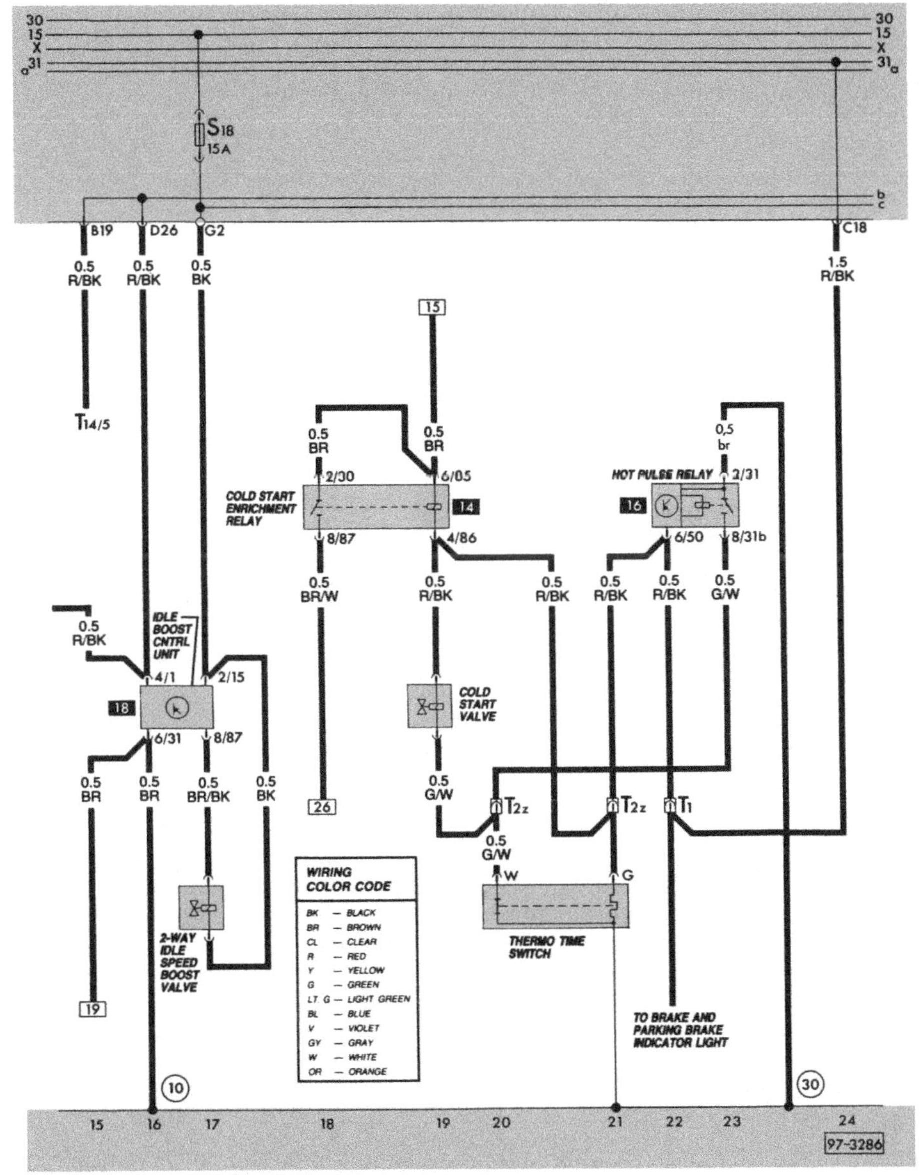

Electronic engine controls – Scirocco (except 16-valve) from July 1987
page 3 of 3

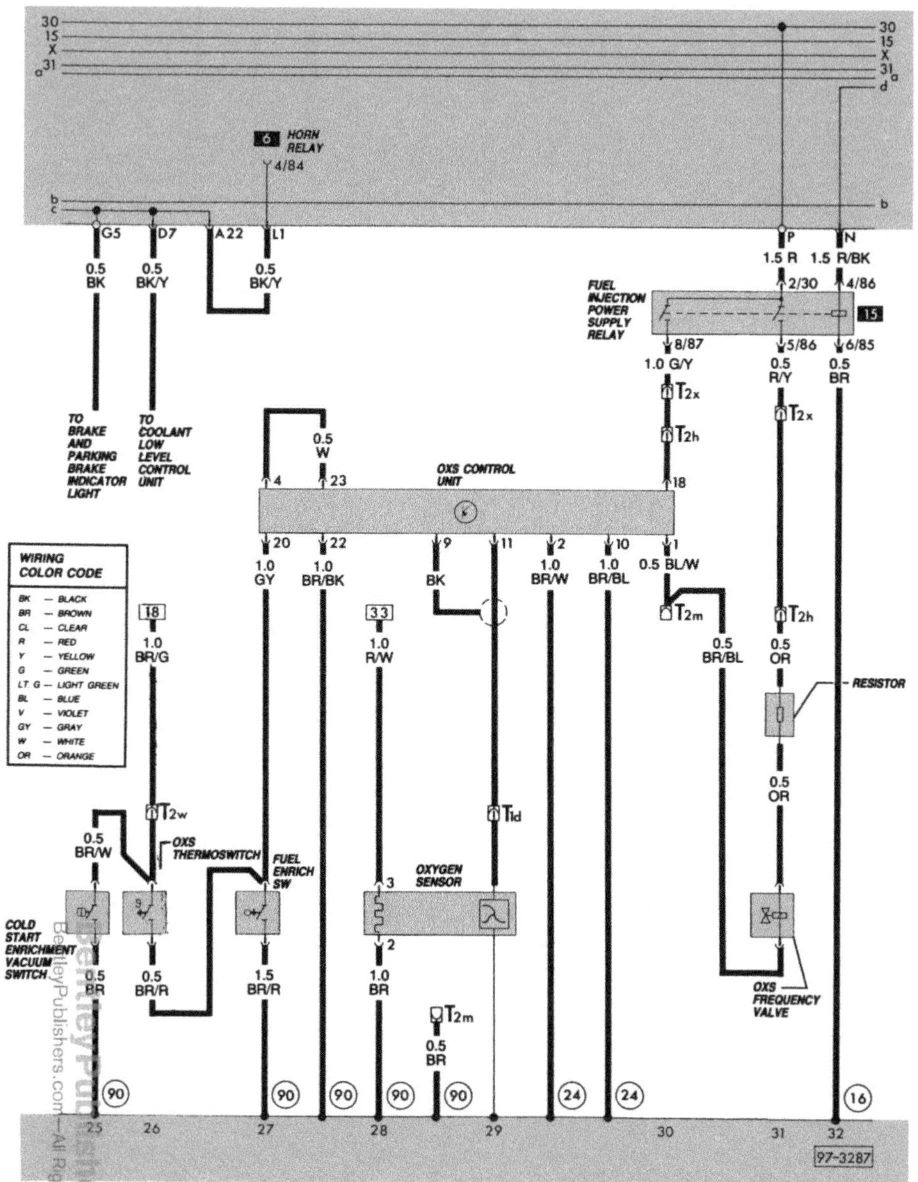

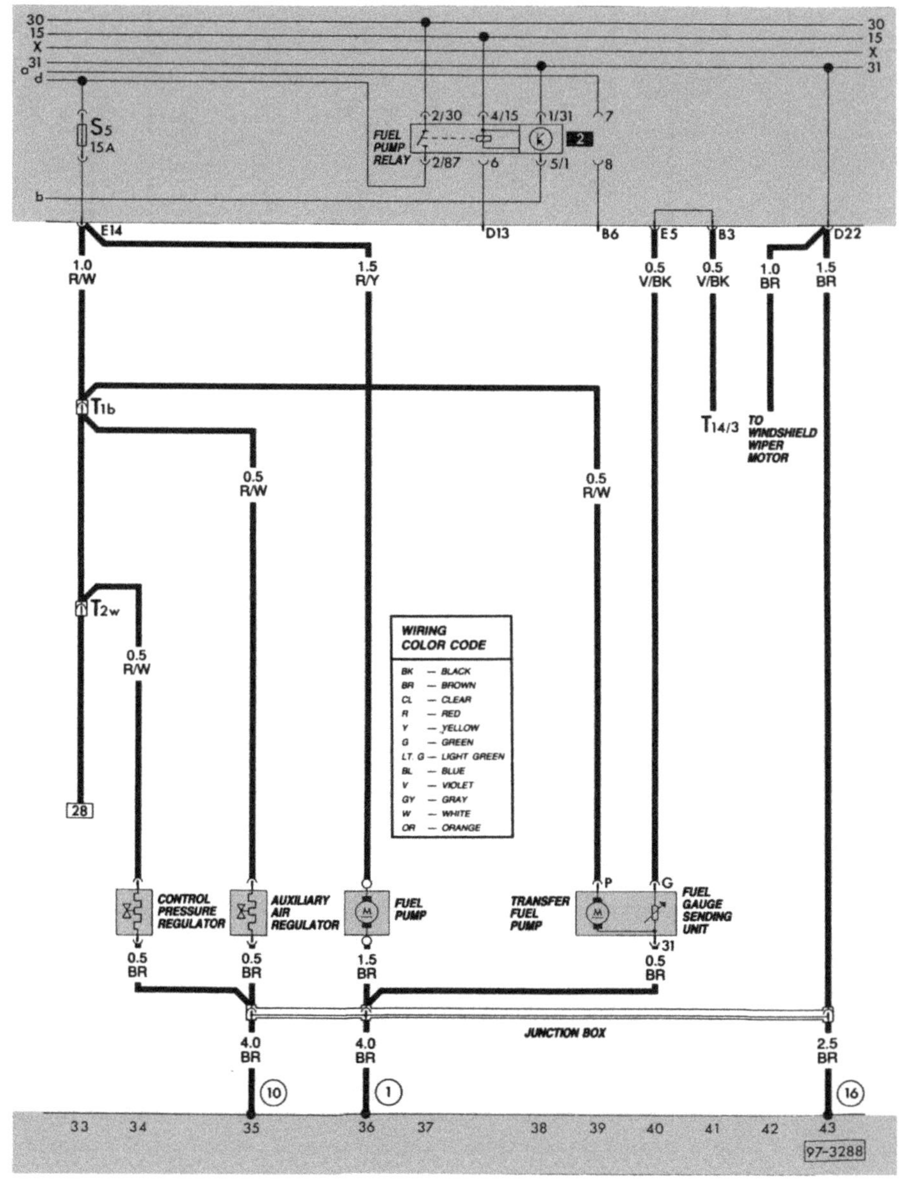

Electronic engine controls – Cabriolet, July 1987-1989

Description	Current track
2-way idle speed boost valve	18
Alternator	1-3
Auxiliary air regulator	46
Battery	5
Cold start enrichment relay	31-35
Cold start enrichment vacuum switch	33
Cold start valve	33
Control pressure regulator	44
Fuel enrichment switch	29
Fuel gauge sending unit	48
Fuel injection power supply relay	23-27
Fuel pump	49
Fuel pump relay	47
Hall generator	9-11
Hot pulse relay	37,38
Idle boost control unit	17,18
Ignition coil	12-14
Ignition control unit	8-13
Ignition distributor	12-14
OXS control unit	24-28
OXS frequency valve	26
OXS thermoswitch	31
Oxygen sensor	26,27
Spark plug connectors	12-14
Spark plugs	12-14
Starter	6,7
Thermo time switch	34-36
Transfer fuel pump	47
Voltage regulator	1-3

Wire connectors

T1a — single, in engine compartment, front left
T1b — single, behind fuse/relay panel
T1d — single, behind fuse/relay panel
T2a — double, near intake manifold
T2b — double, in engine compartment, right
T2c — double, in right plenum, near intake manifold
T2d — double, behind fuse/relay panel
T14 — fourteen-point, on instrument cluster

Ground connections

(1) — ground strap, battery to body
(3) — ground strap, engine to body
(10) — in plenum
(15) — on cylinder head
(16) — on valve cover
(30) — beside fuse/relay panel
(90) — in fuel injection wiring harness

Electronic engine controls – Cabriolet, July 1987-1989
page 2 of 3

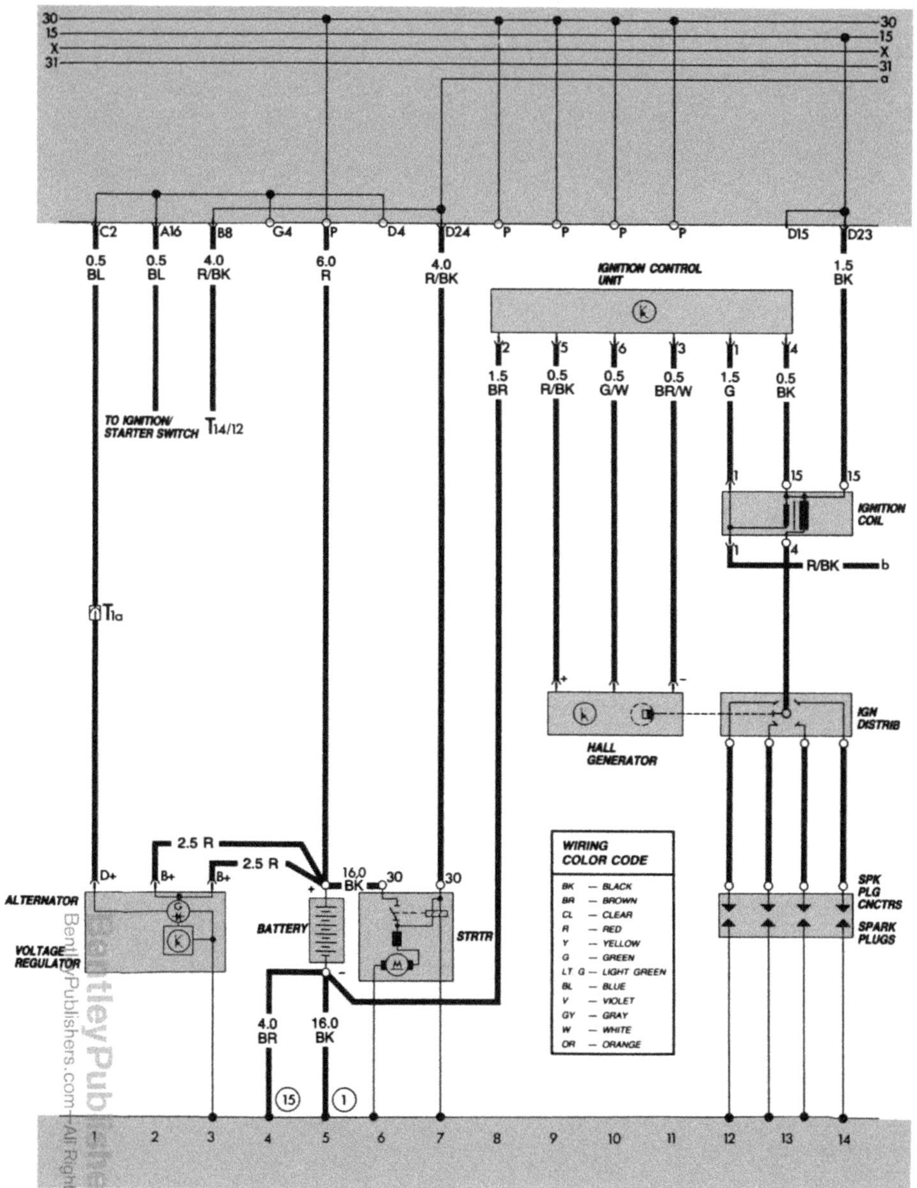

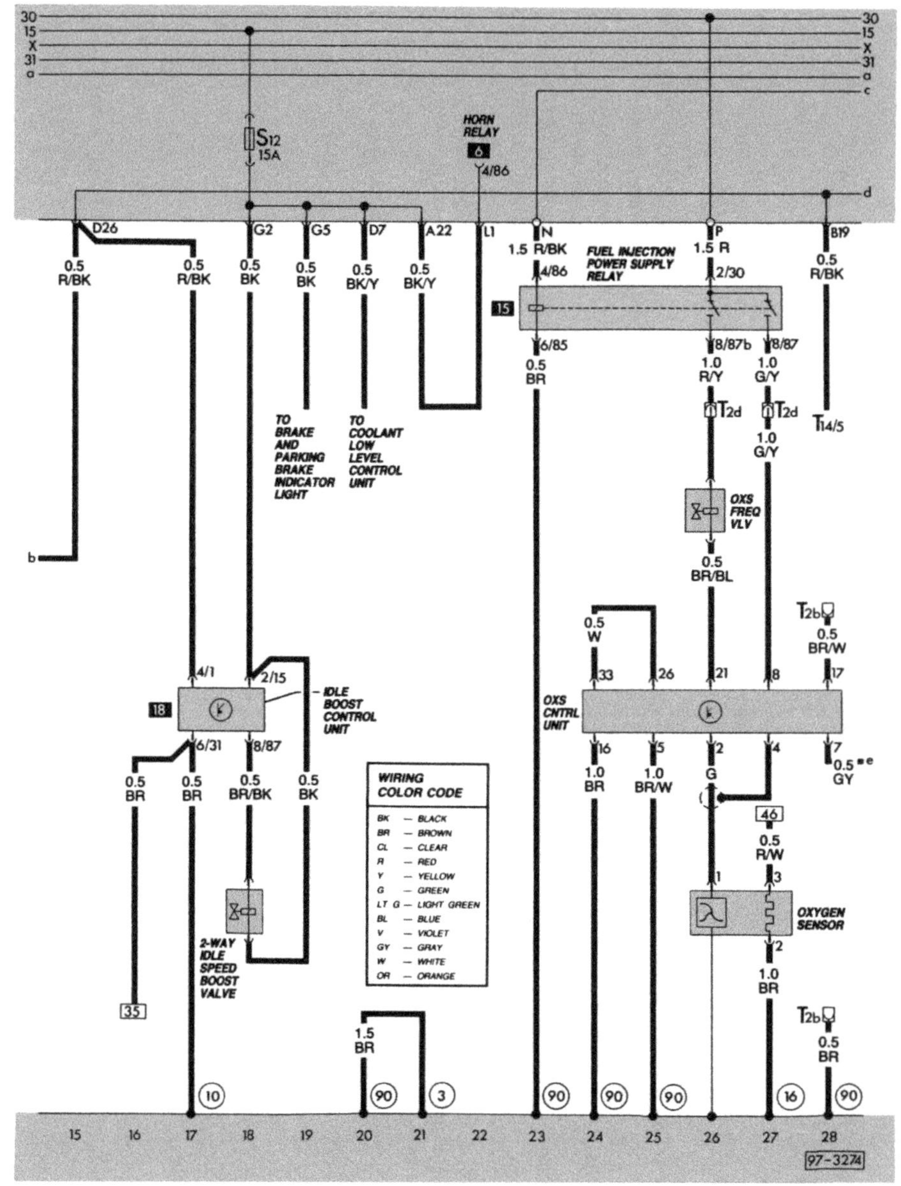

Electronic engine controls – Cabriolet, July 1987-1989
page 3 of 3

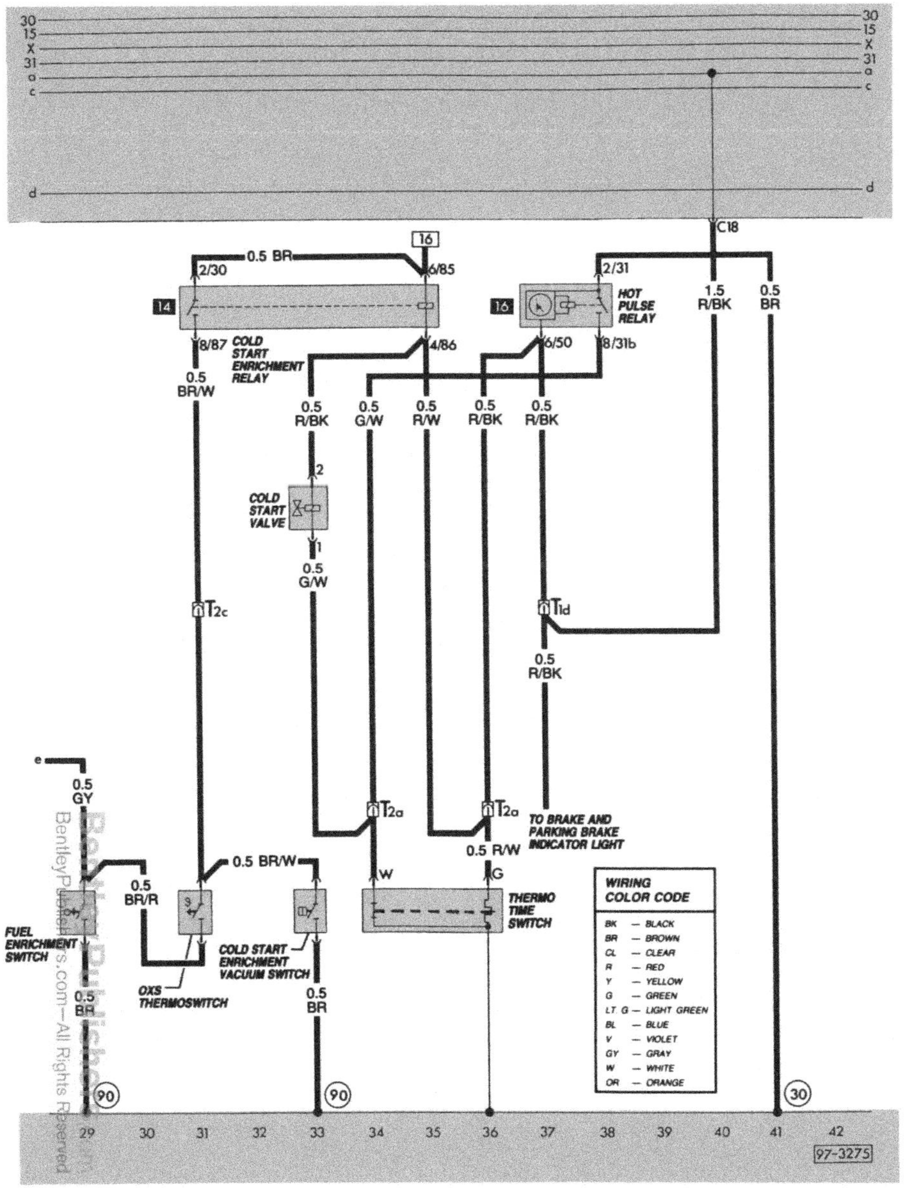

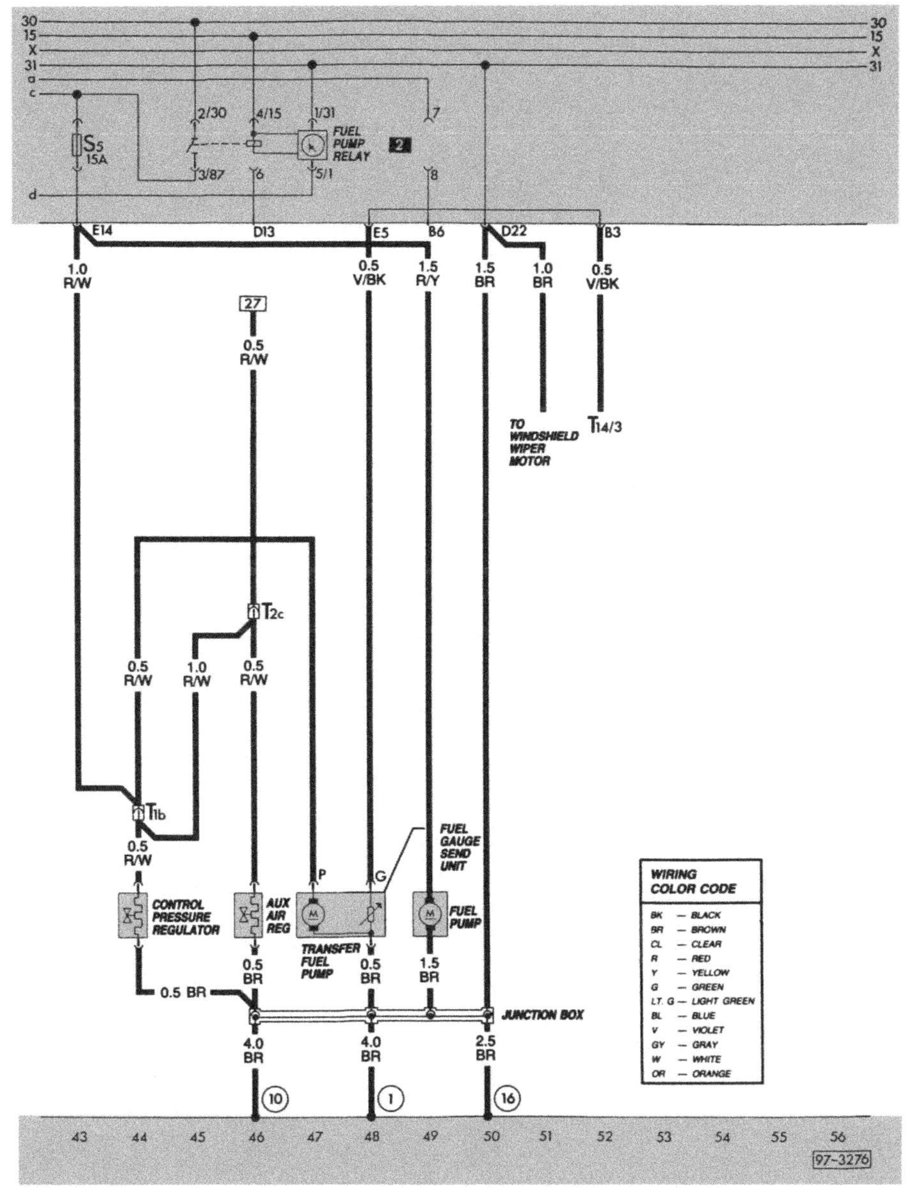

Electronic engine controls, Digifant I – Cabriolet from 1991 (California only)

Description	Current track
Air flow sensor potentiometer, G19	12
Brake and parking brake indicator light, K7	19
Coolant temperature sensor, G62	8
Digifant control unit, J169	5-26
Digifant control unit relay, J176	7-9
Fuel gauge sending unit, G	38
Fuel pump, G6	36
Fuel pump relay, J17	23-26
Fuse, S17	5
Fuse, S18	20
Fuse, S15	29
Fuse, S1	33
Fuse, S5	36
Hall generator, G40	5-7
Idle stabilizer valve, N71	21-22
Ignition coil, N152	1
Ignition coil power stage, N157	3-4
Ignition distributor, O	1-3
Injector, cyl. 1, N30	26
Injector, cyl. 2, N31	25
Injector, cyl. 3, N32	24
Injector, cyl. 4, N33	23
Intake air temperature sensor, G42	11
Knock sensor, G61	11-13
OBD fault warning light, K83	21
Oxygen sensor, G39	19-20
Radiator cooling fan, V7	34-35
Radiator cooling fan after run control unit, J138	29-32
Radiator cooling fan thermoswitch, F18	33-35
Radiator cooling fan thermoswitch, F87	29
Seat belt warning light, K19	20
Spark plug connectors, P	1-3
Spark plugs, Q	1-3
Throttle valve potentiometer, G69	15-16
Transfer fuel pump, G23	39

Wire connectors

T — junction box, behind fuse/relay panel
T1 — single, behind fuse/relay panel
T1a — single, behind fuse/relay panel
T1b — single, behind fuse/relay panel
T1d — single, behind fuse/relay panel
T1f — single, behind fuse/relay panel
T1k — single, behind fuse/relay panel
T1m — single, behind fuse/relay panel
T2 — double, behind fuse/relay panel
T2l — double, near headlight, left
T2x — double, black, diagnostic connector, below transmission shift lever boot
T2z — double, white, diagnostic connector, below transmission shift lever boot
T3 — three-point, on ignition coil power stage
T3b — three-point, in engine compartment, rear
T3g — three-point, near starter
T4 — four-point, behind fuse/relay panel
T5a — five-point, near intake manifold
T5c — five-point, on firewall, left
T6 — six-point, behind fuse/relay panel

Ground connections

(12) — in engine compartment, left
(18) — on engine block
(30) — beside fuse/relay panel
(82) — in left front wiring harness
(94) — in Digifant wiring harness

Welded wiring harness points

(C3) — plus connection (30), in left front wiring harness
(G1) — plus connection, in Digifant wiring harness
(G3) — plus connection, in injector wiring harness

Electronic engine controls, Digifant I – Cabriolet from 1991 (California only) page 2 of 3

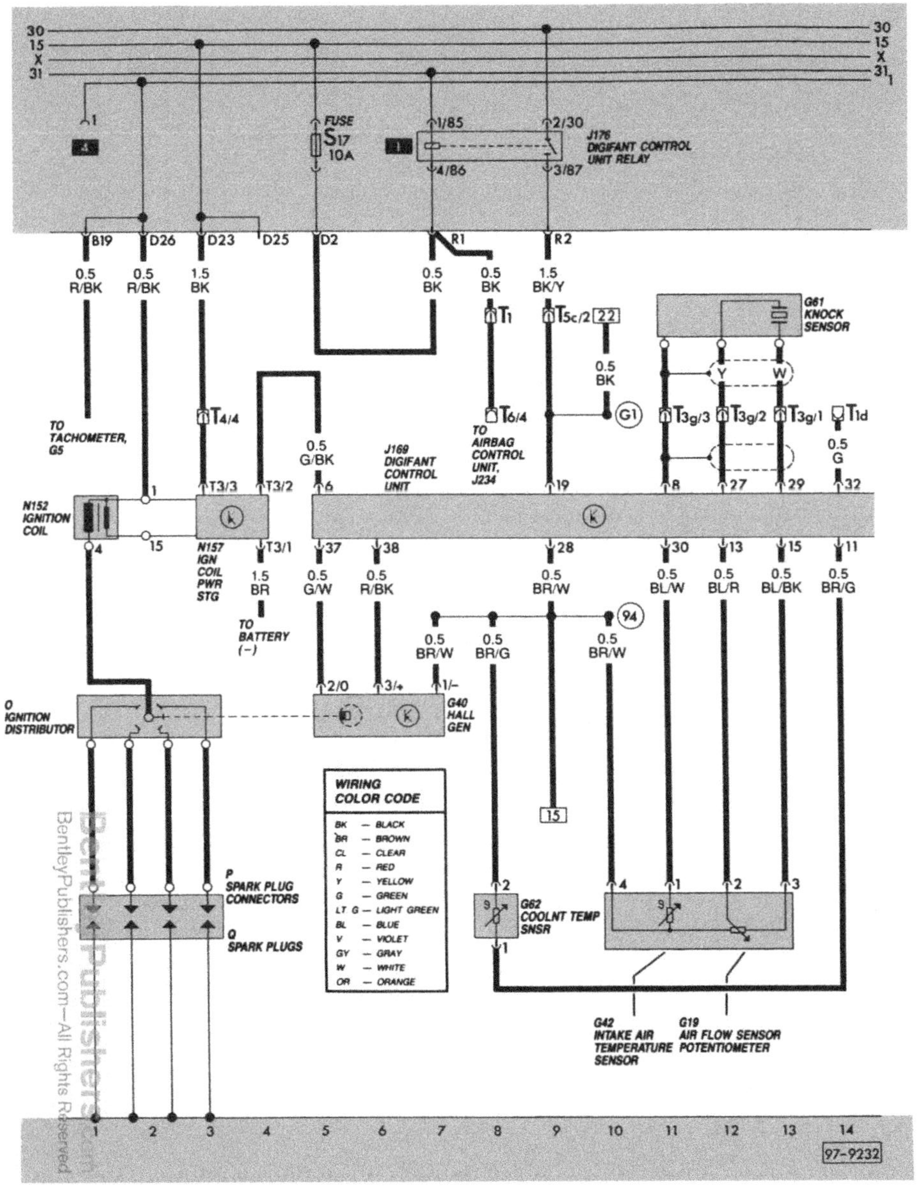

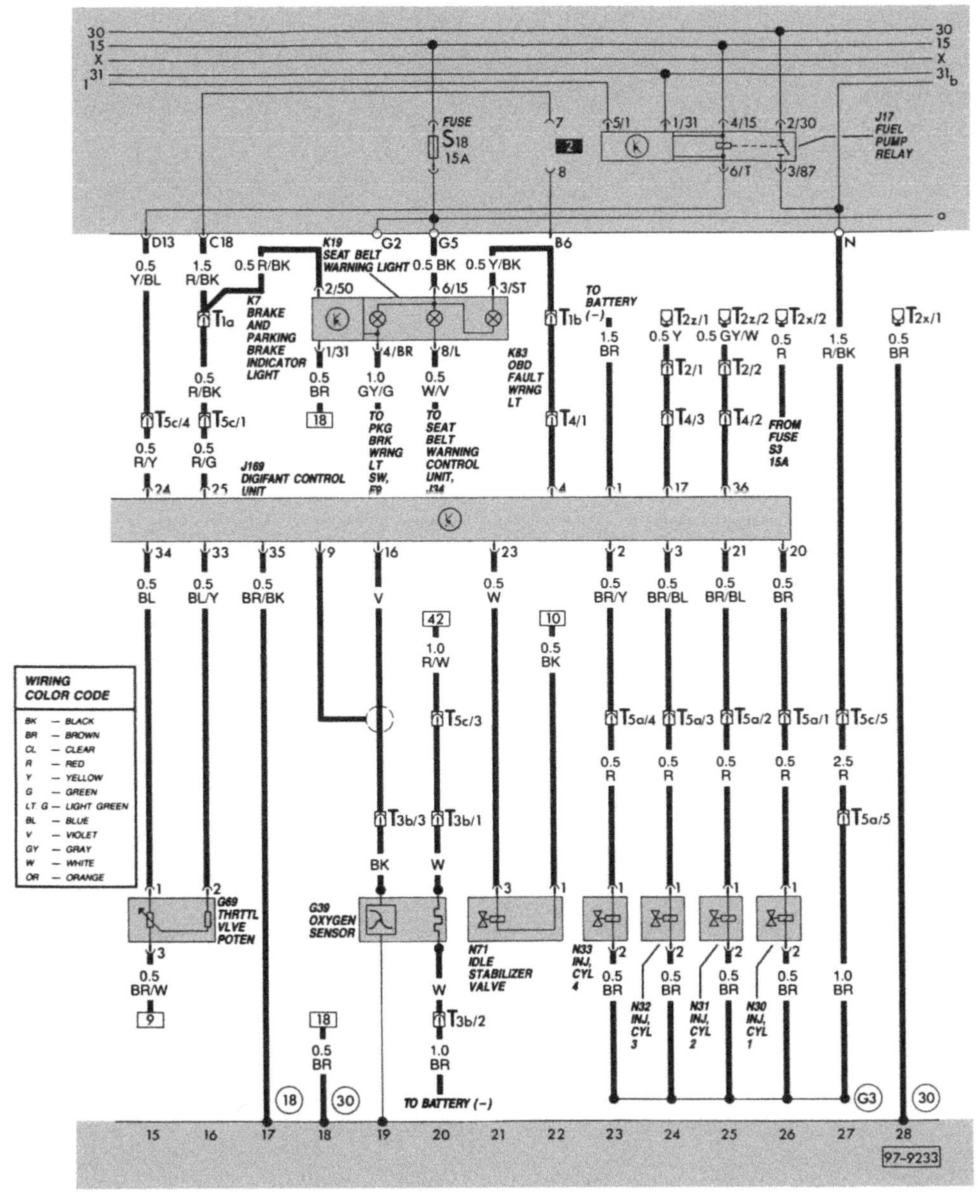

Electronic engine controls, Digifant I – Cabriolet from 1991 (California only) page 3 of 3

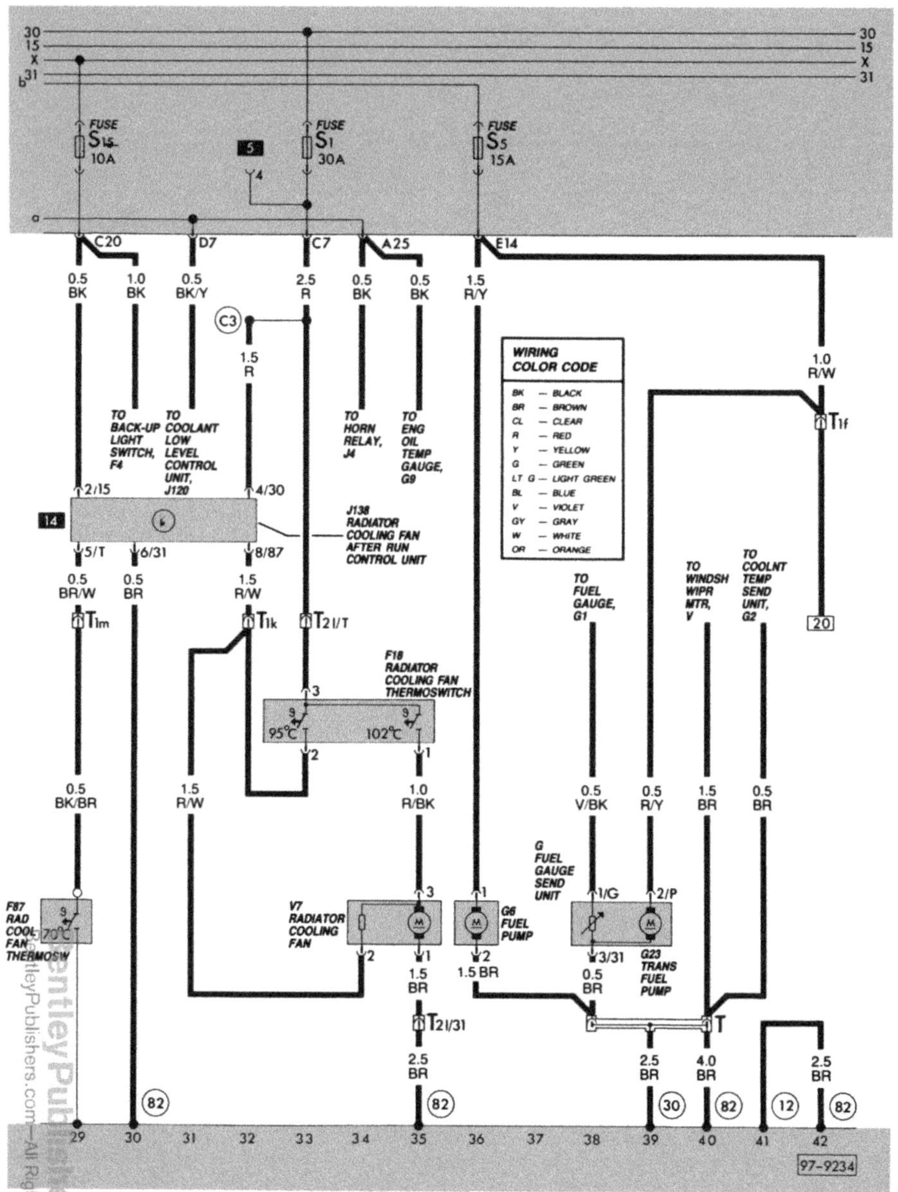

Description	Current track
Driver seat control unit	13-16
Driver seat switch (in console)	10
Driver seat temp. sensor	13
Heat element, driver seat	16
Heat element, driver backrest	16
Heat element, passenger backrest	1
Heat element, passenger seat	1
Passenger seat control unit	1-4
Passenger seat switch (in console)	7
Passenger seat temp. sensor	4
Seat heater indicator lite	17

Wire connectors

T1 — single, behind dashboard
T6a — 6 point, below driver seat
T6b — 6 point, below passenger seat

Ground connectors

(10) — on fuse/relay panel

Heated seats – Scirocco, Cabriolet to 1990
page 2 of 2

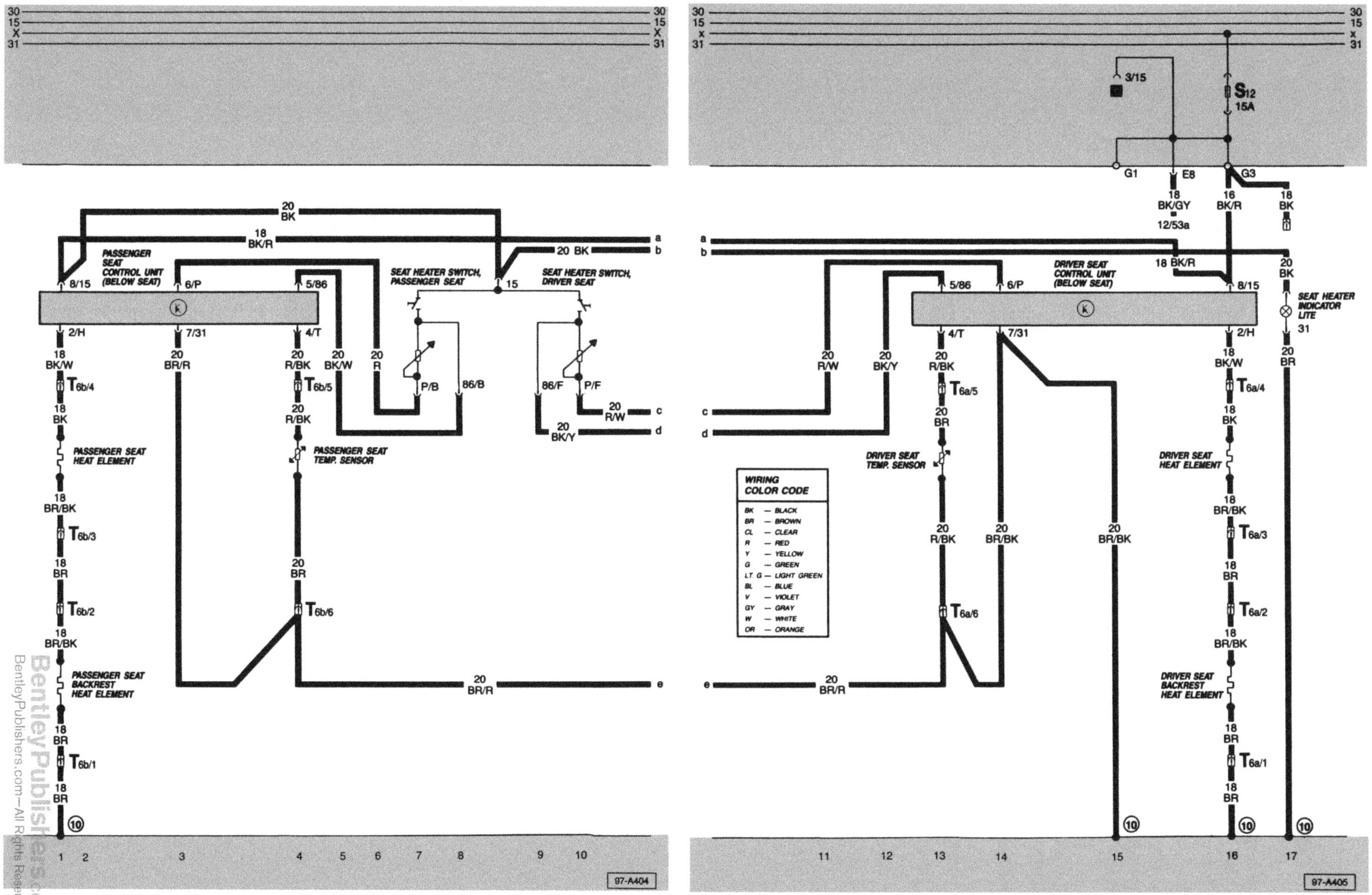

Heated seats – Cabriolet from 1991
page 1 of 1

Description	Current track
Driver seat heater control unit, J131	3-6
Driver seat temperature sensor, G59	4
Fuse, S12	3
Heat element, driver backrest, Z7	1
Heat element, driver seat, Z6	2-3
Heat element, passenger backrest, Z9	14
Heat element, passenger seat, Z8	12-13
Heat regulating switch, driver seat, E94	5
Heat regulating switch, passenger seat, E95	10
Passenger seat heater control unit, J132	9-12
Passenger seat temperature sensor, G60	11
Seat heater switch indicator light, L44	7

Wire connectors

T1 — single, behind fuse/relay panel
T6a — six-point, below driver's seat
T6b — six-point, below passenger's seat

Ground connections

(30) — beside fuse/relay panel

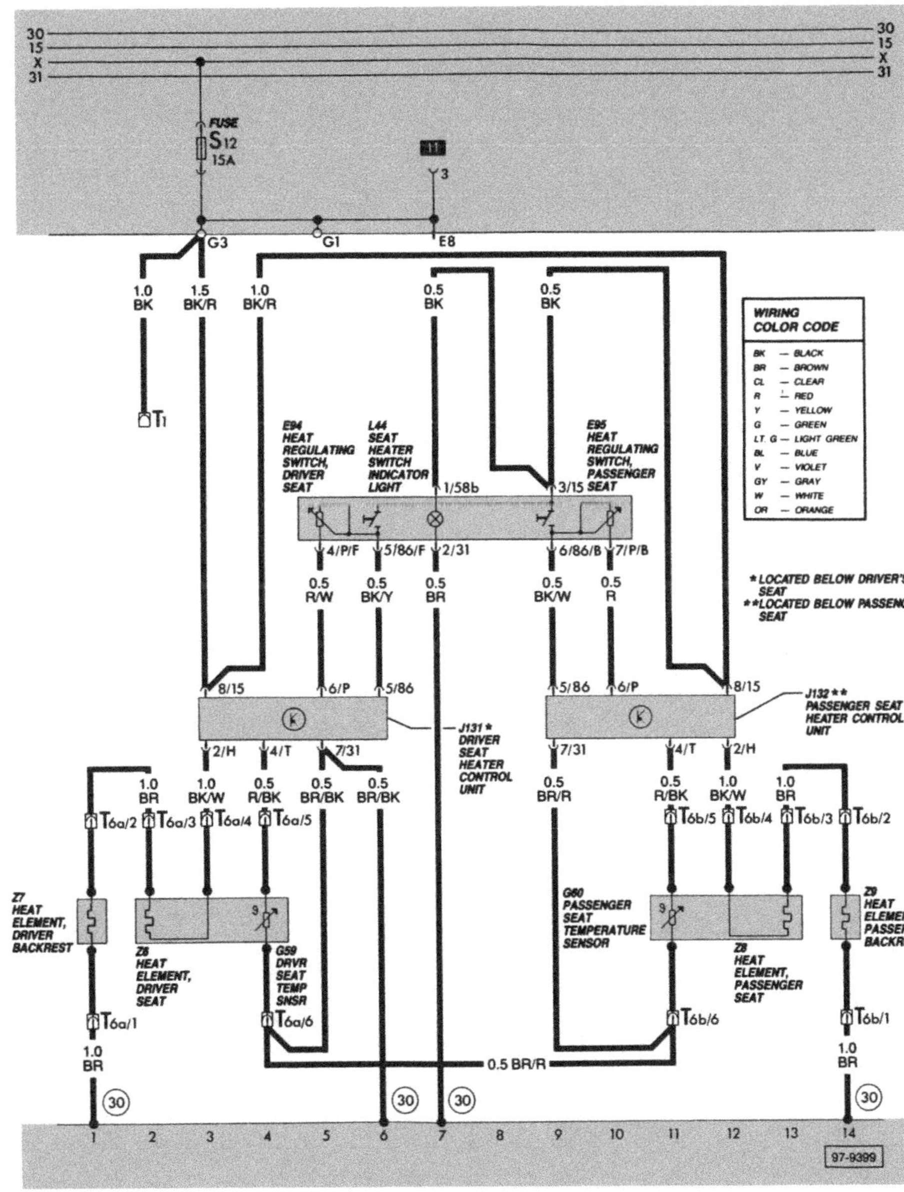

Lights – Scirocco, Cabriolet 1989
page 1 of 2

Description	Current track
Headlight dimmer/flasher switch	21-22
Headlight, left	27-28
Headlight, right	31-32
High-beam headlight, left	25
High-beam headlight, right	29
Instrument panel light dimmer switch	9
License plate light	11-12
Light switch	2-9
Load reduction relay	2-4
Luggage compartment light	13
Luggage compartment light switch	13
Parking light, left	19
Parking light, right	18
Side marker lights, front	17, 20
Tail light, left	15
Tail light, right	14

Wire connectors

T1e — single, in luggage compartment, left
T2 — double, in luggage compartment, left rear

Ground connections

(50) — in luggage compartment, left

(58) — near rear window wiper motor

(82) — in left front wiring harnesss

(108) — in left front wiring harness

Welded wiring harness points

(A11) — plus connection (58 b) in instrument panel wiring harness

Lights – Scirocco, Cabriolet 1989
page 2 of 2

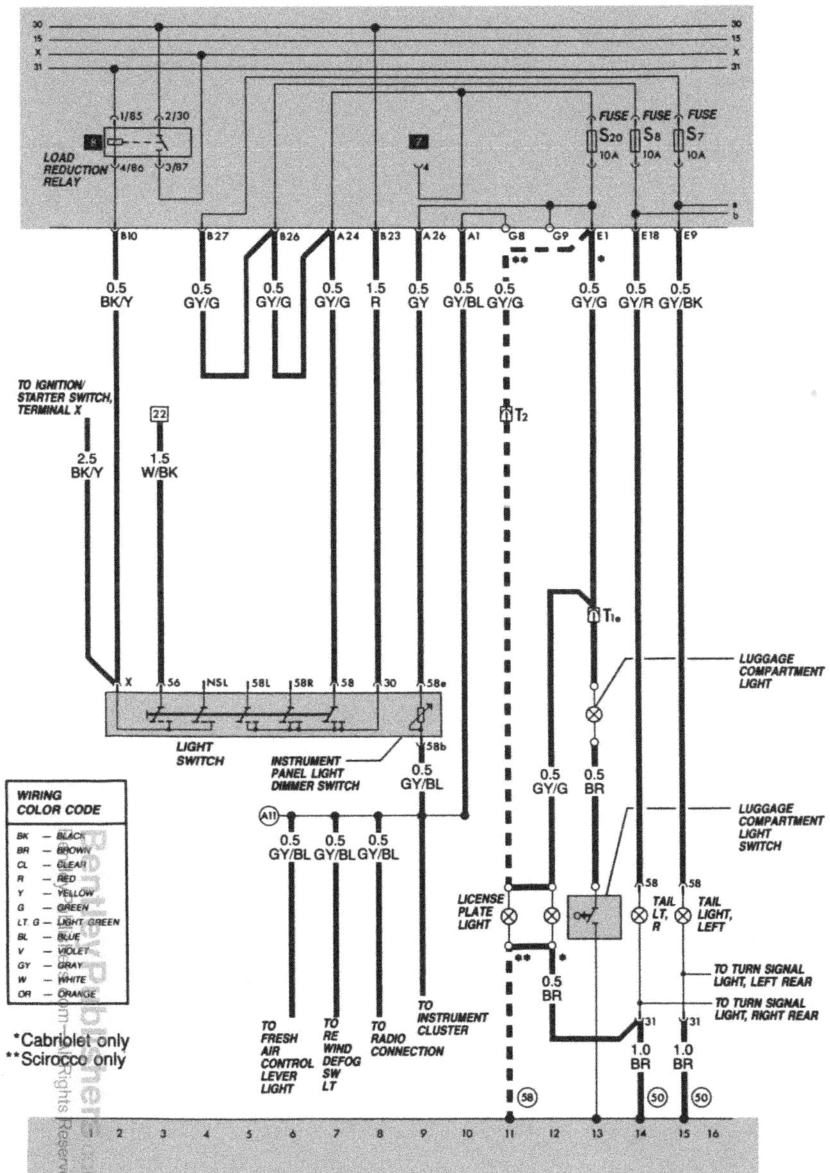

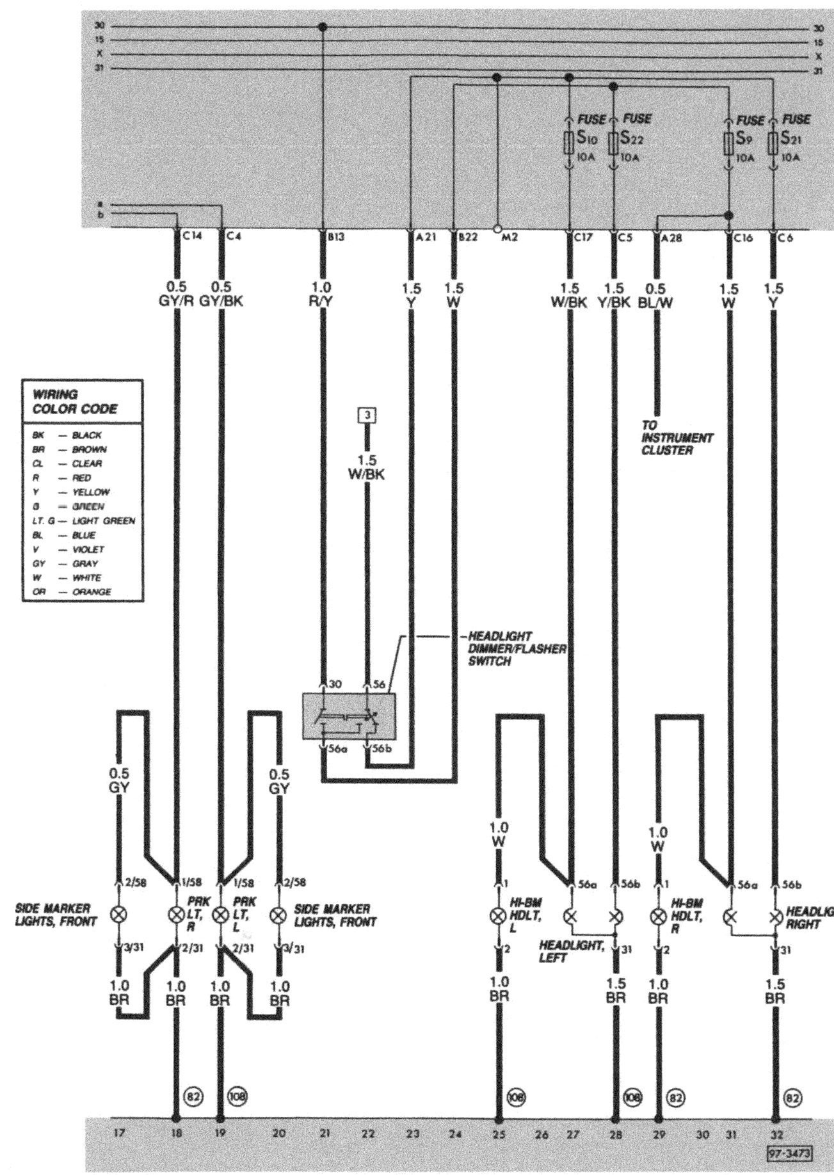

Multifunction indicator – Scirocco 1985-1989

Description	Current Track
Alternator	13
Alternator warn. light	14
Clock light	21
Coolant temp. gauge	22
Coolant temp. light	24
Coolant temp. sender	35
Emerg. flash. relay	15
Eng. oil press. control unit	11, 12
Eng. oil press. light	13
Eng. oil press. sender(0.3 bar)	33
Eng. oil press. sender (1.8 bar)	34
Eng. oil temp. sender	26
Fuel gauge	20
Fuel pump relay	8
Headlight, right	17
Hi-beam warn. light	18
Idle stabilizer relay	9
Ign./start switch	19
Instrument panel light dimmer	23
Interior light, front	29
Luggage compart. light	30
Memory switch	31, 32
Multi-function gauge	25, 32
Outside temp. sensor	27
OXS elapsed mileage counter	3, 4
OXS light	5
Speedometer sender	22
Tachometer	10
Turn sig. indicator	16

Wire connectors

T1 — single, under relay panel
T1a — single, under relay panel
T1f — single, under relay panel
T2a — double, under relay panel
T2c — double, in eng. compart. near trans.
T2d — double, in eng. compart. near trans.
T4a — four point, under steering column cover
T7 — seven point, on instrument cluster
T7a — seven point, on instrument cluster
T7b — seven point, on instrument cluster
T7c — seven point, on instrument cluster
T8 — eight point, on transmission

Ground Connections

(10) — near relay panel

(12) — on cylinder head cover

Multifunction indicator – Scirocco 1985-1989
page 2 of 2

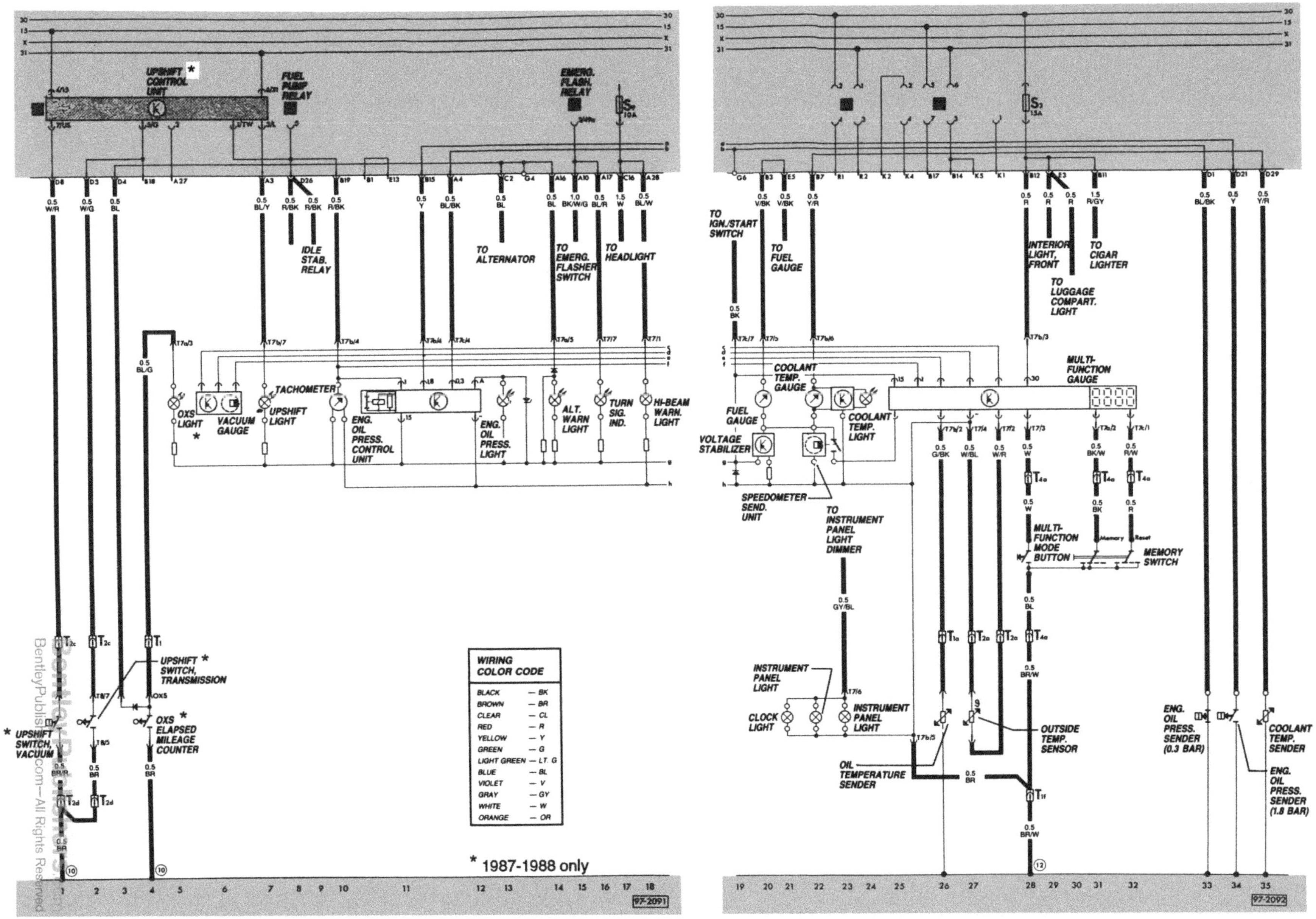

Power convertible top – Cabriolet from 1991
page 1 of 1

Description	Current track
Contact, left, (cover installed), F155	3
Contact, right, (cover installed), F156	6
Convertible top circuit breaker, S68	8
Convertible top control unit, J256	3-11
Convertible top motor, V82	9
Convertible top switch, E137	3-5
Convertible top switch light, L72	2

Wire connectors

T1 — single, behind fuse/relay panel
T1a — single, behind fuse/relay panel
T2 — double, on convertible top motor
T5 — five-point, on convertible top switch
T17 — seventeen-point, in luggage compartment, left

Ground connections

(30) — beside fuse/relay panel
(50) — in luggage compartment, left
(51) — in luggage compartment, right

Welded wiring harness points

(A11) — plus connection, (58b) in instrument panel wiring harness

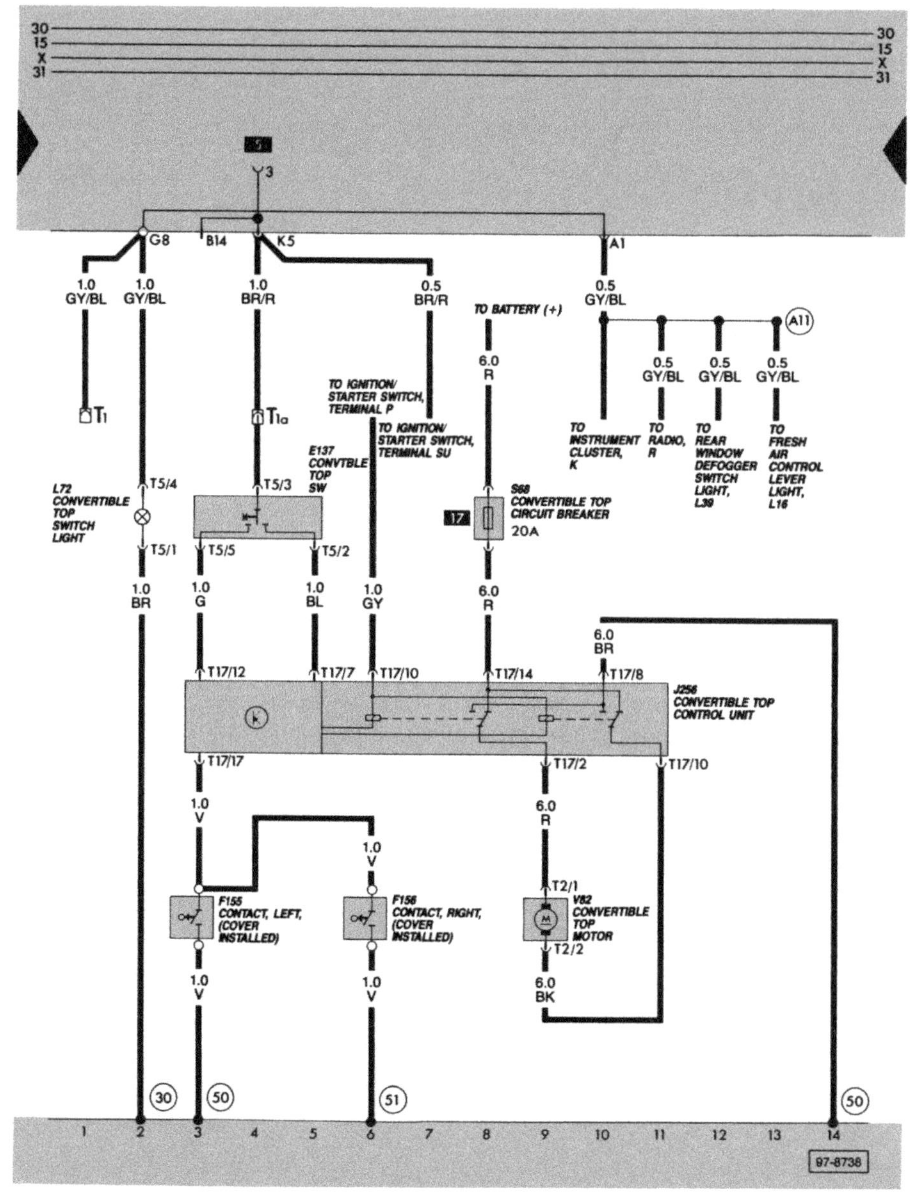

Power/heated mirrors – Scirocco 1985-1988

Description	Current Track
Fuse on fuse/relay panel (15A)	2
Left mirror heating element	7
Left mirror motor	3
Magnetic clutch for left mirror	4
Magnetic clutch for right mirror	5
Mirror adjusting switch	2-5
Mirror selector switch	2-5
Rear window defogger switch	6
Right mirror heating element	6
Right mirror motor	2

Wire Connectors

T1 — single, behind dash
T3 — 3-point, in right door
T3a — 3-point, in right door
T3b — 3-point, in left door
T3c — 3-point, in left door
T6 — 6-point, below dash, left
T6a — 6-point, below dash

Ground Connectors

(10) — on bracket of fuse/relay panel

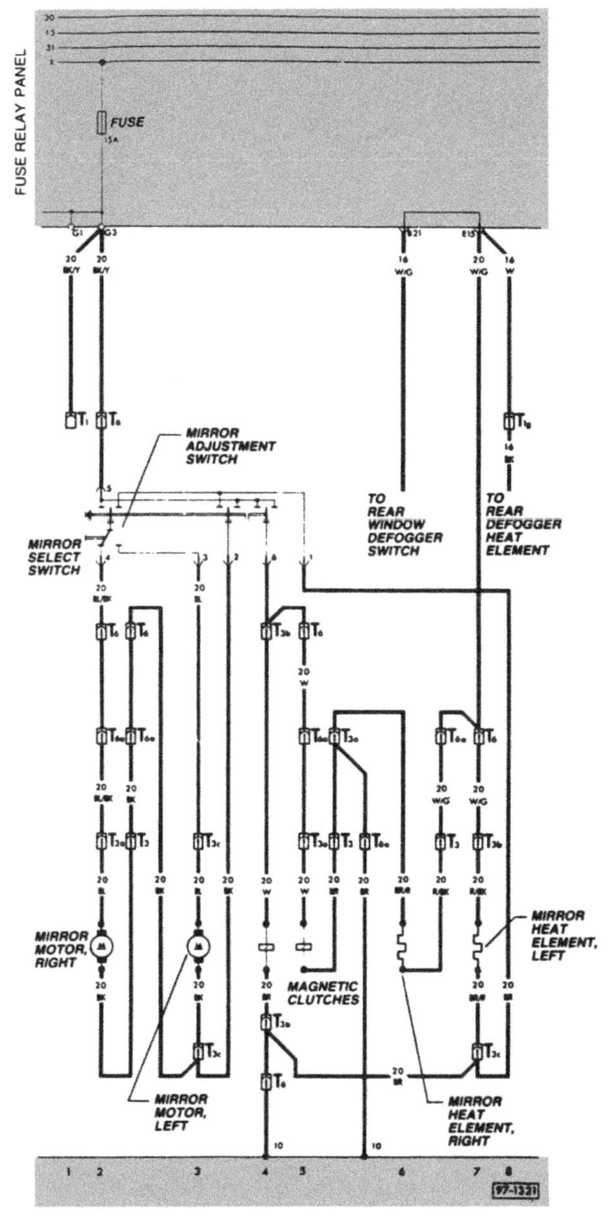

Power windows – Scirocco 1985-1989
page 1 of 1

Description	Current Track
Window motor, LF	3
Window motor, RF	10
Window relay	1, 2
Window switch, left	2-5
Window switch, right	6-8, 9-12

Wire Connectors

T1a — single, under inst. panel, right
T2a — double, behind driver's door panel
T2b — double, under inst. panel, left
T2c — double, near fuse/relay panel
T2d — double, under inst. panel, right
T2e — double, behind passenger's door panel
T5a — five-point, behind driver's door panel
T5b — five-point, behind driver's door panel
T5c — five-point, behind passenger's door panel

Ground Connections

(10) — near fuse/relay panel

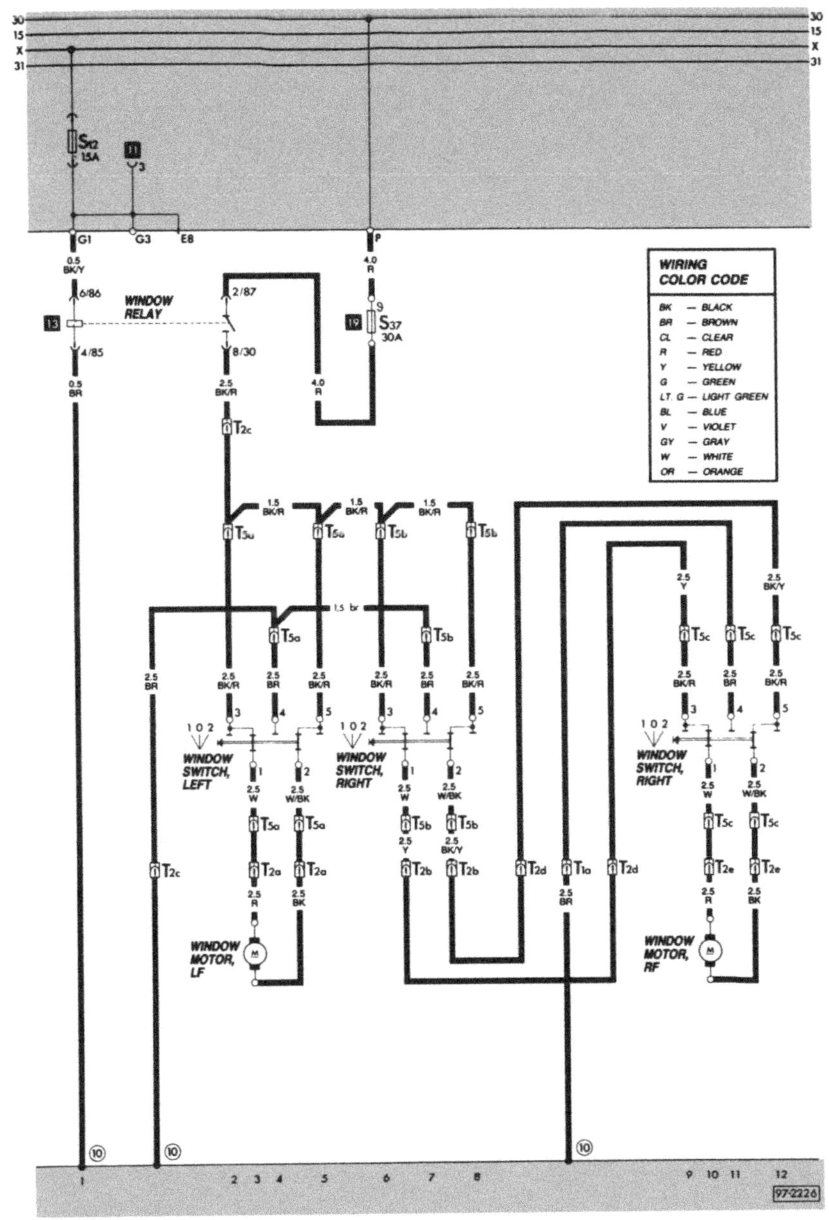

Power windows – Cabriolet from 1990
page 1 of 2

Description	Current track
Door contact switch, left front, F2	12
Power window control unit, J139	1-15
Power window fuse, S37	11
Power window switch light, L53	4, 5
Window motor, left front door, V14	13
Window motor, right front door, V15	14
Window switch, left front, E40	1-4
Window switch, right front, E41	5-8

Wire connectors

T2a — double, in driver's door
T2b — double, in passenger's door
T8a — eight-point, on power window control unit
T8b — eight-point, on power window control unit
T8c — eight-point, on power window control unit

Ground connections

(40) — below rear seat, right

(89) — in power window wiring harness

Welded wiring harness points

(Q9) — plus connection, in power window wiring harness

(Q13) — plus connection, in power window wiring harness

(Q15) — plus connection, in power window wiring harness (left up)

(Q16) — plus connection, in power window wiring harness (left down)

Power windows – Cabriolet from 1990
page 2 of 2

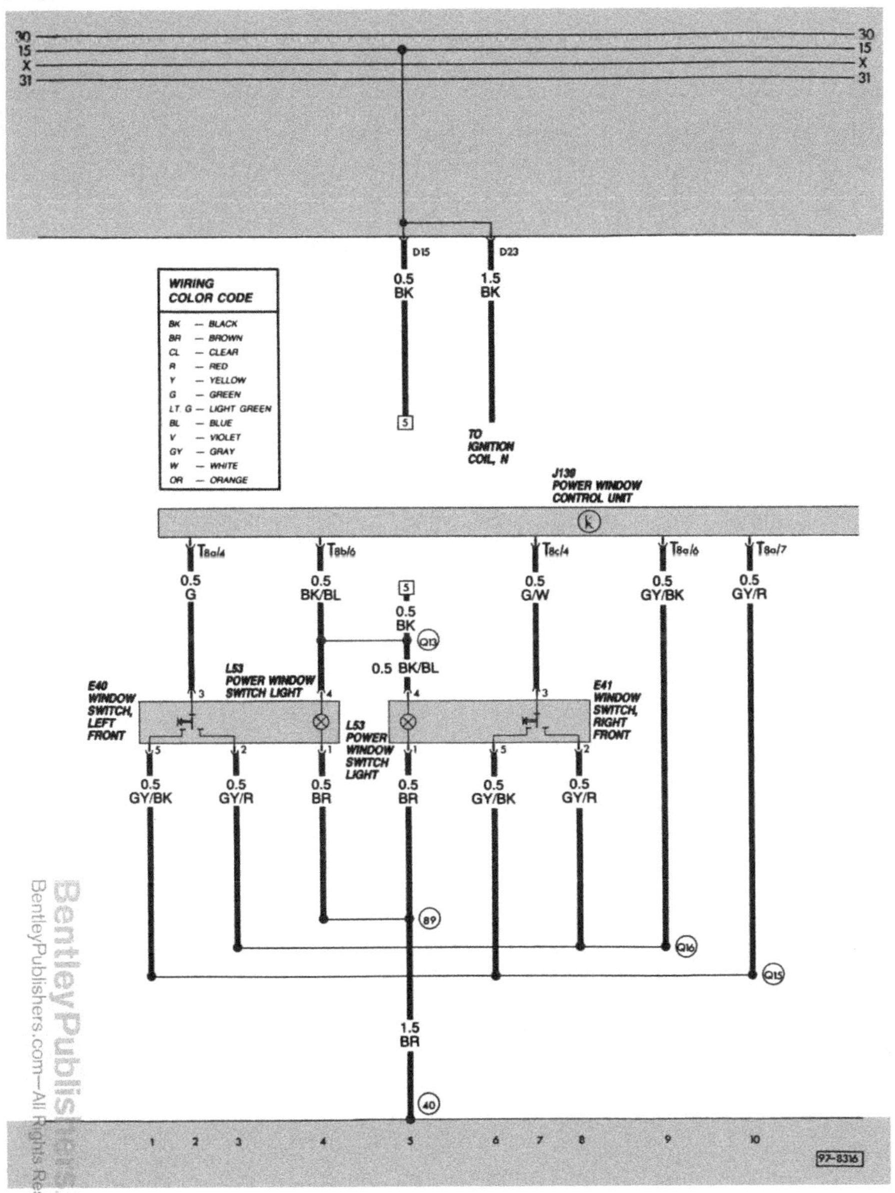

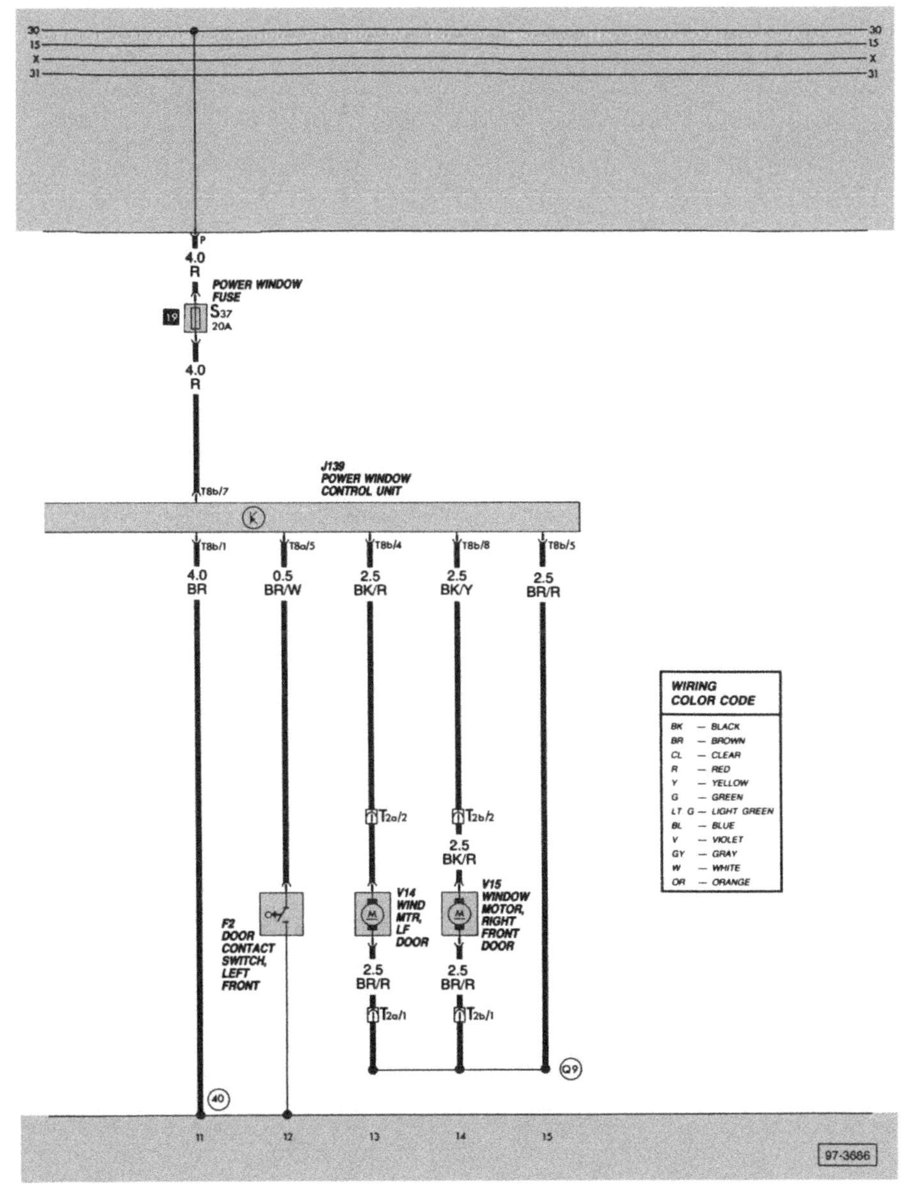

Stereo radio – Scirocco, Cabriolet 1985-1989
page 1 of 1

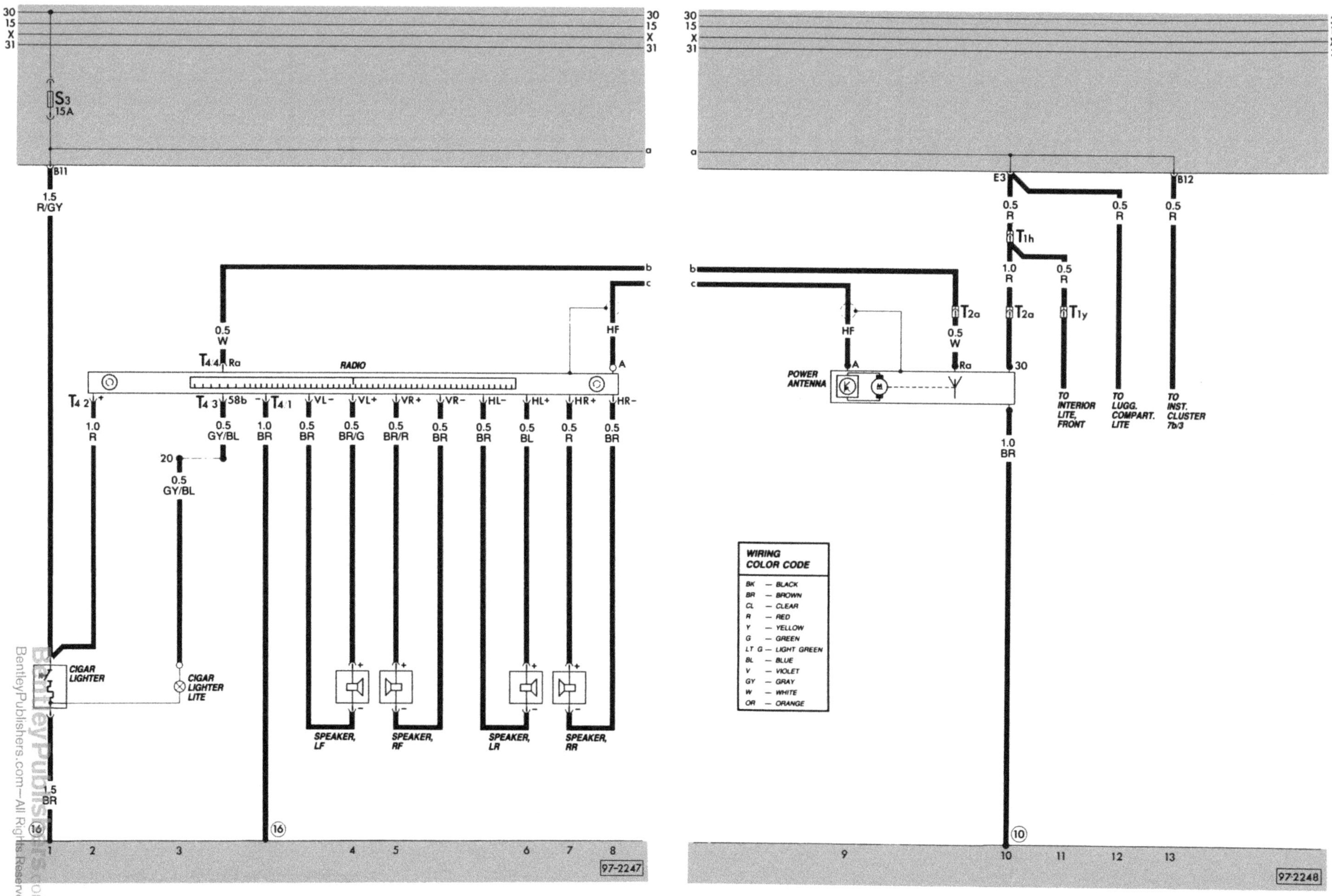

Stereo radio with roof-mounted antenna – 16 valve Scirocco
page 1 of 1

Description	Current track
Amplifier/radio	9,10
Antenna	9,10
Cigarette lighter lite	4
Radio connection	2-10
Refrigerator 12V connector	1
Speaker, left front	6
Speaker, left rear	8
Speaker, right front	7
Speaker, right rear	9

Wire connectors

T2b — double, in left front footwell
T2c — double, in right front footwell
T4/ — four-point, under instrument panel, middle
T7b/— seven-point, on instrument cluster (long white housing)
T8/ — eight-point, on radio
T8a/— eight-point, on radio

Ground connections

(81)— in instrument panel wiring harness

Plus (+) connections

(A4)— (58b) in instrument panel wiring harness

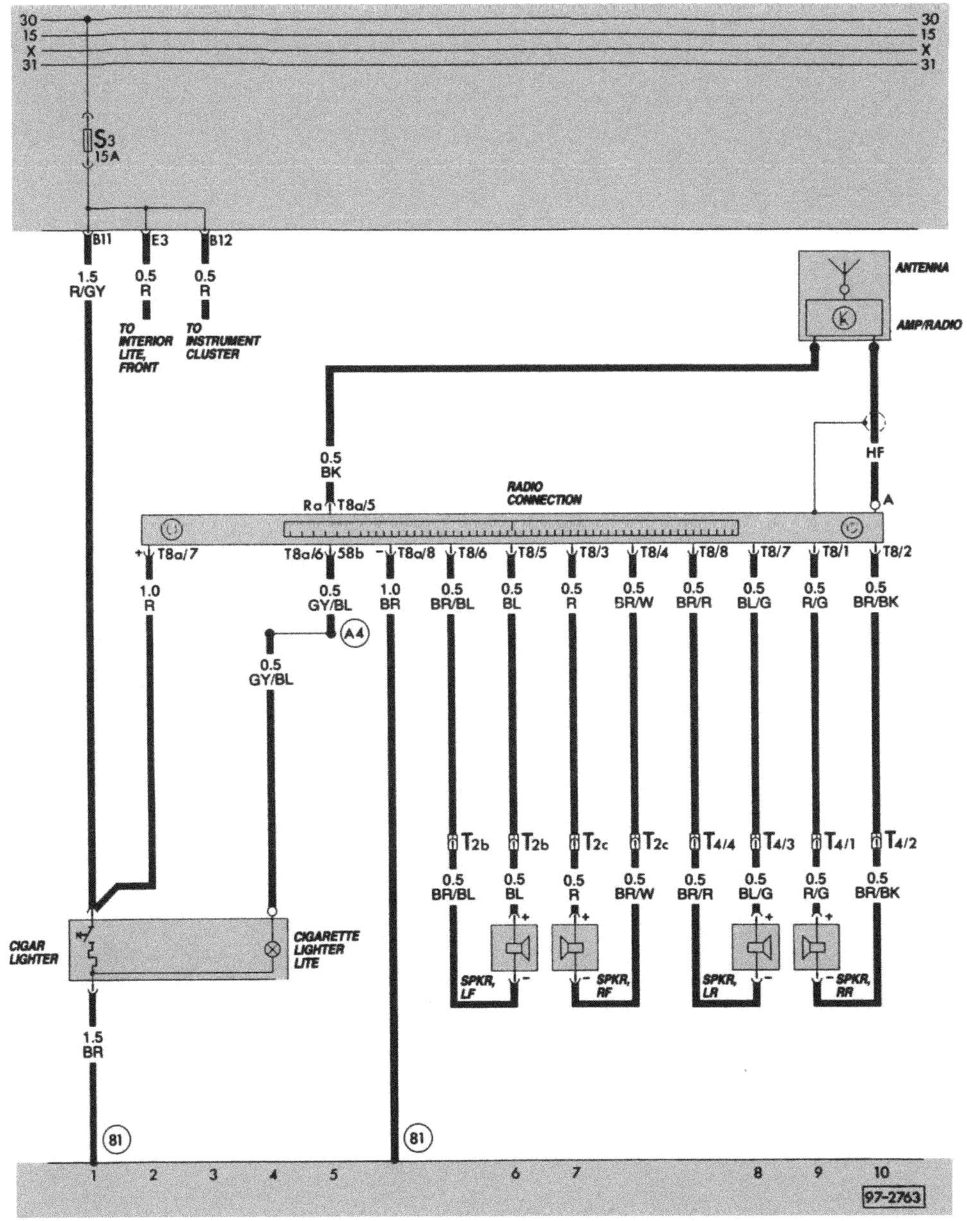

Stereo radio with power antenna – Scirocco, Cabriolet 1985-1989
page 1 of 2

Description	Current track
Cigarette lighter lite	3
Electrical antenna	9,10
Radio connection	2-8
Refrigerator 12V connector	1
Speaker, left front	4
Speaker, left rear	6
Speaker, right front	5
Speaker, right rear	7

Wire connectors

T1h — single, under fuse/relay panel
T1y — single, under fuse/relay panel
T2a — double, under instrument panel, middle
T2b — double, left front footwell
T2c — double, right front footwell
T4/ — four-point, under instrument panel, middle
T7a/— seven-point, on instrument cluster (long white housing)
T8/ — eight-point, on radio
T8a/— eight-point, on radio

Ground connections

(30)— near fuse/relay panel

(81)— in instrument panel wiring harness

Plus (+) connections

(A4)— (58b) in instrument panel wiring harness

Stereo radio with power antenna—Scirocco, Cabriolet 1985-1989
page 2 of 2

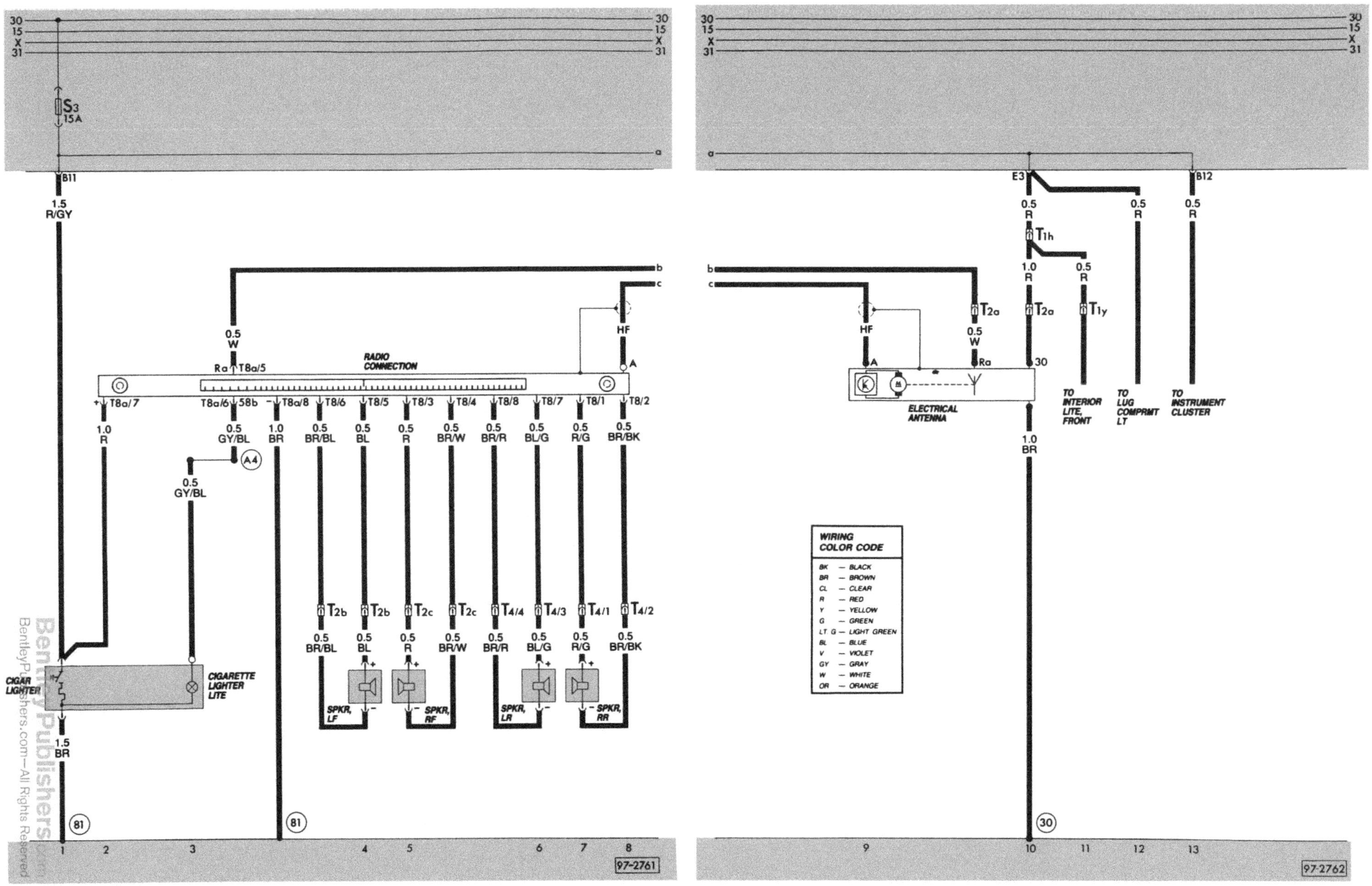

Stereo radio – Cabriolet 1990
page 1 of 2

Description	Current track
Cigarette lighter lite	4
Electrical antenna	11,12
Fresh air control lever lite	5
Radio connection	2-10
Refrigerator 12V connector	1
Speaker, left front	6
Speaker, left rear	8
Speaker, right front	7
Speaker, right rear	9

Wire connectors

T1h — single, under fuse/relay panel
T1y — single, under fuse/relay panel
T2a — double, under instrument panel, middle
T2b — double, in left front footwell
T2c — double, in right front footwell
T3a — three-point, under instrument panel, middle
T4/ — four-point, in right front footwell
T8/ — eight-point, on radio
T8a/— eight-point, on radio
T14 — fourteen-point, on instrument cluster

Ground connections

(30) — near fuse/relay panel

(81) — in instrument panel wiring harness

Plus (+) connections

(A4) — (58b) in instrument panel wiring harness

Stereo radio – Cabriolet 1990
page 2 of 2

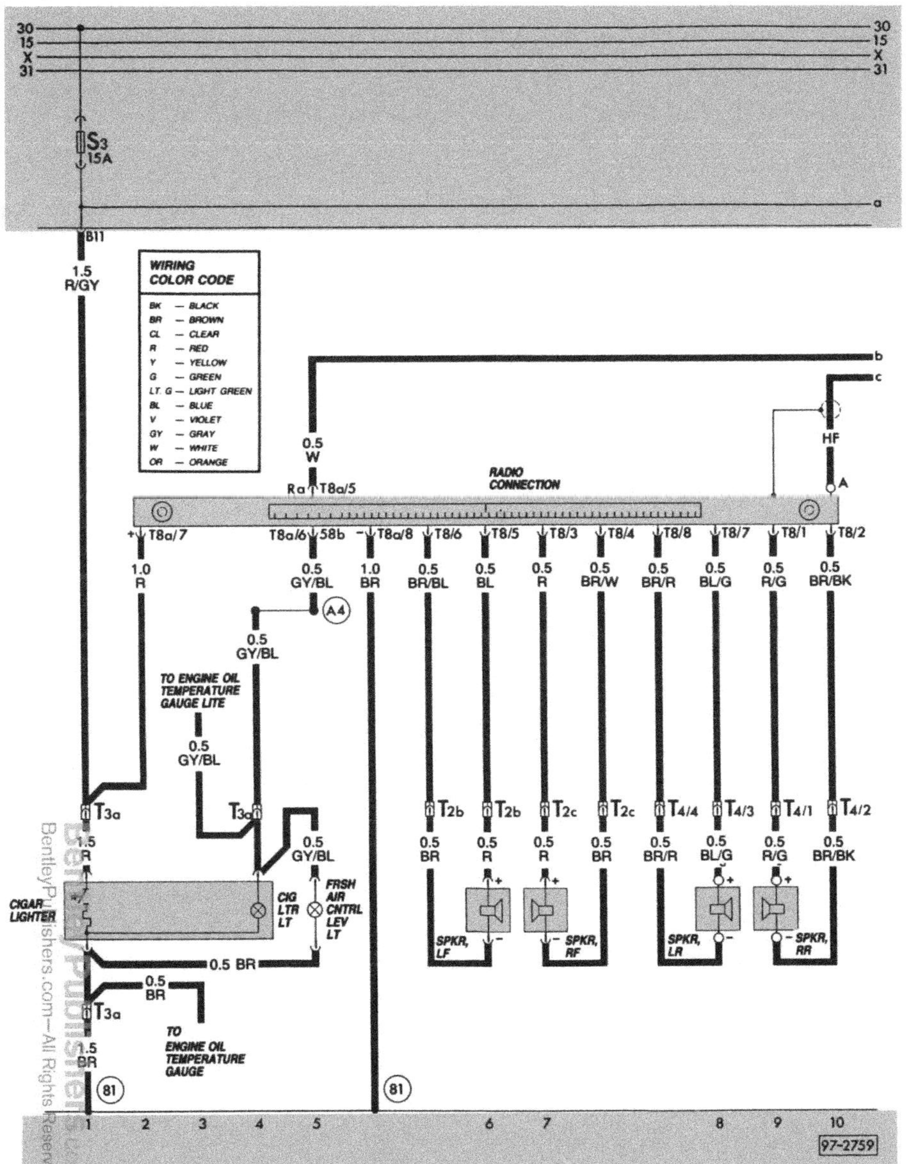

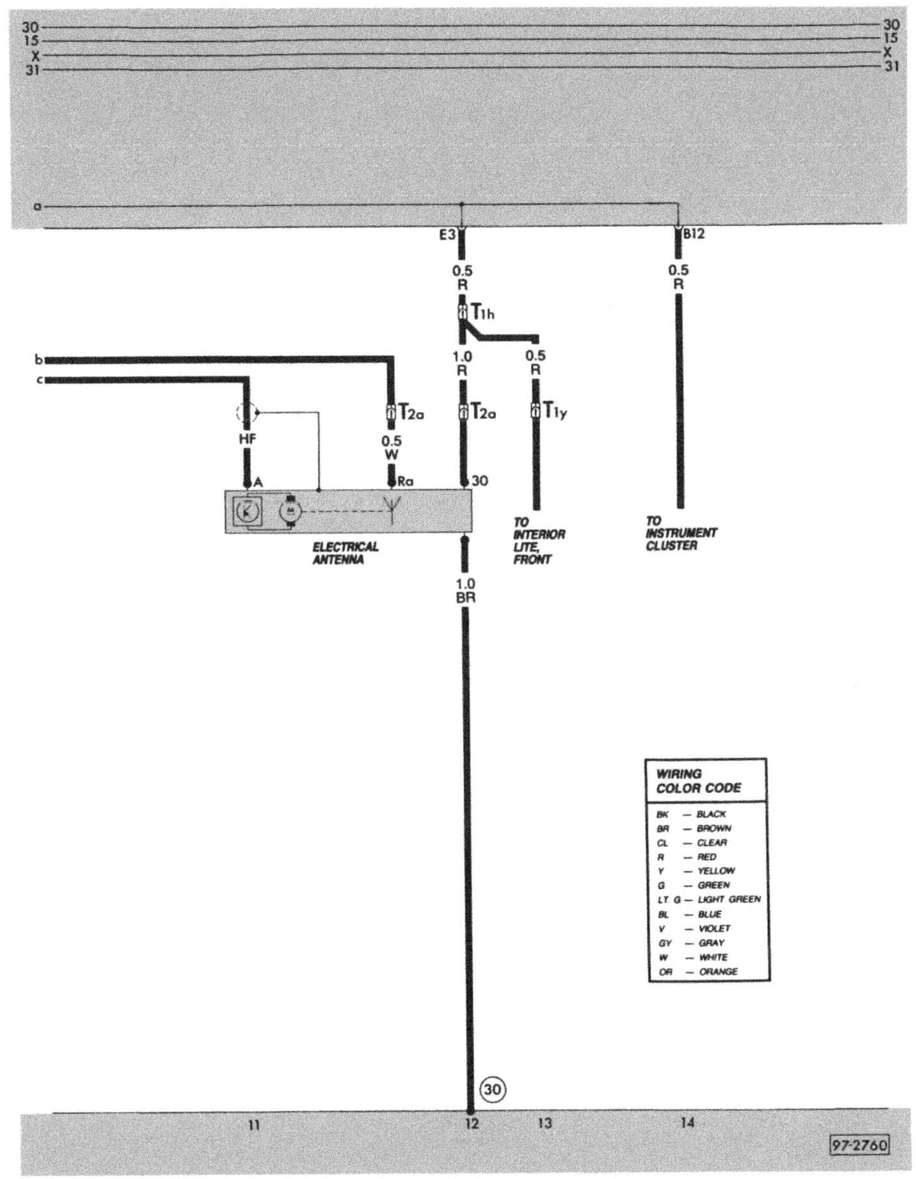

Stereo radio with six amplified speakers – Cabriolet from 1991
page 1 of 2

Description	Current track
Amplified speaker (bass), left, R32	5-13
Amplified speaker (bass), right, R33	15-21
Antenna, R11	27
Fuse, S3	7
High frequency speaker, left front, R20	3
High frequency speaker, right front, R22	24
Midrange speaker, left rear, R34	1
Midrange speaker, right rear, R35	26
Radio, R	9-28

Wire connectors

T1 — single, behind fuse/relay panel
T2 — double, behind fuse/relay panel
T3b — three-point, below console
T4 — four-point, on lower right A-pillar
T4a — four-point, on lower right A-pillar
T4b — four-point, on lower right A-pillar
T4c — four-point, on lower left A-pillar
T4d — four-point, on lower left A-pillar
T4e — four-point, in passenger's door
T4f — four-point, in driver's door
T6 — six-point, in driver's door
T6a — six-point, in passenger's door
T8 — eight-point, on radio
T10 — ten-point, on radio

Ground connections

(30) — beside fuse/relay panel

(43) — on lower right A-pillar

Welded wiring harness points

(A4) — plus connection (58b), in instrument panel wiring harness

Stereo radio with six amplified speakers – Cabriolet from 1991 page 2 of 2

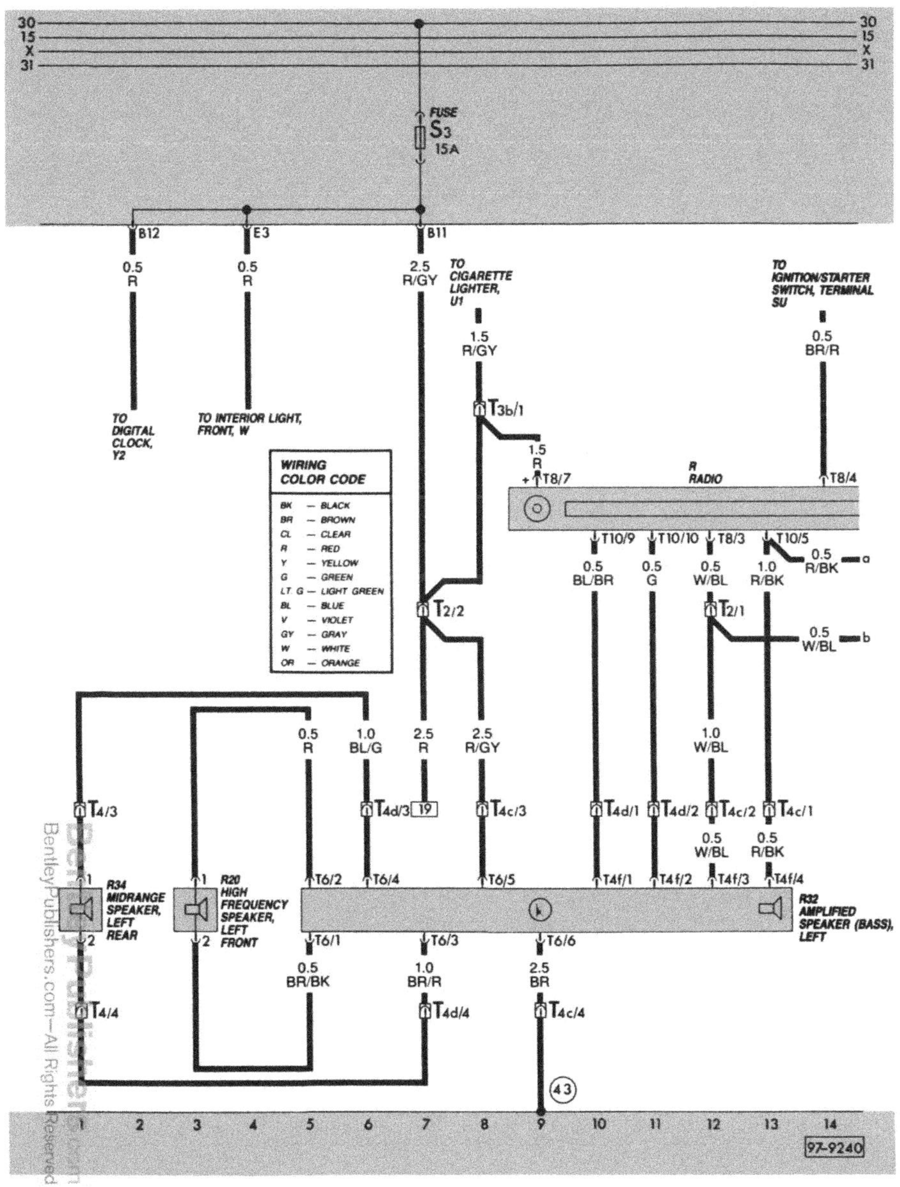

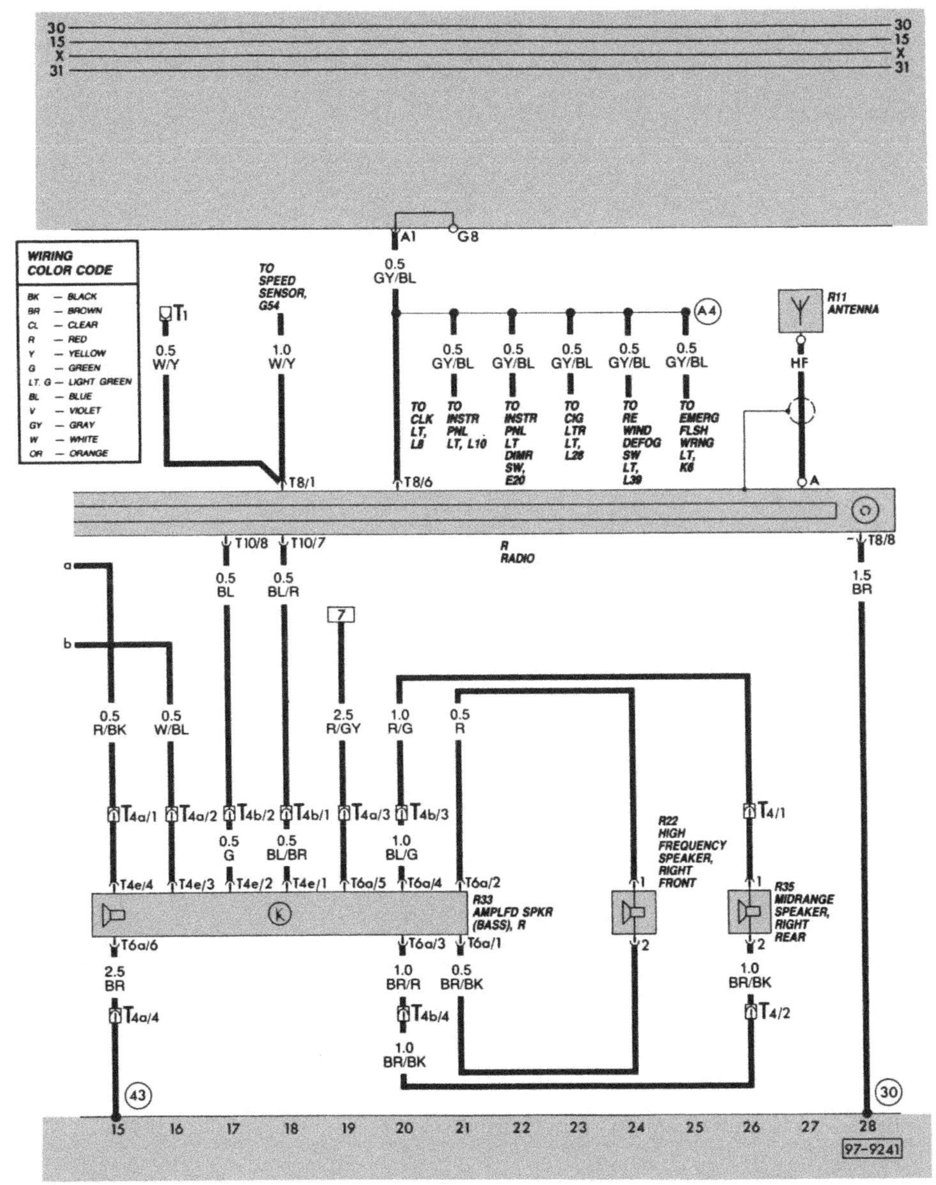

Stereo radio with six passive speakers – Cabriolet from 1991
page 1 of 2

Description	Current track
Antenna, R11	26
Fuse, S3	1
High frequency speaker, left front, R20	7
High frequency speaker, right front, R22	9
Low frequency speaker, left front, R21	5
Low frequency speaker, right front, R23	11
Radio, R	2-28
Speaker, left rear, R4	20
Speaker, right rear, R5	23

Wire connectors

T1 — single, behind fuse/relay panel
T2 — double, on lower left A-pillar
T2a — double, on lower right A-pillar
T3b — three-point, below console
T4 — four-point, on lower right A-pillar
T8 — eight-point, on radio
T8a — eight-point, on radio

Ground connections

(30) — beside fuse/relay panel

Welded wiring harness points

(A4) — plus connection (58b), in instrument panel wiring harness

Stereo radio with six passive speakers – Cabriolet from 1991
page 2 of 2

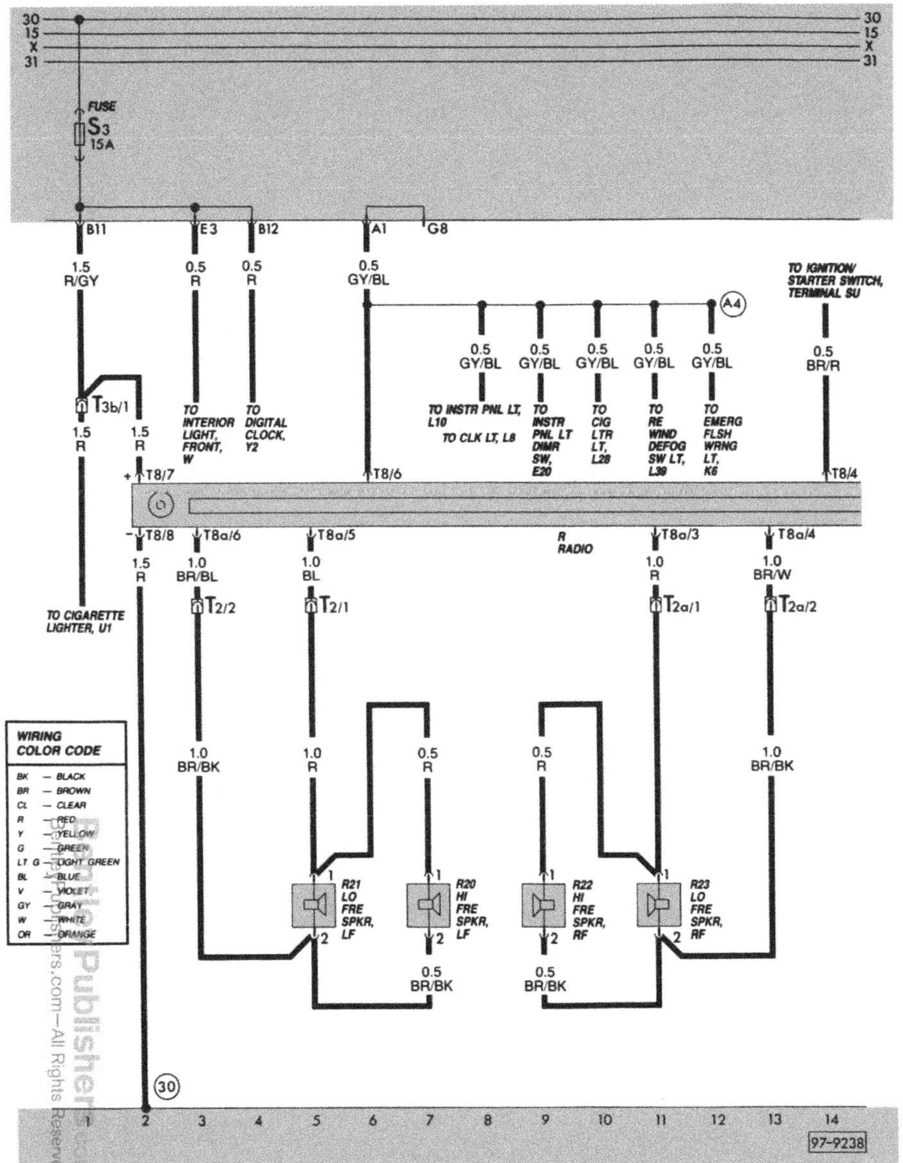

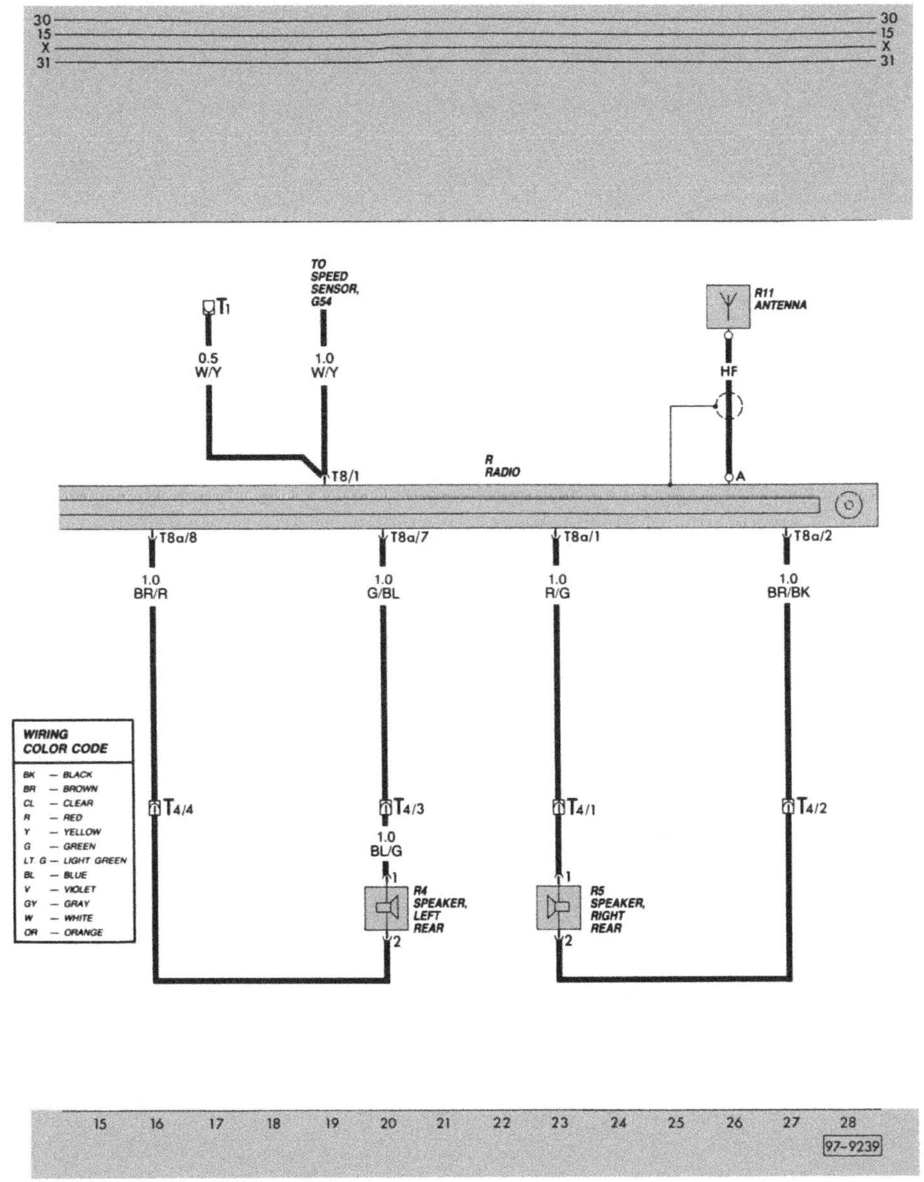

WARNING —

- *Automotive service and repair is serious business. You must be alert, use common sense, and exercise good judgement to prevent personal injury and complete the work safely.*
- *Before beginning any work on your Volkswagen, thoroughly read all the Cautions and Warnings listed near the front of this manual.*
- *Always read the complete procedure before you begin the work. Pay special attention to any Cautions and Warnings that accompany that procedure, or other information on a specific topic.*

A

B

C

D

E

> **WARNING —**
> • *Automotive service and repair is serious business. You must be alert, use common sense, and exercise good judgement to prevent personal injury and complete the work safely.*
>
> • *Before beginning any work on your Volkswagen, thoroughly read all the Cautions and Warnings listed near the front of this manual.*
>
> • *Always read the complete procedure before you begin the work. Pay special attention to any Cautions and Warnings that accompany that procedure, or other information on a specific topic.*

WARNING —

- *Automotive service and repair is serious business. You must be alert, use common sense, and exercise good judgement to prevent personal injury and complete the work safely.*
- *Before beginning any work on your Volkswagen, thoroughly read all the Cautions and Warnings listed near the front of this manual.*
- *Always read the complete procedure before you begin the work. Pay special attention to any Cautions and Warnings that accompany that procedure, or other information on a specific topic.*

R

S

T

> **WARNING —**
> • *Automotive service and repair is serious business. You must be alert, use common sense, and exercise good judgement to prevent personal injury and complete the work safely.*
>
> • *Before beginning any work on your Volkswagen, thoroughly read all the Cautions and Warnings listed near the front of this manual.*
>
> • *Always read the complete procedure before you begin the work. Pay special attention to any Cautions and Warnings that accompany that procedure, or other information on a specific topic.*

U

V

W